ULTRAFAST OPTICS

ULTRAFAST OPTICS

ULTRAFAST OPTICS

ANDREW M. WEINER
Purdue University

WILEY

A JOHN WILEY & SONS, INC., PUBLICATION

Published by John Wiley & Sons, Inc., Hoboken, New Jersey.
Published simultaneously in Canada.

Library of Congress Cataloging-in-Publication Data:

Weiner, Andrew Marc
Ultrafast optics / Andrew M. Weiner.
p. cm.
Includes bibliographical references and index.
ISBN 978-0-471-41539-8 (cloth)
1. Laser pulses, Ultrashort. 2. Laser pulses, Ultrashort–Industrial applications. 3. Mode-locked lasers. 4. Nonlinear optical spectroscopy. 5. Nonlinear optics. I. Title.

QC689.5.L37W45 2009
621.36'6–dc22 2008052027

In memory of

Jason Weiner, my father

Prof. Hermann Haus

and

Bill Drake, Jr.

CONTENTS

PREFACE

This book deals with the optics of picosecond and femtosecond light pulses, primarily at wavelengths in the visible range and longer. Research in ultrafast optics started roughly forty years ago, and although this field is now tremendously active, in many aspects it has also reached a level of maturity. However, relatively few broad treatments of ultrafast optics are available. It is hoped that this book, which is both detailed and comprehensive, will be a valuable resource not only for graduate students and researchers seeking to enter ultrafast optics but also for colleagues already engaged in this fascinating field.

I would like to mention a few pertinent points about my perspective in this book. First, in keeping with my training as an electrical engineer, the signals aspect of ultrafast optics is emphasized. That is, I often attempt to capture the detailed form of ultrashort pulses as they are transformed in various optical systems or evolve inside mode-locked lasers. Similarly, the detailed form of measurement data, as in ultrashort pulse characterization and ultrafast spectroscopy, is analyzed when possible.

Second, although a detailed theoretical treatment is often presented, I strive to balance this with an experimental perspective. Accordingly, many examples of data from the literature are included, especially in the later chapters. These examples are selected to provide concrete illustration of material that otherwise might remain abstract, to provide evidence of certain important phenomena, and sometimes to illustrate applications.

Third, although the suite of applications of ultrafast optics is now very rich, this book is concerned primarily with fundamental principles. No attempt is made to cover the applications space comprehensively. Two applications are covered in some depth: ultrafast spectroscopy and ultrafast electromagnetic pulse generation and measurement. Both of these are subjects of individual chapters. Certain other applications, such as the application of optical frequency combs for precision frequency metrology, are discussed briefly within appropriate sections of related text.

Fourth, the book is focused on ultrafast optics in visible and lower-frequency spectral bands, on time scales down to femtoseconds, and at intensities for which perturbative non-linear optics applies. Under these conditions the motions of bound electrons that mediate important laser–matter interactions may usually be viewed as instantaneous. Extreme nonlinear optics phenomena, arising from high-intensity laser–matter interactions for which the laser field reaches or exceeds the interatomic field, are not covered. One important example of extreme nonlinear optics is high harmonic generation, in which visible wavelength femtosecond pulses result in emission of photons in the vacuum ultraviolet (XUV) and soft x-ray bands. The use of high harmonic generation to realize attosecond pulses has become an active research topic within the last few years. Attosecond time scales and XUV and x-ray frequencies bring in entirely new physics that are beyond the scope of this book. Attosecond technology and science are in a stage of rapid evolution and will undoubtedly be the subject of future treatises.

The structure of the book is as follows:

- Chapter 1 begins with a brief overview and motivation, discussing key attributes of ultrashort laser pulses and some application examples. Important background material, including simple electromagnetics and laser essentials, is reviewed. The chapter continues with a phenomenological introduction to short pulse generation via mode-locking and concludes with a review of Fourier transforms, a mathematical tool essential to much of our treatment.
- Chapter 2 covers basic principles of laser mode-locking in some depth. The intent is not only to cover one of the most interesting topics at the outset, but to use the discussion on mode-locking as a physical context in which to introduce a variety of important ultrafast optical effects (e.g., dispersion, filtering, self-phase modulation), many of which are themselves treated in detail in subsequent chapters.
- Measurement of pulses on the femtosecond time scale is an important issue, since the speed required is considerably faster than that of existing photodetectors and oscilloscopes. In Chapter 3 we discuss methods for characterizing ultrashort pulses. Included are historical techniques dating back to the early years of ultrafast optics (these offer only partial information but remain in widespread use) as well as more powerful techniques offering full waveform information. The field of ultrashort pulse characterization has continued to grow, and new techniques continue to be introduced. I have not attempted to cover all the interesting measurement techniques that have been invented. My hope is that the discussion accompanying those methods that are included will prepare the reader to quickly grasp additional methods that may interest him or her.
- Dispersion is often a key limiting effect in ultrafast systems. Accordingly, in Chapter 4 we focus on dispersion and its compensation. After defining key concepts, the discussion covers material dispersion, then temporal dispersion arising from angular dispersion (including important grating and prism pair setups), and finally, dispersion effects in mirror structures. The effect of dispersion in the focusing of light by lenses is also discussed, as are methods for measurement of dispersion.
- Chapters 5 and 6 deal with ultrafast nonlinear optics. Chapter 5 emphasizes second-order nonlinear effects. After an introduction to nonlinear optics and a review of continuous-wave second-harmonic generation (SHG), new effects arising in ultrashort pulse SHG, sum and difference frequency generation, and optical parametric generation are discussed. Such effects are of primary interest in frequency conversion and pulse measurement applications. Chapter 6 focuses on refractive index (third-order) nonlinearities, which have seen very wide applications in ultrafast optics. Topics

include self-phase modulation, pulse compression, solitons, continuum generation, and propagation equations, including propagation equations relevant for pulses down to a few optical cycles.

- Chapter 7 takes advantage of material developed in earlier chapters to continue the discussion of mode-locking at a more advanced level. Included are soliton nonlinear optics phenomena observed in the mode-locking of fiber lasers, stretched pulse lasers operating in the normal dispersion regime, and soliton mode-locking of solid-state lasers with slow saturable absorbers. Important aspects of sub-10-fs laser design and stabilized frequency combs important for precision frequency metrology are also discussed.
- Chapter 8 focuses on the manipulation of ultrashort pulses. The chapter begins with detailed coverage of ultrafast Fourier optics methods that enable ultrashort pulse shaping and arbitrary waveform generation. It then treats various chirped pulse approaches for waveform manipulation and measurement, including interesting time lens approaches. Finally, femtosecond pulse amplification techniques, leading to realization of pulses with unparalleled peak power, are discussed.
- Ultrafast time-resolved spectroscopy is possibly the most widely practiced application of ultrashort light pulses. This field is highly interdisciplinary, and the number of different physical systems probed using ultrafast techniques is large. In Chapter 9 I present and analyze selected important concepts in femtosecond time-resolved spectroscopy. No attempt is made to cover all the applications or all the experimental variations. A pedagogical challenge is that students studying ultrafast optics may have very different degrees of preparation in quantum mechanics, which is needed for a detailed microscopic understanding of many of the systems studied via ultrafast techniques. Therefore, this chapter takes a two-pronged approach. First, techniques such as pump-probe, which principally probe incoherent (phase insensitive) population relaxation processes, are treated classically, although quantum concepts such as energy levels do appear phenomenologically. Experiments on population relaxation in organic dye molecules and in direct-gap semiconductors such as GaAs are discussed for illustration. Second, after a necessarily brief (and possibly inadequate) introduction to relevant quantum mechanics, spectroscopies sensitive to coherent (phase sensitive) phenomena are discussed. These subjects are treated with the help of quantum mechanics. Topics include wave packet phenomena in semiconductors and molecules, coherent polarization effects, and measurement of dephasing. A final topic, impulsive stimulated Raman scattering, is treated in a largely classical framework.
- The final chapter deals with another important application of ultrafast optics: the generation and measurement of picosecond and subpicosecond electrical and electromagnetic transients. Both electrical signals propagating on-chip on transmission-line structures and terahertz (THz) electromagnetic radiation freely propagating in space are considered. Finally, THz time-domain spectroscopy, a technique that provides exciting capability for materials characterization and sensing in a spectral region that is difficult to access by either direct electronic or optical means, is discussed.

Several problems are provided at the end of each chapter, ranging from simple theoretical questions to practical exercises requiring numerical computation. As an example of the latter, in the Chapter 2 problems the student is asked to simulate pulse evolution through many round trips in a mode-locked laser cavity, finally arriving at the self-consistent pulse solution. I regularly assign such numerical problems in my own course on ultrafast optics at Purdue University. Although in my experience such problems require substantial effort on the part of the student, they result in a much better understanding of the phenomena involved, not to mention improved skill in applying numerical tools such as the fast Fourier transform. For homework on pulse measurement I have frequently synthesized FROG or other data on a computer; I then distribute the data file to the class with the assignment to process the data to extract the pulse shape. (I have not included such problems in the current book, as I deemed it more expedient to let instructors generate their own pulses and corresponding data files.) The numerical problems included in the book may be used as is or may simply serve to inspire instructors to invent their own numerically oriented problems.

Authoring this book has been a project of nearly ten years. I began formal writing while on sabbatical during the 1999–2000 academic year at the Max Born Institute (MBI) for Nonlinear Optics and Ultrashort Pulse Spectroscopy in Berlin, Germany. Work continued for many years, but in a fragmented way, at my home institution. I made substantial progress toward completion during a second sabbatical during the 2006–2007 academic year in Boulder, Colorado, where I split my time at the National Institute of Standards of Technology (NIST) and at JILA, a joint enterprise of NIST and the University of Colorado. I owe great thanks to my sabbatical hosts, Prof. Thomas Elsaesser of the MBI, Dr. Leo Hollberg of NIST, and Prof. Steve Cundiff of JILA, for making these stays possible. I would also like to thank the MBI and the Alexander von Humboldt Foundation for assistance with funding during my stay in Berlin and NIST and JILA for assistance with funding during my stay in Boulder.

I would like to thank many persons who generously provided input and assistance in various aspects of this project. Giullo Cerullo, Steve Cundiff, Alex Gaeta, and Franz Kaertner provided helpful comments and clarification on various technical topics (in some cases, on multiple topics). Virginia Lorenz made available a copy of her University of Colorado Ph.D. thesis, which provided a very helpful overview of dephasing. At Purdue University, Dee Dee Dexter provided invaluable logistical and secretarial assistance throughout the course of this project. Dan Leaird was always willing to lend an ear when I wanted to voice ideas about the book project; Dan also deserves great thanks for his unflagging attention to our ultrafast optics and fiber communications research laboratory, even when I sometimes became distracted by the burdens of authorship. Many students deserve credit for identifying errors in preliminary versions of the manuscript, which were used over several iterations of my graduate course. Prof. Dongsun Seo, a sabbatical visitor from Korea, also pointed out several items in need of correction. A number of graduate students kindly agreed to carry out numerical work, generating data that resulted in a number of figures. These students include Jung-Ho Chung, Ehsan Hamidi, Zhi Jiang, Houxun Miao, Bhaskaran Muralidharan, Ninad Pimparkar, Haifeng Wang, Mark Webster, and Shang-Da Yang. V. R. Supradeepa checked several equations on my behalf. Zhi Jiang was especially helpful in proofing the typeset manuscript.

This book includes well over 200 figures, many of which were composed especially for this project. Although many others were taken from the literature, almost all of these were modified or redrawn to ensure readability and to achieve consistency of appearance and notation. I am tremendously grateful to Bill Drake, Jr., for fulfilling this responsibility with great skill from the inception of this project until it neared completion. Tragically, Bill succumbed to cancer at an early age. He continued to contribute to this book even as he struggled against the disease that ultimately killed him. Michael Black took over technical illustration responsibilities during the final period of this project; he also deserves much gratitude.

Finally, I would like to thank my parents, Jason and Geraldine Weiner, for fostering in me a love of learning; my graduate advisors, Profs. Hermann Haus and Erich Ippen, for attracting me to the field of ultrafast optics; and my wife, Brenda, and children, Roberta, Steven, and Gabriela, for their love and patience.

ANDREW M. WEINER

West Lafayette, Indiana
September 2008

1

INTRODUCTION AND REVIEW

1.1 INTRODUCTION TO ULTRASHORT LASER PULSES

This book is about ultrafast laser pulses: what they are, linear and nonlinear optical effects which they experience, methods by which they are generated and measured, and how they can be used for measurement of ultrafast physical processes. Let us begin with a definition of the relevant time units.

$$
\begin{aligned}
&1 \text{ nanosecond (ns)} &&= 10^{-9}\ \text{s} &&= 0.000000001\,\text{s}\\
&1 \text{ picosecond (ps)} &&= 10^{-12}\ \text{s} &&= 0.000000000001\,\text{s}\\
&1 \text{ femtosecond (fs)} &&= 10^{-15}\ \text{s} &&= 0.000000000000001\,\text{s}\\
&1 \text{ attosecond} &&= 10^{-18}\ \text{s} &&= 0.000000000000000001\,\text{s}
\end{aligned}
$$

To put these very short time units in perspective, it is useful to consider their spatial equivalent. If we could take a snapshot of a 1-s light pulse, this pulse would stretch over a distance of 186,000 miles (or 300,000 km), equal to the speed of light multiplied by 1s. This is roughly three-fourths of the distance from the Earth to the moon, a distance we will consider very slow! Now skipping over milliseconds and microseconds, we arrive at nanoseconds. One nanosecond has a spatial extent of 30 cm (ca. 1 ft). Although still rather slow by the standards of ultrafast optics, the nanosecond is the approximate time scale for high-speed electronic chips and computers. The word *ultrafast* is usually applied to the picosecond time scale and below. A picosecond has an extent of 0.3 mm, roughly the thickness of a business card. Given that typical garden-variety laser beams have beam diameters on the order of a few millimeters, we should perhaps envision pulses a picosecond and shorter not as pencils of light but as pancakes of light! In the visible and near-infrared spectral regions, pulses as

Ultrafast Optics, By Andrew M. Weiner

short as a few femtoseconds can now be generated. The spatial extent of even a 10-fs laser pulse is only 3 μm, much less than the diameter of a human hair.

Pulse durations of a few femtoseconds in the visible are approaching the fundamental pulse-width limitation of roughly one optical cycle (roughly one wavelength in spatial extent). Research into attosecond pulse generation is also under way [1]. One key theme in attosecond pulse generation is the use of highly nonlinear optical frequency-conversion methods to produce radiation at much higher frequencies (much shorter wavelengths), corresponding to extreme ultraviolet (XUV) and x-ray spectral regions. At such frequencies the duration of a single optical cycle (and hence the attainable pulse-width limit) is reduced, making attosecond pulses possible.

In this book we specifically focus on ultrafast optics in visible and lower-frequency spectral bands and on time scales down to femtoseconds. Within this time scale the motions of bound electrons that mediate important laser–matter interactions may usually be viewed as instantaneous. Conversely, attosecond time scales and XUV and x-ray frequencies bring in entirely new physics in which laser–matter interactions are sensitive to the noninstantaneous dynamics of bound electron motions. Attosecond technology and science are in a stage of rapid evolution and will undoubtedly be the subject of future books.

Ultrashort pulses have several related characteristics which make them useful for applications. These include the following:

- *High time resolution*. By definition, the pulse duration is in the picosecond or femtosecond range (or below). This provides very high time resolution for excitation and measurement of ultrafast physical processes in solid-state, chemical, and biological materials.
- *High spatial resolution*. The spatial extent of a short light pulse is given by the pulse duration multiplied by the speed of light. As noted above, for very short pulse durations, the spatial pulse length can be on the order of micrometers. This makes ultrashort pulses useful for some microscopy and imaging applications.
- *High bandwidth*. By the uncertainly principle, the product of the pulse-width times the optical bandwidth must be of order unity (or larger). As the pulse duration decreases, the bandwidth increases correspondingly. Pulses of 100 fs have bandwidths on the order of 10 terahertz (THz), and the shortest visible laser pulses contain so much of the visible spectrum that they appear white. This high-bandwidth feature can be important for optical communications as well as other applications.
- *Potential for high intensity*. For a given pulse energy, the peak power and peak intensity are inversely proportional to the pulse duration. Because the size (hence cost) of high-power lasers usually scales with pulse energy, femtosecond pulse technology can be used to obtain ultrahigh peak intensities at moderate energy levels. Amplified femtosecond pulses have produced peak powers up to the petawatt level (1 petawatt = 10^{15} W) and peak intensities exceeding 10^{20} W/cm^2.

The field of ultrafast optics has traditionally been a highly interdisciplinary one, with a wide range of applications areas. To give a flavor for the nature of application areas, we comment below on a few of the research applications.

Ultrafast Spectroscopy Time-resolved spectroscopy is a very successful and probably the most widespread application of picosecond and femtosecond laser technology. The idea is that ultrashort laser pulses can be used to make "stop-action" measurements of ultrafast physical processes, just as high-speed (microsecond) electronic flashes have been used starting several decades ago to make such stop-action photographs of bullets traveling through apples and milk droplets splashing into milk bowls [2]. On the femtosecond time scale, macroscopic objects such as bullets and milk droplets are motionless, and therefore ultrafast spectroscopy is best applied to study microscopic processes. Examples include investigations of femtosecond interactions of photoexcited electrons and holes with each other and with lattice vibrations in semiconductor crystals, ultrafast laser-induced melting, photodisassociation and ultrafast solution dynamics of chemical species, and ultrafast internal rearrangements of the large organic molecule bacteriorhodopsin as photons absorbed in the retina initiate the first biochemical steps in the process of vision. The principles of ultrafast spectroscopy are covered in Chapter 9 with examples.

Laser-Controlled Chemistry In a research area closely related to ultrafast spectroscopy, researchers are using specially engineered femtosecond laser waveforms to try to influence the course of photoinduced chemical reactions. In addition to observing ultrafast chemical motions as in time-resolved spectroscopy, the added idea here is to control the motions that take place. Since the intrinsic time scale for nuclear motions in chemical systems is tens to hundreds of femtoseconds, femtosecond laser pulses are a natural tool in pursuing the challenging goal of laser-controlled chemistry.

Frequency Metrology Ultrashort pulses are usually emitted from lasers in the form of periodic trains, which under certain conditions can exhibit very high timing stability and long-term coherence. The spectrum of such a periodic train is a comb of up to hundreds of thousands of discrete spectral lines, which may be stabilized to permit precision measurements of optical frequencies with sub-hertz uncertainties across the optical spectrum. Such stabilized frequency combs are now widely adopted for high-precision frequency metrology and for investigations of precision optical clocks. Related topics are discussed in Section 7.5.

High-Speed Electrical Testing Testing is a key issue in the development of high-speed electronic devices and circuits. Electronic test instrumentation based on established technology is usually slower than advanced high-speed research devices. However, since even the very fastest electronic devices only reach into the picosecond range, ultrafast laser technology offers speed to spare. Thus femtosecond optical pulses have been applied to generate subpicosecond electrical pulses and to measure operation of the highest-speed electronic devices. Ultrafast electrical pulse generation and measurement are discussed in Chapter 10.

Laser–Plasma Interactions Lasers with intensities of 10^{13} W/cm^2 and above (easily achieved using amplified femtosecond pulses) directed onto solid targets are sufficient to strip electrons from their nuclei, resulting in a laser-generated plasma. On the 100-fs time scale, the resulting free electrons do not have enough time to separate from the ionized nuclei. This provides the opportunity to study solid-density plasmas at temperatures as high as 1 million degrees.

Short-Wavelength Generation High-intensity ultrashort pulses at visible wavelengths can also be used to generate coherent short-wavelength radiation in the vacuum ultraviolet and x-ray ranges through highly nonlinear harmonic generation processes or by pumping x-ray lasers. Coherent short-wavelength radiation may be important, for example, for imaging microscopic structures such as DNA.

Optical Communications The low-loss transmission window of optical fibers has a bandwidth comparable to that of a 100-fs pulse, and therefore ultrashort-pulse technology may play an important role in optical communications. Subpicosecond pulses have already been used for laboratory experiments demonstrating fiber optic transmission of data at Tbit/s (10^{12} bit/s) rates. Here ultrafast optics technology is important not only for pulse generation but also for signal processing, for data detection, and for the advanced metrology necessary for characterizing and optimizing ultrashort-pulse transmission [3,4]. Ultrashort pulses may also prove important in wavelength-division-multiplexing (WDM) systems in which the fiber bandwidth is carved up into different wavelength bands or channels. For WDM applications it is the large bandwidth of the ultrashort pulse (not the short duration) which is useful, since a single pulse contains enough bandwidth to produce a number of wavelength channels.

Biomedical Applications Ultrashort pulses are finding substantial applications in biomedical imaging. Attractive features include the ability to perform optical imaging within scattering media (e.g., most tissues) and to obtain high-resolution depth information. An example of such an application is discussed in Section 3.3.3. In confocal microscopy significantly improved spatial resolution has been demonstrated by relying on two-photon excitation. The ability of ultrashort pulses to provide high intensity without high pulse energy is important in the use of this technique with sensitive biological samples. In laser-assisted surgical procedures ultrashort pulses may in some cases reduce collateral tissue damage by reducing heat deposition.

Materials Processing High-power lasers are used for a variety of industrial applications, such as cutting and drilling. With continuous-wave or "long"-pulse (nanoseconds) lasers, the minimum feature size and the quality of the cut are limited by thermal diffusion of heat to areas neighboring the laser focus. With femtosecond lasers, materials processing is possible using lower pulse energies, due to the very high peak powers, which lead to new physical mechanisms. This reduces the heat deposited into the sample during the laser machining process and leads to a much cleaner cutting or drilling operation.

1.2 BRIEF REVIEW OF ELECTROMAGNETICS

Since ultrashort laser pulses are made up of light, and light is a form of electromagnetic radiation, we very briefly review Maxwell's equations, which describe all forms of electromagnetic radiation, including light. We use MKS (SI) units here and throughout the book. It is assumed that the reader is already familiar with vector calculus. For a more detailed treatment of electromagnetics, the reader is directed to textbooks on this subject [5,6].

1.2.1 Maxwell's Equations

Maxwell's equations are a set of relationships between the electric field **E** and magnetic field **H** (boldface symbols denote vectors). Inside a medium we must also consider the

Table 1.1 Names and Units of Symbols in Maxwell's Equations

Symbol	Name	Units or Numerical Value
E	Electric field	$\mathrm{V\,m^{-1}}$
D	Electric flux density	$\mathrm{C\,m^{-2}}$
H	Magnetic field	$\mathrm{A\,m^{-1}}$
B	Magnetic flux density	T (or $\mathrm{V{\cdot}s\,m^{-2}}$)
P	Polarization density	$\mathrm{C\,m^{-2}}$
M	Magnetization density	$\mathrm{A\,m^{-1}}$
J	Current density	$\mathrm{A\,m^{-2}}$
ρ	Charge density	$\mathrm{C\,m^{-3}}$
ε_0	Permittivity of free space	$8.85 \times 10^{-12}\ \mathrm{F\,m^{-1}}$ (or $\mathrm{C\,V^{-1}\,m^{-1}}$)
μ_0	Permeability of free space	$4\pi \times 10^{-7}\ \mathrm{H\,m^{-1}}$ (or $\mathrm{V\,s^2\,m^{-1}\,C^{-1}}$)

charge density ρ, current density **J**, polarization density **P**, and magnetization density **M**, and in order to include the effect of the fields on the matter, the electric and magnetic flux densities, **D** and **B**, are also introduced. Units for these quantities are given in Table 1.1. Maxwell's equations are then written as follows:

$$\nabla \cdot \mathbf{D} = \rho \tag{1.1}$$

$$\nabla \cdot \mathbf{B} = 0 \tag{1.2}$$

$$\nabla \times \mathbf{E} = \frac{-\partial \mathbf{B}}{\partial t} \tag{1.3}$$

$$\nabla \times \mathbf{H} = \mathbf{J} + \frac{\partial \mathbf{D}}{\partial t} \tag{1.4}$$

The relations defining **D** and **B** are

$$\mathbf{D} = \varepsilon_0 \mathbf{E} + \mathbf{P} \tag{1.5}$$

$$\mathbf{B} = \mu_0 (\mathbf{H} + \mathbf{M}) \tag{1.6}$$

The constants ε_0 and μ_0 are known as the *permittivity* and *permeability* of free space, with the numerical values and units given in Table 1.1. Note also that the symbol ρ refers to the free charge density (i.e., any bound charge density associated with the polarization is not included). Similarly, the current density **J** does not include any currents associated with the motion of bound charges (changes in polarization). In free space we would have $\rho = \mathbf{J} = \mathbf{P} = \mathbf{M} = 0$.

For now we specialize to the case of a linear, isotropic, and source-free medium. By *source-free* we mean that the charge and current densities are zero ($\rho = 0$ and $\mathbf{J} = 0$). By *linear* we mean that the medium response (i.e., the polarization and magnetization) is linear in the applied fields. For the case of the electric field, we write

$$\mathbf{P} = \varepsilon_0 \chi_e \mathbf{E} \tag{1.7}$$

where χ_e is known as the electric susceptibility (dimensionless). Inserting into eq. (1.5), one obtains

$$\mathbf{D} = \varepsilon_0(1 + \chi_e)\mathbf{E} = \varepsilon\mathbf{E} \tag{1.8}$$

The proportionality constant ε is termed the dielectric constant, with

$$\varepsilon = (1 + \chi_e)\,\varepsilon_0 \tag{1.9}$$

Other common terms include the relative dielectric constant ($\varepsilon/\varepsilon_0$) and the index of refraction n, which is commonly used in optics, where

$$n^2 = \frac{\varepsilon}{\varepsilon_0} \tag{1.10}$$

For the case of the magnetic field, we write

$$M = \chi_m H \tag{1.11}$$

where χ_m is the magnetic polarizability. Using eq. (1.6), we obtain

$$\mathbf{B} = \mu_0\,(1 + \chi_m)\mathbf{H} = \mu\mathbf{H} \tag{1.12}$$

In most cases in ultrafast optics, one is interested in nonmagnetic materials, for which $\mathbf{M} = 0$. In this case of zero magnetization, one has

$$\mathbf{B} = \mu_0\mathbf{H} \tag{1.13}$$

Equations (1.7) and (1.11) are examples of constitutive laws, which specify the response of the material to the fields. The form of these equations as written arises because we have assumed both linear and isotropic media (for nonisotropic media, one would need to replace the assumed scalar susceptibilities with tensors). We note that there are many situations in ultrafast optics where these assumptions are not valid. For example, nonlinear optical effects, which we discuss in later chapters, require by definition that **P** be a nonlinear function of **E**.

1.2.2 The Wave Equation and Plane Waves

We now consider electromagnetic wave propagation in linear, isotropic, source-free media. To derive the wave equation, we take the curl of eq. (1.3) and insert eq. (1.4), which, using the stated assumptions and a well-known vector identity,[1] gives the following:

$$\nabla \times \nabla \times \mathbf{E} = \nabla(\nabla \cdot \mathbf{E}) - \nabla^2\mathbf{E} = -\mu\varepsilon\frac{\partial^2\mathbf{E}}{\partial t^2} \tag{1.14}$$

[1] The identity is $\nabla \times \nabla \times \mathbf{A} = \nabla(\nabla \cdot \mathbf{A}) - \nabla^2\mathbf{A}$. Note that in Cartesian coordinates ∇^2 has a very simple form, namely $\nabla^2\mathbf{A} = (\partial^2/\partial x^2 + \partial^2/\partial y^2 + \partial^2/\partial z^2)\mathbf{A}$.

Since $\nabla \cdot \mathbf{E} = 0$ also under our conditions, we obtain the wave equation

$$\nabla^2 \mathbf{E} = \mu\varepsilon \frac{\partial^2 \mathbf{E}}{\partial t^2} \tag{1.15}$$

One situation of special interest is the case where the field varies in only one direction, which without loss of generality we take as the z direction. Then the wave equation becomes

$$\frac{\partial^2 \mathbf{E}}{\partial z^2} = \mu\varepsilon \frac{\partial^2 \mathbf{E}}{\partial t^2} \tag{1.16}$$

The general solution takes the form

$$\mathbf{E}(z, t) = \mathbf{E}_0 \left(t - \frac{z}{v}\right) \tag{1.17}$$

where $\mathbf{E}_0$ is a vector in the x–y plane [eq. (1.1) precludes E from having a z-component] and $v = 1/\sqrt{\mu\varepsilon}$. The solution can be verified by plugging back into the wave equation. Equation (1.17) is called a *plane-wave solution*, since the field does not vary in the transverse (x–y) plane. It also represents a traveling wave, since the field propagates in the z direction without changing its form. In the case of a pulsed field, $\mathbf{E}_0(t)$ represents the pulse shape. The propagation velocity is given by v. Note that

$$\frac{1}{\sqrt{\mu_0\varepsilon_0}} = c \cong 2.998 \times 10^8 \text{ m s}^{-1} \tag{1.18}$$

is the velocity of light in free space. Therefore, for the case most common in optics where $\mu = \mu_0$, the velocity of propagation within a medium is given by

$$v = \frac{c}{n} \tag{1.19}$$

where n is the refractive index according to eq. (1.10). Note also that in deriving eqs. (1.14) to (1.17), we have assumed implicitly that the refractive index n is independent of frequency. When n does have a frequency dependence, this can change the propagation velocity or cause the pulse to distort during propagation. These effects are discussed in Chapter 4.

The case of a sinusoidal solution to the wave equation will be of special importance. Then eq. (1.17) takes the form

$$\mathbf{E}(z, t) = \mathbf{E}_0 \cos(\omega t - kz + \phi) \tag{1.20}$$

where $\mathbf{E}_0$ is now a constant vector, ω is the angular frequency, and the propagation constant k must satisfy the dispersion relation

$$k = \omega\sqrt{\mu\varepsilon} \tag{1.21}$$

or again, assuming that $\mu = \mu_0$,

$$k = \frac{\omega n}{c} \tag{1.22}$$

The wave has a temporal oscillation period equal to $2\pi/\omega$ and a spatial period or wavelength *in the medium* given by $\lambda = 2\pi/k$. The wavelength in free space is denoted λ_0 and is given by

$$\lambda_0 = \frac{2\pi c}{\omega} \tag{1.23}$$

Equation (1.20) represents the ideal case of single frequency or monochromatic laser radiation. It can also be written in the equivalent form

$$\mathbf{E}(z, t) = \mathrm{Re}\left\{\tilde{\mathbf{E}}_0 e^{j(\omega t - kz)}\right\} \tag{1.24}$$

where $\mathrm{Re}\{\cdots\}$ denotes the real part and the phase ϕ has been incorporated into the complex vector $\tilde{\mathbf{E}}_0$. We refer to this form as complex notation. As we will see shortly, ultrashort light pulses are conveniently described as superpositions of sinusoidal solutions of the form (1.20) or (1.24) with different frequencies.

Finally, we note that similar solutions can be written for propagation in directions other than along z, as follows:

$$\mathbf{E}(\mathbf{r}, t) = \mathrm{Re}\left\{\tilde{\mathbf{E}}_0 e^{j(\omega t - \mathbf{k}\cdot\mathbf{r})}\right\} \tag{1.25}$$

Here $\mathbf{k}$ is the propagation vector; it points along the direction of propagation and its magnitude $k = |\mathbf{k}|$ still satisfies the dispersion relation (1.21).

1.2.3 Poynting's Vector and Power Flow

We also review the expressions for energy flow with electromagnetic waves. To arrive at the required formulas, we form the dot product of eq. (1.3) with $\mathbf{H}$ and subtract from this the dot product of eq. (1.4) with $\mathbf{E}$. Using another vector identity,[2] we find that

$$\nabla \cdot (\mathbf{E} \times \mathbf{H}) + \mathbf{H} \cdot \frac{\partial \mathbf{B}}{\partial t} + \mathbf{E} \cdot \frac{\partial \mathbf{D}}{\partial t} + \mathbf{E} \cdot \mathbf{J} = 0 \tag{1.26}$$

We also make use of the divergence theorem,

$$\int \nabla \cdot \mathbf{A}\, dV = \int \mathbf{A} \cdot \hat{\mathbf{n}}\, dS \tag{1.27}$$

which states that the surface integral of a vector $\mathbf{A}$ over a closed surface is equal to the volume integral of $\nabla \cdot \mathbf{A}$ over the volume bounded by that surface. $\hat{\mathbf{n}}$ is the unit vector

[2] The identity is $\nabla \cdot \mathbf{A} \times \mathbf{B} = \mathbf{B} \cdot \nabla \times \mathbf{A} - \mathbf{A} \cdot \nabla \times \mathbf{B}$.

normal to the surface and pointing outward. The result is

$$\int (\mathbf{E} \times \mathbf{H}) \cdot \hat{\mathbf{n}}\, dS + \int dV \left\{ \mathbf{H} \cdot \frac{\partial \mathbf{B}}{\partial t} + \mathbf{E} \cdot \frac{\partial \mathbf{D}}{\partial t} + \mathbf{E} \cdot \mathbf{J} \right\} = 0 \tag{1.28}$$

Finally, assuming a linear medium and substituting for **D** and **B** using eqs. (1.8) and (1.12), we obtain

$$\int (\mathbf{E} \times \mathbf{H}) \cdot \hat{\mathbf{n}}\, dS + \int dV \left\{ \frac{\partial \left(\frac{1}{2}\varepsilon\, |\mathbf{E}|^2\right)}{\partial t} + \frac{\partial \left(\frac{1}{2}\mu\, |\mathbf{H}|^2\right)}{\partial t} + \mathbf{E} \cdot \mathbf{J} \right\} = 0 \tag{1.29}$$

Equations (1.28) and (1.29) are representations of Poynting's theorem, which describes conservation of energy in electromagnetic systems. We can identify specific meanings for each of the terms. Look at eq. (1.29), for example:

- $\int (\mathbf{E} \times \mathbf{H}) \cdot \hat{\mathbf{n}}\, dS$ is the net rate of energy flow out of the closed surface. It has units of power (watts). $\mathbf{E} \times \mathbf{H}$ is called the *Poynting vector* and has units of intensity ($\mathrm{W/m^2}$). It gives the power density carried by an electromagnetic wave and the direction in which power is carried.
- $\frac{1}{2}\varepsilon\, |\mathbf{E}|^2$ and $\frac{1}{2}\mu\, |\mathbf{H}|^2$ are the local energy densities ($\mathrm{J/m^3}$) associated with the electric and magnetic fields, respectively. $\partial/\partial t \int \frac{1}{2}\varepsilon\, |\mathbf{E}|^2\, dV$ and $\partial/\partial t \int \frac{1}{2}\mu\, |\mathbf{H}|^2\, dV$ represent the time rate of change of electric and magnetic field energy stored within the volume, respectively.
- $\int \mathbf{E} \cdot \mathbf{J}\, dV$ represents power dissipation or generation within the volume (in watts). When $\mathbf{E} \cdot \mathbf{J}$ is positive, this term represents power dissipation due, for example, to ohmic losses. Energy is transferred out of the fields and into the medium, typically as heat. When $\mathbf{E} \cdot \mathbf{J}$ is negative, this term represents power supplied by the currents and fed into the electromagnetic fields.

Overall, Poynting's theorem is a power balance equation, showing how changes in stored energy are accounted for by power dissipation and energy flow.

It is worth specializing once more to the case of single-frequency sinusoidal fields, with **E** given by eq. (1.25). The **H** field is obtained using eq. (1.3), with the result

$$\mathbf{H} = \mathrm{Re}\left\{ \sqrt{\frac{\varepsilon}{\mu}} \frac{\mathbf{k} \times \tilde{\mathbf{E}}_0}{k} e^{j(\omega t - \mathbf{k}\cdot\mathbf{r})} \right\} \tag{1.30}$$

Thus **H** is perpendicular to both **E** and **k**, and its magnitude is equal to $\sqrt{\varepsilon/\mu}\, |\mathbf{E}|$. The factor $\sqrt{\mu/\varepsilon}$ is termed the *characteristic impedance* of the medium, and $\sqrt{\varepsilon/\mu}$ is therefore the *admittance*.

In optics one is usually interested in the time-average power flow. This is calculated in complex notation as follows. First consider scalar functions $f(t)$ and $g(t)$, where

$$f(t) = \mathrm{Re}\left\{\tilde{f}e^{j\omega t}\right\} \quad \text{and} \quad g(t) = \mathrm{Re}\left\{\tilde{g}e^{j\omega t}\right\} \tag{1.31}$$

The time average of $f(t)g(t)$ is given by

$$\langle fg \rangle = \frac{1}{2}\mathrm{Re}\{\tilde{f}\tilde{g}^*\} \tag{1.32}$$

Here $\langle \cdots \rangle$ denotes the time average and * indicates a complex conjugate. Similarly, if $\mathbf{f(t)}$, $\mathbf{g(t)}$, $\tilde{\mathbf{f}}$, and $\tilde{\mathbf{g}}$ now denote vectors, the time average of $\mathbf{f} \times \mathbf{g}$ is given by

$$\langle \mathbf{f} \times \mathbf{g} \rangle = \frac{1}{2}\mathrm{Re}\{\tilde{\mathbf{f}} \times \tilde{\mathbf{g}}^*\} \tag{1.33}$$

Using these relations, the time-average Poynting vector for the plane waves of eqs. (1.25) and (1.30) becomes

$$\langle \mathbf{E} \times \mathbf{H} \rangle = \frac{1}{2}\sqrt{\frac{\varepsilon}{\mu}}\left|\tilde{\mathbf{E}}_\mathbf{0}\right|^2 \frac{\mathbf{k}}{k} \tag{1.34}$$

where we have assumed that ε and μ are real. Power is carried along the direction of $\mathbf{k}$. In the case of a nonmagnetic material, we can write the magnitude of the time-average Poynting vector, commonly called the *intensity* I, in the following useful form:

$$I = |\langle \mathbf{E} \times \mathbf{H} \rangle| = \frac{1}{2}\varepsilon_0 cn \left|\tilde{\mathbf{E}}_\mathbf{0}\right|^2 \tag{1.35}$$

1.3 REVIEW OF LASER ESSENTIALS

We will shortly discuss in some detail methods by which lasers can be made to produce ultrashort light pulses. First, however, we give a brief and simple review of lasers in general. More detail can be found in texts on lasers, such as [7,8].

1.3.1 Steady-State Laser Operation

Schematic drawings of two simple laser geometries are shown in Fig. 1.1. Both lasers consist of a set of mirrors and a gain medium. The gain medium is an optical amplifier which coherently amplifies light passing through it. The mirrors may be curved or planar and together make up the laser cavity or resonator. The cavity is aligned so that light reflects back and forth again and again, passing along the same path every time. If we imagine even a very weak light intensity in the cavity (due to spontaneous emission from the gain medium), then for sufficiently high gain, the intensity increases from one round trip through the laser to the next, eventually resulting in an intense beam. In steady state the gain per round trip must equal the loss. Part of the light passes through the partially transmissive output coupler, and this forms the output laser beam, which can be used for experiments.

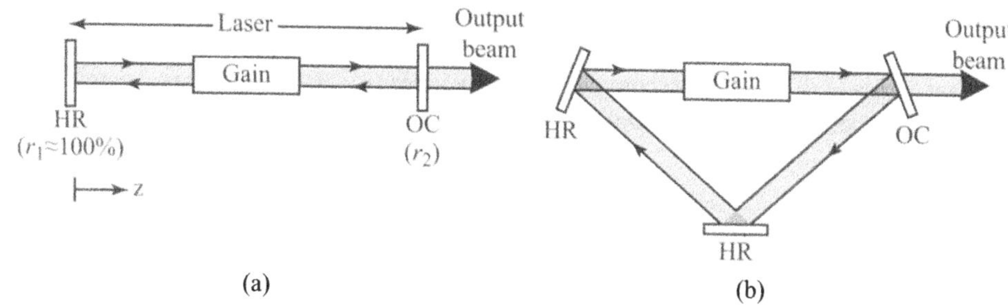

Figure 1.1 Schematic of (a) a linear cavity laser and (b) a ring laser. HR, high reflector; OC, output coupler.

In the linear or Fabry–Perot laser cavity shown in Fig. 1.1a, light passes along the same path, traveling from left to right and right to left. The light makes two passes through the gain medium per round trip. In the ring cavity shown in Fig. 1.1b, light passes through the gain medium only once per round trip. In the diagram we have assumed unidirectional operation (i.e., light is traveling around the ring in only one direction). In real ring lasers either unidirectional or bidirectional operation is possible. Both linear and ring geometries have been used in femtosecond laser design.

For our purposes we usually consider the gain medium as a black box. The physics of laser gain media is usually analyzed using quantum mechanics in courses on laser fundamentals. Here we usually stick to a classical description, although we will use the indispensable energy-level concept, which does come from quantum mechanics. Common laser media used for ultrafast lasers include impurity-doped crystals or glasses such as Ti:sapphire, Nd:YAG or Nd:glass, organic dyes, doped optical fibers, and semiconductor heterojunction diodes. Note that in thermodynamic equilibrium, materials can absorb light but cannot amplify it. To achieve gain, power must be supplied to the medium to promote electrons into excited-state energy levels. When the electrons are "pumped" to the excited state at a sufficiently high rate (i.e., when enough power is supplied), a *population inversion*, in which the population of electrons in an excited energy level exceeds that in a lower level, can be achieved. A population inversion is a necessary condition to achieve optical gain. Power can be supplied to the laser medium in many different ways. Optical pumping, in which absorption of pump photons from a flashlamp or an external laser promotes the electrons to the excited state, is often used in ultrashort-pulse lasers. Other pumping methods include current injection in semiconductor diode lasers or electric discharges in gas lasers. The need for a pump source to obtain optical gain can be likened to the need to plug in an electronic amplifier.

To achieve steady-state laser operation, the electric field must repeat itself after a round trip through the laser cavity. Therefore, let us now consider a single round trip of a monochromatic field through a cavity. To be specific, we consider the linear cavity of Fig. 1.1a. The optical path length between mirrors is denoted l; the total optical path length is therefore $2l$. The electric field amplitude of a monochromatic plane wave propagating the $+z$ direction can be written

$$E(z,t) = \mathrm{Re}\left\{E_0 e^{j(\omega t - kz)}\right\} \tag{1.36}$$

where the field is now taken as a scalar for convenience. E_0 is the field amplitude just to the right of the high reflector at $z = 0$, ω is the angular frequency, $k = \omega n/c$ is the propagation constant, and $n = n' + jn''$ is the complex refractive index (n' and n'' are both real numbers). The field just before the output coupler is written

$$E = \mathrm{Re}\left\{ E_0 e^{(\omega/c)n'' l_g} e^{j\omega t} e^{-j(\omega/c)(n' l_g + l_a)} \right\} \tag{1.37}$$

In eq. (1.37) n' and n'' refer to the gain medium; the regions outside the gain medium are taken as air, with $n = 1$. l_g and l_a are the physical lengths of the gain medium and air region, respectively. We can now identify $l = n' l_g + l_a$ as the optical path length. We also see that for $n'' > 0$, the field has been amplified.

The field at $z = 0$, corresponding to a round-trip path through the resonator, is obtained by reflecting off the output coupler with amplitude r_2, passing back through the cavity, and then reflecting off the high reflector with amplitude r_1. Although ideally we would have $r_1 = 100\%$, we keep the variable r_1 in our expressions to account for any imperfections of the high reflector as well as any other losses in the laser cavity besides those arising from the output coupler. The resulting expression for the field is as follows:

$$E = \mathrm{Re}\left\{ r_1 r_2 e^{2(\omega/c)n'' l_g} E_0 e^{j\omega t} e^{-j(2\omega/c)l} \right\} \tag{1.38}$$

To satisfy the steady-state requirement, eq. (1.38) must be equal to the initial field, eq. (1.36), evaluated at $z = 0$: $E = \mathrm{Re}\left[E_0 e^{j\omega t} \right]$. This leads to two conditions: the gain condition and the phase condition. The *gain condition*,

$$r_1 r_2 e^{(2\omega/c)n'' l_g} = 1 \tag{1.39}$$

states that the round-trip gain exactly balances the round-trip loss. The *phase condition*,

$$\frac{2\omega l}{c} = 2m\pi \tag{1.40}$$

requires that the the round-trip phase shift is equal to an integer m times 2π. This means that the laser is only allowed to oscillate at certain discrete angular frequencies, given by

$$\omega_m = \frac{m\pi c}{l} \tag{1.41}$$

These frequencies are known as the *longitudinal modes* of the cavity. In terms of the frequency $f = \omega/2\pi$, the mode frequencies are

$$f_m = \frac{mc}{2l} \tag{1.42}$$

and the mode spacing (taking n' as frequency independent) is

$$\Delta f = f_m - f_{m-1} = \frac{c}{2l} \tag{1.43}$$

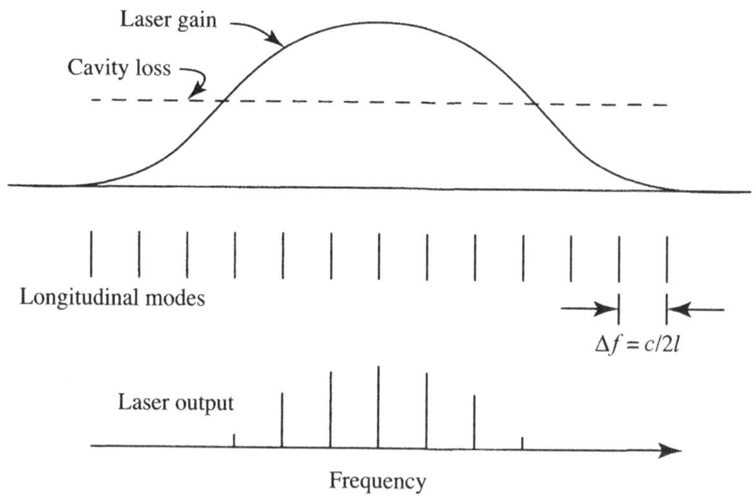

Figure 1.2 Laser gain and cavity loss spectra, longitudinal mode locations, and laser output for multimode laser operation.

A frequency-domain view of basic laser operation is pictured schematically in Fig. 1.2. A comparison of the gain vs. loss is shown at the top, with the locations of the longitudinal modes shown below. The resonator loss is taken as frequency independent, while the gain is assumed to have a bandpass spectral response. Laser oscillation occurs only for those modes where the gain lies above the loss line. In the situation shown, gain exceeds loss for several longitudinal modes, and multiple output frequencies appear simultaneously. This is called *multimode operation.*

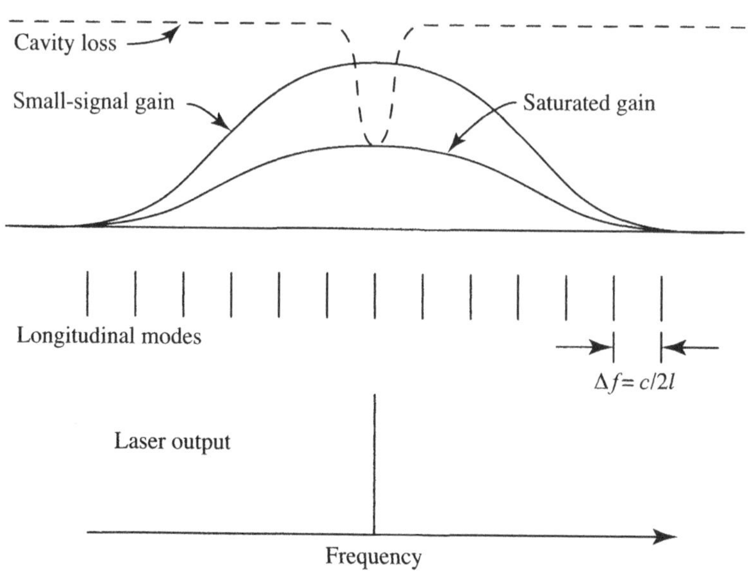

Figure 1.3 Gain and loss spectra, longitudinal mode locations, and output of a homogeneously broadened laser under single-mode operation.

To obtain monochromatic or single-mode laser radiation, it is usually necessary to insert a frequency-dependent loss element (a filter) to ensure that gain exceeds loss for only a single longitudinal mode. This situation is sketched in Fig. 1.3. Note, however, that the steady-state laser gain condition [eq. (1.39)] requires that the gain exactly equal rather than exceed the loss. For this reason we now distinguish between small-signal gain and saturated gain. *Small-signal gain* is gain available under conditions of zero or at least very weak light intensity and depends only on the properties of the laser medium and the pump level. For weak spontaneous emission to build up to produce a significant laser intensity, the small-signal gain must indeed exceed the loss. However, as the intracavity intensity increases, the small-signal condition is violated. As laser photons are amplified in the laser medium through stimulated emission, at the same time electrons in the excited energy state are stimulated back down to the lower-energy state. The intracavity field extracts energy that was stored in the gain medium by the pump and reduces the number of excited-state electrons available for amplification. As a result, the gain is reduced. The actual gain that results, known as the *saturated gain*, depends on the properties of the laser medium, the pump level, and the intracavity laser intensity. Note that saturation phenomena are quite familiar in the context of electronic amplifiers, where the full amplifier gain is available only for input signals below a certain voltage level; for higher voltage levels the output may appear to be clipped.

To clarify the role of gain saturation in lasers, let us imagine that the pump intensity is increased slowly from zero in a single-mode laser cavity. As long as the small-signal gain remains below the loss, the laser intensity is zero. When the pump intensity is sufficient to raise the small-signal gain to equal the loss exactly, the laser reaches *threshold*. The laser power is still zero at this point. When the pump is increased above threshold, the laser power increases, and this saturates the gain. For a given pump level, the laser intensity builds up just enough to maintain the saturated gain at exactly the loss level, as pictured in Fig. 1.3. Thus, pump power above the threshold value is converted into stimulated emission. As a result, the laser intensity increases linearly with pump power above threshold.

In our discussion we have implicitly assumed that the gain medium is broadened homogeneously. This means that all the excited-state electrons have identical gain spectra, so that the overall gain is simply the gain spectrum per electron times the population difference between upper and lower laser states. Gain saturation in a homogeneously broadened medium results from a decrease in this population difference, and therefore the saturated gain has the same spectral shape as the small-signal gain.

Inhomogeneously broadened media and inhomogeneously broadened lasers, in which different excited-state electrons have different spectra, are also possible. This may result, for example, due to the fact that different impurity ions in a glass laser experience different local environments. As shown in Fig. 1.4a, the overall absorption or gain spectrum is then determined by both the individual homogeneous lines (associated with electrons with identical resonance frequencies) and the inhomogeneous distribution function $G(\omega - \Omega_0)$. When the inhomogeneous distribution is wide compared to the homogeneous linewidth, the net absorption or gain spectrum is determined mainly by $G(\omega - \Omega_0)$ and peaks at the center of the inhomogeneous distribution Ω_0. As shown in Fig. 1.4b, the saturation behavior of an inhomogeneously broadened medium interacting with a narrow-bandwidth laser field is quite different than saturation behavior for homogeneous broadening. In particular, saturation induces a "spectral hole" in the vicinity of the laser field, while the spectrum is essentially unchanged for frequencies much more than a homogeneous linewidth away from the laser frequency. Because inhomogeneously broadened lasers are not in common use for ultrashort

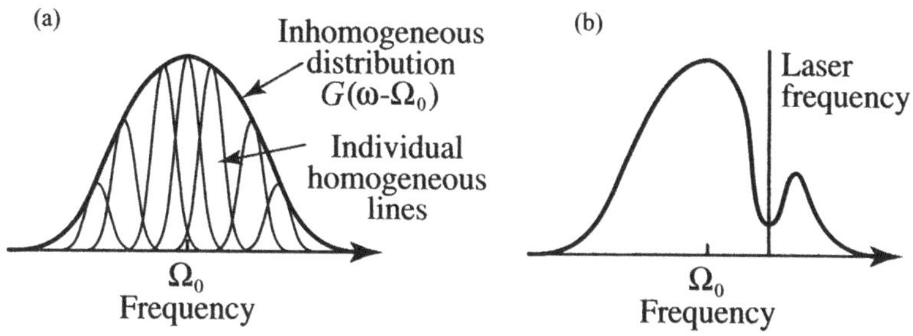

Figure 1.4 (a) Absorption or gain spectrum of a medium with an inhomogeneous distribution $G(\omega - \Omega_0)$ of transition frequencies; (b) spectral hole burning associated with saturation of an inhomogeneously broadened medium by a narrowband laser field.

pulse generation, we do not discuss them further at this time. However, we return to this topic in Chapter 9, where we discuss line-shape theory and ultrafast spectroscopy techniques for probing physics associated with homogeneous and inhomogeneous broadening.

1.3.2 Gain and Gain Saturation in Four-Level Atoms

To provide further insight, we discuss gain saturation for the four-level atom. The four-level atom approximation, sketched in Fig. 1.5, is commonly used to model important mode-locking media such as Ti:sapphire or dye molecules. Here we discuss continuous-wave (CW) saturation; saturation in response to pulses is covered in Chapter 2. Note that we do not mean the word *atom* in the term *four-level atom* literally. Rather, this word refers to whatever entity (molecule, impurity complex, etc.) is active in the laser gain process.

Referring to Fig. 1.5, electrons are promoted from the lowest state, denoted level 1, up to level 2 (e.g., via optical pumping). Electrons in level 2 are usually assumed to relax very rapidly to level 3, so that the population of electrons in level 2 remains close to zero. Physically, this relaxation usually arises from vibronic rearrangement of the nuclei within the crystal-impurity or molecular system, which typically occurs on a subpicosecond time scale following the electronic transition from 1 to 2. The transition from 3 to 4 is the lasing transition, and levels 3 and 4 are the upper and lower laser levels, respectively. Electrons from level 3 can undergo stimulated emission down to level 4, giving up a photon to the laser field in the process, or they can also relax spontaneously down to level 4 with rate τ_G^{-1}, in which case their energy is not available to the laser field. τ_G is the energy storage time of the gain medium, which may range from nanoseconds to milliseconds in materials used for femtosecond pulse generation. Electrons in level 4 are again assumed to relax very rapidly

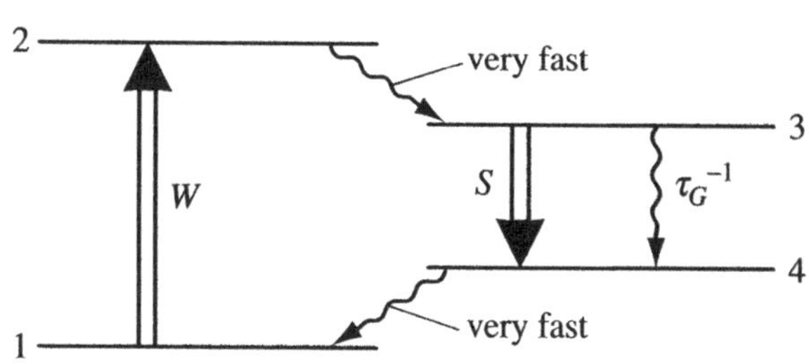

Figure 1.5 Energy-level structure for a four-level atom.

back to level 1, so that the lower laser level remains nearly empty (this is the principal advantage of four-level atoms).

Mathematically, we denote the total density of atoms (in m^{-3}) as N_G, and the number density in each of the four levels as N_1, N_2, N_3, and N_4, respectively. Naturally, we have

$$N_1 + N_2 + N_3 + N_4 = N_G \tag{1.44}$$

The level populations are described by the following rate equations:

$$\dot{N}_1 = -W(N_1 - N_2) + \frac{N_4}{\tau_{41}} \tag{1.45}$$

$$\dot{N}_2 = W(N_1 - N_2) - \frac{N_2}{\tau_{23}} \tag{1.46}$$

$$\dot{N}_3 = -S(N_3 - N_4) - \frac{N_3}{\tau_G} + \frac{N_2}{\tau_{23}} \tag{1.47}$$

$$\dot{N}_4 = S(N_3 - N_4) - \frac{N_4}{\tau_{41}} + \frac{N_3}{\tau_G} \tag{1.48}$$

Here W is the pumping rate per atom (in s^{-1}) from level 1 to 2, and its strength is controlled via the external excitation of the laser medium. For example, in the case of optical pumping, W is proportional to the intensity of the pump laser. S is the stimulated emission rate per atom (in s^{-1}), which is proportional to intensity and given in physical units by

$$S = \frac{\sigma_{34} I\,(\omega_{34})}{\hbar\omega_{34}} \tag{1.49}$$

Here σ_{34} is the "cross section" of the laser transition, which characterizes the strength of the laser–matter interaction, $I\,(\omega_{34})$ is the laser intensity inside the resonator, and $\hbar\omega_{34}$ gives the photon energy of the $3 \rightarrow 4$ transition. In the case of optical pumping, a similar expression applies for W. τ_{23} and τ_{41} are the $2 \rightarrow 3$ and $4 \rightarrow 1$ relaxation times, respectively. The gain is proportional to the population difference between the laser levels:

$$g = \frac{\sigma_{34}}{2}\,(N_3 - N_4)\,l_g \tag{1.50}$$

The factor of $1/2$ arises because we have written the gain coefficient for the *field* in eq. (1.50); this factor is not present when writing the gain coefficient for the *intensity*. In steady state all the time derivatives in the rate equations are set to zero. Furthermore, since the relaxation from levels 2 and 4 is very fast, we can also approximate $N_1 - N_2 = N_1$ and $N_3 - N_4 = N_3$ in eqs. (1.45 to 1.48). Solving for the upper-laser-level population under these conditions yields the following expression for the gain:

$$g = \left(\frac{1}{2}\right) \frac{\sigma_{34} W N_G \tau_G l_g}{1 + (W + S)\,\tau_G} \tag{1.51}$$

The gain is a function of both W and S. In the small-signal regime ($S = 0$), the small-signal gain g_0 is given by

$$g_0 = \left(\frac{1}{2}\right) \frac{\sigma_{34} W N_G \tau_G l_g}{1 + W\tau_G} \tag{1.52}$$

g_0 starts at zero, then increases linearly with pump rate W at first but saturates for large W when N_3 approaches N_G. In the large-signal regime (S is finite), the gain can be rewritten

$$g = \frac{g_0}{1 + S/S_{\text{sat}}} \tag{1.53}$$

where

$$S_{\text{sat}} = W + \frac{1}{\tau_G} \tag{1.54}$$

is a saturation parameter. Within a laser, the gain first increases with increasing pump until threshold is reached. Above threshold the small-signal gain g_0 continues to increase as W increases, but the actual gain g is clamped or saturated at the value needed for threshold, which we denote g_{th}. The laser intensity (inside the laser) is found by setting eq. (1.53) to g_{th}, with the result

$$S = S_{\text{sat}} \left(\frac{g_0}{g_{\text{th}}} - 1 \right) \tag{1.55}$$

Lasing can occur in three-level systems as well as in the four-level systems discussed above. For further discussion, see standard laser texts.

1.3.3 Gaussian Beams and Transverse Laser Modes

Real laser beams have a finite transverse extent; they are not plane waves. In most cases of interest to us, however, laser beams may be considered paraxial. This means that they are made up of a superposition of plane waves with propagation vectors close to a single direction (which we take as z). Equivalently, the variation of the field in the transverse (x–y) direction must be much weaker than in the z direction. Under these conditions the electric field vector still lies mainly in the x–y plane. In the following we give a brief summary of paraxial laser beams and the related transverse mode structure of lasers. More detail can be found in most texts on lasers (e.g., [7–9]).

For a monochromatic, paraxial wave it is useful to write

$$E(z, t) = \text{Re}\left\{ \tilde{E}_0 u(x, y, z) e^{j(\omega t - kz)} \right\} \tag{1.56}$$

where k is given by eq. (1.22) and E refers to a single transverse polarization component. In this form the most rapid wavelength-scale variation of the field is carried by the e^{-jkz}

term; $u(x, y, z)$ is a slowly varying envelope. Substitution into the wave equation, eq. (1.15), yields

$$\nabla_T^2 u + \frac{\partial^2 u}{\partial z^2} - 2jk\frac{\partial u}{\partial z} = 0 \tag{1.57}$$

where $\nabla_T^2 = \partial^2/\partial x^2 + \partial^2/\partial y^2$. Since u is assumed to be slowly varying, $\left|\partial^2 u/\partial z^2\right| \ll 2k\,|\partial u/\partial z|$, and therefore $\partial^2 u/\partial z^2$ can be neglected. The resulting paraxial wave equation is

$$\nabla_T^2 u - 2jk\frac{\partial u}{\partial z} = 0 \tag{1.58}$$

The paraxial wave equation has Gaussian beam solutions that provide a good description of laser beams both inside and outside the laser cavity. A particularly useful solution is written as follows:

$$u_{00}(x, y, z) = \frac{w_0}{w(z)} e^{-(x^2+y^2)/w^2(z)} e^{-jk(x^2+y^2)/2R(z)} e^{j\phi(z)} \tag{1.59}$$

where we define

$$w^2(z) = w_0^2\left[1 + \left(\frac{z}{z_0}\right)^2\right] \tag{1.60a}$$

$$\frac{1}{R(z)} = \frac{z}{z^2 + z_0^2} \tag{1.60b}$$

$$\phi(z) = \tan^{-1}\left(\frac{z}{z_0}\right) \tag{1.60c}$$

$$z_0 = \frac{\pi w_0^2 n}{\lambda} \tag{1.60d}$$

This solution, sketched in Fig. 1.6, describes a beam that comes to a focus at $z = 0$ with a radius w_0 (measured at $1/e$ of the on-axis field). The beam radius $w(z)$ changes only slowly within the region $|z| < z_0$; far outside this region the beam spreads with a half angle equal to $\lambda/\pi w_0 n$. The depth b over which the beam radius remains less than $\sqrt{2}\,w_0$ is given by $b = 2z_0$, where b is termed the depth of focus or the confocal parameter. $R(z)$ is the radius of curvature of the phase fronts. $R = \infty$ at the beam waist ($z = 0$), corresponding to a planar phase front; for $|z| \gg z_0$, the phase fronts become spherical with a center of curvature at $z = 0$. Equation (1.60c) indicates a phase shift that accompanies propagation through the focal region. This Gaussian beam solution describes both laser beam propagation in free space as well as the spatial behavior of the fundamental transverse modes that are allowed inside a laser resonator with flat or spherical mirrors. The particular transverse mode of this

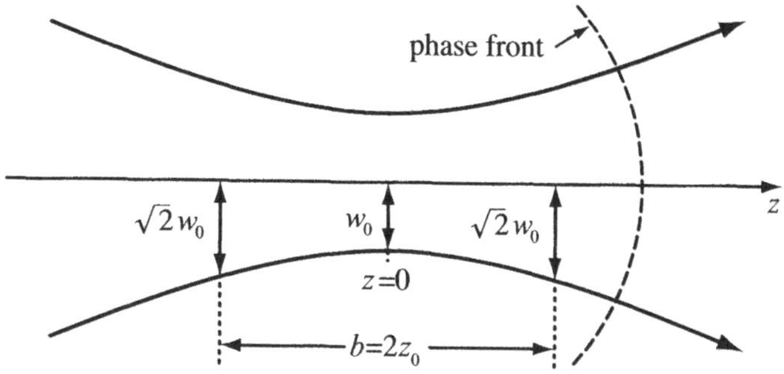

Figure 1.6 Gaussian beam solution of the paraxial wave equation, showing evolution of the beam radius and representative phase front.

type allowed within a given laser resonator is selected by the requirement that $R(z)$ must match the radii of curvature of the mirrors within the laser, which determines the location of the beam waist and the value of w_0; for free propagation outside a laser, any solution of this type is allowed.

It is useful to know the power P carried by a Gaussian beam, which is obtained by integrating the intensity over its cross-sectional area. The result is

$$\begin{aligned} P &= \int dA \, \frac{\varepsilon_0 nc}{2} \left|\tilde{E}_0\right|^2 |u|^2 = \frac{\varepsilon_0 nc}{2} \left|\tilde{E}_0\right|^2 \frac{w_0^2}{w^2(z)} \int_0^\infty 2\pi r \, dr \, e^{-2r^2/w^2(z)} \\ &= \frac{1}{2}\varepsilon_0 nc \left|\tilde{E}_0\right|^2 \frac{\pi w_0^2}{2} \end{aligned} \tag{1.61}$$

The power is equal to the on-axis intensity at the focus, multiplied by the area of a circle of radius $w_0/\sqrt{2}$, which is the radius of the intensity at the e^{-1} point.

Equation (1.59) is the lowest-order member of an entire family of Hermite–Gaussian solutions to the paraxial wave equation, given by

$$\begin{aligned} u_{mn}(x, y, z) \sim & \frac{1}{w(z)} H_m\left(\frac{\sqrt{2}\,x}{w(z)}\right) H_n\left(\frac{\sqrt{2}\,y}{w(z)}\right) \\ & \times e^{-(x^2+y^2)/w^2(z)} e^{-jk(x^2+y^2)/2R(z)} e^{j(m+n+1)\phi(z)} \end{aligned} \tag{1.62}$$

The $H_\nu(\xi)$ with ν a nonnegative integer are Hermite polynomials [10], where for example, $H_0(\xi) = 1$, $H_1(\xi) = 2\xi$, and $H_2(\xi) = 4\xi^2 - 2$. In general, the H_ν are polynomials of order ν, which become wider and exhibit a greater number of zero crossings as ν increases. The expressions for $w(z)$, $R(z)$, and $\phi(z)$ are still as given above. Although the fundamental Gaussian beam is by far the most important in most ultrafast optics applications, the higher order ($m \neq 0$ or $n \neq 0$) Hermite–Gaussian solutions will be relevant in our study for several reasons. For example, all the u_{mn} in eq. (1.62) are allowed transverse modes of laser resonators, termed TEM$_{mn}$ *modes*, although the fundamental TEM$_{00}$ mode usually has the lower loss and is therefore favored. We shall see however that higher-order modes were significant in the discovery of the important Kerr lens mode-locking technique for short pulse generation with solid-state lasers. Additionally, the Hermite–Gaussians form a complete orthogonal basis set, which means that an arbitrary spatial beam profile can be decomposed into a superposition of these functions.

To calculate the propagation of Gaussian beams through linear optical systems, it is convenient to introduce a complex q parameter, such that

$$\frac{1}{q(z)} = \frac{1}{R(z)} - \frac{j\lambda}{\pi w^2(z)n} \tag{1.63}$$

The effect of the optical system can then be characterized by a bilinear transformation of q: namely,

$$q_{\text{out}} = \frac{Aq_{\text{in}} + B}{Cq_{\text{in}} + D} \tag{1.64}$$

The coefficients appearing in eq. (1.64) are usually written in matrix form. The matrices for a few common systems are as follows:

- A length d of a homogeneous medium:

$$\begin{pmatrix} A & B \\ C & D \end{pmatrix} = \begin{pmatrix} 1 & d \\ 0 & 1 \end{pmatrix} \tag{1.65a}$$

- A lens with focal length f (this matrix also applies to a spherical mirror of radius R if we set $f = R/2$):

$$\begin{pmatrix} A & B \\ C & D \end{pmatrix} = \begin{pmatrix} 1 & 0 \\ -1/f & 1 \end{pmatrix} \tag{1.65b}$$

- A planar interface, perpendicular to the propagation direction, from a medium with index n_1 to a medium with index n_2:

$$\begin{pmatrix} A & B \\ C & D \end{pmatrix} = \begin{pmatrix} 1 & 0 \\ 0 & n_1/n_2 \end{pmatrix} \tag{1.65c}$$

An important property of such matrices is that the effect of cascaded systems is computed via matrix multiplication. If $M_1, M_2, \ldots, M_N$ represent the matrices for N cascaded systems, with the beam entering system 1 first and system N last, the effect of the cascaded system is given by $M_N \cdots M_2M_1$. For example, the matrix for a homogeneous slab of thickness d and index n and surrounded on either side by free space is

$$\begin{pmatrix} A & B \\ C & D \end{pmatrix} = \begin{pmatrix} 1 & 0 \\ 0 & n \end{pmatrix} \begin{pmatrix} 1 & d \\ 0 & 1 \end{pmatrix} \begin{pmatrix} 1 & 0 \\ 0 & 1/n \end{pmatrix} = \begin{pmatrix} 1 & d/n \\ 0 & 1 \end{pmatrix} \tag{1.66}$$

The matrices given above are also useful in the ray optics description of optical systems. Here one writes

$$\begin{pmatrix} x_{\text{out}} \\ x'_{\text{out}} \end{pmatrix} = \begin{pmatrix} A & B \\ C & D \end{pmatrix} \begin{pmatrix} x_{\text{in}} \\ x'_{\text{in}} \end{pmatrix} \tag{1.67}$$

where x_{in} and x'_{in} are the position and slope of the incoming ray, respectively, and similarly for the outgoing ray. For a given optical system, the matrixes for the ray optics description and the Gaussian beam (q parameter) description are identical. The ray optics description is sometimes helpful in identifying or interpreting the *ABCD* matrices. For example, the matrix in eq. (1.66) says that the slab of physical thickness d has an effective thickness of d/n. From the ray optics description, it is clear that this occurs due to bending of the rays toward the normal as they enter the medium.

The q parameter and the *ABCD* matrices are useful in solving for the transverse mode of laser resonators. Assume that we have calculated the matrix for one complete round trip through the resonator, starting, for example, at a particular mirror $\mathcal{M}$. Then one condition for an acceptable transverse mode is that the field pattern must exactly reproduce itself after one round trip, which can be formulated as

$$q = \frac{Aq + B}{Cq + D} \tag{1.68}$$

A further condition is that for a stable mode, the beam radius must be finite, which means that q must have a nonzero imaginary part. From the ray optics perspective, this stability condition means that incoming rays remain confined after many passes through the system. Multiplying through by the denominator in eq. (1.68) yields a quadratic polynomial equation for q. The root of this equation with a negative imaginary part gives q at mirror $\mathcal{M}$, and q in turn yields the mode size within the laser resonator. Not all resonators have stable solutions. A well-known and important example is the simple two-mirror cavity (Fig. 1.7), consisting of mirrors with radii of curvature R_1 and R_2 which are separated by distance d. After calculating the round-trip *ABCD* matrix and requiring that the roots of eq. (1.68) have nonzero imaginary part, one obtains the stability relation

$$0 < \left(1 - \frac{d}{R_1}\right)\left(1 - \frac{d}{R_2}\right) < 1 \tag{1.69a}$$

or simply

$$0 < g_1 g_2 < 1 \tag{1.69b}$$

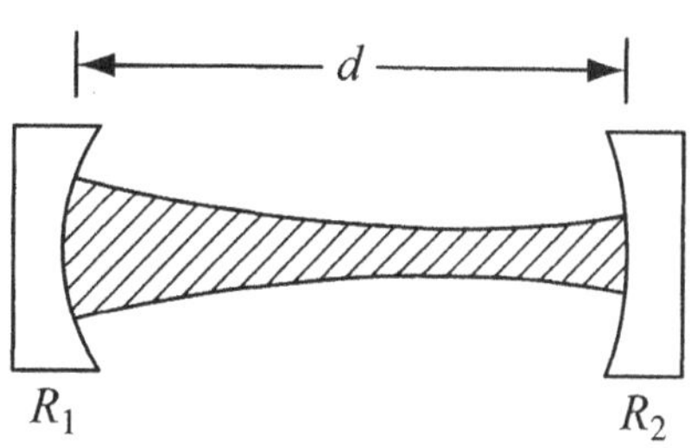

Figure 1.7 Simple two-mirror optical resonator.

where $g_1 = 1 - d/R_1$ and $g_2 = 1 - d/R_2$ are symbols often used to express the stability condition succinctly. We make use of this stability condition when we discuss Kerr lens mode-locking in Chapter 2.

1.4 INTRODUCTION TO ULTRASHORT PULSE GENERATION THROUGH MODE-LOCKING

The single-mode laser discussed in Section 1.3 can have a very narrow optical frequency spectrum. This results in the well-known monochromaticity property of lasers. However, short pulse generation requires a broad optical bandwidth and hence multiple longitudinal mode operation. Even with the bandwidth-limiting filter removed, however, the saturated gain in a homogeneously broadened laser is below the loss level except for those modes near the peak of the gain spectrum, and this also limits the number of oscillating modes. Methods for forcing a great number of modes to oscillate to obtain broad bandwidths and ultrashort pulses are covered in detail later. In the following we analyze the time-domain characteristics of a laser which we assume already has multimode operation.

For a multimode laser we can write the electric field as follows:

$$e(z,t) = \mathrm{Re}\left\{\sum_m E_m e^{j(\omega_m t - k_m z + \phi_m)}\right\} \tag{1.70a}$$

where

$$\omega_m = \omega_0 + m\,\Delta\omega = \omega_0 + \frac{2m\pi c}{L} \tag{1.70b}$$

and

$$k_m = \frac{\omega_m}{c} \tag{1.70c}$$

Henceforth we use lowercase letters [e.g., $e(z, t)$] for the time-domain representation of the field and capital letters (e.g., E_m) to refer to the frequency domain. Equation (1.70a) most accurately models a unidirectional ring laser, where the z coordinate refers to travel around the ring in the direction of laser oscillation, although the same essential results will also hold for a linear (Fabry–Perot) laser. The round-trip distance around the ring is denoted L (for a Fabry–Perot laser we would use $L = 2l$, where l is the Fabry–Perot mirror separation). In this formulation we are assuming that the laser is oscillating in a single transverse mode, so that the spatial profile may be dropped.

Equation (1.70a) can be rewritten as follows:

$$e(z,t) = \mathrm{Re}\left\{e^{j\omega_0(t-z/c)}\sum_m E_m e^{j[m\Delta\omega(t-z/c)+\phi_m]}\right\} \tag{1.71a}$$

$$= \mathrm{Re}\left\{a\left(t - \frac{z}{c}\right) e^{j\omega_0(t-z/c)}\right\} \tag{1.71b}$$

where

$$a\left(t-\frac{z}{c}\right)=\sum_{m} E_m e^{j[m\Delta\omega(t-z/c)+\phi_m]} \tag{1.71c}$$

Thus, the electric field is the product of a complex envelope function $a(t-z/c)$ with the optical carrier $e^{j\omega_0(t-z/c)}$.

This terminology is convenient because in many cases the carrier term, which oscillates with a period of just a few femtoseconds (for visible wavelengths), varies much more rapidly than the envelope function. (However, for the shortest pulses available today, comprising just a few optical cycles, this distinction is blurred.) Both carrier and envelope functions travel around the laser cavity at the speed of light. Furthermore, for a specific location in the cavity (z fixed), the envelope function is periodic with period

$$T=\frac{2\pi}{\Delta\omega}=\frac{2l}{c}=\frac{L}{c} \tag{1.72}$$

This corresponds to the time required for light to make one round trip around the resonator.

If we are to say anything further about the shape of the envelope function, we need to specify the mode amplitudes E_m and phases ϕ_m. If we assume that there are N oscillating modes all with equal amplitudes E_0 and with the phases identically zero, eq. (1.71a) becomes easy to evaluate. The process (discussed later) by which the modes are held with fixed relative phases is known as *mode-locking*, and as we shall see, having all the phases equal is a particularly useful form of mode-locking. Equation (1.71a) now becomes

$$e(z,t)=\mathrm{Re}\left\{E_0 e^{j\omega_0(t-z/c)}\sum_{-(N-1)/2}^{(N-1)/2} e^{j[m\Delta\omega(t-z/c)]}\right\} \tag{1.73}$$

To evaluate this we substitute $m'=m+(N-1)/2$ and use the summation formula

$$\sum_{m=0}^{q-1} a^m=\frac{1-a^q}{1-a} \tag{1.74}$$

Upon some further simplification, the result is

$$e(z,t)=\mathrm{Re}\left\{E_0 e^{j\omega_0(t-z/c)}\frac{\sin(N\,\Delta\omega/2)(t-z/c)}{\sin(\Delta\omega/2)(t-z/c)}\right\} \tag{1.75}$$

The laser intensity (I), averaged over an optical cycle, is proportional to $|e(z,t)|^2$. At a specific cavity location, say $z=0$, we have

$$I(t)\sim|E_0|^2\,\frac{\sin^2(N\,\Delta\omega t/2)}{\sin^2(\Delta\omega t/2)}. \tag{1.76}$$

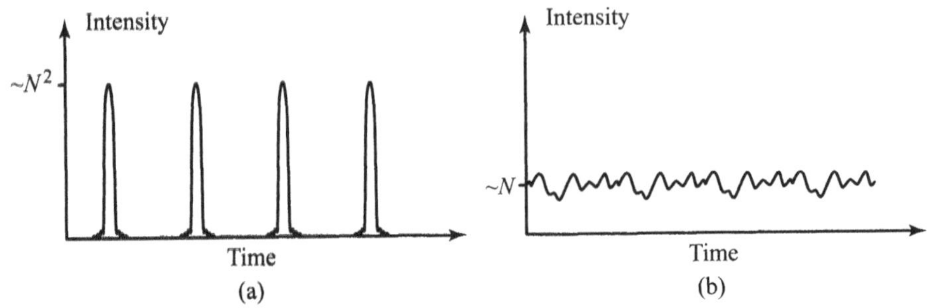

Figure 1.8 (a) Mode-locked laser output with constant mode phase; (b) laser output with randomly phased modes.

The resulting intensity profile is sketched in Fig. 1.8a. Some of the key points are as follows:

- The output consists of a periodic series of short pulses, with period $T = 1/\Delta f = L/c$.
- The pulse duration is approximately $\Delta t = 2\pi/N\,\Delta\omega = 1/N\,\Delta f$. Thus, the pulse duration is equal to the periodicity divided by the number of modes. Equivalently, the pulse width is equal to the inverse of the total laser bandwidth.
- The peak intensity is proportional to $N^2\,|E_0|^2$. In comparison, the average intensity (averaged over the pulse period), given by the number of modes times the power per mode, is proportional to $N\,|E_0|^2$. Thus, the peak intensity of a mode-locked pulse is enhanced by a factor N.

A very different situation arises if we assume that the mode phases are random, as in a garden-variety multimode laser. In this case the intensity profile takes on a random appearance, as shown in Fig. 1.8b. Key points are as follows:

- The intensity fluctuates randomly about the average intensity value. The average intensity is $\sim N\,|E_0|^2$, which is the same as in the mode-locked case, although the peak intensity is a factor of N lower.
- The time scale over which the fluctuations vary (also known as the *correlation time*) is still $\Delta t \approx 1/N\,\Delta f$.
- The intensity fluctuations are still periodic with period $T = 1/\Delta f$.

In addition, in the garden-variety multimode laser, the random phases are likely to vary slowly with time as well. This introduces yet an additional degree of randomness to the multimode laser output. To avoid such fluctuations and obtain a well-defined output, one can either utilize single-mode lasers with narrow spectra, as in Section 1.3, or mode-locked lasers with narrow pulses and broad spectra.

It is also worth noting that the carrier frequency ω_0 is left arbitrary in this very simple treatment. Indeed, in many mode-locked lasers, the carrier frequency is allowed to drift. On the other hand, for some purposes it is important to stabilize and know the carrier frequency precisely. These issues are discussed in Section 7.5.

1.5 FOURIER SERIES AND FOURIER TRANSFORMS

We shall have many occasions where we wish to describe pulses in terms of a frequency-domain representation. We encountered one example in Section 1.4, where we wrote a mode-locked periodic pulse train as a summation over the longitudinal cavity modes. In this section we briefly review Fourier series and Fourier transforms, which are the mathematical tools used to convert between time- and frequency-domain representations. For more detail, see texts such as [11,12]. The key point is that any time-dependent signal can be written as a superposition or summation of sines and cosines (or complex exponentials) with different frequencies.

1.5.1 Analytical Aspects

If a function $f(t)$ is *periodic* with period T, we can write it as a Fourier series:

$$f(t) = \sum_{n=-\infty}^{\infty} F_n e^{jn\Delta\omega t} \tag{1.77}$$

where $\Delta\omega = 2\pi/T$ is the [angular] frequency. The F_n are the Fourier amplitudes or Fourier coefficients and can be obtained from the time-domain signal $f(t)$ as follows[3]:

$$F_n = \frac{1}{T}\int_0^T f(t)e^{-jn\Delta\omega t}\,dt \tag{1.78}$$

In the case of an aperiodic time-domain function, we can consider that the period T goes to ∞. We then have the frequency spacing $\Delta\omega \to 0$, and therefore we replace the discrete variable $n\,\Delta\omega$ with a continuous variable ω. This results in the Fourier transform:

$$f(t) = \frac{1}{2\pi}\int_{-\infty}^{\infty} F(\omega)e^{j\omega t}\,d\omega \tag{1.79a}$$

and

$$F(\omega) = \int_{-\infty}^{\infty} f(t)e^{-j\omega t}\,dt \tag{1.79b}$$

$F(\omega)$ is known as the Fourier transform of $f(t)$, and $f(t)$ is obtained by performing the inverse Fourier transform of $F(\omega)$.

We now review a few useful properties of Fourier transforms, most of which we state without proof. For all of the following we assume that $f(t)$ and $F(\omega)$ are a Fourier transform pair.

- *Reality condition.* If $f(t)$ is a real function, which it will be whenever it represents an actual physical observable, $F(-\omega) = F^*(\omega)$.

[3] Instead of integrating from 0 to T, one can get the same results by integrating over any other interval of duration T (e.g., $-T/2$ to $T/2$).

- *Scaling formula:*

$$\text{If } h(t) = f(at), \quad \text{then } H(\omega) = \frac{1}{a}F\left(\frac{\omega}{a}\right) \tag{1.80}$$

- *Time-delay formula:*

$$\text{If } h(t) = f(t-\tau), \quad \text{then } H(\omega) = F(\omega)e^{-j\omega\tau} \tag{1.81}$$

- *Frequency-offset formula:*

$$\text{If } h(t) = f(t)e^{j\omega_0 t}, \quad \text{then } H(\omega) = F(\omega - \omega_0) \tag{1.82}$$

- *Convolution formula.* The convolution of two functions $f(t)$ and $g(t)$ is denoted $f(t) * g(t)$ and is defined by

$$f(t) * g(t) = \int dt'\, f(t')g(t-t')$$

  The convolution formula states that

$$\text{if } h(t) = f(t) * g(t) \qquad \text{then } H(\omega) = F(\omega)G(\omega) \tag{1.83}$$

- *Parseval's theorem.* We know that the intensity $I(t)$ of a pulse whose electric field profile is $f(t)$ is proportional to $|f(t)|^2 = f(t)f^*(t)$. Recall that the units of intensity are W/m^2. The pulse energy is given by the time-integrated intensity integrated over the cross-sectional area. Parseval's theorem says that

$$\int f(t)f^*(t)\,dt = \frac{1}{2\pi}\int F(\omega)F^*(\omega)\,d\omega \tag{1.84}$$

  Therefore, the time-integrated intensity is equal to the frequency-integrated intensity, except for a multiplicative factor. For this reason the quantity $|F(\omega)|^2 = F(\omega)F^*(\omega)$ is called the *power spectral density.*

The use of these formulas will be illustrated as they are needed. We do provide a few examples of the Fourier transform:

1. $f(t) = \delta(t)$, where $\delta(t)$ is the delta function or unit impulse function. Recall that $\delta(t) = 0$ for $t \neq 0$, $\delta(t) = \infty$ for $t = 0$, and $\int_{-\infty}^{\infty} dt\, \delta(t) = 1$. By substitution into eq. (1.79b), we find that $F(\omega) = 1$.
2. If $f(t) = e^{j\omega_0 t}$, then $F(\omega) = 2\pi\delta(\omega - \omega_0)$. This is verified by plugging into eq. (1.79a).

3. $f(t)$ is a Gaussian, $f(t) = e^{-t^2/t_p^2}$. Then

$$F(\omega) = \int dt\, e^{-t^2/t_p^2}\, e^{-j\omega t} = \int dt\, e^{-\left[(t+j\omega t_p^2/2)^2/t_p^2\right]}\, e^{-\omega^2 t_p^2/4}$$

Using the substitution $u = (t + j\omega t_p^2/2)/t_p$ and $\int_{-\infty}^{\infty} du\, e^{-u^2} = \sqrt{\pi}$, we obtain

$$F(\omega) = t_p\sqrt{\pi}\, e^{-\omega^2 t_p^2/4}$$

The procedure we have used above, called *completing the square*, is very useful for evaluating Fourier transforms of Gaussian functions.

Finally, we frequently write a pulse in terms of a slowly varying envelope function times an optical carrier term:

$$e(t) = \mathrm{Re}\left\{a(t)\; e^{j\omega_0 t}\right\} = \frac{1}{2}\left[a(t)\, e^{j\omega_0 t} + a^*(t)\, e^{-j\omega_0 t}\right] \tag{1.85}$$

The spectrum $E(\omega)$ is then given by

$$E(\omega) = \frac{1}{2}\left[A(\omega - \omega_0) + A^*(-\omega - \omega_0)\right] \tag{1.86a}$$

where

$$A(\omega) = \int_{-\infty}^{\infty} a(t) e^{-j\omega t}\, dt \tag{1.86b}$$

is the Fourier transform of the envelope function $a(t)$. Equation (1.86a) is sketched in Fig. 1.9. Separating the field into envelope and carrier terms is most useful when (as in Fig. 1.9) the bandwidth of $A(\omega)$ is much less than the carrier frequency ω_0. In the time domain, this means that the envelope is much longer than the optical cycle.

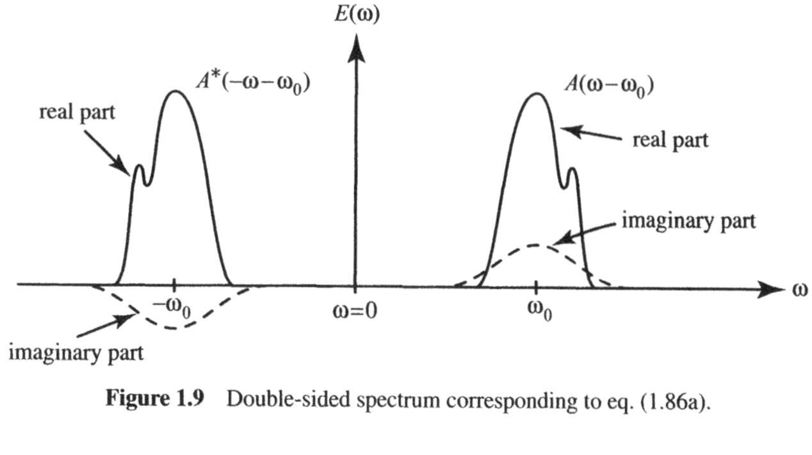

Figure 1.9 Double-sided spectrum corresponding to eq. (1.86a).

The pulse $e(t)$ can be obtained directly by performing the inverse Fourier transform of the double-sided spectrum described by eq. (1.86a) or from the single-sided spectrum using the formula

$$e(t) = \mathrm{Re}\left\{\frac{1}{2\pi}\int A(\omega - \omega_0)\, e^{j\omega t}\, d\omega\right\} \tag{1.87}$$

1.5.2 Computational Aspects

Fourier transform methods are frequently used for computations involving ultrashort pulses. Accordingly, we introduce briefly some relevant points relating to numerical computations of Fourier transforms. For detailed discussion, see [13].

Usually computations involving a continuous-time function $f(t)$ involve only samples of $f(t)$ at a set of discrete times t_n. Let us assume that we have N evenly spaced samples with time spacing Δ, so that $t_n = n\Delta$. We designate the sample values as $f_n = f(t_n)$ for $n = 0, 1, \ldots, N-1$. The Fourier transform of $f(t)$ is then computed numerically as

$$F_k = \sum_{n=0}^{N-1} f_n e^{-2\pi jnk/N} \tag{1.88}$$

Equation (1.88) is formally called the *discrete Fourier transform*. The inverse discrete Fourier transform is given by

$$f_n = \frac{1}{N}\sum_{k=0}^{N-1} F_k e^{2\pi jnk/N} \tag{1.89}$$

Indices n and k take on integer values in the range $0, 1, \ldots, N-1$.

Discrete Fourier transform routines are widely available in numerical computing software. Practically, it is preferred to choose the vector length N to be a power of 2. This allows the numerical package to evaluate the discrete Fourier transform using the fast Fourier transform (FFT) algorithm, which dramatically reduces computing time. Although the relationship to the continuous-time Fourier transform may appear to be straightforward, several points must be kept in mind if one is to obtain sensible computational results. These points include the following:

- Although the f_n vector contains a sequence of samples of the time-domain function $f(t)$, it does not explicitly provide information on the sampling times t_n or the sampling interval. Similarly, the vector of Fourier amplitudes F_k does not explicitly stipulate the frequencies at which the spectrum is sampled. Hence, time and frequency vectors must be tracked by the user as auxiliary information. Note, however, that time and frequency vectors are related. If the sampling interval is known to be Δ, the time vector represents a total time span $N\Delta$; the frequency vector ν_k represents a total frequency span $1/\Delta$, with $\nu_k = k/N\Delta$ and spacing $1/N\Delta$ between frequency samples. Here frequency refers to $\nu = \omega/2\pi$ and is measured in hertz (as opposed to angular frequency ω measured in rad/s).

- Inspection of eqs. (1.88) and (1.89) reveals that if indices n and k were allowed to extend beyond the stipulated range, f_n and F_k would be periodic sequences with period N. f_n and F_k for n and k in the range $0, 1, \ldots, N-1$ would each then represent one period of the corresponding periodic sequences. For signals that are time-limited to duration less than $N\Delta$, or frequency-limited to bandwidth less than $1/\Delta$, meaning that the signal is nonzero only over a range of samples of length less than N, this is of little consequence. However, for signals that are not time- or frequency-limited in this way, signal content appearing at the end of the time (or frequency) vector effectively wraps around to appear at the beginning of the vector.
- Consider a continuous-time signal $f(t)$ that is frequency-limited such that its spectrum is zero for all frequencies outside the region $-B \leq \nu \leq B$. A well-known theorem states that all information about $f(t)$ is retained in its sampled representation provided that the sampling interval satisfies $\Delta \leq 1/2B$. This requirement, called the *Nyquist criterion*, is equivalent to specifying a minimum of two samples within each period of the highest-frequency component of the frequency-limited signal. Conversely, when the Nyquist criterion is not satisfied, information is lost. In the context of our discussion on the discrete Fourier transform, we note that if F_k is frequency-limited to a bandwidth less than $1/\Delta$, we have $2B < 1/\Delta$, and the Nyquist criterion is satisfied. Accordingly, in setting up an F_k vector for a computation, it is usually advisable to ensure that the nonzero values of F_k fit comfortably within the length of the vector. If necessary, this can be accomplished by zero padding the F_k vector, which is equivalent in the time domain to decreasing the sample interval (increasing the sample rate).
- For similar reasons it is usually advisable to make sure that the nonzero values of f_n fit comfortably within the vector length as well. Again this can be accomplished by zero padding if needed.
- Inspection of eqs. (1.88) and (1.89) also shows that time $t = 0$ of the physical continuous-time function $f(t)$ and frequency $\nu = 0$ of the continuous-frequency function $F(\omega)$ correspond to the first point in the f_n and F_k vectors, respectively (i.e., $t_0 = 0$ and $\nu_0 = 0$). For a signal with spectrum centered at $\nu = 0$, the "right" half of the spectrum appears at the beginning of the vector; the "left" half of the spectrum wraps around and appears at the end of the vector. A similar statement is true for a signal centered at $t = 0$ in the time domain.
- If, instead, one inadvertently positions an input spectrum so that it is centered in the frequency vector, then upon performing the inverse transform, one will notice that the expected time-domain signal is multiplied by a mysterious $(-1)^n$ function. That is, data points have alternating signs, which is equivalent to a π phase shift between points. The explanation is that the discrete Fourier transform "interprets" a shift in the positioning of the input spectrum as an actual frequency shift. Centering the input spectrum amounts to a frequency shift of $1/2\Delta$ in real units. This corresponds to a period of 2Δ in time units, equivalent to two sample intervals.
- Similarly, if one inadvertently positions an input time-domain signal so that it is centered in the time vector, then upon performing the forward transform, one will notice that the expected spectrum is multiplied by a $(-1)^k$ function. The explanation here is that a time shift in the input signal gives rise to a sampled linear phase variation in its frequency representation (F_k). For a time shift of $N\Delta/2$, the linear spectral phase amounts to a π phase shift per frequency sample.

PROBLEMS

1.1. As a review of complex notation, prove eq. (1.32).

1.2. Derive expressions for the saturated and small-signal gain of a three-level atom. Your analysis should parallel the analysis given in Section 1.3.2 for a four-level atom, but with level 4 missing. Assume that stimulated emission occurs between levels 3 and 1 and that the population in level 3 relaxes spontaneously to level 1 with time constant τ_G. Plot a family of curves for the saturated gain as a function of W (different curves represent different values of S) for both three- and four-level atoms. Comment on the main differences in gain behavior of three- and four-level atoms and on the implications for laser operation.

1.3. Analyze the gain of a two-level atom. Now there is only one laser field, denoted W. What is the gain as a function of W and τ_G, and how does this differ from three- and four-level atoms? Explain why lasers are never based on two-level atoms.

1.4. In this problem we explore the steady-state laser oscillation condition by analyzing an optical transmission resonator with gain. The setup is similar to Fig. 1.1a, but with a plane-wave field E_{in} with frequency ω incident on mirror r_1 from the left and with r_1 allowed to take on arbitrary reflectivity. The output field emerging from mirror r_2 may be written

$$E_{\text{out}} = E_{\text{in}} t_1 t_2 \sum_{n=0}^{\infty} (r_1 r_2)^n \, e^{(2n+1)g} \, e^{-(2n+1)j\omega l/c}$$

where e^g and l are the single-pass gain and optical path length, respectively, and $t_i = \sqrt{1 - r_i{}^2}$ is a field transmission coefficient.

(a) Assuming zero gain ($g = 0$), work out a formula for the intensity transmission coefficient $|E_{\text{out}}/E_{\text{in}}|^2$. [*Hint:* Equation (1.74) may come in handy.] Assuming a symmetric resonator ($r_1 = r_2$), plot $|E_{\text{out}}/E_{\text{in}}|^2$ as a function of ω for several values of r_1, ranging from 90 to 99%. Comment on key features in your plots, including linewidths and trends with increasing r_1.

(b) Repeat part (a) but now include gain in the resonator. Plot $|E_{\text{out}}/E_{\text{in}}|^2$ vs. ω for various values of g below the laser oscillation condition. Comment on the behavior of your plots as the gain increases toward the oscillation condition. Also explain how the phase condition shows up in your plots.

1.5. Verify by direction substitution that the Gaussian beam given by eq. (1.59) is a solution of the paraxial wave equation.

1.6. A mode-locked laser generates pulses with 10^5 W of peak power. Spatially the laser output is a Gaussian beam. If the beam is focused in air to a diameter of 10 μm (at e^{-2} points of the intensity), give the peak intensity and the corresponding peak electric field amplitude.

1.7. A mode-locked laser usually has a smooth frequency spectrum. Consider a Gaussian spectrum given by

$$A(\tilde{\omega}) = e^{-\tilde{\omega}^2/(\Delta\Omega)^2} \sum_m \delta(\tilde{\omega} - m\,\Delta\omega)\, e^{j\phi_m}$$

where $\tilde{\omega} = \omega - \omega_0$.

(a) Assuming constant spectral phase, $\phi_m = 0$, work out the expression for the time-domain complex envelope function $a(t)$. (*Hint:* The Fourier transform of a periodic train of evenly spaced, equal-amplitude delta functions is a periodic train of evenly spaced, equal-amplitude delta functions. This is shown most easily using Fourier series.)

(b) Use a computer and an FFT algorithm to evaluate the time-domain envelope function $a(t)$ for $\Delta\Omega/2\pi = 10$ GHz, $\Delta\omega/2\pi = 100$ MHz, and $\phi_m = 0$. Plot the spectral amplitude function $A(\tilde{\omega})$ and the temporal intensity and phase. Give the pulse duration (intensity full width at half maximum). Discuss how the setup of your array representing $A(\tilde{\omega})$ determines the number of pulses in the time-domain plot.

(c) Now obtain the phases ϕ_m from a random number generator. Plot two examples of the temporal intensity and phase and comment on all the key differences compared to the uniform phase case.

2

PRINCIPLES OF MODE-LOCKING

2.1 PROCESSES INVOLVED IN MODE-LOCKING

The single-mode lasers reviewed in Chapter 1 were relatively simple and could be adequately modeled including only the laser gain, the cavity loss, and the effects of feedback. However, such lasers do not generate short pulses! To obtain a mode-locked laser generating ultrashort pulses, one must incorporate either active or nonlinear pulse-forming elements (modulators) into the laser cavity. Furthermore, one must also consider new processes affecting the short pulse as it propagates within the laser cavity. The following nonexhaustive list describes some of the processes that can be important in mode-locked lasers:

- *Gain.* Gain is required in mode-locked lasers just as it is in single-mode lasers. As we have seen, gain saturation, in which the gain decreases for increasing laser power, is important in any laser. In single-mode lasers, the gain is sensitive only to the average laser power. In some (not all) mode-locked lasers, however, the gain saturates dynamically on the time scale of the pulse and can be an important mode-locking process.
- *Linear loss.* Linear loss, which is also present in any laser, refers to that portion of the loss that is independent of laser power.
- *Bandwidth limitations.* Passive filtering effects can limit the bandwidth that can be transmitted through the laser cavity. Filtering can arise either from frequency-dependent loss elements or from the finite bandwidth of the laser gain medium. Bandwidth limiting effects are often undesirable in that they limit how short a pulse can be obtained.
- *Dispersion.* This refers to passive, frequency-dependent phase variations encountered by the pulse as it travels through the laser cavity. Dispersion usually leads to pulse broadening and can be an extremely important limitation for femtosecond pulse

Ultrafast Optics, By Andrew M. Weiner

generation. Consequently, dispersion compensation devices are commonly included within femtosecond laser cavities.

- *Active modulation.* This refers to an externally driven optical modulator that modulates either the amplitude or the phase of the circulating pulse. For mode-locking the modulation frequency is usually chosen to coincide with the cavity round-trip time. The modulator may be either an acousto-optic or electro-optic loss or phase modulator, or in some cases the gain element itself is modulated.
- *Saturable loss* (*self-amplitude modulation*). The cavity loss may be a function of pulse intensity or pulse energy. The loss changes dynamically in response to the pulse, which is itself modulated by the dynamically changing loss. This nonlinear process leads to a self-induced modulation with a period that is automatically equal to the cavity round-trip time. Saturable loss elements, also known as *saturable absorbers*, are classified as either fast or slow, depending on how rapidly they recover with respect to the pulse duration.
- *Self-phase modulation.* The phase can also vary nonlinearly with the time-dependent pulse intensity. This leads to a nonlinear self-phase modulation process, which again occurs at the pulse round-trip time. In mode-locked lasers the nonlinear phase shift usually has a very rapid response time and thus depends only on the instantaneous laser intensity, but in some types of lasers slower nonlinear phase processes have also been important. Self-phase modulation can interact in a very interesting and useful way with dispersion, leading to *soliton* pulses which are stable against perturbations. Such solitons are important not only in femtosecond solid-state lasers, but also in optical fiber communications.

Modulation plays a key role in initiating and maintaining mode-locked laser operation. *Active mode-locking* refers to the case where the modulator is externally driven. *Passive mode-locking* refers to situations in which the pulse forms its own modulation through nonlinearities; both amplitude and phase nonlinearities can be important. The shortest pulses have been generated using passively mode-locked lasers. In this chapter we cover the more basic concepts of active and passive mode-locking. More advanced mode-locking concepts are covered in Chapter 7 after we have provided prerequisite information on dispersion and ultrafast nonlinear optics in the necessary depth.

One can imagine many different combinations of the pulse-shaping processes listed above, corresponding to many different flavors of mode-locked lasers. In our treatment we concentrate on a few of the main possibilities that are of historical or current importance. In doing so, we will not only learn about basic mode-locking principles but also learn more about processes affecting ultrafast optical signals and about lasers in general.

2.2 ACTIVE MODE-LOCKING

We begin by discussing active mode-locking, one of the first mode-locking techniques demonstrated and conceptually one of the simplest. As sketched in Fig. 2.1a, mode-locking is achieved with the aid of an externally driven intracavity loss modulator. The modulator is driven at a frequency $\omega_m = 2\pi/T$ corresponding to the longitudinal mode spacing (where T is the round-trip transit time). In the time domain this means that the modulator acts like a periodic shutter which opens once per pulse round-trip time (Fig. 2.1b). We can envision

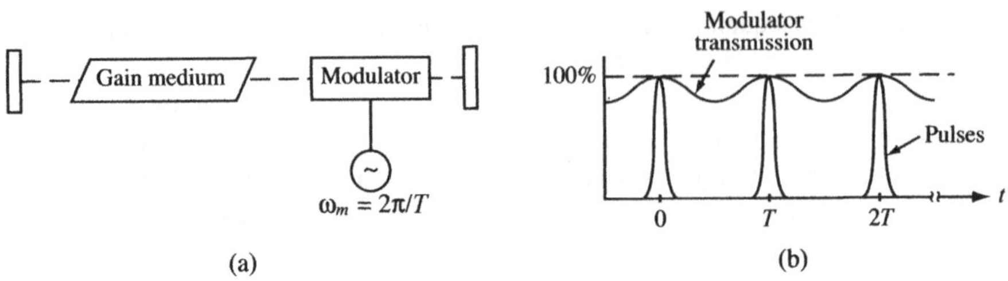

Figure 2.1 (a) Actively mode-locked laser arrangement, with an intracavity modulator driven at the cavity round-trip period; (b) periodic modulator transmission and resulting mode-locked pulses.

that the initial laser radiation is shortened again and again every time it passes through the modulator. Thus, to minimize the intracavity losses, one would expect laser emission to occur in the form of short pulses synchronized with positions of maximum shutter opening. Also in the steady state, we will find that the per-pass pulse shortening by the modulator must be balanced exactly by the (unwanted) per-pass pulse-broadening mechanisms. In the following section we analyze active mode-locking from this time-domain point of view. Subsequently, we present an alternative analysis from a frequency-domain perspective.

2.2.1 Time-Domain Treatment

In our analysis we will consider propagation of an ultrashort pulse around an actively mode-locked laser cavity [8,14]. Our model for the laser cavity is shown in Fig. 2.2. The principal elements in our model are the laser cavity itself, the gain medium, and the modulator. To arrive at a stable mode-locked solution, we require that the pulse reproduce itself after one round trip through the laser.

We start off by assuming a Gaussian laser pulse of the form

$$e(t) = \mathrm{Re}\left\{a(t)e^{j\omega_0 t}\right\} = \mathrm{Re}\left\{E_0 e^{-\Gamma t^2} e^{j\omega_0 t}\right\} \tag{2.1}$$

where $a(t) = E_0 e^{-\Gamma t^2}$ is the envelope function and Γ is a pulse-width parameter. From our results in Section 1.5, the spectrum can be written

$$E(\omega) = \frac{1}{2}\left[A(\omega - \omega_0) + A^*(-\omega - \omega_0)\right] \tag{2.2}$$

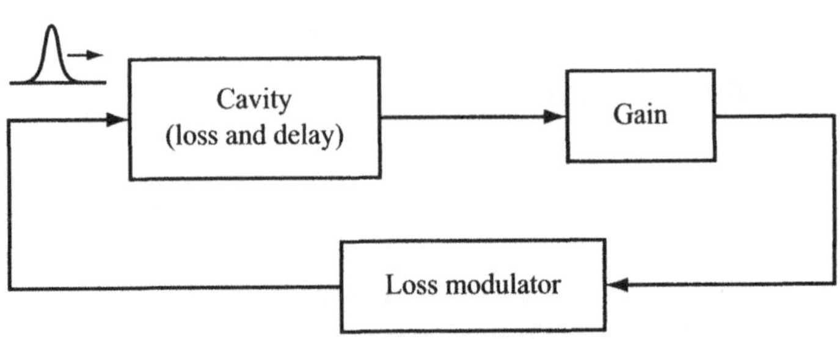

Figure 2.2 Model of an actively mode-locked laser.

with

$$A(\omega) = E_0 \left(\frac{\pi}{\Gamma}\right)^{1/2} e^{-\omega^2/4\Gamma}$$

We now consider each laser element in turn.

The cavity has two effects: It introduces loss and a round-trip time delay. We can represent these effects by the multiplicative factor

$$e^{-\ell} e^{-j\omega L/c}$$

Here $e^{-\ell}$ is the round-trip attenuation of the field accounting for everything except the loss modulator. The term $e^{-j\omega L/c}$ corresponds to a delay by the round-trip propagation time L/c, where $L = l_a + n'_g l_g + n'_m l_m$ is the round-trip optical path length, l_a, l_g, and l_m are the physical lengths of air regions, gain medium, and modulator, respectively, and n'_g and n'_m are the real parts of the refractive index of the gain medium and the modulator. Note that n'_g does not include the part of the gain medium refractive index due to the laser transition (see below).

The purpose of the gain medium is, of course, to amplify the field. We make the following assumptions:

- The gain medium is homogeneously broadened and has a finite optical bandwidth. The finite-gain bandwidth acts to limit how short a pulse can be attained.
- The laser spectrum is centered at the peak of the gain spectrum.
- The relaxation time of the gain medium is much longer than the pulse repetition period. Consequently, the gain saturates only in response to the average laser power. The gain can be considered constant during the passage of a single laser pulse.

Specifically, we express the action of the gain medium by the multiplicative factor

$$\exp\left[\frac{g}{1 + 2j[(\omega - \omega_0)/\omega_G]}\right]$$

Here g refers to the *saturated* gain, and the particular gain line-shape function

$$G(\omega) = \frac{g}{1 + 2j[(\omega - \omega_0)/\omega_G]} \tag{2.3}$$

is known as a *Lorentzian* and corresponds to our assumption of a homogeneously broadened atomic-like gain medium. Introducing the notation $\tilde{\omega} = \omega - \omega_0$, we can rewrite eq. (2.3) as

$$G(\omega) = \frac{g(1 - 2j\tilde{\omega}/\omega_G)}{1 + \left(4\tilde{\omega}^2/\omega_G^2\right)} \tag{2.4}$$

$G(\omega)$ can be related to the complex refractive index $n = n' + jn''$ by identifying $G(\omega) = -j\omega n l_g/c$ [see eq. (1.37)]. Consequently, we have

$$n' = \frac{c}{\omega l_g} \frac{2g\,(\tilde{\omega}/\omega_G)}{1 + \left(4\tilde{\omega}^2/\tilde{\omega}_G^2\right)} \tag{2.5a}$$

and

$$n'' = \frac{c}{\omega l_g} \frac{g}{1 + \left(4\tilde{\omega}^2/\tilde{\omega}_G^2\right)} \tag{2.5b}$$

n'' is associated with gain, and n' is associated with real refractive index (i.e., phase) effects. Figure 2.3 shows a plot of n' and n''. The gain term (n'') has a symmetric peak centered at ω_0 with a full width at half maximum (FWHM) of ω_G. n' is an antisymmetric function with a zero at ω_0 and with $n' < 0$ below resonance ($\omega < \omega_0$).

Before proceeding, we note that the value of the peak gain g is proportional to the population difference $(N_u - N_l)$ between upper and lower laser energy levels. For gain to occur, the population must be inverted $(N_u - N_l > 0)$, and n' and n'' are as shown in Fig. 2.3. If the pump is turned off, the population is not inverted $(N_u - N_l < 0)$, and the signs of n' and n'' are opposite to those shown in the figure. Therefore, for unpumped media, $n'' < 0$, signifying absorption, and $n' > 0$ for $\omega < \omega_0$.

In actively mode-locked lasers, the pulse usually has a narrow bandwidth compared to the gain bandwidth ω_G. Therefore, we can approximate eq. (2.3) using a second-order Taylor series expansion. Using

$$\frac{1}{1 + \Delta} \approx 1 - \Delta + \Delta^2 \tag{2.6}$$

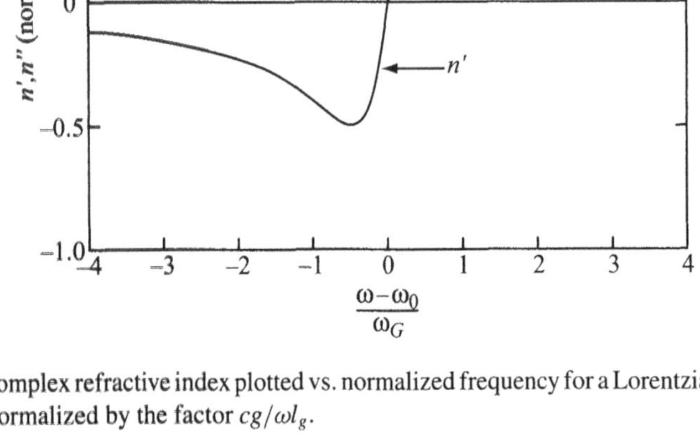

Figure 2.3 Complex refractive index plotted vs. normalized frequency for a Lorentzian gain medium. n' and n'' are normalized by the factor $cg/\omega l_g$.

which is valid for small Δ, we obtain

$$G(\omega) \approx g\left(1 - \frac{2j\tilde{\omega}}{\omega_G} - \frac{4\tilde{\omega}^2}{\omega_G^2}\right) \tag{2.7}$$

Now by using eqs. (2.2) and (2.7), we can obtain an expression for the spectrum as affected by the cavity and the gain medium:

$$A(\tilde{\omega}) = E_0\left(\frac{\pi}{\Gamma}\right)^{1/2} e^{-j\omega_0 L/c} e^{g-\ell} e^{-j\tilde{\omega}(L/c+2g/\omega_G)} e^{-\tilde{\omega}^2/4\Gamma'} \tag{2.8a}$$

where

$$\frac{1}{\Gamma'} = \frac{1}{\Gamma} + \frac{16g}{\omega_G^2} \tag{2.8b}$$

If the fractional change in Γ (or equivalently, in the pulse width) per pass through the gain medium is small, we can rewrite eq. (2.8b) as

$$\Gamma' \approx \Gamma - \frac{16g\Gamma^2}{\omega_G^2} \tag{2.9}$$

At this point we Fourier-transform back into the time domain. Assuming that $\omega_0 L/c = 2m\pi$ (i.e., ω_0 is equal to one of the cavity mode frequencies), we obtain the following for the envelope function $a(t)$:

$$a(t) = E_0 e^{g-\ell}\left(\frac{\Gamma'}{\Gamma}\right)^{1/2} e^{-\Gamma'[t-(L/c+2g/\omega_G)]^2} \tag{2.10}$$

We see that Γ' is decreased by passage through the gain medium. This means that the gain spectral profile reduces the bandwidth of the circulating pulse, resulting in an increase in pulse duration. Additionally, the pulse is delayed by a time

$$T = \frac{L}{c} + \frac{2g}{\omega_G} = \frac{L}{c}\left(1 + \frac{g}{\pi}\frac{\Delta\omega}{\omega_G}\right) \tag{2.11}$$

where $\Delta\omega = 2\pi c/L$ is the *cold cavity* (i.e., the cavity without including the effect of the laser transition) longitudinal mode spacing. The round-trip delay time is increased over that of the cold cavity (L/c) by an amount $Lg\,\Delta\omega/c\pi\omega_G$. This extra delay arises due to the spectral variation of n' due to the laser transition. When n' is a function of frequency, the *pulse propagation velocity* (also known as the *group velocity* v_g) differs from the *propagation velocity of the carrier* (also known as the *phase velocity* $v_p = c/n'$). We discuss group and phase velocities further in a later chapter.

We will now account for the loss modulator by multiplying eq. (2.10) by its transmission function $\eta(t)$, where

$$\eta(t) = \exp\left[-\Delta_m(1 - \cos\omega_m t)\right] \tag{2.12}$$

Thus, $\eta(t)$ varies periodically with period $2\pi/\omega_m$. For best mode-locking the modulator frequency is tuned to

$$\omega_m = \frac{2\pi}{T} \tag{2.13}$$

where T is the actual round-trip time given by eq. (2.11). The peak transmission is unity at $t = 0, T, 2T, \ldots$, and the minimum transmission is $e^{-2\Delta_m}$. Δ_m is called the *modulation depth*. If the pulse is much shorter than the modulation period (as indeed we desire for a mode-locked pulse!) and is centered near the times of peak transmission, we can approximate $\cos\omega_m t \approx 1 - \frac{1}{2}\omega_m^2 t^2$ near $t = 0$, $\cos\omega_m t \approx 1 - \frac{1}{2}\omega_m^2 (t-T)^2$ near $t = T$, and so on. Then the transmission for the pulse in eq. (2.10) becomes

$$\eta(t) = e^{-\frac{1}{2}\Delta_m \omega_m^2 (t-T)^2} \tag{2.14}$$

Multiplying (2.10) by (2.14) and using (2.11), we obtain an expression for the pulse after one round trip through the laser:

$$a(t) = E_0 e^{g-\ell} \left(\frac{\Gamma'}{\Gamma}\right)^{1/2} e^{-\Gamma''(t-T)^2} \tag{2.15a}$$

where

$$\Gamma'' = \Gamma' + \frac{1}{2}\Delta_m \omega_m^2 \approx \Gamma - \frac{16g\Gamma^2}{\omega_G^2} + \frac{1}{2}\Delta_m \omega_m^2 \tag{2.15b}$$

From eq. (2.15b) the modulator decreases the pulse width.

A steady-state mode-locked solution requires that the pulse reproduce itself exactly after one round trip. Therefore, we require that eq. (2.15a) be identical to our starting point, eq. (2.1), except for the time delay. As a result, we have

$$\Gamma'' = \Gamma \tag{2.16}$$

and

$$\left(\frac{\Gamma'}{\Gamma}\right)^{1/2} e^{g-\ell} = 1 \tag{2.17}$$

Equation (2.16) says that the pulse width must reproduce itself after a round trip. Substituting in eq. (2.15b), we obtain

$$\frac{1}{2}\Delta_m \omega_m^2 - \frac{16g\Gamma^2}{\omega_G^2} = 0 \tag{2.18}$$

and

$$\Gamma = \frac{1}{4\sqrt{2}} \left(\frac{\Delta_m}{g} \right)^{1/2} \omega_m \omega_G \tag{2.19}$$

Equation (2.18) leads to the physical interpretation that the per-pass pulse shortening due to the modulator exactly balances the per-pass broadening due to the gain medium's finite bandwidth.

Equation (2.19) is the desired solution for the pulse-width parameter Γ. It is customary to express the pulse width in terms of the full width at half maximum (FWHM) of the intensity, which we denote Δt.

$$\Delta t = 2 \left(\frac{\ln 2}{2\Gamma} \right)^{1/2} = \frac{2^{3/4}\sqrt{\ln 2}}{\pi} \left(\frac{g}{\Delta_m} \right)^{1/4} (f_m f_G)^{-1/2}$$

$$\approx 0.45 \left(\frac{g}{\Delta_m} \right)^{1/4} (f_m f_G)^{-1/2} \tag{2.20}$$

Here $f_m = \omega_m / 2\pi$ and $f_G = \omega_G / 2\pi$. In obtaining the shortest pulses, one prefers a large modulation index, a large modulation frequency, and a large gain bandwidth. The pulse width depends only weakly on these parameters, however, and usually relatively long pulses in the range 10 to 100 ps are generated.

Equation (2.17) is the gain condition for the actively mode-locked laser. The peak saturated gain (g) is slightly greater than the loss, in order to balance the losses in the modulator and the fact that the overall gain is lower than g due to the frequency spread within the mode-locked pulse and the finite gain bandwidth. The average power P circulating in the laser can be obtained from the gain saturation equation,

$$g = \frac{g_0}{1 + P/P_{\text{sat}}} \tag{2.21}$$

where g_0 is the small-signal gain and P_{sat} is the saturation power of the gain medium. In evaluating P, g can be found exactly from eq. (2.17) or can be approximated by $g \approx \ell$.

It should be noted that in our analysis, we have assumed that the modulation is exactly synchronous with the cavity transit time. Under this condition, the mode-locked pulse is positioned exactly at the peak transmission point of the applied modulation, which was also assumed. In practice, the modulation frequency must match the mode spacing very precisely, with a tolerance typically in the range of a few hundred hertz to a few kilohertz. For a typical cavity length corresponding to a repetition rate on the order of 100 MHz, this means that the frequency must be set correctly within 1 part in 10^5 or better, and the cavity length must remain stable to better than a few tens of micrometers. When the modulation frequency is detuned (within this narrow range), the mode-locked pulse will be offset from the time of peak transmission. The nonsymmetrical variation of the modulator transmission during the duration of the mode-locked pulse will then act to push the pulse slightly back toward the transmission peak, which provides a timing restoration mechanism. Stable mode-locked operation can be maintained only when the amount of timing restoration per laser transit is enough to fully compensate the difference between the modulation period and the laser transit time.

2.2.2 Frequency-Domain Treatment

We now consider an alternative frequency-domain theory of active mode-locking. As before, we keep track of the effects induced by each laser element and require that the net change per round trip be zero. Here we do not assume a Gaussian pulse at the outset. Instead, by assuming that the change per element per round trip is small, we arrive at a differential equation for the mode-locked pulse. We shall follow a similar course in analyzing other types of mode-locking in later sections. A principal merit of the treatment in this section is to illustrate the locking of cavity modes leading to the term *mode-locking*. Our discussion most closely follows [15]; related treatments include [16,17].

We start with the effects of the cavity. We assume that the longitudinal modes of the cold cavity are at frequencies $\Omega_q = \omega_0 + q\,\Delta\omega$, where $\Delta\omega$ is the cold-cavity-mode spacing. For a low-loss (high-Q) resonator, the time variation of the field resonating with the qth cavity mode E_q can be written

$$\frac{dE_q}{dt} - j\Omega_q E_q + \frac{E_q}{\tau_0} = \text{source terms} \tag{2.22}$$

The identity of the source terms in our mode-locking problem will become clear soon. As before, we assume an intracavity loss modulator driven at frequency ω_m. For steady-state mode-locked operation, the laser output should be periodic with period $2\pi/\omega_m$. Hence, the laser emission can be written as a Fourier series with harmonics $\omega_n = \omega_0 + n\omega_m$, where n is an integer. Here we have assumed that the center frequency is still at ω_0, which will be true when the central cavity mode exactly coincides with the peak of the gain spectrum. As long as ω_m nearly matches the mode-spacing $\Delta\omega$, each of the driven frequencies $\omega_n = \omega_0 + n\omega_m$ interacts strongly only with the corresponding cavity resonance at $\Omega_n = \omega_0 + n\Delta\omega$. The equation for the nth harmonic, written $E_n = \tilde{E}_n e^{j\omega_n t}$, is

$$\frac{d\tilde{E}_n}{dt} + (jn\omega_m\delta)\tilde{E}_n + \frac{\tilde{E}_n}{\tau_0} = \text{source terms} \tag{2.23a}$$

where

$$\delta = 1 - \frac{\Delta\omega}{\omega_m} \tag{2.23b}$$

Note also that we identify the loss per round trip, which was taken as $e^{-\ell}$ in the preceding section, with the factor e^{-T/τ_0} arising from eq. (2.22). T is the round-trip propagation time, as before. Thus, we can equate

$$\frac{1}{\tau_0} = \frac{\ell}{T} \tag{2.24}$$

Equivalently, the loss coefficient (τ_0^{-1}) in the rate equations for the harmonic amplitudes is obtained by taking the natural logarithm of the single-pass loss, then dividing by the round-trip time.

In a similar way, the gain is incorporated into eq. (2.23a) by adding the term $G(\omega)\tilde{E}_n/T$ to the right-hand side of the equation. Here $G(\omega)$ is approximated by eq. (2.7), as in Section 2.2.1.

Now we consider the modulator. Using eq. (2.12) for the modulator transmission function $\eta(t)$, the source term representing the effect of the modulator can be written

$$\frac{\ln[\eta(t)]}{T}E_n = \frac{-\Delta_m}{T}\tilde{E}_n e^{j\omega_n t} + \frac{\Delta_m}{2T}\tilde{E}_n\left(e^{j\omega_{n-1}t} + e^{j\omega_{n+1}t}\right) \tag{2.25}$$

This expression is inserted into the right-hand side of eq. (2.23a). We see that as a result of the modulation, each harmonic acquires sidebands that act as source terms for the fields resonating with the adjacent cavity modes. Thus, the modulator couples the cavity modes together, which results in mode-locking.

Putting all these terms together, the equation for the nth harmonic is written

$$\frac{d\tilde{E}_n}{dt} = \left(-\frac{\ell}{T} + \frac{g}{T}\right)\tilde{E}_n - jn\omega_m\left(\delta + \frac{2g}{\omega_G T}\right)\tilde{E}_n - \frac{4gn^2\omega_m^2}{T\omega_G^2}\tilde{E}_n + \frac{\Delta_m}{2T}\left[\tilde{E}_{n+1} - 2\tilde{E}_n + \tilde{E}_{n-1}\right] \tag{2.26}$$

For steady-state mode-locking the changes affecting any particular harmonic $\tilde{E}_n$ should cancel out after one round trip. Therefore, in eq. (2.26) we set $d\tilde{E}_n/dt = 0$. We also assume that $\delta = -2g/\omega_G T$; this corresponds to a modulation frequency matched to the actual round trip as modified by the laser transition [see eq. (2.11)]. Finally, we replace the discrete harmonic spectrum $\tilde{E}_n$ by a continuous spectrum $A(\omega)$, where $\omega = n\omega_m$. The last term in eq. (2.26) representing the mode coupling is replaced by a second derivative,

$$\frac{\Delta_m}{2T}\left[\tilde{E}_{n+1} - 2\tilde{E}_n + \tilde{E}_{n-1}\right] \rightarrow \frac{\Delta_m\omega_m^2}{2T}\frac{d^2A(\omega)}{d\omega^2} \tag{2.27}$$

These approximations are valid when the spectrum is broad compared to the mode spacing and varies little from one mode to the next. This corresponds to the condition that the mode-locked pulses are short compared to the period and hence well separated—a condition that is almost always valid. Equation (2.26) now becomes

$$\frac{-\Delta_m\omega_m^2}{2}\frac{d^2A(\omega)}{d\omega^2} + \frac{4g\omega^2}{\omega_G^2}A(\omega) = (g - \ell)A(\omega) \tag{2.28}$$

Equation (2.28) is identical to the time-independent Schrödinger equation for a particle in a harmonic well, which is usually written

$$\frac{-\hbar^2}{2m}\frac{\partial^2\psi}{\partial x^2} + \frac{1}{2}Cx^2\psi = E\psi \tag{2.29}$$

This has the well-known solution for the ground state (see any quantum mechanics text)

$$\psi \sim H_\nu(u)e^{-u^2/2} \tag{2.30a}$$

where

$$u = \frac{(Cm)^{1/4}}{\hbar^{1/2}} x \qquad \text{and} \qquad E = \left(\nu + \frac{1}{2}\right) \hbar \sqrt{\frac{C}{m}} \tag{2.30b}$$

The $H_\nu(u)$ are the Hermite polynomials, which were introduced briefly in Section 1.3.3. The ground state of the quantum mechanical problem corresponds to $\nu = 0$, for which the corresponding Hermite polynomial is $H_0(u) = 1$. By identifying $\Delta_m \omega_m^2 \leftrightarrow \hbar^2/m$ and $8g/\omega_G^2 \leftrightarrow C$, we obtain solutions for the mode-locked spectrum. For the ground-state solution ($\nu = 0$), the result is a Gaussian, identical to eq. (2.2), with Γ given by eq. (2.19). Our choice of the pure Gaussian solution is justified since this is the only stable solution to the mode-locking equation, as determined by a stability analysis published in [15]. From a simple physical point of view, this is consistent with the observation that the $\nu = 0$ solution requires the lowest gain, as can be obtained through inspection of eq. (2.30b) with the identification $g - \ell \leftrightarrow E$. This result is also expected since the higher-order Hermite solutions have a broader temporal duration and therefore experience a greater loss from the modulator.

The key lesson of this section is that from a frequency-domain perspective, the effect of the modulator is to generate sidebands that couple adjacent cavity modes together. This coupling stabilizes the relative phases of the modes and leads to mode-locking. This process is summarized by Fig. 2.4, which shows the small-signal gain spectrum as well as the saturated gain of the mode-locked and free-running (i.e., continuous-wave) laser. As we have seen, for continuous-wave operation of a homogeneously broadened laser, saturated gain just equals loss at line center; ideally, all modes other than the central mode are below threshold. In the case of active mode-locking, the sideband creation process redistributes energy from modes near the line center to modes in the wings of the gain spectrum. As a result, the saturated gain must exceed the loss at line center. Furthermore, even modes where

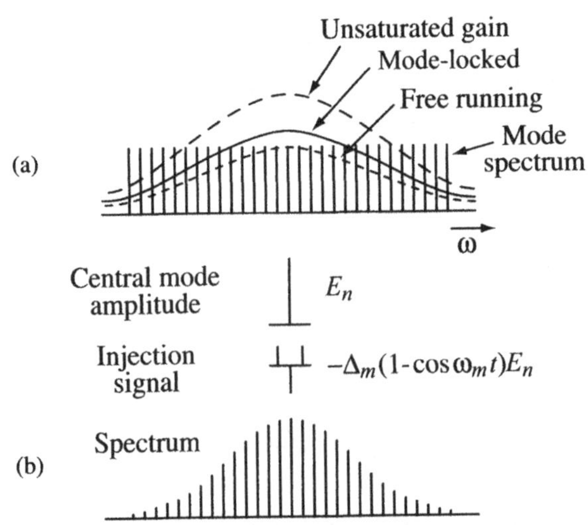

Figure 2.4 Frequency-domain view of active mode-locking: (a) unsaturated gain spectrum and saturated gain spectrum for free-running and mode-locked laser operation; (b) resulting mode-locked spectrum. From [51].

the gain is below the loss line can oscillate if sufficient power is injected by the modulator through sideband generation. From this picture the role of mode-locking in inducing a large number of modes to oscillate should be clear.

Finally, it is worth noting that the differential equation for the spectrum, eq. (2.28), can be Fourier-transformed into a differential equation in the time domain. We use the Fourier transform identities

$$\frac{d}{dt} \leftrightarrow j\omega \qquad \frac{d^2}{dt^2} \leftrightarrow -\omega^2$$
$$\frac{d}{d\omega} \leftrightarrow -jt \qquad \frac{d^2}{d\omega^2} \leftrightarrow -t^2 \tag{2.31}$$

The interpretation of eq. (2.31) is that we can replace the factor $-\omega^2$ appearing in the frequency-domain equation with the operator d^2/dt^2 in the time-domain equation, and so on. The result of the Fourier transform is

$$\frac{-4g}{\omega_G^2}\frac{d^2a(t)}{dt^2} + \frac{\Delta_m\omega_m^2}{2}t^2a(t) = (g - \ell)a(t) \tag{2.32}$$

The solution is obtained once again by comparing to the Schrödinger equation for a particle in a harmonic well. The result is identical to eq. (2.1), with Γ given by eq. (2.19). Of course, this solution for $a(t)$ is just the Fourier transform of the solution for $A(\omega)$ from eq. (2.28).

2.2.3 Variations of Active Mode-Locking

Our treatment of active mode-locking via synchronous loss modulation covers the most common form of active mode-locking. A very similar variant of this technique is synchronous gain modulation, which is important in semiconductor diode lasers, since the gain can conveniently be modulated by modulating the current drive to the laser. Our active mode-locking analysis applies to this case also with only minor modifications. In the following we briefly enumerate several other variations of active mode-locking, although detailed analysis is not included.

- Active mode-locking can be achieved using *synchronous phase modulation* [14,18]. The analysis given above still applies, provided that the modulation depth Δ_m takes on an imaginary value. In the circulating pulse picture, a time derivative of the phase gives the pulse a frequency shift, which reduces its gain due to the finite gain bandwidth. For light sitting exactly at the phase extremum, the frequency shift goes to zero, and therefore the gain is higher. This provides the pulse-shortening mechanism needed for successful mode-locking.
- In *regenerative mode-locking* the modulation signal is derived by using a photodiode to detect the beat note between the initially unlocked longitudinal modes of the laser [19]. The beat note is amplified and then used to drive an intracavity modulator. The modulation frequency is therefore adjusted automatically to match the cavity mode spacing, and provided that the beat note is properly phase shifted so that the modulation applied will enhance the initial mode beating fluctuation, mode-locking results. This relaxes the requirements on careful adjustment of the cavity length in active

mode-locking. In some sense regenerative mode-locking is a cross between active and passive mode-locking; the modulation signal originates due to the laser signal itself as in passive mode-locking but is applied via an electrically driven modulator as in active mode-locking. Regenerative mode-locking has also been used as a starting mechanism for solid-state lasers where passive mode-locking dominates the steady-state mode-locked dynamics.

- In *harmonic mode-locking* the modulator is driven at a harmonic of the cavity mode spacing. The more rapid variation of the applied modulation function can lead to shorter pulses; note the modulation frequency dependence in eq. (2.20). Harmonic mode-locking allows generation of pulse trains with the pulses spaced more closely than the cavity round-trip time. This is especially useful in fiber lasers for optical communications applications, where the round-trip time can be very long (often corresponding to hundreds of nanoseconds), while pulse trains may be desired at multiple gigahertz rates [20]. It should be noted, however, that there is no guarantee that a pulse will be present in every transmission peak in harmonic mode-locking; therefore, if this is important, other stabilizing mechanisms must be incorporated into the laser.
- In another form of mode-locking most appropriate for optically pumped systems, the laser is *synchronously pumped* by the pulses from another mode-locked laser [21,22]. This technique was commonly applied to dye lasers, where relatively long pulses (ca. 100 ps) from an actively mode-locked argon ion or Nd:YAG laser were used to pump the much broader bandwidth dye laser. Under proper conditions, this could lead to considerably shorter output pulses (ca. picoseconds) at the wavelength of choice (determined by the choice of dye). In these systems the gain is built up by the arrival of the pumping pulse, and then turned off rapidly due to dynamic saturation induced by the resulting dye laser pulse. Such transient gain saturation is discussed in Section 2.3. Similar to other actively mode-locked systems, cavity-length matching between the two lasers is critical.
- In *hybrid mode-locking* active modulation is combined with passive mode-locking, most commonly using a slow saturable absorber. The active modulation is commonly in the form of electrically driven gain modulation, as in the case of a semiconductor diode laser [23], or optically driven gain modulation, as in the case of synchronously pumped dye lasers. Typically in hybrid mode-locking, the active modulation assists in pulse formation and helps to stabilize the mode-locking process, while the saturable absorber is responsible for significant reduction of the final pulse duration. Much of the insight we develop in the next section on saturable absorber mode-locking is also relevant to the hybrid mode-locking case.

2.3 PASSIVE MODE-LOCKING USING SATURABLE ABSORBERS

Although active mode-locking is a well-established technique, it suffers from two disadvantages:

1. An externally driven modulator is required, and the modulation frequency must be matched precisely to the cavity mode spacing. (However, this can be an advantage for applications where synchronization to an external clock is required!)

2. Pulse shortening due to the modulator becomes ineffective for very short pulses, and this limits the attainable pulse width.

These drawbacks can be overcome by using passive mode-locking, in which the externally driven modulator is replaced by a nonlinear optical element whose loss depends on the laser pulse intensity. For mode-locking the loss should decrease with increasing intensity. An ultrashort pulse circulating in the laser will modulate the intracavity loss, which in turn will modulate the circulating pulse. As a result, the loss modulation is synchronized automatically with the laser pulses. Furthermore, if the response time of the nonlinearity is sufficiently fast, the optically driven modulation function gets faster as the pulse becomes shorter. Hence, the pulse-shortening action can remain effective even for very short pulses.

We begin our treatment of passive mode-locking by discussing saturable absorbers as the nonlinear loss element. By *saturable absorber* we mean an optically absorbing medium whose absorption decreases with intensity or energy. Throughout most of the 1970s and 1980s and into the early 1990s, the shortest pulses on record were generated using organic dye lasers and saturable absorber mode-locking. Furthermore, the theory of saturable absorber mode-locking provides important background for today's solid-state lasers, which are mode-locked very successfully using artificial saturable absorbers implemented using nonlinear phase modulation effects. A full understanding of such mode-locked solid-state lasers requires consideration not only of nonlinear amplitude modulation, but also of nonlinear phase modulation and dispersion effects, which we discuss subsequently. In the current section we concentrate on nonlinear amplitude modulation.

Our general model of saturable absorber mode-locking is shown in Fig. 2.5. As usual, we consider the changes encountered by a pulse during a single round trip through the laser cavity. For steady-state mode-locking, the original pulse should be reproduced after a single round trip.

The initial pulse envelope is written $a(t)$ with Fourier transform $A(\omega)$. We take the cavity into account by writing

$$A'(\tilde{\omega}) = e^{-\ell_0} e^{-\tilde{\omega}^2/\omega_c^2} A(\tilde{\omega}) \tag{2.33}$$

where $\tilde{\omega} = \omega - \omega_0$, $e^{-\ell_0}$ is the linear time-independent cavity loss, and $e^{-\tilde{\omega}^2/\omega_c^2}$ is a filtering term related to the finite bandwidth of the optical cavity. If there are no other bandwidth-limiting elements, the bandwidth may be determined by the gain bandwidth, as previously. In many cases, however, an intracavity etalon or other filter may limit the cavity bandwidth. In general, there is also a phase term that is linear in frequency. As we have seen, this linear phase term corresponds to a time delay T, where T is the total linear cavity round-trip-time, including the effects of the filter function. Although we have omitted writing the linear phase term here, one should remember that one round trip through the cavity leads to a delay T.

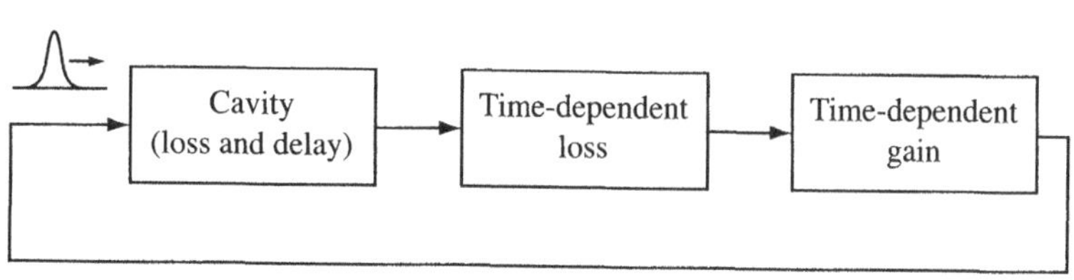

Figure 2.5 Model of a passively mode-locked laser.

To develop an analytic theory [24,25], one assumes that the change wrought by each element (loss, gain, filter) per pass is small, and that the mode-locked spectrum is narrow compared to the filter bandwidth ω_c. Consequently, eq. (2.33) can be approximated using first-order Taylor expansions as

$$A'(\tilde{\omega}) \approx \left(1 - \ell_0 - \frac{\tilde{\omega}^2}{\omega_c^2}\right) A(\tilde{\omega}) \tag{2.34}$$

We now return to the time domain by Fourier transforming eq. (2.34). The result is

$$a'(t) = \left(1 - \ell_0 + \frac{1}{\omega_c^2}\frac{d^2}{dt^2}\right) a(t) \tag{2.35}$$

Now we can take into account the nonlinear loss and gain. The loss $\ell(t)$ and [sometimes] the gain $g(t)$ are taken to be time dependent and controlled via the optical nonlinearity. $\ell(t)$ refers to the nonlinear time-dependent part of the loss only. Assuming again that the nonlinear loss and the gain per pass are small, we can write

$$a''(t) = e^{-\ell(t)} e^{g(t)} a'(t) \approx \left(1 - \ell_0 - \ell(t) + g(t) + \frac{1}{\omega_c^2}\frac{d^2}{dt^2}\right) a(t) \tag{2.36}$$

Both nonlinear loss and gain are taken to be frequency independent, since bandwidth limitations are already modeled as cavity effects.

We arrive at an equation for the mode-locked pulse by requiring that $a''(t)$ be equal to the initial pulse, possibly shifted in time by an amount δT. Here δT is a time shift (not included in the linear cavity delay T) arising due to the nonlinear pulse shaping action of $\ell(t)$ and $g(t)$ and is usually small compared to the steady-state pulse width. $\delta T > 0$ corresponds to a shift forward in time. Mathematically, this is expressed as

$$a''(t) = a(t + \delta T) \approx a(t) + \frac{da}{dt}\delta T \tag{2.37}$$

Putting everything together, we arrive at

$$\frac{1}{\omega_c^2}\frac{d^2 a(t)}{dt^2} + \left[g(t) - \ell(t) - \ell_0\right] a(t) - \delta T \frac{da(t)}{dt} = 0 \tag{2.38}$$

To proceed further we will need to specify the forms of $g(t)$ and $\ell(t)$. However, we do note that the second-derivative term represents pulse broadening due to the bandwidth-limiting filter, which must be compensated by pulse shortening arising from $g(t)$ and $\ell(t)$.

Equation (2.38) is one example of Haus's *master equation* for mode-locking [24,25]. The main assumption of the master equation approach is that the per-pass change wrought by each laser element is small, so that these changes can be expressed by the leading terms in the Taylor series expansions. It is also assumed that the order in which the laser elements are placed has no effect. The master equation approach provides analytical insight for a wide range of mode-locking systems.

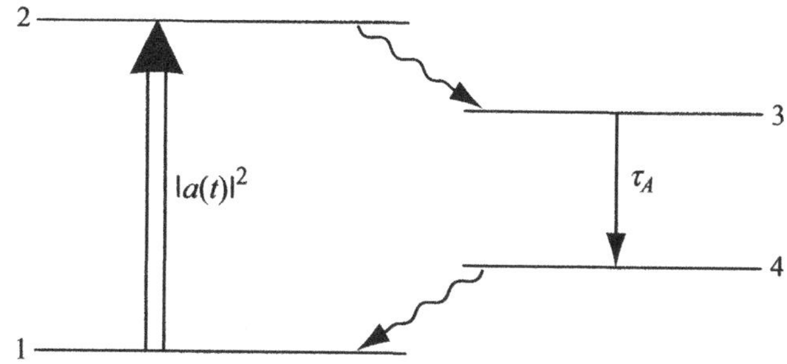

Figure 2.6 Model of a four-level saturable absorber.

2.3.1 Saturation Model

Absorber The most common saturable absorber media which have been used for mode-locking are organic dye solutions and semiconductors. These can be modeled as four-level systems (see also Section 1.3), as shown in Fig. 2.6. The $1 \rightarrow 2$ transition is a resonant absorption for the laser radiation, and the absorption strength is proportional to the population densities $N_1 - N_2$ (where N_j is the density in units of m^{-3} of absorbers in level j). The total density of absorbers is N_A. The $2 \rightarrow 3$ and $4 \rightarrow 1$ relaxations are taken to be very fast. The $3 \rightarrow 4$ relaxation time is finite and is denoted τ_A. We assume that the laser radiation does not interact with the $3 \rightarrow 4$ transition, which is red-shifted compared to the $1 \rightarrow 2$ transition. We also assume that the absorption spectrum is homogeneously broadened and can be considered constant within the mode-locked bandwidth. This assumption is equivalent to ignoring coherent laser–absorber interactions.

With these assumptions we can describe the absorber using a simple rate equation [24,25]:

$$\frac{\partial N_1}{\partial t} = \frac{N_3}{\tau_A} - \frac{\sigma_A |a(t)|^2}{\hbar\omega_0 A_A}(N_1 - N_2) \tag{2.39a}$$

with

$$N_1 + N_3 = N_A \qquad \text{and} \qquad N_2 \approx N_4 \approx 0 \tag{2.39b}$$

The first term on the right is relaxation out of level 3 and the second term represents stimulated absorption. The pulse is normalized so that $|a(t)|^2$ gives the time-dependent power carried by the pulse. σ_A is the $1 \rightarrow 2$ absorption cross section, $\hbar\omega_0$ is the photon energy, and A_A is the beam cross-sectional area in the absorber. Using eq. (2.39b) we can rewrite eq. (2.39a) as

$$\frac{\partial N_1}{\partial t} = \frac{N_A - N_1}{\tau_A} - \frac{|a(t)|^2 N_1}{P_A \tau_A} \tag{2.40a}$$

where

$$P_A = \frac{\hbar\omega_0 A_A}{\sigma_A \tau_A} \tag{2.40b}$$

is the absorber saturation power. Assuming a small loss per pass, the time-dependent loss term $\ell(t)$ in eq. (2.38) is proportional to the ground-state absorber density N_1:

$$\ell(t) = \frac{\sigma_A}{2} N_1(t) l_a \tag{2.41}$$

where l_a is the length of the absorber medium.

We now use eq. (2.40a) to determine $N_1(t)$ in two important limiting cases. These cases are distinguished by the magnitude of the relaxation time τ_A compared to the mode-locked pulse width (which we designate t_p). For $\tau_A \ll t_p$ we speak of a *fast saturable absorber.* In the opposite case, $t_p \ll \tau_A$, we speak of a *slow saturable absorber.*

Fast Saturable Absorber When $\tau_A \ll t_p$, we can set $\partial N_1/\partial t = 0$ in eq. (2.40a). We then solve easily for N_1, with the result

$$N_1(t) = \frac{N_A}{1 + |a(t)|^2/P_A} \tag{2.42}$$

$N_1(t)$ and hence $\ell(t)$ vary instantaneously with the laser power $|a(t)|^2$. The absorption decreases with increasing laser power; hence, the peak of a mode-locked pulse will see a lower loss than will the wings of the pulse.

Slow Saturable Absorber When $t_p \ll \tau_A$, we can solve for the absorber dynamics during the mode-locked pulse by setting $(N - N_1)/\tau_A \approx 0$. With this proviso, eq. (2.40a) becomes

$$\frac{\partial N_1}{\partial t} = \frac{-|a(t)|^2}{P_A \tau_A} N_1 \tag{2.43}$$

The solution is found by integration, with the result

$$N_1(t) = N_1^{(i)} e^{-\int^{t} dt |a(t)|^2/P_A\tau_A} = N_1^{(i)} e^{-U(t)/U_A} \tag{2.44}$$

where

$$U(t) = \int^{t} dt\, |a(t)|^2 \qquad \text{and} \qquad U_A = P_A \tau_A \tag{2.45}$$

Here $N_1^{(i)}$ is the initial absorber population in level 1 just before the laser pulse, $U(t)$ is the pulse energy up to time t, and U_A is the absorber saturation energy. $N_1(t)$ and $\ell(t)$ decrease monotonically during the pulse. The degree of saturation depends on cumulative pulse energy. Note that it is the saturation *energy* that is important in the slow absorber case, whereas for the fast absorber the saturation *power* is important.

After the end of the laser pulse the absorber population relaxes exponentially back to its equilibrium condition. Thus, after the laser pulse we have

$$N_1(t) = N_A + \left(N_1^{(i)} e^{-U/U_A} - N_A\right) e^{-t/T_A} \tag{2.46}$$

where U is the total laser pulse energy, and the pulse is centered at $t = 0$.

Gain Medium We analyze the gain medium using a four-level model (see Fig. 1.5) similar to that for the absorber. The main differences are that the laser radiation is resonant with the $3 \rightarrow 4$ transition, and a pump W drives the $1 \rightarrow 2$ transition out of the ground state. With assumptions similar to these stated previously, the gain medium rate equation is

$$\frac{\partial N_3}{\partial t} = W(N_G - N_3) - \frac{N_3}{\tau_G} - \frac{|a(t)|^2}{P_G \tau_G} N_3 \tag{2.47a}$$

where

$$P_G = \frac{\hbar\omega_0 A_G}{\sigma_G \tau_G} \tag{2.47b}$$

Here N_G is the total population density of ions (etc.) responsible for gain, and N_3 is the population density in level 3, the upper laser level. The gain is given by $g(t) = \sigma_G N_3(t) l_g/2$, where l_g is the length of the gain medium.

Only the case of a *slow* saturable gain medium ($t_p \ll \tau_G$) is of practical importance for mode-locking dynamics. In this case the gain is written[1]

$$g(t) = g^{(i)} e^{-U(t)/U_G} \tag{2.48a}$$

where

$$U_G = P_G \tau_G \tag{2.48b}$$

$g^{(i)}$ is the gain just before the laser pulse. After the pulse the gain recovers exponentially to its small-signal value g_0, given by eq. (1.52):

$$g(t) = \left(g^{(i)} e^{-U/U_G} - g_0\right) e^{-t/\tau_G} + g_0 \tag{2.49}$$

Note that dynamic gain saturation is not always significant. For media such as semiconductors or dye molecules, where τ_G is on the order of the pulse repetition time (typically, nanoseconds), dynamic gain saturation per eq. (2.48a) can play an important role in mode-locking. However, when τ_G is much larger than the pulse period (e.g., doped solid-state media such as Ti:sapphire, Nd:YAG, or Er-doped glass fibers), dynamic (per pulse) gain saturation is very small, even though the gain medium does saturate in response to the average power. This is analogous to a resistor–capacitor low-pass filter, which exhibits small response for drive frequencies exceeding $1/RC$. In the laser gain medium, τ_G plays the role of the RC time.

[1] In addition to $t_p \ll \tau_G$, we also assume that $t_p \ll W^{-1}$. This assumption is justified since according to eq. (1.52), the small-signal gain saturates for $W\tau_G \gg 1$. Therefore, for most passively mode-locked lasers, there is usually no point in setting the pump rate (W) too far above τ_G^{-1}.

2.3.2 Slow Saturable Absorber Mode-Locking

In the early 1970s, subpicosecond pulse generation from passively mode-locked dye lasers was demonstrated using a nonlinear absorbing dye as the mode-locking element [26–28]. In the early 1980s, an improved passively mode-locked dye laser in a ring cavity, termed the *colliding pulse mode-locked* (CPM) *ring dye laser*, resulted in the first generation of sub-100-fs pulses [29,30]. The CPM laser became the dominant tool for femtosecond pulse generation until 1990, when new solid-state laser mode-locking techniques were invented. A very interesting feature of such passively mode-locked dye lasers is that the relaxation times of the amplifying and absorbing dyes were typically on the order of nanoseconds, much longer than the femtosecond pulse widths obtained. The following question then naturally arises: How is it that one can generate pulses a thousand times faster than the response time of the saturable dyes responsible for mode-locking? In the following we outline a theory of slow saturable absorber mode-locking put forth by Haus in the mid-1970s which explains this puzzle [25]. A related treatment is given in [31].

The time-dependent net (or total) gain function $g_T(t) = g(t) - \ell(t) - \ell_0$ plays an important role in the theory of slow saturable absorber mode-locking. $g_T(t)$ must be positive near the center of the pulse to provide amplification. Before and after, the pulse $g_T(t)$ must be negative to suppress the wings of the pulse and provide pulse shortening. For steady-state mode-locking, the pulse shortening per pass must balance pulse broadening, and the integrated gain must just balance the loss so that the pulse energy is kept constant.

Figure 2.7 shows schematically the pulse-shortening mechanism in mode-locking with a slow saturable absorber. Before the pulse the loss exceeds the gain. As the pulse arrives, it saturates the absorber so that the loss line drops below the gain. At this point the pulse experiences net amplification. Later in the pulse the gain begins to saturate, and as a result, the gain drops below the loss. The combined saturation dynamics of $g(t)$ and $\ell(t)$ lead to a net gain window during the pulse, with loss before and after the pulse. Thus, the absorber suppresses the leading edge of the pulse, then saturates to pass the center and trailing edge. The gain amplifies the center of the pulse but then saturates to suppress the trailing edge.

To treat this process analytically, the results of previous sections are used to describe the saturation dynamics. Our treatment closely follows [25]. The saturable loss $\ell(t)$ is written

$$\ell(t) = \ell^{(i)} e^{-U(t)/U_A} \approx \ell^{(i)} \left[1 - \frac{U(t)}{U_A} + \frac{1}{2}\frac{U^2(t)}{U_A^2}\right] \tag{2.50}$$

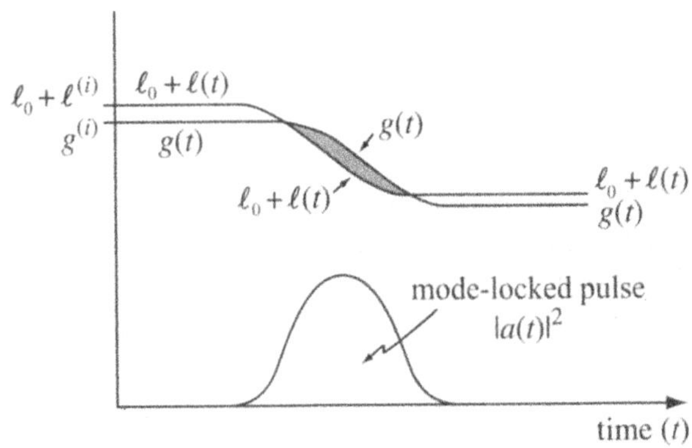

Figure 2.7 Pulse-shortening process in slow saturable absorber mode-locking. The shaded region indicates net positive gain.

where we assume the pulse energy is small enough that a second-order Taylor series expansion is adequate to describe the saturation dynamics. Following the pulse, the nonlinear loss recovers exponentially with time constant τ_A. The initial saturable loss $\ell^{(i)}$ follows the relation

$$\ell^{(i)} = \ell_{\text{sat}}^{(0)} + \left(\ell^{(i)} e^{-U/U_A} - \ell_{\text{sat}}^{(0)}\right) e^{-T/\tau_A} \tag{2.51}$$

Here $\ell_{\text{sat}}^{(0)}$ is the small-signal value of the nonlinear loss, U is the total pulse energy, and T is the round-trip time. The initial nonlinear loss may be less than the small-signal value if the absorber does not fully recover between pulses. In Haus's theory, however, it is assumed that the saturable absorber recovers essentially fully, and therefore

$$\ell^{(i)} \approx \ell_{\text{sat}}^{(0)} \tag{2.52}$$

This was a reasonable assumption for most subpicosecond dye lasers.

Similarly, the gain is given by

$$g(t) = g^{(i)} e^{-U(t)/U_G} \approx g^{(i)} \left[1 - \frac{U(t)}{U_G}\right] \tag{2.53}$$

Here a first-order Taylor series expansion is used, since according to Fig. 2.7, the gain saturates less deeply than the loss and since a first-order expression for the gain is sufficient to describe a net gain window. The initial gain $g^{(i)}$ is related to the small-signal gain g_0 by

$$g^{(i)} = g_0 + \left(g^{(i)} e^{-U/U_G} - g_0\right) e^{-T/\tau_G} \tag{2.54}$$

We can make several important observations right away:

- First, the net gain $g_T(t)$ must be less than zero both before and after the pulse for stability. Therefore,

$$g^{(i)} < \ell_0 + \ell^{(i)} \approx \ell_0 + \ell_{\text{sat}}^{(0)} \tag{2.55a}$$

and

$$g^{(i)} e^{-U/U_G} < \ell_0 + \ell_{\text{sat}}^{(0)} e^{-U/U_A} \tag{2.55b}$$

- For the laser to self-start, the small-signal gain must exceed the small-signal loss:

$$g_0 > \ell_0 + \ell_{\text{sat}}^{(0)} \tag{2.56}$$

Therefore, by comparing with (2.55a), we see that $g^{(i)} < g_0$. This means that the gain must not recover completely between pulses.

- To achieve a net gain window, the absorber must saturate before the gain. This means that

$$\frac{g^{(i)}}{U_G} < \frac{\ell_{\text{sat}}^{(0)}}{U_A} \tag{2.57}$$

Provided that the gain and absorber cross sections (σ_G and σ_A) are comparable, this condition can be achieved by focusing more tightly into the absorber.

Although satisfying all of these conditions simultaneously is possible, careful selection and matching of gain and absorber media are clearly required.

We now proceed to a more analytical treatment. The mode-locking equation is written

$$\frac{1}{\omega_c^2}\frac{d^2a(t)}{dt^2} - \delta T\frac{da(t)}{dt} + g_T(t)a(t) = 0 \tag{2.58}$$

where

$$g_T(t) = g(t) - \ell(t) - \ell_o \approx g^{(i)}\left(1 - \frac{U(t)}{U_G}\right) - \ell_{\text{sat}}^{(0)}\left(1 - \frac{U(t)}{U_A} + \frac{1}{2}\frac{U^2(t)}{U_A^2}\right) - \ell_0 \tag{2.59}$$

and $U(t)$ is the pulse energy up to time t as given by eq. (2.45). The net gain is plotted vs. energy $U(t)$ in Fig. 2.8. With the approximations made, g_T is a quadratic function of $U(t)$. This results in a net gain parabola which provides the pulse shortening necessary for mode-locking. Note that the net gain parabola cannot be selected arbitrarily. The net gain is a function of the mode-locked pulse, which is itself a function of the net gain function. One must solve for the net gain and the pulse shape and intensity self-consistently, unlike the case of active mode-locking, where the modulation is driven under external control.

We can most easily solve eq. (2.58) in the wings of the pulse, where g_T becomes a constant with $g_T < 0$. The solution is an exponential of the form

$$a(t) \sim e^{st}$$

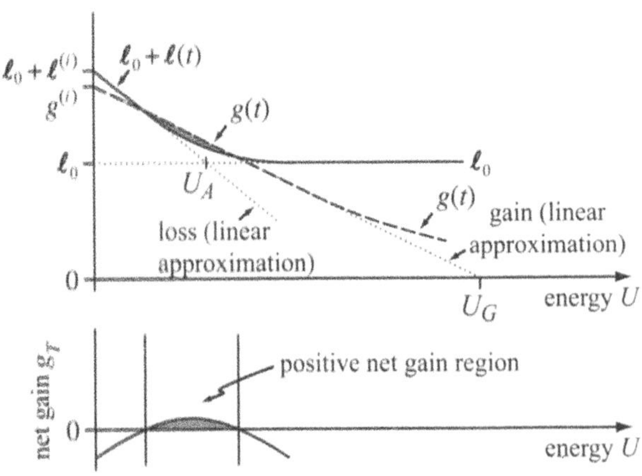

Figure 2.8 Time-dependent loss $\ell(t)$, gain $g(t)$, and net gain $g_T(t)$ plotted as a function of pulse energy $U(t)$. The shaded regions indicate net positive gain.

where

$$s = \frac{\omega_c^2 \delta T}{2} \pm \omega_c \sqrt{\left(\frac{\omega_c \delta T}{2}\right)^2 - g_T} \tag{2.60}$$

The plus sign in eq. (2.60) is taken for the leading edge of the pulse ($t < 0$), and the minus sign is taken for the trailing edge. This ensures that the pulse decays exponentially. Note that since the values of g_T before and after the pulse need not be the same, the exponential decay constants also do not have to be the same. The exponential rather than Gaussian decay is one clear difference compared to active mode-locking.

A symmetric solution of eq. (2.58) is given by [25]

$$a(t) = \frac{a_o}{\cosh(t/t_p)} = a_0 \operatorname{sech}\left(\frac{t}{t_p}\right) \tag{2.61}$$

(Asymmetric solutions also exist but are not discussed here.) The time-dependent energy $U(t)$ becomes

$$U(t) = \int^t |a(t)|^2 \, dt = \frac{U}{2}\left[1 + \tanh\left(\frac{t}{t_p}\right)\right] \tag{2.62}$$

where $U = 2a_0^2 t_p$ is the total pulse energy (we assume that a_0 is real). At this point there are four unknowns that must be solved for: the pulse width parameter t_p, the pulse amplitude a_0 (or equivalently, the pulse energy U), the time shift δT, and the initial gain $g^{(i)}$. All other parameters in the equation are inputs. Substituting eq. (2.61) and (2.62) into eq. (2.58) leads to three distinct sets of terms, proportional to $\operatorname{sech}(t/t_p)$, $\operatorname{sech}(t/t_p)\tanh(t/t_p)$, and $\operatorname{sech}^3(t/t_p)$, respectively. Requiring that each distinct set of terms sum to zero independently leads to three characteristic equations relating t_p, U, δT, and $g^{(i)}$. The fourth equation necessary to obtain a unique solution is the gain recovery equation (2.54).

The most successful example of a laser mode-locked using a slow saturable absorber was the colliding pulse mode-locked (CPM) ring dye laser, sketched in Fig. 2.9 [29]. The gain is provided by a flowing stream of rhodamine 6G dye dissolved in a suitable solvent, pumped

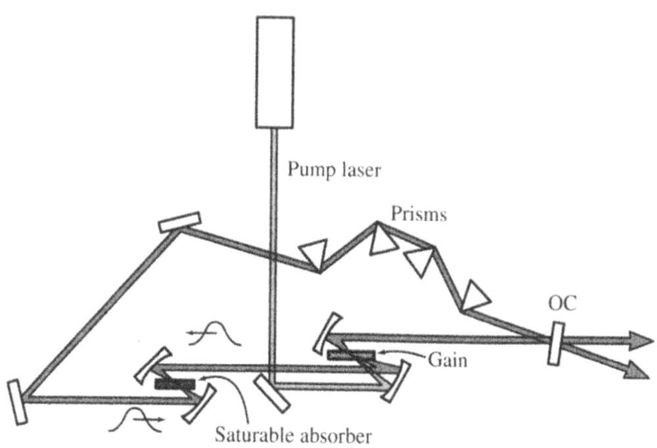

Figure 2.9 Colliding-pulse mode-locked ring dye laser. OC, output coupler.

by several watts of light from a continuous-wave argon ion laser. The saturable absorber is a flowing stream of an absorbing dye solution known as DODCI. The ring geometry leads to an additional mode-locking mechanism that improves the efficiency of the pulse-shortening process in slow absorber passive mode-locking. A ring resonator can support two pulses at the same time, one traveling clockwise and the other counterclockwise. It is most favorable energetically for these two pulses to meet, or collide, in the absorber stream. The standing-wave interference pattern generated when the pulses overlap in the absorber minimizes the energy lost because the absorber saturation is greatest where the optical field is most intense and weakest at the nulls of the interference pattern, where there is no optical energy. To best utilize this effect, the nozzle used to produce the absorber stream is squeezed to yield a dye thickness below a few tens of micrometers (compared to several hundred micrometers for the gain stream), which matches the absorber thickness to the spatial extent of the pulse collision region. This colliding pulse geometry enhances the saturable absorber mechanism, leading to shorter pulses and increased stability. The arrangement of four Brewster angle prisms (see Chapter 4) is used to adjust the sign and magnitude of group velocity dispersion in the resonator. Optimization of the CPM laser eventually led to pulse durations as short as 27 fs; in this mode of operation, intracavity pulse compression due to dispersion acting in concert with self-phase modulation in the dye appeared to supplement the pulse-shortening mechanisms caused by saturation [30]. A master equation analysis of passive mode-locking, including slow saturation as well as both group velocity dispersion and energy- and intensity-dependent self-phase modulation in the dye streams, was published in [32]. However, we defer treatment of such phase terms until we discuss solid-state laser mode-locking, where these phase effects are much more prominent.

2.3.3 Fast Saturable Absorber Mode-Locking

Fast saturable absorber mode-locking using real absorbers did not achieve the widespread use that was enjoyed by slow saturable absorber mode-locked systems. One reason for this may be that it is difficult to find real absorbers whose relaxation is fast compared to femtosecond pulses! Fast saturable absorbers have been used mainly for picosecond pulse generation with relatively narrowband solid-state laser systems, either (a) by using fast absorber dyes for simultaneous passive mode-locking and Q-switching [33] or (b) by using semiconductor saturable absorbers for steady-state mode-locking [34]. The theory presented in this section provides excellent background for Section 2.4, which covers today's leading mode-locking technology based on passive mode-locking of broadband solid-state lasers using artificial fast saturable absorbers implemented using the optical Kerr effect.

To analyze fast saturable absorber mode-locking, we begin with eq. (2.38) (for more details, see [24]). Since this technique is usually applied to solid-state lasers with low-gain cross sections and long relaxation times (τ_G typically on the order of microseconds to milliseconds), dynamic gain saturation during the pulse is very small. Therefore, we replace $g(t)$ with a constant value for the saturated gain, g, which is a function of the small-signal gain g_0 and the time-average laser power. The time-dependent loss $\ell(t)$ is given by

$$\ell(t) = \frac{\ell^{(i)}}{1 + |a(t)|^2/P_A} \approx \ell^{(i)}\left[1 - \frac{|a(t)|^2}{P_A}\right] \tag{2.63}$$

where $\ell^{(i)} = \ell_{\text{sat}}^{(0)}$ is the small-signal value of the saturable loss, which is expanded to first order in the power $|a(t)|^2$. This is justified if the mode-locked power remains sufficiently below the saturation power P_A. One difference in the treatment here compared to that in [24] is that we assume that the gain is frequency independent, whereas [24] assumes that the finite gain bandwidth plays the main role in limiting the bandwidth.

The resulting mode-locking equation is as follows:

$$\left(\frac{1}{\omega_c^2}\frac{d^2}{dt^2} - \delta T\frac{d}{dt} + \left(g - \ell_0 - \ell^{(i)}\right) + \ell^{(i)}\frac{|a(t)|^2}{P_A}\right) a(t) = 0 \tag{2.64}$$

The last term proportional to $|a(t)|^2 a(t)$ is referred to as *self-amplitude modulation* (SAM). In the wings of the pulse when $|a(t)|^2$ is very small, this SAM term is nearly zero, and the pulse decays exponentially, as in the preceding section. Also, as before, a secant hyperbolic pulse is a solution to the mode-locking equation, with

$$a(t) = a_0 \operatorname{sech}\left(\frac{t}{t_p}\right) \tag{2.65}$$

Substituting eq. (2.65) into (2.64) and requiring that the coefficients of the resulting $\operatorname{sech}(t/t_p)$, $\operatorname{sech}(t/t_p)\tanh(t/t_p)$, and $\operatorname{sech}^3(t/t_p)$ terms independently sum to zero yields three characteristic equations:

$$\frac{1}{\omega_c^2 t_p^2} + g - \ell_0 - \ell^{(i)} = 0 \tag{2.66a}$$

$$\frac{-2}{\omega_c^2 t_p^2} + \frac{\ell^{(i)} a_0^2}{P_A} = 0 \tag{2.66b}$$

$$\delta T = 0 \tag{2.66c}$$

Equation (2.66c) says that there is no time shift arising from the fast SAM process. To satisfy eq. (2.66a), we see that the net gain before and after the pulse must be less than zero. This is, in fact, a stability condition, since if the net gain were positive before or after the pulse, perturbations before or after the pulse would grow in amplitude. Figure 2.10 shows the net gain curve corresponding to these observations.

From eq. (2.66b) we see that the peak mode-locked power is inversely proportional to the pulse width squared. The pulse energy $\left(2a_0^2 t_p\right)$ is inversely proportional to t_p. To actually determine the pulse width and power, eq. (2.66b) is substituted in eq. (2.66a), with the result

$$\ell_0 + \ell^{(i)} - g = \frac{\ell^{(i)}}{2}\frac{a_0^2}{P_A} \tag{2.67}$$

Finally, by using the gain saturation equation

$$g = \frac{g_0}{1 + \langle P\rangle / P_G} \tag{2.68}$$

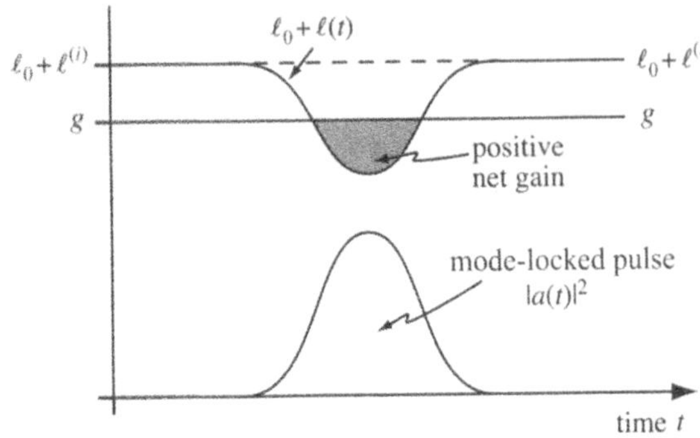

Figure 2.10 Gain and loss dynamics in fast saturable absorber mode-locking. The shaded region corresponds to positive net gain.

we obtain

$$\ell_0 + \ell^{(i)} - \left\{1 + \left(\frac{2P_A}{\ell^{(i)}}\right)^{1/2} \frac{2a_0}{\omega_c P_G T}\right\}^{-1} g_0 = \frac{\ell^{(i)} a_0^2}{2P_A} \tag{2.69}$$

where we have used $\langle P\rangle = 2a_0^2 t_p/T$ for the average laser power. A graphical solution to (2.69) is sketched in Fig. 2.11. The figure assumes that $g_0 > \ell_0 + \ell^{(i)}$, which is required for the laser to self-start even in continuous-wave operation. The *saturated gain* still satisfies $g < \ell_0 + \ell^{(i)}$, as required. Depending on the parameter values, eq. (2.69) has either zero or two solutions. No solutions are found when g_0 is too large (for fixed ℓ_0, $\ell^{(i)}$, etc.). In this case, even though there is no single-pulse mode-locked solution, solutions with two or

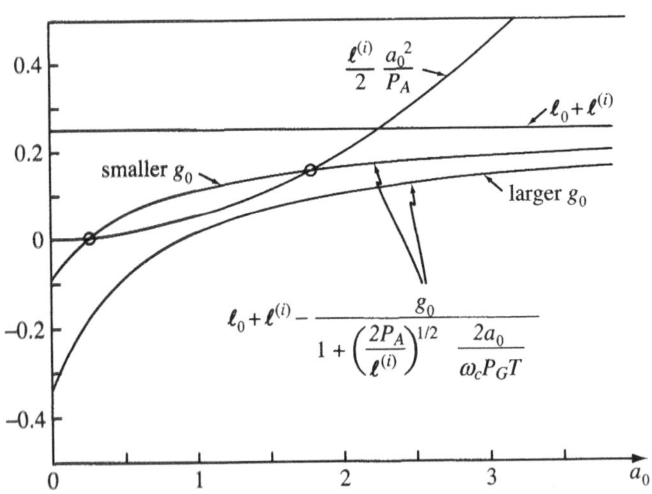

Figure 2.11 Graphical solution of eq. (2.69), in which the left- and right-hand sides of the equation are plotted vs. pulse amplitude a_0. Two cases are plotted: the case of a relatively large g_0 where no single-pulse solutions are found, and the case of a smaller g_0 where two solutions are found (marked by circles). Of these, only the longer-pulse, lower-peak-power solution is stable.

more separated pulses coexisting within a single round-trip time are still possible. When g_0 is reduced, two distinct solutions to eq. (2.69) are found, as indicated in the figure. Of the two solutions, the solution closer to the origin corresponds to the lower mode-locked power and hence the longer pulse. Haus has shown that only this longer pulse solution is stable.

Finally, we note that the strength of the SAM term responsible for pulse shortening and mode-locking is proportional to the pulse power $|a(t)|^2$. As long as the pulse energy remains roughly constant, we can conclude that the strength of the SAM is inversely proportional to the pulse width. As pulses get shorter and shorter during startup of mode-locked operation, the pulse shortening becomes more and more effective. As a result, fast absorber mode-locking can in principle support extremely short pulses. By the same token, when the pulses are still very long, the mode-locking process is very weak. This means that initiating mode-locking may be difficult. This corresponds to the experimental observation that passively mode-locked solid-state lasers are often not self-starting.

2.4 SOLID-STATE LASER MODE-LOCKING USING THE OPTICAL KERR EFFECT

During the early 1990s, passively mode-locked dye lasers were largely supplanted by passively mode-locked solid-state lasers. Today, such lasers produce femtosecond pulses at a variety of wavelengths, with pulses down to roughly 5 fs [35,36] and with average powers up to several watts (compared to at most tens of milliwatts from mode-locked dye lasers). This solid-state laser mode-locking is usually achieved using artificial saturable absorbers based on the *nonlinear refractive index*, also known as the *optical Kerr effect*. In addition to fast amplitude modulation, phase effects are very important in such lasers. In the following, we first briefly introduce the nonlinear refractive index and the pertinent new effects. We then discuss in some depth specific examples of solid-state laser geometries where the Kerr effect is converted into a fast amplitude modulation. At this point we write down the mode-locking equation and discuss important features of its solution. Finally, we comment briefly regarding self-starting of solid-state mode-locked lasers.

2.4.1 Nonlinear Refractive Index Changes

The nonlinear refractive index is usually written

$$n = n_0 + n_2 I(t) \tag{2.70}$$

where $I(t)$ is the pulse intensity, given by $I(t) = |a(t)|^2 / A_{\text{eff}}$ with $|a(t)|^2$ normalized to give the time-dependent power associated with the pulse as before, and A_{eff} is the effective area of the beam in the Kerr medium. Thus, the pulse induces a change in refractive index proportional to the instantaneous intensity. In most laser materials $n_2 > 0$, and therefore the laser pulse increases the refractive index momentarily. This is valid for nonresonant nonlinear index changes associated with intensity-dependent distortions of the electron clouds (electronic wave functions) in the material. The response time of the nonresonant optical Kerr effect is not known exactly but is usually estimated to be 1 to 2 fs. The optical Kerr effect is allowed for gases, liquids, and solids regardless of their symmetry. The size of the nonlinear refractive index (n_2) is material dependent, but a typical (perhaps somewhat small) value is that of fused silica, $n_2 = 3 \times 10^{-16}\,\text{cm}^2/\text{W}$ [37]. Ultrafast nonlinear effects arising due to the optical Kerr effect are discussed in detail in Chapter 6.

The nonlinear phase shift observed after propagating in a nonlinear medium of length L is given by

$$\Delta\phi(t) = \frac{-\omega}{c} n_2 I(t) L = \frac{-2\pi}{\lambda} n_2 I(t) L \tag{2.71}$$

For mode-locking, this fast time-dependent phase modulation must somehow be converted into a fast amplitude modulation to stabilize the wings of the mode-locked pulse. There are, in fact, numerous schemes for achieving this self-amplitude modulation. Some of the more popular schemes included *Kerr lens mode-locking* (KLM), in which the nonlinear index leads to self-focusing in the laser cavity; *additive pulse mode-locking* (APM), in which the nonlinear index modifies the interference between two coupled laser cavities; and *nonlinear polarization rotation*. This list is by no means exhaustive; a number of other schemes have also been demonstrated. In addition to self-amplitude modulation, which is similar to the fast saturable absorber action discussed in Section 2.3.3, the nonlinear refractive index also leads to new effects due to self-phase modulation. The interplay of self-phase modulation with group velocity dispersion is also important. The influence of these effects on laser operation is discussed initially in this chapter and then in more detail in Chapter 7.

2.4.2 Self-Amplitude Modulation, Self-Phase Modulation, and Group Velocity Dispersion

Self-Amplitude Modulation Similar to the fast saturable absorption studied previously, *self-amplitude modulation* (SAM) can be modeled by writing the loss terms in the mode-locking equation as

$$\ell_0 + \ell(t) \approx \ell - \gamma\,|a(t)|^2 \tag{2.72}$$

Here we have lumped both the constant cavity loss ℓ_0 and the small-signal value of the nonlinear loss (this would be $\ell^{(i)} = \ell_{\text{sat}}^{(0)}$ in the fast saturable absorber theory) into the loss constant ℓ, and γ (equal to $\ell^{(i)}/P_A$) is the SAM coefficient. Since $|a(t)|^2$ is normalized to give the time-varying power $P(t)$ carried by the laser beam [i.e., $P(t) = |a(t)|^2$], γ is the (negative) derivative of the *field loss coefficient* with respect to power:

$$\gamma = \frac{-\partial\ell}{\partial P} \tag{2.73}$$

In terms of the loss coefficient for the power, $\mathcal{L}$, which for small loss is given by $\mathcal{L} \approx 2\ell$, one has

$$\gamma = \frac{-1}{2}\frac{\partial\mathcal{L}}{\partial P} \tag{2.74}$$

Calculating γ requires a calculation specific to the particular laser system being considered and can usually be controlled over some range by adjusting the laser parameters. Here we use γ to characterize the strength of the SAM without specifying a particular laser system.

Self-Phase Modulation SAM is based on a time-varying nonlinear phase shift which itself modifies the pulse, called *self-phase modulation* (SPM). Assuming as usual that the SPM per pass is small, we can account for this effect by writing

$$a'(t) = e^{j\Delta\phi(t)}a(t) \approx \left[1 + j\Delta\phi(t)\right] a(t) \tag{2.75}$$

where $a(t)$ and $a'(t)$ represent the pulse envelope before and after the nonlinear refractive index medium. From eq. (2.71) the nonlinear phase shift is proportional to the intensity. We introduce an SPM coefficient δ where $\Delta\phi(t) \approx -\delta\,|a(t)|^2$ with $\delta > 0$ for positive nonlinear index media ($n_2 > 0$). We can then write

$$a'(t) \approx \left[1 - j\delta\,|a(t)|^2\right] a(t) \tag{2.76}$$

The principal effects of SPM can be understood by recognizing that the instantaneous frequency ω_{inst} of the pulse is simply the time derivative of the total phase:

$$\omega_{\text{inst}} = \frac{d}{dt}\left(\omega_0 t + \Delta\phi(t)\right) = \omega_0 + \frac{d\Delta\phi(t)}{dt} \tag{2.77}$$

Thus, the time-varying nonlinear phase shift gives rise to time-varying frequency shifts away from the initial carrier frequency. For a medium with $n_2 > 0$, SPM gives rise to red shifts (lower frequency, longer wavelengths) on the front edge of the pulse and blue shifts (higher frequencies, shorter wavelengths) on the trailing edge. This variation of instantaneous frequency as a function of time is called a *chirp*—in this case, an *up-chirp*, since the frequency increases with time. SPM creates new frequency components which can contribute to the mode-locking process. It turns out that the spectral broadening is small compared to the input bandwidth for $\Delta\phi_{\max} \ll \pi$. For $\Delta\phi_{\max} \geq \pi$, the output spectrum is roughly $\Delta\phi_{\max}/\pi$ times wider than the input bandwidth [37,38]. SPM is discussed in more detail in Chapter 6.

Group Velocity Dispersion We have also noted that group velocity dispersion (GVD) is important in femtosecond solid-state lasers (because the path lengths in material are relatively large and the pulses are very short). *Dispersion* refers to the dependence of group velocity on frequency. In the case of nonzero GVD, different frequencies have slightly different cavity round-trip times, and this can be an important pulse broadening mechanism. Mathematically, we can treat dispersion by considering the frequency dependence of the propagation constant β (also called k in Chapter 1). Using a second-order Taylor series expansion, we write

$$\beta(\omega) = \beta(\omega_0) + \beta_1(\omega - \omega_0) + \frac{\beta_2}{2}(\omega - \omega_0)^2 \tag{2.78}$$

Here $\beta_1 = \partial\beta/\partial\omega$ and $\beta_2 = \partial^2\beta/\partial\omega^2$, with both derivatives evaluated at $\omega = \omega_0$. We account for the dispersion in a medium of length L by writing

$$A''(\tilde{\omega}) = e^{-j\beta_0 L} e^{-j\beta_1 \tilde{\omega} L} e^{-j(\beta_2/2)\tilde{\omega}^2 L} A(\tilde{\omega}) \tag{2.79}$$

where $A(\tilde{\omega})$ and $A''(\tilde{\omega})$ are the spectrum before and after the dispersive medium, respectively, and $\tilde{\omega} = \omega - \omega_0$. For our purposes we can ignore the first two factors, since $e^{-j\beta_0 L}$ represents a constant phase shift and $e^{-j\beta_1 \tilde{\omega} L}$ represents a constant delay. We identify β_1 as the inverse of the group velocity. Assuming that the dispersion per pass is small, we obtain

$$A''(\tilde{\omega}) \approx \left(1 - j\frac{\beta_2}{2}\tilde{\omega}^2 L\right) A(\tilde{\omega}) \tag{2.80}$$

Taking the Fourier transform yields

$$a''(t) \approx \left(1 + jD\frac{d^2}{dt^2}\right) a(t) \tag{2.81a}$$

where

$$D = \frac{\beta_2 L}{2} \tag{2.81b}$$

The physical meaning of these equations is easily identified. We know that the Fourier transform of $a(t - \tau)$ is $A(\tilde{\omega})e^{-j\tilde{\omega}\tau}$. For a spectrum with an arbitrary spectral phase variation, written $A(\tilde{\omega}) = |A(\tilde{\omega})|\, e^{j\psi(\omega)}$, we can define a frequency-dependent delay $\tau(\omega)$ given by

$$\tau(\omega) = -\frac{\partial\psi(\omega)}{\partial\omega} \tag{2.82}$$

From eq. (2.79), we find that

$$\tau(\omega) = \beta_1 L + \beta_2 \tilde{\omega} L \tag{2.83}$$

Thus, for $\beta_2 > 0$ (called *normal dispersion*), higher frequencies (shorter wavelengths) travel more slowly and are displaced toward the trailing edge of the pulse. This leads to an up-chirp. For $\beta_2 < 0$ (*anomalous dispersion*), higher frequencies move faster and are displaced to the leading edge of the pulse, leading to a down-chirp.

We discuss dispersion in more detail in Chapter 4.

2.4.3 Additive Pulse Mode-Locking

The first widely exploited method for producing a self-amplitude modulation via the optical Kerr effect used the nonlinear phase shift to modify the interference between two coupled laser cavities. This method is called *additive pulse mode-locking* (APM) or *coupled cavity mode-locking*. A schematic setup is shown in Fig. 2.12. The gain medium sits in the main cavity, which is coupled to an auxiliary cavity through a partially reflective mirror. The lengths of the two cavities are very close to equal. A pulse traveling to the right in the main cavity is partially reflected by the coupling mirror. Part of the pulse is transmitted into the auxiliary cavity. Since the lengths of the two cavities are nearly equal, the pulses transmitted from the auxiliary cavity back into the main resonator arrive in synchronism and interfere with the main cavity pulses reflected from the coupling mirror. The overall reflectivity seen by the main cavity depends on the phase between these two sets of pulses.

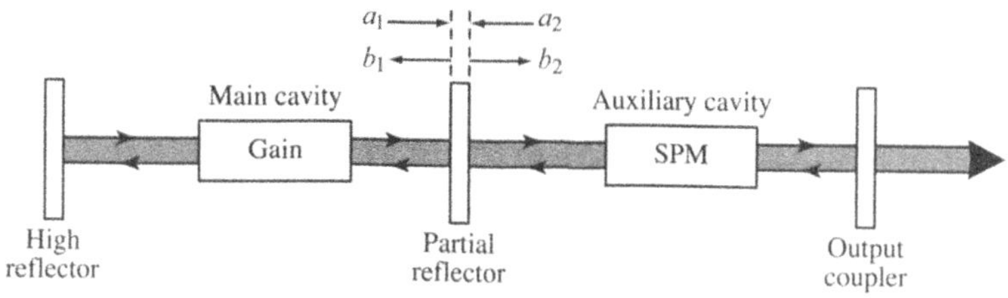

Figure 2.12 Layout of an additive pulse mode-locked or coupled cavity mode-locked laser. The lengths of the main and auxiliary cavities are closely matched, although the exact lengths may differ slightly to provide the required phase bias.

The highest reflectivity occurs when the round-trip phases of the two cavities are identical (modulo 2π). To produce self-amplitude modulation, a medium with a nonlinear refractive index is inserted into the auxiliary cavity. If the two cavities are initially out of phase and the nonlinear phase shift brings them closer to being in phase, the overall reflectivity seen by the main cavity increases. This results in a favorable self-amplitude modulation, which gives the highest reflectivity for the peak of the pulse. Of course, if the initial phase mismatch is such that the nonlinear phase shift puts the cavities more out of phase, the reflectivity will decrease with intensity and mode-locking will be suppressed. For this reason the relative cavity lengths in APM lasers should be interferometrically stabilized.

A forerunner of APM was the *soliton laser*, which also used a nonlinear coupled cavity principle to shorten the pulses from a synchronously pumped color-center laser [39]. However, a significant difference compared to the APM principle is that the fiber in the auxiliary cavity was sufficiently long that both dispersion and self-phase modulation were important. In fact, these effects interacted to form solitons, and soliton propagation was thought to be the main pulse-shortening mechanism for this laser. Although this was correct for the soliton laser as originally implemented, several papers in the late 1980s showed that neither soliton formation nor substantial dispersion were required to obtain effective passive mode-locking in coupled cavity systems [40–44]. The understanding that nonlinear coupled cavities provided a general mechanism for achieving an artificial rapidly saturable absorber without the need for soliton formation led within a few years to the application of APM in a variety of laser systems (e.g., [45–50]).

In one of the earliest clear examples of APM [43], a color-center laser with a broad gain bandwidth around 1.5 μm was modified to include an auxiliary cavity. The main cavity was synchronously pumped by 100-ps pulses from an actively mode-locked Nd:YAG laser. With the main cavity round-trip time adjusted to match the pump laser periodicity, synchronous pumping was sufficient to generate 23-ps pulses. With the addition of the auxiliary cavity containing a short (40-cm) fiber as the nonlinear index medium, the laser produced pulses as short as 127 fs via APM. Figure 2.13 shows data that resulted from examining the APM laser operation as the auxiliary cavity length was slowly ramped to bring the auxiliary cavity repeatedly in and out of phase with the main cavity [43]. For the laser configuration studied, the power in the main cavity and the output power were at a maximum and a minimum, respectively, when the two cavities were in phase. These quantities vary periodically, repeating when the relative phases of the two cavities change by 2π. The curve marked "SHG" indicates the power in the second-harmonic beam generated by focusing the mode-locked pulse onto an appropriate nonlinear crystal. Since the average

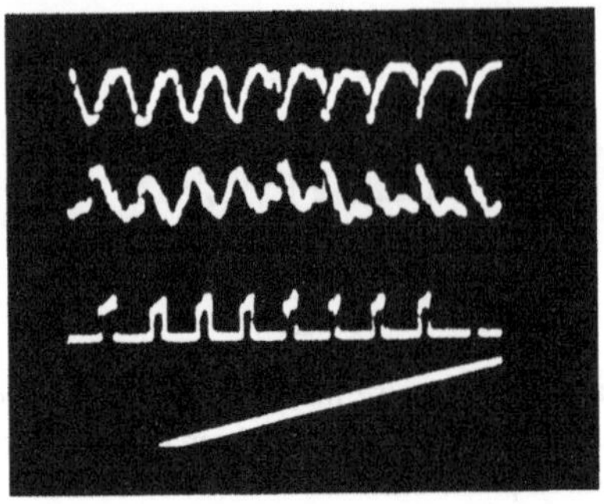

Figure 2.13 Variations of internal power, output power, and second-harmonic generation (SHG) as a function of the length of the fiber cavity. From [43].

SHG power is inversely proportional to the pulse width, the periods when the SHG is high indicate femtosecond pulse operation. The important finding is that the laser mode-locks only for certain relative cavity phases such that the nonlinear phase shift reduces the cavity loss.

The operation of an APM system can be understood by analyzing the reflectivity seen by the main cavity, taking the nonlinear auxiliary cavity into account. Following [43,44,51], we start with the scattering matrix [9] of the mirror separating the main and external cavities. This matrix relates the complex amplitudes of the waves a_i incident on the mirror and the waves b_i reflected from the mirror as follows:

$$\begin{pmatrix} b_1 \\ b_2 \end{pmatrix} = \begin{pmatrix} r & \sqrt{1-r^2} \\ \sqrt{1-r^2} & -r \end{pmatrix} \begin{pmatrix} a_1 \\ a_2 \end{pmatrix} \tag{2.84}$$

Here $i = 1$ and $i = 2$ refer to waves in the main cavity (left side of the mirror) and the auxiliary cavity (right side of the mirror), respectively. The mirror has field reflectivity r, which may be either positive or negative.[2] a_2 and b_2 are also related, since a_2 results simply from b_2 after one transit through the auxiliary cavity. Therefore,

$$a_2 = \eta b_2 e^{-j\phi} \tag{2.85}$$

where η represents the amplitude transmission factor through the auxiliary cavity and ϕ is the phase shift in the auxiliary cavity relative to that of the main cavity. The phase ϕ is

[2] The definition of the scattering matrix requires that $|a_i|^2$ and $|b_i|^2$ represent intensity, so the a_i and b_i actually represent the field amplitudes in medium i times the square root of the refractive index in that medium. Furthermore, the specific form of the scattering matrix given here applies only for certain choices of the reference planes where the a_i and b_i are measured. Transformations of the scattering matrix as the reference planes are translated with respect to the mirror are discussed in [9]. However, the results given below for the SAM and SPM coefficients are independent of the choice of reference planes.

given by $\phi = \phi_b + \Phi$, where ϕ_b is the phase bias of the auxiliary cavity ($\phi_b > 0$ when the auxiliary cavity is longer than the main cavity) and Φ is the nonlinear phase shift due to the double pass through the fiber, which can be written

$$\Phi = \frac{4\pi n_2 \left|a_f\right|^2 L_f}{\lambda A_{\text{eff}}} \tag{2.86}$$

where n_2, L_f, and A_{eff} are the Kerr coefficient, length, and effective area of the fiber, respectively, and a_f is the normalized amplitude of the pulse in the fiber. It is worth noting that although this formalism is most obviously applicable to continuous-wave light, it also works for mode-locked pulses, due to the assumption that the round-trip times in the main and external cavities are equal to within a small phase difference. The composite reflection coefficient $\Gamma = b_1/a_1$ seen by the main cavity can be obtained exactly by combining eqs. (2.84) and (2.85), which yields

$$\Gamma = \frac{b_1}{a_1} = \frac{r + \eta e^{-j\phi}}{1 + \eta r e^{-j\phi}} \tag{2.87}$$

In the usual case where an optical fiber is responsible for the nonlinearity, transmission through the auxiliary cavity is typically small, due to coupling losses into the fiber ($\eta \ll 1$). In this limit Γ can be expanded to first order in η, which gives

$$\Gamma \approx r + \eta(1 - r^2)e^{-j\phi} \tag{2.88}$$

This is equivalent to considering only a single pass through the external cavity. If we further assume a small nonlinear phase shift ($\Phi \ll 1$), this can be rewritten as

$$\Gamma \approx r + \eta(1 - r^2)e^{-j\phi_b}(1 - j\Phi) \tag{2.89}$$

Positive self-amplitude modulation occurs when the nonlinear phase shift brings the second term in eq. (2.89) closer to zero phase, since this results in the greatest reflectivity for the main cavity. Since the Kerr coefficient n_2 is positive, mode-locking requires $\phi_b < 0$ (i.e., the auxiliary cavity shorter than the main cavity). This is consistent with the experimental observations of [43] in Fig. 2.13.

Using eq. (2.89), we can determine the SAM and SPM coefficients. If we identify the field amplitude a_1 in the main cavity as the quantity of interest for the mode-locking equation and assume that $\left|a_f\right| = \eta\sqrt{1 - r^2}\,|a_1|$, which is appropriate if coupling into the fiber dominates the loss in the external cavity, the SAM coefficient is

$$\gamma = \frac{-4\pi\eta^3\left(1 - r^2\right)^2 n_2 L_f}{\lambda A_{\text{eff}}} \sin\phi_b \tag{2.90}$$

γ is maximized by setting $\phi_b = -\pi/2$. For $\phi_b = 0$ or π (linear reflectivity either a maximum or minimum, respectively), the SAM coefficient is zero. The SPM coefficient is given by

$$\delta = \frac{4\pi\eta^3\left(1 - r^2\right)^2 n_2 L_f}{\lambda A_{\text{eff}}} \cos\phi_b \tag{2.91}$$

Except for special settings of ϕ_b, both the SAM and SPM coefficients are nonzero. Furthermore, setting the phase bias of the auxiliary cavity allows the relative strengths of γ and δ to be varied.

It is worth noting that large nonlinear phase shifts are not unusual in APM systems. As a simple estimate, consider a laser at a 1.5-μm wavelength with a steady-state pulse width of 125 fs, a 100-MHz repetition rate, and 5 mW of average power coupled into the fiber. (Considering typical average powers of several hundred milliwatts to several watts for solid-state lasers, 5 mW in the fiber is rather conservative!) If we further assume a 80-cm round-trip distance in the fiber and a fiber effective area of 80 μm^2, we estimate a nonlinear phase shift of roughly 0.5 rad. When large nonlinear phase shifts Φ occur, the assumption of a SAM simply proportional to the intensity is no longer valid, since from eq. (2.88) the composite reflectivity Γ varies sinusoidally with the phase shift. Since the peak intensity and therefore the nonlinear phase shift increase with decreasing pulse width, the nonmonotonic variation of Γ with intensity can lead to a saturation of the pulse-shortening process, as can dispersion effects (not included above), which also become more important for shorter pulses. Nevertheless, this can generally occur only when the laser is already successful at generating pulses that are very short!

Due to the need for interferometric stabilization and the development of other successful implementations of fast absorber solid-state laser mode-locking, lasers mode-locked via APM using nonlinear coupled cavities are no longer common. However, several schemes widely used for mode-locking of fiber lasers are closely related to APM. Although these schemes do not use nonlinear coupled cavities directly, like APM they obtain self-amplitude modulation through the effect of nonlinear phase shifts on interfering waves. Examples include nonlinear polarization rotation, where the nonlinear phase shifts change the coherent summation of the two polarization modes in a fiber, and the nonlinear optical loop mirror, where the interference takes place between counterpropagating waves in a single loop of fiber. Both these implementations have the advantage of a stable geometry where active interferometric stabilization is not necessary. For further discussion, see Chapter 7.

2.4.4 Kerr Lens Mode-Locking

In current practice the most important method for pulse generation from broadband solid-state lasers is usually called *Kerr lens mode-locking* (KLM). KLM lasers based on titanium:sapphire (Ti:S) are particularly widespread. Sub-100-fs pulse generation from Ti:S lasers was first reported in 1991 [52]. Initially, these lasers were described simply as "self-mode-locked," since the mode-locking mechanism was not immediately identified. It was subsequently determined that a nonlinear lensing (or "*Kerr lensing*") *effect* associated with the nonlinear refractive index provides a fast self-amplitude modulation useful for mode-locking [53,54]. This Kerr lensing effect is important in most of today's shortest pulse Ti:S lasers, emitting pulses in the sub-10-fs range, and has also been applied for mode-locking of a number of other solid-state laser materials.

General Description The four-mirror laser cavity used for mode-locking of Ti:S in [52] is shown in Fig. 2.14a. Gain is provided by a 2-cm-long Ti:S crystal pumped through a dichroic mirror by an argon ion laser. Ti:S's gain spectrum is extremely broadband, with lasing possible from below 700 nm to longer than 1 μm (the lasing wavelength was around 880 nm in [52]). In some experiments a single-plate birefringent filter was used to select the

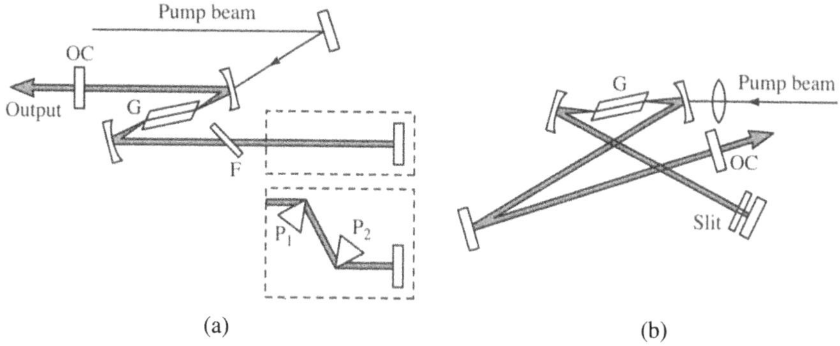

Figure 2.14 (a) Cavity configuration for the original self mode-locked Ti:S laser (the inset shows the intracavity prism sequence for dispersion compensation); (b) sub-10-fs Ti:S laser in which the prism pair is replaced by chirped, dispersion-compensating mirrors. G, gain medium (Ti:S crystal); OC, output coupler; F, birefringent filter. (a) Adapted from [52]; (b) adapted from Fig. 5 of [55], with permission from Springer Science and Business Media.

center wavelength. In other experiments the birefringent filter was removed, and instead, a prism pair (see Chapter 4) was used to compensate for the group velocity dispersion of the Ti:S rod. In the case of the prism pair, wavelength tuning could be accomplished using a vertical slit placed in front of the cavity end mirror, where the various optical frequencies are slightly displaced spatially. When the laser was properly aligned for a single-transverse mode (TEM_{00}) and lowest threshold continuous-wave operation, no mode-locking occurred. The remarkable observation was that when this laser was misaligned for a mixture of TEM_{00} and a higher-order transverse mode such as TEM_{05} (corresponding to a higher threshold), mode-locking could be induced by a suitable external perturbation, such as tapping one of the cavity mirrors. Although the mode-locking was not self-starting, once mode-locking was initiated, it was self-sustaining. The cavity containing the birefringent filter and no prisms produced 2-ps pulses that were highly chirped, with an optical bandwidth greatly exceeding that required to support a 2-ps pulse. The cavity containing the prisms generated pulses as short as 60 fs, with no appreciable chirp. The dramatic improvement seen when using prisms to compensate dispersion clearly indicates the importance of phase effects in such femtosecond lasers, as we discuss presently.

One arrangement for Ti:S lasers generating pulses in the range 10 fs and below is shown in Fig. 2.14b. The layout is similar to that of the original KLM laser geometry in Fig. 2.14a, with the differences that an "X" rather than a "Z" arrangement is used for the spherical mirrors, and the placement of the mirrors and gain rod are optimized. Another important difference [35,36,56–60] is that it is essential to minimize the intracavity dispersion over a very broad bandwidth, including higher-order dispersion terms, such as those where phase is proportional to $(\omega - \omega_0)^3$, which have not been included in our simple treatment so far. This is achieved in part by using shorter, more highly doped gain crystals. Furthermore, the intracavity dispersion-compensation elements must be optimized. This can be realized by choosing the prism material carefully; by replacing the prisms with specially designed chirped, dispersion-compensating mirrors; or by using a combination of prisms and chirped mirrors. Chirped mirrors also have the advantage of a very broad reflectivity bandwidth, which is an important issue in sub-10-fs laser design. A detailed treatment of dispersion and dispersion compensation is the subject of Chapter 4.

To explain the nonlinear lensing responsible for the self-mode-locking observed in the very simple laser cavities of Fig. 2.14, we observe that the mode-locked pulse is a function of both time and space. Assuming that the time and space coordinates are independent, the field can be written most simply as a circular Gaussian beam in the manner of eq. (1.59). The intensity $I(r, t)$ is given by

$$I(r,t) = \frac{2}{\pi w^2} |a(t)|^2 e^{-2r^2/w^2(z)} = \frac{2}{\pi w^2} P(t) e^{-2r^2/w^2(z)} \tag{2.92}$$

where $w(z)$ is the beam radius, which varies along the longitudinal cavity coordinate z; r is the normal radial distance in cylindrical coordinates; and the pulse shape $a(t)$ is normalized such that $|a(t)|^2$ gives the instantaneous power $P(t)$, as usual. It is clear that the intensity and hence the nonlinear index change induced in the Ti:S crystal vary with r. The largest index change occurs at the center of the beam. This radially varying index (hence, phase) change is equivalent to a self-induced lens which can modify the beam radius $w(z)$ of the lasing mode. Furthermore, the strength of the self-induced lensing effect is time dependent and follows the pulse temporal intensity profile. Under suitable conditions this lensing effect can reduce the loss or increase the gain with increasing power. When this occurs, the Kerr lens acts as a fast saturable absorber.

There are several ways in which a nonlinear change in mode structure or beam radius can reduce the loss. For example:

- The misaligned laser containing higher-order transverse mode content such as TEM_{05} may experience loss due to beam clipping on the edge of one of the end mirrors. Self-lensing may improve the mode quality or decrease the beam size on the end mirrors, hence reducing loss.
- A similar effect may arise with respect to any other optical element in the cavity. For example, the beam may be clipped by the apex of one of the dispersion-compensating prisms. In one study [54] a self-mode-locked Ti:S laser was aligned so that light leaking past one of the prisms accounted for a 2% loss in continuous-wave operation. After mode-locking was initiated, the measured prism loss was reduced tenfold. Measurements showing a substantial change in beam profile with mode-locking were also performed and are reproduced in Fig. 2.15. Data of this sort give direct evidence of the importance of nonlinear mode profile changes in the mode-locking process.

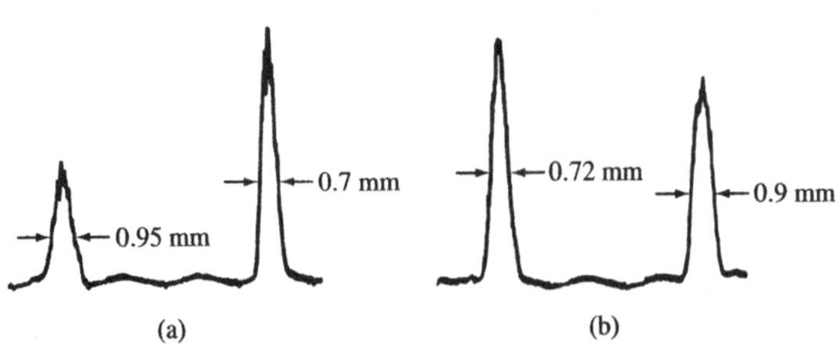

Figure 2.15 Measured horizontal mode profiles from a KLM Ti:S laser: (a) CW operation; (b) mode-locked. The profiles were recorded at the prism closest to the laser rod. For both (a) and (b) the right peak corresponds to the beam coming from the crystal, and the left peak is the beam returning toward the crystal after reflecting from the end mirror. From [54].

- To accentuate and control the beam clipping effects, an appropriately sized slit may be placed near one of the end mirrors. This can increase the beam clipping effects and the size of the loss modulation resulting when the mode size is modified through self-lensing. This approach is termed *hard-aperture KLM*.
- Self-lensing and self-focusing may affect the overlap of the lasing mode with the spatial profile of the pump laser beam in the Ti:S crystal. If spatial overlap improves with power, the net gain can be increased, creating a situation similar to that of a fast saturable absorber [61]. This approach is termed *soft-aperture KLM*.

It is important to note that depending on the cavity configuration, the nonlinear lens can either decrease or increase the net loss. When the loss is decreased, mode-locking is possible; when the loss is increased, mode-locking is suppressed.

Properties of KLM Resonators and the SAM Coefficient We have seen that the SAM and SPM coefficients are important in modeling laser behavior. In the following we outline how these are calculated, given knowledge of the (power-dependent) spatial mode structure for a specific laser configuration.

For hard-aperture KLM, the fraction T of the power transmitted through an aperture with power transmission function $S(x, y)$ is given by [62]

$$T = \frac{2}{\pi w^2} \int_{-\infty}^{\infty} \int_{-\infty}^{\infty} dx\, dy\, S(x, y) e^{-2\left(x^2+y^2\right)/w^2} \tag{2.93}$$

where a circular Gaussian beam is assumed and w is the beam radius at the aperture. For an aperture with circular symmetry, this can be written

$$T = \frac{2}{\pi w^2} \int_0^{\infty} 2\pi r\, dr\, S(r) e^{-2r^2/w^2} \tag{2.94}$$

The loss for the power is given by $\mathcal{L} = 1 - T$.

In the case of a circular aperture which transmits only within a radius w_s, we get

$$T = \frac{2}{\pi w^2} \int_0^{w_s} 2\pi r\, dr\, e^{-2r^2/w^2} = 1 - e^{-2w_s^2/w^2} \tag{2.95}$$

and

$$\mathcal{L} = e^{-2w_s^2/w^2} \tag{2.96}$$

To calculate the linear loss, one uses this formula with w given by its value at low power. The SAM coefficient is given by

$$\gamma = \frac{-1}{2} \frac{\partial \mathcal{L}}{\partial P} = \frac{-2w_s^2}{w^2} e^{-2w_s^2/w^2} \left[\frac{1}{w(0)} \frac{\partial w(0)}{\partial P} \right] = \frac{-2w_s^2}{w^2} \left[\frac{1}{w(0)} \frac{\partial w(0)}{\partial P} \right] \mathcal{L} \tag{2.97}$$

Considerable effort has gone into optimizing cavity geometries to obtain strong SAM, which means from eq. (2.97) that $[1/w(0)][\partial w(0)/\partial P]$, the fractional change in spot size with respect to power evaluated at $P = 0$, should be large. We also note from eq. (2.97) that γ is proportional to the linear loss introduced by the aperture. To avoid excessive linear loss, the beam radius w should be slightly smaller than the aperture radius w_s.

Similar calculations can be applied for a slit aperture, which is the geometry most commonly employed in real hard-aperture KLM lasers. For example, if the slit has width $2w_s$ centered at $x = 0$, we obtain from eq. (2.93)

$$T = \operatorname{erf}\left(\frac{\sqrt{2}\, w_s}{w}\right) \tag{2.98}$$

where the error function is defined as $\operatorname{erf}(x) = (2/\sqrt{\pi}) \int_0^x e^{-t^2} dt$ [10]. The SAM coefficient γ is given by

$$\gamma = \frac{-\sqrt{2}}{\sqrt{\pi}} \frac{w_s}{w} e^{-2w_s^2/w^2} \left[\frac{1}{w(0)} \frac{\partial w(0)}{\partial P}\right] \tag{2.99}$$

In the limit of low loss ($w_s \gg w$), the error function in eq. (2.98) can be simplified by using the asymptotic expression $\operatorname{erf}(z) \approx 1 - (1/\sqrt{\pi} z) \exp(-z^2)$ for $z \to \infty$ [10], which yields

$$\mathcal{L} = \frac{1}{\sqrt{2\pi}} \frac{w}{w_s} e^{-2w_s^2/w^2} \tag{2.100}$$

and

$$\gamma = \frac{-2w_s^2}{w^2} \left[\frac{1}{w(0)} \frac{\partial w(0)}{\partial P}\right] \mathcal{L} \tag{2.101}$$

For small $\mathcal{L}$ the relation between γ and $\mathcal{L}$ is exactly the same as for a circular aperture.

Note that the calculations above neglect any diffraction caused by the aperture. Therefore, these formulas somewhat underestimate both the loss and the SAM coefficient.

For soft-aperture KLM, the SAM coefficient depends on the power-dependent single-pass gain, which can be calculated by performing a three-dimensional overlap integral between the laser mode in the crystal and the position-dependent saturated gain coefficient. Omitting proportionality constants, this integral has the form

$$g_{sp} = \int dV \frac{\left|E_{pump}(x, y, z)\right|^2 \left|E_{laser}(x, y, z)\right|^2}{1 + 2\langle P(x, y, z)\rangle / P_G} \tag{2.102}$$

Here g_{sp} and $E_{laser}(x, y, z)$ refer to the single-pass gain experienced by a specific time slice within the laser pulse and the corresponding spatial profile corresponding to that time slice. The local small-signal gain is proportional to the local pump intensity, but the actual gain is saturated by the average laser power $\langle P(x, y, z)\rangle$ at that point in the crystal. The factor of 2 arises in a linear cavity because the pulse makes a double pass through the crystal. A detailed discussion of this gain integral may be found in [63], for example, Calculation of

g_{sp} and its variation with instantaneous power is usually performed numerically for KLM lasers (for a discussion, see, e.g., [54,64]). For mode-locking to be possible, $\partial g_{sp}/\partial P$ must be positive.

The SPM coefficient δ reflects the nonlinear phase shift caused by the focused beam in the laser crystal. It can be estimated from the on-axis nonlinear phase shift, which for a double pass through a crystal of length L is [65]

$$-\delta\,|a(t)|^2 = \frac{-4\pi n_2}{\lambda}\int_{-L/2}^{L/2}\frac{2P(t)\,dz}{\pi w^2(z)} \tag{2.103}$$

where we have used eq. (2.92) for the on-axis intensity $I(0, t)$. The beam radius varies in the crystal according to the Gaussian beam law,

$$w^2(z) = w_0^2\left[1+\frac{z^2}{z_0^2}\right] \tag{2.104}$$

where $z_0 = \pi n w_0^2/\lambda$ and we have assumed that the beam waist occurs in the center of the crystal. In the limit of weak focusing ($L \ll z_0$), $w(z) = w_0$ is a constant, and

$$\delta = \frac{8n_2 L}{\lambda w_0^2} \qquad \text{for} \qquad L \ll z_0 \tag{2.105}$$

In the opposite, tight-focusing extreme ($L \gg z_0$), the integral can be extended to infinity, with the result

$$\delta = \frac{8\pi^2 n n_2}{\lambda^2} \qquad \text{for} \qquad L \gg z_0 \tag{2.106}$$

Note that calculation of δ for a given laser configuration only requires knowledge of the linear mode properties (i.e., the beam radius in the crystal).

To evaluate γ, from eq. (2.97) for example, one must know the variation in mode size with power. This requires analysis of the laser resonator. Let us first discuss the stability conditions for the resonator in the low-power (linear) regime. A simplified diagram for analysis of the resonator is shown in Fig. 2.16a. Compared to Fig. 2.14, the pair of spherical mirrors is represented by a pair of internal lenses (with lens focal length equal to half the mirror radius of curvature). Usually, the two spherical mirrors have identical radii of curvature, typically 10 cm. The distances s_1 and s_2 from the lenses to the planar end mirrors are typically between 50 and 150 cm. Prism pairs, which may be used for dispersion compensation, are not included in Fig. 2.16, since they do not affect the stability analysis. From eq. (1.66) the effective distance between the curved mirrors is given by $d' = d - (n-1)d_{\text{crystal}}/n$, where d is the physical distance and d_{crystal} is the length through the laser crystal with refractive index n. The resonator analysis is simplified by recognizing that each lens–planar mirror combination can be replaced by an equivalent curved mirror as shown in Fig. 2.16b [66,67]. The equivalent curved mirrors sit in the planes where the planar mirrors are imaged, that is,

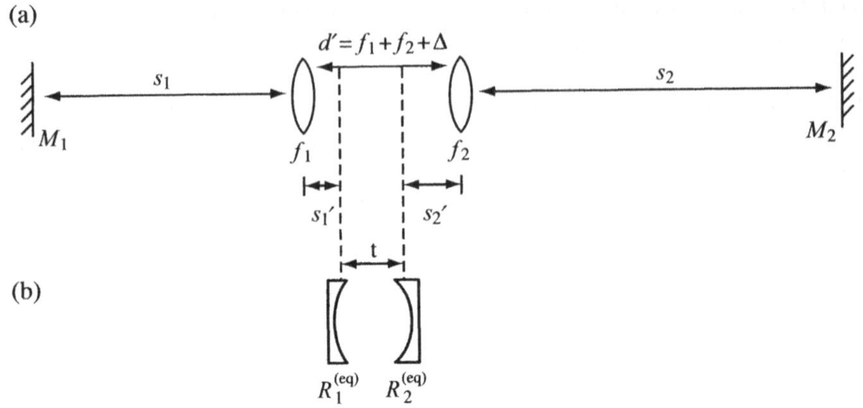

Figure 2.16 (a) Internal lens model for analysis of a KLM laser resonator; (b) equivalent two-mirror resonator.

$$\frac{1}{s_i'}+\frac{1}{s_i}=\frac{1}{f_i} \qquad \text{for} \quad i=1,2 \tag{2.107}$$

and the equivalent radii of curvature[3] are given by

$$R_i^{(eq)}=\frac{-f_i^2}{s_i-f_i} \qquad \text{for} \quad i=1,2 \tag{2.108}$$

If we also introduce a parameter Δ which characterizes the range of stable operation, defined by

$$d'=f_1+f_2+\Delta \tag{2.109}$$

we find that the equivalent distance t between the two equivalent curved mirrors is

$$t=R_1^{(eq)}+R_2^{(eq)}+\Delta \tag{2.110}$$

This result is useful because the simple two-mirror resonator of Fig. 2.16b has been analyzed thoroughly. For example, if we substitute eq. (2.110) into the stability condition for the two-mirror resonator, eq. (1.69a), we find that the resonator is stable only for values of Δ satisfying

$$0<\Delta<-R_2^{(eq)} \qquad \text{or} \qquad -R_1^{(eq)}<\Delta<-\left(R_1^{(eq)}+R_2^{(eq)}\right) \tag{2.111}$$

where without loss of generality we have taken $|R_2^{(eq)}|\leq|R_1^{(eq)}|$. Figure 2.17 shows the stability regions and the beam diameters calculated at the planar end mirrors for a cavity with $f_1=f_2=5\,\text{cm}$, $s_1=68\,\text{cm}$, and $s_2=118\,\text{cm}$ as a function of the mirror spacing parameter [55]. There are two distinct stability zones, each with a calculated width of 2.2 mm, and separated by 1.8 mm. At the edges of the stability zones, the calculated beam

[3] In most mode-locked Ti:S lasers, $s_i>f_i$, and therefore the equivalent radii of curvature are negative.

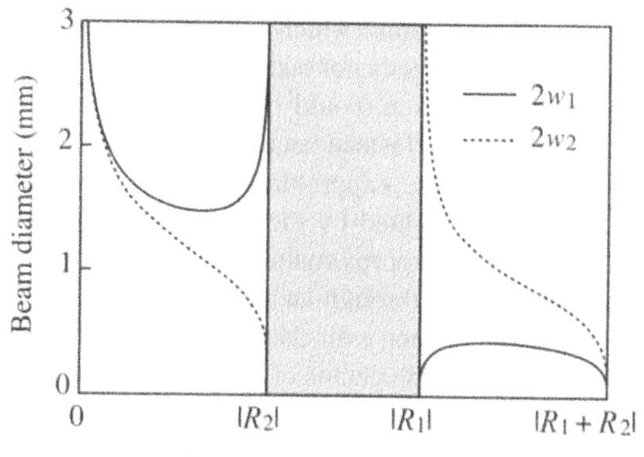

Figure 2.17 Stability regions and calculated low-power beam diameters for a KLM resonator, where $2w_1$ and $2w_2$ are the beam diameters at mirrors M_1 and M_2 respectively, in Fig. 2.16(a). R_1 and R_2 refer to the equivalent radii of curvature of eq. (2.108). Cavity parameters are given in the text. Adapted from Fig. 2 of [55], with permission from Springer Science and Business Media.

diameters vary very rapidly, either diverging or collapsing to zero. It turns out that the largest SAM coefficient in hard-aperture KLM also occurs near the edges of a stability zone. Intuitively, this makes sense. Since the mode parameters are very sensitive to the spacing of the curved mirrors, it is also reasonable that they should be very sensitive to a nonlinear lens placed between these curved mirrors.

To calculate the power-dependent mode profile, one must model the propagation in the laser crystal as influenced by the nonlinear lens. Such nonlinear propagation has been studied in nonlinear optics for many years under the term *self-focusing* [68,69]. In KLM lasers self-focusing has been handled by both numerical and analytical methods. In the numerical approaches (e.g., [54]), the crystal is typically divided into many thin slices, and in each slice the effect of linear diffraction and of the nonlinear lens are calculated essentially independently. For thin enough slices, this accurately models nonlinear propagation through the entire crystal. The remainder of the laser resonator can be modeled via conventional diffraction theory. This calculation is performed iteratively for many passes through the laser until a self-consistent mode profile is obtained.

In the analytical methods self-focusing is treated under the *aberrationless approximation*, in which the nonlinear lens is assumed to have a parabolic index profile [70–72]. This approximation is useful because a Gaussian beam propagating through a medium with a parabolic index profile remains Gaussian. Therefore, the entire laser resonator can be treated using the formulas for Gaussian beam propagation or the *ABCD* matrices, introduced in Chapter 1. This simplifies the calculation and allows analytic results to be derived. The index profile is taken as

$$n(r) = n_0 + \Delta n(0)e^{-2r^2/w^2(z)} = \Delta n(0)\left(1 - \frac{2r^2}{a_{SF}w^2(z)}\right) \tag{2.112a}$$

where $\Delta n(0)$ is the on-axis index change given by

$$\Delta n(0) = n_2 I(t) = \frac{2n_2\,|a(t)|^2}{\pi w^2(z)} \tag{2.112b}$$

Here $w^2(z)$ is the Gaussian beam radius (which varies as the beam propagates through the crystal). Note that the index change is not taken as simply proportional to the second-order expansion of the intensity, which would overestimate the effect of the nonlinear lens on the wings of the beam profile. Instead, an adjustable parameter a_{SF} is introduced to improve the accuracy of the aberrationless approximation [72,73]. Adjusting a_{SF} to minimize the mean-square error in $\Delta n(r)$ weighted by the spatial intensity profile yields $a_{SF} = 4$ [74]. Furthermore, the aberrationless approximation has been compared with experimental studies of the nonlinear transmission through an aperture placed in the far field of a thick self-focusing medium; good agreement with data was achieved when a_{SF} was adjusted between 3.77 and 6.4, depending on the details of specific experiments [72].

The nonlinear propagation within the laser rod has been modeled analytically by several authors (e.g., [67,73,74]). The behavior is found to depend on the ratio P/P_{crit}, where P_{crit} is the critical power for self-focusing, given by

$$P_{crit} = \frac{a_{SF}\lambda^2}{8\pi n n_2} \tag{2.113}$$

The significance of the critical power in three-dimensional self-focusing theory is that in a long enough medium, a collimated or initially focused beam will eventually always spread due to diffraction for $P < P_{crit}$, whereas for $P > P_{crit}$, the beam will always self-focus catastrophically to a very small spot (where, for example, damage may ensue). The distance at which the beam self-focuses (assuming $P > P_{crit}$) is shorter for higher P/P_{crit}. It is also informative to note that the nonlinear phase shift which would be predicted using eq. (2.106) for the case of Gaussian beam in a long medium is $a_{SF}\pi P/P_{crit}$ (this equation, of course, neglects nonlinear beam reshaping and focusing). Thus, we have the intuitive result that large self-focusing effects are expected when the integrated nonlinear phase shifts approach or exceed π. The parameters in a KLM laser are, of course, chosen to remain below the range for catastrophic self-focusing within the crystal; nevertheless, the ratio P/P_{crit} is still the important parameter.

As long as the laser crystal can be described analytically, the rest of the resonator can be analyzed using the conventional *ABCD* matrices. Although detailed coverage of these calculations is beyond the scope of our treatment, there are several general results that emerge from the analysis [75] that we enumerate here. These results are expressed in terms of a coefficient,

$$\delta_w = \frac{1}{w(0)} \frac{\partial w(0)}{\partial (P/P_{crit})} \tag{2.114}$$

which characterizes the variation in mode size with respect to normalized power:

- For resonators with planar end mirrors, as is usually the case for KLM lasers, the maximum attainable value of δ_w occurs at one of the end mirrors and at positions internal to the resonator corresponding to image or Fourier transform planes of that end mirror. This means that placing an aperture near an end mirror is optimum for achieving large γ (assuming, of course, that δ_w is negative at that end mirror).

- The maximum first-order variation of the normalized spot size with respect to normalized power is given by

$$|\delta_w| \leq \frac{1}{4\sqrt{g_1 g_2 (1 - g_1 g_2)}} \tag{2.115}$$

where g_1 and g_2 are the resonator stability parameters from eq. (1.69b). Recall that the resonator is stable only for $0 < g_1 g_2 < 1$. Therefore, the largest γ coefficient can be attained close to the stability limits.

- To reach the limit given by eq. (2.115), the resonator, including the position of the gain medium, must be designed optimally. Convenient formulas expressing δ_w in terms of the placement of the resonator elements are given in [75]. The most critical parameters are the distance between the spherical mirrors, which determines the stability of the resonator, and the longitudinal position of the laser rod with respect to the spherical mirrors, which largely determines the nonlinear variation of the far-field spot size for a given resonator stability. Calculated values of δ_w are plotted in Fig. 2.18 as a function of spherical mirror spacing and the position of the laser rod. Two different cases, corresponding to two different degrees of resonator asymmetry, are shown. The contour plots, which include only negative values of δ_w, demonstrate that (1) the largest negative values of δ are achieved near stability limits, and (2) negative δ_w is obtained only when the laser rod is positioned asymmetrically with respect to the midpoint between the two spherical mirrors. Figure 2.18 also indicates (filled circles) the resonator configurations for which self-sustaining hard-aperture KLM was observed in corresponding experiments. The close correspondence between these data and the δ_w contours provides strong evidence of the need for a sufficiently large SAM coefficient for successful KLM.

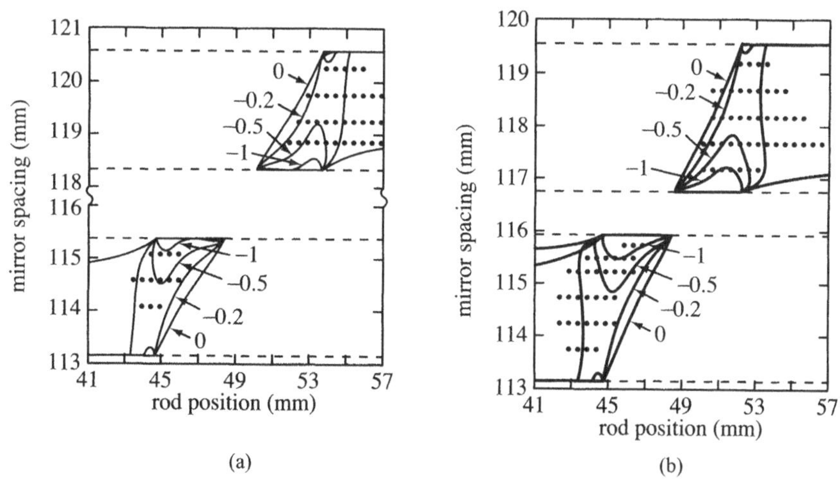

Figure 2.18 Contour plots showing calculated values of δ_w as a function of spherical mirror spacing and laser rod position. (a) $s_1 = 50$ cm, $s_2 = 110$ cm; (b) $s_1 = 70$ cm, $s_2 = 90$ cm. Filled circles indicate resonator configurations for which self-sustaining hard-aperture KLM was observed experimentally, and dashed horizontal lines indicate stability limits. In both cases, $f_1 = f_2 = 5$ cm and the slit for hard-aperture KLM was placed in the short arm of the laser cavity. Adapted from [76].

The dependence of δ_w on the position of the laser rod can be understood by noting that a nonlinear self-focusing medium (i.e., $n_2 > 0$) placed prior to the focus of a converging beam tends to enhance the focusing, leading to greater divergence and reduced transmission through an aperture in the far field; conversely, a self-focusing medium placed after the focus tends to make the beam more collimated, increasing transmission through a far-field aperture. Consequently, the nonlinear transmission through the aperture changes sign when the nonlinear medium is scanned through the focus and is small or zero when the nonlinear medium is placed symmetrically with respect to the focus [72,77].

At this point we are ready to discuss the size of the nonlinear effects in KLM lasers. As an example, we consider the power levels in the original self-mode-locked laser of [52], which are fairly typical for KLM Ti:S lasers. From the 450-mW output power emerging from a 3.5% output coupler, we calculate an average intracavity power of 12.8 W. To estimate the size of the SAM effect, we use eq. (2.99). We take $\delta_w = 1$, which according to Fig. 2.18 is a relatively large value, and we assume that $w/w_s = 0.8$, which gives a linear loss of 1.4% for the power. This yields $\gamma = 0.044 P_{crit}^{-1}$. For Ti:S the critical power is usually taken as 2.6 MW [73], which yields $\gamma \approx 0.02\ \text{MW}^{-1}$. Let us assume that before the laser is mode-locked, the initial power fluctuations due to random-mode beating between multiple unlocked longitudinal laser modes are on the order of the average continuous-wave power (i.e., 12.8 W). Then we have $\gamma P \approx 2 \times 10^{-7}$ for the self-amplitude modulation term, which is quite weak. For this reason many KLM lasers are not self-starting. On the other hand, once the laser is mode-locked, the nonlinear pulse shaping effects can be quite strong. For mode-locked operation the peak pulse power is enhanced by the ratio of the pulse round-trip time to the pulse width, so for 60-fs pulses with a 12-ns round-trip time, the peak power is about 2.6 MW, which is roughly P_{crit} in this particular laser. We estimate a SAM term $\gamma P \approx 0.05$. Since the analysis we are using consists of a first-order expansion in P, this estimate for γP may become inaccurate at the high peak powers encountered in mode-locked operation; nevertheless, the basic conclusion that SAM is a large effect does not change! Since $P \approx P_{crit}$ in this laser, from our previous discussion we conclude that the nonlinear phase shift is itself on the order of π, and therefore SPM can also be a strong effect. Thus, even in the 60-fs regime, both the nonlinear amplitude modulation and the nonlinear phase modulation are becoming large. For even shorter pulses (e.g., 6 fs), the peak powers that drive the nonlinearities can be an order of magnitude higher.

In Section 2.4.5 we describe the solutions to the master equation governing mode-locking of such solid-state lasers. Although the master equation is derived with the assumption of small changes per element per round trip, the solutions provide important insight into mode-locked laser operation, especially for pulses of a few tens of femtoseconds and above. Additional phenomena, associated with violation of the assumption of small changes per round trip, are discussed in Chapter 7.

Before concluding this section, we discuss briefly the role of astigmatism in KLM lasers, which we have neglected so far. By *astigmatism* we mean that the Gaussian beam radius, and the position of the beam waist, may be different for the x and y transverse directions. KLM lasers as depicted in Fig. 2.14 contain two sources of astigmatism: the spherical mirrors used away from normal incidence and the Brewster angle cut laser rod [78]. In astigmatic systems one distinguishes the tangential plane, which lies in the plane of Fig. 2.14, from the sagittal plane, which lies perpendicular to the figure. Thus, the beam folding at the spherical mirror and the change of direction due to refraction at the surface of the Brewster cut rod both lie within the tangential plane. In astigmatic resonators the stability zones calculated for tangential and sagittal planes can be displaced and may not overlap at all. Fortunately, astigmatically compensated cavities can be realized in which the angles of incidence on the spherical mirrors are chosen so that the mirror astigmatism compensates for that of the laser rod; such astigmatically compensated cavities, which were analyzed in [78], are generally used for KLM lasers. Note that although with astigmatic compensation the stability regions are the same for the tangential and sagittal planes, the positions of the beam waists and the beam radii in the laser rod may still be different for the two planes. As a result, astigmatism may still influence KLM even with an astigmatically compensated laser cavity.

Many of the analyses of KLM resonators account for astigmatism by calculating the mode properties and δ_w independently for tangential and sagittal planes. That is, the calculation is performed for a cylindrically symmetric beam; but different values for the focal lengths and other parameters are used to predict the different behavior for the two planes. The results indicate that for most KLM resonators, the astigmatism enhances (decreases) δ_w in the tangential (sagittal) plane [67,76]. This is the reason that a vertical slit is generally used for hard-aperture KLM. In reality, however, the beam radii in the two directions are not strictly independent, since a change in beam radius along one direction, x for example, will change the intensity and therefore the nonlinearity seen by the entire beam [79]. An analytical treatment of astigmatic Gaussian beams in KLM resonators which accounts for the x–y coupling is given in [80]. Using this treatment, the δ_w contour plots, which were shown in Fig. 2.18 for the tangential plane using a cylindrical beam treatment, were recalculated with the coupling included. The results showed that in most cases the overall trends observed from the δ_w contour plots in Fig. 2.18 remain valid.

2.4.5 Mode-Locking Solutions

We now write down the mode-locking equation and discuss its solution, following the treatment of [81]. The main bandwidth-limiting effect is assumed to arise from the gain itself, so that the $(1/\omega_c^2)(d^2/dt^2)$ term in eq. (2.38) is replaced with $(g/\Omega_g^2)(d^2/dt^2)$.[4] The changes in the pulse in one round trip through the laser, caused by gain, loss, SAM, SPM, and dispersion, are given by

$$a_{n+1}(t) = \left\{1 + g\left(1 + \frac{1}{\Omega_g^2}\frac{d^2}{dt^2}\right) - \left(\ell - \gamma\,|a(t)|^2\right) - j\delta\,|a(t)|^2 + jD\frac{d^2}{dt^2}\right\} a_n(t) \tag{2.116}$$

where $a_n(t)$ and $a_{n+1}(t)$ represent the pulse after n and $n+1$ passes through the laser, respectively, and g is, as usual, the saturated gain. For steady-state mode-locking we require that the pulse reproduce itself to within a phase shift ψ:

$$a_{n+1}(t) = a_n(t)e^{j\psi} \tag{2.117}$$

Note that previously we implicitly assumed that any such phase shift could be ignored in the equations for active mode-locking and passive mode-locking using saturable absorbers. This is exactly correct when the gain lineshape and any bandwidth-limiting filters are centered exactly on one of the longitudinal cavity modes. However, the assumption is also

[4] Here our notation is that the gain is approximated by $G(\tilde{\omega}) \approx g\left(1 - \tilde{\omega}^2/\Omega_g^2\right)$. For a Lorentzian Ω_g would be the half width at half maximum of the gain line. In contrast, in Section 2.2 we used the notation ω_G, which was the full width at half maximum of the gain line. Hence, $\Omega_g = \omega_G/2$.

approximately correct whenever the cavity modes are closely spaced relative to the gain or filter bandwidth, which is usually satisfied in practice. Here, however, we must include ψ due to the possibility of contributions from self-phase modulation and dispersion. If we assume that the phase shift per pass is small, then $\exp(j\psi) \approx 1 + j\psi$, and we can combine eq. (2.116) and (2.117) to obtain

$$\left\{ (D_g + jD) \frac{d^2}{dt^2} + (g - \ell - j\psi) + (\gamma - j\delta)\,|a(t)|^2 \right\} a(t) = 0 \tag{2.118a}$$

where

$$D_g = \frac{g}{\Omega_g^2} \tag{2.118b}$$

A number of new phenomena can occur in mode-locked lasers described by this equation. These include the following:

- SPM can create new frequency components, which can broaden the spectrum and sometimes lead to shorter output pulses.
- Due to the action of SPM, GVD, or both, the mode-locked pulses may be chirped. In the case of chirped pulses, a chirp compensator located outside the laser may be used to compress the output pulse duration. Pulses discussed in previous sections (e.g., active mode-locking, saturable absorber passive mode-locking) were always chirp-free.
- SPM and GVD can interact strongly. The character of the interaction depends on the sign of the dispersion, which using prism pairs or other devices, can be tuned to be positive, negative, or approximately zero.

The solution of eq. (2.118a) is a chirped secant hyperbolic [81], written as follows:

$$a(t) = a_0 \left[\operatorname{sech}\left(\frac{t}{t_p}\right) \right]^{1+j\beta} = a_0 \operatorname{sech}\left(\frac{t}{t_p}\right) e^{j\beta \ln[\operatorname{sech}(t/t_p)]} \tag{2.119}$$

For convenience we take a_0 to be real. The new feature is the time-dependent phase term, which leads to a time-dependent frequency shift $\Delta\omega(t)$, given by

$$\Delta\omega(t) = \frac{\partial\phi(t)}{\partial t} = \frac{-\beta}{t_p} \tanh\left(\frac{t}{t_p}\right) \tag{2.120}$$

The frequency shift is equal to β/t_p before the pulse, zero at the center of the pulse, and $-\beta/t_p$ after the pulse. $\beta < 0$ ($\beta > 0$) corresponds to an up-chirp (down-chirp). Since the bandwidth of an unchirped secant hyperbolic would be on the order of t_p^{-1}, the chirp corresponds to large additional spectral broadening when $|\beta| \gg 1$ and small spectral broadening when $|\beta| \ll 1$. Roughly, the spectral width is increased by a factor $\sqrt{1+\beta^2}$ compared to the bandwidth in the unchirped ($\beta = 0$) case.

Inserting eq. (2.119) into the mode-locking equation (2.118a) and requiring that the coefficients of $\operatorname{sech}^{1+j\beta}(t/t_p)$ and $\operatorname{sech}^{3+j\beta}(t/t_p)$ independently sum to zero yields two complex characteristic equations:

$$g - \ell - j\psi + \frac{(1 + j\beta)^2}{t_p^2}\left(D_g + jD\right) = 0 \tag{2.121a}$$

$$\left(D_g + jD\right)\frac{2 + 3j\beta - \beta^2}{t_p^2} = (\gamma - j\delta)\, a_0^2 \tag{2.121b}$$

These equations (equivalent to four real equations) are supplemented by the gain saturation equation (2.68), which gives g in terms of the small-signal gain and average laser power. This system of nonlinear equations can be solved simultaneously to yield the five unknowns: a_0, t_p, β, g, and ψ.

This calculation can be simplified by noting that the saturated gain is not usually too far below the loss line, and therefore to first order, one can approximate $g \approx \ell$. Under this approximation the average power and the pulse energy $U = 2a_0^2 t_p$ can be obtained directly from eq. (2.68). It is then convenient to introduce a normalized (dimensionless) dispersion parameter

$$D_n = \frac{D}{D_g} \tag{2.122}$$

which describes the relative strength of dispersive pulse spreading compared to pulse spreading due to bandwidth filtering. It is also useful to define a normalized (not dimensionless!) pulse width

$$t_{p,n} = \frac{U t_p}{2D_g} \tag{2.123}$$

Equation (2.121b) can be rewritten in terms of these normalized variables as:

$$(1 + jD_n)(2 + 3j\beta - \beta^2) = (\gamma - j\delta)t_{p,n} \tag{2.124}$$

This single complex equation can be solved to yield the chirp parameter β and the normalized pulse-width parameter $t_{p,n}$.

Results are shown in Fig. 2.19, in which β, $t_{p,n}$, a normalized mode-locked bandwidth $(1/t_{p,n})\sqrt{1+\beta^2}$, and a stability parameter (discussed below) are all plotted as a function of D_n. This figure assumes a fixed SAM coefficient ($\gamma = 1$) with varying degrees of SPM ($\delta = 0, 1, 2, 4$). Due to the approximation $g \approx \ell$ and the use of the normalized variables, the small-signal gain does not appear explicitly, and therefore a single set of curves is able to represent solutions for various g_0. The value of g_0 and U do enter when eq. (2.123) is used to convert the normalized pulse width back to real pulse-width units.

When there is no SPM ($\delta = 0$), the curves are symmetric. The shortest pulses and highest bandwidth occur at $D = 0$, for which the pulses are chirp-free. (With no dispersion or SPM, the behavior is the same as for fast absorber mode-locking, considered previously.) For $D \neq 0$ the bandwidth decreases while the pulses grow longer and pick up a chirp. The sign of the chirp changes when the sign of the dispersion changes.

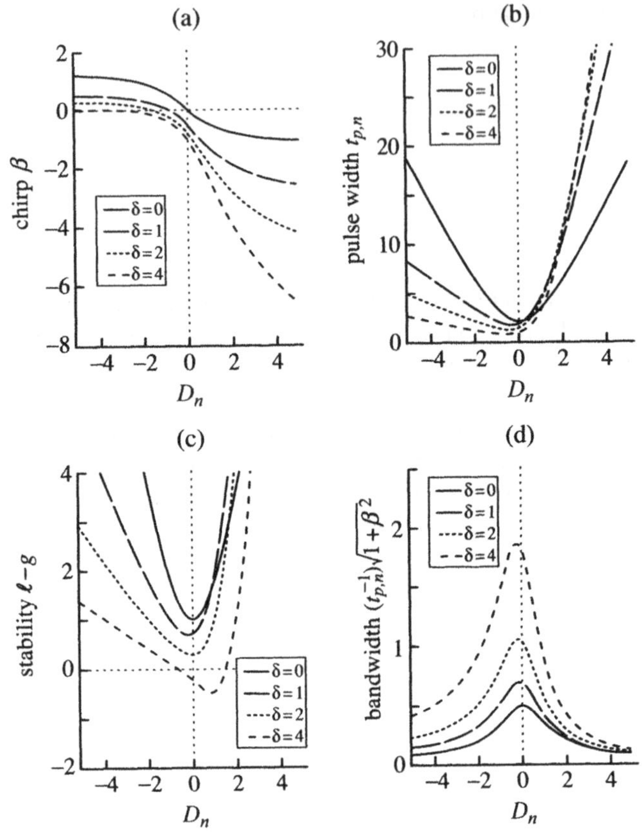

Figure 2.19 Pulse parameters vs. normalized dispersion D_n for $\gamma = 1$ and various SPM parameters δ: (a) chirp parameter β; (b) normalized pulse width $t_{p,n}$; (c) stability parameter $\ell - g$; (d) normalized bandwidth $(1/t_{p,n})\sqrt{1+\beta^2}$. Adapted from [81].

When SPM is turned on ($\delta > 0$), the behavior depends strongly on the sign of the dispersion. Even at zero dispersion, the pulses are up-chirped, with shorter pulses and higher bandwidth compared to $\delta = 0$. These effects are attributed to the new frequency components generated through SPM.

For $D > 0$ (normal dispersion), SPM and GVD acting individually both lead to an up-chirp. The down-shifted frequency components that SPM creates on the leading edge of the pulse are shifted to even earlier times by GVD, while the up-shifted frequency components created on the trailing edge of the pulse are shifted to even later times. Thus, under the combined influence of SPM and GVD, pulses spread faster and experience greater chirping than they would under the influence of GVD alone. In Fig. 2.19 this manifests itself as a rapid increase in the pulse width and in the magnitude of the chirp parameter, either for increasing D_n with δ fixed or for increasing δ with D_n fixed. For fixed dispersion the bandwidth also increases with increasing SPM, so chirp compensation outside the laser can lead to compressed pulses shorter than could be obtained without SPM but with the same intracavity dispersion. Finally, it is interesting to note that when a highly chirped pulse passes through a bandwidth-limiting filter (e.g., due to the finite gain bandwidth), it can actually be narrowed in time as well as in frequency. This is in contrast to chirp-free pulses considered earlier, where filters were always associated with pulse-broadening effects.

For $D < 0$ (anomalous dispersion), GVD leads to a down-chirp while SPM creates an up-chirp. If the strengths of these two effects are matched, they compensate each other, and no net chirping or pulse spreading occurs. In nonlinear fiber optics, such compensation results in pulses called *solitons*, which can propagate without temporal or spectral distortion [37,82,83]. Solitons have been used in optical communication systems for distortionless propagation of short pulses over extremely long distances (e.g., pulses a few tens of picoseconds long over 10,000 km of fiber, corresponding to trans-Pacific distances!) [84]. The operation of mode-locked lasers with $D_n < 0$ can be heavily influenced by soliton-like effects, and such effects can lead to new pulse-compression mechanisms which contribute strongly to pulse shortening in the mode-locked laser. These effects manifest themselves in Fig. 2.19 in several ways. First, for fixed δ, the shortest pulses and the largest bandwidth actually occur for D_n slightly less than zero. Compared to the positive dispersion case, the pulse width increases rather slowly as the negative dispersion is increased beyond the optimum point, especially for large SPM. For a given D_n, the pulses become shorter with increasing δ. At the same time, the pulses remain remarkably chirp-free over a substantial range of negative dispersion.

Mode-locked lasers operating in a solitonlike regime exhibit definite relations between pulse intensity, pulse width, and dispersion (since the SPM effect, whose strength is proportional to peak intensity, must balance dispersive effects, whose strength turns out to depend inversely on pulse width squared). This can be seen by inserting $\beta = 0$, which we have seen is approximately true within much of negative dispersion range, into eq. (2.124). The imaginary part of this equation yields

$$t_{p,n} = \frac{2|D_n|}{\delta} \tag{2.125}$$

In terms of the original rather than the normalized variables, this gives

$$\frac{\delta U t_p}{4|D|} = \frac{\delta a_0^2 t_p^2}{2|D|} = 1 \tag{2.126}$$

As we discuss in further detail in Chapter 6, this is exactly the relationship that applies for ideal soliton propagation.

Figure 2.19c shows the stability parameter plotted vs. D_n. The stability parameter is taken as $\ell - g$, which gives the net loss before and after the pulse. As in the case of saturable absorber mode-locking, loss must exceed gain before and after the pulse to suppress amplification of perturbations on the pulse's leading or trailing edge. Therefore, $\ell - g > 0$ for stability (see Fig. 2.10). Since this stability parameter cannot be evaluated using the approximation $g \approx \ell$, it is obtained from the full solution of eqs. (2.121a) and (2.121b). The key point to be learned from Fig. 2.19c is that although the mode-locking solution is stable over most of the parameter range investigated, there are unstable ranges, particularly for large SPM and when one approaches positive dispersion. Experimentally, one can encounter the instability when using an intracavity prism sequence to tune close to zero dispersion or into the positive dispersion range. At this point one may observe, for example, either multiple pulses or chaotic behavior in which the laser power, pulse width, and so on, exhibit strong fluctuations on a time scale long compared to the laser round-trip time.

the random beating between these modes. Initiation of mode-locking depends on whether the nonlinearities are strong enough (over a large number of round trips through the laser) to amplify the energy within the fluctuation with the highest peak intensity and suppress the remainder of the laser radiation. Once a single fluctuation is selected, the nonlinearities must act on it repeatedly before reaching the mode-locked steady state.

One simple rule that has been advanced provides a criterion for energy growth of an intense fluctuation as affected by two competing processes: self-amplitude modulation, which enhances the fluctuation, and dynamic gain saturation, which occurs during the fluctuation and reduces its gain [86]. The fluctuation can grow only if the increase in energy due to SAM exceeds the decrease in energy due to dynamic gain saturation, which is satisfied when

$$\frac{\gamma}{g} \gtrsim \frac{\sigma_g \tau_f}{\hbar\omega A_g} \tag{2.127}$$

Here σ_g, A_g, $\hbar\omega$, and τ_f are the gain cross section, the beam area within the gain medium, the photon energy, and the duration of the fluctuation, respectively, and a pulse shape–dependent numerical factor of order unity has been omitted. This relation was cited [86] as one reason why APM with Ti:S lasers ($\sigma_g = 2.7 \times 10^{-19}\ \text{cm}^2$) and other solid-state lasers with lower gain cross sections was observed to be self-starting, whereas APM with color-center lasers ($\sigma_g \sim 10^{-17}\ \text{cm}^2$) was not.

Additional analyses [87–89] provide an explanation for the observation of a threshold power for mode-locking, which is not predicted by eq. (2.127). In one time-domain formulation, one requires that not only must the initial fluctuation grow in energy, but this growth must occur faster than a correlation time τ_c, which is taken as the lifetime of the fluctuation in the absence of nonlinearities [87]. This leads to the criterion

$$\gamma \Delta P \gtrsim \frac{T}{\tau_c} \tag{2.128a}$$

or

$$\gamma P \gtrsim \frac{1}{\ln(N)} \frac{T}{\tau_c} \tag{2.128b}$$

where P is the average intracavity power, $\Delta P \approx P \ln(N)$ gives a typical value for the peak intensity of the strongest fluctuation, N is the number of modes that are oscillating in the free-running laser, T is the cavity round-trip time, and the effect of dynamic gain saturation has been assumed to be small. This criterion says that mode-locking can be initiated only when the change in loss experienced by the initial strong fluctuation per round trip, multiplied by the number of round trips within the correlation time τ_c, is of order unity or larger. Although a few theoretical studies give insight into physical mechanisms leading to a finite τ_c [88,89], it is most practical to determine τ_c experimentally. This can be achieved by measuring the laser intensity noise with a radio-frequency spectrum analyzer and identifying $\tau_c \approx 1/(\pi \Delta\nu)$, where $\Delta\nu$ is the full width at half maximum of the first beat note. Measurements yield values for $\Delta\nu$ typically in the several kilohertz range for various solid-state lasers, corresponding to τ_c on the order of 100 μs. Self-starting thresholds have been measured for several lasers and seem to agree at least qualitatively with the predictions of eq. (2.128a) (e.g., [87,90,91]). Additionally, eq. (2.128a) also explains the greater propensity for APM lasers to self-start (compared to KLM with the same laser materials), since the self-amplitude modulation coefficient can be substantially higher in APM lasers using nonlinearities in optical fibers than in KLM lasers using bulk nonlinearities.

PROBLEMS

2.1. A mode-locked laser generates Gaussian pulses with a 1.06-μm center wavelength, 10-ns cavity round-trip time, 1-W average power, and 20-ps FWHM duration. Give values for the number of cavity modes that are oscillating, the optical bandwidth, the peak power, and the pulse energy.

2.2. In the time-domain treatment of active mode-locking, we analyzed changes to Gaussian pulses incurred in passing through cavity elements. Define pulse-broadening velocity as the fractional change in pulse width in passing once through the gain medium. Similarly, define pulse-shortening velocity as the fractional change in pulse width in passing once through the modulator. Plot pulse-broadening and pulse-shortening velocities as a function of pulse width for $\omega_m = 2\pi \times 80$ MHz, $\omega_G = 2\pi \times 200$ GHz, $g = 5\%$, and $\Delta_m = 25\%$. Indicate the position of the steady-state pulse width on your plot, and comment on your results.

2.3. In this problem we consider active mode-locking with the modulation frequency detuned from the optimum. For small detuning, mode-locking is still possible, but the timing of the steady-state pulse does not coincide with the peak transmission of the modulator.

(a) Consider a delayed Gaussian pulse of the form $a(t) \sim e^{-\Gamma(t-\tau)^2}$, which passes through a loss modulator with transmission given by eq. (2.12). What is the delay of the pulse after it passes through the modulator? How much has the delay changed, and has it moved toward or away from $t = 0$? (*Hint:* Expand the cosine near $t = 0$.)

(b) Give an expression for the difference between the modulation period and the actual laser round-trip time as a function of the detuning. You may assume that the detuning is small.

(c) For steady-state operation, the change in delay caused by the modulator must equal the change in delay caused by the frequency detuning. Use this fact to obtain an expression for τ.

(d) Assuming that the maximum permissible value for τ is 1 ns, what is the maximum allowable frequency detuning? Evaluate your result using the laser parameters from the preceding problem. Also give the time shift resulting from a single pass through the modulator.

(e) Evaluate the maximum frequency detuning as above, but with the laser gain bandwidth increased by a factor of 10. Is the modulator more or less effective at shifting the pulses toward $t = 0$? Why?

2.4. Consider an actively mode-locked laser with a phase modulator instead of a loss modulator. The modulator multiplies the field by the factor $e^{j\Delta_m \cos \omega_m t}$.

(a) Repeat the recirculating Gaussian pulse analysis of Section 2.2.1 for the case of the phase modulator. You may assume that pulses are short compared to the modulation period and occur near $t = 0, 2\pi/\omega_m, \ldots$. Comment on the character of the mode-locked pulse solution. What is the time–bandwidth product, and how does it compare to active mode-locking with a loss modulator?

(b) Can you think of a physical explanation for how a phase modulator could lead to short-pulse formation? That is, how does a phase modulator give a net gain window?

(c) Suppose that the pulses occur at the negative peaks of the cosine at $t = \pi/\omega_m$, $3\pi/\omega_m, \ldots$ rather than at the positive peaks ($t = 0, \ldots$). What if anything will be different about the solution? What implications does this have for practical laser operation?

2.5. Consider passive mode-locking using a fast saturable absorber. The linear loss and initial value of the saturable loss are $\ell_0 = 5\%$ and $\ell^{(i)} = 10\%$, respectively; the steady-state pulse duration is 1 ps, and the pulse energy is 1 nJ. Assuming that dynamic gain saturation may be ignored, give bounds for the value of saturated gain consistent with stable steady-state mode-locking.

2.6. Modify eq. (2.64) to describe fast saturable absorber mode-locking, including the possibility of dynamic gain saturation (take the gain recovery time to be fast compared to the pulse width). What new condition must be satisfied to achieve stable, steady-state mode-locking? Give a brief interpretation of this condition in words.

2.7. Consider fast saturable absorber mode-locking with the following parameters: $\hbar\omega_0 = 2.5 \times 10^{-19}$ J, $\ell_0 = 5\%$, $\ell^{(i)} = 20\%$, $T = 10$ ns, $A_A = A_G = 10^{-4}$ cm^2, $\sigma_A = 10^{-16}$ cm^2, $\sigma_G = 10^{-19}$ cm^2, $\tau_A = 100$ fs, $\tau_G = 1$ μs, $\omega_c = 2\pi \times 10^{12}$ rad s^{-1}.

(a) Determine the minimum and maximum values of small-signal gain g_0 that yield a stable single-pulse solution. Setting g_0 to the average of these values, what are the values of t_p and a_0 for the (stable) longer-pulse solution? Give numbers for the peak power, pulse energy, average power, and FWHM pulse width.

(b) How would you change the analysis in Section 2.3.3 to handle solutions with two identical pulses (assumed to be separated and noninteracting)? No need to carry out the analysis; just explain what must be changed.

(c) Using the same g_0 as in part (a), write a program to simulate the pulse evolution as it runs repeatedly through the laser cavity. As a suggestion, start your simulation with a noise burst roughly 100 times longer than the steady-state pulse width from above. Do you get a steady-state solution, and does it agree with the solution from theory? Plot pulse parameters (e.g., pulse width, peak power, energy) vs. number of iterations through the laser. Also show a few representative plots of the temporal intensity at different stages during your simulation. [*Hint:* After each pass through the laser you should recalculate the pulse energy and the gain to be used for the next iteration. The gain may be calculated, for example, from eq. (2.54).]

2.8. (a) Demonstrate that the laser cavity with internal lenses of Fig. 2.16a is equivalent to the two-mirror laser cavity of Fig. 2.16b. That is, verify eq. (2.108) through explicit calculation.

(b) As noted in the text, the equivalent radii of curvature in the two-mirror cavity of Fig. 2.16b are usually negative. Normally, a cavity consisting of a pair of convex reflectors is unstable. Explain how the equivalent cavity discussed in connection with Fig. 2.16 can be stable.

3

ULTRAFAST-PULSE MEASUREMENT METHODS

Up until now we have only discussed how to generate ultrashort pulses, not how to measure them. Measurement of pulses on the femtosecond time scale is an important issue, since the speed required is considerably faster than existing photodetectors and oscilloscopes. In this chapter we cover several measurement techniques capable of resolving femtosecond pulses. We begin by discussing correlation measurements, which are widespread experimentally but which offer only partial information about the optical pulse. We then cover several approaches that can reveal complete intensity and phase information, including the frequency-resolved optical gating approach based on time–frequency analysis concepts as well as techniques based on frequency filtering and on interferometry. Finally, we discuss methods for characterizing the intensity noise and timing jitter of mode-locked lasers based on radio-frequency spectral analysis.

3.1 TERMINOLOGY AND DEFINITIONS

We start with some definitions. As usual we write the field as

$$e(t) = \mathrm{Re}\left\{a(t)e^{j\omega_0 t}\right\} \tag{3.1}$$

with $a(t)$ normalized so that the power averaged over an optical cycle is given by

$$P(t) = |a(t)|^2 \tag{3.2}$$

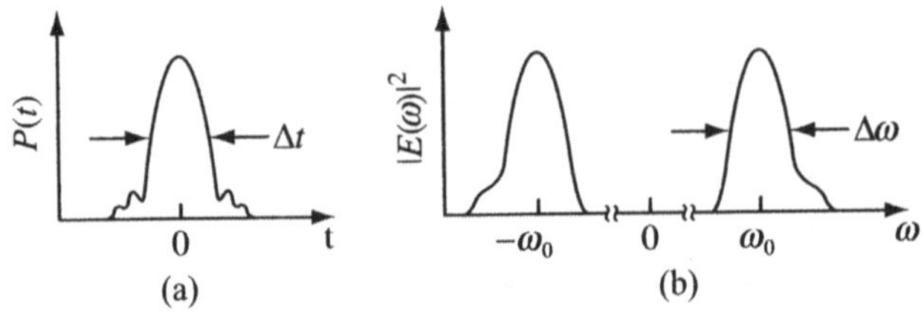

Figure 3.1 Time profile of (a) power $P(t)$ and (b) power spectrum. Δt and $\Delta\omega$ are the full width at half maximum pulse width and bandwidth.

The intensity $I(t)$ is, of course, simply proportional to $P(t)$. The Fourier transform and the power spectrum are given by [see eq. (1.86a)]

$$E(\omega) = \frac{1}{2}\left[A(\omega - \omega_0) + A^*[-(\omega + \omega_0)]\right] \tag{3.3}$$

and

$$|E(\omega)|^2 = \frac{1}{4}\left[|A(\omega - \omega_0)|^2 + |A[-(\omega + \omega_0)]|^2\right] \tag{3.4}$$

where $A(\omega)$ is the Fourier transform of $a(t)$. In writing the power spectrum, we have assumed that the positive and negative frequency components of the spectrum do not overlap. The power and the power spectrum are sketched in Fig. 3.1.

The most widely used measure of the pulse duration in ultrafast optics is the intensity full width at half maximum (FWHM), which we denote Δt. The bandwidth is usually measured as the FWHM of the positive frequency part of the power spectrum, denoted $\Delta\nu$ (in hertz), where $\Delta\nu = \Delta\omega/2\pi$ and $\Delta\omega$ is the FWHM in angular frequency units. Δt and $\Delta\omega$ are sensitive only to the central portion of a pulse or its spectrum. Root-mean-square (rms) pulse widths or bandwidths have also been introduced for use in ultrafast optics [94]. The rms pulse width is defined

$$\Delta t_{\text{rms}} = 2\sqrt{\langle t^2\rangle - \langle t\rangle^2} \tag{3.5}$$

where

$$\langle f(t)\rangle = \frac{\int_{-\infty}^{\infty} dt\, f(t)\,|a(t)|^2}{\int_{-\infty}^{\infty} dt\, |a(t)|^2} \tag{3.6}$$

Similarly, the rms bandwidth is defined

$$\Delta\nu_{\text{rms}} = \frac{1}{\pi}\sqrt{\langle\omega^2\rangle - \langle\omega\rangle^2} \tag{3.7}$$

where

$$\langle G(\omega)\rangle = \frac{\int_0^{\infty} d\omega\, G(\omega)\,|A(\omega - \omega_0)|^2}{\int_0^{\infty} d\omega\, |A(\omega - \omega_0)|^2} \tag{3.8}$$

Table 3.1 Time–Bandwidth Products for Three Transform-Limited Pulse Shapes[a]

$a(t)$	$\|A(\omega)\|^2$	Δt	Δt_{rms}	$\Delta\nu\,\Delta t$	$\Delta\nu_{\text{rms}}\,\Delta t_{\text{rms}}$
$e^{-(t/t_p)^2}$	$e^{-(\omega^2 t_p^2/2)}$	$1.177t_p$	t_p	0.441 $(2\ln 2/\pi)$	0.318 $(1/\pi)$
$\text{sech}\left(\frac{t}{t_p}\right)$	$\text{sech}^2\left(\frac{\pi\omega t_p}{2}\right)$	$1.763t_p$	$1.814t_p$ $(\pi t_p/\sqrt{3})$	0.315	$\frac{1}{3}$
$\frac{1}{1+(t/t_p)^2}$	$e^{-2\|\omega t_p\|}$	$1.287t_p$	$2t_p$	0.142	0.450 $(\sqrt{2}/\pi)$

[a] Δt and $\Delta\nu$ are full widths at half maximum of $|a(t)|^2$ and $|A(\omega)|^2$, respectively. Δt_{rms} and $\Delta\nu_{\text{rms}}$ are the corresponding rms pulse width and spectral bandwidth. $|A(\omega)|^2$ is normalized to one.

In eq. (3.8) only the positive frequency part of the power spectrum is included in the integrals; this is necessary to give a result for $\langle\omega\rangle$ that is in the vicinity of ω_0. The rms pulse width (bandwidth) has sensitivity to energy in the wings of a pulse (spectrum) and therefore provides complementary information to the FWHM widths.

In measuring ultrashort pulses, one is interested not only in the intensity profile but also in any phase variations in either the frequency or time domains. Conversely, one is often interested in knowing when a pulse is free of such phase variations. We use $\phi(t)$ and $\psi(\omega)$, respectively, to define the temporal and spectral phases, as follows:

$$a(t) = |a(t)|\, e^{j\phi(t)} \tag{3.9}$$

and

$$A(\omega) = |A(\omega)|\, e^{j\psi(\omega)} \tag{3.10}$$

When $\psi(\omega)$ is a constant, we speak of the pulse as *bandwidth limited.*[1] When $\phi(t)$ is a constant, we say the pulse is free of *phase or frequency modulation.*[2] When $\phi(t) = 0$, $a(t)$ is real, and the Fourier transform gives $A(\omega) = A^*(-\omega)$. If, in addition, $a(t)$ is even, one finds $A(\omega) = A^*(\omega)$, which means that the pulse is bandwidth limited. Conversely, when $A(\omega)$ is real, one finds that $a(t) = a^*(-t)$. If, in addition, $A(\omega)$ is even, then $a(t) = a^*(t)$ and the pulse is free of phase modulation. We note that for a pulse to be simultaneously bandwidth limited and free of phase modulation [i.e., $\psi(\omega) = 0$ and $\phi(t) = 0$], it must also be an even function of both time and frequency. Therefore, it is possible for a bandwidth-limited pulse to show a phase modulation or a pulse free of any phase modulation to show a spectral phase variation. We speak of a pulse as *chirped* when it has a nontrivial variation in both $\phi(t)$ and $\psi(\omega)$.

One simple metric for characterizing the degree of chirp of a pulse is the time–bandwidth product $\Delta\nu\,\Delta t$, which is written in terms of the FWHMs. Values of $\Delta\nu\ \Delta t$ are given in Table 3.1 for three different simple chirp-free pulse shapes [94,95]. For a given pulse shape, an increased value of $\Delta\nu\,\Delta t$ is an indicator of chirp. The rms time–bandwidth product, written $\Delta\nu_{\text{rms}}\ \Delta t_{\text{rms}}$, is also useful [94]. Unlike the time–bandwidth product written in

[1] A pulse with $\psi(\omega)$ linear in ω is also considered bandwidth limited, since a linear spectral phase shift corresponds to a simple delay in the time domain.

[2] A pulse with $\phi(t)$ linear in t may also be considered as free of phase modulation, since a linear temporal phase is simply a shift of center frequency.

terms of FWHMs, it has the important property that it can be given a lower bound without specifying the pulse shape. Specifically, the *uncertainty relation*

$$\Delta\nu_{\rm rms}\,\Delta t_{\rm rms} \geq \frac{1}{\pi} \tag{3.11}$$

is always satisfied, independent of pulse shape. The minimum value occurs when the pulse shape is Gaussian [11].

3.2 ELECTRIC FIELD AUTOCORRELATION MEASUREMENTS AND THE POWER SPECTRUM

Since the ultrashort pulse itself is often the shortest event available, a common measurement strategy is to try to let the pulse sample itself. A simple setup for doing this is sketched in Fig. 3.2, where a Michelson interferometer is used to split the input pulse into two identical pulses which are then recombined with a time delay τ. The delay can be varied simply by moving one of the stages forward or backward, often using a computer-controlled translation stage with submicrometer step sizes. The simplest measurement consists of recording the average power (using a slow, time-averaging detector) emerging from the interferometer as a function of delay τ. As we shall see, this linear (in power) measurement technique is actually not capable of determining ultrashort pulse intensity or phase profiles. However, before proceeding to related nonlinear techniques which can yield the needed information, it is first important to understand what the linear methods tell us.

We now analyze the measurement outlined above [9,96]. The output field from the interferometer can be written

$$e_{\rm out}(t) = \frac{1}{2}{\rm Re}\left\{a(t-t_1)e^{j\omega_0(t-t_1)} + a(t-t_2)e^{j\omega_0(t-t_2)}\right\} \tag{3.12}$$

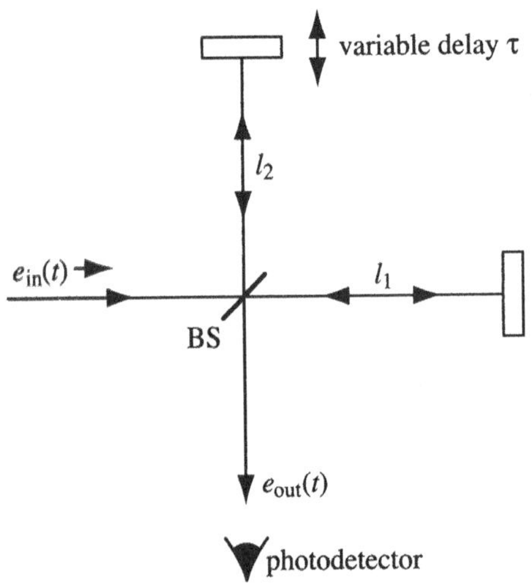

Figure 3.2 Michelson interferometer setup for measuring the electric field autocorrelation function.

The delays t_1 and t_2 are given by $t_1 = 2l_1/c$ and $t_2 = 2l_2/c$, respectively. The factor of 1/2 takes into account one reflection from and one pass through a 50–50 splitter. (The field amplitude is multiplied by the factor $1/\sqrt{2}$ after a single reflection from or transmission through the beamsplitter, resulting in 50% power reflected and 50% transmitted.) The output power is written

$$P_{\text{out}}(t) = \frac{1}{4}\Big\{ |a(t-t_1)|^2 + |a(t-t_2)|^2 + \big[a(t-t_1)a^*(t-t_2)e^{j\omega_0(t_2-t_1)} + \text{c.c.}\big]\Big\} \tag{3.13}$$

where "c.c." stands for *complex conjugate*. If we assume that the detector is slow compared to the pulse, so that it measures the total time-integrated intensity (i.e., the pulse energy), the measured quantity is given by

$$\int dt\, P_{\text{out}}(t) = \frac{1}{2}\int dt\, |a(t)|^2 + \left[\frac{e^{j\omega_0\tau}}{4}\int dt\, a(t)a^*(t-\tau) + \text{c.c.}\right] \tag{3.14}$$

Here we have introduced the delay $\tau = t_2 - t_1$ and assumed a *time-stationary process* (i.e., the time-integrated intensity is independent of the pulse delay, and the integral of the cross terms depends only on the relative delay between the two pulses).

At this point we are ready to define the first-order correlation function $\Gamma_a(\tau)$ of the complex electric field amplitude $a(t)$, given by

$$\Gamma_a(\tau) = \frac{1}{\Delta T}\int_{-\Delta T/2}^{\Delta T/2} dt\, a(t)a^*(t-\tau) = \langle a(t)a^*(t-\tau)\rangle \tag{3.15}$$

where ΔT is assumed to approach ∞, and the angular brackets $\langle\cdots\rangle$ denote the time average. In the ultrafast optics community, the quantity given in eq. (3.15) is often called simply the *field* or *amplitude autocorrelation*. The term *autocorrelation* means that the two pulses being correlated are identical. In the case of an *ergodic process*, the time average is the same as the ensemble average (a concept that is particularly useful in the case of a random or noise input electric field). Unless otherwise noted, we assume ergodicity, and therefore we do not need to distinguish between time and ensemble averages.

The output of the photodetector in Fig. 3.2 is proportional to the time-average power, which we now rewrite using eqs. (3.14) and (3.15) as follows:

$$\langle P_{\text{out}}\rangle = \frac{1}{2}\Gamma_a(0)\left\{1 + \frac{1}{2}\left(\frac{\Gamma_a(\tau)e^{j\omega_0\tau} + \Gamma_a^*(\tau)e^{-j\omega_0\tau}}{\Gamma_a(0)}\right)\right\} \tag{3.16}$$

Since $a(t)$ is in general complex, $\Gamma_a(\tau)$ is also complex, and therefore we write $\Gamma_a(\tau) = |\Gamma_a(\tau)|\, e^{j\Phi(\tau)}$. Equation (3.16) now becomes

$$\langle P_{\text{out}}\rangle = \frac{1}{2}\Gamma_a(0)\,[1 + G_1(\tau)] \tag{3.17}$$

where we have introduced the first-order correlation function of the electric field, $G_1(\tau)$, which is related to $\Gamma_a(\tau)$ by

$$G_1(\tau) = \frac{|\Gamma_a(\tau)|}{\Gamma_a(0)} \cos(\omega_0 \tau + \Phi(\tau)) \tag{3.18}$$

To interpret these equations, we first note that $\Gamma_a(\tau) = \Gamma_a^*(-\tau)$, as one can easily verify by substitution into eq. (3.15). Consequently, $G_1(\tau)$ is even. We also note that $|\Gamma_a(\tau)| \,/\, |\Gamma_a(0)| \le 1$. The maximum value occurs at $\tau = 0$. In the case of pulses, this corresponds to perfect temporal overlap between the recombined pulses emerging from the interferometer. For nonzero delays the overlap decreases and the value of the correlation function is reduced. It is typical to define a correlation time τ_c as the delay for which the correlation function is reduced to half of the peak value. A different way of defining τ_c uses a weighted average of τ defined by

$$\tau_c = \frac{\int_0^\infty \tau \, |\Gamma_a(\tau)| \, d\tau}{\int_0^\infty |\Gamma_a(\tau)| \, d\tau} \tag{3.19}$$

With either definition the overlap is very small for large delays ($|\tau| \gg \tau_c$), and the correlation approaches zero.

As one simple example, consider a Gaussian pulse with a complex amplitude given by $a(t) = \exp(-t^2/t_p^2)$. By direct integration, one can easily show that $\Gamma_a(\tau)/\Gamma_a(0) = \exp(-\tau^2/2t_p^2)$. An example of the time-average output power for a Gaussian pulse is sketched in Fig. 3.3. The power oscillates rapidly as a function of τ, with a period corresponding to one optical wavelength path-length change. Near $\tau = 0$ the power swings between zero and $\Gamma_a(0)$, which is the full input power. Thus, for phase differences of 0 or π, the power emerges completely in the output direction indicated in Fig. 3.2 or is reflected completely back toward the source, respectively. The average power (averaged over a few cycles of delay τ) is $\frac{1}{2}\Gamma_a(0)$ in either direction. The oscillations are damped as the

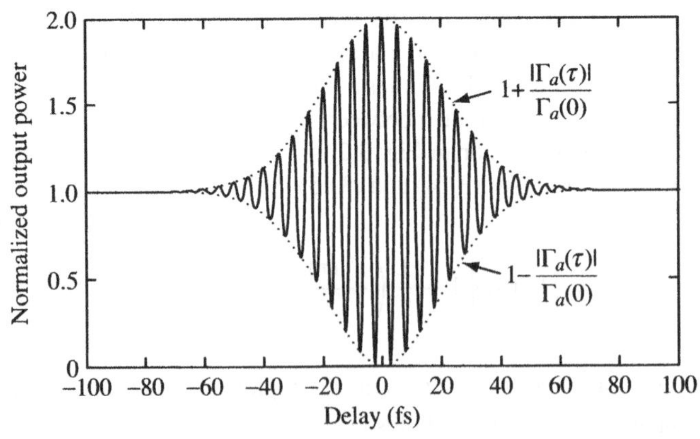

Figure 3.3 Example of time-average interferometer output power as a function of delay τ. The power is normalized by the factor $\frac{1}{2}\Gamma_a(0)$. This figure corresponds to a Gaussian input pulse with $t_p = 20$ fs and an optical period equal to 5 fs.

delay is increased, and the envelope of the oscillations, normalized by the average power, is $1 \pm |\Gamma_a(\tau)|$.

We see from this example that in the case of a coherent short-pulse input, the width of the electric field autocorrelation function is related in a simple way to the actual pulse width. The following question then arises: Can we determine the pulse width by measuring $G_1(\tau)$? In general, the answer is NO! To see this, we calculate the Fourier transform of $\Gamma_a(\tau)$ as follows:

$$\mathcal{F}\{\Gamma_a(\tau)\} \sim \int d\tau\, e^{-j\omega\tau} \int dt\, a(t)a^*(t-\tau) \tag{3.20}$$

By substituting $\tau = t - \eta$, we obtain

$$\mathcal{F}\{\Gamma_a(\tau)\} \sim \int dt\, a(t)e^{-j\omega t} \int d\eta\, e^{j\omega\eta} a^*(\eta) = |A(\omega)|^2 \tag{3.21}$$

Similarly, we can calculate the Fourier transform of $G_1(\tau)$, with the result

$$\mathcal{F}\{G_1(\tau)\} \sim |E_{\text{in}}(\omega)|^2 \tag{3.22}$$

These results say that the electric field correlation function is the inverse Fourier transform of the power spectrum, which can be measured, for example, by using a spectrometer. This has several important implications:

- The information obtained from measuring $\Gamma_a(\tau)$ and measuring the optical power spectrum is identical.
- The correlation time τ_c is roughly the inverse of the optical bandwidth.
- The correlation measurement gives no information about the spectral phase.
- Therefore, on the basis of these measurements, one cannot distinguish a bandwidth-limited pulse from a longer chirped pulse with the same bandwidth. In fact, a coherent ultrashort pulse and continuous-wave incoherent light (i.e., noise) give the same result, provided again that the optical spectra are the same.

On the other hand, if we somehow know a priori that the spectrum has a constant phase, we can determine the input pulse uniquely. We would start with the power spectrum, measured either directly using a monochromator or indirectly by measuring $\Gamma_a(\tau)$ and then doing the Fourier transform. Assuming for simplicity that the spectral phase is zero, we would take the square root of the power spectrum to get $A(\omega)$. Performing an inverse Fourier transform would then yield the input pulse $a(t)$.

3.3 ELECTRIC FIELD CROSS-CORRELATION MEASUREMENTS AND SPECTRAL INTERFEROMETRY

The situation changes when the correlation measurement involves two different input pulses, especially when one of the pulse shapes is known. This situation typically arises when the shape of the original optical signal is manipulated or distorted as the result of some linear or nonlinear process occurring in the experiment. We refer to the known pulse as the *reference* and to the unknown pulse as the *signal*. If one wishes to characterize such pulse shape changes, one can perform a cross-correlation measurement by splitting off a portion of the original light from the laser to use as a reference. In some cases the shape of the reference pulse is measured independently using one of the methods described later in the chapter.

3.3.1 Electric Field Cross-Correlation

We denote the reference and signal pulses as $a_r(t)$ and $a_s(t)$, respectively, and we assume that the reference pulse is delayed by time τ relative to the signal. The total electric field after the pulses are combined by a beamsplitter is given by

$$e_{\text{out}}(t) = \frac{1}{\sqrt{2}} \operatorname{Re}\left\{ \left[a_s(t) + a_r(t-\tau)e^{-j\omega_0\tau}\right] e^{j\omega_0 t} \right\} \tag{3.23}$$

The time-integrated output power becomes

$$\int dt\, P_{\text{out}}(t) = \frac{1}{2} \int dt\, \left\{ |a_r(t)|^2 + |a_s(t)|^2 + \left[a_s(t)a_r^*(t-\tau)e^{j\omega_0\tau} + \text{c.c.}\right] \right\} \tag{3.24}$$

or

$$\langle P_{\text{out}}(t) \rangle = \frac{1}{2} \left\{ U_r + U_s + \left[e^{j\omega_0\tau} \left\langle a_s(t)a_r^*(t-\tau) \right\rangle + \text{c.c.} \right] \right\} \tag{3.25}$$

where U_r and U_s are the energies of the reference and signal pulses, respectively. The factor of 1/4 in eq. (3.13) is changed to 1/2 since we assume that the reference and signal are already separate at the input to the cross-correlation setup; therefore, the beamsplitter is needed only once (to combine the beams). In the case where the reference pulse is known to be much shorter than the signal pulse, it can be approximated as a delta function. The time-integrated output power then reduces to

$$\langle P_{\text{out}}(t) \rangle = \frac{1}{2} U_r + \frac{1}{2} U_s + \operatorname{Re}\left\{ a_s(\tau) e^{j\omega_0\tau} \right\} \tag{3.26}$$

Thus, recording the oscillatory part of the output signal as a function of τ gives a direct measurement of the signal field. The envelope of the oscillations gives $|a_s(t)|$; measuring the phase of the fringes yields the phase of $a_s(t)$. When $a_r(t)$ is not truly a delta function but is still much narrower than $a_s(t)$, the cross-correlation yields an approximate but somewhat broadened measurement of the signal field.

It is also useful to analyze the electric field cross-correlation measurement in the frequency domain. The Fourier transform of eq. (3.24) with respect to τ is given by

$$\mathcal{F}\left\{ \int dt\, P_{\text{out}}(t) \right\}$$
$$= \cdots + \frac{1}{2} \left[A_s(\omega-\omega_0)A_r^*(\omega-\omega_0) + A_s^*(-\omega-\omega_0)A_r(-\omega-\omega_0) \right] \tag{3.27}$$

where the noninterferometric terms have been omitted. Provided that the reference spectrum is known and is broad compared to that of the signal, the signal spectrum and therefore the signal pulse can be obtained by deconvolution [i.e., by dividing the first term on the right-hand side of eq. (3.27) by $A_r^*(\omega - \omega_0)$]. The requirement of a broad reference spectrum is equivalent to the requirement that the reference pulse be short compared to the signal pulse.

From an experimental perspective, measuring the phase of the fringes requires a great deal of care. In particular, an online calibration of the sweep in the length of the interferometer is usually required in order to account for the possibly irregular motion of the delay stage. This can be achieved by counting the fringes made by a narrowband reference laser, such as a He–Ne laser, which is simultaneously passed through the interferometer [97,98].

3.3.2 Spectral Interferometry

In another measurement technique called *spectral interferometry*, τ is kept fixed, and a spectrometer is used to measure the output power spectrum [99–101]. In addition to its application in ultrafast optics, this method has also been applied for characterization of short optical fibers [102,103]. In terms of the envelope functions, the power spectrum can be written

$$|A_{\text{out}}(\tilde{\omega})|^2 = \frac{1}{2}\left\{|A_s(\tilde{\omega})|^2 + |A_r(\tilde{\omega})|^2\right\} + \frac{1}{2}\left\{\left[A_s(\tilde{\omega})A_r^*(\tilde{\omega})e^{j(\omega_0+\tilde{\omega})\tau} + \text{c.c.}\right]\right\} \tag{3.28}$$

Here $|E_{\text{out}}(\omega)|^2$ is related to $|A_{\text{out}}(\tilde{\omega})|^2$ through eq. (3.4) and we use the symbol $\tilde{\omega} = \omega - \omega_0$ to emphasize that $A_{\text{out}}(\tilde{\omega})$ is a baseband function peaked around zero frequency. If we now write $A_s(\tilde{\omega}) = |A_s(\tilde{\omega})|\, e^{j\psi_s(\tilde{\omega})}$ and $A_r(\tilde{\omega}) = |A_r(\tilde{\omega})|\, e^{j\psi_r(\tilde{\omega})}$, eq. (3.28) becomes

$$|A_{\text{out}}(\tilde{\omega})|^2 = \frac{1}{2}\left\{|A_s(\tilde{\omega})|^2 + |A_r(\tilde{\omega})|^2\right\} + |A_s(\tilde{\omega})|\;|A_r(\tilde{\omega})|\cos\left((\omega_0 + \tilde{\omega})\tau + \psi_s(\tilde{\omega}) - \psi_r(\tilde{\omega})\right) \tag{3.29}$$

The overall power spectrum exhibits oscillations with period $2\pi/\tau$, which are modulated by the spectral phase variations of the pulses. If one knows the spectral amplitude and phase of the reference pulse, one can derive the spectral phase of the signal pulse from the oscillatory part of the power spectrum. The spectral amplitude of the signal pulse $|A_s(\tilde{\omega})|$ can also be recovered from the spectral interferometry signal; however, it is often simpler to determine $|A_s(\tilde{\omega})|$ from the signal power spectrum $|A_s(\tilde{\omega})|^2$, which can be measured directly by blocking the reference beam.

Thus, spectral interferometry can be used to determine the complete amplitude and phase of the signal pulse, provided that a well-characterized reference pulse is available. For accurate results it is important that the reference spectrum be as broad or broader than that of the pulse being measured (otherwise, the oscillatory cross-term in the power spectrum will vanish at some frequencies of interest). This requirement is consistent with the requirement that the reference pulse must be shorter than the signal pulse, which was already mentioned in connection with time-domain electric field cross-correlation measurements.

One important form of spectral interferometry arises when the signal pulse is related to the reference pulse through a linear filtering operation. As shown in Fig. 3.4, both pulses

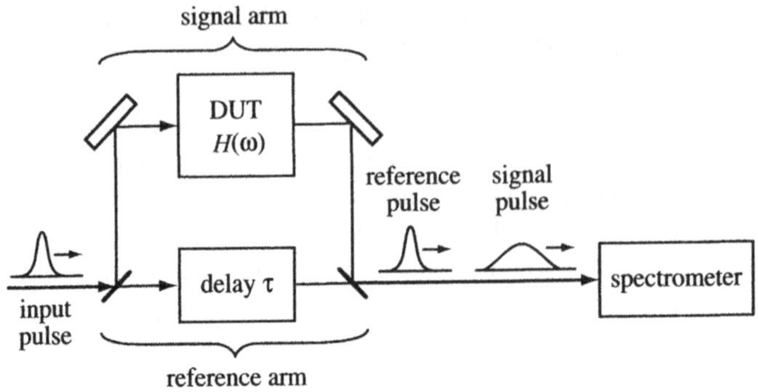

Figure 3.4 Spectral interferometry setup for characterizing the frequency response of linear optical elements. DUT, device under test.

are created from a common input pulse $a_{\text{in}}(t)$ and recombined after passing through an imbalanced interferometer in which the signal arm contains an optical component or device under test. Let us call the frequency response of the device under test $H(\omega)$, defined by $E_{\text{out}}(\omega) = H(\omega)E_{\text{in}}(\omega)$. Equivalently, the device under test has an impulse response function $h(t) = \mathcal{F}^{-1}\{H(\omega)\}$. Then the reference and signal pulses are given, respectively, by

$$A_r(\tilde{\omega}) = \frac{1}{\sqrt{2}}A_{\text{in}}(\tilde{\omega}) \tag{3.30a}$$

and

$$A_s(\tilde{\omega}) = \frac{1}{\sqrt{2}}H(\omega_0 + \tilde{\omega})A_{\text{in}}(\tilde{\omega}) \tag{3.30b}$$

Equation (3.28) for the output power spectrum now becomes

$$|A_{\text{out}}(\tilde{\omega})|^2 = \frac{1}{4}\left\{\left[1 + |H(\omega)|^2\right] + H(\omega)e^{j\omega\tau} + H^*(\omega)e^{-j\omega\tau}\right\}|A_{\text{in}}(\tilde{\omega})|^2 \tag{3.31}$$

If we separate the phase and amplitude response of the component under test, $H(\omega) = |H(\omega)|\, e^{j\psi(\omega)}$, the total power spectrum becomes

$$|A_{\text{out}}(\tilde{\omega})|^2 = \frac{1}{4}\left\{1 + |H(\omega)|^2 + 2\,|H(\omega)|\cos\left(\omega\tau + \psi(\omega)\right)\right\}|A_{\text{in}}(\tilde{\omega})|^2 \tag{3.32}$$

Thus, the oscillatory part of the spectral interferometry signal allows one to determine the spectral phase and amplitude response of the component under test. This makes spectral interferometry a valuable tool for measuring the dispersion of fibers, mirrors, prism sequences, and many other optical elements.

We briefly outline one algorithm for recovering the transfer function $H(\omega)$ based on Fourier transform processing of the spectral interferometry data. The result of taking the inverse Fourier transform of eq. (3.31) is sketched schematically in Fig. 3.5 for the case $\tau > 0$. The feature centered at $t = 0$, marked a, arises from the term proportional to $\left[1 + |H(\omega)|^2\right]|A_{\text{in}}(\tilde{\omega})|^2$ in eq. (3.31) and is simply the incoherent superposition of the field

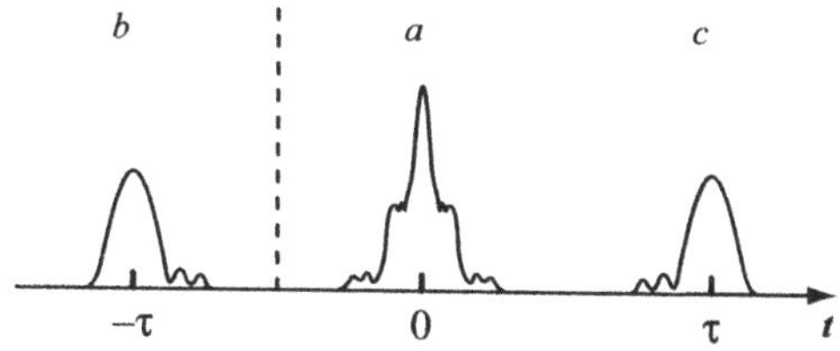

Figure 3.5 Fourier transform of spectral interferometry data for the case $\tau > 0$. Features marked a, b, and c correspond to the first, second, and final terms, respectively, in eq. (3.31). Feature b contains the desired information.

autocorrelation envelopes of the reference and signal pulses. This term does not contain any phase information. The term marked b is located at $t = -\tau$ and has the form

$$\Gamma_a(t) * \left[h(t+\tau)e^{-j\omega_0 t}\right]$$

which represents a baseband version of the impulse response function convolved with the autocorrelation of $a(t)$. $h(t)$ and $H(\omega)$ can be extracted from this term provided that the autocorrelation is sufficiently short (this is equivalent to the requirement of a sufficiently broad spectrum). The third term, marked c and located at $t = \tau$, involves a time-reversed version of $h(t)$ but is otherwise similar to b. The algorithm proceeds as follows:

1. The spectral interferometry data is first inverse-Fourier-transformed.
2. Terms a and c are set to zero. Provided that $\tau > 0$, as in our example, this means that everything to the right of the dashed vertical line in the figure is replaced by zero. Note that one needs to know the sign of τ in order to distinguish correctly between terms b and c.
3. Now the result is Fourier transformed back into the frequency domain and the delay is suppressed. There are two equivalent procedures for this:
   (a) Term b in the time-domain data can simply be shifted to $t = 0$ before inverse Fourier transforming back to the frequency domain.
   (b) Alternatively, term b can be Fourier-transformed back into the frequency domain without any temporal shifting, yielding $H(\omega)\,|A_{\text{in}}(\tilde{\omega})|^2\,e^{j\omega\tau}$. The linear spectral phase term (which can be obtained by curve fitting) is then subtracted off.
4. Finally, the result is divided by $|A_{\text{in}}(\tilde{\omega})|^2$. This leaves $H(\omega)$, as desired.

It should be clear from Fig. 3.5 that this algorithm places a constraint on τ; namely, $|\tau|$ should be sufficiently large to avoid overlap between terms a and b. In the frequency domain, this constraint ensures that the period of the spectral fringes is finer than the fastest spectral variation in $H(\omega)$ and $|E_{\text{in}}(\omega)|^2$ (i.e., the spectral fringes are able to adequately sample the desired spectral phase information). As an extreme case, consider the result of setting $\tau = 0$. In this case the last term in eq. (3.32) becomes proportional to $\cos(\psi(\omega))$. Since the cosine is an even function, information is clearly lost. On the other hand, $H(\omega)$ can be recovered even for $\tau = 0$ provided that both $\sin(\psi(\omega))$ and $\cos(\psi(\omega))$ are available. The sine function can be obtained by repeating the spectral interferometry measurement after introducing an additional $\pi/2$ relative phase shift between the signal and reference beams. From an experimental perspective, however, it is usually preferred to acquire a single spectral interferometry trace at sufficiently large τ.

A few additional points are worth mentioning:

- Since electric field cross-correlation and spectral interferometry are linear techniques, they can be applied to low power signals (unlike some of the nonlinear optical methods discussed later in the chapter).
- The phase of the input signal does not appear in eq. (3.32). Therefore, in order to characterize the complex spectral response of a linear optical system, as shown in Fig. 3.4, the input field need not be a short pulse. Incoherent light can also be used to characterize the complex spectral response of the component under test as long as the input spectrum covers the spectral range of interest. On the other hand, if one wishes to extract the signal field waveform itself, this can be done only if the reference pulse is known. These remarks also apply to time-domain electric field cross-correlation measurements.
- In spectral interferometry there is no need for an accurate calibration of the lengths of the interferometer arms; the measurement itself yields the relative delay. This makes spectral interferometry somewhat simpler to implement than electric field cross-correlation, where an independent online length calibration is required.

3.3.3 Application: Optical Coherence Tomography

In addition to waveform measurement, interferometry techniques have important applications in imaging and depth profiling. Here we briefly introduce one such application, *optical coherence tomography* (OCT), which has found extensive application for imaging within biological samples [104,105].

A basic setup for OCT is sketched in Fig. 3.6a. The apparatus is based on electric field cross-correlation and is usually implemented using fiber optic components. A broadband light source is directed to a 50–50 coupler which splits the light into a reference arm equipped with a retroreflecting mirror and a sample arm. Back-scattered light from the sample and retroreflected reference light are recombined in the coupler and measured using a slow detector as the reference mirror is scanned longitudinally. The measurement principle

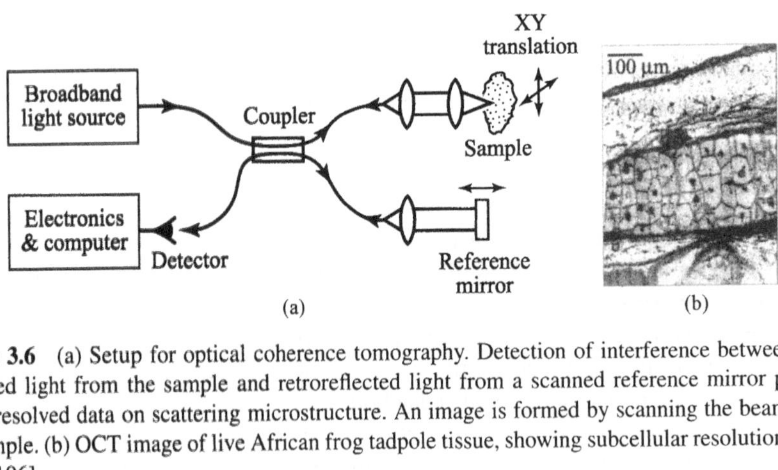

Figure 3.6 (a) Setup for optical coherence tomography. Detection of interference between back-scattered light from the sample and retroreflected light from a scanned reference mirror provides depth-resolved data on scattering microstructure. An image is formed by scanning the beam across the sample. (b) OCT image of live African frog tadpole tissue, showing subcellular resolution. Image from [106].

is that interference between reference and sample arms is observed only when the delay difference is less than the coherence time. Consequently, detecting the interference acts as a coherence gate. The depth into the sample corresponding to zero delay depends on the delay of the reference arm. As a result, the interferogram recorded as the reference mirror is scanned provides a profile of scattering strength vs. depth. Images are built up by using a focused beam and recording a sequence of depth scans at different transverse positions on the sample. The transverse position on the sample may be varied either by translating the sample, as depicted in the figure, or by deflecting the beam.

OCT has been used extensively to image microstructure in both transparent and non-transparent tissue. The eye is the prime example of transparent tissue, whereas most other biological tissues are highly scattering. In either case the ability to select time slices of the back-scattered field through coherence gating is the key factor. The dynamic range of such measurements can reach 110 dB, which means that the system is sensitive to reflected or scattered powers as low as 10^{-11} of the input power [106]. In highly scattering media where attenuation is very strong, this allows imaging to millimeter depths, which is not possible using standard optical microscopy techniques. An example of an OCT image demonstrating subcellular spatial resolution is shown in Fig. 3.6b.

Usually, very good spatial resolution is desired. An advantage of OCT is that transverse resolution and depth resolution are essentially decoupled: Transverse resolution is determined by the focused spot size, while depth resolution is determined by the coherence time of the light source. In particular, depth resolution Δz is given by

$$\Delta z = \frac{v_g \tau_c}{2} \tag{3.33}$$

where τ_c here is the FWHM of the electric field autocorrelation function, v_g is the group velocity in the sample, and a factor of 2 arises due to the double pass through the sample in a back-scattering geometry. Because coherence time is inversely proportional to optical bandwidth, sources with broad bandwidth are preferred. Note that as in other electric field correlation measurements, coherent short pulses are not required. Temporally incoherent sources are also acceptable as long as they have broad bandwidth and good spatial properties. Furthermore, dispersion compensation is not required. As long as the individual arms of the interferometer are adjusted to have identical dispersion, there is no effect on the interferometry measurement. It is this point that makes OCT compatible with fiber optic implementations.

These points are illustrated in Fig. 3.7, which shows optical spectra and interference fringes obtained in an autocorrelation geometry (sample replaced by a second retroreflecting mirror) [106]. Data are shown for two different light sources: a mode-locked Ti:sapphire (Ti:S) laser with a 260-nm bandwidth and a superluminescent diode (SLD) with a 32-nm bandwidth. The SLD is a semiconductor diode device similar to a semiconductor injection laser but with lasing suppressed. Hence, the SLD provides light that is spatially coherent but temporally incoherent. The interference fringes give the electric field autocorrelation functions, which are equal to the inverse Fourier transforms of the optical power spectra. The depth resolutions are 11.5 and 1.5 μm for the SLD and Ti:S laser, respectively, in inverse proportion to the optical bandwidths.

A few additional points are worthy of comment. First, unlike the simulated data of Fig. 3.3, the interferometry data of Fig. 3.7 have zero mean value. This occurs because the results shown in Fig. 3.7 were acquired using a differential measurement scheme, not

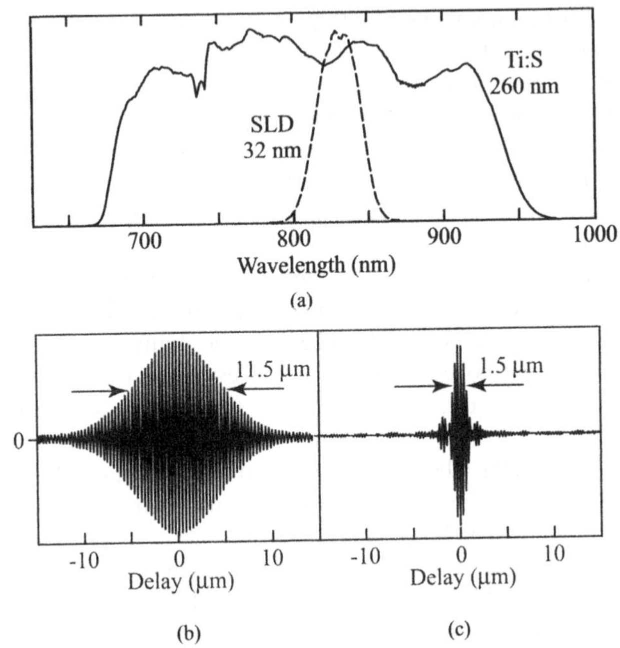

Figure 3.7 Comparison of optical power spectra and resulting OCT spatial resolution for superluminescent diode (SLD) and mode-locked Ti:S laser light sources: (a) power spectra; (b,c) electric field autocorrelation interference fringes from (b) SLD and (c) Ti:S laser. Adapted from [106].

depicted in Fig. 3.2 or 3.6. Referring to Fig. 3.2, there are actually two interferometer outputs, one in the direction of the photodetector, as portrayed, the other in a direction back toward the source. By detecting both outputs with a pair of photodetectors and differencing the result, one obtains a zero-mean signal in which only the fringe terms remain. This differential scheme has the practical advantage of suppressing intensity noise for increased sensitivity. A second point is that although autocorrelations are supposed to be symmetric, close inspection of the Ti:S correlation signal in Fig. 3.7 reveals a slight asymmetry. Such asymmetry may arise from small differences in the dispersion of the two interferometer arms. Thus, although compensation of dispersion common to both interferometer arms is not required, for very broadband signals great care must be exercised to equalize the dispersion experienced by the interfering signals.

A number of different modalities exist for processing of OCT signals. Arguably, the most basic difference between various schemes is the choice of coherent vs. incoherent demodulation. These two main approaches are described briefly below.

- In *incoherent demodulation* approaches, one is concerned only with the envelope of interference fringes. The positions of peaks in the interferogram envelope reveal the depths of strongly back-scattering structures, much as delays of return signals in a pulsed radar system yield a target range. Phase information associated with the interference fringes is not used.
- In *coherent demodulation* approaches, one acquires and analyzes the full interferometric signal, including phase and frequency information associated with the fringes themselves. In this mode, OCT measurements become fully analogous to electric field

cross-correlation measurements described in Section 3.3.1. With full fringe data one may extract new information through Fourier transform analysis. For example, in Doppler OCT one assigns shifts in interferogram spectra to shifts in the mean frequency of back-scattered light [107,108]. Such frequency shifts provide information on scatterer velocities, projected in the direction of light propagation, through the Doppler effect. Furthermore, by analyzing Doppler OCT data in terms of joint time–frequency distributions (see Section 3.5), it becomes possible to acquire depth-resolved velocity images. This approach has enabled mapping of flow in subsurface blood vessels in living tissues.

3.4 INTENSITY CORRELATION MEASUREMENTS

3.4.1 Correlation Measurements Using Second-Harmonic Generation

We have already seen that in the case of a single unknown pulse, a common measurement strategy is to arrange for the pulse to sample itself. Mathematically, this means that one measures a correlation function of the pulse. We have also seen that measurement of the electric field autocorrelation function using a linear Michelson interferometer does not provide any new information about the pulse. To perform a more useful measurement, one can insert a nonlinear element into the interferometer. The most common approach involves second-harmonic generation (SHG), in which a nonlinear crystal is used to generate light at twice the input optical frequency (i.e., at the second harmonic $2\omega_0$ of the input light at the fundamental frequency ω_0). The measurement procedure is to record the time-averaged second-harmonic power as a function of the relative delay τ between the two identical versions of the input pulse [109–112]. Due to the nonlinearity, the total energy in the second-harmonic pulse is greater when the two pulses incident on the nonlinear crystal overlap in time. Therefore, the peak in second-harmonic power plotted as a function of τ contains information about the pulse width. For an excellent discussion of SHG and other nonlinear correlation techniques, see [113].

We now analyze this SHG measurement in the collinear geometry shown in Fig. 3.8. We assume (1) that the intensity is low enough to avoid saturation associated with pump

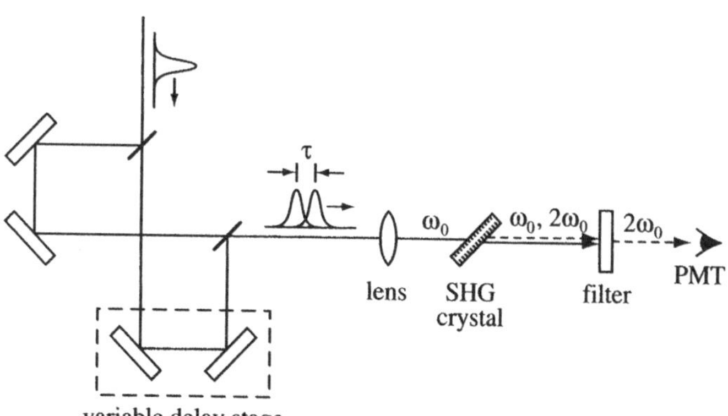

Figure 3.8 Collinear second-harmonic generation geometry for measurement of the intensity autocorrelation function. PMT, photomultiplier tube.

depletion and (2) that the bandwidth of the SHG process is large enough to accommodate the entire spectrum. The first assumption is easily met in practice. The second criterion requires the use of thin nonlinear crystals, especially for very short pulses, so that second-harmonic generation is *phase matched* over a sufficiently broad spectrum. The concept of phase matching and its implications for the bandwidth of the SHG process are discussed in Chapter 5.

The field emerging out of the interferometer and incident on the SHG crystal is written

$$e_{\text{out}}(t) = \frac{1}{2}\text{Re}\left\{a(t)e^{j\omega_0 t} + a(t-\tau)e^{j\omega_0(t-\tau)}\right\} \tag{3.34}$$

The field at the second-harmonic frequency, which is proportional to the square of input field, is given by

$$\begin{aligned} e_{\text{SHG}}(t) &\sim e_{\text{out}}^2(t) \\ &\sim \text{Re}\left\{\left[a^2(t) + a^2(t-\tau)e^{-2j\omega_0\tau} + 2a(t)a(t-\tau)e^{-j\omega_0\tau}\right]e^{2j\omega_0 t}\right\} \end{aligned} \tag{3.35}$$

The actual measurement records the time-integrated (or equivalently, the time-averaged) second-harmonic power (not the instantaneous power, since this would require an ultrafast detector, which we have assumed is not available). After some algebra one obtains the following expression:

$$\begin{aligned} \langle P_{\text{SHG}}(t)\rangle &\sim 2\langle|a(t)|^4\rangle + 4\langle|a(t)|^2|a(t-\tau)|^2\rangle + \left\{\left\{2\langle\left(|a(t)|^2 + |a(t-\tau)|^2\right)\right.\right. \\ &\quad \left.\left.\times\, a(t)a^*(t-\tau)\rangle e^{j\omega_0\tau} + \langle\left[a(t)a^*(t-\tau)\right]^2\rangle e^{2j\omega_0\tau}\right\} + \text{c.c.}\right\} \end{aligned} \tag{3.36}$$

In arriving at this expression, we have assumed a time-stationary process, as previously. The first term in eq. (3.36) is independent of τ, and the second term varies relatively slowly with τ. The third and fourth terms represent interferometric contributions at frequencies ω_0 and $2\omega_0$, respectively, which oscillate rapidly as τ is varied over one optical period.

Fringe-Averaged Autocorrelation Often, one suppresses the interferometric terms by averaging the SHG signal over a few optical periods. The measured time-average second-harmonic power then becomes

$$\langle P_{\text{SHG}}(t)\rangle \sim \left\langle|a(t)|^4\right\rangle\left(1 + \frac{2\left\langle|a(t)|^2\,|a(t-\tau)|^2\right\rangle}{\left\langle|a(t)|^4\right\rangle}\right) \tag{3.37}$$

At this point we define the normalized intensity autocorrelation function $G_2(\tau)$:

$$G_2(\tau) = \frac{\langle I(t)I(t-\tau)\rangle}{\left\langle|I(t)|^2\right\rangle} \tag{3.38}$$

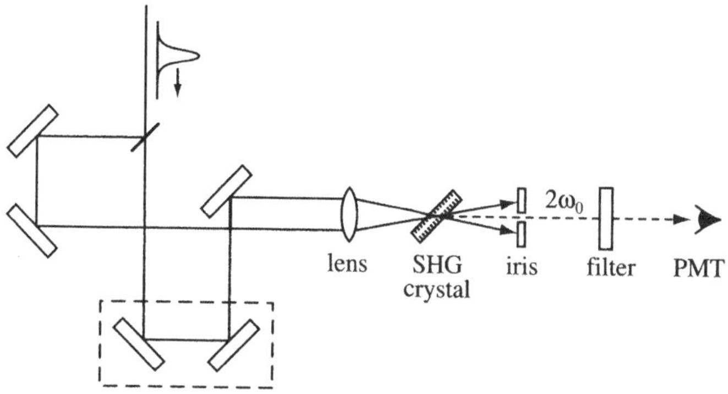

Figure 3.9 Noncollinear second-harmonic generation geometry for background-free intensity autocorrelation function measurement. PMT, photomultiplier tube.

where $I(t) \sim |a(t)|^2$ as usual. Equation (3.37) for the time-average second harmonic power in a collinear measurement geometry then becomes

$$\langle P_{\mathrm{SHG}}(t) \rangle \sim 1 + 2G_2(\tau) \tag{3.39}$$

The measured second-harmonic power is the sum of a constant background term (which represents the SHG from each pulse individually) and twice the intensity autocorrelation function (which arises from the interaction of the two time-delayed pulses).

Another common experimental arrangement uses a noncollinear geometry (Fig. 3.9), in which the second harmonic due to the interaction between the time-delayed pulses is emitted in a different direction than that due to the individual pulses. In this case, the constant background terms can be suppressed, both by using an iris and by adjusting the SHG phase-matching conditions. The output signal in the background-free noncollinear geometry becomes

$$\langle P_{\mathrm{SHG}}(t) \rangle \sim G_2(\tau) \tag{3.40}$$

Note that the noncollinear geometry automatically averages out the fringe terms in eq. (3.36).

It is useful at this point to comment on the properties of $G_2(\tau)$. Several important points are listed below.

- The normalized intensity correlation function assumes its maximum value of unity at $\tau = 0$.
- In the case of finite-duration pulses, the pulses exhibit no overlap for delays much greater than the pulse width, and the correlation function goes to zero. The width of the correlation peak gives information about the pulse width.
- The correlation function is an even function of τ, independent of the symmetry of the actual pulse. Therefore, one cannot uniquely recover the pulse intensity profile from $G_2(\tau)$.

- If one somehow knew a priori that the intensity profile were symmetric, one could indeed recover $I(t)$ from $G_2(\tau)$. We note that the Fourier transform of $G_2(\tau)$ is $\left|\tilde{I}(\omega)\right|^2$, where $\tilde{I}(\omega)$ is the Fourier transform of $I(t)$. If $I(t)$ is symmetric, $\tilde{I}(\omega)$ must be real; therefore, $\tilde{I}(\omega)$ can be obtained by taking the square root of $\left|\tilde{I}(\omega)\right|^2$.
- Conversely, in the usual case where $I(t)$ is not known to be symmetric, $\tilde{I}(\omega)$ cannot be assumed real. Therefore, $G_2(\tau)$ only gives $\left|\tilde{I}(\omega)\right|$ and not its phase, which would be needed to reconstruct $I(t)$.
- By definition, the intensity autocorrelation function is also insensitive to the phase of the electric field itself. Therefore, the fringe-averaged intensity autocorrelation $G_2(\tau)$ cannot be used to characterize chirps or phase modulations on the pulse.

It is also worth discussing the behavior of $G_2(\tau)$ in the case of continuous-wave noise [113]. The intensity correlation function still takes on its maximum value of unity at $\tau = 0$. However, for sufficiently large values of τ, the two relatively delayed versions of the same noise signal are uncorrelated. Since the intensity is always nonnegative, the correlation never drops to zero, but the value is reduced below unity. This results in a coherence peak in the intensity autocorrelation function. It is important to note that this coherence peak has nothing to do with the actual pulse width; the width of the peak is governed by the correlation time of the intensity fluctuations, which can be as short as the inverse of the optical bandwidth. For the specific common case of a Gaussian random field (i.e., the real and imaginary parts of the electric field are independent, zero-mean random variables described by Gaussian probability distributions with equal variances), the probability distribution $p_I(I)$ for the intensity is given by

$$p_I(I) = \frac{1}{\langle I \rangle} e^{-I/\langle I \rangle} \tag{3.41}$$

where $\langle I \rangle$ is the average intensity. At delays large enough so that the two noise signals are independent, we get

$$G_2(\tau) = \frac{\langle I(t) \rangle^2}{\langle I^2(t) \rangle} = \frac{1}{2} \tag{3.42}$$

where we have used $\langle f(I) \rangle = \int dI\, f(I) p_I(I)$.

These findings are summarized in Fig. 3.10. In the collinear measurement geometry, the correlation function for a single coherent pulse decays smoothly to a constant background level, with a 3 : 1 contrast ratio between the peak at $\tau = 0$ and the background. This theoretical contrast ratio can be used as a check for proper experimental alignment. In the case of continuous-wave noise, the only feature is the coherence spike centered near $\tau = 0$, with a contrast ratio of 3 : 2. The intermediate case of a finite-duration noise burst has a coherence spike centered on top of a finite-duration pedestal, with a contrast ratio of 3 : 2 : 1. It is important not to use the duration of the coherence spike to estimate the pulse duration, and it also important to note that the intensity autocorrelation of a pulse with a smooth intensity profile but containing excess bandwidth in the form of a chirp will not exhibit a coherence spike, which arises only from substructure in the intensity. One can estimate the pulse duration from the full width at half maximum (FWHM) of the correlation trace (excluding coherence spike), denoted $\Delta\tau$ in the figure.

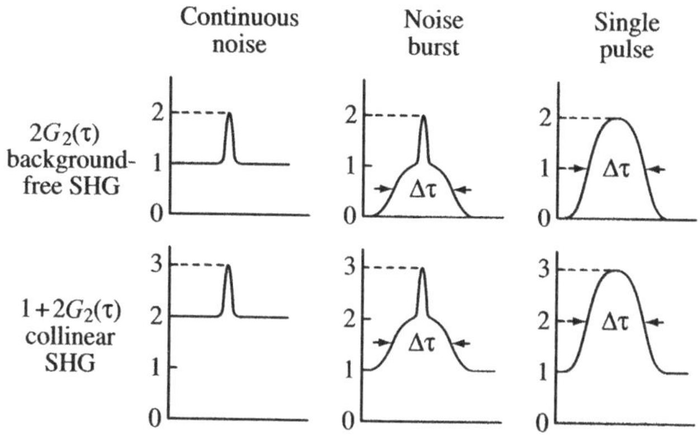

Figure 3.10 Theoretical traces for collinear and noncollinear SHG autocorrelation measurements. From Fig. 3.3 of [113], with permission from Springer Science and Business Media.

The situation is similar in the noncollinear measurement geometry, except that there is no constant background for finite-duration pulses. The zero background makes it easier to identify small amounts of energy in the wings of the pulse. The contrast ratio is 2 : 1 for continuous-wave noise and 2 : 1 : 0 for a finite-duration noise burst.

To estimate the intensity FWHM pulse width from the autocorrelation, one must assume a specific pulse shape. The intensity FWHM (Δt) is then determined by dividing the intensity autocorrelation FWHM ($\Delta\tau$) by a pulse-shape-specific deconvolution factor. See Table 3.2 for autocorrelation functions and deconvolution factors for a few different pulse shapes [113,114]. For smoothly varying pulse shapes (e.g., Gaussian or sech^2 pulses), the variation in the deconvolution factor is of order 10%, and therefore the pulse duration can be obtained approximately, even though the pulse shape is not, in fact, known. For less-well-behaved pulses (e.g., rectangular or single-sided exponential pulses), the deconvolution factor can vary by a larger amount. To do a consistency check, one can fit the autocorrelation data to the functional form of $G_2(\tau)$ corresponding to the assumed pulse shape. It is also useful to compare the measured time–bandwidth product $\Delta\nu\,\Delta t$ with that corresponding to the assumed pulse shape. A discrepancy between the experimental and theoretical time–bandwidth product indicates either a deviation from the assumed pulse shape or the presence of chirp. Note, however, that these consistency checks do not

Table 3.2 Autocorrelation Functions and Deconvolution Factors for Four Pulse Shapes[a]

$I(t)$	$G_2(\tau)$	$\dfrac{\Delta\tau}{\Delta t}$
sq (t)	$1-\lvert\tau\rvert$ for $\lvert\tau\rvert \le 1$, 0 otherwise	1
e^{-2t^2}	$e^{-\tau^2}$	$\sqrt{2}$
$\mathrm{sech}^2\,(t)$	$\dfrac{3\,(\tau\cosh(\tau)-\sinh(\tau))}{\sinh^3(\tau)}$	1.543
e^{-t} for $t \ge 0$, 0 otherwise	$e^{-\lvert\tau\rvert}$	2

[a] Δt and $\Delta\tau$ are full widths at half maximum of the intensity $I(t)$ and the intensity autocorrelation function $G_2(\tau)$, respectively. sq(u) is defined as a unit square pulse such that sq$(u) = 1$ for $\lvert u\rvert \le \frac{1}{2}$ and 0 otherwise; and t and τ are in normalized units.

assure uniqueness (i.e., it is still possible that the actual pulse shape deviates from the shape assumed).

Additional information is available using the root-mean-square (rms) pulse width Δt_{rms}, as discussed in [94]. The definition of the rms pulse width was given in Section 3.1. A very useful property of the rms pulse width is that it can be found directly from the intensity autocorrelation function without any assumption about the pulse shape using the following formula:

$$(\Delta t_{\text{rms}})^2 = \frac{2\int d\tau\, \tau^2 G_2(\tau)}{\int d\tau\, G_2(\tau)} \tag{3.43}$$

It was also demonstrated that for a given power spectrum $|E(\omega)|^2$, Δt_{rms} assumes its minimum value $\Delta t_{\text{rms}}^{(0)}$ when $E(\omega)$ has a constant spectral phase [i.e., when $E(\omega)$ is bandwidth-limited]. A similar relation is often assumed for the FWHM pulse width Δt, but there is no proof of this assumption. Based on this minimum rms pulse width theorem, Sorokin et al. [94] also introduced an rms chirp parameter

$$C = \sqrt{\langle \tau^2(\omega) \rangle - \langle \tau(\omega) \rangle^2} \tag{3.44}$$

which they showed could also be expressed in terms of the rms pulse widths:

$$C = \frac{1}{2}\sqrt{(\Delta t_{\text{rms}})^2 - \left(\Delta t_{\text{rms}}^{(0)}\right)^2} \tag{3.45}$$

In these expressions $\tau(\omega)$ is the frequency-dependent delay as defined in eq. (2.82), and eq. (3.8) is used. An increase in Δt_{rms} over the bandwidth-limited value $\Delta t_{\text{rms}}^{(0)}$ gives a direct indication of spectral chirp, as quantified through the C parameter. An attractive feature of this definition is that both Δt_{rms} and $\Delta t_{\text{rms}}^{(0)}$ can be obtained directly from commonly available experimental diagnostics: Δt_{rms} from intensity autocorrelation through eq. (3.43) and $\Delta t_{\text{rms}}^{(0)}$ by taking the square root of the power spectrum, assuming constant phase, and Fourier transforming to get $I(t)$.

Interferometric Autocorrelation We now return to the full expression describing the SHG pulse-width measurement including interferometric terms, eq. (3.36). Detailed information on interpreting interferometric autocorrelation data is given in [95,114]. The interferometric terms include some phase and coherence information that is lacking in $G_2(\tau)$ itself. These terms may be recorded by using a collinear measurement geometry and exercising care to avoid averaging over the rapid interferometric oscillations. As an example, consider the data plotted in Figs. 3.11 and 3.12, which correspond to pulses from a self-mode-locked Ti:S. Figure 3.11 corresponds to pulses obtained when a prism pair is used for intracavity dispersion compensation. The intensity autocorrelation function is smooth and yields a 60-fs intensity FWHM pulse duration assuming a sech^2 intensity profile. The width of the power spectrum is such that the time–bandwidth product is $\Delta\nu\, \Delta t \approx 0.33$, in close agreement with the theoretical value of 0.315 for secant hyperbolic pulses. Thus, these pulses appear to be close to *bandwidth limited*, meaning that (1) the pulse duration is as short as it can be given the mode-locked bandwidth, and (2) there is little or no chirp. The most notable

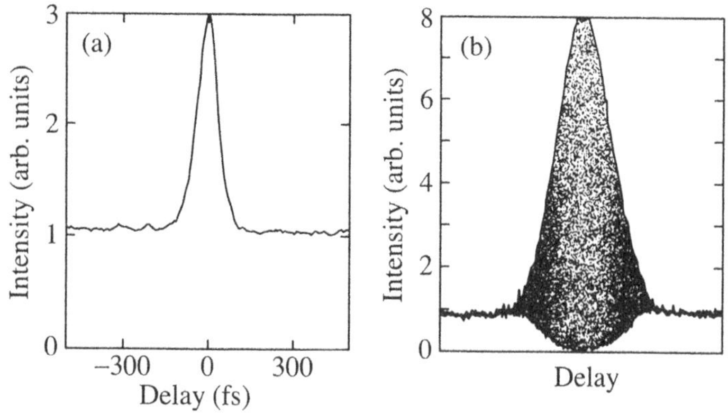

Figure 3.11 (a) Intensity autocorrelation and (b) interferometric autocorrelation for nearly bandwidth-limited pulses from a mode-locked Ti:S laser. A 60-fs pulse width and a time–bandwidth product $\Delta\nu\,\Delta t = 0.33$ are calculated assuming a sech^2 intensity profile. For (b), the delay axis is zoomed by roughly a factor of 3 compared to (a). From [52].

feature of the interferometric autocorrelation is that the fringes last throughout the entire pulse duration. This is another sign that the pulse is bandwidth limited.

The situation is rather different in Fig. 3.12, which corresponds to data obtained when the prism pair is removed from the laser. In this case the dispersion in the laser cavity is dominated by the Ti:S crystal and is large and positive. Comparing with Figs. 2.19 and 2.20, one expects relatively long mode-locked pulses with a strong chirp. This expectation is confirmed by the measurements. The intensity autocorrelation remains smooth (no coherence spike), but with the intensity FWHM pulse duration broadened to 2 ps (again assuming sech^2 pulses). At the same time the spectral width is narrowed from 14 nm to 3.5 nm. The calculated time–bandwidth product increases to 2.7, indicating that the pulses are far from the bandwidth limit. The appearance of the interferometric autocorrelation is changed dramatically: The fringes now occur only near the center of the trace and die off

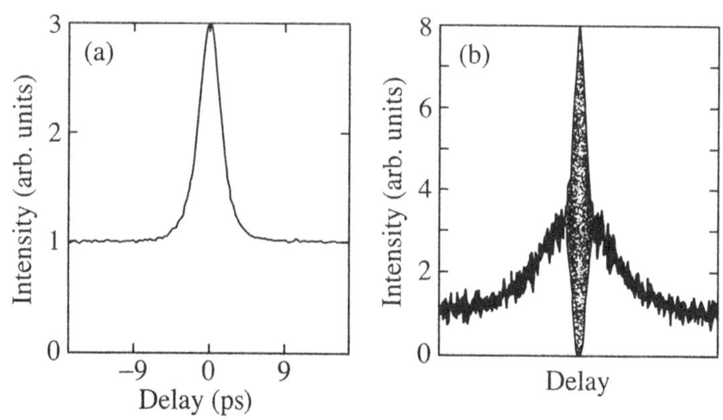

Figure 3.12 (a) Intensity autocorrelation and (b) interferometric autocorrelation for highly chirped pulses from a mode-locked Ti:S laser. A 2-ps pulse width and a time–bandwidth product $\Delta\nu\,\Delta t = 2.7$ are calculated assuming a sech^2 intensity profile. For (b), the delay axis is zoomed by roughly a factor of 3 compared to (a). From [52].

much faster than the nonoscillating portion of the experimental trace. The envelope of the fringes is related to the inverse of the spectral bandwidth (but in a more complicated way than in electric field autocorrelation). The data indicate that the coherence time of the pulses is much shorter than the duration of the intensity profile, consistent with the large $\Delta\nu\ \Delta t$ product and a strong chirp. This behavior should not be confused with coherence spikes which can be observed in measurements of $G_2(\tau)$, which arise due to noise substructure on the intensity profile and are insensitive to chirp.

It is also worth discussing the contrast ratio observed in interferometric autocorrelation measurements. The upper envelope of the fringes results when the two time-delayed fields are exactly in phase and add to produce a maximum total field; the lower envelope results when the two fields are exactly out of phase and cancel to produce the minimum total field. For the case where the coherence time is essentially equal to the pulse width (i.e., the temporal phase is essentially constant), the upper and lower envelopes are proportional to

$$\text{upper envelope} \sim \left\langle \left[|a(t)| + |a(t-\tau)|\right]^4 \right\rangle \tag{3.46a}$$

$$\text{lower envelope} \sim \left\langle \left[|a(t)| - |a(t-\tau)|\right]^4 \right\rangle \tag{3.46b}$$

where the two pulses are also assumed to have equal amplitudes. When there is significant phase modulation (i.e., the coherence time is shorter than the pulse width), this equation applies only for $|\tau| \ll \tau_c$; for $|\tau| \gg \tau_c$ the fringes disappear and only the nonoscillating terms given by eq. (3.39) remain. In all cases the background level observed for large delays when the pulses no longer overlap is proportional to

$$\text{background} \sim 2\left\langle |a(t)|^4 \right\rangle \tag{3.47}$$

From these formulas we can easily see that the contrast ratio between the peak of the upper fringe envelope and the background level should be 8 : 1, as observed in the data shown above. The lower envelope is at the zero level near $\tau = 0$, also as observed. If one averages over the fringes, one gets back to $1 + 2G_2(\tau)$ as indicated in eq. (3.39), with a contrast ratio of 3 : 1 between the peak and the background. (This last statement is not obvious from Figs. 3.11 and 3.12 but is true nevertheless!)

Note that the presence or absence of strong chirp in Figs. 3.11 and 3.12 could be deduced from the time–bandwidth product alone, without using the interferometric autocorrelation. The interferometric autocorrelation is useful because it can serve as a clear visual indicator of moderate to large chirp and because it provides additional data for quantitative pulse-shape analysis. It has been shown that if one measures the term varying as $\exp\{2j\omega_0\tau\}$ in the interferometric autocorrelation, that is,

$$\Xi(\tau) = \left\langle \left[a(t)a^*(t-\tau)\right]^2 \right\rangle e^{2j\omega_0\tau}$$

then together with the fringe-averaged autocorrelation $G_2(\tau)$ and the power spectrum of the input pulse $|A(\omega - \omega_0)|^2$, this is enough information to determine the complete electric field profile uniquely[3] [115]. Reconstruction of the electric field from these data has been

[3] The electric field is determined uniquely except for a time-reversal ambiguity; $a(t)$ and $a^*(-t)$ are both allowed solutions. Furthermore, the proof assumes that the field has finite support (i.e., it is strictly zero outside a finite time window).

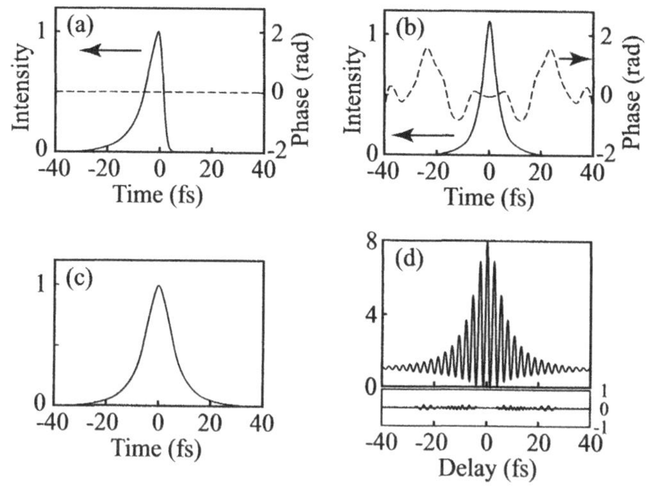

Figure 3.13 Calculation results for a pair of pulses constructed to have identical fringe-averaged autocorrelation traces and power spectra: (a) asymmetric pulse; (b) symmetric pulse; (c) fringe-averaged autocorrelation trace, identical for both pulses; (d) interferometric autocorrelation of asymmetric pulse (top) and difference in asymmetric and symmetric pulse interferometric autocorrelations (bottom). Adapted from [116]. Copyright ©2001, IEEE.

demonstrated by using iterative phase retrieval techniques related to those described later in Section 3.6.5. Practically, this technique has seen limited use, possibly because of the need for high measurement accuracy to ensure valid reconstruction of $a(t)$. It is also interesting that in the case of a pulse free of frequency modulation, the term multiplying $\exp\{2j\omega_0\tau\}$ in $\Xi(\tau)$ is identical to $G_2(\tau)$. Therefore, by carefully comparing these two terms, one can directly check whether $a(t)$ is real. Finally, we note that $\Xi(\tau)$ is simply the electric field autocorrelation of the second-harmonic pulse $a^2(t)$ produced in response to a single pulse illuminating the second-harmonic crystal. Therefore, measurement of the power spectrum of a single second-harmonic pulse gives equivalent information.

These points are illustrated in Fig. 3.13, which shows calculation results for a pair of pulses constructed to have identical fringe-averaged autocorrelation traces and power spectra. One of the pulses (Fig. 3.13a) has an asymmetric intensity profile with a 28-fs FHWM pulse width and a flat temporal phase, while the other (Fig. 3.13b) has a symmetric intensity profile with a 37-fs FWHM pulse width and phase modulation. $G_2(\tau)$ is shown in Fig. 3.13c and is identical for both pulses, as stated. This demonstrates the ambiguity inherent in fringe-averaged intensity autocorrelation measurements. Figure 3.13d shows the interferometric autocorrelation of the asymmetric pulse. The interferometric autocorrelation of the symmetric pulse is almost identical and is not included in the main frame of Fig. 3.13d. Instead, the difference between the two interferometric autocorrelations is plotted in the lower part of Fig. 3.13d. The rms difference between the two traces is less than 1%. Thus, although the interferometric autocorrelations of the two pulses are in fact distinct, as predicted [115], the differences are subtle. The conclusion is that interferometric autocorrelation data may be only weakly sensitive to rather significant changes in pulse shape and may therefore be quite challenging to apply for unambiguous pulse-shape retrieval in a practical experimental context.

Intensity Cross-Correlation In addition to intensity autocorrelation measurements, one can also perform intensity cross-correlation measurements in which the two pulses being correlated have different intensity profiles. If we denote the intensity profiles of reference and signal pulses as $I_r(t)$ and $I_s(t)$, respectively, one measures

$$\langle P_{\mathrm{SHG}}(\tau)\rangle \sim \langle I_s(t)I_r(t-\tau)\rangle \sim \int dt\, I_s(t)I_r(t-\tau) \tag{3.48}$$

where we have assumed a background-free noncollinear measurement geometry. When the reference pulse is much shorter than the signal pulse, the measurement returns $I_s(t)$ directly, including any asymmetry. To estimate the amount of broadening due to the finite duration of the reference pulse, we consider the case of Gaussian pulses [i.e., $a_r(t) \sim e^{-t^2/t_r^2}$ and $a_s(t) \sim e^{-t^2/t_s^2}$]. The intensity cross-correlation measurement then results in

$$\langle P_{\mathrm{SHG}}(\tau)\rangle \sim \exp\left(-\frac{2\tau^2}{t_s^2+t_r^2}\right) \tag{3.49}$$

The measured pulse width t_s' is proportional to

$$t_s' = \sqrt{t_s^2 + t_r^2} \tag{3.50}$$

If the reference pulse duration is 50% that of the signal pulse, the cross-correlation trace is broadened by only 11%. The broadening rapidly becomes very small as the reference pulse duration is further decreased relative to the signal pulse. Intensity cross-correlation measurements are particularly useful when the signal pulse is known to be broader than the reference, for example, in the case when the signal is generated from the reference pulse using a pulse-shaping apparatus (see Chapter 8). If, in addition, the intensity profile of the reference pulse is already known, $I_s(t)$ can be extracted from the cross-correlation data by deconvolution.

3.4.2 Experimental Procedures

Repetitive Pulse Measurements In terms of experimental procedure, with CW mode-locked systems [113] one commonly uses a chopper to modulate one of the interferometer arms and a lock-in amplifier to detect the component of the photomultiplier (PMT) current modulated at that frequency. A stepper motor–driven delay stage is stepped under computer control (typically, with a 0.1- or 1-μm step size) in order to vary the delay, and the lock-in output is recorded as a function of delay. The delay range can be scanned many times to allow signal averaging for improved sensitivity. In this procedure the measurement is the average over a large number (millions or billions) of laser pulses. This is valid provided that all the pulses in the ensemble are essentially identical, which is usually the case for modern mode-locked sources. However, when the measurement averages over pulses with widely varying pulse durations, it has been shown that the experimental correlation trace assumes a double-sided exponential shape, independent of the actual pulse shape [117].

Single-Shot Measurements In high-power amplified laser systems with low pulse repetition rates, one often prefers a different measurement setup, one allowing single-shot

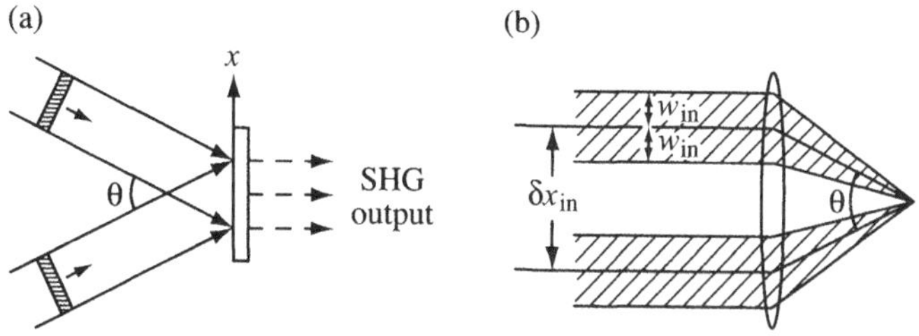

Figure 3.14 (a) Transverse variation of the delay in noncollinear intensity autocorrelation measurements; (b) focusing geometry for noncollinear intensity autocorrelation.

measurement capability. This is usually achieved by crossing the two beams (which typically are not focused) at a relatively large angle in the second-harmonic crystal [118]. For a sufficiently large beam size and crossing angle, this leads to an appreciable variation in delay across the transverse beam profile, where the delay is given by

$$\tau = \frac{2x\sin(\theta/2)}{c} \tag{3.51}$$

As illustrated in Fig. 3.14, x is the lateral spatial coordinate on the surface of the crystal and θ is the full angle between the incident beams (in air). The single-shot measurement is accomplished by recording the spatial profile of the generated second harmonic (e.g., using a CCD camera).

Transverse Delay Effects in Noncollinear Repetitive Pulse Measurements This same transverse variation of the delay, which is desirable for single-shot measurements, can give rise to an undesirable temporal broadening in noncollinear autocorrelation measurements for high-repetition-rate laser systems, in which the entire second-harmonic beam is incident on a single photodetector. From eq. (3.51), the range of delay is clearly on the of order $\theta d/c$, where d is the beam diameter and θ has been assumed small. To make this more precise, we follow a procedure similar to [119] and write down an expression for the measured SHG power in the noncollinear experiment in the case of a Gaussian beam profile [i.e., $E(r,t) \sim a(t)e^{-r^2/w^2}$]. The measured second-harmonic power in the direction of the detector, integrated over both time and the spatial profile, takes the form

$$\begin{aligned} \langle P_{\mathrm{SHG}}\rangle \sim & \iint dx\,dt\,\left|a\left(t+\frac{x}{c}\sin(\theta/2)\right)\right|^2 \left|a\left(t-\tau-\frac{x}{c}\sin(\theta/2)\right)\right|^2 \\ & \times \exp\left(\frac{-4x^2\cos^2(\theta/2)}{w^2}\right) \end{aligned} \tag{3.52}$$

These integrals can be evaluated analytically when the pulse shape is also Gaussian [i.e., $a(t) \sim e^{-t^2/t_p^2}$]. After some algebra one obtains the result:

$$\langle P_{\mathrm{SHG}}\rangle \sim \exp\left(-\frac{\tau^2}{t_p^2 + w^2\tan^2(\theta/2)/c^2}\right) \tag{3.53}$$

Thus, the measurement yields a pulse width proportional to t'_p, where

$$\left(t'_p\right)^2 = t_p^2 + (\delta T)^2 \qquad \text{with} \qquad \delta T = \frac{w\tan(\theta/2)}{c} \tag{3.54}$$

To evaluate this expression, we consider Fig. 3.14b, which depicts the typical beam geometry used for focusing into the SHG crystal in noncollinear autocorrelations. The input pulses are assumed to have Gaussian beams with radii w_{in} and center-to-center separation δx_{in}. Assuming that the input beams are parallel prior to the lens (focal length f), the angle between the focused beams is given by $\tan(\theta/2) = \delta x_{\text{in}}/2f$. The beam radius at the focus of the lens is given by $w = f\lambda/\pi w_{\text{in}}$, which can be verified[4] by using the Gaussian beam formulas and *ABCD* matrices summarized in Section 1.3.3. Inserting into eq. (3.54) gives

$$\delta T = \frac{\lambda}{2\pi c}\frac{\delta x_{\text{in}}}{w_{\text{in}}} \tag{3.55}$$

Since $\delta x_{\text{in}} \geq 2w_{\text{in}}$ is required in order that the input beams are actually separated, $\delta T \geq \lambda/\pi c$. The same result is obtained if one requires instead that the beam-crossing angle at the crystal exceeds the far-field beam divergence angle (which can also be verified using the Gaussian beam formulas). The minimum value for the time-broadening parameter is the optical period divided by π, which gives about 0.9 fs for $\lambda = 0.8$ μm. Thus the time broadening inherent in noncollinear SHG is quite small, even for pulses in the 10-fs regime, provided that the crossing angle is kept near the minimum.

3.4.3 Correlation Measurements Using Two-Photon absorption

In addition to SHG, other nonlinear effects, such as two-photon fluorescence (TPF) and two-photon absorption (TPA), can also be used to measure the intensity autocorrelation. In TPA the measurement depends on an intensity-dependent absorption, so that there is increased absorption when the pulses are overlapped in time. The absorption coefficient α for a TPA process is given by

$$\alpha = \beta I(t) \tag{3.56}$$

where β is the TPA coefficient in cm/W. It is assumed here that the TPA process occurs essentially instantaneously (or at least very fast compared to the pulse being measured), which is valid for electronic two-photon transitions with no intermediate single-photon resonance. The linear absorption is also assumed to be negligible. The absorbed energy ΔU can then be written

$$\Delta U \sim \int dt\, \alpha I(t) = \int dt\, \beta I^2(t) \tag{3.57}$$

If one uses the field of a delayed pulse pair from eq. (3.34) to compute the intensity in this equation, one finds that the absorbed energy as a function of delay, $\Delta U(\tau)$, has the same

[4] This formula applies exactly when the input beams have their waists a distance f in front of the lens.

form as eq. (3.36). Therefore, measurement of $\Delta U(\tau)$ yields the same results as the SHG measurement.

In the TPF technique [120,121], which was introduced in the early days of ultrafast optics, the pulses are passed through a medium with fluorescence proportional to the square of the intensity (but ideally, with no fluorescence possible via single-photon excitation). The two pulses are usually arranged in a counterpropagating geometry, so that the relative delay varies along the sample thickness. Recording the fluorescence signal as a function of position yields an intensity autocorrelation measurement (with background). In this technique the time-integrated fluorescence provides a measure of the energy absorbed. TPF is no longer widely used in current ultrafast optics practice, although it can be useful for autocorrelation measurements at very short wavelengths where efficient second-harmonic crystals are not available.

TPA correlation measurements are usually performed in semiconductor materials [122–127], typically using photodetector structures. The material composition and optical wavelength are chosen such that the light is below bandgap, so that linear absorption is not allowed, but above half-bandgap, so that TPA is allowed. TPA leads to the generation of charge carriers which are collected with the aid of the electric field present in the photodetector structure. The resulting time-average photocurrent is a direct measure of the energy absorbed. TPA correlators have two attractive features compared to SHG: (1) for some wavelengths it is possible to use commercially available photodetectors, which can be very inexpensive compared to an SHG crystal and the associated photomultiplier; and (2) the TPA process alleviates some of the difficulties associated with measurement of very short pulses. With respect to the latter point, two-photon detectors intrinsically have a quite broad optical bandwidth, which can only be obtained in SHG using very thin crystals; furthermore, the electrically active region of conventional surface-normal TPA photodetector structures usually has a depth on the order of micrometers, so that pulse broadening due to dispersion within the nonlinear material is also excluded. Another interesting experimental implementation of a TPA correlator uses a two-photon optical waveguide photodetector as the nonlinear medium [123,127]. The waveguide structure keeps the light confined to a diameter of a few micrometers over a device length on the order of 1 mm, which gives high sensitivity, enabling measurement of very low intensity pulses (e.g., from semiconductor diode lasers with pulse durations on a time scale of tens of picoseconds.

3.4.4 Higher-Order Correlation Techniques

Finally, we mention briefly higher-order correlation measurements which depend on nonlinearities stronger than the SHG or TPA used for second-order correlations. Because these methods depend on higher powers of the electric field, they offer greater pulse shape sensitivity but at the cost of higher peak power requirements. The possibility of higher-order correlation measurements has been known for some time [113,128] and has seen some use for the characterization of pulses from high-power amplifier systems. It has also been demonstrated that third-order correlations can be performed with high-repetition-rate femtosecond oscillator systems using either third-harmonic generation (THG) [129] or three-photon absorption (3PA) [130]. Both sets of experiments were performed with 130-fs, 1.5 to 1.6-μm pulses from an optical parametric oscillator system with average powers from a few tens to a few hundreds of milliwatts at an 80-MHz repetition rate.

In the THG experiments, the pulses were focused by a high-numerical-aperture microscope objective onto the surface of a glass plate. Theory shows that in the case of tight focusing, efficient THG occurs only when the input beam is focused at an interface [131]. This means that the THG interaction is confined to within a few micrometers of the surface, which eliminates the requirement for phase matching, resulting in a broad optical bandwidth for the harmonic generation process. The time-integrated third-harmonic power is given by

$$\langle P_{\text{THG}}(t)\rangle \sim \int dt \left| \left[a(t) + \eta a(t-\tau)e^{-j\omega_0\tau}\right]^3 \right|^2 \tag{3.58}$$

where we have introduced η to allow for the possibility that the time-delayed pulse pair have different amplitudes. Meshulach et al. [129] give the full functional form of the THG correlation signal and demonstrate that when $\eta = 1$, the fringe-averaged third-order autocorrelation gives a contrast ratio of 10 : 1, while the interferometric (fringe-resolved) autocorrelation gives a 32 : 1 contrast ratio.

The 3PA experiments used a photodiode with a bandgap near 680 nm, so that both single- and two-photon absorption were disallowed [130]. The expression for the three-photon absorption photocurrent is given by

$$\langle P_{\text{3PA}}(t)\rangle \sim \int dt \left\{ \left|a(t) + \eta a(t-\tau)e^{-j\omega_0\tau}\right|^2 \right\}^3 \tag{3.59}$$

Since eqs. (3.58) and (3.59) are easily shown to be equal, we conclude that THG and 3PA measure the same third-order correlation function, just as SHG and TPA yield the same second-order correlation function. Experimentally, the 3PA measurements yielded the same fringe-averaged and fringe-resolved contrast ratios as for THG. Furthermore, it was also demonstrated in the 3PA experiments that by setting $\eta \ll 1$, the correlation measurements can reveal asymmetries in the intensity profile (when $\eta = 1$, the correlation trace remains symmetric).

3.5 CHIRPED PULSES AND MEASUREMENTS IN THE TIME–FREQUENCY DOMAIN

We now begin to discuss methods for determining completely the phase and intensity profiles of ultrashort pulses. These methods are particularly important in the case of chirped or phase-modulated pulses, since intensity autocorrelation methods do not provide much chirp information. Before covering these pulse measurement methods, let us first review the character of chirped pulses and introduce powerful time–frequency techniques for visualizing such pulses.

Consider a Gaussian pulse written in the time and frequency domains as follows:

$$e(t) = \text{Re}\left\{a(t)e^{j\omega_0 t}\right\} \tag{3.60a}$$

with

$$a(t) = e^{-\Gamma t^2} \tag{3.60b}$$

and

$$A(\omega) = \left(\frac{\pi}{\Gamma}\right)^{1/2} e^{-\omega^2/4\Gamma} \tag{3.60c}$$

This represents a linearly chirped pulse if we choose a complex pulse-width parameter Γ. Let us write

$$\frac{1}{\Gamma} = \alpha + j\beta \tag{3.61a}$$

with

$$\Gamma = \frac{\alpha - j\beta}{\alpha^2 + \beta^2} = \Gamma_r + j\Gamma_i \tag{3.61b}$$

Here α and β are both real numbers; α affects the spectral amplitude, and β is a chirp parameter. The time-domain pulse envelope can now be rewritten

$$a(t) = e^{-\Gamma_r t^2} e^{-j\Gamma_i t^2} \tag{3.62}$$

The pulse duration is determined by Γ_r, while Γ_i is related to the chirp. For a fixed spectral width (fixed α), the minimum pulse width occurs for $\beta = 0$. The pulse width for arbitrary β is given by

$$\frac{\Delta t(\alpha, \beta)}{\Delta t(\alpha, \beta = 0)} = \sqrt{1 + \frac{\beta^2}{\alpha^2}} \tag{3.63}$$

Let us examine this chirped pulse in more detail. We can first calculate the frequency-dependent delay $\tau(\omega)$, given by

$$\tau(\omega) = \frac{-\partial\psi(\omega)}{\partial\omega} = \frac{\beta(\omega - \omega_0)}{2} \tag{3.64}$$

The spectral phase $\psi(\omega)$ and the frequency-dependent delay $\tau(\omega)$ are sketched in Fig. 3.15a for $\beta > 0$. $\psi(\omega)$ is a parabola and $\tau(\omega)$ is a straight line with positive slope. Since positive frequencies experience greater delays, this represents a positive linear chirp (or an up-chirp).

Similarly, we can calculate the instantaneous frequency $\omega_{\text{inst}}(t)$, given by

$$\begin{aligned}\omega_{\text{inst}}(t) = \omega_0 + \frac{\partial\phi(t)}{\partial t} &= \omega_0 - 2\Gamma_i t \\ &= \omega_0 + \frac{2\beta t}{\alpha^2 + \beta^2}\end{aligned} \tag{3.65}$$

The temporal phase $\phi(t)$ and the instantaneous frequency are sketched in Fig. 3.15b, again assuming that $\beta > 0$. This plot is similar to Fig. 3.15a: $\phi(t)$ is a parabola and $\omega_{\text{inst}}(t)$ is again a straight line with positive slope. Larger times correspond to higher frequencies, so this still represents a positive chirp.

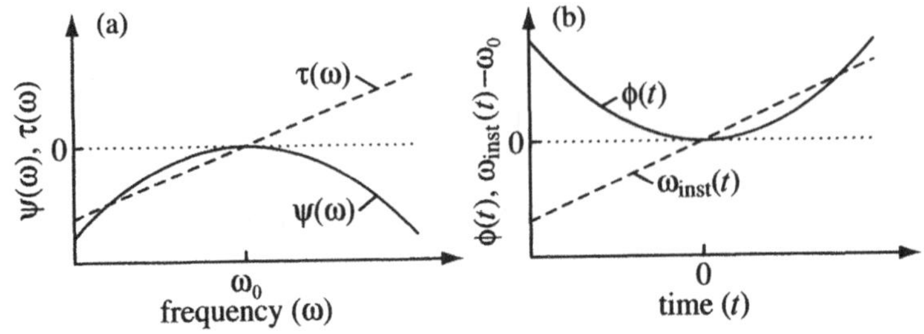

Figure 3.15 (a) Phase $\psi(\omega)$ (solid line) and frequency-dependent delay $\tau(\omega)$ (dashed line) vs. frequency for $\beta > 0$; (b) Phase $\phi(t)$ (solid line) and instantaneous frequency $\omega_{\text{inst}}(t)$ (dashed line) vs. time for $\beta > 0$. All units are arbitrary.

Such plots of frequency-dependent delay and instantaneous frequency give a good visual indication of the chirp, and they are useful not only for linear chirps as in this example but for nonlinear chirps as well. Note, however, that these plots give no direct indication of how close to or far from bandwidth limited the pulse is. It turns out that when $\beta \gg \alpha$, the pulses are highly chirped and far from the bandwidth limit. In this limit the pulse duration is proportional to the bandwidth:

$$\Delta t \approx \frac{1}{2}\beta\,\Delta\omega \tag{3.66}$$

Equation (3.66) clearly breaks down when the chirp is small ($\beta \leq \alpha$). The pulse width does not approach zero as $\beta \to 0$; rather, it approaches a limiting value proportional to $\alpha^{1/2}$. In the case of small chirp, the $\Delta\nu\,\Delta t$ product is close to the bandwidth limit (0.44 for a Gaussian). Plots of $\tau(\omega)$ and $\omega_{\text{inst}}(t)$ as in Fig. 3.15 are most useful for the case of relatively large chirp.

Let us consider one further example: a pulse with a cubic spectral phase corresponding to a nonlinear chirp. Here we assume that the spectrum is given by

$$A(\omega) = e^{-\omega^2/4\Gamma}e^{j\gamma\omega^3} \tag{3.67}$$

where γ represents the size of the cubic phase variation and Γ is now real. The frequency-dependent delay is

$$\tau(\omega) = -3\gamma(\omega - \omega_0)^2 \tag{3.68}$$

The spectral phase and the frequency-dependent delay are sketched in Fig. 3.16 for $\gamma < 0$ and Γ real. $\tau(\omega)$ is a parabola, and all the frequencies have a delay greater than or equal to zero. The highest frequencies have the highest delay and lead to an asymmetric tail (for $t > 0$) in the time-domain intensity profile. The central frequencies contribute to the main peak, which is somewhat broadened and slightly delayed. Figure 3.16 also shows the calculated intensity profile of a pulse with a significant amount of cubic spectral phase. The shift of the central peak and the long asymmetric tail are clearly evident, as predicted. In addition, rapid oscillation is observed. The reason for this oscillation can also be identified with the help of Fig. 3.16a. We see from the figure that for every time $t > 0$,

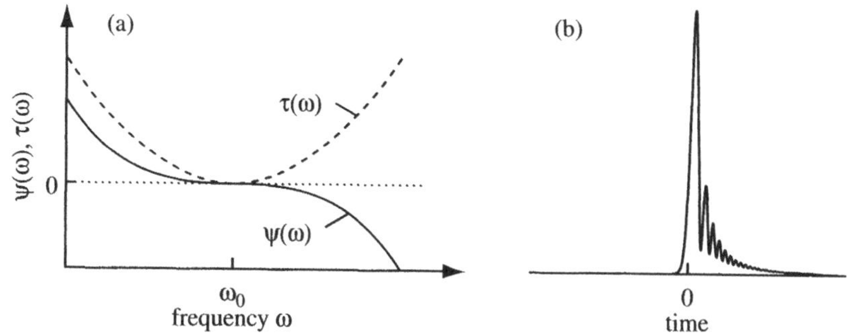

Figure 3.16 (a) Phase $\psi(\omega)$ (solid line) and frequency-dependent delay $\tau(\omega)$ (dashed line) vs. frequency for a cubic spectral phase with $\gamma < 0$; (b) calculated intensity profile. All units are arbitrary.

two different optical frequencies contribute to the signal. When these two frequencies are in phase, they add constructively to give a large field at the corresponding time instant, but when they are out of phase, they interfere destructively to give a weak field or zero. The phase difference is a cubic function of frequency, leading to the oscillations observed.

Let us now discuss how one might measure $\omega_{\text{inst}}(t)$ or $\tau(\omega)$. In the case of the instantaneous frequency, one could multiply the actual field by a suitable gating function $g(t-\tau)$ centered at $t=\tau$, resulting in a signal field given by

$$e_{\text{sig}}(t,\tau) = e(t)g(t-\tau) \tag{3.69}$$

One would then determine the instantaneous frequency by measuring the power spectrum of the time-gated pulse. This procedure, depicted in Fig. 3.17, yields a two-dimensional function $S_e(\omega,\tau)$, given by

$$S_e(\omega,\tau) = \left|\int dt\, e_{\text{sig}}(t,\tau)e^{-j\omega t}\right|^2 = \left|\int dt\, e(t)g(t-\tau)e^{-j\omega t}\right|^2 \tag{3.70}$$

$S_e(\omega,\tau)$ is known as the *spectrogram* of the electric field $e(t)$. For a constant τ, $S_e(\omega,\tau)$ yields a one-dimensional function of ω which provides information about the instantaneous frequency distribution at that τ. For a constant ω, $S_e(\omega,\tau)$ yields a one-dimensional function of τ which provides information about the times at which frequency ω appears.

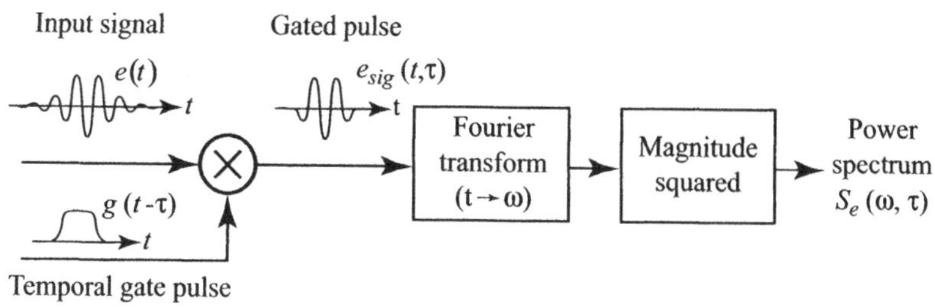

Figure 3.17 Spectrogram time–frequency distribution function.

Spectrograms are not unique to the field of ultrafast optics. They are used extensively to characterize acoustic signals (e.g., speech [132]). Musical scores, which specify a series of musical notes as a function of time, may also be regarded as a type of spectrogram. Spectrograms are an example of a joint time–frequency distribution, an analysis tool used to characterize signals whose spectral content is varying in time. Time–frequency distributions are themselves the subject of a rich and still evolving field, and many different types of time–frequency distributions have been investigated. For a review, see [133]. Here we discuss only the spectrogram and the closely related sonogram, which are the time–frequency distributions that have played the most important role in the measurement of ultrafast optical signals. We note, however, that the Wigner function has also seen some use for the description of ultrafast pulses in the time–frequency domain [134–136].

In Section 3.6 we discuss ultrafast optical methods for generating a gate function and measuring the spectrogram as well as methods for retrieving the actual pulses shape from the spectrogram data. For now we simply note the importance of choosing a proper gate function! If the $g(t)$ is too short, the gated function $e_{sig}(t, \tau)$ will have a spectrum much wider than the original spectrum. In this case any variations in the instantaneous frequency will be much smaller than the spectrum of $e_{sig}(t, \tau)$ and will therefore be difficult to detect. Similarly, if $g(t)$ is much longer than the input pulse, $e_{sig}(t, \tau)$ will be identical to $e(t)$, and the spectrogram will reveal only the original power spectrum. Therefore, one should generally select a gate function whose duration is on the order of the inverse bandwidth of the pulse to be characterized.

Instead of measuring $\omega_{inst}(t)$ as in eqs. (3.69) and (3.70), one can also imagine measuring $\tau(\omega)$. In this case one would first use a narrowband filter to select a specific frequency band. The frequency-dependent delay is determined by measuring the temporal intensity profile of the filtered waveform and then repeating as the passband of the narrowband filter is slid across the spectrum of the pulse. This procedure is sketched in Fig. 3.18. Mathematically, it turns out that such a measurement is equivalent to a spectrogram. This can be seen by substituting $e(t) = (1/2\pi) \int d\omega\, E(\omega)e^{j\omega t}$ and $g(t) = (1/2\pi) \int d\omega\, G(\omega)e^{j\omega t}$ into eq. (3.70), where $E(\omega)$ and $G(\omega)$ are the Fourier transforms of $e(t)$ and $g(t)$, respectively. The result is

$$\hat{S}_e(\omega, t) = \left| \frac{1}{2\pi} \int d\omega'\, E(\omega')H(\omega' - \omega)e^{j\omega' t} \right|^2 \tag{3.71}$$

which describes the temporal intensity profile of the original waveform after filtering by a sliding filter of the form $H(\omega) = G(-\omega)$. $\hat{S}_e(\omega, t)$ is sometimes called a *sonogram*.

The sonogram and the spectrogram of a given signal are identical provided that the sonogram filter function is equal to the frequency-reversed Fourier transform of the spectrogram

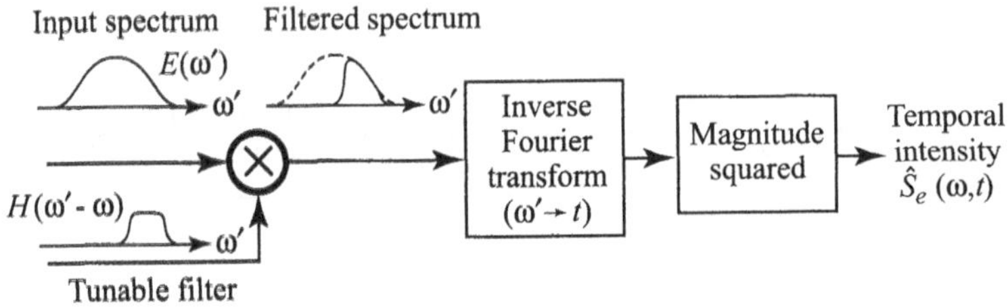

Figure 3.18 Sonogram time–frequency distribution function.

gate function. Furthermore, due to the Fourier transform relationship between the spectrogram gate function and the sonogram filter function, the rms widths of these functions in the time domain (Δt_{rms}) and in the frequency domain ($\Delta\omega_{\text{rms}}$) must satisfy the uncertainty relationship, eq. (3.11). This helps to quantify our discussion about selecting the duration of the gate function. High time resolution in the spectrogram can be obtained by using a narrow gate function, but unavoidably at the cost of spectral resolution.

We illustrate these concepts by discussing the form of the spectrogram for the case where the input pulse $s(t)$ and the gate pulse $g(t)$ are a chirped Gaussian centered at frequency ω_0 and a baseband chirped Gaussian, respectively, written as

$$s(t) = e^{-a_s t^2/2} e^{jb_s t^2/2} e^{j\omega_0 t} \tag{3.72}$$

$$g(t) = e^{-a_g t^2/2} e^{jb_g t^2/2} \tag{3.73}$$

$s(t)$ has the same form as the chirped Gaussian considered earlier in this section if we identify $\Gamma_r = a_s/2$ and $\Gamma_i = -b_s/2$ from eq. (3.61b). The resulting spectrogram of $s(t)$ can be written in two equivalent forms, as follows [133]:

$$S_s(\omega, \tau) \sim P_\tau(\tau) e^{-[\omega - \langle\omega\rangle_\tau]^2/2\sigma_\omega^2} \tag{3.74}$$

and

$$S_s(\omega, \tau) \sim P_\omega(\omega) e^{-[\tau - \langle\tau\rangle_\omega]^2/2\sigma_\tau^2} \tag{3.75}$$

Here $P_\tau(\tau)$ and $P_\omega(\omega)$ are known as the *marginal delay distribution* (delay marginal) and *marginal frequency distribution* (frequency marginal), obtained by integrating the spectrogram over either frequency or delay, respectively, and are given by

$$P_\tau(\tau) = \int d\omega\, S_s(\omega, \tau) \sim e^{-a_g a_s \tau^2 / (a_g + a_s)} \tag{3.76}$$

$$P_\omega(\omega) = \int d\tau\, S_s(\omega, \tau) \sim e^{-a_g a_s (\omega - \omega_0)^2 / \left[a_s\left(a_g^2 + b_g^2\right) + a_g\left(a_s^2 + b_s^2\right)\right]} \tag{3.77}$$

with

$$\langle\omega\rangle_\tau = \frac{1}{P_\tau(\tau)} \int d\omega\, \omega S_s(\omega, \tau) = \omega_0 + \frac{(a_g b_s - b_g a_s)\tau}{a_g + a_s} \tag{3.78}$$

$$\langle\tau\rangle_\omega = \frac{1}{P_\omega(\omega)} \int d\tau\, \tau S_s(\omega, \tau) = \frac{(a_g b_s - b_g a_s)(\omega - \omega_0)}{a_s\left(a_g^2 + b_g^2\right) + a_g\left(a_s^2 + b_s^2\right)} \tag{3.79}$$

$$\sigma_\omega^2 = \frac{(a_g + a_s)^2 + (b_g + b_s)^2}{2(a_g + a_s)} \tag{3.80}$$

$$\sigma_\tau^2 = \frac{(a_g + a_s)^2 + (b_g + b_s)^2}{2\left[a_s\left(a_g^2 + b_g^2\right) + a_g\left(a_s^2 + b_s^2\right)\right]} \tag{3.81}$$

$\langle\omega\rangle_\tau$ and $\langle\tau\rangle_\omega$ are known as the *first conditional moments*, which provide an *estimate* of the instantaneous frequency and frequency-dependent delay, respectively.

Several points are worth discussing:

- Writing the time–frequency distribution in two different forms emphasizes the equivalence of the spectrogram and the sonogram. Equation (3.74) may be considered the spectrogram form, while eq. (3.75) may be considered the sonogram form.
- The spectrogram takes the form of an ellipse, with an inclination that depends on the chirp of both the input pulse and the gate pulse. For a real gate pulse ($b_g = 0$), the inclination depends only on the input pulse chirp, and unchirped input pulses have major and minor axes aligned with the τ and ω axes. Thus, an intuitive interpretation of the spectrogram is more easily accomplished in the case of a real gate pulse, although both chirped or unchirped gate pulses are permissible.
- The area of the ellipse remains finite, even when the duration of the input pulse or the gate pulse becomes extremely short. This is a manifestation of the uncertainty relationship, eq. (3.11).
- From eq. (3.74), for a fixed delay the distribution is centered at $\langle\omega\rangle_\tau$. For an accurate estimation of the instantaneous frequency, one should use a very short gate pulse. For an unchirped gate pulse with $a_g \gg a_s$, one obtains $\langle\omega\rangle_\tau = \omega_0 + b_s\tau$. This is identical to eq. (3.65) for ω_{inst} after we substitute in $b_s = -2\Gamma_i$. Furthermore, $P_\tau(\tau)$ reduces exactly to the intensity profile $|s(t)|^2$ in this case. However, at the same time the frequency spread parameter σ_ω^2 becomes large.
- Conversely, from eq. (3.75), for a fixed frequency the distribution is centered at $\langle\tau\rangle_\omega$. In order that $\langle\tau\rangle_\omega$ give an accurate estimate of the frequency-dependent delay, one should use a very long gate pulse with a narrow spectrum. For an unchirped gate pulse with $a_g \ll a_s$, one obtains $\langle\tau\rangle_\omega = b_s(\omega - \omega_0)/(a_s^2 + b_s^2)$, which reduces to eq. (3.64) for $\tau(\omega)$ after we substitute in for a_s and b_s in terms of Γ_r and Γ_i. At the same time the temporal spread parameter σ_τ^2 becomes large, while $P_\omega(\omega)$ reduces exactly to the power spectrum.
- The need for a short gate pulse to portray temporal information (e.g., ω_{inst}) accurately and a long gate pulse to portray spectral information [e.g., $\tau(\omega)$] accurately is another manifestation of the uncertainty relation, which dictates an inverse relation between the temporal resolution and the spectral resolution of the information displayed by a spectrogram.

Pulse measurement techniques based on spectrograms and sonograms are discussed in the following two sections, respectively. For simplicity of understanding, in ultrafast optics we generally use the term *spectrogram* to refer to measurements involving multiplication by a relatively short time gate and reserve the term *sonogram* to refer to measurements involving the use of a series of relatively narrow filters. However, from a mathematical point of view, this distinction is not necessary.

3.6 FREQUENCY-RESOLVED OPTICAL GATING

In this section we discuss methods for measuring spectrograms in ultrafast optics and for retrieving complete amplitude and phase information about the electric field from the spectrogram. The techniques we discuss are called *frequency-resolved optical gating*, also known by the acronym FROG [137–140]. The main experimental requirement in measuring ultrafast optical spectrograms is in implementing an ultrafast gate function. In FROG, as in other ultrashort-pulse measurement techniques, the pulse is used to gate itself via a nonlinear optical interaction. A number of nonlinear interactions, including polarization gating, self-diffraction, transient gratings, and second- and third-harmonic generation, have been demonstrated for FROG measurements. In the following we first describe three of the more important FROG geometries and discuss general features of the FROG spectrograms. We then discuss the procedure for recovering a complete electric field profile from the FROG spectrograms.

3.6.1 Polarization-Gating FROG

One common FROG geometry, called *polarization gating FROG* (PG-FROG) [138,139], is shown in Fig. 3.19. Ultrafast gating is achieved through the optical Kerr effect, in which the phase modulation and nonlinear birefringence induced by the gating pulse are used to rotate the probe pulse polarization. The probe pulse is linearly polarized and is passed through a pair of crossed polarizers. In the absence of the gating pulse, the probe pulse is rejected by the second polarizer and does not pass through to the rest of the measurement apparatus. The gate pulse, which has its polarization oriented at an angle of 45° with respect to the probe, intersects the probe pulse in a thin nonlinear medium with an ultrafast third-order optical nonlinearity (the optical Kerr effect). We discussed previously nonlinear refractive index changes induced by an intense ultrashort pulse. Here we note that the optical Kerr effect also results in a nonlinear birefringence, since the nonlinear index change seen by a probe pulse with polarization parallel to that of a linearly polarized gate pulse is different than that seen by a probe pulse with polarization perpendicular to that of the gate pulse. For an electronic Kerr effect, the nonlinear birefringence is proportional to the instantaneous intensity of the gate pulse. This time-dependent birefringence leads to a time-dependent polarization rotation of the probe pulse, which is therefore transmitted in part through the second polarizer. The power spectrum of the probe pulse transmitted is measured using a spectrometer. By recording the spectrogram as a function of delay τ between gate and probe pulses, one obtains the spectrogram.

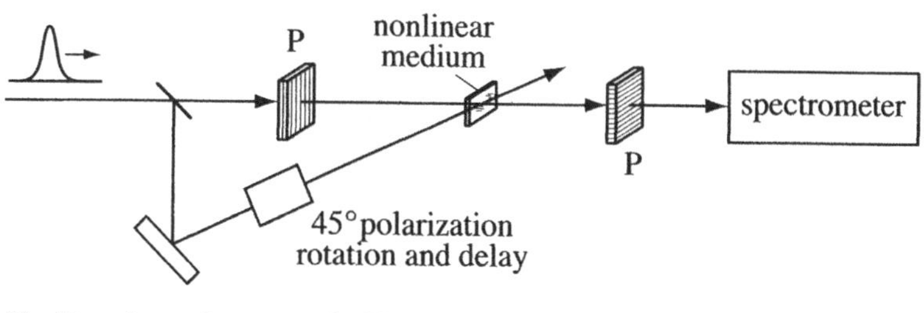

Figure 3.19 Experimental geometry for FROG using polarization gating. Usually, this measurement is performed in a single-shot geometry, where the delay varies across the nonlinear medium. The output of the spectrometer is captured using a CCD camera, which displays a two-dimensional delay–frequency plot. P, polarizer.

We can analyze the nonlinear polarization rotation by writing

$$n_{\|} = n_0 + \Delta n_{\|}(t) = n_0 + n_{2,\|} I_{\text{gate}}(t) \tag{3.82a}$$

and

$$n_{\perp} = n_0 + \Delta n_{\perp}(t) = n_0 + n_{2,\perp} I_{\text{gate}}(t) \tag{3.82b}$$

Here $n_{\|}$ and $n_{\perp}$ are the index of refraction seen by light polarized parallel or perpendicular to the linearly polarized gate pulse, respectively, and $I_{\text{gate}}(t)$ is the gate pulse intensity. For the electronic Kerr effect in an isotropic medium, $n_{2,\perp} = \frac{1}{3} n_{2,\|}$ [37]. We assume that the input probe and gate pulses can be written

$$\begin{aligned} \mathbf{e}_{\text{probe}}(t) &= E_{\text{pr}} e(t) \hat{\mathbf{x}} \\ \mathbf{e}_{\text{gate}}(t) &= \frac{1}{\sqrt{2}} E_g e(t-\tau)(\hat{\mathbf{x}} + \hat{\mathbf{y}}) \end{aligned} \tag{3.83}$$

E_{pr} and E_g are the amplitudes of the probe and gate pulses, respectively, and $e(t)$ is the electric field temporal profile (assumed the same for both pulses). The probe pulse after the nonlinear medium is written

$$\mathbf{e}_{\text{probe}}(t) = \frac{1}{\sqrt{2}} E_{\text{pr}} e(t) \left\{ \frac{\hat{\mathbf{x}} + \hat{\mathbf{y}}}{\sqrt{2}} e^{-j(\omega/c)\Delta n_{\|}(t) L} + \frac{\hat{\mathbf{x}} - \hat{\mathbf{y}}}{\sqrt{2}} e^{-j(\omega/c)\Delta n_{\perp}(t) L} \right\} \tag{3.84}$$

where the linear phase shift corresponding to passage through the nonlinear medium of length L has been neglected. After some algebra, eq. (3.84) can be rewritten

$$\begin{aligned} \mathbf{e}_{\text{probe}}(t) &= E_{\text{pr}} e(t) e^{-j(\omega/2c)(\Delta n_{\|} + \Delta n_{\perp})L} \\ &\quad \times \left\{ \hat{\mathbf{x}} \cos\left[\frac{\omega}{2c}\left(\Delta n_{\|} - \Delta n_{\perp}\right) L\right] - j\hat{\mathbf{y}} \sin\left[\frac{\omega}{2c}(\Delta n_{\|} - \Delta n_{\perp})L\right]\right\} \end{aligned} \tag{3.85}$$

Finally, after passage through a polarizer passing $\hat{\mathbf{y}}$-polarized light, the gated probe field is

$$\mathbf{e}_{\text{probe}}(t) = -jE_{\text{pr}} e(t) \hat{\mathbf{y}} e^{-j(\omega/2c)(\Delta n_{\|} + \Delta n_{\perp})L} \sin\left[\frac{\omega}{2c}(\Delta n_{\|} - \Delta n_{\perp})L\right] \tag{3.86}$$

The probe transmitted is gated according to the time-dependent birefringence $\Delta n_{\|}(t) - \Delta n_{\perp}(t)$ and also experiences a phase modulation induced by the gate pulse. The appearance of the phase modulation together with the amplitude modulation desired is related to the appearance of both self-amplitude-modulation and self-phase-modulation terms in our treatment of solid-state laser mode-locking in Section 2.4. If we assume that the induced birefringence in eq. (3.86) remains small, we can expand the sine to first order and neglect the SPM term, which will also be small. Substituting in for the nonlinear birefringence, we obtain

$$\mathbf{e}_{\text{probe}}(t) \sim \hat{\mathbf{y}} E_{\text{pr}} e(t) \left| e(t-\tau) \right|^2 \tag{3.87}$$

This gives us the gated field $e_{sig}(t, \tau)$ used in calculating the spectrogram. Thus, we make the identification

$$e_{sig}(t, \tau) \sim e(t)\,|e(t-\tau)|^2 \tag{3.88}$$

The apparatus measures the power spectrum, resulting in a spectrogram that we call $I_{FROG}(\omega, \tau)$, written as

$$I_{FROG}(\omega, \tau) = \left| \int dt\, e_{sig}(t, \tau) e^{-j\omega t} \right|^2 = \left| \int dt\, e(t)\,|e(t-\tau)|^2\, e^{-j\omega t} \right|^2 \tag{3.89}$$

It is important to remember that we assumed small birefringence in our derivation. This results in the specific functional form of eq. (3.89) and also justifies neglect of the gate pulse–induced phase modulation term, which would otherwise modify the probe pulse phase profile. Of course, this means that the efficiency of the Kerr gating process must be low. We have also assumed that the probe pulse intensity is low enough that there is negligible self-phase modulation and that the nonlinear medium is thin enough that we can ignore dispersion.

From the experimental point of view, we note that PG-FROG is used mainly with amplified femtosecond laser systems with sufficient intensity to perform single-shot measurements. For single-shot measurements the probe and gate pulses, which are at most weakly focused, cross at a relatively large angle in the nonlinear material. This leads to a variation in the relative delay across the beam cross section, as given by eq. (3.51). The spectrometer disperses the frequency spectrum in a direction perpendicular to the beam-crossing direction, resulting in a two-dimensional image (one axis for frequency, one axis for delay) which is recorded using a CCD camera. In principle, PG-FROG can also be used with CW mode-locked lasers, but due to the lower peak powers provided by such sources, PG-FROG signal levels are usually too low for practical measurements.

PG-FROG traces calculated for unchirped and linearly chirped Gaussian pulses are shown in Fig. 3.20. Following our earlier discussion of conventional spectrograms with independent gate functions, the FROG traces are ellipses rather than lines, reflecting the minimum time–bandwidth product property of the Fourier transform. The tilt (or lack of tilt) of the plots is an obvious visual indication of the magnitude and sign of the chirp. Another example of a FROG trace, this time calculated for a pulse with cubic spectral phase, is also shown in Fig. 3.20. The shape of the trace reveals a roughly parabolic delay vs. frequency dependence, as expected. Thus, even though FROG is a time-domain measurement technique, it is able to provide information such as $\tau(\omega)$ which is more intuitively connected with sonogram measurements. This illustrates once again the duality between spectrograms and sonograms.

In each of these examples, the FROG traces give a qualitative visual indication of the type of chirp present. Note, however, that the instantaneous frequency or frequency-dependent delay cannot be taken directly from the FROG trace (without first retrieving the electric field profile itself). As a simple illustration, consider the PG-FROG distribution for the case of a linearly chirped Gaussian, according to eq. (3.72). Since the gate pulse is $|e(t)|^2$, the form of the FROG trace is then given by eqs. (3.74) to (3.81) with $a_g = 2a_s$ and $b_g = 0$. From eq. (3.78) we obtain $\langle\omega\rangle_\tau = \omega_0 + \frac{2}{3}b_s\tau$. Comparing our earlier discussion on spectrograms, we find that $\langle\omega\rangle_\tau - \omega_0 = \frac{2}{3}(\omega_{inst} - \omega_0)$. The slope of the FROG trace is only two-thirds of the

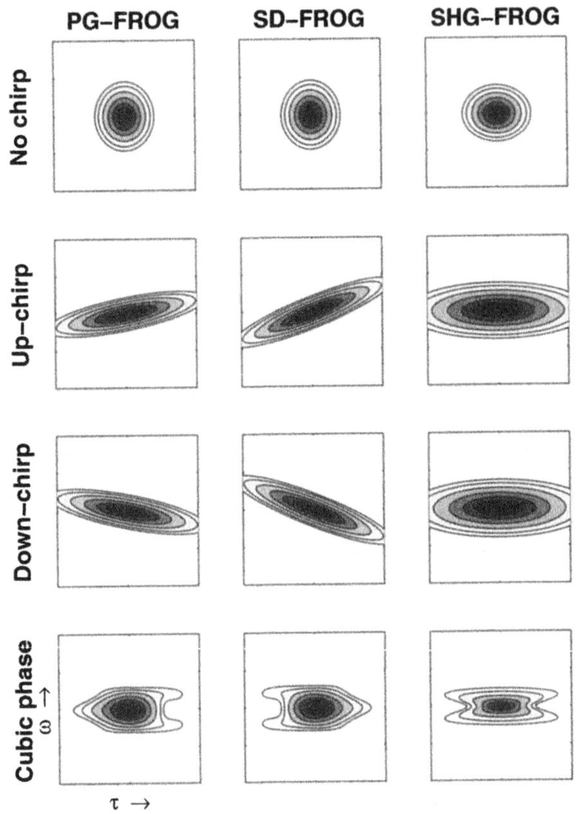

Figure 3.20 Calculated spectrogram traces for PG-FROG (left column), SD-FROG (middle), and SHG-FROG (right). The FROG traces are shown as shaded contour plots, with darker regions indicating higher intensity. All traces correspond to identical Gaussian power spectra, but with different spectral phase profiles. Row one, bandwidth-limited; row two, linear up-chirp; row three, linear down-chirp; row four, cubic spectral phase ($\gamma < 0$ case).

slope in a corresponding $\omega_{\text{inst}}(t)$ plot. The reason for this can readily be understood: The quantity $e(t)\,|e(t-\tau)|^2$ is centered at $t = 2\tau/3$, and therefore the PG-FROG spectrogram with delay τ samples the frequency content around $2\tau/3$, which accounts for the difference between $\langle\omega\rangle_\tau$ and ω_{inst} [139]. Nevertheless, since the gate function is strictly real, PG-FROG still provides relatively intuitive traces.

3.6.2 Self-Diffraction FROG

Another common FROG geometry for amplified pulse measurement is based on self-diffraction in a thin nonlinear refractive index medium [139,141]. The geometry is sketched in Fig. 3.21. Two identical and usually co-polarized pulses with a relative time delay τ intersect at an angle in the nonlinear material. The interference between the two pulses induces a transient refractive index grating varying in the transverse direction of the form

$$\Delta n_{\text{grating}} = n_{2,\|}\left[a(t)a^*(t-\tau)e^{-j(\mathbf{k}_1-\mathbf{k}_2)\cdot\mathbf{r}} + \text{c.c.}\right] \tag{3.90}$$

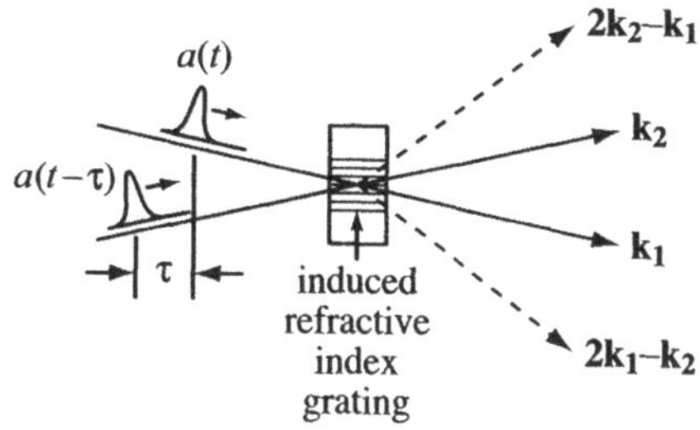

Figure 3.21 Geometry used in self-diffraction experiments.

where $\mathbf{k_1}$ and $\mathbf{k_2}$ are as shown in the figure and $a(t)$ is the complex amplitude function. Each of the pulses can diffract off this grating, giving light in the new directions $2\mathbf{k_1} - \mathbf{k_2}$ and $2\mathbf{k_2} - \mathbf{k_1}$. This provides the self-diffraction FROG (SD-FROG) signal fields

$$e_{\text{sig}(1)}(t, \tau) \sim a^2(t)a^*(t - \tau)e^{j[\omega_0 t-(2\mathbf{k_1}-\mathbf{k_2})\cdot\mathbf{r}]} \tag{3.91a}$$

and

$$e_{\text{sig}(2)}(t, \tau) \sim a^2(t - \tau)a^*(t)e^{j[\omega_0 t-(2\mathbf{k_2}-\mathbf{k_1})\cdot\mathbf{r}]} \tag{3.91b}$$

where low diffraction efficiency is assumed. The FROG trace is obtained by measuring the power spectrum of the signal field as a function of τ. The FROG trace for the $2\mathbf{k_1} - \mathbf{k_2}$ direction is written

$$I_{\text{FROG}}(\omega, \tau) = \left|\int dt\, e_{\text{sig}(1)}(t, \tau)e^{-j\omega t}\right|^2 = \left|\int dt\, a^2(t)a^*(t - \tau)e^{-j\omega t}\right|^2 \tag{3.92}$$

The FROG trace using direction $2\mathbf{k_2} - \mathbf{k_1}$ is easily shown to have the same form with the substitution $\tau \to -\tau$ and therefore carries no additional information.

From a practical point of view, one advantage of SD-FROG is that no polarizer is needed, which may be helpful for measurements at very short wavelengths where high-quality polarizers are difficult to realize. On the other hand, self-diffraction is not phase matched, which means that the angle between the two beams and the thickness of the nonlinear medium must be kept small. This limits the sensitivity of SD-FROG and in principle can introduce an undesirable wavelength dependence to the generation process. However, these difficulties can be avoided with appropriate experimental care.

Calculated SD-FROG traces for unchirped and linearly chirped pulses (Fig. 3.20) show behavior similar to those for PG-FROG. The trace for a bandwidth-limited pulse is an untilted ellipse. The chirped pulses are tilted ellipses, where the sign of the tilt indicates the sign of the chirp. The magnitude of the tilt is different than for PG-FROG. The form of the SD-FROG trace for a complex Gaussian pulse, eq. (3.72), is given by eqs. (3.74) to (3.81), provided that we make the following substitutions: $a_s \to 2a_s$; $b_s \to 2b_s$; $a_g \to a_s$; $b_g \to -b_s$. As a result, we find that $\langle\omega\rangle_\tau - \omega_0 = \frac{4}{3}b_s\tau = \frac{4}{3}(\omega_{\text{inst}} - \omega_0)$. This can be understood by noting that the FROG signal field peaks at $\tau/3$, and therefore the contribution to $\langle\omega\rangle_\tau - \omega_0$ from the three factors involved in $e_{\text{sig}(1)}(t, \tau)$ is $2\omega_{\text{inst}}(\tau/3) - \omega_{\text{inst}}(-2\tau/3) = 4b_s\tau/3$. Thus, the slope of SD-FROG traces for linearly chirped pulses is twice that of PG-FROG.

Note that in SD-FROG the gate function is not purely real; it also carries phase information about the pulse. As a result, the phase information in the signal and gate functions can interact, leading to spectrogram traces that are in many cases more difficult to interpret than for PG-FROG. One example is the case of a spectral cubic phase. As shown in Fig 3.20, the trace is still qualitatively consistent with a parabolic $\tau(\omega)$; however, the direction of the trace with respect to positive τ is reversed compared to the PG-FROG case and compared to the actual frequency-dependent delay.

3.6.3 Second-Harmonic-Generation FROG

PG-FROG and SD-FROG are difficult to apply with CW mode-locked sources, since the combination of low peak powers together with a third-order optical nonlinearity leads to polarization gating or self-diffraction efficiencies which are too low for convenient operation. By using second-harmonic generation (SHG), which is a second-order nonlinearity, as the gating mechanism, one can obtain sufficient FROG signal levels with CW mode-locked systems. The geometry for SHG-FROG is similar to the noncollinear SHG intensity autocorrelation setup shown in Fig. 3.9, except that a spectrometer is placed in front of the detector. The cost of this approach is that SHG-FROG traces are less amenable than PG-FROG traces to simple qualitative interpretation. The SHG-FROG approach has been described, for example, in [138,142,143].

In SHG-FROG the gated field and the FROG trace are given, respectively, by

$$e_{\text{sig}}(t,\tau) \sim e(t)\, e(t-\tau) \tag{3.93}$$

and

$$I_{\text{FROG}}(\omega,\tau) = \left| \int dt\, e(t) e(t-\tau) e^{-j\omega t} \right|^2 \tag{3.94}$$

By making the substitution $t' = t - \tau$, one can easily show that

$$I_{\text{FROG}}(\omega,\tau) = I_{\text{FROG}}(\omega,-\tau) \tag{3.95}$$

That is, the SHG-FROG trace is a symmetric function of τ. This is not surprising, since the pulse and the gating function are identical, and therefore one cannot tell whether the pulse is ahead of the gating function or the gating function is ahead of the pulse. In contrast, the gating function in PG-FROG, $|e(t-\tau)|^2$, is distinguishable from the pulse itself, and therefore FROG traces that are not symmetric in delay are possible. A further consequence of eqs. (3.93) to (3.95) is that SHG-FROG has a time-reversal ambiguity not present in PG-FROG. In particular, the fields $e(t)$ and $e(-t)$, corresponding to complex amplitude functions $a(t)$ and $a^*(-t)$, respectively, have identical SHG-FROG traces. To determine the pulse shape uniquely, one must perform additional measurements. One possibility that can be used to resolve the ambiguity is to first measure the FROG trace of the initial pulse and then measure the FROG trace of the pulse after transmission through a known dispersive medium.

One other nontrivial ambiguity has been identified for SHG-FROG. In the case of two isolated pulses, relative phases between the pulses of ϕ and $\phi + \pi$ lead to identical $I_{\text{FROG}}(\omega, \tau)$. Therefore, these two cases cannot be distinguished on the basis of the FROG trace.

The third column of Fig. 3.20 shows calculated SHG-FROG traces for transform-limited and linearly chirped Gaussian pulses. The trace for the transform-limited pulse is a horizontal ellipse and is qualitatively similar to FROG traces shown earlier. The SHG-FROG traces for the linearly chirped pulses, however, show qualitative differences compared to the earlier results. As the chirp is increased, the trace grows wider in the horizontal (delay) direction, but the major and minor axes of the ellipse remain in the horizontal and vertical directions. This is a consequence of the symmetry of the FROG trace with respect to τ. Identical traces are obtained for equal-magnitude positive and negative linear chirps; this further illustrates one of the disadvantages of SHG-FROG: the inability to determine the sign of a chirp. Additionally, one clearly sees that the area contained under the SHG-FROG trace increases substantially for the chirped pulse. This is a consequence of the increased time–bandwidth product of the chirped pulse.

The formula for the SHG-FROG trace for a linearly chirped Gaussian pulse is given by eqs. (3.74) to (3.81), with the assignments $a_g \rightarrow a_s$ and $b_g \rightarrow b_s$, resulting in the following particularly simple form:

$$I_{\text{FROG}}(\omega, \tau) \sim e^{-a_s\tau^2/2} e^{-\omega^2/2\sigma_\omega^2} \tag{3.96}$$

where $\sigma_\omega^2 = a_s + b_s^2/a_s$. These expressions conform to our discussion above: The trace retains its untilted orientation independent of the chirp parameter (b_s), while the width of the trace in the frequency direction increases with increasing frequency modulation.

An example of a more complicated SHG-FROG trace, corresponding to a pulse with cubic spectral phase, is also shown in the figure. Compared to corresponding PG- and SD-FROG plots, the symmetrization of the SHG-FROG trace as a function of τ is clearly evident.

It is interesting to consider the marginal frequency distribution $P_\omega(\omega)$ of the SHG-FROG trace. It turns out that $P_\omega(\omega)$ is given by

$$P_\omega(\omega) = \int d\tau\, I_{\text{FROG}}(\omega, \tau) = \frac{1}{2\pi} \int d\omega' \left|E(\omega')\right|^2 \left|E(\omega - \omega')\right|^2 \tag{3.97}$$

This can be derived [132] by replacing $I_{\text{FROG}}(\omega, \tau)$ in the above eq. (3.97) with the sonogram form of the SHG-FROG trace, eq. (3.71), where $g(t)$ is set equal to $e(t)$. $P_\omega(\omega)$ is given by the autoconvolution of the optical power spectrum at the fundamental wavelength. This offers a useful consistency check for measurement data: The SHG-FROG trace can be compared to the independently measured input power spectrum using eq. (3.97). Since the FROG data and input power spectrum are acquired in different wavelength ranges and often using different equipment, it is likely that any data sets corrupted through systematic measurement errors will fail to satisfy the frequency marginal. Therefore, this procedure can be used to test the reliability of the FROG data.

3.6.4 Frequency-Resolved Optical Gating Using Temporal Phase Modulation

In principle, FROG can also use a phase-only gate function. In one early approach the spectra of a pulse before and after *self-phase modulation* (SPM) in a nonlinear refractive index material were measured and analyzed. Reconstruction of electric field profiles was attempted using only a single spectrum after SPM, but with limited success due to the one-dimensional nature of this phase-retrieval problem [144]. Later, temporal modulation on the input signal was imposed using *cross-phase modulation* (XPM) from a pump pulse with a variable time delay. Analysis of the resulting two-dimensional data set, consisting of XPM spectra corresponding to different delays, yielded more consistent success in electric field reconstruction [145]. FROG has also been demonstrated using SPM and XPM of a delayed pulse pair in optical fibers; like SHG-FROG, this technique offers the possibility of using relatively low pulse energies [146]. At present, such pulse measurements based on temporal phase modulation are not used extensively in ultrafast optics; therefore, they are not considered further.

3.6.5 Signal Recovery from FROG Traces

Let us now discuss how to retrieve electric field pulse shapes from FROG traces [138,140,143]. Note that if $e_{\text{sig}}(t,\tau)$ is known, $e(t)$ can be obtained by direct integration:

$$\int d\tau\, e_{\text{sig}}(t,\tau) = \int d\tau\, e(t)g(t-\tau) \sim e(t) \tag{3.98}$$

where the proportionality constant is not important. We therefore ask the question: is it possible to determine $e_{\text{sig}}(t,\tau)$?

To answer this, we express the FROG trace in terms of $\tilde{E}_{\text{sig}}(t,\Omega)$, the Fourier transform of $e_{\text{sig}}(t,\tau)$ with respect to τ:

$$I_{\text{FROG}}(\omega,\tau) \sim \left| \iint dt\, d\Omega\, \tilde{E}_{\text{sig}}(t,\Omega) e^{-j\omega t} e^{j\Omega\tau} \right|^2 \tag{3.99}$$

where

$$\tilde{E}_{\text{sig}}(t,\Omega) = \int d\tau\, e_{\text{sig}}(t,\tau) e^{-j\Omega\tau} \tag{3.100}$$

Equation (3.99) represents a phase-retrieval problem: We wish to recover $\tilde{E}_{\text{sig}}(t,\Omega)$ [and hence $e_{\text{sig}}(t,\tau)$] from its two-dimensional power spectrum. We have already seen that in one dimension this problem does not have a unique solution; that is why we cannot determine $e(t)$ from its power spectrum $|E(\omega)|^2$. It turns out, however, that in the two-dimensional phase-retrieval problem [i.e., eq. (3.99)], one can almost always recover a unique solution.[5] This fact has been the focus of considerable activity in the image science community since the early 1980s [147–150] and is the basis for pulse retrieval in FROG. We can rationalize why unique phase retrieval is possible by noting that what we actually wish to recover is the one-dimensional complex function $e(t)$. If we represent $e(t)$ by a complex vector of length N, the FROG data can be represented by an $N \times N$ two-dimensional array with N^2

[5] Existing proofs of such uniqueness depend on the assumption that the signal to be retrieved, here e_{sig}, has compact support (i.e., it is nonzero only within a bounded region of space). Although in a strict sense this is generally not true in ultrafast optics, in a practical sense $e(t)$ and $e_{\text{sig}}(t,\tau)$ generally decay rapidly and disappear into the noise. Also, the mathematical form constraint provides extra information not available in the usual phase-retrieval problem, which helps enforce uniqueness. Experience strongly suggests that except for certain known ambiguities, FROG returns unique solutions.

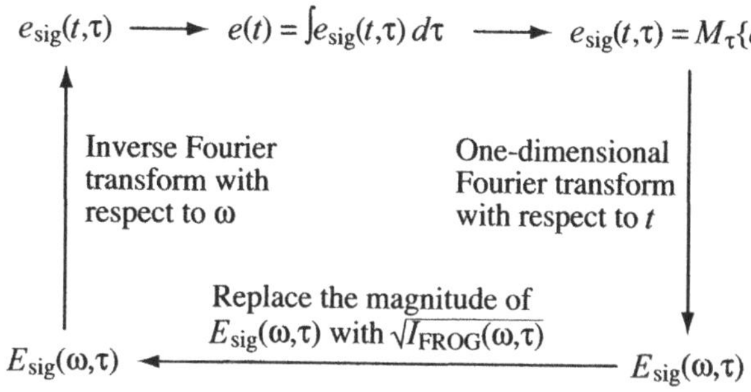

Figure 3.22 Iterative Fourier transform algorithm for retrieving the ultrashort-pulse amplitude and phase from FROG data. $M_\tau\{e(t)\}$ stands for the mathematical form constraint [e.g., for PG-FROG, $M_\tau\{e(t)\} = e(t)\,|e(t-\tau)|^2$]. Adapted from [138].

data values. Since $e(t)$ has only $2N$ degrees of freedom (real and imaginary parts for each element of the $e(t)$ vector), there is considerable redundancy in the FROG data, and this makes phase retrieval possible.

Trebino and co-workers have published several algorithms for recovering the electric field profile from FROG data [138]. Their original iterative Fourier transform algorithm is indicated in Fig. 3.22. One starts at the top center with a guess for $e(t)$. Moving in a clockwise direction, the next step is to use this guess for $e(t)$ to generate $e_{sig}(t, \tau)$ using

$$e_{sig}(t, \tau) = M_\tau\{e(t)\} \tag{3.101}$$

where $M_\tau\{\cdots\}$ represents the mathematical form of the nonlinear process being utilized [e.g., eq. (3.88), (3.91a), or (3.93) for PG-, SD-, or SHG-FROG, respectively] [139]. This is called a *mathematical form constraint*. By performing a series of one-dimensional Fourier transforms with respect to t, one obtains $E_{sig}(\omega, \tau)$. At this point one applies the experimental data (the intensity constraint) by replacing the magnitude of $E_{sig}(\omega, \tau)$ with the magnitude (square root) of the FROG data while leaving its phase unchanged; that is,

$$E_{sig}(\omega, \tau) \rightarrow \frac{E_{sig}(\omega, \tau)}{|E_{sig}(\omega, \tau)|}\sqrt{I_{FROG}(\omega, \tau)} \tag{3.102}$$

The algorithm continues by inverse Fourier transforming with respect to ω to get a new estimate of $e_{sig}(t, \tau)$ and then integrating with respect to τ to obtain an updated estimate of $e(t)$. This completes one iteration. The algorithm continues by using the updated version of $e(t)$ as the input for the mathematical form constraint on the second iteration; and so on. Additional iterations are performed until the solution converges. This algorithm is related to iterative Fourier transform algorithms used previously in image science [148,151], with the extra feature of the mathematical form constraint.

Let $e^{(k)}(t)$ and $E^{(k)}_{sig}(\omega, \tau)$ denote the computed arrays representing $e(t)$ and $E_{sig}(\omega, \tau)$ during iteration number k. One obvious way to check for convergence is to compare these results to the corresponding results from the previous iteration. Another way is to compute the error between the computed and experimental values. Since $e(t)$ is not known directly from the experiments, one cannot compute the error in $e^{(k)}(t)$. However, one can use $E^{(k)}_{sig}(\omega, \tau)$ for comparison with actual FROG data. In particular, it is useful to compute the rms FROG error, defined by

$$\varepsilon_{\text{FROG}}^{(k)} = \left\{ \frac{1}{N^2} \sum_{i=1}^{N} \sum_{j=1}^{N} \left[I_{\text{FROG}}^{(k)}(\omega_i, \tau_i) - I_{\text{FROG}}(\omega_i, \tau_i) \right]^2 \right\}^{1/2} \tag{3.103}$$

where

$$I_{\text{FROG}}^{(k)}(\omega, \tau) = \left| E_{\text{sig}}^{(k)}(\omega, \tau) \right|^2 \tag{3.104}$$

The algorithm should be run until the FROG error cannot be made any smaller. Note that $\varepsilon_{\text{FROG}}$ will not generally reach zero, due to noise on the experimental trace.[6] Note also that because of the redundancy built into the FROG trace, in general there is no $e(t)$ that corresponds precisely to the noisy experimental FROG data. As long as the noise in the input data is not excessive, the final computer value for $E_{\text{sig}}^{(k)}(\omega, \tau)$ may be a better estimate of the actual FROG trace than the experimental data [152]!

It is important to distinguish between convergence and uniqueness [148]. Convergence means that the algorithm reduces the FROG error to a very small value, ideally limited only by experimental or computer noise, whichever is larger. Upon convergence the electric field waveform retrieved is consistent with the FROG trace. Uniqueness means that the waveform retrieved is the only waveform consistent with the FROG trace. Due to the uniqueness of the two-dimensional phase-retrieval problem, convergence of the FROG algorithm implies that a unique electric field solution has been obtained (except for some known ambiguities, as noted below). However, it is important to remember that for other problems, such as one-dimensional phase-retrieval, convergence need not imply uniqueness. Therefore, as new pulse measurement techniques are developed, results from such techniques should be taken with caution until their uniqueness is established.

In addition to the iterative Fourier transform algorithm, other more powerful FROG algorithms which tend to give more robust convergence have also been developed. One of the most widely used of these is called *generalized projections* [140,143]. The idea is pictured in Fig.3.23, in which the upper and lower elliptical regions represent the sets of all possible $e_{\text{sig}}(t, \tau)$ functions that satisfy the mathematical form constraint and the intensity constraint, respectively. According to the uniqueness of the two-dimensional phase retrieval problem, these regions intersect at a single point, which is the solution for $e_{\text{sig}}(t, \tau)$. The algorithm starts with an initial guess and then projects onto (moves to the closest point on) one of the constraint sets. At this point the algorithm iteratively projects from one constraint set onto the other until the solution is reached. Operationally, the algorithm has a flowchart that looks very much like Fig. 3.22, with one exception: At the top of the flowchart, the new estimate for $e(t)$ in each iteration is obtained by projecting onto the mathematical form constraint. This is achieved at the $(k+1)$th iteration by finding the new field $e^{(k+1)}(t)$ which minimizes the difference (Z) between $M_\tau\{e^{(k+1)}(t)\}$ and the most recent (kth) estimate of

[6] The FROG error must obviously be computed prior to the magnitude update of eq. (3.102).

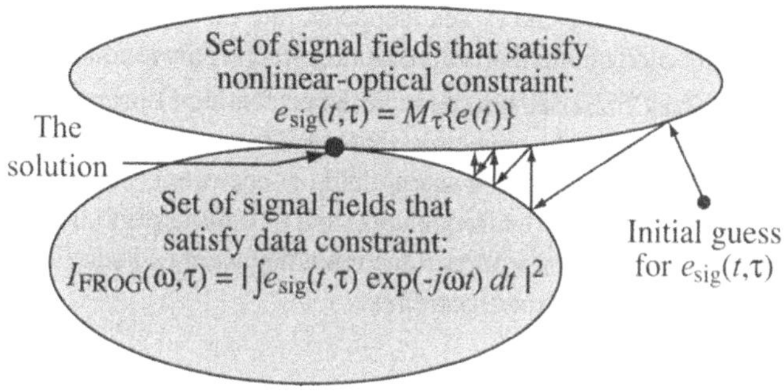

Figure 3.23 Generalized projections algorithm for retrieving the ultrashort-pulse amplitude and phase from FROG data. Adapted from [140], with permission. Copyright ©1997, American Institute of Physics.

$e_{sig}(t, \tau)$; that is,

$$Z = \frac{1}{N^2}\sum_{i=1}^{N}\sum_{j=1}^{N}\left|e_{sig}^{(k)}(t_i, \tau_j) - M_{\tau_j}\{e^{(k+1)}(t_i)\}\right|^2 \tag{3.105}$$

In practice, minimization of Z has been performed approximately, usually by using a simple gradient-based approach [140]. More recently, generalized projections has been implemented by using an outer product matrix method [153]. For more details the reader is directed to the references.

There are a number of practical issues that must keep track of in implementing FROG pulse recovery algorithms [152]. Some of these are listed below.

- The original guess for $e(t)$ is a matter of choice and can range from an intelligent guess based perhaps on the power spectrum and any additional information available to random noise. Using noise as the starting point is often regarded as a favorable choice.
- It is useful to normalize the experimental FROG data to a peak magnitude of 1, and to do the same for the computer value of $|E_{sig}^{(k)}(\omega, \tau)|^2$ during each iteration. This facilitates computation of the rms FROG error ε_{FROG}. As a slight improvement, which makes a difference in the case of real data with noise present, one could instead adjust the normalization of $|E_{sig}^{(k)}(\omega, \tau)|^2$ to minimize the FROG error.
- Although as stated, the shape of the electric field $e(t)$ is essentially uniquely determined from the FROG trace, it is important to note that there are some simple ambiguities: The measurement is not sensitive to the time delay of the pulse or to a constant multiplicative phase factor. In the computer algorithm this means that the time delay or the phase of the pulse may wander as subsequent iterations are performed. In addition, it has already been pointed out that SHG-FROG has some additional ambiguities: a time-reversal ambiguity and a twofold ambiguity in the relative phases of a pair of well-separated pulses.
- It is important to avoid truncation of the FROG trace in either the ω or τ dimensions. This means that the spectral range should be sufficiently broad and the delay scan

sufficiently long that the FROG trace disappears into the noise. The matrix containing the FROG data should contain the valid data completely surrounded by zeros (or noise).

- As noted in Section 1.5.2, one should recall in performing Fourier transforms that FFT algorithms typically have the convention that $t = 0$ ($\omega = 0$) corresponds to the leftmost (first) point in the data arrays. If the actual field or spectrum is centered in the array, the FFT interprets this as a time (or frequency) shift of half a period, resulting in a $(-1)^n$ phase factor in the Fourier transform. One must be aware of such FFT idiosyncracies in implementing pulse retrieval algorithms.

It is worth repeating that the redundancy inherent in the FROG trace substantially overdetermines the electric field $e(t)$. As a result, when the phase retrieval algorithm converges, it is virtually guaranteed that the correct $e(t)$ has been found (except for known ambiguities). Conversely, for the more robust FROG algorithms, lack of convergence can be taken as an indicator of measurement error (excessive noise, incorrect calibration, truncation, etc.) [152]. The marginal distributions can also serve as a check in the consistency of the FROG data. The utility of comparing the SHG-FROG frequency marginal with a curve independently derived from the input power spectrum has already been discussed. Similar marginals exist for other forms of FROG [139]. An important point is that many one-dimensional pulse measurement schemes, such as intensity autocorrelation, lack the redundancy or internal consistency checks of FROG. Therefore, the experimentalist must exercise greater care with such one-dimensional techniques to ensure that any important sources of error are eliminated.

Pulse-shape recovery using FROG has been explored most extensively with relatively simple pulse shapes, such as those emitted from femtosecond modelocked lasers. Although FROG has also demonstrated success for some more highly structured pulses, such as those generated using femtosecond pulse shapers or resulting from nonlinear propagation in fibers, there is also the following alternative [154]. FROG can be used, for example, to completely characterize the relatively simple pulses from the mode-locked laser which provides the input to the pulse shaping apparatus. These characterized pulses can then be used as the reference for measurement of the shaped pulses via spectral interferometry or electric field cross-correlation.

3.7 PULSE MEASUREMENTS BASED ON FREQUENCY FILTERING

We have already seen that the sonogram of a short pulse could be obtained by first filtering the pulse to select a certain frequency band and then measuring the temporal intensity profile of the filtered pulse as the filter is set to different center frequencies. We have also seen that sonograms and spectrograms are equivalent mathematically. Therefore, it should be possible to determine ultrashort-pulse electric field profiles from sonogram measurements, just as FROG retrieves complete pulse information from spectrogram measurements. One important difference is that tunable filters can be constructed independent of the pulse to be measured, while in FROG the gate function is always derived from the pulse. This leads to the new possibility of determining the spectral phase without the need for iterative calculations.

A block diagram depiction of the measurement approach is shown in Fig. 3.24. The input pulse first passes through a filter characterized by its frequency response $H(\omega - \Omega)$,

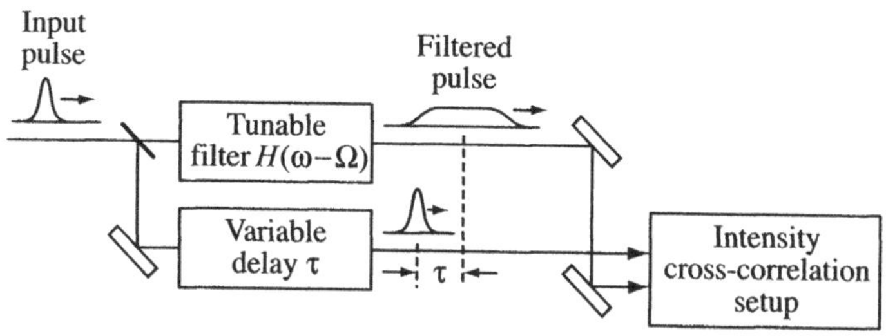

Figure 3.24 Sonogram approach to ultrashort-pulse measurement.

where the filter center frequency Ω is tunable. Often, the filter is implemented using a femtosecond pulse-shaping apparatus [155] constructed from a pair of diffraction gratings and lenses. The first grating and lens serve to separate different optical frequency components, as in a spectrometer, and the second grating and lens reassemble all the optical frequencies into a single collimated beam. This operation is performed in such a way that the output pulse shape is identical to that of the input, without distortion or dispersion. Let us use $\hat{\mathbf{x}}$ to denote the direction along which the optical frequencies are spread. A mask with a transparency function spatially patterned along $\hat{\mathbf{x}}$ is placed at the plane where the frequencies are separated. To a good approximation this yields an optical filter function which is simply a scaled version of the spatial mask pattern. The output pulse from the filter is broadened and reshaped in a manner related to the Fourier transform of the spatial masking pattern. Measurement of the output intensity profile, usually performed via intensity cross-correlation, comprises the final step in the measurement scheme. Either an unfiltered version of the input pulse, or any other sufficiently short pulse that may be available, may be used as the reference for the cross-correlation. Note that femtosecond pulse shaping is discussed in detail in Chapter 8. For the pulse measurement application it provides a means of conveniently tuning the shape and center frequency of the filter function.

3.7.1 Single-Slit Approaches

Single-Slit Sonogram Approach Let us first consider the case where the spatial mask is a simple slit, so that $H(\omega)$ is approximately a rectangle function with spectral width $\delta\omega$ and constant spectral phase. $\delta\omega$ is determined by the width of the slit, and the center frequency Ω can be adjusted by translating the slit along $\hat{\mathbf{x}}$. Let us also assume that the following three approximations are valid:

1. The power spectrum of the input pulse remains essentially constant over a frequency range of width $\delta\omega$.
2. The spectral phase $\psi(\omega)$ of the input pulse varies sufficiently slowly over a frequency $\delta\omega$ that it may be expanded to first order.
3. The duration of the filtered pulse, which is proportional to $\delta\omega^{-1}$, is sufficiently broad that the intensity cross-correlation yields a good estimate of the filtered intensity profile.

These conditions can all be satisfied by choosing a sufficiently narrow filter width. We find then that measurement of the output temporal profile provides direct information about $\psi(\omega)$ [156,157]. This is demonstrated by using these approximations to simplify the expression for the filtered intensity profile, eq. (3.71), which gives

$$\hat{S}_e(\Omega, t) \approx |E(\Omega)|^2 \left|h\left[t + \psi'(\Omega)\right]\right|^2 \tag{3.106}$$

where $h(t)$ is the inverse Fourier transform of $H(\omega)$ and $\psi'(\Omega) = \partial\psi(\omega)/\partial\omega$ evaluated at $\omega = \Omega$. The output intensity profile is shifted in time by $-\psi'(\Omega)$, which is exactly equal to the frequency-dependent delay $\tau(\omega)$! Thus, $\tau(\omega)$ is obtained directly by recording the shift as a function of Ω, which can either be read directly from the data, or determined by calculating the first temporal moment:

$$\tau(\omega = \Omega) = \frac{\int dt\, t\, \hat{S}_e(\Omega, t)}{\int dt\, \hat{S}_e(\Omega, t)} \tag{3.107}$$

where $\hat{S}_e(\Omega, t)$ refers to the actual data. Integrating $\tau(\omega)$ gives $\psi(\omega)$.

This result is independent of the shape of the filter function, provided that $\delta\omega$ is narrow enough to satisfy our assumptions. The case of a Gaussian filter function with a linearly chirped Gaussian input pulse was worked out exactly in Section 3.5, where it was shown that the first conditional moment of the delay, $\langle\tau\rangle_\omega$, approaches $\tau(\omega)$ in the limit of a narrow filter function. However, as was pointed out at the same time, it is undesirable to make the filter much narrower than necessary; the reduced filter throughput and increased output pulse duration both lower the signal-to-noise ratio of the measurement.

This method has the advantage that the spectral phase information desired is obtained directly without the need for iterative phase retrieval. A disadvantage is that the accuracy of the phase information obtained depends on the degree to which the assumptions enumerated above are satisfied and is difficult to assess quantitatively. On the other hand, it is also possible to use the full information inherent in the sonogram, eq. (3.106), together with phase-retrieval algorithms to recover $e(t)$ [158]. This is essentially a sonogram version of FROG. One new feature which this provides is the possibility of substituting a two-photon absorption photodetector in place of the second-harmonic crystal in the intensity correlator, which is not possible in SHG-FROG.

Spectrally and Temporally Resolved Upconversion Technique An alternative measurement approach is depicted in Fig. 3.25. Compared to the previous technique, this spectrally and temporally resolved upconversion technique (STRUT) [159,160] uses a fixed filter which we denote $F(\omega - \Omega)$, where Ω is fixed. Since tunability is not required, an interference filter can be utilized instead of a pulse shaper, if desired. The broadened filtered pulse is controlled with a variable delay and then interacts with a replica of the input pulse in a second harmonic crystal to produce an upconverted signal, which is spectrally resolved using a spectrometer. This procedure produces a two-dimensional data set as a function of delay τ and upconverted frequency ω_{up}. The measurement can be performed in either a repetitively scanned or a single-shot geometry. The arrangement is similar to SHG-FROG, except that the gate is now given by a filtered version of the input pulse. As we shall see, due to this difference, the measurement yields direct information on $\psi(\omega)$ without the need for iterative phase retrieval.

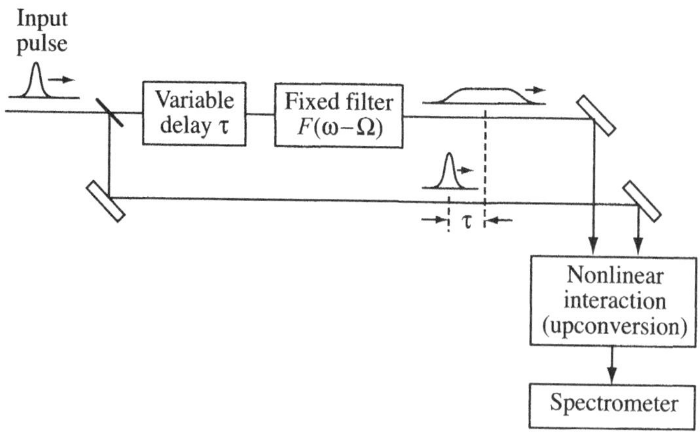

Figure 3.25 Spectrally and temporally resolved upconversion technique (STRUT) for ultrashort-pulse measurement.

We now analyze STRUT assuming that the filter bandwidth is sufficiently narrow that approximations 1 and 2 from earlier in this section still hold. Condition 3 is not needed, since the measurement does not require knowledge of the actual intensity profile of the filtered pulse. The quantity measured by the apparatus is given by

$$I_{\text{STRUT}}(\omega_{\text{up}}, \tau) = \left| \int dt\, e(t) g(t-\tau) e^{-j\omega_{\text{up}} t} \right|^2 \tag{3.108}$$

where $g(t)$ is the filtered version of the pulse. Using the equality between eqs. (3.70) and (3.71), we can rewrite this as

$$I_{\text{STRUT}}(\omega_{\text{up}}, \tau) = \left| \frac{1}{2\pi} \int d\omega'\, E(\omega') E(\omega_{\text{up}} - \omega') F(\omega_{\text{up}} - \omega' - \Omega) e^{j\omega'\tau} \right|^2 \tag{3.109}$$

where we have used $G(\omega) = E(\omega)F(\omega - \Omega)$. This can be simplified if we introduce a new variable $\delta\omega'$ such that $\omega' = \omega_{\text{up}} - \Omega + \delta\omega'$ and expand the spectral phase $\psi(\omega)$ to first order in $\delta\omega'$ (the power spectra are written only to zero order in $\delta\omega'$). After some algebra, this gives the result

$$I_{\text{STRUT}}(\omega_{\text{up}}, \tau) \approx \left| E(\Omega) E(\omega_{\text{up}} - \Omega) \right|^2 \left| f\left[\psi'(\Omega) - \psi'(\omega_{\text{up}} - \Omega) - \tau \right] \right|^2 \tag{3.110}$$

where $f(t)$ is the inverse Fourier transform of $F(\omega)$. The curve shifts along the delay coordinate by an amount given by $\psi'(\omega_{\text{up}} - \Omega)$. Measuring the temporal shift with the spectrometer set for upconverted frequency ω_{up} gives $\tau(\omega)$ for frequency $\omega_{\text{up}} - \Omega$ within the spectrum of the original pulse. The $\psi'(\Omega)$ term is a constant that does not influence the measurement. The fact that $f(\cdots)$ appears in a time-reversed form in eq. (3.110) arises simply because of the sign convention chosen for τ in our analysis and is not consequential.

As described earlier, STRUT has the advantage that the spectral phase information desired is obtained directly, at least approximately, with the concomitant disadvantage that the accuracy of the phase information obtained is not directly known. Of course, it should be possible to run phase retrieval algorithms on the full STRUT data set, but this would offset the main benefit of this technique.

3.7.2 Double-Slit Approach

Let us now consider another technique based on the scheme of Fig. 3.24, but with the single-slit filter of Section 3.7.1 replaced by a double slit. This technique was first used for measurement of picosecond pulses in optical pulse compression experiments [161] and then developed more fully for femtosecond pulse characterization, where it was termed *direct optical spectral phase measurement* (DOSPM) [162,163]. The frequency filter is now written $H(\omega - \omega_1) + H(\omega - \omega_2)$. We assume that the following approximations are valid:

1. Both the power spectrum and the spectral phase of the input pulse remain essentially constant within the bandwidth $\delta\omega$ of a single slit.
2. The reference pulse is sufficiently well behaved that it can measure the intensity profile of the filtered pulse with good accuracy.

Under these conditions, the measured output intensity contains a beat note of the form

$$I_{\mathrm{DOSPM}} \sim |h(t)|^2 \cos^2\left(\frac{\Delta\omega t + \psi(\omega_2) - \psi(\omega_1)}{2}\right) \tag{3.111}$$

where $\Delta\omega = \omega_2 - \omega_1$. The phase of the $\cos^2$ function gives directly the phase difference between the two optical frequencies selected. For example, if the relative phase is zero, an intensity peak occurs at $t = 0$, while for a relative phase of π, there is an intensity minimum at $t = 0$.

There are several strategies for exploiting this temporal interference phenomenon to determine the full spectral phase profile $\psi(\omega)$:

1. The measurement is repeated several times, first for frequencies ω_1 and $\omega_1 + \Delta\omega$, then for frequencies $\omega_1 + \Delta\omega$ and $\omega_1 + 2\Delta\omega$, and so on. Taking $\psi(\omega_1)$ as an arbitrary constant, the first measurement yields $\psi(\omega_1 + \Delta\omega) - \psi(\omega_1)$, the second measurement gives $\psi(\omega_1 + 2\Delta\omega) - \psi(\omega_1 + \Delta\omega)$, and so on. The overall spectral phase function is obtained by adding (or integrating over) the series of phase differences thus obtained. This procedure is a proper sonogram method, since the two-slit frequency filter is slid across the spectrum.
2. Instead of keeping the slit separation fixed, one can instead keep the position of one slit fixed while the second slit is slid across the spectrum. Thus, ω_1 is held constant while $\Delta\omega$ and hence ω_2 are varied. Each measurement yields $\psi(\omega_2)$ directly for the corresponding value of ω_2. This method is no longer a proper sonogram, since the variation of the two-slit filter function from one trace to the next is no longer a simple frequency translation.
3. Instead of a two-slit filter, a $2N$-slit filter can be utilized for simultaneous measurement of $2N - 1$ independent spectral phase values [163]. For example, $2N - 1$ of the slits can be arranged with a constant frequency separation $\Delta\omega$ [i.e., at frequencies $\omega_0 - (N-1)\Delta\omega, \omega_0 - (N-2)\Delta\omega, \ldots, \omega_0 + (N-2)\Delta\omega,$

$\omega_0 + (N-1)\Delta\omega]$. The last slit is deliberately offset (e.g., at $\omega_0 + \Delta\omega/3$). The filtered pulse has a complicated intensity profile in which $N(2N-1)$ individual interference terms are superimposed. However, with the slit positions given above, the terms corresponding to the interference between the single offset slit with each of the other $2N-1$ regularly spaced slits all occur at unique beat frequencies. Hence, each of these terms can be separated by Fourier analysis of the measured intensity profile, which yields the individual spectral phases.

It is interesting to note that although this DOSPM technique is clearly related to sonogram concepts, unlike the single-slit approaches it also makes obvious use of interference phenomena. This brings us to a discussion of interference-based pulse characterization techniques.

3.8 SELF-REFERENCING INTERFEROMETRY

In Section 3.3 we discussed temporal and spectral interferometry techniques in which the complete electric field could be determined via interference with a *completely characterized* reference pulse. Here we discuss two self-referencing interferometric methods in which the need for a fully characterized reference pulse is removed. Like the spectral filtering techniques discussed in Section 3.7, these interferometry-based methods provide a direct measurement of the phase profile without the need for iterative inversion and therefore permit rapid reconstruction of the electric field.

3.8.1 Time-Domain Interferometry of Chirped Pulses

We first describe a time-domain interferometry method applicable to highly chirped pulses when a much shorter (but not fully characterized) reference pulse is available [164]. This technique was developed for characterization of chirped pulses that had been spectrally broadened via self-phase modulation in an optical fiber and subsequently compressed in the context of fiber and grating pulse compression experiments (see Chapter 6). In this context the compressed pulses serve as the short reference pulses needed for measurement of the longer chirped pulses at the fiber output.

The experimental setup is shown schematically in Fig. 3.26. The chirped input pulse is split and sent into the arms of an interferometer, one of which contains a narrowband filter $H(\omega - \omega_0)$ of bandwidth $\delta\omega$ centered at frequency ω_0. The light from each of the two arms is recombined at the output of the interferometer. $\delta\omega$ should be sufficiently small that the spectral phase is essentially constant within the filter bandwidth and the filtered pulse is at least as long as the input pulse; the relative delay of the interferometer should also be set so that the filtered pulse and the input pulse have good temporal overlap. Physically, the output intensity profile exhibits a temporal interference pattern reflecting the time-varying phase difference between the essentially monochromatic (constant phase) filtered pulse and the chirped input pulse. This temporal interference pattern is recorded by means of an intensity cross-correlation using a short reference pulse. In our subsequent analysis we assume that the reference pulse is sufficiently short that the cross-correlation provides a good estimate of the actual intensity profile.

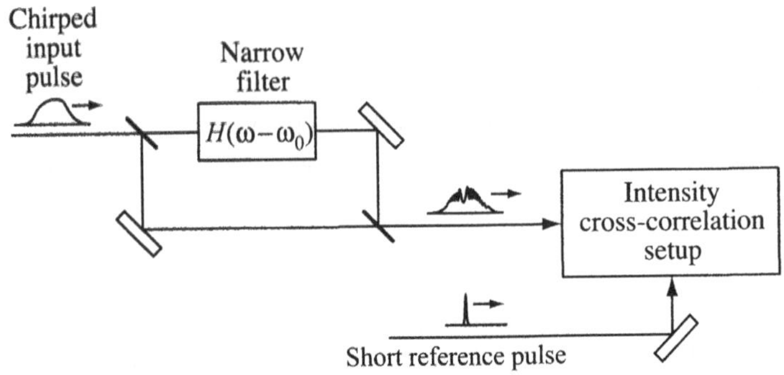

Figure 3.26 Time-domain interferometry of chirped pulses.

Mathematically, the output intensity profile can be written

$$I_{out}(t) \sim \left| |a(t)| e^{j[\omega t+\phi_{in}(t)]} + \alpha |h(t)| e^{j[\omega_0 t+\phi_h(t)+\phi_r]} \right|^2 \tag{3.112}$$

where $\phi_h(t)$ is the temporal phase associated with the impulse response function of the filter, ϕ_r is a constant reference phase which can be controlled by fine tuning the interferometer, and α (assumed real) accounts for the different amplitudes of the two terms. We now assume that $\phi_h(t)$ is a constant, which for simplicity we set to zero. This is valid provided that $H(\omega) = H^*(-\omega)$ in addition to our earlier assumption that $\delta\omega$ is sufficiently narrow. The resulting intensity profile is given by

$$I_{out}(t) \sim |a(t)|^2 + \alpha^2 |h(t)|^2 + 2 |a(t)h(t)| \cos [\phi_{in}(t) - \phi_r] \tag{3.113}$$

The desired information is extracted by subtracting $|a(t)|^2$ and $\alpha^2 |h(t)|^2$; these terms can be measured by performing cross-correlations from one arm of the interferometer at a time, with the other arm blocked. Let us call the result I'_{out}. The results for ϕ_r values of 0 and $\pi/2$ are given, respectively, as follows:

$$I'_{out}(t; \phi_r = 0) \sim 2\alpha |a(t)h(t)| \cos [\phi_{in}(t)] \tag{3.114a}$$

$$I'_{out}(t; \phi_r = \frac{\pi}{2}) \sim 2\alpha |a(t)h(t)| \sin [\phi_{in}(t)] \tag{3.114b}$$

The temporal phase $\phi_{in}(t)$ is obtained by dividing eq. (3.114b) by eq. (3.114a) and taking the arctangent of the result.

Experimental demonstrations of this technique used an etalon to implement $H(\omega)$. An interference filter or a pulse shaper could clearly also be used. It is interesting that in earlier experiments [165], passage of a single beam through a resonant atomic medium was used instead of the interferometer and filter shown in Fig. 3.26. The temporal reshaping caused by a single sharp absorption resonance can be described as arising from the temporal interference between the original input pulse and the nearly monochromatic free induction decay signal emitted by the atomic medium subsequent to its excitation. Thus, these two techniques are very similar. However, the earlier experiments based on atomic media suffered the disadvantages that ϕ_r could not conveniently be varied and that the temporal interference began only partway through the pulse when the input frequency swept through the absorption line.

3.8.2 Self-Referencing Spectral Interferometry

We now discuss a self-referencing spectral interferometry technique based on the interference between two spectrally and temporally shifted but otherwise identical versions of the pulse to be measured. Although this approach has much in common with the spectral interferometry method discussed in Section 3.3, the introduction of a spectral shift (also called a *spectral shear*) allows the pulse shape to be determined without the need for a separate characterized reference pulse. This technique is called *spectral phase interferometry for direct electric field reconstruction* (SPIDER) [166,167].

We write the two pulses that interfere in the SPIDER apparatus as

$$e_1(t) = \mathrm{Re}\{a(t)e^{j\omega_0 t}\} \tag{3.115a}$$

and

$$e_2(t) = \mathrm{Re}\{a(t-\tau)e^{j(\omega_0+\Omega)(t-\tau)}\} \tag{3.115b}$$

We discuss later how the spectral shear by Ω is imposed. In terms of the envelope functions, the output power spectrum can be written

$$\begin{aligned} |A_{\text{out}}(\tilde{\omega})|^2 \sim\ & |A(\tilde{\omega})|^2 + |A(\tilde{\omega}-\Omega)|^2 \\ & + \left[A(\tilde{\omega})A^*(\tilde{\omega}-\Omega)e^{j(\omega_0+\tilde{\omega})\tau} + \text{c.c.}\right] \end{aligned} \tag{3.116}$$

We are using the symbol $\tilde{\omega} = \omega - \omega_0$ since $A(\tilde{\omega})$ is a baseband function. If we now write $A(\tilde{\omega}) = |A(\tilde{\omega})|\, e^{j\psi(\tilde{\omega})}$, eq. (3.116) becomes

$$\begin{aligned} |A_{\text{out}}(\tilde{\omega})|^2 \sim\ & |A(\tilde{\omega})|^2 + |A(\tilde{\omega}-\Omega)|^2 \\ & + |A(\tilde{\omega})|\ |A(\tilde{\omega}-\Omega)| \cos\left[(\omega_0+\tilde{\omega})\tau + \psi(\tilde{\omega}) - \psi(\tilde{\omega}-\Omega)\right] \end{aligned} \tag{3.117}$$

For a flat spectral phase the overall power spectrum exhibits oscillations with period $2\pi/\tau$. The period is modulated by the spectral phase difference $\psi_{\text{diff}}(\tilde{\omega},\Omega) = \psi(\tilde{\omega}) - \psi(\tilde{\omega} - \Omega)$. Provided that the power spectrum is measured with sufficient resolution, this allows determination of the spectral phase.

The algorithm for reconstructing the spectral phase is similar to spectral interferometry. The SPIDER data is first inverse Fourier transformed into the time domain, which leads to terms located around $t = 0$ and $t = \pm\tau$ (refer back to Fig. 3.5). The term at $t = -\tau$ is extracted by setting the terms at $t = 0$ and $t = \tau$ to zero. The result is then Fourier transformed back into the frequency domain. The phase in the frequency domain is equal to $(\omega_0 + \tilde{\omega})\tau + \psi_{\text{diff}}(\tilde{\omega},\Omega)$. Subtracting the term linear in τ gives $\psi_{\text{diff}}(\tilde{\omega},\Omega)$. After assigning a single arbitrary phase to one frequency, say ω_0, the measured phase differences then specify the spectral phase at frequencies $\omega_0 \pm m\Omega$, where m is a positive integer.

In order that this procedure provide an accurate reconstruction of $\psi(\omega)$ and hence $e(t)$, two conditions must be satisfied:

1. The delay τ must be sufficiently large that the terms at $t = 0$ and $t = \pm\tau$ can be clearly separated in the Fourier-transformed SPIDER data. This implies that the $2\pi/\tau$ period of the fringes in the power spectrum is fast compared to the variation of $\psi_{\text{diff}}(\tilde{\omega}, \Omega)$ and of the power spectrum $|A(\tilde{\omega})|^2$. This also implies that the duration of the pulse to be measured, t_p, satisfies $t_p \ll \tau$.
2. The spectral shear Ω must be sufficiently small that it adequately samples the power spectrum and $\psi_{\text{diff}}(\tilde{\omega}, \Omega)$. The latter condition can be restated as $|\psi_{\text{diff}}(\tilde{\omega}, \Omega)| \ll 1$. This implies that $\Omega \ll 2\pi/t_p$. In the unusual case of a pulse known to have finite support (i.e., the intensity is strictly zero outside a region $-T/2 < t < T/2$), this condition would be relaxed to $\Omega \leq 2\pi/T$, which is the familiar sampling theorem of electrical engineering [168]. On the other hand, one should not make Ω so small that $\psi_{\text{diff}}(\tilde{\omega}, \Omega)$ becomes too small to measure with good signal-to-noise ratio.

A schematic SPIDER apparatus is shown in Fig. 3.27. The pulse to be measured is split into two beams. The first is transformed into a pulse doublet with delay τ, which can be performed by passing through either an etalon or a Michelson interferometer. The relative phase between the two pulses in the pulse pair must remain constant during the course of the measurement. The second beam passes through a pulse stretcher, which broadens the pulse and gives it a chirp. To lowest order, pulse stretchers operate by imposing a quadratic spectral phase onto the pulse; for a description of a common pulse stretcher based on a pair of diffraction gratings, see Chapter 4. The pulse pair and stretched pulse then interact in a second harmonic crystal. Each pulse in the pulse pair interacts with a different frequency from within the chirped stretched pulse. This results in a pair of upconverted pulses, still separated by τ but with the required spectral shift. Practically the experiment is usually performed using a phase-matching geometry where the second harmonic is generated only by the interaction between the pulse pair and the stretched pulse, not by either the stretched pulse (or the pulse pair) interacting with itself alone. Finally, the upconverted power spectrum is measured using a spectrometer.

In the case of a Gaussian pulse, the action of the pulse stretcher is described by the formalism of Section 3.5. If we write the spectrum after stretching as $A(\omega) \sim \exp\left\{-(\alpha + j\beta)\omega^2/4\right\}$, the instantaneous frequency is given by eq. (3.65), which reduces to $\tilde{\omega} \approx 2t/\beta$ in the limit $|\beta| \gg \alpha$. This is the limit where the stretched pulse is much longer than the inverse of the bandwidth of the input pulse. The spectral shear is related to the

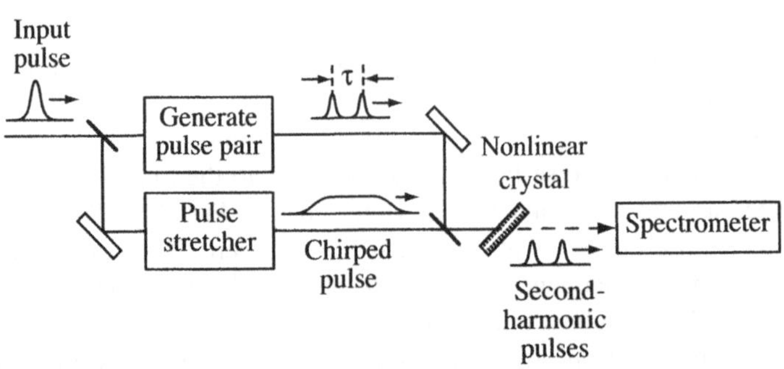

Figure 3.27 Self-referencing spectral shearing interferometry apparatus (SPIDER). Although not depicted explicitly, a background-free geometry is assumed where second harmonic is generated only through the interaction of the pulse pair with the stretched pulse.

delay through $\Omega \approx 2\tau/\beta$. The chirp parameter β should be chosen to satisfy the following constraints:

1. The stretched pulse duration should exceed the pulse pair delay τ.
2. The instantaneous frequency of the stretched pulse should be essentially constant within the duration of a single input pulse. This ensures that the upconversion process faithfully produces a pair of spectrally shifted but otherwise undistorted versions of the complex input spectrum. If the input pulse is close to bandwidth limited, the condition $|\beta| \gg \alpha$ is sufficient to enforce this constant instantaneous frequency constraint. Adherence to this constraint can also be checked experimentally by observing the upconverted spectrum resulting when only one of the pulses from the delayed pulse pair is present (the other pulse is blocked).

Finally, some calibrations are necessary to use the SPIDER technique. The most important is to determine τ accurately, since a linear phase term proportional to τ must be subtracted from the SPIDER data. An error in τ leads to an erroneous quadratic phase term in $\psi(\tilde{\omega})$, even for pulses without any chirp. One calibration method is to use spectral interferometry to determine τ. Spectral interferometry can be performed, with the stretched pulse blocked, either on the delayed pulse pair itself in the fundamental frequency band, or on the pair of second-harmonic pulses produced individually by each of the delayed pulses. In the latter case the second-harmonic crystal orientation is adjusted to permit upconversion of each pulse individually. The accuracy with which τ must be known may be especially demanding for measurement of pulses in the few-cycle regime. A variation of spectral shearing interferometry designed for robust measurement of few-cycle pulses is reported in [169].

3.9 CHARACTERIZATION OF NOISE AND JITTER

The characterization methods described so far in this chapter focus on measurement of isolated pulses. In principle, a perfectly mode-locked train of such pulses would be periodic, with all pulses having the same shape, amplitude, frequency, and timing. In practice, however, these quantities can fluctuate. In this section we discuss characterization of amplitude and timing fluctuations based on radio-frequency (RF) spectral analysis of the mode-locked pulse train after detection via a fast (subnanosecond) photodetector [170]. Using this technique, picosecond timing jitters can be measured using subnanosecond photodetectors!

Consistent with von der Linde's original treatment, we analyze this technique in the context of active mode-locking. Subtle differences that would arise in the passive mode-locking case are mentioned at the end of this section. We write the power of the mode-locked pulse train as

$$P(t) = P_0(t) + \delta P(t) \tag{3.118}$$

Here, $P_0(t)$, written as

$$P_0(t) = \sum_m f_0(t - mT) \tag{3.119}$$

represents the ideal, fluctuation-free pulse train, $f_0(t) = |a(t)|^2$ is the power of a single pulse, T is the period of the pulse train, and $\delta P(T)$ represents the fluctuations. To take into account amplitude and timing variations, we write

$$P(t) = [1 + \mathcal{A}(t)] \sum_m f_0[t - mT - \delta T(t)] \tag{3.120}$$

Here $\mathcal{A}(t)$ and $\delta T(t)$ represent fluctuations in the pulse energy and timing jitter, respectively. Note that changes in the pulse shape itself are not accounted for in this analysis; therefore, pulse amplitude and pulse energy are considered to be equivalent variables. We rewrite this expression in terms of a first order Taylor series expansion:

$$P(t) = [1 + \mathcal{A}(t)]\, P_0(t) - T\mathcal{J}(t)\dot{P}_0(t) \tag{3.121}$$

where $\mathcal{J}(t) = \delta T(t)/T$ is the normalized timing jitter and $\dot{P}_0(t)$ is the first derivative of $P_0(t)$. The first-order series expansion is strictly valid only when δT is much less than the pulse width. However, in the actual experiment, the optical pulse train is detected by a photodetector. The photodetector current response $\mathcal{I}_{\mathrm{PD}}$ can be written

$$\mathcal{I}_{\mathrm{PD}}(t) = P(t) * g_{\mathrm{PD}}(t) \tag{3.122}$$

$g_{\mathrm{PD}}(t)$ is the photocurrent impulse response function. Although $g_{\mathrm{PD}}(t)$ should be chosen to be fast compared to the mode-locked period T, it is usually quite slow compared to the duration of a pulse in ultrafast optics. Once the photocurrent response time is accounted for, the first-order expansion is valid as long as δT is much less than the photodetector impulse response, a much less restrictive requirement.

To relate the fluctuations to the RF spectrum, we first form the correlation function $\langle P(t)P(t-\tau)\rangle$. In contrast to our discussion earlier in this chapter, we are now interested in the behavior of the correlation function for $\tau \geq T$, which is the time scale in which the amplitude and timing fluctuations become important. Assuming that $\mathcal{A}(t)$ and $\delta T(t)$ are independent, zero-mean random variables, we obtain

$$\begin{aligned}\langle P(t)P(t-\tau)\rangle &= \langle P_0(t)P_0(t-\tau)\rangle\, [1 + \langle \mathcal{A}(t)\mathcal{A}(t-\tau)\rangle] \\ &\quad + T^2 \left\langle \dot{P}(t)\dot{P}(t-\tau)\right\rangle \langle \mathcal{J}(t)\mathcal{J}(t-\tau)\rangle\end{aligned} \tag{3.123}$$

The power autocorrelation function depends on the autocorrelation functions of the amplitude and timing fluctuations. Using our earlier results on the Fourier transform relationship between the autocorrelation and the power spectrum, eq. (3.21), we obtain

$$\begin{aligned}\left|\tilde{\mathcal{I}}_{\mathrm{PD}}(\omega)\right|^2 &= \left|\tilde{G}_{\mathrm{PD}}(\omega)\tilde{P}_0(\omega)\right|^2 * \left[2\pi\delta(\omega) + \left|\tilde{\mathcal{A}}(\omega)\right|^2\right] \\ &\quad + \omega^2 T^2 \left|\tilde{G}_{\mathrm{PD}}(\omega)\tilde{P}_0(\omega)\right|^2 * \left|\tilde{\mathcal{J}}(\omega)\right|^2\end{aligned} \tag{3.124}$$

where $\tilde{\mathcal{I}}_{\mathrm{PD}}$, $\tilde{G}_{\mathrm{PD}}$, $\tilde{P}_0$, $\tilde{\mathcal{A}}$, and $\tilde{\mathcal{J}}$ are the Fourier transforms of the corresponding variables without tildes. Substituting for $\tilde{P}_0(\omega)$ using the Fourier transform of eq. (3.119), we obtain

$$\left|\tilde{\mathcal{I}}_{\mathrm{PD}}(\omega)\right|^2 = \left(\frac{2\pi}{T}\right)^2 \sum_k \left|\tilde{G}_{\mathrm{PD}}(\omega_k)\tilde{F}_0(\omega_k)\right|^2 \times \left[2\pi\delta(\omega-\omega_k) + \left|\tilde{\mathcal{A}}(\omega-\omega_k)\right|^2 + \omega_k^2 T^2 \left|\tilde{\mathcal{J}}(\omega-\omega_k)\right|^2\right] \quad (3.125)$$

where $\tilde{F}_0$ is the Fourier transform of f_0 and $\omega_k = 2\pi k/T$.

The photocurrent power spectrum, which can be measured using an RF spectrum analyzer, is pictured in Fig. 3.28. The spectrum consists of a series of lines at DC and at the harmonics ω_k of the mode-locking frequency. The sharp central lines occurring right at the ω_k correspond to the ideal mode-locked train, while the broader low-intensity pedestals are related to the fluctuations. The width of the sharp central lines is limited by the resolution bandwidth of the RF spectrum analyzer. Although not indicated in the figure, the peaks begin to roll off at high frequencies, due to the limited photodetection bandwidth. To analyze the fluctuations, we note the k^2 factor, which multiplies $|\tilde{\mathcal{J}}(\omega)|^2$ but not $|\tilde{\mathcal{A}}(\omega)|^2$. Therefore, we can determine the amplitude noise spectrum $|\tilde{\mathcal{A}}(\omega)|^2$ from the low intensity pedestal centered at DC (or at frequencies ω_k with k sufficiently low that $|\tilde{\mathcal{A}}(\omega)|^2$ dominates). The timing jitter term grows quadratically with k and therefore is the dominant contribution to the noise pedestal at sufficiently large k. This allows determination of $|\tilde{\mathcal{J}}(\omega)|^2$. This behavior is evident in data for the mode-locked argon ion laser shown in Fig. 3.29.

Once $|\tilde{\mathcal{A}}(\omega)|^2$ and $|\tilde{\mathcal{J}}(\omega)|^2$ have been determined, they can be inverse-Fourier-transformed to yield the correlation functions of the amplitude noise and of the timing jitter. The correlation functions characterize the size of the amplitude and timing variations as well as the time scale over which the amplitude and timing fluctuate. The size of the variations is characterized by the variances, which are equal to the correlation functions evaluated at $\tau = 0$. Thus, we have

$$\left\langle \mathcal{A}^2(t)\right\rangle = \frac{1}{2\pi}\int d\omega \left|\tilde{\mathcal{A}}(\omega)\right|^2 \quad (3.126)$$

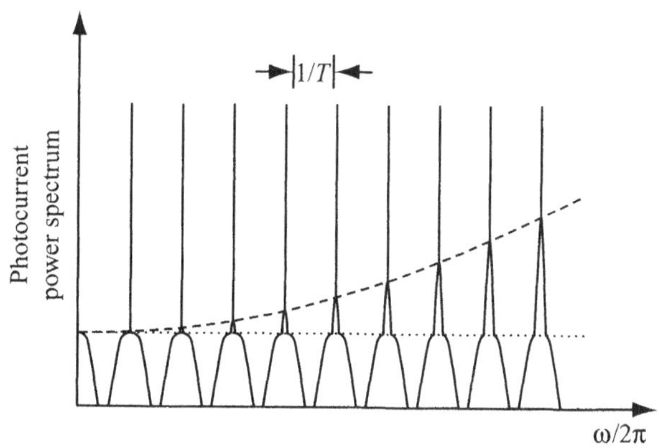

Figure 3.28 Radio-frequency spectrum of a fluctuating mode-locked train. The photocurrent response is assumed to be flat over the frequency range pictured. From Fig. 1 of [170], with permission from Springer Science and Business Media.

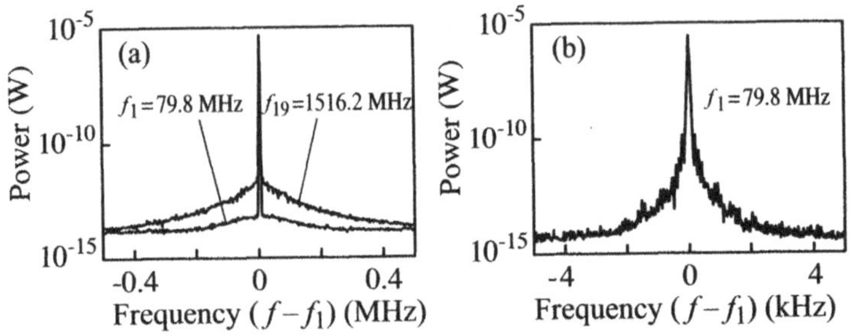

Figure 3.29 (a) RF photocurrent power spectra at the fundamental mode-locking frequency and nineteenth harmonic of a mode-locked argon laser (1-kHz resolution bandwidth); (b) photocurrent power spectrum at fundamental mode-locking frequency with a 30-Hz resolution bandwidth. From Figs. 5 and 6 of [170], with permission from Springer Science and Business Media.

and

$$\left\langle \left(\frac{\delta T(t)}{T} \right)^2 \right\rangle = \frac{1}{2\pi} \int d\omega \left| \tilde{\mathcal{J}}(\omega) \right|^2 \tag{3.127}$$

Practically, in evaluating the area under the noise pedestal, one usually integrates only over a certain range of frequency offsets, $\delta\omega_{\min} \leq |\omega - \omega_k| \leq \delta\omega_{\max}$. There are two reasons for this. First, one may be interested only in the fluctuations occurring within a certain frequency range. Second, there is the practical problem that at sufficiently low frequency offset, the noise pedestal merges with the spike corresponding to the ideal mode-locked train. The frequency at which the spike and pedestal merge depends on the resolution bandwidth $\Delta\omega_{\mathrm{RB}}$ of the RF spectrum analyzer. This effect can have a very significant effect on the measurement, as illustrated by Fig. 3.29, which shows RF spectra taken from the same mode-locked laser, with resolution bandwidths of 1 kHz and 30 Hz, respectively [170]. The higher-resolution trace clearly reveals the structure of additional noise energy not visible in the lower-resolution data. Consequently, it is important to specify the frequency range over which the broad pedestal is integrated. The amplitude variance can be computed using

$$\left\langle \mathcal{A}^2(t) \right\rangle = \frac{\left[\int_{-\delta\omega_{\max}}^{-\delta\omega_{\min}} + \int_{\delta\omega_{\min}}^{\delta\omega_{\max}} \right] d\tilde{\omega} \left| \tilde{\mathcal{I}}_{\mathrm{PD}}(\omega_k + \tilde{\omega}) \right|^2}{\int_{-\Delta\omega_{\mathrm{RB}}}^{\Delta\omega_{\mathrm{RB}}} d\tilde{\omega} \left| \tilde{\mathcal{I}}_{\mathrm{PD}}(\omega_k + \tilde{\omega}) \right|^2} \tag{3.128}$$

The integrals involve only a single harmonic (single k value), with k chosen sufficiently small that the amplitude fluctuations dominate. The integral over the central peak [denominator of eq. (3.128)] should be carried out over a range sufficient to determine its full area (i.e., over a range at least equal to the resolution bandwidth of the spectrum analyzer). Clearly, $\delta\omega_{\min} \geq \Delta\omega_{\mathrm{RB}}$ should be satisfied. The resulting rms amplitude noise given by the square root of eq. (3.128) should be specified in terms of both the percentage value and the corresponding frequency range ($\delta\omega_{\min}$, $\delta\omega_{\max}$). Computation of the timing jitter follows a similar procedure, but using a higher k value such that the timing jitter dominates the area under the pedestal. The timing variance is then given by

$$\left\langle (\delta T(t))^2 \right\rangle = \frac{T^2}{4\pi^2 k^2} \frac{\left[\int_{-\delta\omega_{\max}}^{-\delta\omega_{\min}} + \int_{\delta\omega_{\min}}^{\delta\omega_{\max}}\right] d\tilde{\omega} \left[\left|\tilde{\mathcal{I}}_{\mathrm{PD}}(\omega_k + \tilde{\omega})\right|^2 - \left|\tilde{A}(\omega_k + \tilde{\omega})\right|^2\right]}{\int_{-\Delta\omega_{\mathrm{RB}}}^{\Delta\omega_{\mathrm{RB}}} d\tilde{\omega} \left|\tilde{\mathcal{I}}_{\mathrm{PD}}(\omega_k + \tilde{\omega})\right|^2} \tag{3.129}$$

For sufficiently large k, the $|\tilde{A}(\omega)|^2$ term subtracted in the upper integral is only a small correction, which can be ignored without much loss of accuracy. The rms timing jitter must also be specified together with the frequency range. The photodetector frequency response does not appear in eqs. (3.128) and (3.129) and therefore has no direct effect on the measurement, provided, of course, that the response is sufficiently broadband to allow measurement of the desired harmonics with the desired signal-to-noise ratio. Note that truncation of the RF spectrum for frequency offsets below $\omega_{\min}$ means that information about the full correlation functions $\langle \mathcal{A}(t)\mathcal{A}(t-\tau)\rangle$ and $\langle \mathcal{J}(t)\mathcal{J}(t-\tau)\rangle$ is lost for $|\tau| \gtrsim 2\pi/\omega_{min}$. Similarly, truncation for offsets above $\delta\omega_{\max}$ means that information about the correlation functions is lost for $|\tau| \lesssim 2\pi/\omega_{\max}$.

In some experiments one is interested in the relative timing jitter between two pulses emitted from the laser at a specified delay offset. For example, it is the relative timing jitter that would determine broadening of an experimental cross-correlation trace involving interaction of the pulses emitted from a mode-locked laser with delayed earlier pulses emitted from the same laser. The quantity of interest here would be the variance of $[\delta T(t) - \delta T(t-\tau)]$, which is easily shown to be

$$\left\langle [\delta T(t) - \delta T(t-\tau)]^2 \right\rangle = 2\left\langle [\delta T(t)]^2 \right\rangle \{1 - G_{\mathcal{J}}(\tau)\} \tag{3.130a}$$

where

$$G_{\mathcal{J}}(\tau) = \frac{\langle \mathcal{J}(t)\mathcal{J}(t-\tau)\rangle}{\langle [\mathcal{J}(t)]^2\rangle} = \frac{\int d\omega \left|\tilde{\mathcal{J}}(\omega)\right|^2 e^{j\omega\tau}}{\int d\omega \left|\tilde{\mathcal{J}}(\omega)\right|^2} \tag{3.130b}$$

is the autocorrelation of $\mathcal{J}$, which can be computed using the inverse Fourier transform of $|\tilde{\mathcal{J}}(\omega)|^2$ extracted from the RF spectrum. For delays less than the correlation time of $\mathcal{J}$, this relative timing jitter can be much less than the absolute timing jitter given by eq. (3.127).

Studies of fundamental amplitude noise and timing jitter properties of actively and passively mode-locked lasers may be found in [171–173]. One key difference between actively and passively mode-locked lasers is that in the actively mode-locked laser, the externally driven modulator acts as a restoring force which damps timing fluctuations of the mode-locked pulses with respect to the external timing source. As a result, in the actively mode-locked case, $\left\langle [\delta T(t) - \delta T(t-\tau)]^2 \right\rangle$ approaches a constant for large τ; the size of the timing fluctuations is bounded. In contrast, for the passively mode-locked case, there is no timing reference or restoring force. The timing of the mode-locked pulses is governed by a random walk process. Therefore, $\left\langle [\delta T(t) - \delta T(t-\tau)]^2 \right\rangle$ grows asymptotically as τ. The size of the timing variations with increased observation time is unbounded; therefore, the first-order Taylor series expansion used to describe the timing jitter in eq. (3.121) is not valid for very large observation times. As a consequence, the form of the photocurrent spectrum is modified for passive mode-locking. In particular, the delta function, which is the signature of a perfectly mode-locked pulse train in eq. (3.125), is replaced by a central feature of finite width and amplitude. For passively mode-locked lasers, then, it is fundamentally not possible to distinguish the noise and jitter pedestal from the central feature in the RF spectrum at very low frequency offsets. However, except at very low frequencies, eq. (3.125) remains a good description.

Finally, we note that although independent timing and energy fluctuations have been assumed, in principle these fluctuations can have correlations. For example, energy fluctuations lead to gain changes, which imply changes in the refractive index and hence the cavity round-trip time. One consequence of correlations between $\mathcal{A}$ and δT is the possibility of photocurrent power spectra which are not symmetric with respect to the various ω_k. Pulse-width fluctuations, which we have not considered here, are also possible. For further discussion of these effects, see [172,174,175].

PROBLEMS

3.1. Either numerically or analytically, construct a pulse whose time–bandwidth product in terms of FWHM duration and bandwidth is very small. Can you realize a $\Delta\nu\,\Delta t$ value of less than 0.1? Less than 0.01? Also compute the product of the rms pulse duration and bandwidth for your pulse. Comment on your results.

3.2. **(a)** Spectral interferometry is used to characterize a lossless optical device which imparts a quadratic spectral phase $\Delta\psi(\omega) = A(\omega - \omega_0)^2$ onto the signal field, where $A = 1.5 \times 10^{-26}\ \mathrm{s}^{-2}$. The input field is a bandwidth-limited Gaussian pulse with center frequency $\omega_0/2\pi = 2 \times 10^{14}$ Hz and FWHM power spectral bandwidth $\Delta\nu = 5 \times 10^{12}$ Hz. Plot the spectral interferometry signal assuming that the reference field is delayed relative to the signal field by (i) 2 ps and (ii) −2 ps. Comment on the main features in your data.

(b) Repeat assuming a chirped Gaussian input, with spectral phase $-A(\omega - \omega_0)^2/2$.

3.3. Figure 3.7 shows power spectra and electric field autocorrelation data for superluminescent diode and mode-locked Ti:S laser light sources. Compare the correlation widths to those expected from the spectra. To relate delay in micrometers, as shown in the figure, to delay in femtoseconds, you may assume that $v_g = c$ (i.e., the group velocity is equal to the speed of light in free space).

3.4. Derive eq. (3.43) relating the rms pulse width to the autocorrelation. [*Hint*: Use the substitution $\eta = t - \tau$ in evaluating the numerator on the right-hand side of eq. (3.43).]

3.5. Consider a frequency-modulated Gaussian pulse with center frequency ω_0 and complex envelope function $a(t) = \exp(-t^2/t_p^2)\exp(jt^2/t_c^2)$. Produce plots of the power spectrum, electric field autocorrelation envelope, (fringe-averaged) intensity autocorrelation, and interferometric intensity autocorrelation, for $t_p = 200$ fs, $\omega_0/2\pi = 5 \times 10^{14}$ Hz, and $t_c = 40$, 200, and 800 fs. Comment on the trends in your plots. For the interferometric autocorrelation, it is sufficient to plot the envelopes of the fringes.

3.6. **(a)** Verify the expression for the time broadening of the autocorrelation trace for a Gaussian pulse in a noncollinear geometry, eq. (3.53).

(b) By evaluating eq. (3.55) in light of the requirement that the input beam separation (at the lens) exceed the input beam diameter, $\delta x_{\text{in}} \geq 2w_{\text{in}}$, the text finds a minimum temporal broadening of $\lambda/\pi c$. Referring to the Gaussian beam formulas of Chapter 1, show that the same result is obtained if one instead imposes the requirement that the angular separation of the beams exceed the beam divergence (both evaluated at the nonlinear crystal).

3.7. Using eq. (3.58) for the correlation function measured by third-harmonic generation, give an expression for the contrast ratio of the interferometric correlation trace as a function of η for $0 \leq \eta \leq 1$ [i.e., give the ratios of the maximum (minimum) value at the top (bottom) of a fringe near $\tau = 0$ to the value of the power for large delays where the pulses do not overlap].

3.8. Demonstrate that spectrogram and sonogram time–frequency distributions, eqs. (3.70) and (3.71), are identical, provided that the sonogram filter function and spectrogram gate function satisfy $H(\omega) = G(-\omega)$.

3.9. Consider a pulse whose spectral amplitude function is given by $A(\tilde{\omega}) \sim \exp(-\tilde{\omega}^2 t_p^2/4)\exp(jA\tilde{\omega}^2)\exp(jB\tilde{\omega}^4)$, where $t_p = 100$ fs and (a) $A = \pi \times 10^4$ fs^2, $B = 0$; (b) $A = 0$, $B = (-\pi/16) \times 10^8$ fs^4; (c) $A = \pi \times 10^4$ fs^2, $B = (-\pi/16) \times 10^8$ fs^4. For each case use a real Gaussian gate function to generate a spectrogram, choosing the duration of the gate function to bring out interesting information in your spectrogram plot. Comment on the relationship between your spectrograms and the instantaneous frequency and frequency-dependent delay functions.

3.10. In PG-FROG there is a trade-off between efficiency and nonlinear phase modulation (the latter is usually kept low to avoid distortion of the pulse to be measured). Define efficiency as the fraction of the input power that is passed by the analyzer after the nonlinear index material. How large an efficiency is possible if we restrict the peak nonlinear phase shift to less than 0.1π? To less than 0.01π? To keep the problem simple, you may calculate the instantaneous efficiency at the peak of the gate pulse.

3.11. Derive the formula for the SHG-FROG frequency marginal, eq. (3.97).

3.12. In self-referencing spectral interferometry (SPIDER), an unknown input pulse is split into a pair of identical pulses with relative delay τ and spectral shear Ω. As described in the text, the algorithm for extracting the spectral phase from the SPIDER data requires exact knowledge of τ.

(a) Although the actual delay is τ_0, the experimentalist erroneously believes the delay to be $\tau_0 + \delta\tau$, where $\delta\tau$ is a small error. The erroneous delay value $\tau_0 + \delta\tau$ is used to extract the spectral phase from the SPIDER data. Assuming that all other conditions outlined in the text are satisfied, derive a formula for the error in the estimated spectral phase.

(b) The input pulse is a bandwidth-limited Gaussian with duration Δt (intensity FWHM). Using your result from above, obtain a formula for the estimated pulse duration.

(c) Plot the estimated pulse duration as a function of delay error for $\Delta t = 50$ fs, $\tau = 500$ fs, and $\Omega/2\pi = 400$ GHz. What is the maximum permissible delay error if the pulse duration is to be determined to better than 5% accuracy?

3.13. Again considering self-referencing spectral interferometry, assume that the experimentalist has perfect knowledge of the delay ($\delta\tau = 0$ in the nomenclature of Problem 3.12), but believes the spectral shear to be $\Omega + \delta\Omega$, where Ω and $\delta\Omega$ are, respectively, the actual spectral shear and a small error. Discuss the error introduced into the estimated spectral phase as a result of processing the SPIDER data using the erroneous value of the spectral shear.

4

DISPERSION AND DISPERSION COMPENSATION

4.1 GROUP VELOCITY DISPERSION

We have already introduced the concept of group velocity dispersion (GVD): in Section 2.4.2 on mode-locking of lasers and then in Section 3.5 on chirped pulses and spectrograms. In this section we discuss GVD a little more formally, prior to discussing methods for compensating GVD later in the chapter.

4.1.1 Group Velocity Definition and General Dispersion Relations

Consider an input pulse $e_{\text{in}}(t)$ with spectrum $E_{\text{in}}(\omega)$. After passing through a dispersive system that adds spectral phase $\psi(\omega)$, the output pulse is given by

$$e_{\text{out}}(t) = \frac{1}{2\pi} \int d\omega \, E_{\text{in}}(\omega) e^{j\omega t} e^{j\psi(\omega)} \tag{4.1}$$

When the phase arises due to passage through a bulk dispersion medium of length L, we can write

$$\psi(\omega) = -\beta(\omega) L \tag{4.2}$$

where $\beta(\omega)$ is the propagation constant in the medium. ψ and β are commonly written in terms of their Taylor series expansions, as follows:

$$\begin{aligned}
\psi(\omega) &= \psi(\omega_0) + \frac{\partial\psi}{\partial\omega}(\omega-\omega_0) + \frac{1}{2}\frac{\partial^2\psi}{\partial\omega^2}(\omega-\omega_0)^2 + \frac{1}{6}\frac{\partial^3\psi}{\partial\omega^3}(\omega-\omega_0)^3 + \cdots \\
&= \psi_0 + \psi_1(\omega-\omega_0) + \frac{\psi_2}{2}(\omega-\omega_0)^2 + \frac{\psi_3}{6}(\omega-\omega_0)^3 + \cdots
\end{aligned} \tag{4.3}$$

Ultrafast Optics, By Andrew M. Weiner

$$\begin{aligned}\beta(\omega) &= \beta(\omega_0) + \frac{\partial\beta}{\partial\omega}(\omega-\omega_0) + \frac{1}{2}\frac{\partial^2\beta}{\partial\omega^2}(\omega-\omega_0)^2 + \frac{1}{6}\frac{\partial^3\beta}{\partial\omega^3}(\omega-\omega_0)^3 + \cdots \\ &= \beta_0 + \beta_1(\omega-\omega_0) + \frac{\beta_2}{2}(\omega-\omega_0)^2 + \frac{\beta_3}{6}(\omega-\omega_0)^3 + \cdots \end{aligned} \tag{4.4}$$

Here $\psi_n = \partial^n\psi/\partial\omega^n$, $\beta_n = \partial^n\beta/\partial\omega^n$, and all the derivatives are evaluated at $\omega = \omega_0$. We can identify $\psi_n = -\beta_n L$.

Now if we introduce our normal notation for envelope functions, where the positive frequency part of $E_{\text{in}}(\omega)$ is replaced by $A(\tilde{\omega})$ and $\tilde{\omega} = \omega - \omega_0$, we can rewrite the output pulse as follows:

$$e_{\text{out}}(t) = \text{Re}\left\{e^{j(\omega_0 t - \beta_0 L)} a_{\text{out}}(t)\right\} \tag{4.5}$$

where

$$a_{\text{out}}(t) = \frac{1}{2\pi}\int d\tilde{\omega}\, A(\tilde{\omega})\, e^{j[\tilde{\omega}t - (\beta_1\tilde{\omega} + (\beta_2/2)\tilde{\omega}^2 + (\beta_3/6)\tilde{\omega}^3 + \cdots)L]} \tag{4.6}$$

Thus $e_{\text{out}}(t)$ is the product of a carrier term $e^{j(\omega_0 t - \beta_0 z)}$ and the envelope function $a_{\text{out}}(t)$. The carrier term propagates at the phase velocity

$$v_p = \frac{\omega_0}{\beta_0} \tag{4.7}$$

and is unaffected by the variation of β or ψ with ω. The envelope function $a_{\text{out}}(t)$, however, clearly is affected by the form of $\beta(\omega)$. One very important case occurs when $\beta(\omega)$ is a linear function of ω:

$$\beta(\omega) = \beta_0 + \beta_1\tilde{\omega} \tag{4.8}$$

In this case, we find that

$$a_{\text{out}}(t) = a_{\text{in}}(t - \beta_1 L) \tag{4.9}$$

Thus, the output pulse is an undistorted replica of the input pulse, which travels with velocity β_1^{-1}. This velocity is called the *group velocity* v_g, where

$$v_g = \beta_1^{-1} = \left(\frac{\partial\beta}{\partial\omega}\bigg|_{\omega=\omega_0}\right)^{-1} \tag{4.10}$$

The group velocity is the velocity at which energy is transported through the medium. Its value may differ from the phase velocity v_p, although v_g will be independent of frequency as long as eq. (4.8) holds.

We can also look at the concept of group velocity using simple Fourier transform identities. The Fourier transform of a delayed pulse $a(t-\tau)$ is given by $A(\omega)e^{-j\omega\tau}$, so

$\tau = -\partial\psi(\omega)/\partial\omega$. Using eqs. (4.2) and (4.8), we find that $\tau = \beta_1 L = L/v_g$, which is consistent with our discussion above.

When $\beta(\omega)$ is not a linear function of ω [i.e., eq. (4.8) does not hold], $a_{\text{out}}(t)$ is changed compared to $a_{\text{in}}(t)$. The part of $\beta(\omega)$ that is not linear in ω is called *dispersion*. The $\partial^2\beta/\partial\omega^2$ term contributes a quadratic spectral phase variation, which leads to a linear variation in delay with frequency. This imparts a linear chirp to the output pulse (see Section 3.5). The $\partial^3\beta/\partial\omega^3$ term contributes a cubic spectral phase leading to a quadratic variation in delay with frequency. This results in an asymmetric pulse distortion with an oscillatory structure, as shown in Fig. 3.16. These terms are sufficient to describe femtosecond pulse propagation in most dispersive media in most circumstances. However, in some very critical applications, such as sub-10-fs laser design and chirped pulse amplifiers for very short pulses, additional higher-order terms must also be retained.

We now proceed to give several useful formulas for the group velocity and the group velocity dispersion, specifically the *material dispersion*, encountered during propagation in a bulk medium. Starting with

$$\beta(\omega) = \frac{\omega n(\omega)}{c} \tag{4.11}$$

we find that

$$v_g = \frac{c}{n + \omega(dn/d\omega)} \tag{4.12}$$

It turns out that it is also useful to express v_g in terms of the wavelength λ. Here we use

$$\lambda = \frac{2\pi c}{\omega} \qquad \text{and} \qquad \frac{d\lambda}{d\omega} = \frac{-2\pi c}{\omega^2} = \frac{-\lambda^2}{2\pi c} \tag{4.13}$$

The result using the chain rule is

$$v_g = \frac{c}{n - \lambda(dn/d\lambda)} \tag{4.14}$$

Now we turn our attention to the quadratic dispersion term. For a particular frequency component ω, the time required to propagate through a length L of a dispersive medium is $L/v_g(\omega)$. The propagation time relative to that corresponding to frequency ω_0 is given by

$$\begin{aligned} \tau(\omega) - \tau(\omega_0) = \Delta\tau(\omega) &= \frac{\partial\tau}{\partial\omega}(\omega - \omega_0) \\ &= \frac{\partial\left(v_g^{-1}\right)}{\partial\omega}(\omega - \omega_0) L = \beta_2 (\omega - \omega_0) L \end{aligned} \tag{4.15}$$

Using eq. (4.11) to evaluate eq. (4.15), we obtain

$$\Delta\tau(\omega) = \left(2\frac{dn}{d\omega} + \omega\frac{d^2 n}{d\omega^2}\right)\frac{(\omega - \omega_0) L}{c} \tag{4.16}$$

By using eq. (4.13) and repeated applications of the chain rule, this can be rewritten

$$\Delta\tau(\omega) = \frac{\lambda^3 L}{2\pi c^2}\frac{d^2 n}{d\lambda^2}(\omega - \omega_0) \tag{4.17}$$

In some applications, notably fiber optics, it is customary to express $\Delta\tau$ in terms of the wavelength offset $\Delta\lambda = \lambda - \lambda_0$:

$$\Delta\tau(\lambda) = \frac{\partial\left(v_g^{-1}\right)}{\partial\lambda}(\lambda - \lambda_0)\,L = \frac{-2\pi c\beta_2\,\Delta\lambda L}{\lambda^2} \tag{4.18}$$

Fiber dispersion is usually described in terms of a dispersion parameter D with units $ps\,nm^{-1}\,km^{-1}$, defined by

$$\Delta\tau(\lambda) = D\,\Delta\lambda L \tag{4.19}$$

where

$$D = \frac{\partial\left(v_g^{-1}\right)}{\partial\lambda} = \frac{-2\pi c\beta_2}{\lambda^2} = \frac{-\lambda}{c}\frac{d^2 n}{d\lambda^2} \tag{4.20}$$

For completeness we also give expressions for the second-and third-order spectral phase. Compiling our earlier results, we have

$$\psi_2 = -\beta_2 L = \frac{-\partial\tau(\omega)}{\partial\omega} = \frac{-L}{c}\left[2\frac{dn}{d\omega} + \omega\frac{d^2 n}{d\omega^2}\right] = \frac{-\lambda^3 L}{2\pi c^2}\frac{d^2 n}{d\lambda^2} \tag{4.21}$$

For the third-order dispersion one finds that

$$\psi_3 = -\beta_3 L = \frac{-\partial^2\tau(\omega)}{\partial\omega^2} = \frac{-L}{c}\left[3\frac{d^2 n}{d\omega^2} + \omega\frac{d^3 n}{d\omega^3}\right] = \frac{\lambda^4 L}{4\pi^2 c^3}\left[3\frac{d^2 n}{d\lambda^2} + \lambda\frac{d^3 n}{d\lambda^3}\right] \tag{4.22}$$

We also note that for a Gaussian pulse, the broadening due to pure second-order spectral phase can be expressed very simply. For an unchirped input pulse of the form $a_{\text{in}}(t) \sim \exp\{-t^2/t_p^2\}$, the formulas of Section 3.5 yield

$$a_{\text{out}}(t) \sim \frac{1}{\sqrt{1 - 2j\psi_2/t_p^2}}\exp\left(\frac{-t^2/t_p^2}{1 - 2j\psi_2/t_p^2}\right) \tag{4.23}$$

and

$$|a_{\text{out}}(t)|^2 \sim \frac{1}{\sqrt{1+\left(2\psi_2/t_p^2\right)^2}} \exp\left(\frac{-2t^2/t_p^2}{1+\left(2\psi_2/t_p^2\right)^2}\right) \tag{4.24}$$

The fractional pulse-width broadening is

$$\frac{\Delta t_{\text{out}}}{\Delta t_{\text{in}}} = \sqrt{1+\frac{4\psi_2^2}{t_p^4}} \tag{4.25}$$

In the ultrafast optics community the dispersion associated with β_2 and β_3 is often termed *quadratic* (or second-order) and *cubic* (or third-order) *dispersion*, respectively, in view of the corresponding quadratic and cubic spectral phase variation. In the fiber optics community, this same dispersion associated with β_2 and β_3 is usually called first- and second-order dispersion, respectively, due to the corresponding linear and quadratic variation of delay with frequency or wavelength.

One other term that arises in fiber optics is the *dispersion slope* S, which refers to the derivative of D with respect to wavelength. One can easily show that

$$S = \frac{\partial D}{\partial \lambda} = \frac{4\pi c}{\lambda^3}\beta_2 + \left(\frac{2\pi c}{\lambda^2}\right)^2 \beta_3 \tag{4.26}$$

In cases where $\beta_2 = 0$ (e.g., at the zero-dispersion wavelength λ_0 in fibers), the dispersion slope is proportional to β_3. The time required for a wavelength component λ to propagate through a fiber of length L, relative to the time for wavelength λ_0, is then given by

$$\Delta\tau(\lambda) = \frac{1}{2} S\,(\lambda-\lambda_0)^2 L \tag{4.27}$$

We note that when waveguide dispersion or other sources of dispersion are present in addition to the material dispersion, eq. (4.11) as well as other later equations involving derivatives of the refractive index do not apply unless $n(\omega)$ is interpreted as an effective refractive index *defined* by the relation $n(\omega) = c\beta(\omega)/\omega$. Similarly, the $\beta(\omega)$ picture does not provide a natural description of the dispersion of components based on interference, including multilayer dielectric mirrors. In such situations it is necessary to speak directly in terms of the spectral phase $\psi(\omega)$ acquired in passing through an optical component or medium.

4.1.2 General Aspects of Material Dispersion

Materials that are transparent in the visible part of the spectrum usually exhibit a similar type of dispersion behavior [96,176]. The basics aspects of such *normal dispersion* are obeyed by a wide range of transparent materials and are therefore worth discussion. In the following we refer only to material dispersion; waveguide and other sources of dispersion are not included.

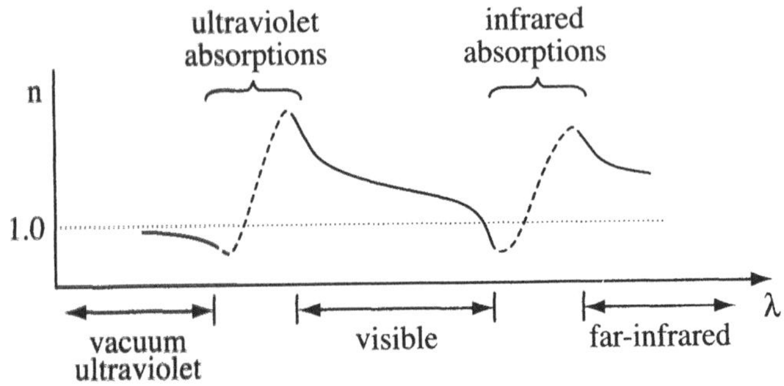

Figure 4.1 Simplified typical refractive index variation for transparent optical materials.

The behavior of transparent materials is usually dominated by electronic absorption resonances, typically at wavelengths below a few hundred nanometers in the ultraviolet spectral region. The typical Lorentzian line shape, which is usually valid in the vicinity of such resonances, was plotted in Fig. 2.3. For regions away from the absorption peak, the refractive index is an increasing function of frequency (decreasing function of wavelength), while close to (within) the absorption peak the converse is true. A simplified version of the resulting typical refractive index behavior is sketched in Fig. 4.1 for wavelengths from the ultraviolet into the infrared. Several of the important points are as follows:

- In the transparency range in the visible, the refractive index and the magnitude of its derivatives increase as the ultraviolet absorptions are approached from the visible. The signs of n and its lowest derivatives satisfy $n > 0$, $dn/d\lambda < 0$ (this means that the group velocity is less than the phase velocity) and $d^2n/d\lambda^2 > 0$. This behavior is termed *normal dispersion*.
- Within each ultraviolet absorption resonance, the index decreases with frequency ($dn/d\lambda > 0$). Note that in general there may be many discrete resonances in the ultraviolet; for simplicity only one is indicated in the sketch. Furthermore, on its short-wavelength side, each resonance gives a negative contribution to $d^2n/d\lambda^2$. This behavior, which is opposite to the normal dispersion observed in the visible, is termed *anomalous dispersion*.
- As the wavelength is increased from the visible into the infrared, normal dispersion behavior remains valid until infrared absorption resonances are approached. Close to such resonances anomalous dispersion may be observed. These absorptions are associated with motions of nuclei (e.g., vibrations) rather than electrons and typically occur somewhere in the range of a few to a few tens of micrometers. We have again indicated only a single infrared absorption in the sketch, although many resonances are usually present.
- Additional resonances not shown in the figure occur in the very short (x-ray) and very long (millimeter-wave) spectral regions. Normal dispersion behavior may again be observed in broad transparency ranges usually found, on the one hand, between the x-ray and ultraviolet resonances and, on the other hand, between the infrared and millimeter-wave resonances.

Some notes with respect to terminology are in order. In the classical optics terminology, normal and anomalous dispersion appear to refer specifically to the sign of $dn/d\lambda$, since this determines the order in which colors are spatially dispersed by a prism. However, in the ultrafast optics community, one is usually more interested in temporal dispersion, which in the case of material dispersion is governed by $d^2n/d\lambda^2$. It is therefore customary to utilize the terms *normal* and *anomalous material dispersion* to mean $d^2n/d\lambda^2 > 0$ or $d^2n/d\lambda^2 < 0$, respectively. The terms *positive* and *negative dispersion* are also often used; however, these can lead to confusion since they are used differently by different communities. Within ultrafast optics these terms usually refer to $\partial\tau/\partial\omega$, whereas in fiber optics they usually refer to D, which is proportional to $\partial\tau/\partial\lambda$.

In this book we use the terms *normal* and *positive dispersion* to refer to $\partial\tau/\partial\omega > 0$; *anomalous* and *negative dispersion* mean $\partial\tau/\partial\omega < 0$. For material dispersion this usage is equivalent to the definition in terms of $d^2n/d\lambda^2$; however, it is more general since it also applies to other sources of dispersion. We also sometimes refer directly to the signs of ψ_2, ψ_3, and so on. Here it is important to note that these signs depend explicitly on the definition of $\psi(\omega)$ in eq. (4.2), which we adopt consistently throughout this book. For normal dispersion our definition leads to $\psi_2 < 0$.

The refractive index of transparent materials in the visible spectral region is usually well approximated by a Sellmeier equation [96] of the form

$$n^2(\lambda) = 1 + \sum_k \frac{A_k\lambda^2}{\lambda^2 - \lambda_k^2} \tag{4.28}$$

Here each term in the summation arises from a different ultraviolet absorption. Usually, the refractive index behavior can be well approximated with only a few terms in the Sellmeier equation, sometimes only one. The specific form of the Sellmeier equation arises by neglecting damping terms associated with the resonances. This approximation, which is equivalent to assuming an infinitely sharp resonance peak, gives good results within the transparency range but is not valid very close to the absorption. One interesting consequence of eq. (4.28) is that for very long wavelengths, the refractive index approaches a constant greater than 1.

The similar form of the refractive index variation for a wide range of transparent materials leads to useful scaling laws connecting different derivatives of the refractive index. These scaling laws do not apply within absorption bands of colored media (e.g., solutions of organic dyes, semiconductor-doped glasses, etc.). As pointed out by [177], in the case of a single-term Sellmeier equation it can be shown that

$$\frac{d^2n}{d\lambda^2} = \frac{-3}{\lambda}\frac{dn}{d\lambda}\left[1 + \lambda\frac{dn}{d\lambda}\frac{3n^2+1}{3n(n^2+1)}\right] \tag{4.29}$$

The last term in the brackets is usually small; when it can be neglected, one obtains the very simple relation

$$\frac{d^2n}{d\lambda^2} \approx \frac{-3}{\lambda}\frac{dn}{d\lambda} \tag{4.30}$$

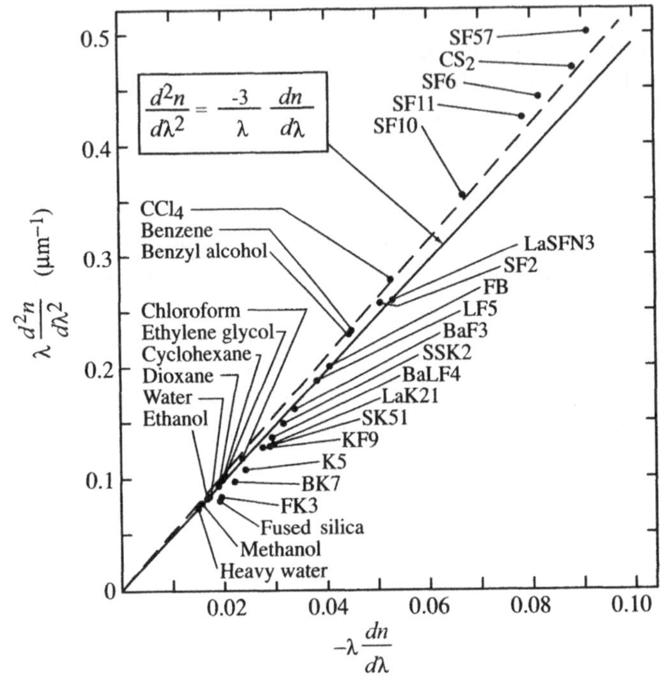

Figure 4.2 Second derivative of refractive index plotted against the first derivative for various materials at 620-nm wavelength. Solid line, eq. (4.30); dashed line, best linear fit. From [177].

The utility of this approximate expression is demonstrated in Fig. 4.2, which plots $\lambda(d^2n/d\lambda^2)$ vs. $\lambda(dn/d\lambda)$ for many different optical glasses and transparent solvents at a wavelength of 620 nm. The linear relation predicted is clearly observed, although the actual slope is slightly (ca. 7%) larger than predicted. The relatively small magnitude of $\lambda(dn/d\lambda)$ is also apparent from the graph, which means that at this wavelength the group velocity differs from the phase velocity by only a few percent for most transparent materials. Finally, the numerical values for $\lambda(d^2n/d\lambda^2)$ permit evaluation of the group velocity dispersion. For example, for fused silica, which is a relatively low dispersion material, $|\psi_2|$ at 620 nm is calculated to be approximately 50 fs^2 mm^{-1} (i.e., 50 fs^2 for a 1-mm piece of material).

Similar approximate expressions were also obtained relating higher-order derivatives of the refractive index. For example, further manipulation of the single-term Sellmeier equation gave the following result for the third derivative:

$$\frac{d^3n}{d\lambda^3} \approx \frac{-4}{\lambda}\frac{d^2n}{d\lambda^2} \tag{4.31}$$

A plot of actual index data confirmed this behavior, although again with a somewhat higher (ca. 13%) slope than predicted by eq. (4.31).

These relations allow one to compare the size of different terms containing derivatives of the refractive index. For example, from eqs. (4.22) and (4.31), one finds that ψ_3 is approximately proportional to $-d^2n/d\lambda^2$, which means that $\psi_3 < 0$ when these relations hold. Also, when eqs. (4.30) and (4.31) hold and $|\lambda(dn/d\lambda)| \ll 1$, one can easily show that

$$\left(\frac{dn}{d\lambda}\right)^2 \ll \frac{d^2n}{d\lambda^2} \quad \text{and} \quad \left|\frac{dn}{d\lambda}\frac{d^2n}{d\lambda^2}\right| \ll \left|\frac{d^3n}{d\lambda^3}\right| \tag{4.32}$$

These inequalities will be helpful in determining the dominant terms in the dispersion associated with prisms, discussed later in the chapter.

4.2 TEMPORAL DISPERSION BASED ON ANGULAR DISPERSION

As noted in Chapter 2, control of dispersion is key to obtaining short pulses from mode-locked solid-state lasers and also for a host of other applications discussed in later chapters, including femtosecond pulse compression and high-power chirped pulse amplifier systems. Two of the most common dispersion control devices—prism sequences and grating pairs—involve optical components with angular dispersion. In this section we first show that angular dispersion leads to temporal dispersion generally, a relationship first pointed out in [178]. We then show that angular dispersion also invariably leads to pulses with tilted intensity fronts. In later sections we examine how the dispersive properties of prism and grating pairs can be understood in terms of their angular dispersion.

4.2.1 Relation Between Angular and Temporal Dispersion

We write the field as a superposition of plane waves at different frequencies which we take to propagate in the x–z plane at angle $\theta(\omega)$ with respect to the z axis:

$$e(x, z, t) = \text{Re}\left\{\frac{1}{2\pi}\int d\tilde{\omega}\, A(\tilde{\omega})e^{j(\omega_0+\tilde{\omega})t}e^{-j\boldsymbol{\beta}(\omega)\cdot\mathbf{r}}\right\} \tag{4.33}$$

where

$$\boldsymbol{\beta}(\omega) = \frac{\omega n(\omega)}{c}\left[\hat{\mathbf{x}}\sin\theta(\omega) + \hat{\mathbf{z}}\cos\theta(\omega)\right] \tag{4.34}$$

The field after propagation to the $z = L$ plane can be rewritten

$$e(x, L, t) = \text{Re}\left\{\frac{e^{j\omega_0 t}}{2\pi}\int d\tilde{\omega}\, A(\tilde{\omega})e^{j\tilde{\omega}t}e^{-j\omega x n(\omega)\sin\theta(\omega)/c}e^{-j\omega L n(\omega)\cos\theta(\omega)/c}\right\} \tag{4.35}$$

For the special cases $(x, z) = (0, 0)$ and $(x, z) = (0, L)$, we have

$$e(0, 0, t) = \text{Re}\left\{\frac{e^{j\omega_0 t}}{2\pi}\int d\tilde{\omega}\, A(\tilde{\omega})e^{j\tilde{\omega}t}\right\} \tag{4.36}$$

and

$$e(0, L, t) = \mathrm{Re}\left\{\frac{e^{j\omega_0 t}}{2\pi}\int d\tilde{\omega}\, A(\tilde{\omega})e^{j\tilde{\omega}t}e^{-j\omega L n(\omega)\cos\theta(\omega)/c}\right\} \tag{4.37}$$

The frequency-dependent phase difference $\psi(\omega)$ between wavefronts passing through (0, 0) and (0, L), respectively, is given by

$$\psi(\omega) = \frac{-\omega L n(\omega)\cos\theta(\omega)}{c} \tag{4.38}$$

The phase depends on the angle of propagation with respect to a reference direction (here taken as $\hat{\mathbf{z}}$). A simple geometrical interpretation, sketched in Fig. 4.3a, is that $\psi(\omega)$ is proportional to the distance between the wavefronts passing through (0, 0) and (0, L), respectively. Equation (4.38) fully accounts for the frequency-dependent phase and hence the temporal dispersion arising from angular dispersion.

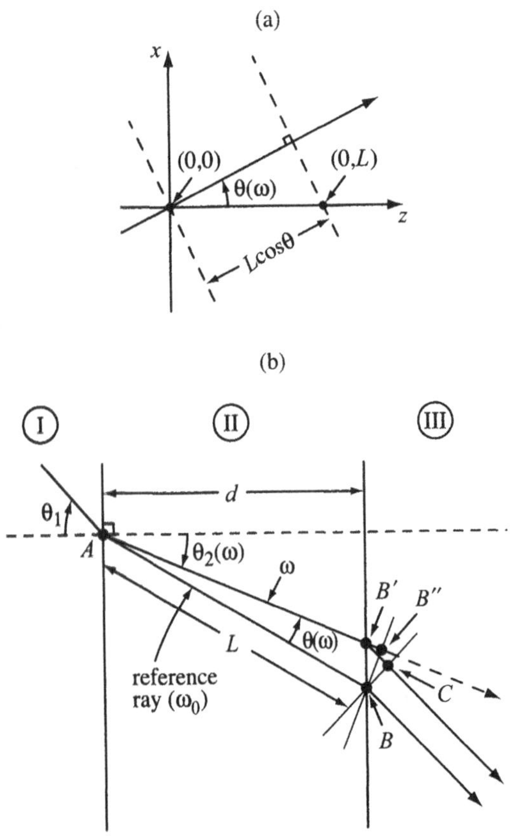

Figure 4.3 (a) Geometrical interpretation of eq. (4.38). $\psi(\omega)$ depends on the distance between the wavefronts passing through (0, 0) and (0, L), respectively. (b) Geometry for calculating the frequency-dependent phase arising in passing through a slab with angular dispersion.

Instead of an infinite medium, as above, let us now consider a uniform slab (Fig. 4.3b), as in the original derivation in [178]. A plane wave incident from an input medium (I) with index $n_1(\omega)$ propagates through the parallel slab (region II) with index $n_2(\omega)$ and then back into a final region (III) with index $n_3(\omega) = n_1(\omega)$ again. All the frequencies are assumed to have the same propagation direction in region I. We consider a ray that enters medium II at point A. The ray bends at this point due to the different refractive indices n_1 and n_2. For some particular reference frequency ω_0, the ray leaves medium II at point B. As the ray enters region III, it refracts again so that it is parallel to the input ray (since regions I and III have the same refractive index). The distance traveled by the reference ray in medium II (i.e., the length of $\overline{AB}$) is denoted L. Now let us discuss light at frequency ω, which travels at an angle $\theta(\omega)$ with respect to the reference ray. The ray leaves region II at point B' and refracts along the $\overline{B'C}$ direction, where it is once again parallel to the input ray. C is chosen so that for frequency ω it lies on a common phase front with point B. Also shown in the figure is a point B'', constructed such that $\overline{AB''}$ is collinear with $\overline{AB'}$ and perpendicular to $\overline{BB''}$. The reason for this construction becomes evident if we momentarily assume region III to be identical to region II [$n_3(\omega) = n_2(\omega)$]. In that case there would be no refraction at the II–III border, and B'' would represent the point along the $\overline{AB'}$ direction on a common phase front with B for frequency ω. The phase accumulated along $\overline{AB''}$ would simply be

$$\psi_{AB''}(\omega) = \frac{-\omega L n_2(\omega)\cos\theta(\omega)}{c} \tag{4.39}$$

Returning now to the actual case where $n_3(\omega) = n_1(\omega)$, we note that since point C is chosen to be on a common phase front with B for frequency ω, and since the length of $\overline{AB}$ remains constant as we vary the index of region III, the phase accumulated along path $AB'C$ must be the same as that accumulated hypothetically along AB'': namely,

$$\psi_{AB'C}(\omega) = \frac{-\omega L n_2(\omega)\cos\theta(\omega)}{c} \tag{4.40}$$

This is of the same form as for eq. (4.38), which means that the dispersion for a finite slab is the same as for the equivalent propagation length in an infinite medium.

We note that eq. (4.40) can also be derived by directly calculating the phases $\psi_{AB'}$ and $\psi_{B'C}$ accumulated along paths AB' and $B'C$. This approach is less elegant but indeed yields the same results (see Section 4.8.1).

We use these results to calculate the dispersion, starting from the relation

$$\frac{\partial\tau}{\partial\omega} = \frac{-\partial^2\psi(\omega)}{\partial\omega^2} \tag{4.41}$$

It is useful to perform the calculation in terms of the wavelength λ and path length P, where

$$\psi(\omega) = \frac{-\omega P(\omega)}{c} \tag{4.42}$$

Using eq. (4.13) and the chain rule, we obtain

$$\frac{\partial^2 \psi}{\partial \omega^2} = \frac{-\lambda^3}{2\pi c^2} \frac{\partial^2 P}{\partial \lambda^2} \tag{4.43a}$$

$$\frac{\partial \tau}{\partial \omega} = \frac{\lambda^3}{2\pi c^2} \frac{\partial^2 P}{\partial \lambda^2} \tag{4.43b}$$

and

$$\frac{\partial \tau}{\partial \lambda} = \frac{-\lambda}{c} \frac{\partial^2 P}{\partial \lambda^2} \tag{4.43c}$$

The dispersion parameter D defined in eq. (4.19) is given by

$$D = \frac{-\lambda}{cL} \frac{\partial^2 P}{\partial \lambda^2} \tag{4.44}$$

Equations (4.41) to (4.44) are true in general.

We now arrive at the dispersion of the slab by identifying the path length as

$$P = n_2 L \cos\theta \tag{4.45}$$

The derivatives are

$$\frac{\partial P}{\partial \lambda} = L\left\{\frac{dn_2}{d\lambda}\cos\theta - n_2 \sin\theta\left(\frac{d\theta}{d\lambda}\right)\right\} \tag{4.46}$$

$$\frac{\partial^2 P}{\partial \lambda^2} = L\left\{\frac{d^2 n_2}{d\lambda^2}\cos\theta - 2\frac{dn_2}{d\lambda}\sin\theta\left(\frac{d\theta}{d\lambda}\right) - n_2\cos\theta\left(\frac{d\theta}{d\lambda}\right)^2 - n_2\sin\theta\left(\frac{d^2\theta}{d\lambda^2}\right)\right\} \tag{4.47}$$

If we evaluate the dispersion along the reference direction, $\omega = \omega_0$ and $\theta = 0$, resulting in

$$\frac{\partial^2 P}{\partial \lambda^2} = L\left\{\frac{d^2 n_2}{d\lambda^2} - n_2\left(\frac{d\theta}{d\lambda}\right)^2\right\} \tag{4.48}$$

and

$$D = \frac{-\lambda}{c}\left\{\frac{d^2 n_2}{d\lambda^2} - n_2\left(\frac{d\theta}{d\lambda}\right)^2\right\} \tag{4.49}$$

The first term in eq. (4.49), proportional to $d^2 n_2/d\lambda^2$, is the material dispersion, which was discussed in Section 4.1. The second term, proportional to $(d\theta/d\lambda)^2$, arises from the angular dispersion. Since $(d\theta/d\lambda)^2$ is always greater than or equal to zero, angular dispersion provides a source of anomalous temporal dispersion.

This anomalous dispersion can in principle be used to compensate for normal material dispersion. However, for most transparent materials in the visible, the angular dispersion term is much smaller than the material dispersion term. Consider, for example, a fused silica slab at $\lambda = 600$ nm, for which $n_2 \approx 1.46$, $dn_2/d\lambda \approx -3 \times 10^{-5}$ nm^{-1}, and $d^2n_2/d\lambda^2 \approx 1.3 \times 10^{-7}$ nm^{-2}. If we assume a fixed incident angle θ_1, then

$$\sin\theta_1 = n_2 \sin\theta_2 \qquad \text{and} \qquad \frac{d\theta_2}{d\lambda} = \frac{-1}{n_2}\frac{dn_2}{d\lambda}\tan\theta_2 \tag{4.50}$$

For the case $\theta_2 = 30°$, $n_2\,(d\theta_2/d\lambda)^2 \approx 2 \times 10^{-10}$ nm^{-2}, which is more than 600 times smaller than $d^2n/d\lambda^2$. Therefore, further tricks are still needed to compensate for the material dispersion.

4.2.2 Angular Dispersion and Tilted Intensity Fronts

Before discussing specific dispersion control devices, we first treat another fundamental aspect of angular dispersion: namely, the introduction of a tilt into the pulse intensity fronts [179–181]. Returning to eq. (4.35), we can write the spectral phase as

$$\psi(x, z, \omega) = \psi_x(x, \omega) + \psi_z(z, \omega) \tag{4.51a}$$

where

$$\psi_x(x, \omega) = \frac{-\omega x n(\omega)\sin\theta(\omega)}{c} \qquad \text{and} \qquad \psi_z(z, \omega) = \frac{-\omega z n(\omega)\cos\theta(\omega)}{c} \tag{4.51b}$$

As before, we assume that the reference ray is along the $\hat{\mathbf{z}}$ direction, so that $\theta(\omega_0) = 0$. Expanding θ to first order, we have $\theta(\omega) \approx \tilde{\omega}\,(\partial\theta/\partial\omega)$, where $\tilde{\omega} = \omega - \omega_0$. Substituting for θ in the expression for ψ_x and retaining only first-order terms in $\tilde{\omega}$, we obtain

$$\psi_x(x, \omega) \approx \frac{-\omega_0 x n(\omega_0)\tilde{\omega}}{c}\frac{\partial\theta}{\partial\omega} \tag{4.52}$$

The frequency-dependent delay includes an x-dependent term given by

$$\tau_x(\omega) = \frac{-\partial\psi_x}{\partial\omega} \approx \frac{\omega_0 x n(\omega_0)}{c}\frac{\partial\theta}{\partial\omega} \tag{4.53}$$

The pulse envelope is delayed by an amount that is proportional to both the transverse position (x) and the angular dispersion.

This effect is illustrated in Fig. 4.4, in which the reference propagation vector is taken to be along the z axis [i.e., $\boldsymbol{\beta}(\omega_0) \sim \hat{z}$]. The contours of constant intensity experience a transverse position-dependent delay with respect to the phase fronts (contours of constant phase), which remain normal to the propagation vector $\boldsymbol{\beta}(\omega_0)$. The constant intensity contour corresponding to the peak of the pulse is usually called the *pulse front*. Thus, the pulse front

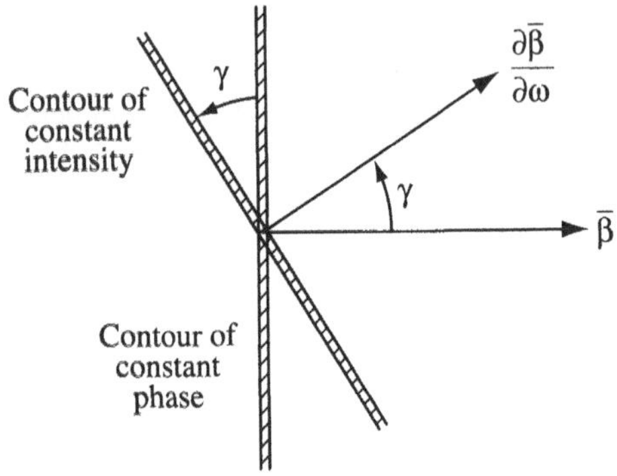

Figure 4.4 Tilt of the pulse fronts compared to the phase fronts.

is tilted with respect to the phase fronts. In free space the tilt angle is given by

$$\tan\gamma = \omega_0 \frac{\partial\theta}{\partial\omega} = -\lambda \frac{\partial\theta}{\partial\lambda} \tag{4.54}$$

whereas in a medium with a frequency-dependent index $n(\omega)$, the tilt angle is slightly modified to [181,182]

$$\tan\gamma = \frac{-\lambda n}{n_g} \frac{\partial\theta}{\partial\lambda} \tag{4.55}$$

Here n_g, the group index, is defined by $v_g = c/n_g$.

Another way of obtaining this result is to generalize equations (4.5) to (4.10) to include angular dispersion. This is achieved by associating each frequency with a plane wave propagating in a frequency-dependent direction. To first order in $\tilde{\omega}$, one finds that

$$e_{\text{out}}(\boldsymbol{r}, t) = \text{Re}\left\{ e^{j\left(\omega_0 t - \boldsymbol{\beta}(\omega_0)\cdot \boldsymbol{r}\right)}\, a_{\text{out}}(\boldsymbol{r}, t) \right\} \tag{4.56a}$$

where

$$a_{out}(\boldsymbol{r}, t) = \frac{1}{2\pi} \int d\tilde{\omega}\, A(\tilde{\omega})\, e^{j\tilde{\omega}\left[t - (\partial\boldsymbol{\beta}/\partial\omega)\cdot \boldsymbol{r}\right]} \tag{4.56b}$$

Clearly $\partial\boldsymbol{\beta}/\partial\omega$ gives the direction in which the pulse envelope propagates. We can obtain $\partial\boldsymbol{\beta}/\partial\omega$ by differentiating eq. (4.34) with respect to ω. For $\boldsymbol{\beta}(\omega_0)$ along $\hat{z}$, the result evaluated at frequency ω_0 is

$$\frac{\partial\boldsymbol{\beta}}{\partial\omega} = \frac{n_g}{c}\hat{z} + \frac{\omega_0 n}{c}\frac{\partial\theta}{\partial\omega}\hat{x} \tag{4.57}$$

The first term (along $\hat{z}$) is the same as that without angular dispersion. The second term arising from the angular dispersion adds a transverse component, which means that the direction in which the pulse front propagates is tilted with respect to the direction in which the phase front propagates. The tilt angle is identical to that given by eq. (4.55).

It is interesting to note that except for pulses with extremely large fractional bandwidths, the magnitude of the tilt angle is much larger than the variation in $\theta(\omega)$ across the optical bandwidth. Furthermore, for a given angular dispersion, the pulse tilt angle does not vary with propagation, as long as the plane-wave approximation holds. For finite-sized beams, this is valid provided that the propagation distance is less than the depth of focus and the spatial separation of the spectral components remains much less than the beam diameter. Additional effects that arise when the spatial separation of the spectral components is not small are analyzed in [180].

4.3 DISPERSION OF GRATING PAIRS

One way to allow the anomalous angular dispersion to overcome the material dispersion is to eliminate the material. This can be done by using a parallel pair of identical diffraction gratings in an arrangement introduced by Treacy [183]. The angular dispersion of the gratings remains, but the free-space region between the gratings contributes no material dispersion.

As shown in Fig. 4.5, the input beam is incident on the first grating at angle θ_i. The diffracted angle θ_D is a function of frequency, and the pulse front after diffraction from the first grating is tilted. After diffracting from the second grating, all the output rays are again parallel to the input beam, independent of frequency, and there is no longer any angular dispersion or pulse tilt. The distance between the gratings measured along the normal is denoted G, and the slant distance measured along the diffracted direction is b.

The dispersion is easily calculated by using eqs. (4.43c) and (4.49), which give

$$\frac{\partial \tau}{\partial \lambda} = \frac{\lambda b}{c}\left(\frac{d\theta_D}{d\lambda}\right)^2 \tag{4.58}$$

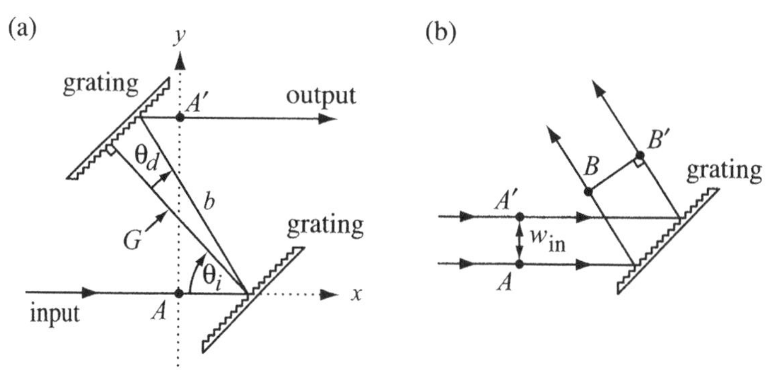

Figure 4.5 (a) Treacy grating pair: a pair of parallel gratings with normal separation G and slant distance b; (b) geometry for evaluating the path-length difference for a finite beam diffracting from a grating. θ_D is the diffraction angle measured with respect to the grating normal.

We have taken the index between the gratings to be a constant ($n_2 = 1$), and the distance L traveled within the "slab" is set to $L = b$. The angular dispersion is obtained from the well-known grating equation

$$\sin\theta_D = \sin\theta_i + \frac{m\lambda}{d} \tag{4.59}$$

where d is the periodicity of the grating rulings and m is the diffraction order. Usually, the gratings are used in the -1 order ($m = -1$), although for generality we retain m in the following expressions. The result obtained by differentiating eq. (4.59) is

$$\frac{d\theta_D}{d\lambda} = \frac{m}{d\cos\theta_D} \tag{4.60}$$

which gives

$$\frac{\partial^2 P}{\partial\lambda^2} = \frac{-m^2 b}{d^2\cos^2\theta_D} \tag{4.61}$$

The resulting dispersion is

$$\frac{\partial\tau}{d\lambda} = \frac{m^2\lambda b}{cd^2\cos^2\theta_D} \tag{4.62}$$

or expressed in terms of the phase,

$$\frac{\partial^2\psi}{\partial\omega^2} = \frac{m^2\lambda^3 b}{2\pi c^2 d^2\cos^2\theta_D} \tag{4.63}$$

Equations (4.62) and (4.63) are a key result which show that a parallel diffraction pair can indeed provide a tunable amount of anomalous dispersion. This device can be used to compensate for material dispersion by adjusting the grating separation to provide a dispersion equal (and opposite!) to that accumulated in propagating through material. For positively chirped input pulses, the grating pair can be used to compress these pulses to the bandwidth limit, provided that the grating separation is chosen properly. For bandwidth-limited input pulses, passage through the grating pair leads to a negative chirp on the output pulse.

It is instructive to note that the dispersion can also be obtained by calculating the delay directly, as in the original Treacy paper. Here we compute the delay between points A and A' where the input and output rays, respectively, cross the $x = 0$ plane. The expression is

$$\tau = \frac{b}{c}\left[1 + \cos\left(\theta_{\text{in}} + \theta_D\right)\right] \tag{4.64a}$$

where

$$b = G\sec\theta_D \tag{4.64b}$$

By careful differentiation of eq. (4.64a), remembering that θ_D and b are both functions of wavelength, one arrives at the same result as in eq. (4.62), although considerably more algebra is required. The main merit of the latter derivation is that it clearly links the dispersion to the different physical paths traveled by different frequencies.

When one wishes to compensate a chirp over a wide bandwidth, the third-order dispersion is also of importance. From eq. (4.43a), we obtain

$$\frac{\partial^3 \psi}{d\omega^3} = \frac{\lambda^4}{4\pi^2 c^3}\left(3\frac{\partial^2 P}{\partial\lambda^2} + \lambda\frac{\partial^3 P}{\partial\lambda^3}\right) \tag{4.65}$$

For our case of parallel diffraction gratings, one finds that

$$\frac{\partial^3 \psi}{\partial\omega^3} = \frac{-3m^2\lambda^4 b}{4\pi^2 c^3 d^2 \cos^2\theta_D}\left[1 + \frac{m\lambda}{d}\frac{\sin\theta_D}{\cos^2\theta_D}\right] \tag{4.66a}$$

$$\frac{\partial^3 \psi}{\partial\omega^3} = \frac{-\partial^2\psi}{\partial\omega^2}\frac{3\lambda}{2\pi c}\left[1 + \frac{m\lambda}{d}\frac{\sin\theta_D}{\cos^2\theta_D}\right] \tag{4.66b}$$

If one assumes a positively chirped input pulse with a purely quadratic spectral phase (i.e., a linearly chirped pulse), the third-order phase term must be small for the gratings to provide high-quality pulse compression. Specifically, we require that

$$\frac{1}{6}\frac{\partial^3\psi}{\partial\omega^3}(\Delta\omega)^3 \ll \pi \tag{4.67}$$

where $\Delta\omega$ is the optical bandwidth. Furthermore, we see from eq. (4.66b) that for fixed $\Delta\omega$, there is a maximum second-order dispersion which the gratings can provide subject to the is third-order dispersion limit.

The third-order dispersion from the gratings always has the opposite sign from that of the second-order dispersion. This is demonstrated by using the grating equation, eq. (4.59), to show that the last factor in eq. (4.66b) can be rewritten as

$$1 + \frac{m\lambda}{d}\frac{\sin\theta_D}{\cos^2\theta_D} = \frac{1 - \sin\theta_i \sin\theta_D}{\cos^2\theta_D} \tag{4.68}$$

which is always positive. On the other hand, the third-order dispersion of transparent materials in the visible usually has the same sign as the (normal) second-order material dispersion. This means that grating pairs cannot compensate simultaneously for both second- and third-order material dispersion; although the second-order dispersions can be made to cancel, the third-order dispersions will add [184,185].

It is also instructive to consider the maximum differential time delay that a grating pair can provide: Given a bandwidth-limited input pulse, what is the longest chirped output pulse that one can produce? Or alternatively, given a positively chirped input pulse that one wishes to compress using the grating pair, what is the longest input pulse duration that the grating pair can accommodate?

The maximum differential time delay is limited by the fact that different optical frequency components separate as they pass through the grating pair. Let us postulate that the largest spatial separation we will tolerate over a bandwidth $\Delta\lambda$ is equal to the radius of the input

Gaussian beam w_{in}. We will calculate the implications of this postulate for the maximum differential time delay $\Delta\tau$. In the following we now specialize to diffraction into the negative first order ($m = -1$).

With respect to Fig. 4.5a, the y coordinate of the output ray is given by

$$y = b \sin(\theta_i + \theta_D) \tag{4.69}$$

where y, b, and θ_D are all functions of frequency (wavelength). The spatial separation Δy between two output rays at different wavelengths separated by $\Delta\lambda$ is

$$|\Delta y| = \left| \frac{\partial y}{\partial \lambda} \Delta\lambda \right| = \frac{b \cos\theta_i \Delta\lambda}{d \cos^2\theta_d} \tag{4.70}$$

where we have used eqs. (4.60) and (4.64b). This equation is valid provided that the wavelength range is sufficiently narrow that the first-order Taylor series expansion is valid. Since we require that $\Delta y \leq w_{\text{in}}$, we have

$$b \leq \frac{w_{\text{in}} d \cos^2\theta_d}{\cos\theta_i \Delta\lambda} \tag{4.71}$$

Using this expression in eq. (4.62), we find the maximum time delay:

$$\Delta\tau \leq \frac{\lambda w_{\text{in}}}{cd \cos\theta_{\text{in}}} \tag{4.72}$$

Remarkably, the bandwidth $\Delta\lambda$ has dropped out of this expression, so the maximum time delay before the spectrum separates by a beam radius depends only on the beam size, the grating period, the wavelength, and the incident angle on the grating.

We can present a simple physical interpretation for this result in terms of the physical delay associated with diffraction of a finite beam from a single grating (see Fig. 4.5b). The delay difference between the two paths $A'B'$ and AB, representing two parallel input rays separated by distance w_{in}, is given by

$$\Delta\tau = \frac{|A'B'| - |AB|}{c} = \frac{w_{\text{in}}}{c} \tan\theta_i - \frac{w_{\text{in}}}{c} \frac{\sin\theta_D}{\cos\theta_i} \tag{4.73}$$

Using the grating equation (4.59), we find that

$$\Delta\tau = \frac{\lambda w_{\text{in}}}{cd \cos\theta_{\text{in}}} \tag{4.74}$$

This yields the same result as eq. (4.72)! Thus, the maximum differential delay that the grating pair can provide without spatial separation of the optical frequencies is set by the physical delay across the output diffracted beam. This physical delay is the same effect as that discussed in Section 4.2.2 leading to pulse front tilt. In fact, since the beam radius w_D

after diffraction from the first grating is related to the input radius by

$$w_D = \frac{w_{\text{in}} \cos\theta_D}{\cos\theta_{\text{in}}} \tag{4.75}$$

the delay can be rewritten

$$\Delta\tau = \frac{\lambda w_D}{cd\cos\theta_D} \tag{4.76}$$

This is exactly the amount of delay across a transverse distance w_D predicted by eq. (4.53) for the special case of a grating.

For another closely related interpretation of eq. (4.72), we note that λ/c is the time associated with one optical period. $w_{\text{in}}/d\cos\theta_{\text{in}}$ is the number of grating periods found within one beam radius on the grating. Therefore, the maximum value for $\Delta\tau$ without the beam physically separating is given by the number of grating periods intercepted by the optical beam multiplied by the optical period.

The discussion above refers to a single pass through a Treacy grating pair. To get around these spatial separation effects, a double-pass arrangement is often used [186]. A flat mirror is used to reflect the output rays of the grating pair in Fig. 4.5a back through the grating pair. This causes all the rays to retrace their paths and recombine into a single collimated output beam with no spatial separation. The output is counter-propagating with respect to the input, from which it is usually separated by a slight vertical tilt in the retroreflecting mirror. The total dispersion is now twice that in eq. (4.63) and is limited by the size of the second grating instead of the beam size as before.

Gratings can also be used profitably together with internal lenses, as sketched in Fig. 4.6 [178,187] . Here a pair of lenses L_1 and L_2 with focal lengths $f_1 = f_2 = f$ are separated by distance $f_1 + f_2 = 2f$ to form a unity magnification telescope. The second grating G_2 is located a distance l_2 beyond the focal plane of lens L_2. Similarly, the first, grating (G_1) can be located a distance l_1 before the front focal plane of lens L_1, although for now we consider $l_1 = 0$ as indicated in the figure. To be clear, the distance between grating G_i and lens L_i is $l_i + f_i$. The inclination of G_1 and G_2 are equal and opposite with respect to the symmetry plane indicated in the figure, which resets the angular dispersion after G_2 to zero.

We can understand the dispersion of this system based on the following simple argument [178]. Consider propagation by a point source from the front focal plane of L_1 to the back focal plane of L_2. We do not consider the gratings yet. Two representative rays 1 and 2

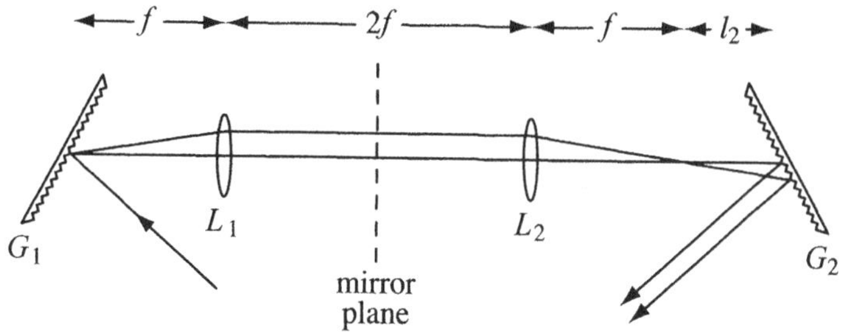

Figure 4.6 Grating pair with internal lenses.

are indicated in the figure. For a given ω and for ideal nondispersive thin lenses, the phase accumulated along paths 1 and 2 are the same, since these paths originate at a common point and end at a common point. Since the phases are the same, the path lengths $P_1(\omega)$ and $P_2(\omega)$ along the two paths are also identical: $P_1(\omega) = P_2(\omega)$. Now consider two frequencies ω_1 and ω_2 incident along a common ray onto G_1, then diffracted so that ω_1 follows path 1 and ω_2 follows path 2. The path lengths, which we have already shown to be identical, do not change as the frequency is changed; therefore, $P_1(\omega_1) = P_2(\omega_2)$. This means that the propagation from focal plane to focal plane is dispersionless. However, beyond the focal planes there is still angular dispersion, which we have learned also gives temporal dispersion. The portion of the grating separation that contributes to temporal dispersion is simply $l_1 + l_2$. The resulting dispersion is equal to

$$\frac{\partial^2 \psi}{\partial \omega^2} = \frac{\lambda^3 (l_1 + l_2)}{2\pi c^2 d^2 \cos^2 \theta_D} \tag{4.77}$$

where we have set $m = -1$.

The dispersion can be controlled by varying the grating separation. For $l_1 + l_2 > 0$, the system gives anomalous dispersion. However, for $l_1 + l_2 < 0$, the dispersion becomes normal, which cannot be achieved with the usual Treacy grating pair. The maximum normal dispersion that can be realized is obtained for $l_1 + l_2 = -2f$ (i.e., the gratings are placed directly adjacent to the lenses). The use of such grating–lens pairs to provide normal dispersion is very important in chirped pulse femtosecond amplifiers, discussed in Chapter 8. For $l_1 + l_2 = 0$, passage through the grating–lens pair imparts zero net dispersion. This configuration forms the basis for the femtosecond pulse shaper (also discussed in Chapter 8), which makes use of the spatial separation of different optical frequencies midway between the two lenses. An alternative treatment of the dispersion properties of such grating–lens pairs, based on a diffraction analysis, is given in Chapter 8 as well.

Finally, we note that similar grating pairs with internal telescopes with nonunity magnification (i.e., $f_1 \neq f_2$) are also possible [188]. The angular dispersion after lens L_2 is increased by the factor f_1/f_2, which means that the dispersion accumulated in a single pass through the system is given by

$$\frac{\partial^2 \psi}{\partial \omega^2} = \frac{\lambda^3 (l_1 + f_1^2 l_2 / f_2^2)}{2\pi c^2 d^2 \cos^2 \theta_D} \tag{4.78}$$

The major merit of this system is that with angular magnification, a large amount of dispersion can be obtained in a system of reduced physical size. Gratings G_1 and G_2 must now have different orientations or ruling periods (d) in order to cancel the angular dispersion after G_2. For practical reasons it is simpler instead to replace G_2 with a corner-type reflector. With the spatial inversion provided by the folding mirror setup, this becomes equivalent to an appropriately oriented pair of identical gratings separated by a double-telescope system.

4.4 DISPERSION OF PRISM PAIRS

Prisms are another angular dispersion device commonly used for dispersion control in ultrafast optics. Compared to gratings, their main advantages are low loss, which makes them

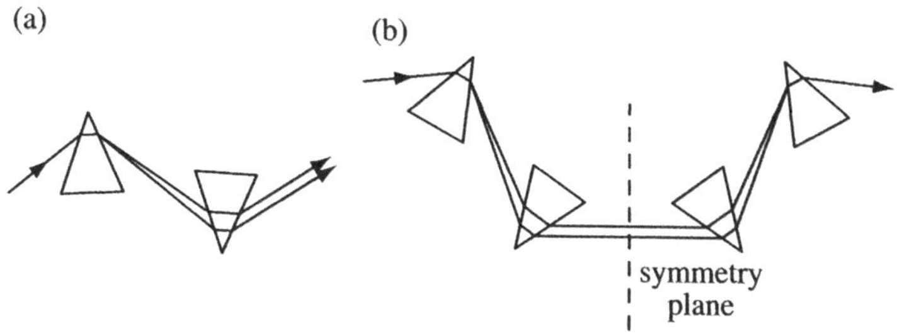

Figure 4.7 Prism sequence for dispersion control: (a) two-prism sequence; (b) four-prism sequence.

suitable for dispersion control inside laser resonators, as well as easy tuning of the sign and magnitude of the dispersion. The main limitation is relatively low angular dispersion, which increases the required prism spacings and restricts the magnitude of temporal dispersion that prism sequences can provide. However, for pulses 100 fs in duration and below, prisms provide sufficient dispersion for most applications.

Figure 4.7 shows the prism configurations most commonly used in ultrafast optics, [178,189]. Figure 4.7a shows a configuration consisting of a pair of anti-parallel prisms. The various frequencies in the collimated input beam emerge from the first prism with different angles but again travel in parallel directions after the second prism. The angular dispersion between the prisms contributes anomalous dispersion, while passage through the prism material is a source of normal dispersion. Unlike the single-slab case, most of the angular dispersion occurs in a material-free region, and therefore the anomalous dispersion can exceed the normal material dispersion. The overall dispersion can be tuned by translating one of the prisms in a direction perpendicular to its base, which varies the material path length without affecting the angular dispersion.

The various optical frequencies experience some transverse spatial separation after the two-prism sequence, which is usually undesirable. This can be rectified by using a four-prism sequence, consisting of two prism pairs arranged about a central symmetry plane (Fig. 4.7b). The second prism pair reverses the transverse separation caused by the first while doubling the temporal dispersion. An equivalent result is obtained with only a single prism pair by placing a mirror at the symmetry plane position, which sends the parallel but spatially dispersed optical frequencies back through the first two prisms.

Before analyzing the dispersion of a prism pair, we briefly discuss the operation of a single prism of index $n(\lambda)$ [96,176]. As depicted in Fig. 4.8, the prism apex angle is α and the incident and output angles are ϕ_1 and ϕ_2. The corresponding interior angles are ϕ_1' and ϕ_2', respectively. The angle of deviation, defined as the angle between the emergent and incident rays, is denoted ϵ. From geometry these angles are related by

$$\phi_1 + \phi_2 = \epsilon + \alpha \qquad \text{and} \qquad \phi_1' + \phi_2' = \alpha \tag{4.79}$$

Furthermore from Snell's law we have

$$\sin\phi_k = n \sin\phi_k' \tag{4.80}$$

where k equals either 1 or 2. From these relations it can be shown that ϵ has a minimum value as a function of ϕ_1, which occurs when $\phi_1 = \phi_2$ (i.e., when the passage of the rays

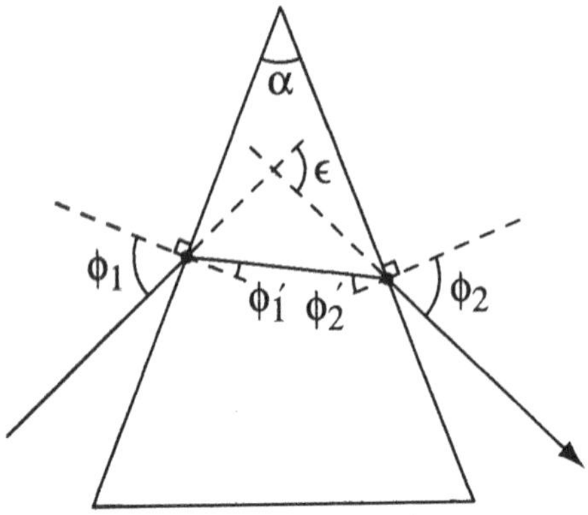

Figure 4.8 Geometry for describing passage of a light ray through a prism.

through the prism is symmetrical). Experimentally, this minimum-deviation phenomenon can be observed by rotating the prism about an axis perpendicular to the plane of the paper; the deviation angle first decreases to a minimum value $\epsilon_{\min}$ and then increases again as the prism is further rotated. Historically, measurements of $\epsilon_{\min}$ performed as a function of optical frequency were used to determine the wavelength-dependent refractive index of prism materials [96]. For the dispersion-compensation application, using the prisms at minimum deviation has the advantage of no astigmatism, since the input and output beam diameters are the same. Furthermore, when $\phi_1 = \phi_2$, the prism apex angle can be chosen to satisfy Brewster's angle for both input and output beams simultaneously, which is important to realize low loss. The Brewster's angle condition is written

$$\tan\phi_1 = \tan\phi_2 = n \tag{4.81}$$

For our dispersion analysis we need to know the derivatives of the output angle (ϕ_2) with respect to wavelength. These are given as follows:

$$\frac{\partial\phi_2}{\partial\lambda} = \frac{\partial\phi_2}{\partial n}\frac{dn}{d\lambda} \tag{4.82}$$

and

$$\frac{\partial^2\phi_2}{\partial\lambda^2} = \frac{\partial^2\phi_2}{\partial n^2}\left(\frac{dn}{d\lambda}\right)^2 + \frac{\partial\phi_2}{\partial n}\frac{d^2n}{d\lambda^2} \tag{4.83}$$

Using eqs. (4.79) and (4.80), we can write

$$\frac{\partial\phi_2}{\partial n} = (\cos\phi_2)^{-1}\left[\sin\phi_2' + \cos\phi_2' \tan\phi_1'\right] \tag{4.84}$$

$$\frac{\partial^2\phi_2}{\partial n^2} = \tan\phi_2\left(\frac{\partial\phi_2}{\partial n}\right)^2 - \frac{\tan^2\phi_1'}{n}\left(\frac{\partial\phi_2}{\partial n}\right) \tag{4.85}$$

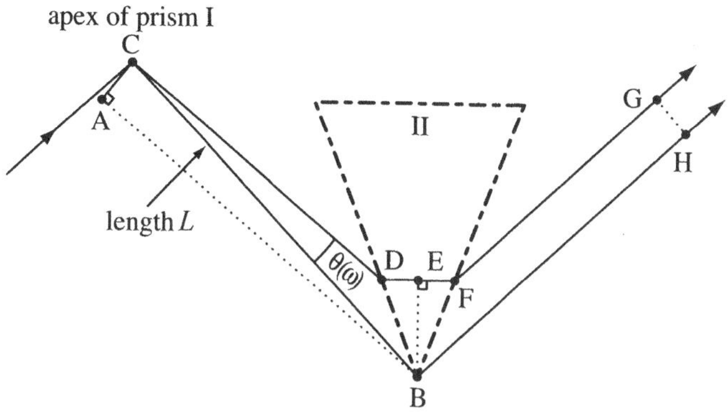

Figure 4.9 Construction for calculating the dispersion of a prism sequence. Adapted from [189].

When the prism is used at minimum deviation and Brewster's angle [i.e., eq. (4.81) applies], these expressions simplify to yield

$$\frac{\partial \phi_2}{\partial n} = 2 \tag{4.86}$$

$$\frac{\partial^2 \phi_2}{\partial n^2} = 4n - \frac{2}{n^3} \tag{4.87}$$

The dispersion is most easily calculated following an elegant construction from [189], which is indicated in Fig. 4.9. Ray CB represents the reference ray of length L, which connects the apices of prisms I and II. The choice of a reference ray wholly in air simplifies the calculation, as we will see below. The center rays of the various optical frequencies make an angle $\theta(\omega)$ with respect to the reference ray. We wish to determine the path length of path CDE, denoted P_{CDE}. The desired result can be found by inspection if we note that both AC and BE are possible wavefronts, where point A is chosen such that AB is parallel to CD and AC is perpendicular to CD.[1] The result is

$$P_{CDE} = P_{AB} = L\cos\theta \tag{4.88}$$

Similarly, since output rays BH and FG are parallel and since BE and HG are possible wavefronts, then P_{EFG} is equal to P_{BH}, which is frequency independent. This result is independent of θ. Therefore, we conclude that there is no temporal dispersion associated with path EFG, and the full dispersion of a single prism pair can be calculated from eq. (4.88). For a four-prism sequence, the result should be multiplied by a factor of 2.

[1] This applies for any single frequency ω and when, for purposes of the construction, prism I is temporarily removed so that AB is entirely in air.

The dispersion is computed from eq. (4.47) with n_2 set equal to 1, which gives the following:

$$\frac{\partial^2 P}{\partial\lambda^2} = -L\left\{\cos\theta\left(\frac{d\theta}{d\lambda}\right)^2 + \sin\theta\left(\frac{d^2\theta}{d\lambda^2}\right)\right\} \tag{4.89}$$

The dispersion follows by substituting in the results from eqs. (4.82) to (4.85) and using (4.43a). It should be noted that θ is defined in the opposite sense compared to ϕ_2, so that

$$\frac{\partial\theta}{\partial\lambda} = \frac{-\partial\phi_2}{\partial\lambda} \tag{4.90}$$

In the special case of minimum deviation and Brewster's angle, the result is

$$\frac{\partial^2\psi}{\partial\omega^2} = \frac{\lambda^3}{2\pi c^2}\left\{4L\cos\theta\left(\frac{dn}{d\lambda}\right)^2 - 2L\sin\theta\left[\frac{d^2 n}{d\lambda^2} + \left(2n - \frac{1}{n^3}\right)\left(\frac{dn}{d\lambda}\right)^2\right]\right\} \tag{4.91}$$

We emphasize that this expression applies to a single prism pair.

To evaluate this result, we note that in the usual case where θ is small, we can set $L\cos\theta \approx L$. Therefore, the first term in eq. (4.91) represents anomalous temporal dispersion due to the angular dispersion of the first prism. Its magnitude can be increased by increasing the apex-to-apex prism spacing. To interpret the second term, we note that $L\sin\theta$ is approximately equal to the distance between point D (the point where the central ray intersects prism II) and the reference ray. Therefore, the minimum value of θ is determined by the requirement that the entire beam should pass through the prisms without clipping. We also note that the distance which the central ray travels in the prism can be shown to satisfy $|DF| = 2L\sin\theta$ (again assuming a Brewster prism at minimum deviation). Therefore, the quantity $-2L\sin\theta(d^2n/d\lambda^2)$, which is usually dominant in the second term in eq. (4.91), is simply the material dispersion arising due to a length $2L\sin\theta$ of material in prism II. Since in our construction we have not yet included the material dispersion of prism I, we can simply increase the value of $L\sin\theta$ used in evaluating eq. (4.91) to account also for the material dispersion of prism I. Thus, the minimum value of $L\sin\theta$ will be about two beam diameters (one beam diameter for prism I, one beam diameter for prism II). The material dispersion term can be varied by translating either or both of the prisms perpendicular to their bases, which changes the amount of material in the beam (i.e., the value of $L\sin\theta$). For suitable values of L, varying the amount of material in the beam allows the overall dispersion to be tuned from the anomalous dispersion regime through zero and into the normal dispersion regime.

It is also important to evaluate the third-order dispersion of the prism pair, which from eq. (4.65) is given by

$$\frac{\partial^3\psi}{d\omega^3} = \frac{\lambda^4}{4\pi^2 c^3}\left(3\frac{\partial^2 P}{\partial\lambda^2} + \lambda\frac{\partial^3 P}{\partial\lambda^3}\right)$$

By differentiating eq. (4.89), we obtain

$$\frac{\partial^3 P}{\partial\lambda^3} = L\left\{\sin\theta\left(\frac{\partial\theta}{\partial\lambda}\right)^3 - 3\cos\theta\left(\frac{\partial\theta}{\partial\lambda}\right)\frac{\partial^2\theta}{\partial\lambda^2} - \sin\theta\left(\frac{\partial^3\theta}{\partial\lambda^3}\right)\right\} \tag{4.92}$$

where

$$\frac{\partial^3\theta}{\partial\lambda^3} = \frac{\partial^3\theta}{\partial n^3}\left(\frac{dn}{d\lambda}\right)^3 + 3\frac{\partial^2\theta}{\partial n^2}\left(\frac{dn}{d\lambda}\right)\frac{d^2n}{d\lambda^2} + \frac{\partial\theta}{\partial n}\left(\frac{d^3n}{d\lambda^3}\right) \tag{4.93}$$

Substituting for the minimum deviation–Brewster case and keeping only the dominant terms (where according to Section 4.1.2 we assume $|d^3n/d\lambda^3| \gg |(dn/d\lambda)(d^2n/d\lambda^2)| \gg |dn/d\lambda|^3$), we obtain

$$\frac{\partial^3 P}{\partial\lambda^3} \approx -12L\cos\theta\left(\frac{dn}{d\lambda}\right)\frac{d^2n}{d\lambda^2} + 2L\sin\theta\left(\frac{d^3n}{d\lambda^3}\right) \tag{4.94}$$

The final expression for the third-order dispersion becomes

$$\frac{\partial^3\psi}{\partial\omega^3} \approx \frac{-\lambda^4}{\pi^2c^3}\left\{3L\cos\theta\left[\left(\frac{dn}{d\lambda}\right)^2 + \lambda\frac{dn}{d\lambda}\frac{d^2n}{d\lambda^2}\right]\right.$$
$$\left. -\frac{3}{2}L\sin\theta\left[\frac{d^2n}{d\lambda^2} + \frac{\lambda}{3}\frac{d^3n}{d\lambda^3}\right]\right\} \tag{4.95}$$

To gain insight into the sign of third-order dispersion, we make use of the approximate expressions from Section 4.1.2 relating different derivatives of the refractive index in the normal material dispersion regime. Using these relations we find that when normal dispersion behavior holds, the third-order dispersion is given very roughly by

$$\frac{\partial^3\psi}{\partial\omega^3} \approx \frac{\lambda^4}{\pi^2c^3}\left\{6L\cos\theta\left(\frac{dn}{d\lambda}\right)^2 - \frac{1}{2}L\sin\theta\left(\frac{d^2n}{d\lambda^2}\right)\right\} \tag{4.96}$$

The key point is that for a sufficiently large prism separation, the signs of both ψ_2 and ψ_3 are positive. The signs of the second- and third-order dispersion for prisms, gratings, and material dispersion are summarized in Table 4.1. Using a combination of grating pairs *and* prism pairs together offers the possibility of compensating for second- and third-order material dispersion simultaneously [190,191]. This can be very important when dealing with pulses below a few tens of femtoseconds, as well as in some cases when longer pulses encounter very large amounts of dispersion.

Measured dispersion data for a pair of SF-glass Brewster angle prisms, obtained using the white light interferometry method described in Section 4.7.1, are shown in Fig. 4.10. Three different curves are shown, corresponding to different amounts of glass in the beam path. Several points consistent with our expectation are evident:

Table 4.1 Typical Signs of Second- and Third-Order Dispersion for Material Dispersion and for Dispersion Arising from Grating Pairs and Prism Pairs

	Material	Gratings	Prisms
$\frac{\partial^2 \psi}{\partial \omega^2}$	−	+	+
$\frac{\partial^3 \psi}{\partial \omega^3}$	−	−	+

1. Each curve exhibits a frequency range where the delay is flat, which occurs when the material dispersion and angular dispersion contributions to ψ_2 are exactly balanced.
2. The large curvature of the delay vs. frequency curves arises from third-order dispersion. The *downward* curvature corresponds to $\partial^2\tau/\partial\omega^2 < 0$ ($\psi_3 > 0$).
3. For all of the curves, anomalous dispersion ($\partial\tau/\partial\omega < 0$, $\psi_2 > 0$) is achieved at sufficiently short wavelengths.
4. For any particular frequency, inserting additional glass increases the net group delay and adds a positive contribution to the slope, which corresponds to additional normal dispersion ($\partial\tau/\partial\omega > 0$) from the extra material.

Finally, we note that as an alternative to the analytical methods used in this section, the dispersion of prisms can also be calculated on a computer using a technique known as *dispersive ray tracing*. This is especially helpful when the approximations and assumptions used above do not hold. As the name implies, in ray tracing the paths of individual rays are traced through the system, with Snell's law applied at every surface. This technique is used heavily in optical design and is discussed in standard optics texts (e.g., [176]). In

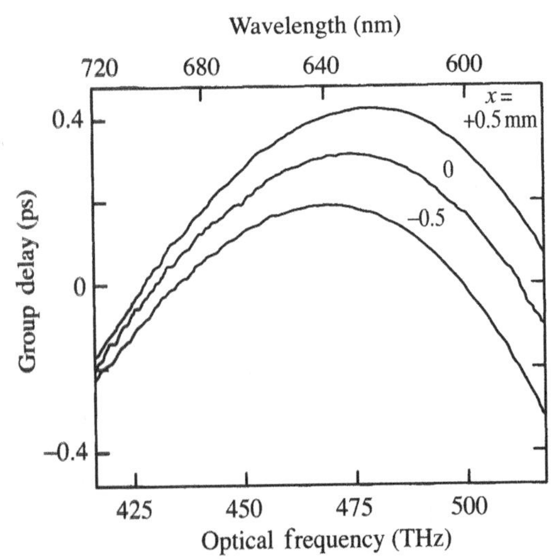

Figure 4.10 Group delay vs. frequency data for an SF-glass prism pair. The prism insertion into the beam is characterized by a relative translation (x) perpendicular to the prism base, with larger x corresponding to more glass in the beam. From [97].

dispersive ray tracing, ray-tracing calculations are repeated individually for many different optical frequencies of interest, keeping track of the optical path length for each frequency. This yields numerical data for $P(\lambda)$ and hence the full dispersion information. Examples of dispersive ray-tracing applied to prism sequences were reported in [192,193]. This technique can also be applied to other dispersive systems, such as those based on gratings.

4.5 DISPERSIVE PROPERTIES OF LENSES

We now turn our attention to the dispersive properties of lenses. In a ray optics picture, a lens achieves focusing by refracting light to different angles as a function of radial position. Since lenses are made of dispersive materials, the angle by which light is refracted is a function of both radius and wavelength. This radially dependent angular dispersion leads to new temporal dispersion effects. The first treatment of the dispersion of a lens was performed by Bor [194], who analyzed the time delay through a lens as a function of transverse position in a ray optics picture. The analysis revealed a radially varying delay connected to the chromatic aberration of the lens. The same conclusion was later obtained by Kempe et al. [195] via an electromagnetic diffraction analysis. Here we derive the dispersive properties of a lens in a different way, drawing on the relationship between angular dispersion and pulse front tilt.

The geometry for our derivation is shown in Fig. 4.11. We assume the case of a thin lens, which means that the input rays are not displaced appreciably in traversing the lens, as well as the paraxial limit. For a planar input phase front, the lens produces a curved output phase front, which then converges to the focus. For this derivation we assume the focus occurs at $z = 0$. Denoting the angle that the ray passing through x to the focus makes with the optical axis ($\hat{\mathbf{z}}$) as ϕ, we have

$$\tan\phi(x, z, \lambda) = -\frac{x}{z} \tag{4.97}$$

Here x and z are the coordinates of a point on the phase front, and the minus sign is valid for $-f < z < 0$ [i.e., in the region between the lens ($z = -f$) and the focus ($z = 0$)]. From

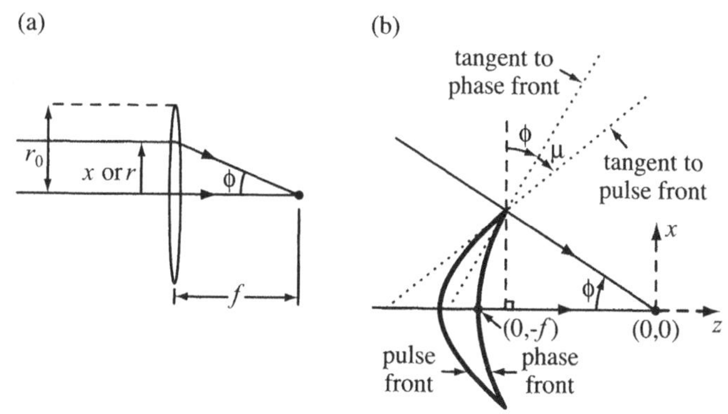

Figure 4.11 (a) Geometry for calculations of short-pulse focusing through a thin lens; (b) phase and pulse fronts after passing through a thin lens.

Fig. 4.11b, the slope of the line tangent to the phase front is $1/\tan\phi$, which means that the phase front is described by

$$\frac{dx}{dz} = \frac{1}{\tan\phi} \tag{4.98}$$

To solve for the phase front at the lens, we set $z = -f$ in eq. (4.97) and insert into eq. (4.98). The solution for the phase front is

$$z = \frac{x^2}{2f} - f \tag{4.99}$$

where we have picked the constant of integration so that the curve passes through $(0, -f)$. This result is true in the paraxial limit ($|x| \ll f$).

We now allow the focal length f and angle ϕ to be wavelength dependent: $f = f(\lambda)$ and $\phi = \phi(\lambda)$. From Section 4.2.2 we know that the pulse front is tilted with respect to the phase front by an angle μ satisfying

$$\tan\mu = -\lambda\frac{\partial\phi}{\partial\lambda} \tag{4.100}$$

In analogy with eq. (4.98), in the paraxial limit the pulse front at the lens ($z = -f$) is described by

$$\frac{dx}{dz} = \frac{1}{\tan(\phi+\mu)} \approx \frac{f}{\left[1 - f\lambda(\partial/\partial\lambda)(1/f)\right] x} = \frac{f^2}{(f + \lambda(\partial f/\partial\lambda))x} \tag{4.101}$$

which has the solution

$$z = \frac{1}{2f}\left[1 - f\lambda\frac{\partial}{\partial\lambda}\left(\frac{1}{f}\right)\right]x^2 + \mathcal{K} = \frac{1}{2f^2}\left(f + \lambda\frac{\partial f}{\partial\lambda}\right)x^2 + \mathcal{K} \tag{4.102}$$

where $\mathcal{K}$ is a constant of integration.

Since the phase difference between input and output phase fronts is independent of transverse position, the relative delay in passing through the lens is proportional to the difference in the z positions of the phase and pulse fronts. Since the problem is cylindrically symmetric, we switch now from x to r to represent the transverse coordinate. The difference in delay between radius r and a reference radius r_0 is

$$T(r) - T(r_0) = \frac{r_0^2 - r^2}{2cf^2}\lambda\frac{\partial f}{\partial\lambda} \tag{4.103}$$

If we choose r_0 to represent the marginal ray passing through the edge of the lens and assume as in Fig. 4.11 that the lens has zero thickness at this point, then the pulse front is coincident with the phase front at r_0, since the phase and group velocities of free space are the same. In this case $T(r_0) = 0$ in eq. (4.103), and $T(r)$ gives directly the delay of the pulse front relative to the phase front.

Equation (4.103), which is valid for single-element as well as compound lenses, says that a pulse experiences a radially varying delay in passing through a thin lens. This situation is

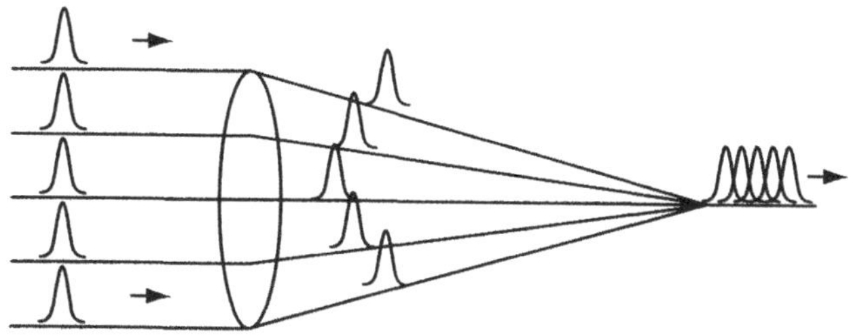

Figure 4.12 Radially varying pulse delay for a lens with chromatic aberration. Material dispersion is not included. From [195].

sketched in Fig. 4.12. The magnitude of the radially varying delay is linked to the derivative of focal length with wavelength, and for an ideal achromatic lens ($\partial f/\partial\lambda = 0$), this effect vanishes. The pulse duration measured on axis near the focus is broadened by roughly

$$T(0) - T(r_0) = \frac{r_0^2}{2cf^2}\lambda\frac{\partial f}{\partial\lambda} \tag{4.104}$$

Equation (4.103) can be confirmed by direct calculation for the case of a singlet lens [194]. Taking R_1 and R_2 as the radii of curvature of the front and rear surfaces of the lens, the thickness of the lens as a function of r is given in the paraxial limit by

$$L(r) = \frac{r_0^2 - r^2}{2}\left(\frac{1}{R_1} + \frac{1}{R_2}\right) \tag{4.105}$$

The group delay exceeds the phase delay by

$$\frac{L(r)}{v_g} - \frac{L(r)}{v_p} = -\lambda\frac{dn}{d\lambda}\frac{L(r)}{c} = -\lambda\frac{dn}{d\lambda}\frac{r_0^2 - r^2}{2c}\left(\frac{1}{R_1} + \frac{1}{R_2}\right) \tag{4.106}$$

On the other hand, the focal length is given by

$$\frac{1}{f} = (n-1)\left(\frac{1}{R_1} + \frac{1}{R_2}\right) \tag{4.107}$$

which yields

$$\lambda\frac{\partial f}{\partial\lambda} = -\lambda f^2\left(\frac{1}{R_1} + \frac{1}{R_2}\right)\frac{dn}{d\lambda} \tag{4.108}$$

By substituting eq. (4.108) into eq. (4.103), one obtains exactly the same result as eq. (4.106). For a positive singlet lens made of normally dispersive material, the delay is greatest at the center of the lens. Furthermore, in some situations the radially dependent delay can be a large effect (> 1 ps), especially for wavelengths in the ultraviolet when $|\lambda(dn/d\lambda)|$ becomes relatively large and for large beam apertures needed for tight focusing [194]. Therefore, for

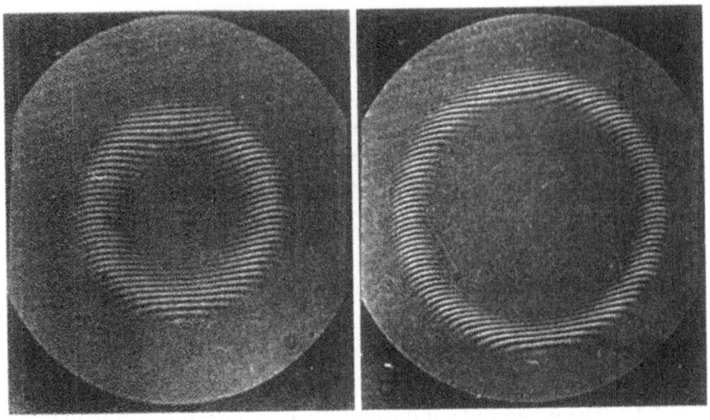

Figure 4.13 Interference fringes observed for an interferometer with a telescope inserted in one arm for two different lengths of the reference arm. The shift in the annuli where interference occurs arises due to the radially dependent pulse-front delay of these lenses. From [196].

focusing of femtosecond pulses, either achromatic lenses or spherical mirrors (which are also achromatic) are often preferred.

The effects described above have been confirmed experimentally. For example, the propagation time through a lens was measured at different radii by placing a pair of singlet lenses forming a telescope[2] into one arm of a Michelson interferometer [196]. As described in Sections 3.2 and 3.3 on electric field correlation measurements, interference occurs only when the delay of light passing through the lenses is matched to the delay of the reference arm of the interferometer to within the coherence time (inverse bandwidth) of the light source. Therefore, in the presence of a radially varying pulse-front delay, interference fringes are observed only within a specific annulus at the output plane of the interferometer, where the delays are matched. This effect is clearly visible in Fig. 4.13, which shows interference fringes recorded using blue light with a coherence time of about 100 fs for two different lengths of the reference arm, corresponding to a delay increment of about 600 fs. Quantitative analysis of data such as these showed that the shift of the annuli vs. reference arm delay closely followed eq. (4.103).

In addition to broadening at the focus due to pulse-front delay, pulses can also be broadened due to material dispersion in the lens. The broadening is most easily calculated for the on-axis rays, which experience no angular dispersion after the lens. From eq. (4.17) the contribution of the material dispersion to the broadening is simply

$$\Delta\tau(r=0) = \frac{\lambda^3 \Delta\omega}{2\pi c^2} \sum_i d_i \frac{d^2 n_i}{d\lambda^2} \tag{4.109}$$

where the summation is over the different elements of a multiple element lens, d_i and n_i are the on-axis thickness and the refractive index of the ith lens element, respectively, and

[2] The pulse front delay from a telescope is simply the sum of the pulse front delays from the individual lenses.

$\Delta\omega$ is the optical bandwidth. For a singlet lens, the broadening is given by

$$\Delta\tau(r = 0) = \frac{\lambda^3 \Delta\omega r_0^2}{4\pi c^2}\left(\frac{1}{R_1} + \frac{1}{R_2}\right)\frac{d^2 n}{d\lambda^2} \tag{4.110}$$

It is useful to compare the magnitude of the pulse broadening due to dispersion to that caused by pulse-front delay [194]. Taking the ratio of eqs. (4.110) and (4.104), and using eq. (4.108), we obtain for a singlet lens

$$\frac{\Delta\tau(r = 0)}{T(0) - T(r_0)} = \frac{\lambda^2(d^2n/d\lambda^2)}{-\lambda(dn/d\lambda)}\frac{\Delta\omega}{\omega_0} \tag{4.111}$$

When normal dispersion holds, we can further simplify this result using eq. (4.30) to obtain the following approximate relationship:

$$\frac{\Delta\tau(r = 0)}{T(0) - T(r_0)} \approx \frac{3\Delta\omega}{\omega_0} \tag{4.112}$$

Although one should recall that the proportionality constant is weakly material and wavelength dependent, it is clear nevertheless that for single-element lenses illuminated over their full aperture, material dispersion is small compared to pulse-front tilting effects except for pulses comprising only a few optical cycles. On the other hand, for achromats $T(0) - T(r_0) = 0$, and therefore broadening arises due to material dispersion alone. The magnitude of the material dispersion in achromats, which are composed of two or more elements, is usually larger than for singlets, which have less overall material. It has been shown [194,195] that for achromatic lenses, the material dispersion effects are uniform across the transverse beam profile. This means that dispersion compensation of the entire beam is possible using elements such as gratings and prisms, which is important for applications requiring tight focusing of very short pulses. This result does not hold for lenses with chromatic aberration.

The effect of chromatic aberration on the beam profile at focus has also been analyzed and has been shown to lead to complicated space–time coupling effects [195,197]. For monochromatic light the profile of the focused beam is determined by the spatial Fourier transform of the beam profile immediately after the lens. However, for very short pulses annuli of light passing through the lens at different radial positions arrive at focus at different times. When this effect is large compared to the input pulse duration, the spatial profile at focus at any one instant is determined (at least qualitatively) by the spatial Fourier transform only of the annulus of light arriving at that instant. As a result, the spatial profile at focus has a complicated shape, is less tightly focused than in the monochromatic case, and is time varying. These effects are absent for ideal achromatic lenses.

4.6 DISPERSION OF MIRROR STRUCTURES

In addition to material dispersion and dispersion arising from angular dispersion, there is a third type of dispersion important in ultrafast optics: the dispersion associated with interference effects. These effects appear primarily in connection with multiple-layer mirrors,

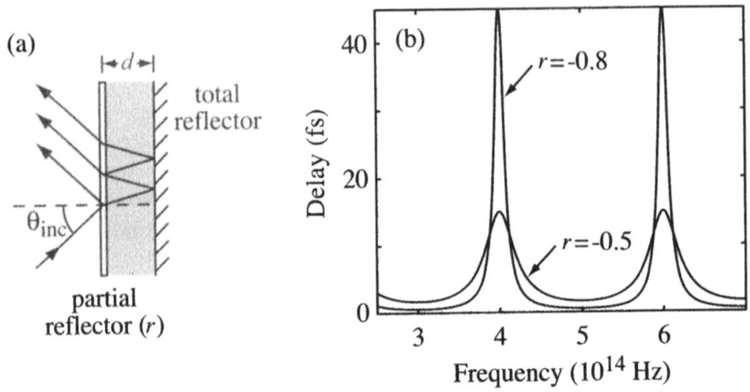

Figure 4.14 (a) Gires–Tournois interferometer; (b) frequency-dependent delay plotted for $r = -0.5$ and $r = -0.8$ and $t_0 = 5$ fs.

which are based on constructive (or destructive) interference of multiple reflections. Next, we consider three important examples.

4.6.1 The Gires–Tournois Interferometer

The Gires–Tournois interferometer (GTI), sketched in Fig. 4.14a, consists of a partially reflective mirror spaced a distance d in front of a total reflector [198]. This forms a simple resonator structure. The spacer between the two mirrors may be either air or a dielectric with index n. The total reflector ensures that all of the light is eventually reflected from the composite mirror system, while interference effects due to partial reflections from the front mirror lead to a frequency-dependent phase. These properties make the GTI useful in some applications as a low-loss dispersion compensator.

The reflectivity Γ of such a two-mirror system as a function of frequency was already derived in Section 2.4.3 on additive pulse mode-locking using the scattering matrix formalism. The result was

$$\Gamma = \frac{r + \eta e^{-j\phi}}{1 + \eta r e^{-j\phi}} \tag{4.113}$$

Here r is the field reflectivity of the partial front reflector, and η represents the transmission coefficient for the field in a single round-trip through the resonator. Unlike the additive pulse mode-locking case, here the round-trip phase shift ϕ results from linear propagation only and can be expressed as

$$\phi(\omega) = \frac{2\omega n d \cos\theta}{c} - \phi_{TR} = \frac{\omega P}{c} - \phi_{TR} \tag{4.114}$$

In this equation P and θ are the round-trip optical path length and propagation angle in the resonator, respectively, and ϕ_{TR} is the phase shift associated with reflection from the total reflector. In the case of an oblique angle of incidence, θ is related to the input angle of incidence θ_{inc} by *Snell's law* : $n \sin\theta = \sin\theta_{inc}$, where the refractive index of the medium in front of the GTI is assumed to be unity. Note that the front reflector may be a bulk optic mirror, a multiple-layer dielectric reflector deposited on top of the spacer layer, or simply the Fresnel reflection from the air–dielectric interface in the case of a bare spacer layer. In the case of the simple air–dielectric interface at normal incidence, the reflectivity of the front reflector measured directly at the interface is given by the well-known formula $r = (1 - n)/(1 + n)$, which gives $r < 0$.

If we assume the ideal case of no loss ($\eta = 1$), the reflectivity becomes

$$\Gamma = \frac{r + e^{-j\phi}}{1 + re^{-j\phi}} \tag{4.115}$$

This satisfies $|\Gamma| = 1$, which is consistent with 100% reflection of the incident power. The spectral phase due to the reflection, $\psi_\Gamma(\omega)$, is given by

$$\tan \psi_\Gamma(\omega) = \frac{-(1 - r^2) \sin \phi(\omega)}{2r + (1 + r^2) \cos \phi(\omega)} \tag{4.116}$$

The frequency-dependent delay $\tau(\omega)$ is found by taking the derivative of $\psi_\Gamma(\omega)$ with respect to ω, keeping in mind that in general r, n, θ, and ϕ_{TR} may all be frequency dependent. After some algebra, the following relatively simple expression may be obtained if we assume that the reflectivity r of the front mirror is frequency independent[3]:

$$\tau(\omega) = -\frac{\partial \psi_\Gamma(\omega)}{\partial \omega} = \frac{(1 - r^2)t_0}{1 + r^2 + 2r \cos \phi(\omega)} \tag{4.117a}$$

with

$$t_0 = \frac{\partial \phi}{\partial \omega} = \frac{P}{c} + \frac{\omega}{c}\frac{\partial P}{\partial \omega} - \frac{d\phi_{TR}}{d\omega} \tag{4.117b}$$

Here t_0 is the group delay corresponding to a single round trip in the cavity. $\tau(\omega)$ is sketched in Fig. 4.14b for $t_0 = 5$ fs and different values of r, under the assumption that t_0 may be taken as frequency independent, which means that dispersive effects due to a single propagation through the cavity are neglected. We see that $\tau(\omega)$ is positive and frequency dependent, with maxima at frequencies where $\phi = 2m\pi$ and minima at frequencies where $\phi = (2m + 1)\pi$. Here m is an integer and we assumed that $r < 0$. If $r > 0$, the positions of the maxima and minima are reversed.

This behavior may be understood intuitively by considering the first two partial reflections, corresponding to zero passes and one pass through the GTI cavity. These partial reflections have amplitudes r and $(1 - r^2)\exp(-j\phi)$, respectively. For $r < 0$ and $\phi = (2m + 1)\pi$, these partial reflections are in phase and reinforce each other; the reflection process begins very promptly. For $\phi = 2m\pi$ the partial reflections are out of phase and interfere destructively. Although all the light must eventually be reflected, the onset of the reflection process is slowed and a larger portion of the field is trapped for some time in the resonator. Thus, the dispersion of a GTI is associated with the frequency dependence of the time that the field is stored in the resonator.

[3] The infinite delay predicted at $|r| = 1$ is not physical, since loss that we have neglected will limit the delay as $|r| \to 1$.

At the delay peaks $\tau(\omega)$ is quadratic in frequency, corresponding to a cubic phase vs. frequency variation. Either positive or negative $\partial^2\psi_\Gamma/\partial\omega^2$ is possible by moving to the appropriate side of a peak. The expression for the dispersion is written

$$\frac{\partial\tau}{\partial\omega} = \frac{(1-r^2)t_0^2 \sin\phi(\omega)}{\left[1+r^2+2r\cos\phi(\omega)\right]^2} + \frac{1-r^2}{1+r^2+2r\cos\phi(\omega)}\frac{\partial t_0}{\partial\omega} \tag{4.118}$$

The first term, which is usually the largest, arises purely from the frequency-dependent storage time, while the second term arises mainly from material dispersion in the spacer layer. Concentrating on the first term, we see that the dispersion can be controlled via ϕ, which can be tuned by varying the frequency, the thickness of the spacer layer, or the angle of incidence. The magnitude of the dispersion available increases quadratically with round trip delay t_0 and can also be enhanced by using a large front reflectivity r. However, as the GTI design is varied to increase the peak dispersion, the higher-order dispersion terms also grow rapidly, and the bandwidth over which the dispersion remains approximately constant decreases. For many applications the dispersion is required to remain approximately constant over the optical bandwidth Δf. For relatively small values of r, it is sufficient that Δf be much less than the frequency spacing between two successive delay maxima, or equivalently, $t_0 \ll \Delta f^{-1}$ (i.e., the round-trip time of the GTI must be much less than the coherence time of the field). To satisfy this condition, GTIs for femtosecond pulses are usually fabricated using thin-film technology [199,200].

4.6.2 Quarter-Wave Stack High Reflectors

Highly reflecting mirrors are usually realized using a stack of thin dielectric films of different refractive indices [201,202]. Analysis methods for determining the reflectivity of arbitrary stacks of dielectric layers are well established. Generally, the reflectivity is calculated using either a matrix method [96], in which the structure is characterized by the product of matrices representing each of the individual layers, or by a closely related impedance transformation method [5,9]. We outline the impedance transformation analysis method briefly in Section 4.8.2.

A typical high reflector design consists of alternating quarter-wave layers of a high- and a low-index material deposited on a substrate, as sketched in Fig. 4.15a. For the particular frequency where the single-pass phase shift in each layer is exactly $\pi/2$, the reflectivity is given by a simple analytic expression, the magnitude of which tends rapidly to unity as the number of layer pairs increases (see Section 4.8.2). This frequency corresponds to the Bragg wavelength

$$\lambda_B = 2(n_H d_H + n_L d_L) \tag{4.119}$$

For wavelengths where the phase shift per layer is other than $\pi/2$, a computer analysis of the reflectivity spectrum is easily implemented. Such computer analyses of quarter-wave stack reflectors were performed by De Silvestri et al. [203], who concentrated on the phase vs. frequency response which determines the dispersion. An example of their results is shown in Fig. 4.15b, which plots both the intensity reflectivity $R = |r|^2$ and the dispersion ψ_2 for the mirror structure described in the figure caption. The intensity reflectivity is very close to 100% over a relatively broad bandwidth centered at λ_B. The width of the

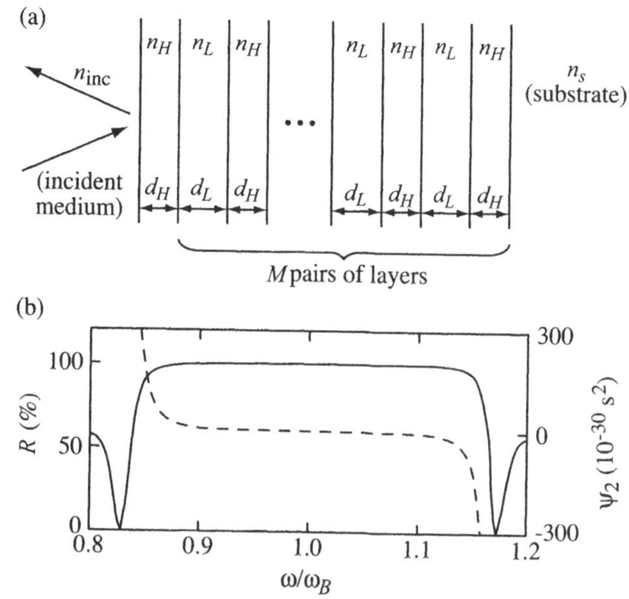

Figure 4.15 (a) Quarter-wave stack-high reflector; (b) calculated frequency-dependent reflectivity ($R = |r|^2$) and dispersion for a quarter-wave dielectric mirror composed of the following layer sequence: (air)$(HL)^M(H)$(substrate). H and L stand for the high- and low-index materials (TiO_2 and SiO_2 with $n_H = 2.28$ and $n_L = 1.45$, respectively), and $M = 9$ is the number of layer pairs. The calculation is for normal incidence and $\lambda_B = 600$ nm. (b) From [203].

high-reflectivity band depends only on n_H/n_L, the ratio of the refractive indices used, and increases with larger refractive index contrast [202]. The bandwidths obtained using available mirror materials are sufficient for most applications with pulses of a few tens of femtoseconds or longer, but may be a significant limitation for very short pulses whose spectra may cover hundreds of nanometers.

The magnitude of the dispersion is zero at λ_B but increases monotonically with increasing frequency detuning within the high-reflectivity band. Qualitatively, in the vicinity of λ_B the field is reflected nearly completely by the top few layer pairs, which means that $\tau(\omega)$ is small and relatively constant in this range. However, for frequencies approaching either edge of the high-reflectivity band, where the mirror begins to transmit, the penetration depth of the field into the mirror, and hence $\tau(\omega)$, increase rapidly with detuning. As a result, $\partial\tau/\partial\omega$ is positive on the high-frequency side of the band and negative on the low-frequency side. The magnitude of ψ_2 for this particular mirror structure varies from 20 to 75 fs^2 for detunings at which the mirror transmission $1 - R$ grows from 1% to 4%. For comparison, the dispersion of 1 mm of fused silica at 600-nm wavelength is approximately 50 fs^2, which is enough to produce significant effects for femtosecond pulses, especially within a laser cavity. Therefore, when using high-reflectivity quarter-wave stacks for femtosecond applications, care is needed to ensure that the spectral range of interest fits comfortably within the high-reflectivity, low-dispersion region.

For very short pulses, mirrors with broader bandwidths as well as controlled dispersion characteristics throughout their high-reflectivity range are needed. Metal mirrors are one possibility; they have very broad bandwidths as well as extremely low dispersion [204,205],

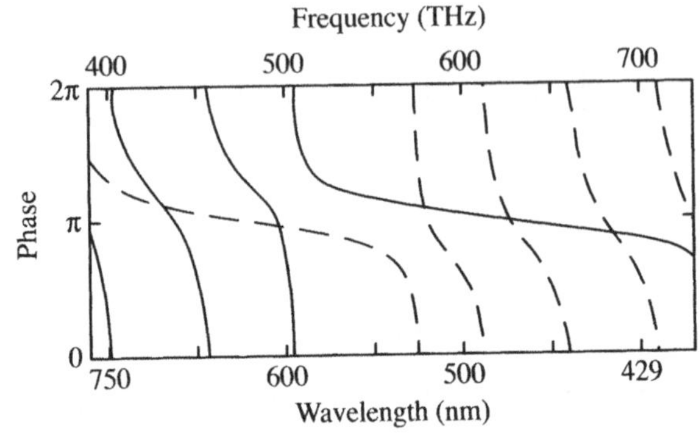

Figure 4.16 Normal incidence frequency-dependent phase calculated for a double-stack dielectric mirror composed of the following layer sequence: (air)$(H_2L_2)^7(H_2)(L_s)(H_1L_1)^7(H_1)$(substrate). H and L stand for ZnS and MgF_2 with $n_H = 2.3$ and $n_L = 1.38$, respectively. Subscripts 1 and 2 refer to the inner or outer stack, which have layer thicknesses corresponding to λ_{B1} and λ_{B2}, respectively. For the blue outer-stack design (solid line), $\lambda_{B1} = 625$ nm and $\lambda_{B2} = 480$ nm; for the red outer-stack design (dashed line), $\lambda_{B1} = 480$ nm and $\lambda_{B2} = 625$ nm. L_s refers to a low-index spacer layer used to suppress a small reflectivity dip between λ_{B1} and λ_{B2}. Adapted from [204].

but typically result in a loss on the order of a few percent per reflection. Chirped mirrors, which are discussed in Section 4.6.3, are another solution. A third possibility, the concatenation of two quarter-wave stack reflectors, each designed for a different center wavelength on a single substrate, turns out to have unfavorable dispersion characteristics [204,206]; however, discussion of this design provides new insights.

Figure 4.16 gives calculated phase spectra for a prototypical double-stack coating designed for normal incidence, with one quarter-wave stack centered at 480 nm and the other at 625 nm. Details of the coating design are given in the caption. Two distinct broadband coatings are possible, since the blue and red stacks can be deposited in either order. In each case the reflectivity is computed to be greater than 99.9% throughout most of the visible. The phase spectra, however, are very different, depending on the order in which the stacks are deposited. For the red outer-stack case, a short input pulse centered at 625 nm would be reflected completely from the outer stack. As a result, the phase variation is almost linear (as long as the optical bandwidth is not too large), and little pulse distortion would occur. In the complementary case of a blue outer stack, an input pulse centered at 625 nm would be strongly distorted. The off-resonant blue stack introduces rapid phase variations as a function of frequency, in part because sidelobes in its reflectivity spectrum lead to frequency-dependent partial reflections, and in part because these partial reflections in conjunction with the inner resonant high reflector form a sort of Gires–Tournois interferometer. A number of experiments have confirmed this behavior [97,204,207]. The extra reflectivity bandwidth provided by the double-stack design is not useful for femtosecond pulse applications, because the extra bandwidth is accompanied by a rapidly varying spectral phase.

4.6.3 Chirped Mirrors

The phenomenon of a frequency-dependent penetration depth can be exploited to design mirrors with altered dispersion characteristics suitable for dispersion-compensation

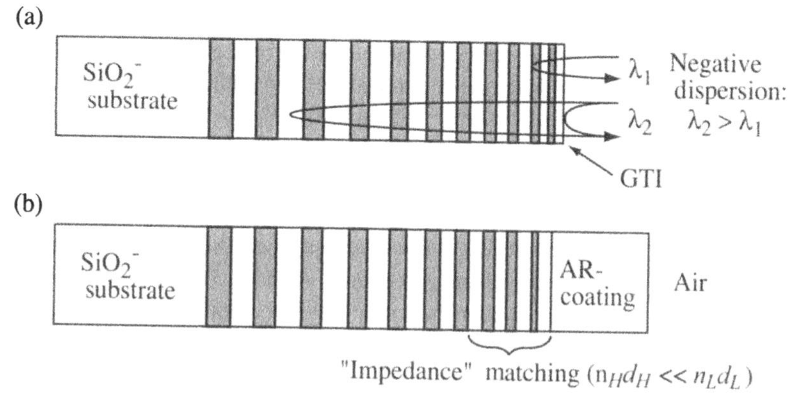

Figure 4.17 (a) Chirped mirror designed to provide negative dispersion; (b) double-chirped mirror. From [208]. Copyright © 1998, IEEE.

applications. This is achieved using a chirped mirror design, in which the thicknesses of the dielectric layers, and hence λ_B, are gradually varied within the mirror stack. Figure 4.17a shows a sketch of a chirped mirror with λ_B increasing toward the substrate. The idea is that each optical frequency component should reflect mainly from that part of the mirror where λ_B matches the free-space optical wavelength. Therefore, different frequencies should be reflected after penetrating to different depths, resulting in a frequency-dependent delay. The structure shown in Fig. 4.17a would be expected to provide negative dispersion, since longer wavelengths would experience greater delay. This would be useful for compensating for positive material dispersion (e.g, in a mode-locked laser cavity).

In reality, the response of a chirped mirror is more complicated [206]. For wavelengths reflected relatively deep within the mirror, interference with the weak partial reflections from the front portion of the mirror leads to GTI effects which disturb the desired dispersion characteristics. Figure 4.18 shows the reflectivity and frequency-dependent delay calculated for a 25-layer-pair linearly chirped mirror (dotted line). The mirror shows high reflectivity

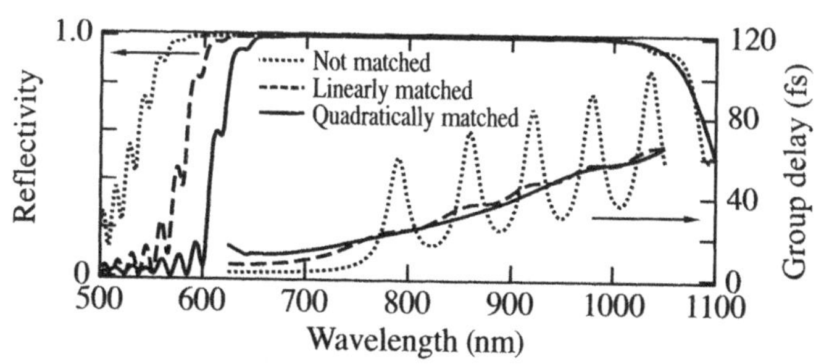

Figure 4.18 Calculated frequency-dependent reflectivity ($R = |r|^2$) and group delay for 25-layer-pair chirped mirrors with $n_H = 2.5$ and $n_L = 1.5$. The Bragg wavenumber $2\pi/\lambda_B$ is linearly chirped from $2\pi/(600\,\text{nm})$ to $2\pi/(900\,\text{nm})$ over the first 20 layer pairs, then held constant. The curves correspond either to a standard chirped mirror, where d_H corresponds to a quarter-wave for all layers, or to DCMs with d_H either linearly or quadratically chirped over the first six layer pairs. From [208]. Copyright © 1998, IEEE.

over a range of more than 400 nm, which is much broader than is possible with standard quarter-wave stack designs. This is one important benefit of chirped mirrors. The group delay does show the desired trend, on average, of delay increasing with wavelength. However, for the longer wavelengths a series of strong oscillations are superimposed onto the group delay curves, corresponding to GTI resonances. Application of chirped mirrors for femtosecond applications requires that these resonances be eliminated.

In one approach to this goal, a mirror structure with a simple chirp in the layer thicknesses is chosen as a starting point for a numerical optimization procedure carried out on a computer [209,210]. The optimization procedure varied the layer thicknesses from the starting values to minimize the squared deviation of the complex reflectivity curve from a predetermined target. This process succeeded in producing chirped mirrors with greatly reduced delay oscillations, suitable for use with sub-10-fs pulses. It was found that the maximum achievable group delay difference was given approximately by

$$\Delta\tau_{\max} = \frac{2(P_{\text{chirped}} - P_{\text{QW}})}{c} \tag{4.120}$$

where P_{chirped} and P_{QW} are the optical path lengths of the entire layer structure and of the penetration depth for the field reflected resonantly from the first few quarter-wave layers, respectively. Equation (4.120) states that the maximum group delay difference for a structure where GTI effects have been eliminated is limited to the single round-trip transit-time difference corresponding to reflection from the front and back layers, respectively. Therefore, the maximum group delay difference is limited by the number of layers in the chirped mirror structure, which in turn is limited by the precision of the coating technology.

It has been shown that the equations governing reflections from a multilayer dielectric mirror can be transformed exactly into equations governing coupling between forward- and backward-propagating waves in a weakly modulated, quasi-periodic medium [208,211,212]. These coupling-of-modes equations [9] are in turn equivalent to the equations describing wave propagation on a transmission line. The case of a chirped mirror corresponds to an inhomogeneous transmission line. This formalism furnishes new insight into the design of chirped mirrors free of GTI delay resonances. In particular, it is well known from transmission-line theory that impedance discontinuities lead to reflections, whereas reflections can be avoided provided that the properties of the line vary adiabatically (i.e., continuously and sufficiently slowly). This idea led to the introduction of the doubly chirped mirror (DCM) structure sketched in Fig. 4.17b, which may be regarded as an impedance-matching method for avoiding unwanted reflections in the front section of the mirror. There are two key aspects to this idea:

1. Impedance matching from the first layer of the coating into the rest of the coating. For the sake of discussion we assume here that this first layer consists of the low-index material. Impedance matching is achieved by varying the duty cycle of the high-index material (i.e., the thicknesses d_H and d_L of the high- and low-index materials are varied, while maintaining the local Bragg wavelength λ_B at the desired value). In the DCM structure, d_H starts near zero and increases gradually over several layer pairs until both d_H and d_L represent $\lambda/4$ optical thicknesses. For regions where $d_H = 0$, the coating would be simply a homogeneous low-index material; clearly, no reflections should originate within this region. Increasing d_H slowly from that point corresponds to adiabatically turning on the reflectivity of the multilayer, which

according to the inhomogeneous transmission-line analogy does not result in any "impedance mismatch" reflections.

2. Suppression of reflections at the interface between the incident medium (usually air) and the top coating layer. At present this is usually achieved via a broadband anti-reflection (AR) coating.

The term *doubly chirped mirror* then refers to independently chirping both the local Bragg wavelength, as in a normal chirped mirror, and the duty cycle of the high-index/low-index layers, which helps to suppress unwanted partial reflections in the front section of the mirror.

Figure 4.18 shows computed reflectivity and group delay curves for a DCM where the optical thickness of the high-index layer is increased from zero at the top of the coating until it reaches a quarter-wave after six layer pairs. Computed results for both a linear and a quadratic chirping function are shown. The group delay oscillations are strongly suppressed, especially for the quadratically chirped case. At the same time, the high-reflectivity band is narrowed on the short-wavelength side of the spectrum. This arises because d_H is not yet fully turned on in the top layers, where λ_B corresponds to the shortest wavelengths. Thus, for a fixed number of layers, the apodization of the layer thicknesses responsible for suppression of GTI delay resonances comes at the expense of the effective number of layers contributing to the high reflectivity and to the maximum group delay variation.

We briefly describe a method for synthesizing DCMs with desired dispersion profiles, which may include a quadratic as well as a higher-order phase [212]. This procedure provides a prescription for the variation of λ_B and assumes that double chirping is performed independently, as required. The starting point is a simple equation for the frequency-dependent delay:

$$\tau(\omega) = \frac{2P_r(\omega)}{c} \tag{4.121}$$

Here $P_r(\omega)$ refers to the optical path length into the mirror where frequency ω is most strongly reflected. This equation is valid in the absence of GTI effects and assuming that the layer parameters vary slowly from one layer pair to the next. This is equivalent to the slowly varying transmission-line picture, for which $P_r(\omega)$ signifies the location where the input traveling wave becomes evanescent. Let us assume that path length $P_r(\omega)$ corresponds to layer pair $m_r(\omega)$, counting from the top of the mirror. We can then write

$$\tau(\omega) = \frac{1}{c}\sum_{m=1}^{m_r(\omega)} \lambda_B(m) \approx \frac{1}{c}\int_0^{m_r(\omega)} dm\, \lambda_B(m) \tag{4.122}$$

This can be written in terms of the desired dispersion profile $\partial\tau/\partial\omega$ by differentiating with respect to ω:

$$c\frac{\partial\tau}{\partial\omega} \approx \lambda_B\left[m_r(\omega)\right]\frac{\partial m_r(\omega)}{\partial\omega} \tag{4.123}$$

To proceed further, we note that the reflection should occur when the local Bragg wavelength is approximately equal to the free-space wavelength, or

$$\lambda_B\left[m_r(\omega)\right] \approx \alpha\lambda = \frac{2\pi c\alpha}{\omega} \tag{4.124}$$

Here α is a numerical factor that takes into account modification of the effective propagation constant in the layered medium due to the coupling between counter-propagating waves [212]. To first order in this coupling, $\alpha \approx 1 - 2|r|/\pi$, where r is the reflectivity between high- and low-index layers. For very weak coupling (small index steps), $\alpha \approx 1$. The correction remains rather small even with a strong index contrast (e.g., $\alpha \approx 0.82$ for $n_H = 2.5$ and $n_L = 1.5$) but should, nevertheless, be taken into account. Combining eqs. (4.123) and (4.124) gives the desired result:

$$m_r(\omega) \approx \frac{1}{2\pi\alpha}\int_{\omega_{\max}}^{\omega} \omega' \frac{\partial\tau}{\partial\omega'}d\omega' \tag{4.125}$$

where we have assumed that the mirror is designed for a negative chirp (the top layers reflect the highest frequency, $\omega_{\max}$). This equation can be inverted (analytically in some cases, numerically or graphically in others) to determine the frequency ω corresponding to each integer value of m_r, which then yields the local Bragg wavelength $\lambda_B(m_r)$ for layer m_r via eq. (4.124). This procedure has been found to yield good initial mirror designs, which are then improved further via computer optimization. The advantage of using eq. (4.125) as a starting point is that only relatively small changes need to be introduced by the optimization procedure.

Chirped and doubly chirped mirrors are widely used for both intracavity and extracavity dispersion compensation applications for pulses in the sub-10-fs regime. They are limited mainly by small fabrication errors and by the difficulty in completely eliminating reflections over a wide bandwidth using an AR coating.

4.7 MEASUREMENTS OF GROUP VELOCITY DISPERSION

To accurately compensate for dispersion, accurate measurements of the group velocity dispersion of optical elements are needed. A number of different measurement techniques have been established, many of which have a strong resemblance to the pulse characterization methods discussed in Chapter 3. Figure 4.19 shows one scheme which illustrates the basis for this relationship. A short test pulse is first characterized in both amplitude and phase, then transmitted through (or reflected from) the optical element of interest and recharacterized. As long as the optical element is linear, comparison of the complex electric field before and after the element yields its complex transmission spectrum, and hence its dispersion, within the spectral range where the test pulse has significant energy. In the following we briefly (and nonexhaustively) survey several dispersion measurement techniques that work without requiring full characterization of the electric field.

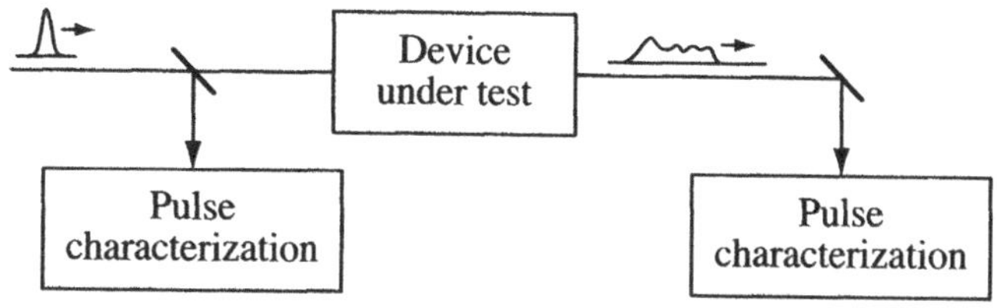

Figure 4.19 Dispersion characterization via phase and amplitude pulse measurement.

4.7.1 Interferometric Methods

In principle, dispersion can be characterized by measuring the time of flight of a frequency-tunable femtosecond pulse through the device under test. However, this approach is technically challenging. Equivalent information can be obtained via several different white light interferometry techniques. These techniques have the significant practical advantage that they can be performed using a simple continuous-wave, broadband incoherent source, where the time resolution is determined by the coherence time of the field. Certain white light interferometry methods have been used for measuring the dispersion of optical fibers since the early 1980s, where they provide a unique capability for characterizing short (ca. meter) fiber lengths [213]. Here we summarize the use of white light interferometry for dispersion measurements in ultrafast optics.

The Fringe Contour Shift Method In a technique known as the *fringe contour shift* (FCS) *method* [207], the device under test is inserted into one arm of a simple interferometer, similar to Fig. 3.2, which is illuminated using a broadband (typically, incoherent) source. A tunable optical filter, which can be placed either before, after, or within the interferometer, selects a particular frequency band. For each frequency band, the delay between the two arms is scanned and the envelope of the resulting interference fringes is recorded (the individual fringes themselves do not need to be resolved). An example of experimental data is shown in Fig. 4.20a for the case of an empty interferometer and a 10-nm filter bandwidth, corresponding to about a 50-fs coherence time.[4] The fringe patterns are all centered at the same time delay. Figure 4.20b shows the data after a 1-mm-thick glass slide is placed into one of the arms. The dramatic shift of the fringe envelopes as a function of wavelength gives a direct measurement of the dispersion. More precisely, for each frequency ω, measuring the delay corresponding to the centroid of the fringe envelope yields the frequency-dependent delay $\tau(\omega)$.

An assumption underlying the use of the FCS method is that the group delay should remain constant within the bandwidth set by the filter. Otherwise, the group delay is not well defined over the bandwidth selected, and the fringe pattern may be strongly distorted. For elements with rapidly varying spectral features, the FCS method has difficulties: Reducing the filter passband broadens the fringe envelope, which makes it more difficult to locate the centroid. This issue is similar to what is encountered in the single-slit sonogram pulse characterization approach of Section 3.7.1, which is also based on measuring the frequency-dependent delay.

[4] Even in the "empty" interferometer, passage through the beamsplitter imbalances the dispersion in the two arms; therefore, an identical beamsplitter is placed into one of the arms as a compensation plate.

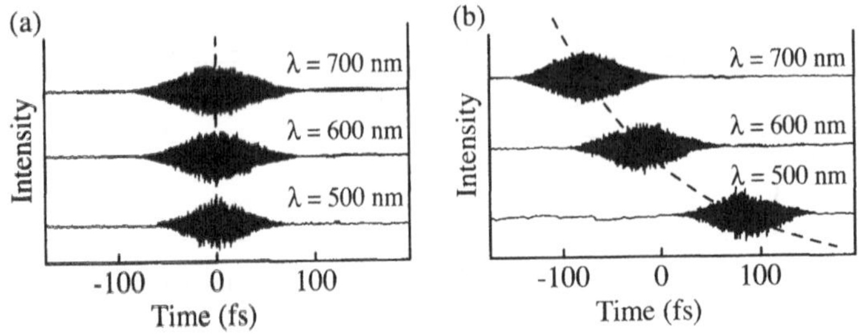

Figure 4.20 Fringe contour shift method for dispersion characterization: (a) data for an empty interferometer; (b) data with 1-mm glass slide inserted into one arm of interferometer. From [207].

Electric Field Cross-Correlation The FCS method uses only the envelope of the fringes, not their phase. One method [97] which does utilize the phase information in the fringes is based on the complete electric field cross-correlation, which was described in Section 3.3.1. The experimental setup is similar to that for the FCS method, with two principal differences:

1. No optical filter is needed—the entire optical bandwidth can be used at one time.
2. An online calibration of the length of the interferometer is required as the interferometer is swept. This is usually achieved by simultaneous recording of the fringes from a narrowband reference laser, which is also sent through the interferometer [97].

According to Section 3.3.1, the rapidly oscillating terms in the power measured at the interferometer output are given by

$$\langle P_{\text{out}}(t)\rangle = \frac{1}{2}\left\{\cdots + e^{j\omega_0\tau}\left\langle a_s(t)a_r^*(t-\tau)\right\rangle + \text{c.c.}\right\} \tag{4.126}$$

where a_r and a_s are the complex electric field amplitudes corresponding to passage through the reference and sample arms of the interferometer, respectively, and ω_0 is the center frequency. Once the data are recorded, they can be Fourier transformed with respect to τ to yield

$$\mathcal{F}\left\{\int dt\; P_{\text{out}}(t)\right\} = \cdots + \frac{1}{2}\left[A_s(\omega-\omega_0)A_r^*(\omega-\omega_0) + A_s^*(-\omega-\omega_0)A_r(-\omega-\omega_0)\right] \tag{4.127}$$

where again the noninterferometric terms have been omitted. If we assume that the effect of the sample is given by a complex frequency response $H(\omega)$, defined by $E_{\text{out}}(\omega) = H(\omega)E_{\text{in}}(\omega)$, eq. (4.127) can be rewritten as follows:

$$\mathcal{F}\left\{\int dt\; P_{\text{out}}(t)\right\} = \cdots + \frac{1}{2}\left[H(\omega)\left|A_r(\omega-\omega_0)\right|^2 + H^*(-\omega)\left|A_r(-\omega-\omega_0)\right|^2\right] \tag{4.128}$$

Thus Fourier analysis of the electric field cross-correlation data yields $H(\omega)$ directly. The phase of $H(\omega)$ gives the dispersion information desired. For an example of experimental results, see Fig. 4.10, where this technique was applied to measure the group delay spectrum of a prism pair.

An interesting variation of the electric field cross-correlation method has been used for characterization of the intracavity dispersion inside laser resonators [98,214]. The laser is kept below threshold so that it produces broadband-amplified spontaneous emission (ASE). The ASE output from the laser is then passed through an empty interferometer, where the relative delay between the arms is set to differ by approximately the laser round-trip time. Scanning the interferometer delay yields the cross-correlation of two different ASE fields, whose difference arises from one extra pass through the laser resonator by one of the ASE fields. This difference may be analyzed using eq. (4.128) to find the cavity response function $H(\omega)$, where the phase of $H(\omega)$ gives the intracavity dispersion. This technique has been applied to a number of laser types and appears to be of special interest for semiconductor lasers, where other dispersion characterization methods may be difficult to apply.

Spectral and Spatial-Spectral Interferometry Spectral interferometry is another method for using the phase information. The delay τ is kept fixed, and the resulting fringes in the output power spectrum are analyzed. The application of spectral interferometry for dispersion measurement was discussed explicitly in Section 3.3.2. Here we consider an extension of spectral interferometry in which a tilt is introduced between the two interfering beams. This results in fringes both in one spatial dimension and in frequency. This spatial-spectral interferometry method provides for a clear real-time visualization of dispersion information [215,216].

Suppose that the delayed reference pulse and the signal pulse from Section 3.3.2 are now propagating in the y–z plane at angles of $\pm\theta/2$ with respect to the z axis, such that the propagation vector of the reference (signal) has a component along the positive (negative) y axis. Then the power spectrum, eq. (3.29), is modified, resulting in the following expression:

$$|A_{\text{out}}(\tilde{\omega})|^2 = \frac{1}{2}\left\{|A_s(\tilde{\omega})|^2 + |A_r(\tilde{\omega})|^2\right\} + |A_s(\tilde{\omega})|\,|A_r(\tilde{\omega})|\cos\left[\left(\tau + \frac{2y\sin(\theta/2)}{c}\right)\omega + \psi_D(\tilde{\omega})\right] \tag{4.129}$$

Here $\tilde{\omega} = \omega - \omega_0$, and $\psi_D(\tilde{\omega}) = \psi_s(\tilde{\omega}) - \psi_r(\tilde{\omega})$ is the spectral phase due to the dispersion of the sample placed in the signal arm. The fringes are defined by the contours in ω–y space where the argument of the cosine is constant. To gain further insight, we expand about the point $\omega = \omega_0$, $y = 0$. This is valid for sufficiently small fractional bandwidths. Substituting in $\omega \to \omega_0 + d\omega$, $y \to dy$, and keeping only the lowest-order terms, the condition that the argument of the cosine remains a constant is approximately

$$\tau + \frac{2\omega_0 \sin(\theta/2)}{c}\frac{dy}{d\omega} + \frac{d\psi_D}{d\omega} \approx 0 \tag{4.130}$$

which has the solution

$$\frac{2\omega_0 \sin(\theta/2)}{c} y + \tilde{\omega}\tau + \psi_D(\tilde{\omega}) - \psi_D(\tilde{\omega} = 0) = 0 \tag{4.131}$$

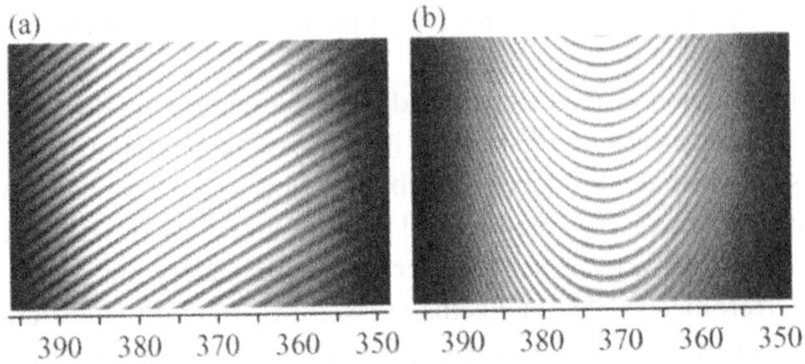

Figure 4.21 Spatial-spectral interference data at an 800-nm center wavelength: (a) empty interferometer, $\tau = 333$ fs; (b) 28 cm of BK-7 glass, $\tau = 0$. From [216].

The fringe contours give direct information about the spectral phase. By using a prism or grating to disperse the optical frequencies along the second transverse direction (x), ω is translated into x, resulting in an intuitive two-dimensional fringe pattern which can be visualized using a CCD camera. Figure 4.21 shows two examples of such data. In one example the delay is set to 333 fs and the interferometer is left empty, which leads to a tilt in the fringe pattern (for $\tau = 0$ the fringes would be untilted within the framework of our approximation). In a second example, 28 cm of BK-7 glass is placed in the signal arm and τ is set to zero. The parabolic form of the fringes, corresponding to the quadratic spectral phase arising from the dispersion of the glass, is evident.

4.7.2 Frequency-Domain Intracavity Dispersion Measurements

Finally, we discuss a very simple measurement technique for determining the total dispersion inside an operating mode-locked laser [205,217]. This frequency-domain dispersion (FDD) technique relies on measuring the change in pulse repetition rate as the optical frequency is tuned. Let f_{ML} denote the mode-locking frequency corresponding to the inverse of the cavity round-trip time T. Then

$$T = \frac{1}{f_{\mathrm{ML}}} \tag{4.132}$$

and

$$\frac{\partial T}{\partial \omega} = \frac{-1}{(f_{\mathrm{ML}})^2} \frac{\partial f_{\mathrm{ML}}}{\partial \omega} \tag{4.133}$$

Equation (4.133) relates the change in cavity round-trip time with optical frequency (i.e., the dispersion) to the change in laser repetition rate with respect to frequency.

Experimentally, this method was originally implemented by inserting a slit into a Ti:S laser between the intracavity prism pair and the back mirror. Since the optical frequencies are slightly separated at this point, the center frequency ω_0 can be tuned by translating the slit laterally. Data are collected by measuring f_{ML} with a frequency counter and ω_0 with a spectrometer as the slit is translated. An example of data obtained from a Ti:S laser is shown in Fig. 4.22 for two different amounts of intracavity glass. Without the extra glass,

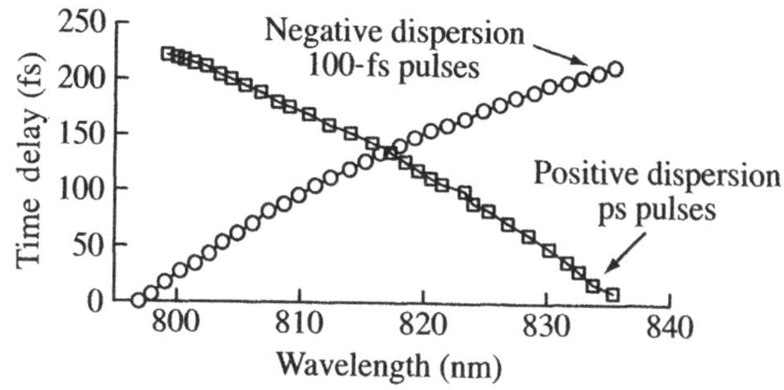

Figure 4.22 Frequency-domain intracavity dispersion data for a mode-locked Ti:S laser. The addition of 2 cm of glass moves the laser from the negative to the positive dispersion regime. From Fig. 22 of [205], with permission from Springer Science and Business Media.

the intracavity dispersion was negative and the laser produced ca. 100-fs pulses. With an additional 2 cm of glass, the intracavity dispersion became positive and the laser produced longer (picosecond) pulses. This behavior is consistent with the predictions of the mode-locking theory discussed in Section 2.4.5.

We comment briefly on issues affecting the applicability of the FDD technique [218]:

1. A tuning mechanism compatible with mode-locked operation is obviously necessary.
2. This technique works best when ω_0 can be tuned over a range equal to many times the mode-locked bandwidth. To satisfy this condition, relatively long-pulse, narrow-bandwidth operation is preferred. Since the group delay in fact varies over the mode-locked bandwidth and since the measurement gives some average of the group delay over the bandwidth, a smooth mode-locked spectrum without significant structure is needed to avoid difficulties of interpretation.
3. The method is based on the premise that the change in f_{ML} with optical frequency is dominated by the intracavity dispersion and that the intracavity nonlinearities needed for mode-locking do not have a significant effect. This assumption can be rationalized by noting that in a laser where self-phase modulation leads to self-amplitude modulation, a 2π nonlinear phase shift normally leads to strong amplitude modulation. At 800 nm this would correspond to an error in the measurement of T of only 3 fs, which is small compared to the dispersion effects typically of interest, and the error in $\partial T/\partial\omega$ would be smaller still.

4.8 APPENDIX

4.8.1 Frequency-Dependent Phase Due to Propagation Through a Slab: Alternative Derivation

With reference to Fig. 4.3b, the phase along path $AB'C$ can be written

$$\psi_{AB'C} = \psi_{AB'} + \psi_{B'C} = \frac{-\omega}{c}(n_2\ell_{AB'} + n_1\ell_{B'C}) = \frac{-\omega P_{AB'C}}{c} \tag{4.134}$$

where $P_{AB'C} = n_2\ell_{AB'} + n_1\ell_{B'C}$ is defined as the optical path length along $AB'C$ and $\ell_{AB'}$ and $\ell_{B'C}$ represent the physical lengths along $\overline{AB'}$ and $\overline{B'C}$, respectively. From trigonometry we can write

$$\ell_{AB'} = \frac{d}{\cos\theta_2} \tag{4.135}$$

and

$$\ell_{B'C} = d\,[\tan(\theta + \theta_2) - \tan\theta_2]\sin\theta_1 \tag{4.136}$$

Here d is the thickness of the slab, θ_1 is the angle with respect to the slab normal in regions I and III, $\theta_2(\omega)$ is the angle of the ray $\overline{AB'}$ in region II from the slab normal, and $\theta(\omega)$ is the angle from $\overline{AB}$ to $\overline{AB'}$, as before.

Substituting, we have

$$P_{AB'C} = \frac{n_2 d}{\cos\theta_2} + n_1 d\,[\tan(\theta + \theta_2) - \tan\theta_2]\sin\theta_1 \tag{4.137}$$

After simplifying the trigonometry, we obtain

$$P_{AB'C} = \frac{n_2 d\cos(\theta + \theta_2) + n_1 d\sin\theta_1\sin\theta}{\cos\theta_2\cos(\theta + \theta_2)} \tag{4.138}$$

Now we substitute in $d = L\cos(\theta + \theta_2)$ and use Snell's law, $n_1\sin\theta_1 = n_2\sin\theta_2$. After further simplification, we obtain

$$P_{AB'C} = n_2 L\cos\theta \tag{4.139}$$

Substituting into eq. (4.134), we obtain the desired result:

$$\psi_{AB'C}(\omega) = \frac{-\omega L n_2(\omega)\cos\theta(\omega)}{c} \tag{4.140}$$

This is the same as eq. (4.40).

4.8.2 Impedance Method for Analysis of Dielectric Mirror Stacks

The geometry of a general multilayer dielectric reflector is pictured in Fig. 4.23. The structure consists of N layers deposited on a substrate of index n_s. The ith layer has index n_i and thickness d_i. The interface between layers i and $i+1$ is located at position z_i, where $z_1 = -d_1$, $z_2 = -(d_1 + d_2)$, and so on. Here we outline the impedance transformation method for calculating the reflectivity of such a structure [5,9].

We begin by considering the reflection of plane waves from the interface at $z = 0$ between layer 1 and the substrate. For the moment we assume normal incidence with the electric field polarized along $\hat{\mathbf{x}}$. The $\mathbf{E}$ and $\mathbf{H}$ fields in layer i can be written as the superposition of

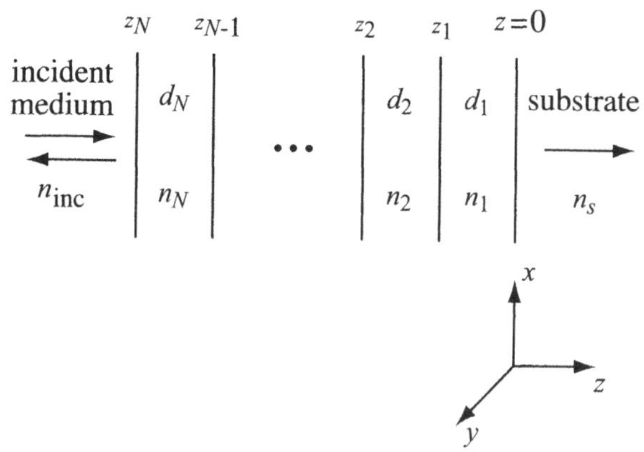

Figure 4.23 General N-layer dielectric film, showing parameters used in calculating the reflectivity.

plane waves traveling in the $+z$ and $-z$ directions:

$$\mathbf{E_i} = \hat{\mathbf{x}} \left[E_i^+ e^{-j\beta_i z} + E_i^- e^{j\beta_i z} \right]$$

$$\mathbf{H_i} = \hat{\mathbf{y}} \left[\frac{E_i^+}{Z_i} e^{-j\beta_i z} - \frac{E_i^-}{Z_i} e^{j\beta_i z} \right] \tag{4.141}$$

As usual, the propagation constant is given by $\beta_i = \omega n_i / c$. The ratio between $|E|$ and $|H|$ for a single plane wave is called the *characteristic impedance*. Assuming nonmagnetic materials, the characteristic impedance of layer i is given by

$$Z_i = \frac{|E|}{|H|} = \sqrt{\frac{\mu_0}{\varepsilon_i}} = \frac{1}{n_i} \sqrt{\frac{\mu_0}{\varepsilon_0}} \tag{4.142}$$

In the substrate there is only a wave traveling in the $+z$ direction, so the fields can be written

$$\mathbf{E_s} = \hat{\mathbf{x}} E_s^+ e^{-j\beta_s z}$$

$$\mathbf{H_s} = \hat{\mathbf{y}} \frac{E_s^+}{Z_s} e^{-j\beta_s z} \tag{4.143}$$

The boundary conditions require that the tangential components of **E** and of **H** must be continuous across the interface. Therefore, the ratio of the tangential components of **E** and **H** must also be continuous. Applying this condition at $z = 0$ gives

$$\frac{E_1^-}{E_1^+} = \frac{Z_s - Z_1}{Z_s + Z_1} \tag{4.144}$$

Rewriting this in terms of the refractive indices yields

$$\frac{E_1^-}{E_1^+} = \frac{n_1 - n_s}{n_1 + n_s} \tag{4.145}$$

which is the well-known formula for the reflectivity of a dielectric interface. The reflection goes to zero when the impedances (or indices) are matched.

Now consider the ratio of $E_x(z)$ and $H_y(z)$ at an arbitrary position z within layer 1, which is called the *wave impedance* $Z(z)$. Inserting eq. (4.144) into eq. (4.141) leads to

$$Z(z) = \frac{E_x(z)}{H_y(z)} = Z_1 \frac{Z_s - jZ_1 \tan \beta_1 z}{Z_1 - jZ_s \tan \beta_1 z} \tag{4.146}$$

In particular, at the interface between regions 1 and 2 located at $z_1 = -d_1$, the wave impedance becomes

$$Z(z_1) = Z_1 \frac{Z(0) + jZ_1 \tan \delta_1}{Z_1 + jZ(0) \tan \delta_1} \tag{4.147}$$

where we have used $Z(0) = Z_s$ and defined $\delta_i = \beta_i d_i$. Now using the wave impedance at z_1, we can solve as above for the ratio of forward- and backward- traveling waves in layer 2, yielding

$$\frac{E_2^-}{E_2^+} = \frac{Z(z_1) - Z_2}{Z(z_1) + Z_2} \tag{4.148}$$

This in turn allows computation of the wave impedance at the next interface at z_2:

$$Z(z_2) = Z_2 \frac{Z(z_1) + jZ_2 \tan \delta_2}{Z_2 + jZ(z_1) \tan \delta_2} \tag{4.149}$$

This procedure can be repeated further, moving one layer at a time away from the substrate until the interface with the incident medium is reached. In general, we find that the wave impedance at the interface between layers i and $i + 1$, which occurs at $z = z_i$, is given by

$$Z(z_i) = Z_i \frac{Z(z_{i-1}) + jZ_i \tan \delta_i}{Z_i + jZ(z_{i-1}) \tan \delta_i} \tag{4.150}$$

The reflectivity of the full structure is expressed as

$$r = \frac{Z(z_N) - Z_{\text{inc}}}{Z(z_N) + Z_{\text{inc}}} \tag{4.151}$$

Consider, for example, the case of a quarter-wave stack-high reflector, consisting of $M = N/2$ identical repetitions of a two-layer sequence with refractive indices n_1 and n_2, respectively. The reflectivity can be calculated very simply for the special case where each layer provides exactly a quarter-wave phase shift ($\delta_i = \pi/2$ for all i). In terms of the impedances, one finds that

$$Z(z_1) = \frac{Z_1^2}{Z_s} \qquad \text{and} \qquad Z(z_2) = \left(\frac{Z_2}{Z_1}\right)^2 Z_s$$

For $M = N/2$ repetitions of this two-layer sequence, the wave impedance at the input end becomes

$$Z(z_N) = \left(\frac{Z_2}{Z_1}\right)^{2M} Z_s$$

which tends to either 0 or ∞ as the number of layer pairs increases, depending on the order of the low- and high-index layers. Therefore, from eq. (4.151) the reflectivity r tends to either -1 or 1.

Calculation of the reflectivity of a general dielectric multilayer is easily extended to waves at oblique incidence. Assume that the angle of incidence with respect to normal is θ_{inc}. By Snell's law, the angle θ_i in layer i is obtained from $n_i \sin\theta_i = n_{\text{inc}} \sin\theta_{\text{inc}}$. The reflectivity is solved using the same procedure as for normal incidence, but with the following two changes:

1. The phase shift parameter in layer i becomes $\delta_i = \beta_i d_i \cos\theta_i$.
2. The characteristic impedances must be modified. Recall that we solved at the boundaries using the ratio of the tangential **E** and **H** fields. For transverse electric (TE) waves (magnetic field in the plane of incidence, **E** perpendicular to the plane of incidence), the tangential component of **H** is $|H|\cos\theta_i$ and hence the effective impedance is increased by $1/\cos\theta_i$. Conversely, for transverse magnetic (TM) waves (electric field in the plane of incidence, **H** perpendicular to the plane of incidence), the effective impedance is decreased by a factor of $\cos\theta_i$. Summarizing, in the equations above we must replace Z_i as follows:

$$\textbf{TE:} \quad Z_i \to \frac{Z_i}{\cos\theta_i} = \sqrt{\frac{\mu_0}{\varepsilon_0}}\frac{1}{n_i\cos\theta_i}$$

$$\textbf{TM:} \quad Z_i \to Z_i\cos\theta_i = \sqrt{\frac{\mu_0}{\varepsilon_0}}\frac{\cos\theta_i}{n_i}$$

PROBLEMS

4.1. Using eq. (2.5a) for the refractive index, evaluate the phase delay and group delay contributed by a resonant gain medium at the line center. Show that the modification in the round-trip delay compared to that of the cold cavity in our analysis of the actively mode-locked laser can be explained precisely by the difference in phase and group velocities.

4.2. Examine the magnitude of $\lambda(dn/d\lambda)$ assuming that the refractive index is described by a single-term Sellmeier equation, written $n^2(\lambda) = 1 + A\lambda^2/(\lambda^2 - \lambda_0^2)$, where $\lambda_0 =$ 150 nm and the refractive index is 1.44 at $\lambda = 1$ μm. Plot $\lambda(dn/d\lambda)$ throughout the visible wavelength range. What is the fractional difference between the group velocity and the phase velocity?

4.3. Consider a chirped pulse propagating in a medium of length L and dispersion β_2. The electric field envelope of the input pulse is given by $a_{\text{in}}(t) = \text{sech}(t/t_p)e^{j\phi(t)}$, where $\Phi(t) = B \ln[\text{sech}(t/t_p)]$. Calculate and plot the intensity profile, temporal phase, and instantaneous frequency of the output pulse for $B = 10$, $t_p = 1$ ps, and $\beta_2 L$ set to 0 or $\pm t_p^2/B$. Comment on your results.

4.4. Consider an optical system with angular dispersion and pulse front tilt. Comment on the direction of energy flow in such a system. Does energy flow perpendicular to the phase fronts, perpendicular to the pulse fronts, or in some other direction? Justify your answer.

4.5. Derive the result in eq. (4.62) for the dispersion $(\partial\tau/\partial\lambda)$ of a grating pair by differentiating eq. (4.64a) directly.

4.6. By direct calculation, confirm the relation between the angular dispersion of a prism and the pulse tilt that it imposes on a pulse passing through it. Assume that the prism is used at minimum deviation.

4.7. Starting from eq. (4.113), investigate the frequency-dependent delay of a Gires–Tournois interferometer in the case of finite loss. Express your results in terms of the group delay t_0 corresponding to a single round trip in the cavity, with t_0 taken to be frequency independent. What is the maximum delay variation that the GTI can provide as a function of reflectivity (r) and transmission factor (η)? Give numerical results for $r = 0.9, 0.95,$ and 0.99 over an appropriate range of the transmission factor.

4.8. The phase response of a double-stack broadband reflector, shown in Fig. 4.16, exhibits strong variations when a blue outer-stack design is used to reflect red light, or conversely, when a red outer-stack design is used to reflect blue light. Compare the period of these strong frequency-dependent phase variations with that expected viewing the structure as a Gires–Tournois interferometer.

4.9. Write a program (see Section 4.8.2) to evaluate the complex reflectivity spectrum of a quarter-wave stack mirror designed for high reflectivity at 800-nm center wavelength and normal incidence. The layer structure is taken as (air)$L(HL)^7H$(substrate), where L and H are low and high index layers with $n_H = 2.4$ and $n_L = 1.45$, respectively. The index of the substrate is also equal to 1.45. Use your results to investigate the temporal profile of short pulses reflected by this mirror. Assume 15-fs (intensity FWHM) Gaussian input pulses with center wavelength at various positions within the mirror reflectivity band. Over what range of center wavelengths is the reflected pulse broadened by less than 20%?

4.10. In this problem you will design and analyze simple chirped mirror structures. The design goal is a mirror with a high-reflectivity band spanning 700 to 1000 nm, with constant group velocity dispersion $\psi_2 = 50$ fs^2 (equivalent to $\partial\tau/\partial\lambda = 0.13$ fs nm^{-1} at 850 nm). Refractive indices for the substrate and high- and low-index layers are as specified in Problem 4.9. Limit your design to ≤ 80 dielectric layers. (*Hint:* You may assume that instead of air, the incident medium is the same as that of the top layer. This avoids the need to design an antireflection coating.)

(a) Design structures both with and without the use of double chirping and calculate their frequency-dependent delay and reflectivity responses. Can you get a smooth group delay variation? How does it compare to the constant GVD target? Comment on the differences in your designs.

(b) Attempt to modify your design to provide a controllable amount of cubic spectral phase while maintaining the specified average group velocity dispersion.

4.11. Using the data in Fig. 4.20, which shows fringe contour shifts for a double pass through a 1-mm glass plate, determine approximate values for $d^2n/d\lambda^2$ and $d^3n/d\lambda^3$. Give estimated error bars as well. Compare your values to numbers indicated for various glasses earlier in the chapter (e.g., Fig. 4.2).

4.12. In this problem we explore the form of the fringes in spatial-spectral interferometry when the assumption of small fractional bandwidth is not valid.

(a) Discuss the form of the fringes when both $\tau = 0$ and $\psi_D(\tilde{\omega}) = 0$. Do not make the small-bandwidth assumption. How does this differ from eq. (4.131)?

(b) Repeat for $\tau = 0$ and $\psi_D(\tilde{\omega}) = A\tilde{\omega}^2$ (i.e., pure quadratic spectral phase). What is the lowest-order correction to eq. (4.131)?

5

ULTRAFAST NONLINEAR OPTICS: SECOND ORDER

5.1 INTRODUCTION TO NONLINEAR OPTICS

Nonlinear optical effects play a very important role in ultrafast optics. We have already seen that nonlinearities form the basis for pulse generation through passive mode-locking and for ultrashort-pulse measurement schemes. Nonlinear effects are also central to time-resolved spectroscopy methods for the observation and investigation of ultrafast physical processes in materials (see Chapter 9). In this Chapter we introduce the formalism of nonlinear optics and then specialize to second-order nonlinear processes, which depend on the second power of the input field. Second-order processes that we consider include second harmonic generation, sum frequency and difference frequency generation, and optical parametric amplification. These processes are important for conversion of short pulses from mode-locked lasers to new frequency ranges and, particularly for the case of second-harmonic generation, for pulse measurement. Third-order nonlinear processes involving nonlinear phase modulation effects are covered in Chapter 6. A number of books devoted solely to nonlinear optics are available for the reader interested in a comprehensive treatment of the field (e.g., [37,69,131,219]). In this treatment we emphasize nonlinear processes of particular interest within the field of ultrafast optics.

Maxwell's equations in vacuum are linear. Nonlinear effects arise in materials, since the material response may be nonlinear in the applied field. The nonlinear material response can couple back to the optical field, giving rise to nonlinear optics. In our treatment we take the optical polarization as the material response of interest. In principle, a nonlinear magnetization response could also drive nonlinear optical effects, but this possibility is encountered only infrequently in practice.

As in Section 1.2, the electric flux density $\mathbf{D}$ is defined by

$$\mathbf{D} = \varepsilon_0 \mathbf{E} + \mathbf{P} \tag{5.1}$$

In a linear medium, the polarization is proportional to the field, $P = \varepsilon_0 \chi_e E$, where for the moment we consider scalar quantities. In a nonlinear medium, the polarization is usually written

$$P = \varepsilon_0 \left[\chi^{(1)} E + \chi^{(2)} E^2 + \chi^{(3)} E^3 + \cdots \right] \tag{5.2}$$

where the terms proportional to E^2, E^3, and so on, represent the nonlinear polarization. The coefficients $\chi^{(n)}$ are termed the nth *order susceptibilities*. Many different physical mechanisms can give rise to optical nonlinearities. For example, nonlinearities can arise due to motions of bound electrons, field-induced vibrational or orientational motions, optically induced acoustic waves, thermal effects, and other interactions. Nonlinearities associated with bound electrons in transparent media are of particular interest for ultrafast optics, since the response is very fast (typically, estimated on the order of 1 to 2 fs). Calculation of the nonlinear susceptibilities due to bound electrons generally requires a quantum mechanical treatment. However, these nonlinearities can also be understood phenomenologically on the basis of an anharmonic oscillator model, where a bound electron is driven by an applied electric field. In this model the Coulomb force binding the electron to the nucleus acts as a spring, and the nonlinearity arises from changes in the spring constant when the spring (i.e., the electron–nucleus separation) is stretched or compressed significantly.

To illustrate the concept of the nonlinear polarization, let us consider the second-order nonlinear polarization that arises with input fields at two different frequencies. The input electric field is taken as

$$E = \frac{1}{2} \left[\tilde{E}(\omega_1) e^{j\omega_1 t} + \tilde{E}(\omega_2) e^{j\omega_2 t} + \text{c.c.} \right] \tag{5.3}$$

The nonlinear polarization arising from the second-order nonlinear susceptibility can be written

$$\begin{aligned} P_{\text{NL}}^{(2)} = \frac{\varepsilon_0}{4} \{ & \chi^{(2)}(2\omega_1 : \omega_1, \omega_1) \tilde{E}^2(\omega_1) e^{j2\omega_1 t} + \chi^{(2)}(2\omega_2 : \omega_2, \omega_2) \tilde{E}^2(\omega_2) e^{j2\omega_2 t} \\ & + 2\chi^{(2)}(\omega_1 + \omega_2 : \omega_1, \omega_2) \tilde{E}(\omega_1) \tilde{E}(\omega_2) e^{j(\omega_1+\omega_2)t} \\ & + 2\chi^{(2)}(\omega_1 - \omega_2 : \omega_1, -\omega_2) \tilde{E}(\omega_1) \tilde{E}^*(\omega_2) e^{j(\omega_1-\omega_2)t} \\ & + \chi^{(2)}(0 : \omega_1, -\omega_1) \tilde{E}(\omega_1) \tilde{E}^*(\omega_1) + \chi^{(2)}(0 : \omega_2, -\omega_2) \tilde{E}(\omega_2) \tilde{E}^*(\omega_2) \} \\ & + \text{c.c.} \end{aligned} \tag{5.4}$$

The second-order nonlinearity mixes the input fields, resulting in second-harmonic generation ($2\omega_1$ and $2\omega_2$ terms), sum and difference frequency generation ($\omega_1 + \omega_2$ and $\omega_1 - \omega_2$), and optical rectification (the zero-frequency terms). In general, the nonlinear susceptibility may be both complex and frequency dependent. Therefore, the notation $\chi^{(2)}(\omega_a + \omega_b : \omega_a, \omega_b)$ is used to keep track of the input and output frequencies involved in a particular interaction.

In a full treatment of nonlinear optics, one must consider both the nonlinear polarization and the field as vectors. The nonlinear susceptibility then takes the form of a tensor. To illustrate, let us consider the case of sum frequency generation. The electric field and the nonlinear polarization are now written

$$\mathbf{E} = \frac{1}{2}\sum_{i=1}^{3} \hat{u}_i \left[\tilde{E}_i(\omega_1)e^{j\omega_1 t} + \tilde{E}_i(\omega_2)e^{j\omega_2 t} + \text{c.c.}\right] \tag{5.5a}$$

and

$$\mathbf{P_{NL}} = \frac{1}{2}\sum_{i=1}^{3} \hat{u}_i \left[\tilde{P}_i^{(2)}(\omega_1 + \omega_2)e^{j(\omega_1+\omega_2)t} + \text{c.c.}\right] \tag{5.5b}$$

where the $\hat{u}_i$ represent unit vectors in Cartesian coordinates. The vector components of the nonlinear polarization can be written in terms of the susceptibility tensor as follows:

$$\tilde{P}_i^{(2)}(\omega_1 + \omega_2) = \frac{\varepsilon_0 D}{2}\sum_{j,k=1}^{3} \chi_{ijk}^{(2)}(\omega_1 + \omega_2 : \omega_1, \omega_2)\tilde{E}_j(\omega_1)\tilde{E}_k(\omega_2) \tag{5.6}$$

Here D is a degeneracy factor equal to the number of distinguishable permutations of the input frequencies. For sum frequency generation, we have $D = 2$, while $D = 1$ for second-harmonic generation, as evident from eq. (5.4).

In general, the second-order nonlinear susceptibility tensor in eq. (5.6) has 27 components ($i, j, k = 1{:}3$) that must be known to fully specify the nonlinear polarization, even for a fixed set of frequencies. Fortunately, the nonlinearity tensor exhibits a number of powerful symmetries that can greatly reduce the number of independent tensor components. A comprehensive discussion of such symmetries is given in [131]. Here we make only a few brief observations that are relevant to our discussion of second-harmonic generation and frequency mixing later in the Chapter, where we are concerned with nonlinearities associated with bound electrons. If we assume that all the optical frequencies of interest are well below the electronic absorption resonances of the nonlinear material, and well above any other resonances (such as those associated with lattice vibrations in the nonlinear crystal), the following conditions apply:

- The material is transparent (lossless). It can be shown that $\chi_{ijk}^{(2)}$ is real.
- The nonlinear susceptibility is essentially dispersionless [i.e., $\chi_{ijk}^{(2)}$ becomes independent of the optical frequencies of the interacting waves]. From a time-domain perspective, the nonlinear polarization may be considered to respond essentially instantaneously.

Symmetries associated with the spatial structure of the nonlinear medium are also important and result in a reduction in the number of independent components of the $\chi_{ijk}^{(2)}$ tensor. In the case of a medium with inversion symmetry, the second-order susceptibility vanishes! As a result, second-harmonic generation and other second-order frequency-mixing processes are performed in nonlinear crystals that lack inversion symmetry.

For second-order nonlinear processes satisfying the conditions outlined above, it is common to use engineering notation, where

$$d_{ijk} = \frac{1}{2}\chi^{(2)}_{ijk} \tag{5.7}$$

The frequency dependence of the nonlinear tensor is suppressed in this notation, since we are assuming frequency independence. The form of the nonlinear tensor and numerical values for the tensor elements for common nonlinear crystals are given in references such as [131,220]. Typically, the d_{ijk} coefficients used in frequency-mixing processes range from a few times 10^{-13} m/V to a few times 10^{-11} m/V.

Equation (5.2) implies a perturbation approach which is usually valid for intensities below about 10^{14} W/cm^2, where the applied electric field is well below the interatomic fields. Our treatment will focus on this perturbational regime of nonlinear optics. Investigation of the laser–matter interaction at higher intensity levels leads to *nonperturbational nonlinear optics* (also called *extreme nonlinear optics*). This is an active research area involving a rich set of physical phenomena, such as multiphoton ionization leading to laser-generated plasmas, generation of extremely high-order harmonics of the incident laser field, attosecond pulse generation, and relativistic nonlinear effects. This nonperturbational regime is beyond the scope of our treatment. Some reviews are given in [131,221–224].

5.2 THE FORCED WAVE EQUATION

Following a standard treatment in nonlinear optics [69,131,219], we consider the nonlinear optical polarization as a source term in the wave equation. Assuming a medium that is uniform, nonmagnetic, and source-free (except for the nonlinear polarization), the wave equation can be written

$$\nabla \times \nabla \times \mathbf{E} = \nabla(\nabla \cdot \mathbf{E}) - \nabla^2\mathbf{E} = -\mu_0 \frac{\partial^2 \mathbf{D}}{\partial t^2} \tag{5.8}$$

The **D** field can be written

$$\mathbf{D} = \varepsilon_0\mathbf{E} + \mathbf{P}_{(1)} + \mathbf{P}_{\mathbf{NL}} = \mathbf{D}_{(1)} + \mathbf{P}_{\mathbf{NL}} \tag{5.9}$$

where $\mathbf{P}_{\mathbf{NL}}$ is the nonlinear polarization, $\mathbf{P}_{(1)}$ is the first-order polarization (linear in **E**), and $\mathbf{D}_{(1)}$ is the first-order portion of the **D** field. If we assume that the medium is isotropic in its linear properties, we can write $\mathbf{D}_{(1)} = \varepsilon_{(1)}\mathbf{E} = n^2\varepsilon_0\mathbf{E}$, where n is the (linear) refractive index. Gauss's law then yields

$$\varepsilon_{(1)}\nabla \cdot \mathbf{E} = -\nabla \cdot \mathbf{P}_{\mathbf{NL}} \tag{5.10}$$

In a nonlinear medium, $\nabla \cdot \mathbf{E}$ is not strictly zero. However, in most cases we will consider, the nonlinear polarization is weak, and therefore we make the approximation

$$\nabla \cdot \mathbf{E} \approx 0 \tag{5.11}$$

The wave equation now simplifies to give

$$\nabla^2\mathbf{E} - \mu_0\varepsilon_{(1)}\frac{\partial^2\mathbf{E}}{\partial t^2} = \mu_0\frac{\partial^2\mathbf{P}_{\mathbf{NL}}}{\partial t^2} \tag{5.12}$$

For now we treat the case of plane waves propagating in the $\hat{\mathbf{z}}$ direction. We will develop two alternative formulations, one in the frequency domain and the other in the time domain. We start with the frequency-domain version.

5.2.1 Frequency-Domain Formulation

We express the electric field and the nonlinear polarization in terms of spectral envelope functions and Fourier integrals, as follows:

$$\mathbf{E}(z,t) = \frac{1}{2}\hat{\mathbf{e}}\left[\frac{1}{2\pi}\int_0^\infty d\omega\, \mathcal{E}(z,\omega)\, e^{j[\omega t - k(\omega)z]} + \text{c.c.}\right]$$
$$\mathbf{P}_{\mathbf{NL}}(z,t) = \frac{1}{2}\hat{\mathbf{p}}\left[\frac{1}{2\pi}\int_0^\infty d\omega\, \mathcal{P}_{\text{NL}}(z,\omega)\, e^{j\omega t} + \text{c.c.}\right] \tag{5.13}$$

We do not yet specify the dependence of $\mathbf{P}_{\mathbf{NL}}$ on the electric field. Note, however, that both $\mathbf{E}$ and $\mathbf{P}_{\mathbf{NL}}$ are written in terms of the same carrier frequency. Our perspective here is that we are given a nonlinear polarization at frequency ω, which is inserted into the *linear* wave equation, and therefore has only the possibility of coupling to or driving a field at the same frequency. $k(\omega)$ is taken as the propagation constant for the electric field plane wave in the absence of nonlinearity.[1] The electric field and nonlinear polarization are assumed to point along unit vectors $\hat{\mathbf{e}}$ and $\hat{\mathbf{p}}$, respectively. Using this notation, we obtain an equation that relates $\mathcal{E}$ and $\mathcal{P}_{\text{NL}}$:

$$\left\{\left(\omega^2\mu_0\varepsilon_{(1)} - k^2\right)\mathcal{E} - 2jk\frac{\partial\mathcal{E}}{\partial z} + \frac{\partial^2\mathcal{E}}{\partial z^2}\right\}\hat{\mathbf{e}} = -\omega^2\mu_0\hat{\mathbf{p}}\mathcal{P}_{\text{NL}}\, e^{jkz} \tag{5.14}$$

This equation can be simplified by noting the dispersion relation

$$k^2(\omega) = \omega^2\mu_0\varepsilon_{(1)} \tag{5.15}$$

We now introduce the important *slowly varying envelope approximation* (SVEA), in which we assume that

$$\left|\frac{\partial^2\mathcal{E}}{\partial z^2}\right| \ll 2k\left|\frac{\partial\mathcal{E}}{\partial z}\right| \ll k^2\,|\mathcal{E}| \tag{5.16}$$

This SVEA is valid when the change of the frequency-dependent electric field envelope function $\mathcal{E}(z,\omega)$ due to the effect of nonlinear propagation is small over distance scales

[1] Both k and β are commonly used as symbols for the propagation constant; in our treatment, k and β are taken to have identical meaning and are used interchangeably.

on the order of a single optical wavelength. This approximation, together with a temporal version of the SVEA that we will introduce shortly, has proven to be valid over a remarkably broad range of conditions in ultrafast nonlinear optics, especially when the nonlinearity remains within the limits of a perturbative treatment and the pulse width exceeds several optical periods. Returning to the propagation equation and applying the SVEA, we obtain

$$\frac{\partial \mathcal{E}(z,\omega)}{\partial z} = \frac{-j\omega\mu_0 c}{2n}\mathcal{P}_{\text{NL}}(z,\omega)(\hat{\mathbf{e}}\cdot\hat{\mathbf{p}})\,e^{jk(\omega)z} \tag{5.17}$$

The result is a frequency-domain version of a forced wave equation widely used in nonlinear optics. The nonlinear polarization shows up as a source term which modifies or excites the complex spectral amplitude of the electric field. One important point is that in this particular formulation, we have not yet made any assumption about pulse duration. Therefore, eq. (5.17) is valid even for extremely broadband fields, where one cannot assume that the fractional bandwidth is small compared to the center frequency. Furthermore, the full effect of dispersion is included through the $\exp[jk(\omega)z]$ term.

5.2.2 Time-Domain Formulation

In the time-domain formulation, we express the field and the nonlinear polarization in terms of slowly varying amplitude functions multiplied by a carrier:

$$\begin{aligned}\mathbf{E}(z,t) &= \frac{1}{2}\hat{\mathbf{e}}\left[a(z,t)e^{j(\omega_0 t - k_0 z)} + \text{c.c.}\right]\\ \mathbf{P}_{\mathbf{NL}}(z,t) &= \frac{1}{2}\hat{\mathbf{p}}\left[\tilde{P}_{NL}(z,t)e^{j\omega_0 t} + \text{c.c.}\right]\end{aligned} \tag{5.18}$$

where $k_0 = \omega_0 n_0/c$. The Fourier transform of the envelope function is written

$$A(z,\tilde{\omega}) = \int a(z,t)e^{-j\tilde{\omega}t}\,dt \tag{5.19}$$

The frequency variable $\tilde{\omega}$, connected to the optical frequency via $\tilde{\omega} = \omega - \omega_0$, is used to emphasize that A is centered at baseband. Substituting these expressions into eq. (5.12), we obtain

$$\begin{aligned}&\frac{\hat{\mathbf{e}}}{2}\int\frac{d\tilde{\omega}}{2\pi}\left\{\left(\omega^2\mu_0\varepsilon_{(1)} - k_0^2\right)A - 2jk_0\frac{\partial A}{\partial z} + \frac{\partial^2 A}{\partial z^2}\right\}e^{j(\tilde{\omega}t - k_0 z)}\\ &\quad = \frac{\mu_0\hat{\mathbf{p}}}{2}\left\{-\omega_0^2\tilde{P}_{\text{NL}} + 2j\omega_0\frac{\partial\tilde{P}_{\text{NL}}}{\partial t} + \frac{\partial^2\tilde{P}_{\text{NL}}}{\partial t^2}\right\}\end{aligned} \tag{5.20}$$

We will again be invoking the slowly varying envelope approximation, assuming here that the envelope functions vary slowly in both space and time. This requires that the duration of the envelope functions in time should be significantly greater than the optical period; equivalently, the spatial extents of the envelope functions should significantly exceed the optical wavelength. Within our time-domain formulation, the SVEA is written (for example)

$$\left|\frac{\partial^2 A}{\partial z^2}\right| \ll \left|2k_0\frac{\partial A}{\partial z}\right| \ll \left|k_0^2 A\right| \quad \text{and} \quad \left|\frac{\partial^2 a}{\partial t^2}\right| \ll \left|2\omega_0\frac{\partial a}{\partial t}\right| \ll \left|\omega_0^2 a\right| \tag{5.21}$$

Similar relations are assumed to hold for the nonlinear polarization envelope function $\tilde{P}_{\mathrm{NL}}$.

We can now simplify eq. (5.20) by using the SVEA and the dispersion relation, eq. (5.15), which gives the result

$$\int \frac{d\tilde{\omega}}{2\pi}\left\{-2jk_0\frac{\partial A}{\partial z} + \left[k^2(\tilde{\omega}) - k_0^2\right] A\right\} e^{j\tilde{\omega}t} = -\mu_0\omega_0^2 \tilde{P}_{\mathrm{NL}}(\hat{\mathbf{e}}\cdot\hat{\mathbf{p}})\, e^{jk_0 z} \tag{5.22}$$

We can further simplify by setting $k(\omega) + k_0 \approx 2k_0$, which is equivalent to neglecting any backward-traveling waves. This is valid within the framework of the slowly varying envelope approximation. We obtain

$$\int \frac{d\tilde{\omega}}{2\pi}\left\{\frac{\partial A}{\partial z} + j\left[k(\tilde{\omega}) - k_0\right] A\right\} e^{j\tilde{\omega}t} = \frac{-j\mu_0\omega_0 c}{2n(\omega_0)} \tilde{P}_{\mathrm{NL}}(\hat{\mathbf{e}}\cdot\hat{\mathbf{p}})\, e^{jk_0 z} \tag{5.23}$$

We now approximate $k(\omega)$ with its first-order Taylor series expansion in the frequency offset $\tilde{\omega} = \omega - \omega_0$:

$$k(\omega) \approx k_0 + \left.\frac{\partial k}{\partial \omega}\right|_{\omega_0} \tilde{\omega} = k_0 + v_g^{-1}\,\tilde{\omega} \tag{5.24}$$

where v_g is the group velocity. Using this approximation for the propagation constant, we can easily carry out the inverse Fourier transform in eq. (5.23). The result is

$$\frac{\partial a(z,t)}{\partial z} + \frac{1}{v_g}\frac{\partial a(z,t)}{\partial t} = \frac{-j\mu_0\omega_0 c}{2n} \tilde{P}_{\mathrm{NL}}(z,t)\,(\hat{\mathbf{e}}\cdot\hat{\mathbf{p}})\, e^{jk_0 z} \tag{5.25}$$

As before, the nonlinear polarization shows up as a source term that drives or modifies the electric field. In the absence of nonlinear polarization, the electric field envelope $a(z,t)$ would propagate at the group velocity without any distortion or change. This formulation in which dispersive effects are ignored is adequate to describe many of the key effects that are practically important in ultrafast second-order nonlinear optics. In Chapter 6 we extend this time-domain formulation to include both dispersion and various terms that were omitted under the slowly varying envelope approximation. These additional effects are frequently important for ultrafast third-order nonlinear optics.

5.3 SUMMARY OF CONTINUOUS-WAVE SECOND-HARMONIC GENERATION

For the remainder of this chapter we focus on second-order nonlinear interactions. We begin with second-harmonic generation (SHG). In this section we consider SHG in lossless nonlinear media with the assumption of both plane-wave and continuous-wave lasers.

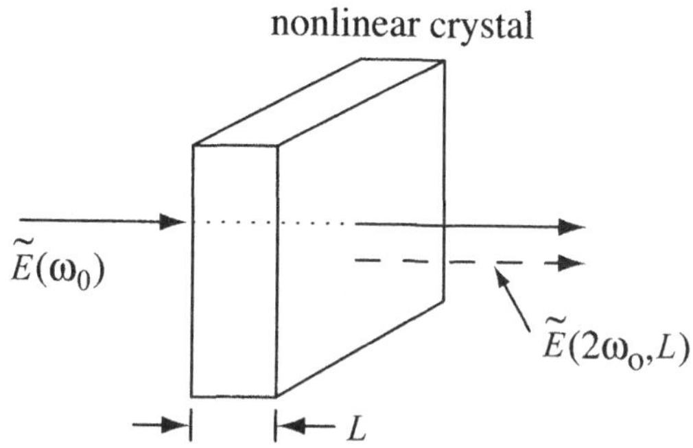

Figure 5.1 Continuous-wave second-harmonic generation setup.

Although this topic is well covered in textbooks on nonlinear optics, we give a brief derivation of the key results which will serve as a foundation for our treatment of pulsed SHG.

The situation is sketched in Fig. 5.1. A monochromatic input field at frequency ω_0 is directed onto a nonlinear crystal of length L and induces a nonlinear polarization at the second-harmonic frequency $2\omega_0$. This nonlinear polarization then excites a new field at the second-harmonic frequency. Following our time-domain formulation, it is convenient to write

$$E(z,t) = \frac{1}{2}\left[\tilde{E}_\omega(z)e^{j(\omega_0 t - k_\omega z)} + \tilde{E}_{2\omega}(z)e^{j(2\omega_0 t - k_{2\omega} z)} + \text{c.c.}\right] \tag{5.26}$$

where we use $\tilde{E}_\omega$, k_ω, $\tilde{E}_{2\omega}$, and $k_{2\omega}$ to represent the complex amplitudes and the propagation constants of the ω_0 and $2\omega_0$ waves, respectively. For continuous-wave (monochromatic) lasers, the complex amplitudes depend only on z. The nonlinear polarization is written

$$P_{\mathrm{NL}}(z,t) = 2\varepsilon_0 d_{\mathrm{eff}} E^2(z,t) \tag{5.27}$$

To keep the notation simple, we have now shifted to scalar notation for the electric field, the nonlinear susceptibility, and the nonlinear polarization. The effective nonlinear coefficient d_{eff} takes into account the tensor products of the fields and the nonlinear susceptibility tensor as well as the projection of the nonlinear polarization onto the fields (the $\hat{\mathbf{e}} \cdot \hat{\mathbf{p}}$ terms). Further information on calculation of d_{eff} starting from the actual nonlinear susceptibility tensor may be found in [220,225]. For lossless nonlinear media, d_{eff} is real. Using eq. (5.26), the nonlinear polarization is now written

$$\begin{aligned} P_{\mathrm{NL}}(z,t) &= \frac{1}{2}\left[\tilde{P}_{\mathrm{NL}}^{(\omega)}(z,t)e^{j\omega_0 t} + \tilde{P}_{\mathrm{NL}}^{(2\omega)}(z,t)e^{2j\omega_0 t} + \text{c.c.}\right] \\ &= \frac{\varepsilon_0 d_{\mathrm{eff}}}{2}\left\{\tilde{E}_\omega^2(z)e^{2j(\omega_0 t - k_\omega z)}\right. \\ &\quad \left. + 2\tilde{E}_{2\omega}(z)\tilde{E}_\omega^*(z)e^{j[\omega_0 t - (k_{2\omega} - k_\omega)z]} + \text{c.c.}\right\} \end{aligned} \tag{5.28}$$

The fact that the same d_{eff} can be used in writing the nonlinear polarization at both ω_0 and $2\omega_0$ is a consequence of the symmetry of the nonlinear susceptibility tensor for lossless media [131]. Note that we have omitted additional terms proportional to $|\tilde{E}_\omega|^2$ and $|\tilde{E}_{2\omega}|^2$, which produce a nonlinear polarization at zero frequency. This optical rectification process does not appreciably couple to either the ω_0 or $2\omega_0$ waves and is therefore neglected.

The nonlinear polarization terms in eq. (5.28) can be inserted directly into the right-hand side of eq. (5.25). Collecting terms at ω_0 and $2\omega_0$, respectively, leads to the following set of coupled equations for the fundamental and second-harmonic fields:

$$\frac{\partial \tilde{E}_\omega}{\partial z} = \frac{-j\omega_0 d_{\text{eff}}}{n_\omega c}\tilde{E}_{2\omega}\tilde{E}_\omega^* e^{-j\Delta k z}$$
$$\frac{\partial \tilde{E}_{2\omega}}{\partial z} = \frac{-j\omega_0 d_{\text{eff}}}{n_{2\omega} c}\tilde{E}_\omega^2 e^{j\Delta k z} \tag{5.29}$$

n_ω and $n_{2\omega}$ are the refractive indices associated with the ω_0 and $2\omega_0$ waves, respectively, and the difference in the propagation constants, $\Delta k = k_{2\omega} - 2k_\omega$, is referred to as the *phase mismatch*. Provided that d_{eff} is real, it can be shown that the total power carried by the fundamental and second-harmonic waves is conserved for arbitrary values of Δk.

Initially we assume that $\Delta k = 0$; this *phase-matching* condition is usually required for efficient harmonic generation. In most of our treatment, we will also assume that the $2\omega_0$ wave remains much weaker than the input wave, $|\tilde{E}_{2\omega}(z)| \ll |\tilde{E}_\omega(0)|$. Thus, we ignore depletion of the input wave, so that $\tilde{E}_\omega$ is taken as a constant throughout the nonlinear medium: $\tilde{E}_\omega(z) \approx \tilde{E}_\omega(0)$. We will also assume that there is no input second-harmonic wave. The propagation equation is easily integrated in this *nondepleted pump approximation*, which yields

$$\tilde{E}_{2\omega}(z) = -j\kappa \tilde{E}_\omega^2(0) z \tag{5.30}$$

where

$$\kappa = \frac{\omega_0 d_{\text{eff}}}{cn}$$

Here we have set $n_\omega = n_{2\omega} = n$, which as we shall see is the condition for phase matching. The second-harmonic field is proportional to the square of the input field and grows linearly with propagation distance in the nonlinear crystal. We can rewrite this result in terms of the intensities. Using eq. (1.35), we obtain

$$I_{2\omega}(L) = \Gamma^2 I_\omega^2 L^2 \qquad \text{where} \qquad \Gamma^2 = \frac{2\kappa^2}{\varepsilon_0 cn} = \frac{2\omega_0^2 d_{\text{eff}}^2}{\varepsilon_0 c^3 n^3} \tag{5.31}$$

The SHG conversion efficiency, $I_{2\omega}/I_\omega$, is proportional to the input intensity and the square of the propagation length. It is important to recall that this result is valid only in the low-conversion-efficiency limit.

It turns out that in the case of perfect phase matching, $\Delta k = 0$, eqs. (5.29) have a simple solution even for arbitrary pump depletion [131]. The result, which can easily be verified by direct substitution, is

$$\tilde{E}_\omega(z) = \tilde{E}_\omega(0)\,\text{sech}(\kappa \tilde{E}_\omega(0) z)$$
$$\tilde{E}_{2\omega}(z) = -j\tilde{E}_\omega(0)\tanh(\kappa \tilde{E}_\omega(0) z) \tag{5.32}$$

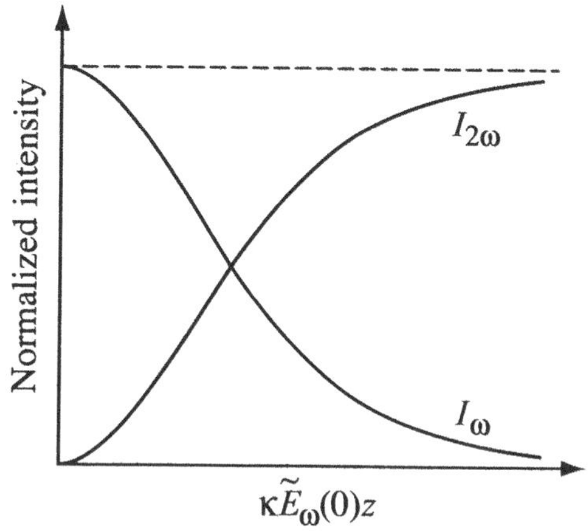

Figure 5.2 Intensities of the fundamental and second-harmonic waves in continuous-wave SHG, assuming plane waves and perfect phase matching, $\Delta k = 0$. The input pump wave is depleted as the second harmonic grows.

where for convenience $\tilde{E}_\omega(0)$ is taken as real. The resulting fundamental and second-harmonic intensities are plotted in Fig. 5.2. To the first order in the input field, eqs. (5.32) give the same result as the nondepleted pump solution. However, for stronger input fields, the second-harmonic intensity grows to a significant fraction of the input intensity. At the same time, the pump intensity is reduced, which leads to a saturation behavior for the second harmonic. In the limit $\kappa\tilde{E}_\omega(0)L \gg 1$, the input is converted completely into second harmonic.

In the following, we confine our treatment to the nondepleted pump approximation. However, it is important to remember that in any situation where the second harmonic reaches an amplitude that is significant compared to the input field amplitude, pump depletion effects must be considered through coupled nonlinear equations such as those in (5.29).

5.3.1 Effect of Phase Matching

Let us consider now the case of nonzero phase mismatch. In terms of the refractive indices, we can write

$$\Delta k = k_{2\omega} - 2k_\omega = \frac{(2\omega_0)n_{2\omega}}{c} - 2\left(\frac{\omega_0 n_\omega}{c}\right) = \frac{2\omega_0}{c}(n_{2\omega} - n_\omega) \tag{5.33}$$

Due to material dispersion (see Chapter 4), we generally have $n_{2\omega} \neq n_\omega$, so that unless we resort to special tricks, $\Delta k \neq 0$.

Even with phase mismatch, eq. (5.29) can still be solved in the nondepleted pump approximation by direct integration, with the result that the second-harmonic field a distance z into the crystal is written

$$\tilde{E}_{2\omega}(z) = \frac{-2j\omega_0 d_{\text{eff}}}{n_{2\omega}c\,\Delta k}\tilde{E}_\omega^2 e^{j\Delta kz/2}\sin\left(\frac{\Delta kz}{2}\right) \tag{5.34}$$

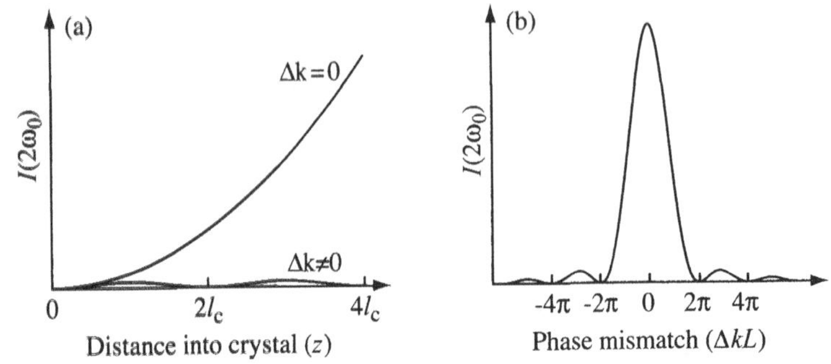

Figure 5.3 Effect of phase matching on continuous-wave second-harmonic generation: (a) second-harmonic intensity in the nondepleted pump approximation, as a function of distance into the nonlinear crystal, both with and without phase matching; (b) output intensity as a function of phase mismatch.

The intensity is given by

$$I_{2\omega} = \Gamma^2 I_\omega^2 \left[\frac{2}{\Delta k} \sin\left(\frac{\Delta k z}{2} \right) \right]^2 \tag{5.35}$$

Figure 5.3a shows the variation of second-harmonic intensity as a function of distance into the nonlinear crystal, both with and without phase matching. For small z both curves increase quadratically with z. However, for the phase-mismatched case, the behavior changes drastically for larger z. The intensity reaches a maximum at distance $l_c = \pi/\Delta k$ and then diminishes to zero, repeating sinusoidally with period $2l_c$. l_c is termed the *coherence length* and indicates the maximum thickness of the nonlinear crystal that can contribute to generating a second harmonic. Often, l_c is as short as tens of micrometers. This behavior clearly demonstrates the need for phase-matching techniques in order to achieve an efficient nonlinear optical interaction.

The need for phase matching can also be understood based on energy transfer as described by Poynting's theorem. Assuming for illustration that we can write $\mathbf{D} = \varepsilon_{(1)}\mathbf{E} + \mathbf{P}_{\mathrm{NL}}$ and $\mathbf{B} = \mu_0 \mathbf{H}$, eq. (1.29) can be rewritten as

$$\int (\mathbf{E} \times \mathbf{H}) \cdot \hat{\mathbf{n}}\, dS + \int dV \left\{ \frac{\partial\left(\frac{1}{2}\varepsilon_{(1)} |\mathbf{E}|^2\right)}{\partial t} + \frac{\partial\left(\frac{1}{2}\mu_0 |\mathbf{H}|^2\right)}{\partial t} \right\} + \int dV \left\{ \mathbf{E} \cdot \mathbf{J} + \mathbf{E} \cdot \frac{\partial \mathbf{P}_{\mathrm{NL}}}{\partial t} \right\} = 0 \tag{5.36}$$

The key points are that the time derivative of the nonlinear polarization, $\partial \mathbf{P}_{\mathrm{NL}}/\partial t$, plays the same role as a current ($\mathbf{J}$), and that both terms contained within the third integral represent energy transfer. When these terms are negative, power is supplied by the current (or nonlinear polarization) and fed into the electromagnetic field. Conversely, if these terms are positive, energy is extracted out of the field and put back into the medium. Thus, the relative phase between the fields and the nonlinear polarization determine the direction of energy transfer. By using eqs. (5.28) and (5.30), it is easily shown that in the case of perfect phase matching, the time average of $\mathbf{E}_{2\omega} \cdot \partial \mathbf{P}_{\rm NL}/\partial t$ is always less than zero. That is, the solution is consistent with energy transfer from the input (ω_0) field into the nonlinear polarization and then from the nonlinear polarization into the second-harmonic field. However, when $\Delta k \neq 0$, the phase relationship between $\tilde{E}_{2\omega}$ and the nonlinear polarization changes with propagation. The phase difference reaches π after a distance l_c into the medium. At this point $\mathbf{E}_{2\omega} \cdot \partial \mathbf{P}_{\rm NL}/\partial t$ changes sign, and the second-harmonic wave now loses its power, returning it back into the ω_0 field via the nonlinear polarization.

In actual experiments one usually measures the second-harmonic output for a nonlinear crystal of fixed length L. It then becomes convenient to rewrite eq. (5.35) as follows:

$$I_{2\omega_0}(L) = \Gamma^2 I^2(\omega_0) L^2 \frac{\sin^2(\Delta k L/2)}{(\Delta k L/2)^2} = \Gamma^2 I^2(\omega_0) L^2 \mathrm{sinc}^2\left(\frac{\Delta k L}{2}\right) \tag{5.37}$$

The output intensity is plotted vs. phase mismatch in Fig. 5.3b. The maximum output intensity is achieved for perfect phase matching ($\Delta k = 0$). For increasing $|\Delta k|$, the intensity falls off as a sinc-function squared. In the next section we discuss how phase matching may be achieved via propagation in a birefringent crystal. Experimentally, Δk is most commonly controlled by varying the propagation angle in the crystal, the temperature, or the optical frequency. In all of these cases, the same sinc-squared dependence on phase mismatch is observed.

5.3.2 Phase Matching in Birefringent Media

We have seen that phase mismatch in second-harmonic generation arises due to the variation of the refractive index with frequency. The most widespread method for achieving phase matching makes use of birefringent media (media with anisotropic *linear* properties) [226,227]. The idea is that the dependence of the refractive index on the polarization and propagation direction in a birefringent nonlinear crystal is an additional degree of freedom that can be exploited to compensate for the variation with frequency. Fortunately, many of the media that have large second-order nonlinearities also possess a linear birefringence that can be used for phase matching. In the following we give a quick overview of wave propagation in linear anisotropic media and then discuss the application of such media for phase matching. Since magnetic effects are not commonly used for phase matching, we confine our discussion to nonmagnetic media ($\mathbf{B} = \mu_0 \mathbf{H}$). A second important phase-matching technique, termed *quasi-phase matching*, is discussed in Section 5.4.3.

The linear anisotropy is described through the relation between the first-order $\mathbf{D}$ and $\mathbf{E}$ fields. Writing the definition of D, eq. (1.5), in terms of the vector components gives

$$D_i = \varepsilon_0 E_i + P_i \tag{5.38}$$

In general, the linear (first-order) polarization can be written

$$P_i = \varepsilon_0 \chi_{ij} E_j \tag{5.39}$$

where χ_{ij} is the linear susceptibility tensor and repeated indices imply summation. Combining these relations yields

$$D_i = \varepsilon_{ij} E_j \qquad \text{where} \qquad \varepsilon_{ij} = \varepsilon_0 \left(\delta_{ij} + \chi_{ij}\right) \tag{5.40}$$

Here δ_{ij} is the identity matrix and ε_{ij} is the dielectric tensor. For media with isotropic linear properties, χ_{ij} and ε_{ij} are both proportional to the identity matrix, and therefore we have the common result that the first-order **D** and **E** fields are parallel. However, in anisotropic media, χ_{ij} and ε_{ij} are not proportional to the identity matrix, and in general we find that **D** and **E** do not point in the same direction.

We now explore the character of plane-wave solutions of Maxwell's equations in anisotropic media [5,96,228]. We write the fields in complex form according to eq. (1.25); for example,

$$\mathbf{E}(\mathbf{r}, t) = \mathrm{Re}\left\{\tilde{\mathbf{E}} e^{j(\omega t - \mathbf{k}\cdot\mathbf{r})}\right\} \tag{5.41}$$

where for plane waves $\tilde{\mathbf{E}}$ is a constant vector. Similar expressions are assumed for the other field vectors. Applying Maxwell's curl equations, (1.3) and (1.4), for a current-free region yields[2]

$$\mathbf{k} \times \tilde{\mathbf{E}} = \mu_0 \omega \tilde{\mathbf{H}} \qquad \text{and} \qquad \mathbf{k} \times \tilde{\mathbf{H}} = -\omega \tilde{\mathbf{D}} \tag{5.42}$$

We find that $\tilde{\mathbf{H}}$ is perpendicular to both $\tilde{\mathbf{E}}$ and $\tilde{\mathbf{D}}$ as well as to the propagation vector **k**. $\tilde{\mathbf{D}}$ and **k** are also perpendicular. These conditions also hold for waves in isotropic media. However, unlike the isotropic plane-wave case, $\tilde{\mathbf{E}}$ need not be perpendicular to **k**.

The orientation of the field vectors is sketched in Fig. 5.4 for the choice of an $\hat{\mathbf{x}}$-polarized **H** field. According to our findings, $\tilde{\mathbf{E}}$, $\tilde{\mathbf{D}}$, and **k** are coplanar and lie in the y–z plane. We use θ to denote the angle that the propagation vector **k** makes with the z axis. As we shall see, phase matching may be achieved under suitable conditions by selecting θ properly. We also note one further peculiarity that arises in anisotropic media: The direction of energy flow, $\mathbf{E} \times \mathbf{H}$, need not be parallel to **k**. For this reason the direction of energy flow is also sketched in Fig. 5.4. $\mathbf{E} \times \mathbf{H}$ lies in the y–z plane but is separated by a *walk-off* angle ρ with respect to **k**; ρ is also the angular separation between the **E** and **D** fields.

We gain further information by applying the wave equation, eq. (5.8). For plane-wave fields of the form (5.41), the wave equation reduces to

$$k^2 \tilde{\mathbf{E}} - \mathbf{k}\left(\mathbf{k} \cdot \tilde{\mathbf{E}}\right) = \omega^2 \mu_0 \tilde{\mathbf{D}} \tag{5.43}$$

This can be rewritten in terms of the vector components as

$$k^2 \tilde{E}_i - k_i k_j \tilde{E}_j = \omega^2 \mu_0 \varepsilon_{ij} \tilde{E}_j \tag{5.44}$$

[2] We are using the vector identity $\nabla \times (\phi \mathbf{A}) = \nabla\phi \times \mathbf{A} + \phi \nabla \times \mathbf{A}$, so that $\nabla \times \tilde{\mathbf{E}} e^{-j\mathbf{k}\cdot\mathbf{r}} = -j\mathbf{k} \times \tilde{\mathbf{E}} e^{-j\mathbf{k}\cdot\mathbf{r}}$.

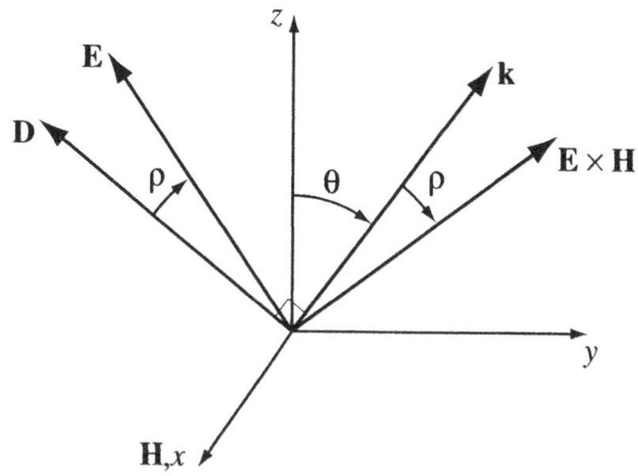

Figure 5.4 Field orientations for propagation in birefringent media. For illustration, **k** is chosen to lie in the $y-z$ plane, and **H** is taken to be polarized along $\hat{\mathbf{x}}$. The **E** and **D** fields, as well as the propagation vector **k** and the direction of energy flow (**E** × **H**), all lie in a plane perpendicular to **H**. In this sketch x, y, and z refer to crystallographic axes of the medium.

or equivalently,

$$\left(k^2\delta_{ij} - k_i k_j - \omega^2\mu_0\varepsilon_{ij}\right)\tilde{E}_j = 0 \tag{5.45}$$

We recognize the last version as a simple system of simultaneous linear equations. The solution is obtained by setting to zero the determinant of the matrix represented by the expression contained within the parentheses. The resulting equation is quadratic in k^2. Hence, for a specified propagation direction in the crystal, there are two solutions for k, the magnitude of the propagation vector. Associated with each solution for k is a distinct solution for the propagating fields.

To proceed further, we now specialize to the special case of *uniaxial* media, for which the dielectric tensor can be written

$$\varepsilon_{ij} = \varepsilon_0 \begin{pmatrix} n_0^2 & 0 & 0 \\ 0 & n_0^2 & 0 \\ 0 & 0 & n_e^2 \end{pmatrix} \tag{5.46}$$

The case $n_e > n_o$ is known as *positive uniaxial*, and conversely, the case $n_e < n_o$ is known as *negative uniaxial*. In either case, from the form of ε_{ij}, we expect that electric fields polarized in the x–y plane will see an index of refraction n_o, while electric fields polarized along z will see index of refraction n_e.[3] Solving eq. (5.45) by setting the determinant to

[3] The dielectric tensor of eq. (5.46) is written in a reference frame defined with respect to the principal axes of the crystal. Hence in this section we are using $\hat{x}$, $\hat{y}$, and $\hat{z}$ to refer to the orientation of the crystal axes, while **k** refers to the propagation direction. This should not be confused with our usage in other sections of the Chapter, where z usually refers to the propagation direction. The distinction between these two usages should be clear from the context.

zero, as suggested above, yields two solutions:

1. An *ordinary wave* (o-wave), with **E** in the x–y plane. The propagation constant is given simply by

$$k = \frac{\omega n_o}{c} \tag{5.47}$$

   For the o-wave, we have no walk-off, $\rho = 0$. Therefore, **E** is parallel to **D**, and **k** is parallel to **E** × **H**. For **k** in the y–z plane, **E** is along $\hat{\mathbf{x}}$ and **H** is in the y–z plane.

2. An *extraordinary wave* (e-wave), where **E** is perpendicular to the o-wave electric field solution. The propagation constant is now given by the relation

$$k = \frac{\omega n_e(\theta)}{c} \tag{5.48a}$$

   where

$$\frac{1}{n_e^2(\theta)} = \frac{\cos^2(\theta)}{n_o^2} + \frac{\sin^2(\theta)}{n_e^2} \tag{5.48b}$$

   The effective index of refraction for the extraordinary wave $n_e(\theta)$ depends on the propagation angle with respect to the z-axis and varies smoothly between n_o for $\theta = 0$ (propagation along $\hat{\mathbf{z}}$) and n_e for $\theta = 90^\circ$ (propagation in the $x - y$ plane). The orientations of the fields associated with the e-wave are as sketched in Fig. 5.4 for the case of a propagation vector in the $y - z$ plane. In particular, **D** and **E** are not mutually parallel, and the walk-off angle ρ is nonzero.

The fact that the e-wave exhibits walk-off while the o-wave does not means that for beams of finite size, the e- and o-waves will eventually separate. As we shall see later, this can have consequences for the second-harmonic generation process. A general formula for spatial walk-off in anisotropic media, obtained using an approach similar to the derivation of group velocity in Chapter 4, is presented in Section 5.6.1. However, we can also compute the walk-off angle in a very simple way. Let us take the case of **k** in the $y - z$ plane, as in Fig. 5.4. For uniaxial crystals this assumption does not limit the generality of our result. The **D** and **E** fields may be written

$$\mathbf{D} = D_0\left[-\cos\theta\hat{\mathbf{y}} + \sin\theta\hat{\mathbf{z}}\right] \tag{5.49a}$$

$$\mathbf{E} = E_0\left[-\cos(\theta+\rho)\hat{\mathbf{y}} + \sin(\theta+\rho)\hat{\mathbf{z}}\right] \tag{5.49b}$$

Since **D** and **E** are related via the dielectric tensor, eq. (5.46), we can also write

$$\mathbf{D} = \varepsilon_0 E_0\left[-n_o^2\cos(\theta+\rho)\hat{\mathbf{y}} + n_e^2\sin(\theta+\rho)\,\hat{\mathbf{z}}\right] \tag{5.50}$$

Combining eqs. (5.49a) and (5.50), we obtain

$$n_o^2\tan\theta = n_e^2\tan(\theta+\rho) \tag{5.51}$$

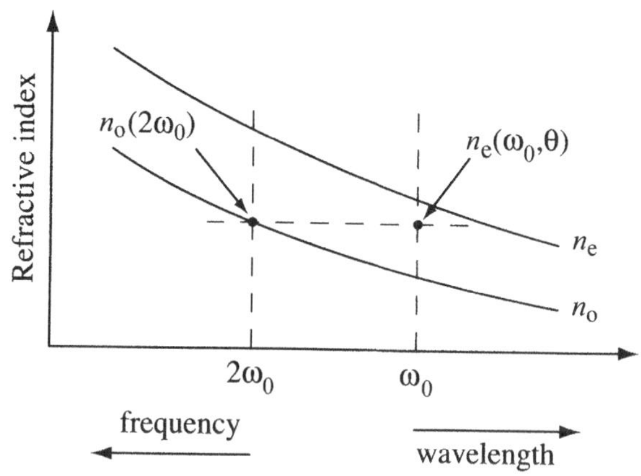

Figure 5.5 Phase-matching diagram for second-harmonic generation in a positive uniaxial medium, for the case of type I phase matching.

which gives ρ implicitly in terms of the propagation angle θ. The walk-off angle goes to zero when $\theta = 0°$ or $\theta = 90°$, as expected for these special cases where the fields are polarized exactly along principal crystal axes.

We are now in a position to understand birefringent phase matching [225,228]. Let us first consider positive uniaxial media. Figure 5.5 shows a sketch of index of refraction data (n_o and n_e) vs. wavelength. There are two schemes to achieve phase matching:

1. An input e-wave at the fundamental frequency drives an o-wave at the second-harmonic frequency. Since the electric field associated with the input e-wave appears twice, we abbreviate this interaction $e + e \to o$. Phase matching occurs when

$$n_o(2\omega_0) = n_e(\omega_0, \theta) \tag{5.52}$$

   This condition can be achieved by adjusting the propagation angle θ in the nonlinear crystal, which tunes the effective index $n_e(\omega_0, \theta)$ for the e-wave to the needed value. This procedure is known as *angle tuning*. Substituting eq. (5.48b) into eq. (5.52) yields the following expression for the phase-matching angle θ_p:

$$\sin^2\theta_p = \left[\frac{1}{n_o^2(2\omega_0)} - \frac{1}{n_o^2(\omega_0)}\right] \Big/ \left[\frac{1}{n_e^2(\omega_0)} - \frac{1}{n_o^2(\omega_0)}\right] \tag{5.53}$$

   For a phase-matching angle to exist, the refractive indices must satisfy $n_e(\omega_0) \geq n_o(2\omega_0)$ (i.e., the crystal must have sufficient birefringence). This scheme in which two identically polarized fundamental waves interact to produce the second harmonic is termed *type I* phase matching and is sketched in Fig. 5.5.

2. One input o-wave and one input e-wave, both at ω_0, interact to drive an o-wave at $2\omega_0$ ($o + e \to o$). This scheme, where both an o-wave and e-wave are present at the input, is known as *type II* phase matching. As above, phase matching is achieved via

angle tuning. The condition for type II phase matching is

$$\frac{2\omega_0 n_o(2\omega_0)}{c} = \frac{\omega_0 n_o(\omega_0)}{c} + \frac{\omega_0 n_e(\omega_0, \theta)}{c} \tag{5.54a}$$

which reduces to

$$2n_o(2\omega_0) = n_o(\omega_0) + n_e(\omega_0, \theta) \tag{5.54b}$$

This equation has a real solution when $n_e(\omega_0) + n_o(\omega_0) \geq 2n_o(2\omega_0)$. Therefore, greater birefringence is required compared to type I phase matching.

Type I and type II phase-matching schemes are also available for second-harmonic generation in negative uniaxial crystals. For type I, two o-waves at ω_0 combine to produce an e-wave at $2\omega_0$ ($o + o \rightarrow e$), whereas for type II, an input o-wave and input e-wave interact to produce a second-harmonic e-wave ($o + e \rightarrow e$). The phase-matching conditions for both positive and negative uniaxial crystals are summarized in Table 5.1. For type I phase matching in a negative uniaxial crystal, the following simple formula may be obtained for the phase-matching angle:

$$\sin^2\theta_p = \left[\frac{1}{n_o^2(\omega_0)} - \frac{1}{n_o^2(2\omega_0)}\right] \Big/ \left[\frac{1}{n_e^2(2\omega_0)} - \frac{1}{n_o^2(2\omega_0)}\right] \tag{5.55}$$

Finally, we comment briefly on the important special case of phase matching for $\theta_p = 90^\circ$, known as noncritical or 90° phase matching. This case is desirable since there is no spatial walk-off. When 90° phase matching is possible, temperature tuning (i.e., exploiting the temperature dependence of the refractive indices) is often used for fine adjustments of the birefringence while maintaining a 90° propagation geometry.

Table 5.1 Summary of Birefringent Phase Matching for Second Harmonic Generation

	Interaction Type	Phase-Matching Condition	Birefringence Requirement
Positive uniaxial ($n_e > n_o$)			
Type I	$e + e \rightarrow o$	$n_o(2\omega_0) = n_e(\omega_0, \theta)$	$n_e(\omega_0) \geq n_o(2\omega_0)$
Type II	$o + e \rightarrow o$	$2n_o(2\omega_0) = n_o(\omega_0) + n_e(\omega_0, \theta)$	$n_o(\omega_0) + n_e(\omega_0) \geq 2n_o(2\omega_0)$
Negative uniaxial ($n_o > n_e$)			
Type I	$o + o \rightarrow e$	$n_e(2\omega_0, \theta) = n_o(\omega_0)$	$n_o(\omega_0) \geq n_e(2\omega_0)$
Type II	$o + e \rightarrow e$	$2n_e(2\omega_0, \theta) = n_o(\omega_0) + n_e(\omega_0, \theta)$	$n_o(\omega_0) + n_e(\omega_0) \geq 2n_e(2\omega_0)$

5.3.3 Focusing Effects in Continuous-Wave SHG

Second-harmonic generation is a nonlinear process whose efficiency depends on optical intensity. Therefore, laser beams are often focused in second-harmonic generation experiments in order to increase intensity. In this section we work out simple approximate formulas for the second-harmonic output intensity in the case of focused, continuous-wave beams in the low-conversion-efficiency limit. An in-depth treatment of this problem may be found in [229].

Neglecting spatial walk-off at this point in the discussion, we write the input (ω_0) and generated ($2\omega_0$) fields as Gaussian beams of the form given in Section 1.3.3:

$$\mathbf{E}_{(\mathrm{i})} = \frac{1}{2}\hat{\mathbf{e}}\left[\tilde{E}_{(i)}(z)\,u_{(i)}(x, y, z)\,e^{j(\omega_i t - k_i z)} + \text{c.c.}\right] \tag{5.56}$$

Here i takes on the values ω or 2ω, which we use as shorthand to keep track of the fields at ω_0 and $2\omega_0$, respectively, and the $u_{(i)}$ are the Gaussian beam functions for the ω_0 and $2\omega_0$ fields, written

$$u_{(i)}(x, y, z) = \frac{w_{0(i)}}{w_{(i)}(z)} \exp\left(\frac{-r^2}{w_{(i)}^2(z)} - \frac{jk_i r^2}{2R_{(i)}(z)} + j\phi_{(i)}(z)\right) \tag{5.57}$$

with

$$w_{(i)}^2(z) = w_{0(i)}^2\left[1 + \left(\frac{z}{z_{0(i)}}\right)^2\right] \tag{5.58a}$$

$$\frac{1}{R_{(i)}(z)} = \frac{z}{z^2 + z_{0(i)}^2} \tag{5.58b}$$

$$\phi_{(i)}(z) = \tan^{-1}\left(\frac{z}{z_{0(i)}}\right) \tag{5.58c}$$

$$z_{0(i)} = \frac{\pi w_{0(i)}^2 n_i}{\lambda_i} \tag{5.58d}$$

The intensity profile of a Gaussian beam is sketched in Fig. 1.6. Note that the depth of focus $b_{(i)} = 2z_{0(i)}$ depends on the minimum beam radius $w_{0(i)}$. The more tightly a beam is focused, the more quickly it spreads due to diffraction. The value of b in relation to the length L of the nonlinear crystal will play an important role in determining the efficiency of the SHG process.

In the nondepleted pump approximation, the amplitude of the ω_0 field remains constant, $\tilde{E}_\omega(z) = \tilde{E}_\omega(0)$. The field is simply a Gaussian beam solution of the paraxial wave equation, eq. (1.58):

$$\nabla_T^2 u_\omega - 2jk_\omega \frac{\partial u_\omega}{\partial z} = 0 \tag{5.59}$$

Note that the nonlinear polarization does not affect the propagation of the ω_0 wave in the nondepleted pump approximation.

The propagation of the generated second harmonic is governed by a forced wave equation, similar to eq. (5.17), but retaining the transverse derivatives of the field as in the paraxial wave equation. The propagation equation for a continuous-wave field can be written

$$\left\{ \frac{j}{2k_{2\omega}} \nabla_T^2 + \frac{\partial}{\partial z} \right\} \left(\tilde{E}_{2\omega}(z) u_{2\omega}(x, y, z) \right) = \frac{-j(2\omega_0)\mu_0 c}{2n_{2\omega}} \tilde{P}_{\rm NL}(\hat{\mathbf{e}} \cdot \hat{\mathbf{p}})\, e^{jk_{2\omega}z} \tag{5.60}$$

Since $u_{2\omega}$ is taken as a solution of the nonforced paraxial wave equation, eq. (1.58), the propagation equation simplifies to yield

$$u_{2\omega} \frac{\partial \tilde{E}_{2\omega}}{\partial z} = \frac{-j\omega_0 \mu_0 c}{n_{2\omega}} \tilde{P}_{\rm NL}(\hat{\mathbf{e}} \cdot \hat{\mathbf{p}})\, e^{jk_{2\omega}z} \tag{5.61}$$

In analogy with eq. (5.28), the nonlinear polarization driving the $2\omega_0$ wave is written

$$\tilde{P}_{\rm NL}^{(2\omega)} = \varepsilon_0 d_{\rm eff} \tilde{E}_\omega^2(0)\, u_\omega^2(x, y, z)\, e^{-2jk_\omega z} \tag{5.62}$$

Inserting into eq. (5.61), we have

$$\frac{\partial \tilde{E}_{2\omega}}{\partial z} = \frac{-j\omega_0 d_{\rm eff}}{n_{2\omega} c} \tilde{E}_\omega^2(0) \frac{u_\omega^2(x, y, z)}{u_{2\omega}(x, y, z)} e^{j\Delta kz} \tag{5.63}$$

Since $\tilde{E}_{2\omega}$ has been defined as a function of z alone, we require that $u_\omega^2/u_{2\omega}$ must also be a function of z alone in order to have a valid Gaussian beam solution for the second-harmonic wave. This is achieved for

$$w_{0(2\omega)} = \frac{w_{0(\omega)}}{\sqrt{2}} \tag{5.64}$$

for which[4]

$$z_{0(\omega)} = \frac{\pi w_{0(\omega)}^2 n_\omega}{\lambda_0} \simeq \frac{\pi w_{0(2\omega)}^2 n_{2\omega}}{(\lambda_0/2)} = z_{0(2\omega)} \equiv z_0 \tag{5.65}$$

[4] The exact equality depends on $n_{2\omega} = n_\omega$, which is true for perfect phase matching, $\Delta k = 0$. However, even for small phase mismatch, $n_{2\omega}$ and n_ω are sufficiently close that the error in eq. (5.65) is very small. We simplify some of the equations that follow by introducing the notation $n \approx n_\omega \approx n_{2\omega}$.

The final propagation equation becomes

$$\frac{\partial \tilde{E}_{2\omega}}{\partial z} = \frac{-j\omega_0 d_{\text{eff}}}{nc} \frac{w_{0(\omega)}}{w_{(\omega)}(z)} \tilde{E}_{\omega}^2(0) e^{j\phi_{(\omega)}(z)} e^{j\Delta k z} \tag{5.66}$$

An important point is that the input Gaussian beam drives a second-harmonic Gaussian beam with a factor of $\sqrt{2}$ smaller beam radius but with the same depth of focus $b = 2z_0$.

Equation (5.66) can be integrated numerically, yielding the second-harmonic output in the case of negligible spatial walk-off for arbitrary focusing and phase-matching conditions. Here we discuss the limiting situations of weak and strong focusing, both with $\Delta k = 0$. In all cases we assume that the beam comes to a focus at the center of the nonlinear crystal.

Weak Focusing, Negligible Walk-off For sufficiently large focused spot sizes, the depth of focus can be made to significantly exceed the nonlinear crystal length, $b \gg L$. This is the weak-focusing regime. If the length scale l_s for spatial walk-off, which we discuss below, also exceeds the crystal length, $l_s \gg L$, walk-off is also negligible. In this regime $w_{(\omega)}(z)$ and $\phi_{(\omega)}(z)$ can be taken as approximately constant, so that eq. (5.66) can then be integrated analytically, with the same result as eq. (5.30):

$$\tilde{E}_{2\omega}(L) = \frac{-j\omega_0 d_{\text{eff}}}{nc} \tilde{E}_{\omega}^2 L \tag{5.67}$$

The power in the second-harmonic beam is calculated by integrating the intensity over the beam cross section:

$$P_{2\omega} = I_{2\omega} \int dx\,dy\, |u_{2\omega}|^2 = (\Gamma I_{\omega} L)^2 \int dx\,dy\, |u_{\omega}|^4 = \frac{(\Gamma P_{\omega} L)^2}{A_{\text{eff}}} \tag{5.68}$$

where $I_{(i)} = \frac{1}{2} c n \varepsilon_0 \left|\tilde{E}_{(i)}\right|^2$, Γ is given in eq. (5.31), and the area term is defined by

$$A_{\text{eff}} = \frac{\left[\int dx\,dy\, |u_{\omega}|^2\right]^2}{\int dx\,dy\, |u_{\omega}|^4} \tag{5.69}$$

For Gaussian beams, A_{eff} reduces to $\pi w_{0(\omega)}^2$ [see eq. (1.61)]. We obtain the result

$$P_{2\omega} = \frac{(\Gamma P_{\omega} L)^2}{\pi w_{0(\omega)}^2} = \gamma P_{\omega}^2 \frac{L^2}{b} \tag{5.70a}$$

where

$$\gamma = \frac{2n\Gamma^2}{\lambda} = \frac{4\omega_0^2 d_{\text{eff}}^2}{n^2 c^3 \varepsilon_0 \lambda} \tag{5.70b}$$

Writing the power in terms of the depth of focus in this way highlights the effect of focusing.

To increase the efficiency, the beam should be focused more tightly until the depth of focus is reduced to approximately the crystal length, $b \simeq L$. If b is further reduced, the weak-focusing approximation fails, and we must consider strong focusing.

Strong Focusing, Negligible Walk-off For sufficiently tight focusing, the depth of focus is considerably less than the crystal length, $b \ll L$. This is the strong-focusing regime. Since most of the second-harmonic power is generated near the focus, we may estimate the output simply by replacing the crystal length (L) in eq. (5.70a) with a value approximately equal to the depth of focus. According to [229], the best result is obtained when L is replaced by $\pi b/2$. This gives

$$P_{2\omega} \approx \left(\frac{\pi}{2}\right)^2 \gamma P_\omega^2 b \tag{5.71}$$

In order to increase the output power, one should increase the depth of focus until $b \approx L$, which is at the limit of the tight-focusing approximation. Combining this result with that from the weak-focusing regime, we find that the overall optimum is obtained near $b \approx L$. The efficiency is reduced if the focusing is either too strong or too weak.

Equation (5.71) actually underestimates $P_{2\omega}$, since SHG continues (at least weakly) outside the focal region. Therefore, the best second-harmonic power should be obtained for slight tighter focusing. The exact treatment of [229] shows that optimum SHG actually occurs for $L = 2.84b$, for which the power generated is

$$P_{2\omega} \approx 1.068\gamma P_\omega^2 L \tag{5.72}$$

Weak Focusing with Walk-off We now consider the case for which walk-off is important, sketched in Fig. 5.6. The second-harmonic energy generated propagates at an angle ρ with respect to the input field and the nonlinear polarization. The second-harmonic field walks off by one beam diameter at a distance of $\sqrt{2}\, w_{0(\omega)}/\rho$. We call this distance l_s, the spatial walk-off length or the aperture length. The analysis in [229] gives a slightly longer aperture length,

$$l_s = \frac{\sqrt{\pi} w_{0(\omega)}}{\rho} \tag{5.73}$$

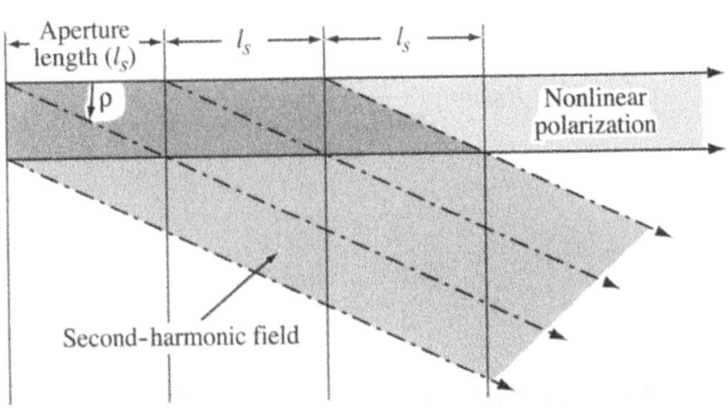

Figure 5.6 Continuous-wave second-harmonic generation with spatial walk-off and weak focusing.

Walk-off becomes significant when the aperture length is less than the length of the nonlinear crystal. Here we consider the case of walk-off with weak focusing, $l_s \ll L \ll b$.

Our strategy for estimating the second-harmonic output power is sketched in Fig. 5.6. The crystal is divided into segments of length l_s. The power generated by the first segment is given approximately by eq. (5.70a), with L replaced by l_s. At the end of the first segment, the second-harmonic is largely separated from the nonlinear polarization. Therefore, the second-harmonic generated in the second segment is spatially separated from and essentially independent of the second-harmonic field generated in the first segment. The total second-harmonic power is approximately equal to the power generated in any individual segment multiplied by the number of independent segments (L/l_s). The result is

$$P_{2\omega} \approx \gamma P_\omega^2 \frac{l_s^2}{b} \frac{L}{l_s} = \gamma P_\omega^2 \frac{L l_s}{b} \tag{5.74}$$

The output second-harmonic beam is spatially elongated in the walk-off direction. Its diameter is approximately $\sqrt{2}\, w_{0(\omega)}$, the diameter in the absence of walk-off, multiplied by L/l_s, the number of walk-off lengths.

Strong Focusing with Walk-off Finally, we consider the case $l_s \ll b \ll L$ (i.e., strong focusing with strong walk-off). As before, most of the second-harmonic generation occurs within a distance $\pi b/2$ near the focus. Therefore, the number of independent segments may be taken as $\pi b/2l_s$. The second-harmonic power is approximately

$$P_{2\omega} \approx \gamma P_\omega^2 \frac{l_s^2}{b} \frac{\pi b}{2 l_s} = \frac{\pi \gamma}{2} P_\omega^2 l_s \tag{5.75}$$

Phase-Matching Effects with Strong Focusing A brief discussion of phase matching is also worthwhile. In the case of weak focusing, the phase-matching behavior is essentially unchanged compared to the plane-wave case (i.e., the second harmonic peaks at $\Delta k = 0$ and exhibits a sinc-function dependence for nonzero phase mismatch). However, for strong focusing, different behavior arises. Figure 5.7a shows a plot of the phase-matching response calculated from the theory in [229] for $L/b = 50$, assuming no spatial walk-off. The second

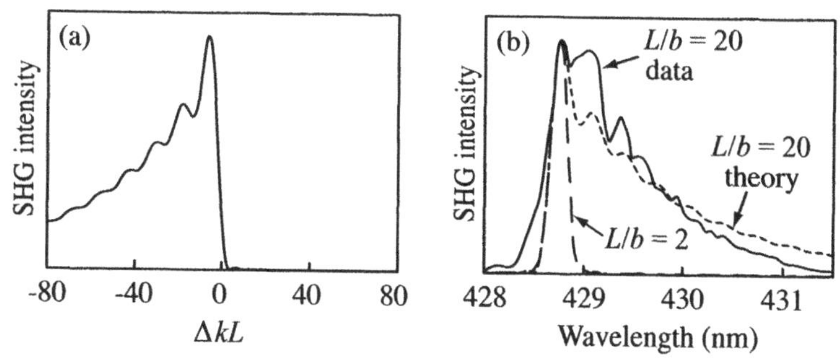

Figure 5.7 (a) Theoretical second-harmonic power as a function of phase mismatch for tight focusing, $L/b = 50$, assuming no spatial walk-off; (b) measured second-harmonic spectra for experiments with 120-fs, 860-nm input pulses in noncritically phase-matched $KNbO_3$, with $|\Delta(v_g^{-1})| L = 3.6$ ps, for relatively weak ($L/b \approx 2$) and strong ($L/b \approx 20$) focusing. The theoretical curve calculated from the expression in [229] is also shown for $L/b = 20$. (b) Adapted from [235].

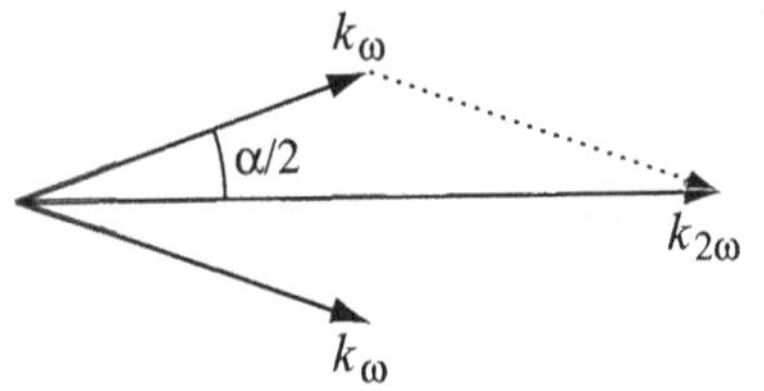

Figure 5.8 Two noncollinear field components from the fundamental interacting to give phase-matched second-harmonic generation.

harmonic exhibits a strongly asymmetric phase-matching response, peaking at $\Delta k < 0$ and with a large tail extending to large values of negative Δk. Very little second harmonic is obtained for positive phase mismatch ($\Delta k > 0$).

We can understand this behavior in two ways. First, we may view this effect as arising from the interaction between the $e^{j\phi_{(\omega)}(z)}$ and the $e^{j\Delta kz}$ terms in eq. (5.66). In particular, the $\phi_{(\omega)}(z)$ term has a positive derivative near focus, so that a negative value of Δk is required to compensate its effect. A second, more physical view is tied to the angular spectrum of a Gaussian beam. Since a Gaussian beam contains a range of **k** vectors at different angles, various noncollinear components of the fundamental beam can interact to generate second harmonic. As sketched in Fig. 5.8, phase matching is achieved in such noncollinear interactions when

$$2\,|k_\omega|\cos\left(\frac{\alpha}{2}\right) = |k_{2\omega}| \tag{5.76}$$

where α is the full angle between the fundamental field components. Phase matching can only be achieved for $2\,|k_\omega| \geq |k_{2\omega}|$ (i.e., for $\Delta k < 0$).

5.4 SECOND-HARMONIC GENERATION WITH PULSES

We are now ready to cover pulsed second-harmonic generation. Before proceeding to new effects that occur with ultrashort pulses, we first discuss the long pulse or quasi-continuous-wave limit.

5.4.1 SHG in the Quasi-Continuous-Wave Limit

For input pulses that are sufficiently long, the formulas that we have already developed for continuous-wave SHG may still be applied on an instant-by-instant basis. How long is long enough will be clarified in the next section. Consider, for example, our result for plane waves with perfect phase matching, eq. (5.31). In the case of long pulses, it is rewritten simply by including a time dependence for the input and output intensities:

$$I_{2\omega}(t) = \Gamma^2 I_\omega^2(t) L^2 \tag{5.77}$$

Due to the quadratic dependence on the input intensity, the second harmonic pulse is generally somewhat narrower than the input pulse.

For pulses we are often interested in the integrated intensity or *fluence* $\mathcal{F}$. The output fluence can be written in terms of the input fluence as follows:

$$\mathcal{F}_{2\omega} = \int dt\, I_{2\omega}(t) = \frac{\Gamma^2 L^2 \mathcal{F}_\omega^2}{\Delta t_{\text{eff}}} \tag{5.78a}$$

where Δt_{eff} is an effective pulse duration defined by

$$\Delta t_{\text{eff}} = \frac{\left[\int dt\, I_\omega(t)\right]^2}{\int dt\, I_\omega^2(t)} \tag{5.78b}$$

We see from this relation that for a given pulse shape, the output fluence is proportional to the square of the input fluence divided by the pulse duration. For a given input fluence, the SHG efficiency is better for shorter pulses.

Finally, for real beams with finite cross-sectional area, we are often interested in the pulse energy, which is obtained by integrating over the transverse coordinates. The result is

$$U_{2\omega} = \frac{\Gamma^2 L^2 U_\omega^2}{A_{\text{eff}} \Delta t_{\text{eff}}} \tag{5.79}$$

where A_{eff} is an effective area defined by eq. (5.69). In the case of a repetitively pulsed laser with repetition period t_{rep}, the average input power is $\langle P_\omega \rangle = U_\omega / t_{\text{rep}}$. The second-harmonic efficiency can be written

$$\frac{U_{2\omega}}{U_\omega} = \frac{\Gamma^2 L^2 t_{\text{rep}}}{A_{\text{eff}}\ \Delta t_{\text{eff}}} \langle P_\omega \rangle \tag{5.80}$$

The efficiency is enhanced compared to a continuous-wave laser by the ratio of the repetition period to the pulse duration, provided of course that the long pulse approximation remains valid. The enhancement results simply from the increase in peak power accompanying pulsed laser operation.

5.4.2 Ultrashort-Pulse SHG

Basic Formulation To analyze new effects that occur with ultrashort pulses [230–233], we follow the frequency-domain formulation given in Section 5.2.1. The input field, second-harmonic field, and nonlinear polarization, respectively, are written[5]

$$\begin{aligned}
E_\omega(t) &= \frac{1}{2}\left[\frac{1}{2\pi}\int_0^\infty d\omega\, \mathcal{E}_\omega(\omega)\, e^{j[\omega t - k_\omega(\omega)z]} + \text{c.c.}\right] \\
E_{2\omega}(t) &= \frac{1}{2}\left[\frac{1}{2\pi}\int_0^\infty d\omega\, \mathcal{E}_{2\omega}(\omega)\, e^{j[\omega t - k_{2\omega}(\omega)z]} + \text{c.c.}\right] \\
P_{\text{NL}}(t) &= \frac{1}{2}\left[\frac{1}{2\pi}\int_0^\infty d\omega\, \mathcal{P}_{\text{NL}}(\omega)\, e^{j\omega t} + \text{c.c.}\right]
\end{aligned} \tag{5.81}$$

[5] Note that although both the time- and frequency-domain field and nonlinear polarization variables are functions of z, in most cases we suppress the z dependence for notational simplicity.

We have implicitly assumed type I phase matching since only one input wave is included in the formulation. For any wave that is an extraordinary wave, the propagation constant (k_ω or $k_{2\omega}$) includes appropriate dependence on polarization and propagation direction in addition to the frequency dependence.

The nonlinear polarization responsible for the second-harmonic field is given by $P_{\rm NL}(t) = 2\varepsilon_0 d_{\rm eff} E_\omega^2(t)$. Substituting for the input field, we have

$$P_{\rm NL}(t) = \frac{\varepsilon_0 d_{\rm eff}}{2} \int_0^\infty \frac{d\omega}{2\pi} \mathcal{E}_\omega(\omega) e^{j\omega t} \int_0^\infty \frac{d\omega'}{2\pi} \mathcal{E}_\omega(\omega') e^{j\omega' t} e^{-j[k_\omega(\omega)+k_\omega(\omega')]z} + \text{c.c.} \quad (5.82)$$

Now we introduce a new frequency variable, $\Omega = \omega + \omega'$. In the following we use Ω to designate second-harmonic frequencies (in the vicinity of $2\omega_0$), while ω is reserved for frequencies in the fundamental band ($\approx \omega_0$). The area element in the integral, $d\omega\, d\omega'$, becomes $d\Omega\, d\omega'$. As a result, the nonlinear polarization becomes

$$P_{\rm NL}(t) = \frac{\varepsilon_0 d_{\rm eff}}{2} \int_0^\infty \frac{d\Omega}{2\pi} e^{j\Omega t} \times \int_0^\infty \frac{d\omega'}{2\pi} \mathcal{E}_\omega(\omega') \mathcal{E}_\omega(\Omega - \omega') e^{-j[k_\omega(\omega')+k_\omega(\Omega-\omega')]z} + \text{c.c.} \quad (5.83)$$

Comparing with eq. (5.81), we identify the nonlinear polarization spectrum as

$$\mathcal{P}_{\rm NL}(\Omega) = \varepsilon_0 d_{\rm eff} \int_0^\infty \frac{d\omega'}{2\pi} \mathcal{E}_\omega(\omega') \mathcal{E}_\omega(\Omega - \omega') e^{-j[k_\omega(\omega')+k_\omega(\Omega-\omega')]z} \quad (5.84)$$

According to this expression, frequencies from the input pulse combine in various pairs (i.e., ω' and $\Omega - \omega'$) to drive frequency components (i.e., Ω) of the nonlinear polarization, and hence of the second-harmonic field. Energy conservation is satisfied automatically. This interaction is sketched from a frequency-domain perspective in Fig. 5.9.

We now solve for the second-harmonic field, assuming plane waves and no pump depletion. In this case the nonlinearity is accounted for completely in the expression for $\mathcal{P}_{\rm NL}(\Omega)$.

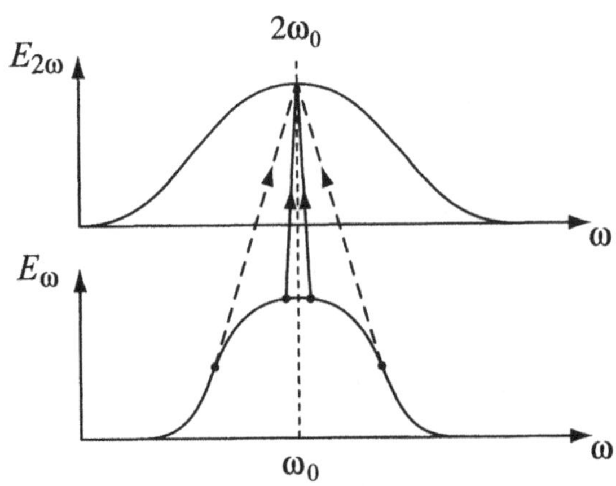

Figure 5.9 Frequency-domain picture of ultrashort-pulse SHG. Frequencies from the input pulse combine in pairs to drive second-harmonic frequency components.

Therefore, we can solve for the Fourier component of the second-harmonic field at Ω by using the forced wave equation, eq. (5.17), with $\mathcal{P}_{\mathrm{NL}}(\Omega)$ as a source term. The resulting equation for positive second-harmonic frequencies Ω is

$$\frac{\partial \mathcal{E}_{2\omega}(\Omega)}{\partial z} = \frac{-j\omega_0 d_{\mathrm{eff}}}{n_{2\omega} c}\left[\frac{1}{2\pi}\int_0^{\infty} d\omega'\, \mathcal{E}_{\omega}(\omega')\mathcal{E}_{\omega}(\Omega-\omega')\, e^{j\Delta k z}\right] \tag{5.85}$$

where

$$\Delta k = k_{2\omega}(\Omega) - k_{\omega}(\omega') - k_{\omega}(\Omega-\omega') \tag{5.86}$$

and we have set $\Omega \approx 2\omega_0$ in the multiplicative factor preceding the integral. The solution is

$$\mathcal{E}_{2\omega}(L,\Omega) = \frac{-j\omega_0 d_{\mathrm{eff}}}{n_{2\omega} c}\int_0^{\infty}\frac{d\omega'}{2\pi}\,\mathcal{E}_{\omega}(\omega')\mathcal{E}_{\omega}(\Omega-\omega')\left[L\exp\left(\frac{j\Delta k L}{2}\right)\mathrm{sinc}\left(\frac{\Delta k L}{2}\right)\right] \tag{5.87}$$

Equation (5.87) is valid for arbitrary variations of $k_{2\omega}(\Omega)$ and $k_{\omega}(\omega)$ (i.e., dispersion is fully accounted for). However, to identify the dominant effects, we expand the phase mismatch to the first order in optical frequency. We find

$$\Delta k = k_{2\omega}(2\omega_0) - 2k_{\omega}(\omega_0) + \left(\left.\frac{\partial k_{2\omega}}{\partial\omega}\right|_{2\omega_0} - \left.\frac{\partial k_{\omega}}{\partial\omega}\right|_{\omega_0}\right)(\Omega - 2\omega_0) \tag{5.88}$$

If we assume phase matching for input frequency ω_0, then $k_{2\omega}(2\omega_0) = 2k_{\omega}(\omega_0)$. This gives the result

$$\Delta k(\Omega) = \left[\left.\frac{\partial k_{2\omega}}{\partial\omega}\right|_{2\omega_0} - \left.\frac{\partial k_{\omega}}{\partial\omega}\right|_{\omega_0}\right](\Omega - 2\omega_0) \tag{5.89}$$

Identifying the derivatives with respect to ω as the inverse group velocities of the input and second-harmonic waves evaluated at frequencies ω_0 and $2\omega_0$, respectively, we have

$$\Delta k(\Omega) = \left[v_g^{-1}(2\omega_0) - v_g^{-1}(\omega_0)\right](\Omega - 2\omega_0) = \Delta(v_g^{-1})(\Omega - 2\omega_0) \tag{5.90}$$

Within the limits of the first-order expansion, the phase mismatch is a function of the second-harmonic frequency only. Its magnitude is related to the mismatch in *inverse group velocities*, $\Delta(v_g^{-1}) = v_g^{-1}(2\omega_0) - v_g^{-1}(\omega_0)$. Note that phase matching does not imply group velocity matching. Except in special cases, the group velocity mismatch is nonzero.

We are now able to make several observations about the second-harmonic field:

- The nonlinear polarization driving the second harmonic is proportional to the autoconvolution of the fundamental field complex spectrum. Therefore, for bandwidth-limited pulses, the spectral width of the nonlinear polarization is slightly broader than that of the input spectrum.
- Not all of the second-harmonic frequencies that could potentially be driven by the nonlinear polarization are phase matched. The effect is that of a filter acting solely

on the second-harmonic field and not on the fundamental, with an effective filter bandwidth, $\Delta\nu = 0.88/\left|\Delta(v_g^{-1})\right| L$, that is inversely proportional to the difference in group delays through the crystal for the fundamental and second-harmonic waves. Therefore, in cases where there is substantial *group velocity mismatch (GVM)*, the second-harmonic spectrum can be significantly narrower than the input spectrum.

- All of the frequencies in the input spectrum can participate in the generation of phase-matched second-harmonic frequencies. Hence, a high-energy conversion efficiency into the second harmonic is possible even in the presence of group velocity mismatch leading to the filtering effect cited above.
- These results are strictly valid only for a first-order expansion in the frequency variables. If second- or higher-order effects are included, Δk is a function of both the fundamental and second-harmonic frequencies [both ω' and Ω in eq. (5.87)]. In the time domain, neglect of the second-order terms means that broadening of the fundamental pulse itself (or the second harmonic) due to group velocity dispersion is omitted. Although the effect of group velocity dispersion does become important for sufficiently short pulses, the lower-order group velocity mismatch considered here is usually the dominant effect.

Let us now return to the time domain. Substituting eq. (5.87) into eq. (5.81) gives

$$E_{2\omega}(L,t) = \frac{1}{2}\left[\int_0^\infty \frac{d\Omega}{2\pi}\left(\frac{-j\omega_0 d_{\text{eff}}}{n_{2\omega}c}\right) F(\Omega)H(\Omega)e^{j\Omega t} + \text{c.c.}\right] \tag{5.91a}$$

where F and H represent, respectively, the forcing term and the effective filter function due to the group velocity mismatch, given by

$$F(\Omega) = \frac{1}{2\pi}\int_0^\infty d\omega'\, \mathcal{E}_\omega(\omega')\mathcal{E}_\omega(\Omega - \omega') \tag{5.91b}$$

and

$$H(\Omega) = \left[Le^{\frac{j\Delta k(\Omega)L}{2}}\operatorname{sinc}\left(\frac{\Delta k(\Omega)L}{2}\right)\right] e^{-jk_{2\omega}(\Omega)L} \tag{5.91c}$$

Since the Fourier transform of the convolution of two functions is equal to the product of the Fourier transforms, we obtain

$$E_{2\omega}(L,t) = \frac{1}{2}\left[\frac{-j\omega_0 d_{\text{eff}}}{n_{2\omega}c} f(t) * h(t) + \text{c.c.}\right] \tag{5.92}$$

where $*$ denotes convolution and $f(t)$ and $h(t)$ are the inverse Fourier transforms of $F(\Omega)$ and $H(\Omega)$, respectively. We find that

$$f(t) = a_\omega^2(0,t)\, e^{2j\omega_0 t} \tag{5.93}$$

where $a_\omega(0, t)$ is the slowly varying amplitude function of the fundamental field evaluated at the input end of the crystal, defined by

$$E_\omega(0, t) = \frac{1}{2}\left[a_\omega(0, t)e^{j\omega_0 t} + \text{c.c.}\right] \tag{5.94}$$

Thus, the driving term for the nonlinear polarization is just the slowly varying amplitude squared. Note that the driving term is not intensity; it is field squared, which includes phase. To evaluate $h(t)$, we recall that the sinc function is the Fourier transform of a square pulse. After some manipulation, we obtain

$$h(t) = \frac{e^{j[2\omega_0 t - k_{2\omega}(2\omega_0)L]}}{\Delta(v_g^{-1})}\,\text{sq}\left(\frac{t - [v_g^{-1}(2\omega_0) + v_g^{-1}(\omega_0)](L/2)}{|\Delta(v_g^{-1})|L}\right) \tag{5.95}$$

where $\text{sq}(u)$ is defined as a unit square pulse, such that $\text{sq}(u) = 1$ for $|u| \le \frac{1}{2}$ and 0 otherwise. Putting these results together, we can now write the output second-harmonic field. The result is

$$E_{2\omega}(L, t) = \frac{1}{2}\left[\frac{-j\omega_0 d_{\text{eff}}}{n_{2\omega}c}\frac{e^{j[2\omega_0 t - k_{2\omega}(2\omega_0)L]}}{\Delta(v_g^{-1})}\int_{t - v_g^{-1}(2\omega_0)L}^{t - v_g^{-1}(\omega_0)L} dt'\, a_\omega^2(0, t') + \text{c.c.}\right] \tag{5.96}$$

Since we are ignoring pump depletion and dispersion, the fundamental field exiting the crystal is the same as the input field with a delay, $a_\omega(L, t) = a_\omega(0, t - v_g^{-1}(\omega_0)L)$. We can rewrite the output second-harmonic field as

$$E_{2\omega}(L, t) = \frac{1}{2}\left[\frac{-j\omega_0 d_{\text{eff}}}{n_{2\omega}c}\frac{e^{j[2\omega_0 t - k_{2\omega}(2\omega_0)L]}}{\Delta(v_g^{-1})}\int_{t - \Delta(v_g^{-1})L}^{t} dt'\, a_\omega^2(L, t') + \text{c.c.}\right] \tag{5.97}$$

We can understand this result as follows. Consider, for example, the case where the second harmonic travels with a lower group velocity than the fundamental. Then the second harmonic generated near the input lags behind the fundamental as both pulses propagate into the crystal. The fundamental continues to generate new contributions to the second-harmonic field, which in turn continue to trail behind the fundamental. The integral in eq. (5.97) keeps track of the different arrival times of the contributions to the second-harmonic originating in different parts of the nonlinear crystal. There are two limiting cases, depending on the magnitude of the group delay difference, $|\Delta(v_g^{-1})|L$, in comparison with the input pulse width. Using Δt_{eff} as a measure of the pulse width, we have:

- $|\Delta(v_g^{-1})|L \ll \Delta t_{\text{eff}}$. The group delay difference is much less than the pulse width and can be neglected. For bandwidth-limited pulses, this condition is equivalent to the requirement that the phase-matching bandwidth be much wider than the input spectrum. The second-harmonic field is proportional simply to the square of the input field and is consequently somewhat shorter than the input pulse. This is the quasi-continuous-wave limit.[6] In the limit where the input field is continuous wave, eq. (5.97) becomes identical to eq. (5.30).

- $|\Delta(v_g^{-1})|L \gg \Delta t_{\text{eff}}$. Now the group delay mismatch is much larger than the pulse width. This is equivalent to the condition that the phase-matching bandwidth is much narrower than the input spectrum. The second-harmonic is essentially square in time, with duration $|\Delta(v_g^{-1})|L$. In this large group delay mismatch regime, the second-harmonic pulse width is much longer than and essentially independent of the input pulse duration.

For second-harmonic generation of visible and near-infrared laser pulses, the group velocity mismatch is typically on the order of tens to hundreds of fs/mm. Group velocity mismatch is both crystal and wavelength dependent and typically increases with decreasing wavelength, especially at wavelengths where the second harmonic approaches an absorption edge. For a specified nonlinear material and wavelength, the total group velocity mismatch can be reduced by using shorter crystals. To avoid broadening of the second-harmonic output, most femtosecond SHG experiments use thin nonlinear crystals where L is on the order of several hundred micrometers or less (tens of micrometers for very short pulses, e.g., in the 10-fs range).

Efficiency and Focusing Effects In Section 5.4.1 we worked out an expression for the second-harmonic efficiency for long pulses in the quasi-continuous-wave limit. We now calculate the second-harmonic pulse energy in the limit of large group delay mismatch. We continue to assume that spatial walk-off is negligible. For a calculation of the second-harmonic efficiency with simultaneous temporal and spatial walk-off, see [234].

Working from eq. (5.97), we first convert from field to intensity. Assuming that $a_\omega(t)$ is real (bandwidth-limited pulses), we obtain

$$I_{2\omega}(L,t) = \frac{\Gamma^2 \mathcal{F}_\omega^2}{[\Delta(v_g^{-1})]^2}\,\mathrm{sq}\left(\frac{t}{|\Delta(v_g^{-1})|L}\right) \tag{5.98}$$

Here, as previously, $\mathcal{F}_\omega$ is the input fluence and we have exploited the fact that in the large group velocity mismatch limit, the integral extends over the entire pulse. For simplicity we have shifted the time frame so that the second-harmonic pulse is centered at $t = 0$; this has no effect on our efficiency calculation. The second-harmonic fluence is

$$\mathcal{F}_{2\omega} = \frac{\Gamma^2 \mathcal{F}_\omega^2 L}{|\Delta(v_g^{-1})|} \tag{5.99}$$

To convert to pulse energies, we must integrate over the cross-sectional area. For now we consider weak focusing, so that the area remains constant throughout the crystal.

[6] For this limit to hold for pulses that are not bandwidth limited, the small group delay difference *and* the large phase-matching bandwidth requirements must be met simultaneously.

The output energy is given by

$$U_{2\omega} = \frac{\Gamma^2 U_\omega^2 L}{A_{\text{eff}} |\Delta(v_g^{-1})|} \tag{5.100}$$

where A_{eff}, defined by eq. (5.69), takes on the value $\pi w_{0(\omega)}^2$. Rewriting in terms of the depth of focus b, we have

$$U_{2\omega} = \frac{\gamma U_\omega^2 L}{|\Delta(v_g^{-1})| b} \tag{5.101}$$

where γ is given by eq. (5.70b).

We can obtain this same result using an alternative approach that closely parallels our treatment of continuous-wave SHG efficiency with spatial walk-off. We introduce a temporal walk-off length l_T,

$$l_T = \frac{t_p}{|\Delta(v_g^{-1})|} \tag{5.102}$$

for which the temporal walk-off is equal to the pulse width t_p. Large group velocity mismatch corresponds to the thick crystal limit, $L \gg l_T$; conversely, thin crystals satisfying $L \ll l_T$ yield the quasi-continuous-wave limit. Similar to Fig. 5.6 for spatial walk-off, we divide the crystal into segments of length l_T. Since the second harmonic separates in time from the fundamental pulse and hence from the nonlinear polarization after length l_T, the energy generated in each segment is essentially independent. Using eq. (5.70a) and assuming simple rectangular pulses, the energy from a single segment is

$$U_{2\omega} \approx \frac{\gamma U_\omega^2 l_T^2}{b t_p} \quad \text{per segment} \tag{5.103}$$

Since the number of segments is L/l_T, the total second-harmonic energy is

$$U_{2\omega} \approx \frac{\gamma U_\omega^2 l_T^2}{b t_p} \frac{L}{l_T} = \frac{\gamma U_\omega^2 L l_T}{b t_p} = \frac{\gamma U_\omega^2 L}{|\Delta(v_g^{-1})| b} \tag{5.104}$$

This is exactly the same result as eq. (5.101), which underscores the close analogy between spatial and temporal walk-off.

Unlike the quasi-continuous-wave case, the second-harmonic output energy scales linearly with crystal length and is proportional to input energy squared, independent of the input pulse width. An important point is that the second-harmonic efficiency continues to grow with increasing crystal length even in the large group velocity mismatch regime, although the generated pulse width increases as well. Thus, the thick crystal (large GVM) regime may be important for applications where high frequency-conversion efficiency is desired but the output pulse duration is not critical [235].

We can easily extend our efficiency analysis to the case of stronger focusing [235]. We proceed by taking A_{eff} in eq. (5.100) to be z dependent according to the

Gaussian beam formula,

$$A_{\text{eff}}(z) = \pi w_{0(\omega)}^2 \left(1 + \frac{z^2}{(b/2)^2}\right) \tag{5.105}$$

and integrating over the length of the crystal. This approach is valid provided that we remain in the large group velocity mismatch limit ($l_T \ll L$) and that the temporal walk-off length remains shorter than the depth of focus ($l_T \ll b$). The resulting expression is

$$U_{2\omega} = \frac{\gamma U_\omega^2}{|\Delta(v_g^{-1})| b} \int_{-L/2}^{L/2} \frac{dz}{1 + z^2/(b/2)^2} = \frac{\gamma U_\omega^2}{|\Delta(v_g^{-1})|} \tan^{-1}\left(\frac{L}{b}\right) \tag{5.106}$$

In the limit of tight focusing and large group velocity mismatch ($l_T \ll b \ll L$), we obtain

$$U_{2\omega} = \frac{\pi}{2} \frac{\gamma U_\omega^2}{|\Delta(v_g^{-1})|} \tag{5.107}$$

This result can be understood on the basis of eq. (5.101) by replacing L with $\pi b/2$, where we use the fact that most of the second-harmonic output is generated near the focus. Within this regime the conversion efficiency is independent not only of the pulse width but also of the focusing, which is again in contrast to the quasi-continuous-wave limit. The focusing dependence described by eq. (5.106) has been verified experimentally [235].

Finally, we discuss the phase-matching response in the large group velocity mismatch limit. Measured second-harmonic spectra are shown in Fig. 5.7b for experiments with 120-fs input pulses from a Ti:S laser upconverted in a noncritically phase-matched $KNbO_3$ crystal with 3.6 ps of group velocity walk-off [235]. For the relatively weakly focused case, the second-harmonic spectrum is symmetric and much narrower than the ca. 10-nm bandwidth of the input pulses. This is explained simply due to the sinc($\Delta kL/2$) phase-matching response in eq. (5.87), with the phase mismatch proportional to the second-harmonic frequency offset. For the tight-focusing case, the second-harmonic spectrum becomes wider and markedly asymmetric. Similar asymmetries that arise for tightly focused Gaussian beams as a function of phase mismatch in continuous-wave SHG were discussed in Section 5.3.3. The pulsed second-harmonic spectrum shown in Fig. 5.7b is, in fact, in good agreement with the tight focusing phase-matching response calculated from the continuous-wave theory of [229], with the phase mismatch taken as proportional to frequency offset according to eq. (5.89). These observations underscore the similarity between phase-matching effects in ultrashort-pulse SHG, where the phase mismatch varies with second-harmonic frequency, and in continuous-wave SHG, where the phase mismatch is usually controlled via crystal angle, temperature, or input frequency.

5.4.3 Quasi-Phase Matching

We have seen that group velocity mismatch can play a key role in determining the duration, bandwidth, and efficiency of the output field in ultrashort-pulse second-harmonic generation. Used in conjunction with a phase-matching technique termed *quasi-phase matching* (QPM), group velocity mismatch effects can also be used to manipulate the *shape* of the

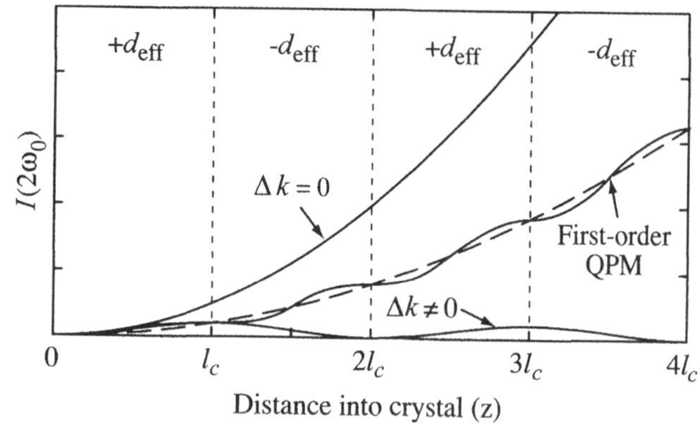

Figure 5.10 Second-harmonic intensity as a function of distance into a nonlinear crystal, for $\Delta k = 0$, $\Delta k \neq 0$, and first-order QPM. The dashed line shows the average growth in the SHG intensity under QPM. Adapted from [238]. Copyright © 1992, IEEE.

output pulses. In this section we first discuss QPM basics and then analyze pulse shaping and spectral patterning techniques based on QPM.

Quasi-phase matching, proposed in [236,237], is an alternative to birefringence phase matching in which a structural periodicity is built into the nonlinear medium, to repeatedly correct the relative phase between the ω_0 and $2\omega_0$ fields. Practically, this periodicity is generally achieved through spatial modulation of the nonlinear coefficient. The most common QPM scheme, where the nonlinear coefficient is assumed to change sign after every coherence length $l_c = \pi/\Delta k$, is depicted in Fig. 5.10. We recall that after one coherence length, the relative phase between $\tilde{E}_{2\omega}$ and $\tilde{P}_{\mathrm{NL}}^{(2\omega)}$ has slipped by π, after which the second-harmonic field begins to lose power back into the ω_0 field. The sign changes in the nonlinear coefficient occur when this π phase slippage has just accumulated, which has the effect of resetting the relative phase back to zero. As a result, the contributions to the second-harmonic field produced by each of the regions of material of length l_c all add. Averaged over a crystal length $L \gg l_c$, the second-harmonic field grows linearly with crystal length, and the intensity grows quadratically, although still less rapidly than the case of perfect phase matching.

To analyze quasi-phase matching, we begin by considering continuous-wave SHG in the nondepleted pump and plane-wave limits, now with the new feature of a spatially varying nonlinear coefficient $d_{\mathrm{eff}}\, g(z)$. Here d_{eff} represents the effective nonlinear coefficient in the absence of spatial modulation as before, and $g(z)$ is the spatial variation, normalized such that $|g(z)| \leq 1$. Following [238,239], a solution for the second-harmonic field can be obtained via direct integration of eq. (5.29), with the result

$$\tilde{E}_{2\omega}(L) = \frac{-j\omega_0 d_{\mathrm{eff}}}{n_{2\omega}c}\, \tilde{E}_{\omega}^2(0)L \left\{ \frac{1}{L} \int_0^L dz\, g(z) e^{j\Delta k z} \right\} \tag{5.108}$$

The output second-harmonic field is proportional to that which would be generated for zero phase mismatch and a spatially uniform nonlinear coefficient, multiplied by the Fourier transform of the spatially varying nonlinear coefficient, evaluated at spatial frequency $-\Delta k$. Note that since $|g| \leq 1$, the magnitude of the multiplicative factor within the braces in eq. (5.108) is also ≤ 1. By comparing with eq. (5.30), we see that the term in braces represents the reduction in the output second-harmonic field amplitude relative to that for the spatially uniform case with no phase mismatch.

Consider now the important case where the nonlinear coefficient has a periodic modulation with period Λ. Using other terminology, we may also say that the nonlinear coefficient forms a periodic grating. In this case the spatial variation may be written

$$g(z) = \sum_q G_q \exp\left(\frac{2\pi jqz}{\Lambda}\right) \tag{5.109}$$

where the Fourier coefficients may be obtained via the Fourier series formula,

$$G_q = \frac{1}{\Lambda}\int_0^{\Lambda} dz\, g(z) \exp\left(\frac{-2\pi jqz}{\Lambda}\right) \tag{5.110}$$

To achieve good quasi-phase matching, one of harmonics of the grating wave vector must nearly equal the phase mismatch; that is, for one of the harmonics m,

$$\Delta k'_m = \Delta k + \frac{2\pi m}{\Lambda} \approx 0 \tag{5.111}$$

The mth harmonic dominates, and all of the other harmonics may be neglected. This situation is termed first-order QPM for $m = \pm 1$, second-order QPM for $m = \pm 2$, and so on. The result for the output field is

$$\tilde{E}_{2\omega}(L) = \frac{-j\omega_0 d_{\text{eff}}}{n_{2\omega}c} \tilde{E}^2_{\omega}(0)L \left[G_m \exp\left(\frac{j\Delta k'_m L}{2}\right) \operatorname{sinc}\left(\frac{\Delta k'_m L}{2}\right)\right] \tag{5.112}$$

This expression, including the sinc function phase-matching response, is similar to our earlier results for the case of a spatially uniform nonlinear medium. The only differences are that the output field is reduced by the factor G_m and the phase mismatch term depends on $\Delta k'_m$, which is simply the actual phase mismatch Δk shifted by the appropriate grating wave vector harmonic.

The spatial variation of $g(z)$ within one spatial period can be tailored to optimize the amplitude of a selected spatial harmonic. For the special case where $g(z) = 1$ for exactly one half of the grating period and -1 for the other, one finds that $G_m = 2/m\pi$ for m odd and zero for m even. The largest efficiency is obtained for first-order QPM. On the other hand, for fixed correlation length l_c, higher-order QPM is less demanding in terms of fabrication, since the spatial period of the QPM grating is larger ($\Lambda = 2|m|l_c$).

The success of QPM for practical applications is enabled by periodic poling of ferroelectric nonlinear optical crystals, such as $LiNbO_3$ and $LiTaO_3$ [240]. During the poling process, a strong DC electric field is used to orient (or pole) the ferroelectric domains in the material. By means of microlithographic methods, a spatial modulation can be imposed onto the applied poling field, which under appropriate conditions is transferred into the crystal as a spatial patterning of the domain orientations. Since flipping the orientation of a crystal domain reverses the sign of the second-order nonlinear coefficient, this method allows fabrication of nonlinear materials with a controlled spatially modulated nonlinearity. Quasi-phase matching offers a number of practical advantages. These include the possibility to utilize ω_0 and $2\omega_0$ polarization combinations for which birefringent phase matching is not possible (which in $LiNbO_3$, for example, allows access to the largest nonlinear coefficients), the possibility to phase-match materials lacking sufficient birefringence, and the ability to engineer nonlinear materials for noncritical phase matching and zero spatial walk-off. Furthermore, the microlithographic patterning process provides the freedom to fabricate aperiodically poled QPM structures, where the QPM grating is itself spatially modulated. As we shall see, this leads to interesting possibilities for shaping of ultrashort pulses.

Let us now consider QPM with broadband ultrashort-pulse fields. Consistent with Section 5.4.2, we use a first-order expansion for the phase mismatch. For an analysis of QPM that includes higher-order terms (i.e., dispersion), see [241]. Working from eq. (5.85), we find that

$$\mathcal{E}_{2\omega}(L,\Omega) = \frac{-j\omega_0 d_{\text{eff}}}{n_{2\omega}c} F(\Omega) \int_0^L dz\, g(z) e^{j\left[\Delta k_0 + \Delta(v_g^{-1})(\Omega - 2\omega_0)\right]z} \tag{5.113}$$

where $F(\Omega)$ is the forcing term given by eq. (5.91b) and $\Delta k_0 = k_{2\omega}(2\omega_0) - 2k_\omega(\omega_0)$. Consider first a periodic QPM grating, with $\Delta k_0 + 2\pi m/\Lambda = 0$. The frequency-dependent phase mismatch for the quasi-phase-matching process becomes

$$\Delta k'_m = \Delta(v_g^{-1})(\Omega - 2\omega_0) \tag{5.114}$$

for which eq. (5.113) simplifies to

$$\mathcal{E}_{2\omega}(L,\Omega) = \frac{-j\omega_0 d_{\text{eff}}}{n_{2\omega}c} F(\Omega) L \left[G_m \exp\left(\frac{j\Delta k'_m L}{2}\right) \operatorname{sinc}\left(\frac{\Delta k'_m L}{2}\right)\right] \tag{5.115}$$

We have assumed here that the same mth harmonic of the QPM grating dominates the entire optical bandwidth. Except for the factor of G_m, this result is nearly identical to those in Section 5.4.2. The output second-harmonic spectrum is proportional to a frequency-domain forcing function, which depends on the input field, multiplied by a filtering function that results from phase-matching effects. In the limit of large GVM, the output pulse in the time domain is simply a square pulse of duration $|\Delta(v_g^{-1})|L$, just as in the case of conventional phase matching.

Let us now generalize to the case of spatially modulated QPM gratings. In the case of poled ferroelectric nonlinear materials, such a spatial modulation has been termed *aperiodic poling*). We represent the normalized nonlinear coefficient as

$$g(z) = \sum_q \tilde{g}_q(z) \exp\left(\frac{2\pi jqz}{\Lambda}\right) = \sum_q \left|\tilde{g}_q(z)\right| \exp\left(\frac{2\pi jqz}{\Lambda} + j\Phi_q(z)\right) \tag{5.116}$$

where the $\tilde{g}_q$ are spatially varying complex amplitude functions. This description is useful provided that the complex amplitude functions are slowly varying relative to l_c. This is equivalent to the condition that the Fourier transform of the nonlinear coefficient should consist of a series of nonoverlapping peaks, each peak corresponding to a specific harmonic q. The spatial modulation broadens the individual peaks and controls their structure, but the peaks remain isolated. Assuming again that $\Delta k_0 + 2\pi m/\Lambda = 0$ and the mth harmonic dominates, the second-harmonic spectrum becomes

$$\mathcal{E}_{2\omega}(L,\Omega) = \frac{-j\omega_0 d_{\text{eff}}}{n_{2\omega}c} F(\Omega)\, H_{\text{QPM}}(\Omega) \tag{5.117}$$

where

$$H_{\text{QPM}}(\Omega) = \int_0^L dz\, \tilde{g}_m(z) e^{j\Delta(v_g^{-1})(\Omega - 2\omega_0)z} \tag{5.118}$$

is an effective filter function for quasi-phase matching which takes into account the spatial modulation of the QPM grating. Following a procedure similar to Section 5.4.2, in the time domain the second-harmonic field generated can be expressed as

$$E_{2\omega}(L,t) = \frac{1}{2}\left[\int_0^{\infty} \frac{d\Omega}{2\pi} \left(\frac{-j\omega_0 d_{\text{eff}}}{n_{2\omega}c}\right) F(\Omega) H'_{\text{QPM}}(\Omega) e^{j\Omega t} + \text{c.c.}\right] \tag{5.119a}$$

where

$$H'_{\text{QPM}}(\Omega) = H_{\text{QPM}}(\Omega)\, e^{-jk_{2\omega}(\Omega)L} \tag{5.119b}$$

The time-domain response is given by

$$E_{2\omega}(L,t) = \frac{1}{2}\left[\frac{-j\omega_0 d_{\text{eff}}}{n_{2\omega}c} \left(a_\omega^2(0,t) e^{2j\omega_0 t}\right) * h'_{\text{QPM}}(t) + \text{c.c.}\right] \tag{5.120}$$

where $h'_{\text{QPM}}(t)$, the inverse Fourier transform of $H'_{\text{QPM}}(\Omega)$, is equal to

$$h'_{\text{QPM}}(t) = \frac{e^{j[2\omega_0 t - k_{2\omega}(2\omega_0)L]}}{\Delta(v_g^{-1})} \tilde{g}_m \left(\frac{-\left(t - v_g^{-1}(2\omega_0)L\right)}{\Delta(v_g^{-1})}\right) \tag{5.121}$$

We see that the filter function is a scaled version of the Fourier transform of the spatial modulation. In the time domain the filter function corresponds to an impulse response function which is a scaled version of the spatial modulation envelope itself.

It is clear from eq. (5.121) that the output pulse shape can be controlled by spatially modulating the QPM grating. The pulse-shaping mechanism arises from an interplay between the GVM, which is intrinsic to the nonlinear material, and the spatial modulation. Each spatial slice of the nonlinear material gives a contribution to the $2\omega_0$ field that arrives at the output with a distinct delay relative to the input ω_0 field. Therefore, changing the amplitude or phase of the second-harmonic contribution from a specific spatial location induces an amplitude or phase change in the output field at the corresponding instant in time [242].

Both the amplitude and the phase of $\tilde{g}_m(z)$ are of interest, as noted explicitly in eq. (5.116), and both may be controlled via the microlithographic patterning process. Modulation of the phase is closely tied to modulation of the QPM grating period. Chirped QPM gratings, where the grating period varies with position, have been applied to compress chirped input pulses during the SHG process [242,243]. The idea is that with chirped gratings, different input

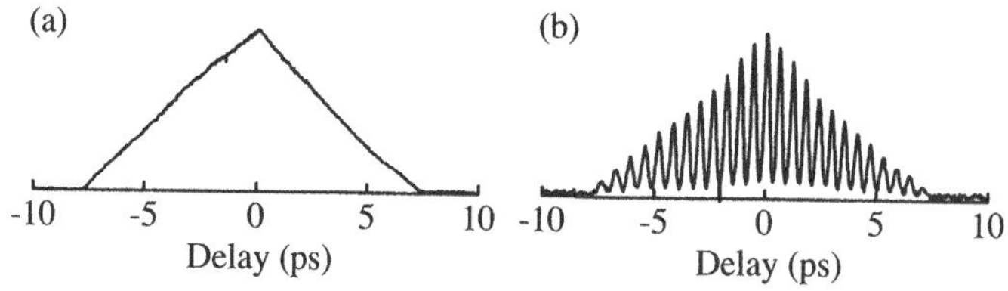

Figure 5.11 Intensity autocorrelation measurements of second-harmonic pulses generated via quasi-phase matched SHG in a 25-mm-long $LiNbO_3$ crystal (input pulses were ca. 250 fs at 1560 nm): (a) uniform QPM grating: the triangular autocorrelation is indicative of a ca. 7.5-ps square pulse; (b) amplitude-modulated QPM grating, consisting of 0.5-mm uniform grating segments alternating with 1.5-mm segments of unmodulated material [i.e., $g(z) = 1$ uniformly, giving $|\tilde{g}_m(z)| = 0$]. The autocorrelation reflects a series of distinct pulses, corresponding to the series of spatially separated grating segments. Both traces are consistent with the 0.3-ps/mm group velocity mismatch. Adapted from [244].

frequencies are quasi-phase matched at different spatial locations, which therefore results in frequency-dependent delays. Modulation of the amplitude can be achieved by varying the structure of the QPM grating within a single grating period, as described by eq. (5.110). For example, for cases where $g(z)$ alternates between $+1$ and -1, $|\tilde{g}_m(z)|$ may be controlled by changing the duty cycle of the grating (i.e., by changing the relative widths of the $+1$ and -1 regions). Examples of data illustrating pulse shaping effects in QPM are shown in Fig. 5.11. In general, the duration of the shortest features in the output field is limited to approximately the inverse bandwidth of the forcing function, $F(\Omega)$, while the maximum duration of the impulse response function is limited to the total GVM, $|\Delta(v_g^{-1})|L$. In the frequency domain, the sharpest features that can be placed onto the effective filter function have a spectral width approximately equal to the inverse of the total GVM, $1/|\Delta(v_g^{-1})|L$. The total bandwidth of $H_{\mathrm{QPM}}(\Omega)$ can be made very large if the QPM grating is given a sufficiently rapid spatial modulation. On the other hand, using Parseval's theorem, we have

$$\frac{1}{2\pi}\int d\Omega\, \left|H_{\mathrm{QPM}}(\Omega)\right|^2 = \frac{1}{|\Delta(v_g^{-1})|}\int dz\, |\tilde{g}_m(z)|^2 \leq \frac{|G_m|^2 L}{|\Delta(v_g^{-1})|} \tag{5.122}$$

G_m is the Fourier coefficient for a perfectly periodic QPM grating, following eq. (5.110), which we assume has been designed to optimize the mth grating harmonic. When the grating is spatially modulated, the power spectral density of the effective filter function must decrease in proportion to the increase in bandwidth.

5.4.4 Effect of Group Velocity Walk-off on SHG-Based Pulse Measurements

Second-harmonic generation plays an important role in ultrashort pulse measurement, specifically in intensity correlation and SHG-FROG. As we have seen, group velocity mismatch in type I SHG acts to narrow the second harmonic spectrally and to broaden it in time. This prompts the following question: How does group velocity mismatch affect ultrashort-pulse measurement techniques based on SHG [245]?

From Chapter 3, intensity autocorrelation measurements via SHG ideally yield the function

$$G_2(\tau) \sim \int dt \, |E_\omega(t)E_\omega(t-\tau)|^2 \tag{5.123}$$

where either fringe averaging or a noncollinear geometry are assumed. SHG-FROG measurements ideally take the form

$$I_{\text{FROG}}(\Omega, \tau) \sim \left| \int dt \, E_\omega(t)E_\omega(t-\tau)e^{-j\Omega t} \right|^2 \tag{5.124}$$

By applying Parseval's theorem, eq. (1.84), one can show that SHG-FROG and intensity autocorrelation are related. In particular, the SHG-FROG delay marginal is equal to the intensity autocorrelation, that is,

$$G_2(\tau) \sim \int d\Omega \, I_{\text{FROG}}(\Omega, \tau) \tag{5.125}$$

For type I phase matching, we include the effect of group velocity mismatch through the effective filter function that acts on the second-harmonic field. This approach is valid both for birefringent phase matching and quasi-phase matching. The second-harmonic field that is produced in either autocorrelation or FROG is

$$E_{2\omega}(t) \sim \int_0^\infty d\Omega \, H(\Omega)e^{j\Omega t} \int dt' \, E_\omega(t')E_\omega(t'-\tau)e^{-j\Omega t'} + \text{c.c.} \tag{5.126}$$

$H(\Omega)$ is the filtering function, which can account for finite phase-matching bandwidth as well as any other second-harmonic filtering effects, such as spectral variation in the detection system. The experimental FROG trace, denoted $I'_{\text{FROG}}(\Omega, \tau)$, follows immediately:

$$I'_{\text{FROG}}(\Omega, \tau) \sim |H(\Omega)|^2 \left| \int dt' \, E_\omega(t')E_\omega(t'-\tau)e^{-j\Omega t'} \right|^2 \tag{5.127}$$

The experimental autocorrelation trace, $G'_2(\tau)$, becomes

$$G'_2(\tau) \sim \int dt \left| \int_0^\infty d\Omega \, H(\Omega)e^{j\Omega t} \int dt' \, E_\omega(t')E_\omega(t'-\tau)e^{-j\Omega t'} \right|^2 \tag{5.128}$$

This expression is considerably simplified by applying Parseval's theorem, which gives

$$G'_2(\tau) \sim \int d\Omega \, I'_{\text{FROG}}(\Omega, \tau) \tag{5.129}$$

This shows that the experimental FROG trace and experimental intensity autocorrelation trace are related in the same way as the ideal FROG and intensity autocorrelation traces.

Within our first-order frequency expansion approximation, eqs. (5.127) and (5.129) completely specify the effect of the group velocity mismatch. We can immediately make two important observations:

- When $|H(\Omega)|$ is sufficiently broadband that it may be taken as constant over the full optical bandwidth, the experimental autocorrelation and FROG trace values are equal to the ideal values. There is no measurement distortion.
- The spectral phase of the filtering function does not play a role. Only the magnitude of $H(\Omega)$ is relevant.

To gain further insight into the effect on autocorrelation measurements, we consider some specific examples [245]. First, consider the special case where

$$E_\omega(t)E_\omega(t-\tau) = \mathcal{E}_1(\tau)\mathcal{E}_2(t - s(\tau)) \tag{5.130}$$

That is, the product of the two time-delayed fields results in a time dependence whose shape is independent of delay, but with a delay-dependent amplitude and time shift. This condition applies both for a complex Gaussian (Gaussian with a linear chirp) and for an unchirped single-sided exponential. In this special case, eq. (5.128) simplifies to yield

$$G_2'(\tau) \sim |\mathcal{E}_1(\tau)|^2 \tag{5.131}$$

For these particular pulse shapes, the experimental intensity autocorrelation trace is independent of group velocity mismatch or other spectral filtering of the second harmonic, and $G_2'(\tau) = G_2(\tau)$. In contrast to the SHG of a single pulse, where the GVM can lead to substantial pulse broadening, group velocity mismatch does not lead to direct broadening of the autocorrelation measurement.

For pulse shapes not satisfying eq. (5.130), the GVM in general does affect the autocorrelation trace. Figure 5.12 shows $G_2'(\tau)$ traces calculated for a square input pulse, $a_\omega(t) \sim \mathrm{sq}(t/t_p)$, for several values of normalized GVM. Perfect phase matching, $\Delta k = 0$, is assumed for second-harmonic frequency $\Omega = 2\omega_0$. The apparent correlation curve is

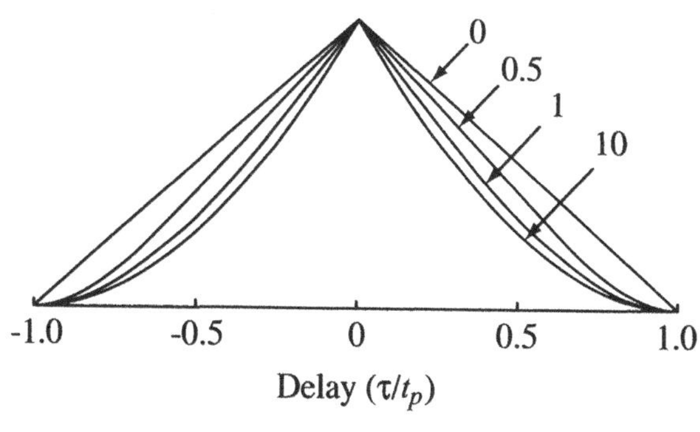

Figure 5.12 Calculated intensity autocorrelation traces as a function of normalized delay, $\tilde{\tau} = \tau/t_p$, taking into account the influence of group velocity mismatch in the second-harmonic crystal. A square input pulse $a_\omega(t) \sim \mathrm{sq}(t/t_p)$ is assumed, with differing amounts of GVM: $\left|\Delta(v_g^{-1})\right| L/t_p = 0, 0.5, 1,$ and 10. From [245]. Copyright © 1983 IEEE.

distorted, and its width decreases with increasing normalized GVM. In the limit of very large GVM, the curve is only 59% as wide as the ideal autocorrelation trace. What is happening can easily be discerned. The product of two time-delayed square pulses becomes shorter and shorter as the delay is increased up to a value of t_p. In the frequency domain, the forcing function for the second harmonic, $F(\Omega)$, becomes broader and broader. Therefore, for fixed GVM, an increase in delay leads to a smaller fraction of the "energy" under $F(\Omega)$ being passed by $H(\Omega)$, the effective filter function. This acts to suppress the correlation signal with increasing delay. On the other hand, for $|\tau| > t_p$, the two time-delayed square pulses have zero overlap, so no second harmonic is generated, independent of the value of the GVM. For this reason, $G_2'(\tau)$ and $G_2(\tau)$ must go to zero at the same delay. As a second example, the second-harmonic trace measured for a secant hyperbolic pulse, $a_\omega(t) \sim \text{sech}(t/t_p)$, can be either slightly narrower or slightly wider than the ideal autocorrelation, depending on the phase mismatch and GVM. In the large GVM limit with $\Delta k = 0$, the trace is broadened by about 10% compared to the ideal autocorrelation.

We observe from these examples that the shape of $G_2'(\tau)$ can indeed be distorted by GVM, while its width can be slightly increased, slightly decreased, or unchanged. However, in all these cases, the width remains on the same order as the width of the ideal intensity autocorrelation function. This behavior is expected for all bandwidth-limited pulses.

Stronger effects are possible for chirped pulses. We know that the low-frequency end of the second-harmonic spectrum arises only when the low-frequency parts of the fundamental spectrum mix with each other, and similarly for the high end of the spectrum. By tuning the effective filter function $H(\Omega)$ (e.g., by varying the phase mismatch), different portions of the second-harmonic spectrum can be selected, which preferentially correspond to different regions of the fundamental spectrum. However, particularly for nonlinearly chirped pulses, different regions of the fundamental frequency spectrum may have very different time behavior, which can lead to strong distortions in the second-harmonic trace. For an example, we can refer to the SHG-FROG trace for a pulse with a moderate level of cubic spectral phase, shown in Fig. 3.20. From eq. (5.125) we know that the autocorrelation trace is simply the integral of the SHG-FROG trace over the frequency variable. However, the filter function arising due to GVM selects out a frequency slice of the FROG trace. We see from the figure that different frequency slices can have significantly different durations!

The effect of GVM or other filtering of the second-harmonic spectrum on SHG-FROG is discussed in [119]. An interesting point is that the redundant information available in the FROG trace allows for some correction of the data. As noted in eq. (3.97), the SHG-FROG marginal frequency distribution is equal to the autoconvolution of the optical power spectrum at the fundamental wavelength:

$$\int d\tau\, I_{\text{FROG}}(\Omega, \tau) \sim \int d\omega\, |\mathcal{E}_\omega(\omega)|^2\, |\mathcal{E}_\omega(\Omega - \omega)|^2 \tag{5.132}$$

This offers a useful consistency check for measurement data: The SHG-FROG trace can be compared to the results from the independently measured input power spectrum. The frequency marginal can also be used to estimate the effective filter function. By integrating eqs. (5.124) and (5.127) and comparing the results, we find that

$$|H(\Omega)|^2 = \frac{\int d\tau\, I'_{\text{FROG}}(\Omega, \tau)}{\int d\tau\, I_{\text{FROG}}(\Omega, \tau)} \sim \frac{\int d\tau\, I'_{\text{FROG}}(\Omega, \tau)}{\int d\omega\, |\mathcal{E}_\omega(\omega)|^2\, |\mathcal{E}_\omega(\Omega - \omega)|^2} \tag{5.133}$$

Finally, by dividing by this estimate of $|H(\Omega)|^2$, the data can be corrected to yield a more accurate estimate of the ideal FROG trace through the formula

$$I_{\text{FROG}}(\Omega, \tau) \sim I'_{\text{FROG}}(\Omega, \tau) \frac{\int d\omega\, |\mathcal{E}_\omega(\omega)|^2\, |\mathcal{E}_\omega(\Omega - \omega)|^2}{\int d\tau\, I'_{\text{FROG}}(\Omega, \tau)} \tag{5.134}$$

This procedure was shown to lead to improved results in measurements on the ca. 10 fs time scale [119]. However, we do note two limitations. First, if $|H(\Omega)|$ has any zeros within the bandwidth of the ideal FROG trace, part of the FROG trace is irretrievably lost. Therefore, this correction procedure is limited to moderate amounts of GVM. Second, the assumption that the filter is a function only of the second-harmonic frequency must remain valid. As we have learned, this is equivalent to a first-order expansion of the wave vectors in the frequency variables; spreading of the individual pulses due to group velocity dispersion in the crystal cannot be compensated.

Finally, we discuss briefly pulse measurements for the case of type II phase matching, in which two nonidentical fundamental waves (one ordinary, one extraordinary) interact to produce the second-harmonic wave. Compared to the preceding analysis, the new feature is that the group velocities of the two fundamental waves are no longer identical. This means that the delay between the two fundamental pulses will vary according to the position in the nonlinear crystal. Because the faster pulse can catch up to and even pass through the slower one, GVM directly broadens the experimental autocorrelation trace. In the large GVM limit, the measurement width becomes simply $|\Delta(v_g^{-1})|L$. Furthermore, because the sign of τ determines whether the two pulses get closer or separate in the medium, the second-harmonic energy need not remain an even function of delay. More details are given in [245]. Because GVM leads to a direct broadening of the experimental trace, type II phase matching can be an unfavorable choice for measurement of very short pulses.

5.5 THREE-WAVE INTERACTIONS

The most general frequency-conversion process supported by second-order nonlinear optics is a three-wave interaction, involving three waves satisfying the energy conservation condition $\omega_1 + \omega_2 = \omega_3$ [7,69,131,246]. We use the convention that ω_3 represents the highest-frequency wave. To have good efficiency, the phase matching or momentum conservation condition should also be satisfied: namely, $\mathbf{k_1} + \mathbf{k_2} = \mathbf{k_3}$. Practically, there are two main cases:

1. In *sum frequency generation*, two input waves at frequencies ω_1 and ω_2 interact to drive a new higher-frequency output wave. Second-harmonic generation is a special case of sum frequency generation. The main difference is that sum frequency generation involves three distinguishable waves, whereas in SHG only two waves are distinguishable (at least for type I phase matching, which we assumed in much of our analysis). Applications of sum frequency generation include generation of coherent ultraviolet radiation and time-resolved gating of weak luminescence signals in ultrafast spectroscopy.
2. In *difference frequency generation*, input waves at ω_3 and ω_1 interact to produce a new wave at ω_2. Closely related is *optical parametric amplification*: In the presence

of a strong pump wave at ω_3, the lower-frequency wave at ω_1 can be amplified, often very strongly. These processes are used to generate tunable femtosecond radiation ranging from the visible to the midinfrared.

We obtain the basic equations governing short pulse three-wave interactions by using the time-domain formulation of the nonlinear wave equation (Section 5.2.2). For plane waves the field is written

$$E(z,t) = \frac{1}{2}\left[\sum_{i=1}^{3} a_i(z,t)e^{j(\omega_i t - k_i z)} + \text{c.c.}\right] \tag{5.135}$$

where the fields are taken as scalars and k_i represents the value of the propagation vector at frequency ω_i. We will also be using the nomenclature n_i and $v_{g,i}$ for the refractive index and the group velocity evaluated at ω_i. The nonlinear polarization is given by eq. (5.27), which yields

$$\begin{aligned} P_{NL}(z,t) = \varepsilon_0 d_{\text{eff}}[&a_1 a_2 e^{j(\omega_3 t - (k_1+k_2)z)} \\ &+ a_3 a_1^* e^{j(\omega_2 t - (k_3-k_1)z)} + a_3 a_2^* e^{j(\omega_1 t - (k_3-k_2)z)} + \text{c.c.}] \end{aligned} \tag{5.136}$$

Here nonlinear polarization terms at frequencies other than ω_1, ω_2, and ω_3 have been omitted. Inserting the nonlinear polarization into eq. (5.25) yields the basic equations for three-wave parametric interactions:

$$\begin{aligned} \frac{\partial a_1}{\partial z} + \frac{1}{v_{g,1}}\frac{\partial a_1}{\partial t} &= -j\kappa_1 a_3 a_2^* e^{-j\Delta k z} \\ \frac{\partial a_2}{\partial z} + \frac{1}{v_{g,2}}\frac{\partial a_2}{\partial t} &= -j\kappa_2 a_3 a_1^* e^{-j\Delta k z} \\ \frac{\partial a_3}{\partial z} + \frac{1}{v_{g,3}}\frac{\partial a_3}{\partial t} &= -j\kappa_3 a_1 a_2 e^{j\Delta k z} \end{aligned} \tag{5.137a}$$

with the coupling constants and the phase-mismatch term given by

$$\kappa_i = \frac{\omega_i d_{\text{eff}}}{2n_i c} \quad \text{and} \quad \Delta k = k_3 - k_1 - k_2 \tag{5.137b}$$

The same phase mismatch Δk governs any three-wave interaction involving the same three frequencies ω_1, ω_2, and ω_3, including both sum and difference frequency generation.

In the case of continuous-wave fields, the time derivative terms in eq. (5.137a) may be dropped. For lossless fields for which d_{eff} is real, it is then easy to show that

$$\frac{1}{\omega_1}\frac{\partial I_1}{\partial z} = \frac{1}{\omega_2}\frac{\partial I_2}{\partial z} = \frac{-1}{\omega_3}\frac{\partial I_3}{\partial z} \tag{5.138}$$

Since $\hbar\omega_i$ is the energy of a photon at frequency ω_i, eq. (5.138), known as the *Manley–Rowe relation*, is a conservation law for the photon numbers [247]. In particular, the *Manley–Rowe*

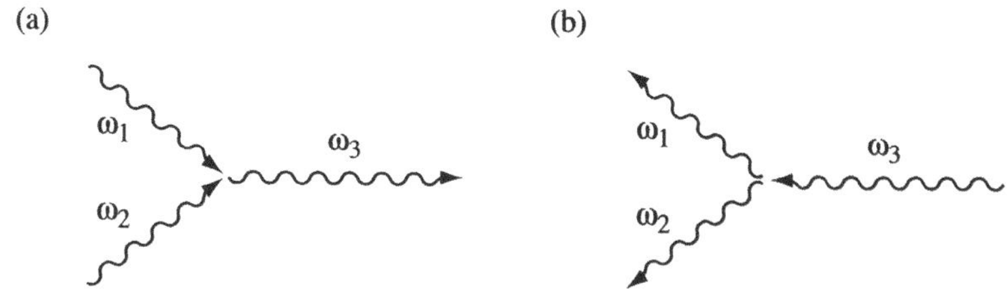

Figure 5.13 (a) Photons at ω_1 and ω_2 combining to create a photon at ω_3; (b) photon at ω_3 splitting to create a pair of photons at ω_1 and ω_2.

relation says that photons at ω_1 and ω_2 combine in pairs to create new photons at ω_3, or conversely photons at ω_3 can be annihilated to form pairs of photons at ω_1 and ω_2. These processes are sketched in Fig. 5.13.

Returning now to pulsed fields, we note that in general the three waves have three different group velocities. Therefore, from the time-domain perspective, there are two different group velocity mismatches that play an important role in determining the character of the interaction. Equivalently, we can also consider the phase-matching condition as a function of frequency. Note that our time-domain formulation already takes into account the frequency dependence of the propagation constants, expanded to first order. Nevertheless, if we explicitly look at the phase mismatch from a frequency-domain perspective, we have, to first order,

$$\Delta k(\Delta\omega_1, \Delta\omega_2, \Delta\omega_3) = \frac{\partial k_3}{\partial\omega_3}\Delta\omega_3 - \left[\frac{\partial k_1}{\partial\omega_1}\Delta\omega_1 + \frac{\partial k_2}{\partial\omega_2}\Delta\omega_2\right] \tag{5.139}$$

Here we replaced ω_i with $\omega_i + \Delta\omega_i$ and have assumed that the crystal is adjusted for zero phase mismatch at $\Delta\omega_1 = \Delta\omega_2 = \Delta\omega_3 = 0$. Now applying the energy conservation condition, which implies that $\Delta\omega_2 = \Delta\omega_3 - \Delta\omega_1$, we obtain

$$\Delta k = \left(\frac{\partial k_3}{\partial\omega_3} - \frac{\partial k_2}{\partial\omega_2}\right)\Delta\omega_3 - \left(\frac{\partial k_1}{\partial\omega_1} - \frac{\partial k_2}{\partial\omega_2}\right)\Delta\omega_1 \tag{5.140}$$

Since $\partial k_i/\partial\omega_i = (v_{g,i})^{-1}$, we see that the mismatches in the inverse group velocities play a determining role in the frequency variation of the phase mismatch. Compared to type I phase-matched SHG, a key difference is that because the low frequency waves may now have different group velocities, the phase mismatch depends on $\Delta\omega_1$, the frequency variation of one of the low-frequency waves, as well as on $\Delta\omega_3$. One consequence of eq. (5.140) is that to maintain phase matching for a fixed crystal orientation, frequency changes in the low-frequency waves must track frequency changes in the high-frequency wave according to the relation

$$\frac{\Delta\omega_1}{\Delta\omega_3} = \frac{\left(v_{g,3}\right)^{-1} - \left(v_{g,2}\right)^{-1}}{\left(v_{g,1}\right)^{-1} - \left(v_{g,2}\right)^{-1}} \tag{5.141}$$

The interaction of three distinguishable waves with three different group velocities gives rise to a rich set of phenomena, which is difficult to catalog exhaustively. Instead, our approach will be to consider certain specific examples, in order to illustrate important concepts.

5.5.1 Sum Frequency Generation

First we discuss sum frequency generation. Input waves at ω_1 and ω_2 are assumed to propagate without change at their respective group velocities in the nonlinear crystal. This fixed-field or nondepleted pump approximation is valid as long as the sum frequency field at ω_3 remains weak compared to both of the input fields. With the additional assumption that $\Delta k = 0$, the equation for a_3 becomes

$$\frac{\partial a_3}{\partial z} + \frac{1}{v_{g,3}}\frac{\partial a_3}{\partial t} = -j\kappa_3 a_1\left(t - \frac{z}{v_{g,1}}\right) a_2\left(t - \frac{z}{v_{g,2}}\right) \tag{5.142}$$

For input pulses sufficiently long that group velocity walk-off may be neglected, the result is closely analogous to our treatment of SHG in the quasi-continuous limit (Section 5.4.1). We find, for example, that the intensity of the sum frequency wave, I_3, is given by

$$I_3(L, t_3') = \frac{\omega_3^2 d_{\text{eff}}^2}{2c^3\varepsilon_0 n_1 n_2 n_3} I_1(t_3') I_2(t_3') L^2 \tag{5.143}$$

where $t_3' = t - z/v_{g,3}$ is the retarded time corresponding to a reference frame moving with the group velocity of a_3. This expression is very similar to the SHG result, eq. (5.77).

More interesting to discuss is the case where group velocity walk-off is significant. Working in the retarded time frame, the propagation equation is written

$$\frac{\partial a_3(z, t_3')}{\partial z} = -j\kappa_3 a_1(t_3' - \eta_{13} z) a_2(t_3' - \eta_{23} z) \tag{5.144}$$

where we have introduced

$$\eta_{ij} = \frac{1}{v_{g,i}} - \frac{1}{v_{g,j}} \tag{5.145}$$

as a shorthand for the various inverse group velocity differences. The solution for the sum frequency wave is

$$a_3(L, t_3') = -j\kappa_3 \int_0^L dz\, a_1(t_3' - \eta_{13} z) a_2(t_3' - \eta_{23} z) \tag{5.146}$$

This is the general result for the case of zero phase mismatch and nondepleted pump waves.

To gain further insight, we examine the case where one of the input waves is very short and can be approximated as a delta function:

$$a_1(0, t) = \mathcal{A}_1 \delta(t - \tau) \qquad \text{where} \qquad \mathcal{A}_1 = \sqrt{\frac{2\mathcal{F}_1 T_1}{c n_1 \varepsilon_0}} \tag{5.147}$$

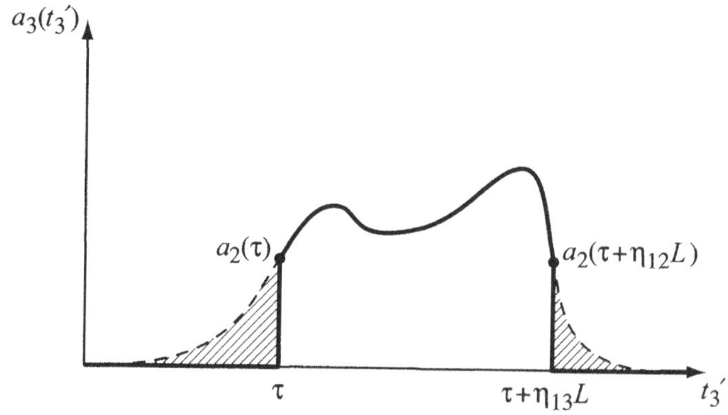

Figure 5.14 Temporal profile of the sum frequency generation output field a_3 when the input a_1 field may be taken as a delta function. The solid line represents a_3, while the dashed line is input field a_2, stretched according to eq. (5.148). For clarity a_3 and a_2 are shown normalized to the same amplitude.

The proportionality constant $\mathcal{A}_1$ is written in terms of the fluence $\mathcal{F}_1$ and the pulse width T_1.[7] After evaluation of the integral, we obtain the following expression for a_3:

$$a_3(L, t_3') = \frac{-j\kappa_3 \mathcal{A}_1}{|\eta_{13}|} \text{sq}\left(\frac{t_3' - \tau}{\eta_{13} L} - \frac{1}{2}\right) a_2\left(\frac{\eta_{12}(t_3' - \tau)}{\eta_{13}} + \tau\right) \tag{5.148}$$

The output field, sketched in Fig. 5.14, is proportional to a temporally windowed and temporally stretched (or compressed) version of the a_2 input field. This result has interesting implications for applications, discussed below.

One straightforward application is in the area of pulse measurement, where sum frequency generation can be used to perform intensity cross-correlation measurements between a sampling pulse a_1 and an unknown pulse a_2 at a different frequency. Similar to autocorrelation measurements, the measurement involves recording the time-integrated sum frequency intensity (i.e., energy) as a function of delay τ. Such a measurement is insensitive to the temporal structure of a_3. However, we infer easily from eq. (5.148) that the measured energy contains contributions from $a_2(t_3')$ for t_3' between τ and $\tau + \eta_{12}L$. Thus, even for a delta-function reference pulse, the measurement gives a minimum time resolution of $|\eta_{12}L|$, which is the total group velocity walk-off between the two input fields. For finite-duration sampling pulses, the minimum time resolution is, of course, increased further. Similar considerations apply for pulse measurements using SHG with type II phase matching [245], since the input waves again have different group velocities, and for ultrafast luminescence spectroscopy, in which a short visible pulse samples weak luminescence signals (often in the infrared) through a sum frequency generation (upconversion) process [248,249]. The use of such ultrafast luminescence spectroscopy to monitor the time- and energy-dependent distribution function of hot carriers in semiconductors is discussed in Chapter 9.

Another application is in the area of short pulse frequency conversion. As in SHG, it is usually desired not only to upconvert an input pulse with good efficiency, but also to

[7] For simplicity we can consider a_1 to be a square pulse, in which case the meaning of T_1 is obvious. For general pulse shapes, T_1 is an effective pulse width given by $T_1 = \left|\int dt\, a_1(t)\right|^2 / \int dt\, |a_1(t)|^2$.

preserve the ultrashort duration of the input pulse. To examine this situation, we consider eq. (5.148) in the long crystal limit, where the group velocity walk-off between the two input pulses is large compared to the duration T_2 of the a_2 field: $|\eta_{12}L| \gg T_2$. Provided that the magnitude and sign of the initial time offset (τ) is chosen so that a_1 walks completely through a_2, the output field simplifies to

$$a_3(L, t_3') = \frac{-j\kappa_3 A_1}{|\eta_{13}|} a_2 \left(\frac{\eta_{12}(t_3' - \tau)}{\eta_{13}} + \tau \right) \tag{5.149}$$

The perhaps surprising result is that the duration of upconverted field becomes independent of the temporal walk-off between the two input fields. a_3 is simply a temporally scaled version of a_2, with the scale factor determined by the ratio of the group velocity differences η_{12} and η_{13}. Either stretching or compression is possible, as well as time reversal, depending on the group velocity mismatches!

We can understand this result as follows. The very short input pulse a_1 overlaps with and upconverts different "temporal slices" of a_2 at different spatial locations within the crystal. Once upconverted, the pulse propagates with velocity $v_{g,3}$ and walks off from the input waves. The total temporal walk-off at the end of the crystal depends on the location in the crystal where the upconverted signal was generated. Thus, the η_{12} group velocity mismatch maps the input temporal profile $a_2(t_3')$ into a temporal–spatial nonlinear polarization within the crystal, which in turn is mapped into the $a_3(L, t_3')$ output temporal profile through the η_{13} group velocity mismatch. This process is analogous to the pulse-shaping mechanism considered in Section 5.4.3, based on the interplay between GVM and spatially patterned QPM gratings, with the difference that in sum frequency generation, the spatially patterning of the nonlinear polarization itself arises due to temporal walk-off between distinct input waves.

Figure 5.15 shows an experimental example involving type II SHG in a 10-mm-thick LBO nonlinear crystal. The a_1 and a_2 waves are a short (ca. 130 fs) reference pulse and a longer shaped signal waveform, respectively, both from a mode-locked Ti:S laser at 800 nm. The signal waveform (Fig. 5.15a) consists of two main pulses separated by 980 fs, as well as several much smaller pulses. The polarizations of a_1 and a_2 are orthogonal; one field propagates as an o-wave while the other propagates as an e-wave. The inverse group velocities of the input fields and the 400-nm sum frequency field are given in the figure caption. Figure 5.15b shows the observed sum frequency waveform obtained with the reference pulse propagating as an o-wave and the signal field propagating as an e-wave. In this configuration $\eta_{12}/\eta_{13} = 0.8$, and a_3 should be stretched by a factor of 1.25 compared to the input signal field. This is consistent with the data, which shows two output pulses separated by 1.22 ps. The smaller pulses evident in Fig. 5.15a do not appear, as these are outside the $|\eta_{12}|L$ time aperture for the interaction (ca. 1.46 ps in this experiment). Figure 5.15c shows the output obtained with the input polarizations interchanged: Now the reference and signal fields propagate, respectively, as an e-wave and an o-wave. In this configuration $\eta_{12}/\eta_{13} \approx -4$, and a_3 should be compressed by a factor of 4 compared to the signal field and time reversed. This is again consistent with the data, where the two output peaks are now separated only by 260 fs. Note that the individual output pulses are much less compressed, as their durations are limited by the finite width of the reference pulse.

The conversion efficiency in sum frequency generation, in terms of the input and output fluences, is obtained in the long crystal limit by integrating eq. (5.149) with respect to time.

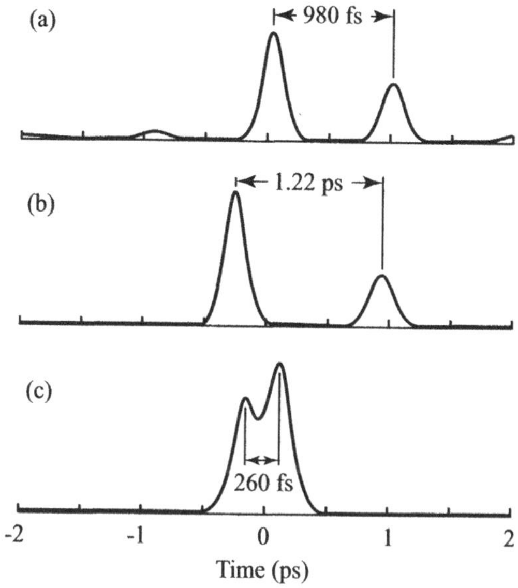

Figure 5.15 Cross-correlation measurements of intensity profiles of (a) input signal field and (b,c) output fields in a sum frequency generation experiment using 10-mm LBO crystal. The duration of the reference pulse is 130 fs (not shown). Inverse group velocities are 5289 fs/mm for the input o-wave, 5435 fs/mm for the input e-wave, and 5472 fs/mm for the sum frequency wave. (b) a_1 is an o-wave; a_2 is an e-wave; (c) a_1 is an e-wave; a_2 is an o-wave. Adapted from [250]. Copyright © 2004, IEEE.

The result is

$$\mathcal{F}_3 = \frac{\omega_3^2 d_{\text{eff}}^2}{2n_1 n_2 n_3 c^3 \varepsilon_0} \frac{\mathcal{F}_1 \mathcal{F}_2 T_1}{|\eta_{12}\eta_{13}|} \tag{5.150}$$

The fluence of the upconverted wave is proportional to the product of the input fluences. This is in analogy to the large temporal walk-off result for SHG, eq. (5.99), where the output fluence was proportional to the square of the input fluence. However, since the interaction length is less than the crystal length, the result is independent of the actual crystal length, unlike the SHG case. To take the analogy further, as in eqs. (5.102) to (5.104), we can divide the nonlinear crystal into segments of length l_T, which we take here as the walk-off length between a_1 and a_3, $l_T = T_1/|\eta_{13}|$. The total interaction length is $L_{\text{eff}} = T_2/|\eta_{12}|$. The total output fluence is equal to the fluence generated per segment, obtained using the quasi-continuous-wave formula for sum frequency generation, eq. (5.143), multiplied by the number of segments within the total interaction length. This gives

$$\mathcal{F}_3 = \left[\frac{\omega_3^2 d_{\text{eff}}^2}{2c^3 \varepsilon_0 n_1 n_2 n_3} I_1(t_3') I_2(t_3') l_T^2 T_1 \right] \times \frac{L_{\text{eff}}}{l_T} \tag{5.151}$$

Substituting the expressions for L_{eff} and l_T and assuming simple rectangular pulses returns eq. (5.150) exactly.[8]

5.5.2 Difference Frequency Generation

By using the close analogy with sum frequency generation, we can quickly obtain results for difference frequency generation for the case $\Delta k = 0$ and with the fixed-field approximation for both of the input waves (a_1 and a_3). The solution for a_2, the new field generated at the difference frequency, is

$$a_2(L, t_2') = -j\kappa_2 \int_0^L dz\, a_1^*(t_2' - \eta_{12}z)a_3(t_2' - \eta_{32}z) \tag{5.152}$$

where $t_2' = t - z/v_{g,2}$ is the retarded time for a frame moving with the group velocity of a_2. According to eq. (5.147), we consider the special case where the a_1 input field is very short and can be approximated as a delta function. The resulting difference frequency field is

$$a_2(L, t_2') = \frac{-j\kappa_2 A_1}{|\eta_{12}|}\mathrm{sq}\left(\frac{t_2' - \tau}{\eta_{12}L} - \frac{1}{2}\right) a_3\left(\frac{\eta_{13}(t' - \tau)}{\eta_{12}} + \tau\right) \tag{5.153}$$

Again the output field is related to a scaled version of the longer of the input fields.

To illustrate these effects, we discuss the short pulse difference frequency generation experiments of [251–253]. Short (ca. 15 fs) input pulses centered at 780 nm with a ca. 70-nm bandwidth are incident on a moderately thick (0.5 to 1.0 mm) GaSe nonlinear crystal. In this simple experimental arrangement, phase-matched difference frequency mixing between higher- and lower-frequency components contained within the same input pulse results in generation of mid-infrared pulses tunable in the range 7 to 20 μm. That is, both a_3 and a_1 are contained within the same 780-nm input pulse. Figure 5.16 shows examples of experimental electric field output waveforms at 13.7 μm, with important parameters given in the caption. For a single input pulse (Fig. 5.16a), the interaction length in the crystal is approximately $\sqrt{2}T_1/|\eta_{13}| \approx 130$ μm, where the $\sqrt{2}$ factor takes into account the fact that the duration of a_3 is the same as a_1. The expected midinfrared output pulse duration is obtained by multiplying by $|\eta_{12}|$, which yields 120 fs, considerably longer than the input pulses and therefore in qualitative agreement with the data. Experiments were also performed with a pair of input pulses separated by 60 fs. The resulting mid-infrared waveform (Fig. 5.16b) now consists of two mid-infrared pulses separated by 400 fs, much larger than the separation of the input pulse pair. This initially surprising behavior is explained by eq. (5.153), which predicts an output pulse separation $|\eta_{12}/\eta_{13}|$ times as large as the input separation.

For the data shown above, the output pulse generated consisted of several midinfrared cycles; and our time-domain formulation, which is limited to a first-order expansion of the propagation constants, is at least approximately valid. However, for very thin nonlinear

[8] The validity of the argument in this paragraph requires that $l_T \ll L_{\mathrm{eff}}$. This condition is satisfied automatically if the duration of the wave assumed to be a delta function, T_1, is much shorter than the duration of the other input wave, T_2, and of the generated output wave, $T_2\, |\eta_{13}/\eta_{12}|$.

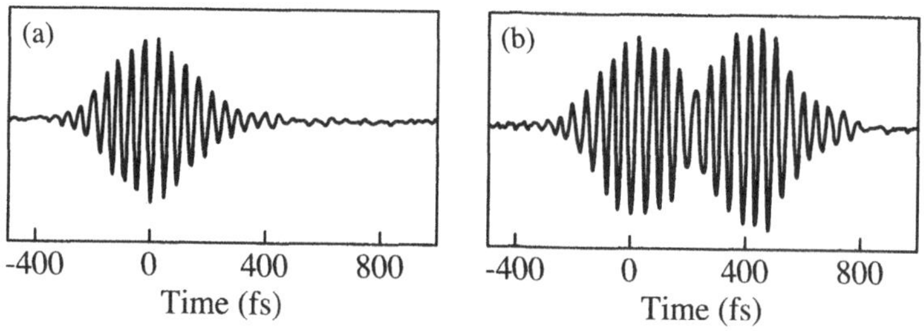

Figure 5.16 Electric field profiles of 13.7-μm pulses generated by short-pulse difference frequency generation in GaSe, using ca. 15-fs input pulses at 780 nm: (a) input: a single pulse; (b) input: a pulse pair separated by 60 fs. The output field profiles are measured by electro-optic sampling (see Chapter 10). The group velocity mismatches are estimated to be $|\eta_{13}| \approx 160$ fs/mm; $|\eta_{12}| \approx 850$ fs/mm. Adapted from [253].

crystals for which the group velocity walk-off is made small, the output can be very broadband, consisting of as few as 1 to 2 midinfrared cycles [254]. In such cases the time-domain slowly varying envelope approximation breaks down, and the full frequency-dependent dispersion must be considered. This situation can be handled using the frequency-domain formulation (Section 5.2.1). Assuming fixed (nondepleted) amplitudes for the input waves, the spectral envelope function for the nonlinear polarization driving the difference frequency field can be written

$$\mathcal{P}_{\mathrm{NL}}(z,\Omega) = \frac{\varepsilon_0 d_{\mathrm{eff}}}{2\pi} \int_0^{\infty} d\omega\, \mathcal{E}_3(\omega)\mathcal{E}_1^*(\omega-\Omega)\, e^{j[k_1(\omega-\Omega)-k_3(\omega)]z} \tag{5.154}$$

where Ω represents the spectral components of the output wave centered near ω_2. Substituting into eq. (5.17), we obtain

$$\frac{\partial \mathcal{E}_2(z,\Omega)}{\partial z} = \frac{-j\Omega d_{\mathrm{eff}}}{2n_2(\Omega)c}\left[\frac{1}{2\pi}\int_0^{\infty} d\omega\, \mathcal{E}_3(\omega)\mathcal{E}_1^*(\omega-\Omega)\, e^{j[k_1(\omega-\Omega)+k_2(\Omega)-k_3(\omega)]z}\right] \tag{5.155}$$

This equation is easily integrated to yield an expression for the output spectral amplitude $\mathcal{E}_2(L,\Omega)$, which is valid even for extremely broadband midinfrared fields [252].

5.5.3 Optical Parametric Amplification

In the preceding section we assumed that the amplitudes of both input waves remained fixed during the interaction. We now extend our treatment of difference frequency generation to allow both low-frequency waves, a_1 and a_2, to vary. The high-frequency wave a_3 is assumed to be strong and to remain undepleted. This results in an amplification process known as *optical parametric amplification* [69,131,228].

Introduction to Parametric Amplification The amplification effect is most easily derived by considering continuous-wave fields. The propagation equations can then be written in the following matrix form:

$$\frac{\partial}{\partial z}\begin{pmatrix} a_1 \\ a_2^* \end{pmatrix} = \begin{pmatrix} 0 & -j\kappa_1 a_3 \\ j\kappa_2 a_3^* & 0 \end{pmatrix}\begin{pmatrix} a_1 \\ a_2^* \end{pmatrix} \tag{5.156}$$

where we have used $\kappa_2^* = \kappa_2$ for lossless media and set $\Delta k = 0$. Since a_3 is taken as a constant, this is a system of linear coupled differential equations, with solutions of the form $a_1 = \tilde{a}_1 \exp(\alpha z)$, $a_2^* = \tilde{a}_2^* \exp(\alpha z)$. Substituting into eq. (5.156) gives

$$\begin{pmatrix} -\alpha & -j\kappa_1 a_3 \\ j\kappa_2 a_3^* & -\alpha \end{pmatrix}\begin{pmatrix} a_1 \\ a_2^* \end{pmatrix} = 0 \tag{5.157}$$

This is solved by setting the determinant to zero, which yields

$$\alpha = \pm\sqrt{\kappa_1\kappa_2}\,|a_3| = \pm\frac{d_{\text{eff}}}{c}\sqrt{\frac{\omega_1\omega_2}{2n_1n_2n_3c\varepsilon_0}I_3} \tag{5.158}$$

Since α is real, the waves can indeed experience exponential growth.

We are most interested in the case, as in difference frequency generation, where there is only one low-frequency field incident on the nonlinear crystal, so that $a_2(0) = 0$. The solution consistent with this boundary condition is

$$\begin{aligned} a_1(z) &= a_1(0)\cosh(|\alpha|z) \\ a_2(z) &= -j\sqrt{\frac{\omega_2 n_1}{\omega_1 n_2}}\frac{a_3}{|a_3|}a_1^*(0)\sinh(|\alpha|z) \end{aligned} \tag{5.159}$$

We see that the phase of the generated a_2 field reflects the phases of the a_1 and a_3 input waves. This is one manifestation of phase sensitivity, which is an interesting property of parametric amplifiers. In terms of intensities, we have

$$\begin{aligned} I_1(z) &= I_1(0)\cosh^2(|\alpha|z) \\ I_2(z) &= \frac{\omega_2}{\omega_1}I_1(0)\sinh^2(|\alpha|z) \end{aligned} \tag{5.160}$$

From eq. (5.160) it follows that the difference in the number of photons at ω_1 and ω_2 remains fixed throughout the amplification process, consistent with the Manley–Rowe relation. Furthermore, if we expand to the lowest order in $|\alpha|z$, we recover the results expected for continuous-wave difference frequency generation within the fixed-field approximation: The intensity I_2 of the new wave grows quadratically, whereas to lowest order the intensity I_1 of the input wave remains constant.

Parametric amplification is applicable both with ultrashort-pulse and quasi-monochromatic sources. However, as illustrated in Table 5.2, the high peak intensities available with short-pulse sources offer the advantage of very high gains. In the following we briefly introduce three main schemes, sketched in Fig. 5.17, in which parametric amplification is utilized.

Table 5.2 Representative Parameters for Blue-Pumped Femtosecond OPOs and OPAs[a]

	Femtosecond OPO	Femtosecond OPA
Pump source	Frequency-doubled femtosecond oscillator	Frequency-doubled femtosecond amplifier
Pump power or energy	1 W (average)	100 μJ
Pump duration	100 fs	100 fs
Pump repetition rate	100 MHz	1 kHz
Pump peak power	10^5 W	10^9 W
Beam diameter	20 μm	250 μm
Pump intensity	3×10^{10} W/cm^2	2×10^{12} W/cm^2
Exponential gain factor ($2\|\alpha\|$)	2.6	21
References	[257,258]	[259–261]

[a] The gain is calculated assuming a nonlinearity of $d_{\text{eff}} = 1$ pm/V, pump, signal, and idler wavelengths of 400, 600, and 1200 nm, respectively, a refractive index of 2, and a 2.0-mm crystal length.

1. In *optical parametric amplifiers* (OPAs), a strong *pump wave* at ω_3 and a weaker *signal wave* at ω_1 are incident on the nonlinear crystal. At the output the signal wave is amplified, and a new *idler wave* at ω_2 is generated. This terminology of pump, signal, and idler waves is in common use in the parametric amplification literature. In conjunction with femtosecond pump pulses from chirped pulse amplifier systems (see Chapter 8) with pulse energies in the microjoule to millijoule range, the parametric gain can be very high. An important application of OPA systems involves seeding with weak but tunable signal pulses generally in the visible or near-infrared range, which results in strong tunable visible/near-infrared pulses at the output, together with new longer-wavelength radiation in the idler beam. For certain crystals the phase-matching bandwidth can be made to be extremely large, which supports generation of amplified signal pulses with durations below 5 fs. We discuss such ultrabroadband OPAs later in the section.
2. *Optical parametric generators* (OPGs) are similar to OPAs, but with no signal input. Provided that the parametric gain is high enough, strong signal and idler outputs can result from amplification of the zero-point fluctuations of the field (i.e., quantum noise). Practically, the gain required is higher than for OPAs. The beam divergence and the spectral content of the output is determined through the phase-matching angular and spectral bandwidth, unlike the OPA case, where the characteristics of

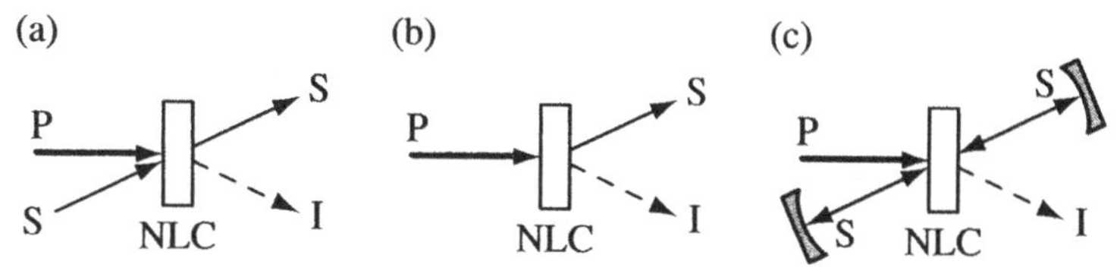

Figure 5.17 Schemes for application of parametric gain in a second-order nonlinear crystal (NLC), pictured here in a noncollinear geometry: (a) optical parametric amplifier (OPA); (b) optical parametric generator (OPG); (c) optical parametric oscillator (OPO). The three interacting waves are the strong *pump field* (P), the *signal field* (S), and the *idler field* (I). Using the terminology adopted in the text, the signal, idler, and pump waves correspond, respectively, to a_1 at ω_1, a_2 at ω_2, and a_3 at ω_3.

the input signal beam also play a key role. This makes it more difficult to obtain diffraction- and bandwidth-limited output fields with an OPG.

3. In *optical parametric oscillators* (OPOs), a resonant cavity reflects the signal wave back and forth in the nonlinear crystal. Provided that the parametric gain exceeds the cavity loss, this results in oscillation of the signal wave. Similar to a laser, a finite signal beam builds up from noise, with a portion of the signal intensity transmitted out of the cavity via an output coupler (not shown in the figure). An idler wave is also generated, although usually it is not fed back into the cavity. Unlike a laser, where the gain is tied to a specific transition or set of transitions, which limits the optical bandwidth and tuning range, in an OPO the dependence of the gain on optical frequency is determined principally via the phase matching condition. This means that OPOs can be tuned over a very broad frequency range, (e.g., by temperature- or angle tuning of the phase matching). Short pulse OPOs are generally synchronously pumped (e.g., using ca. 100-fs pulses from a mode-locked Ti:S laser), resulting in repetition rates typically in the 100-MHz range and pulse energies up to roughly the nanojoule level [255,256]. Like synchronously pumped lasers, the OPO cavity must be kept closely matched to the repetition rate and cavity length of the pump laser. For femtosecond operation group velocity dispersion and self-phase modulation can play a key role in determining the character of the output pulses, with behavior similar to that discussed in Section 2.4.5.

Ultrabroadband Optical Parametric Amplification Figure 5.18 shows schematically an OPA arrangement that allows generation of pulses that can be extremely broadband or widely tunable within the visible spectral region [259–264]. This region is difficult to access directly with mode-locked solid-state lasers. The system starts with a femtosecond laser and amplifier (see Chapter 8), based on Ti:S in current practice, that typically provides ca. 100-fs input pulses with about a 10-nm bandwidth, center wavelength near 800 nm, a 1-kHz repetition rate, and several hundred microjoule pulse energies. Most of the input energy is sent to a SHG stage to yield strong, ca. 400-nm pump pulses for the OPA process, typically with some modest temporal broadening due to group velocity walk-off in the SHG crystal. A small amount of the input energy, typically about 1 μJ, is directed into a setup that broadens the input spectrum into an ultrabroadband white light continuum, with spectral content throughout the visible region and extending into the ultraviolet and infrared. Continuum generation, discussed further in Chapter 6, is based on self-focusing within a transparent dielectric medium, producing extremely high intensities that lead to strong spectral broadening through self-phase modulation and other third-order nonlinear processes. The result is an ultrabroad-bandwidth signal pulse that serves as the second input to the OPA. The pump and signal pulses are adjusted for temporal overlap and directed onto a nonlinear crystal, phase-matched to allow parametric amplification of the signal. As discussed below, an extremely broad phase-matching bandwidth can be obtained by using a noncollinear geometry in conjunction with a suitable nonlinear crystal [257,258]. As a result, the signal pulse can be amplified into the microjoule range while maintaining a bandwidth that far exceeds that available from the input femtosecond laser/amplifier. Figure 5.19a shows an example of a spectrum obtained from such a *noncollinear optical parametric amplifier* (NOPA) that spans most of the visible. By compensating the chirp originating either from the continuum generation process itself or from subsequent propagation through dispersive elements (including the nonlinear crystal), such pulses have been compressed to durations below about 5 fs.

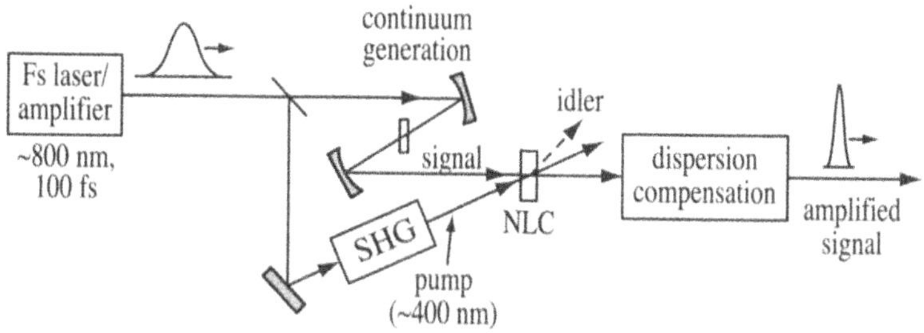

Figure 5.18 Ultrabroadband noncollinear optical parametric amplifier. An extremely broadband signal pulse is obtained via white light continuum generation in a thin dielectric medium and then amplified in a nonlinear crystal (NLC) in the presence of a strong pump pulse. Compensating the chirp of the amplified signal field can lead to compression to durations much shorter than available at the input.

It is worth emphasizing that the gain in an OPA occurs only for a narrow temporal window during which the pump and signal pulses overlap. The extent of this window is determined by the pump duration and the pump–signal temporal walk-off and is limited to at most a few hundred femtoseconds under typical conditions. Therefore, to amplify ultrabroadband spectra as in Fig. 5.19a, the frequency-dependent delay of the white light continuum must be controlled so that all the spectral components of interest arrive at the nonlinear crystal within the temporal gain window. Alternatively, if the frequency-dependent delays are too large, the gain acts only on the slice of spectrum that is properly synchronized. In this situation the amplified spectrum can be tuned by varying the pump–signal delay. This permits broadband tuning across the visible (Fig. 5.19b), with pulse durations that may still be in the sub-20-fs range [259]. This combination of short pulse duration and wide tunability is useful for applications in ultrafast spectroscopy.

Figure 5.20a shows a k-vector diagram for the analysis of noncollinear three-wave mixing [262]. The propagation angle θ of the pump wave with respect to the optic axis is adjusted to achieve birefringent phase matching at the center frequencies of the interacting fields,

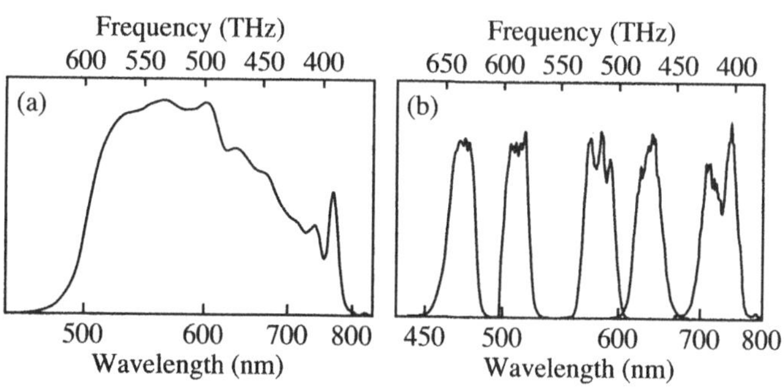

Figure 5.19 Spectra generated by noncollinear OPAs: (a) ultrabroadband spectrum spanning most of the visible, supporting generation of pulses below 5 fs; (b) broadband spectra supporting generation of sub-20-fs pulses with tuning across the visible. (a) Adapted from [263]; (b) adapted from [259].

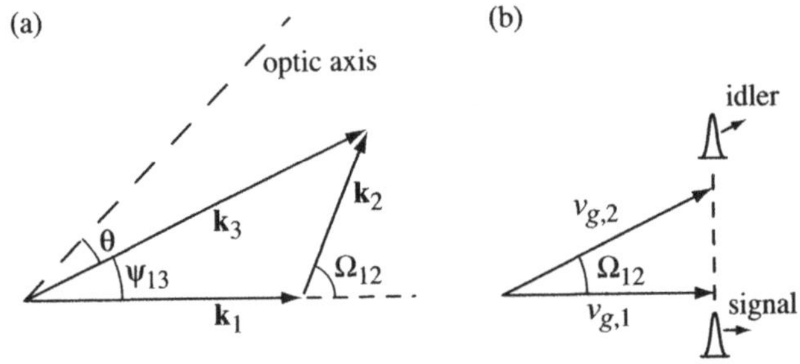

Figure 5.20 (a) k-vector diagram for noncollinear three-wave mixing (all angles refer to angles inside the nonlinear crystal); (b) relation of the group velocities under the broadband phase-matching condition.

as usual. The new idea is to use the angle ψ_{13} between pump and signal fields as an extra degree of freedom, to obtain simultaneous phase matching and group velocity matching, for ultrabroadband operation. The phase-matching condition is now a vector relation, which reads

$$\Delta\mathbf{k} = \mathbf{k_3} - \mathbf{k_1} - \mathbf{k_2} = 0 \tag{5.161}$$

The subscripts 1, 2, and 3 refer to the signal, idler, and pump waves, respectively. In terms of vector components parallel to and perpendicular to $\mathbf{k_1}$, we have

$$\begin{aligned} k_1 + k_2 \cos\Omega_{12} &= k_3 \cos\psi_{13} \\ k_2 \sin\Omega_{12} &= k_3 \sin\psi_{13} \end{aligned} \tag{5.162}$$

For broadband operation $\Delta\mathbf{k}$ should remain at zero as the frequencies of the interacting fields vary. Since the bandwidth of the pump field is narrow compared to the frequency variations of interest here (see Fig. 5.19), we approximate ω_3 as a constant. We can then set $\Delta\omega_2 = -\Delta\omega_1$. To first order in ω_1, the condition for broadband operation becomes

$$\begin{aligned} \frac{\partial k_1}{\partial\omega_1} + \frac{\partial k_2}{\partial\omega_1}\cos\Omega_{12} - k_2\sin\Omega_{12}\frac{\partial\Omega_{12}}{\partial\omega_1} + \frac{\partial k_2}{\partial\Omega_{12}}\cos\Omega_{12}\frac{\partial\Omega_{12}}{\partial\omega_1} &= 0 \\ \frac{\partial k_2}{\partial\omega_1}\sin\Omega_{12} + k_2\cos\Omega_{12}\frac{\partial\Omega_{12}}{\partial\omega_1} + \frac{\partial k_2}{\partial\Omega_{12}}\sin\Omega_{12}\frac{\partial\Omega_{12}}{\partial\omega_1} &= 0 \end{aligned} \tag{5.163}$$

For a given experimental setup, ψ_{13} is taken as fixed; as a consequence, the signal–idler angle Ω_{12} must change as the signal and idler frequencies are varied. This means that the idler wave is created with an inherent angular dispersion in noncollinear three-wave mixing [265]. The terms containing $\partial k_2/\partial\Omega_{12}$ take into account that in cases where the idler is an extraordinary wave, the magnitude of the idler k-vector will vary not only with frequency but also with propagation angle. Now multiplying the first and the second of eqs. (5.163) by $\cos \Omega_{12}$ and $\sin \Omega_{12}$, respectively, and adding, we obtain the following condition:

$$\frac{\partial k_1}{\partial \omega_1} \cos \Omega_{12} + \frac{\partial k_2}{\partial \omega_1} + \frac{\partial k_2}{\partial \Omega_{12}} \frac{\partial \Omega_{12}}{\partial \omega_1} = 0 \tag{5.164}$$

When this condition is satisfied, $\Delta \mathbf{k} = 0$ is satisfied to first order in ω_1, leading to broadband amplification.

It will be useful to obtain an expression for the angular dispersion of the idler wave generated. Referring to Fig. 5.20a, the signal–idler and pump–signal angles are related by

$$\tan \Omega_{12} = \frac{k_3 \sin \psi_{13}}{k_3 \cos \psi_{13} - k_1} \quad \text{and} \quad \tan \psi_{13} = \frac{k_2 \sin \Omega_{12}}{k_1 + k_2 \cos \Omega_{12}} \tag{5.165}$$

The angular dispersion, obtained by differentiating the expression for Ω_{12}, is

$$\frac{\partial \Omega_{12}}{\partial \omega_2} = -\frac{\partial \Omega_{12}}{\partial \omega_1} = -\frac{\sin \Omega_{12}}{k_2} \frac{\partial k_1}{\partial \omega_1} \tag{5.166}$$

where we have simplified using eq. (5.162). This expression is valid for noncollinearly phase matched three-wave mixing interactions in which the pump frequency may be taken as fixed whether or not the broadband condition is satisfied.

We specialize now to type I phase matching in nonlinear crystals such as beta-barium borate (BBO), where the signal and idler fields are o-waves and the pump field is an e-wave. This is an important case in experimental practice. Since the magnitude of the idler k-vector is now independent of propagation angle, the broadband condition becomes

$$\frac{\partial k_1}{\partial \omega_1} \cos \Omega_{12} + \frac{\partial k_2}{\partial \omega_1} = 0 \quad \text{or} \quad v_{g,2} \cos \Omega_{12} = v_{g,1} \tag{5.167}$$

Broadband operation is achieved when the projection of the idler group velocity along the signal direction matches the signal group velocity (Fig. 5.20b). In a collinear geometry with type I phase matching, group velocity matching occurs at the *degeneracy point*, at which the signal and idler have the same frequency, $\omega_1 = \omega_2 = \omega_3/2$. Here the signal and idler waves become indistinguishable, and optical parametric amplification may be regarded simply as SHG running backward. In our analysis of pulsed SHG, we have already seen that for type I phase matching and a fixed second-harmonic frequency, the phase mismatch is independent of variations in the input frequency (to first order). For the same reason the bandwidths of collinear OPAs and OPOs become very large at degeneracy, a finding that is well known in the literature of continuous-wave optical parametric devices. Equation (5.167) shows that group velocity matching and broadband operation can also be achieved away from the degeneracy point, by moving to a noncollinear geometry. Assuming that the nonlinear crystal exhibits normal dispersion, a signal–idler angle satisfying eq. (5.167) can always be found, provided that $\omega_1 \geq \omega_2$.

This noncollinear scheme has been used to shift the group velocity matching point in a number of nonlinear crystals. Figure 5.21a shows calculated signal–idler and pump–signal group velocity mismatch curves for beta-barium borate (BBO). For the collinear geometry, the group velocity mismatches fall roughly in the range 50 to 200 fs/mm over much of the visible, although the signal–idler mismatch drops to zero at the degeneracy point. For

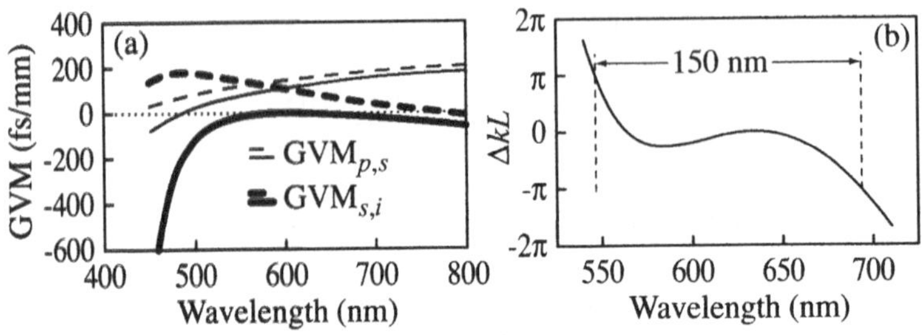

Figure 5.21 Broadband phase matching in BBO. (a) Group velocity mismatch vs. signal wavelength for a 395-nm pump. Heavy lines; Signal–idler GVM; light lines; pump–signal GVM; solid lines; noncollinear geometry, $\psi_{13} = 3.7°$; dashed lines; collinear geometry. (b) Phase mismatch vs. signal wavelength for a 400-nm pump, $\psi_{13} = 3.7°$ noncollinear angle, and a 2-mm nonlinear crystal. (a) From [260]. Copyright © 1998 IEICE. (b) Adapted from [258].

a noncollinear angle $\psi_{13} = 3.7°$, the signal-idler group velocity mismatch drops to zero, not only at a single visible wavelength as predicted by our previous discussion, but over a broad range of wavelengths spanning much of the visible. This fortuitous property of BBO makes it a particularly favorable crystal for ultrabroadband OPA. Figure 5.21b shows the calculated phase mismatch in a 2-mm length of BBO under similar noncollinear phase-matching conditions. The shape of the curve is approximately cubic. If group velocity matching were achieved at a single wavelength only, we would expect the phase mismatch curve to vary quadratically with signal frequency; the observed cubic shape arises because both the signal–idler group velocity and its derivative with respect to signal frequency are approximately equal to zero simultaneously. The result is a huge region of spectrum over which $\Delta kL \leq \pi$, with the consequence of extremely large parametric bandwidths.

The picture of group velocity matching developed above is in fact quite general. The broadband phase-matching condition, eq. (5.163), can be written in a very succinct form as

$$-\frac{\partial \mathbf{k_2}}{\partial \omega_1} = \frac{\partial \mathbf{k_2}}{\partial \omega_2} = \frac{\partial \mathbf{k_1}}{\partial \omega_1} \tag{5.168}$$

The physical significance of vectors of the form $\partial \mathbf{k}/\partial \omega$ is discussed in Section 4.2.2. According to eq. (4.56b), the pulse envelope propagates with a velocity and direction given by $\partial \mathbf{k}/\partial \omega$. The pulse fronts are normal to $\partial \mathbf{k}/\partial \omega$. Also from Section 4.2.2, angular dispersion gives rise to a tilt in the pulse front with respect to the phase fronts. For the case of the idler wave, the pulse front tilt angle γ_2 is given by

$$\tan \gamma_2 = k_2 v_{g,2} \frac{\partial \Omega_{12}}{\partial \omega_2} \tag{5.169}$$

Using eq. (5.166), the tilt angle becomes

$$\tan \gamma_2 = \frac{-v_{g,2}}{v_{g,1}} \sin \Omega_{12} \tag{5.170}$$

Finally, using the broadband condition for type I phase matching, eq. (5.167), we obtain

$$\tan\gamma_2 = -\tan\Omega_{12} \tag{5.171}$$

The interesting result is that when broadband phase matching is achieved, the idler pulse front tilt angle exactly compensates for the inclination of the idler phase fronts. As a result, the pulse fronts of the signal and idler are parallel and propagate with the same velocity [262]. This leads to the picture shown in Fig. 5.20b. Similar results can be derived for type II phase matching; details are given in Section 5.6.2.

5.6 APPENDIX

5.6.1 Spatial Walk-off and Pulse Fronts in Anisotropic Media

In this section we present a general formula for spatial walk-off in anisotropic media, using an approach similar to the derivation of group velocity in Chapter 4. Here, however, the field is taken to be a function of both space and time. We write the input field as

$$E(\mathbf{r}_\mathrm{T}, z=0, t) = \frac{1}{2}\left[a_\mathrm{in}(\mathbf{r}_\mathrm{T}, t)e^{j\omega_0 t} + \mathrm{c.c.}\right] \tag{5.172a}$$

where

$$a_\mathrm{in}(\mathbf{r}_\mathrm{T}, t) = \left(\frac{1}{2\pi}\right)^3 \int d\tilde{\omega} \int d^2\mathbf{k}_\mathrm{T}\, A_\mathrm{in}(\tilde{\omega}, \mathbf{k}_\mathrm{T})e^{j\tilde{\omega}t}e^{-j\mathbf{k}_\mathrm{T}\cdot\mathbf{r}_\mathrm{T}} \tag{5.172b}$$

Here $\mathbf{r}_\mathrm{T}$ and $\mathbf{k}_\mathrm{T}$ represent the transverse components of the position and k vectors (i.e., in the x and y directions), and the beam is assumed to propagate nominally in the z direction. $a_\mathrm{in}(\mathbf{r}_\mathrm{T}, t)$ gives the field at $z = 0$, just inside the medium. The field at an arbitrary position inside the medium is given by

$$E(\mathbf{r}, t) = \frac{1}{16\pi^3}\left\{e^{j\omega_0 t}\int d\tilde{\omega}\int d^2\mathbf{k}_\mathrm{T}\, A_\mathrm{in}(\tilde{\omega}, \mathbf{k}_\mathrm{T})e^{j\tilde{\omega}t}e^{-j\mathbf{k}_\mathrm{T}\cdot\mathbf{r}_\mathrm{T}}e^{-jk_z z} + \mathrm{c.c.}\right\} \tag{5.173}$$

In an anisotropic medium, we take k_z to be a function of both frequency and transverse k vector (equivalent to propagation angle). Expanding k_z to first order gives

$$k_z(\omega, \mathbf{k}_\mathrm{T}) = \bar{k}_z + \frac{\partial k}{\partial\omega}\tilde{\omega} + \frac{\omega_0}{c}\left(\nabla_{\mathbf{k}_\mathrm{T}} n \cdot \mathbf{k}_\mathrm{T}\right) \tag{5.174}$$

with

$$\bar{k}_z = \frac{\omega_0}{c}n(\omega_0, \mathbf{k}_\mathrm{T} = 0) \qquad \text{and} \qquad \frac{\partial k}{\partial\omega} = \frac{n(\omega_0, \mathbf{k}_\mathrm{T} = 0)}{c} + \frac{\omega_0}{c}\frac{\partial n}{\partial\omega}$$

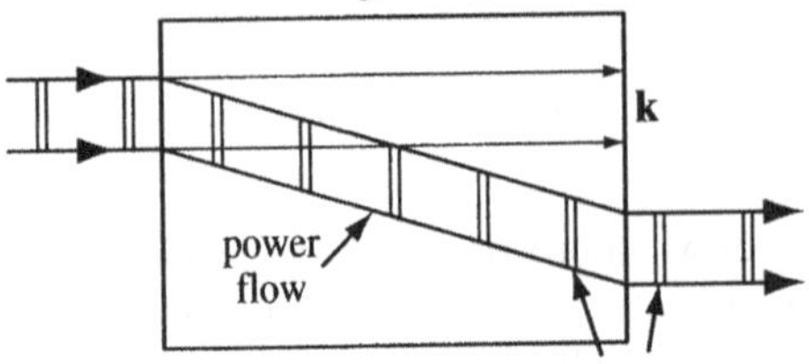

Figure 5.22 Spatial walk-off of an e-wave propagating in an anisotropic medium. The orientation of the pulse fronts are also shown.

Substituting into eq. (5.173), we obtain

$$E(\mathbf{r}_\mathbf{T}, t) = \frac{1}{2}\left[e^{j(\omega_0 t - \bar{k}_z z)}\, a_{\text{in}}\left(\mathbf{r}_\mathbf{T} + \frac{\omega_0}{c}\left(\nabla_{\mathbf{k}_\mathbf{T}} n\right) z\,,\, t - \frac{\partial k}{\partial \omega} z\right) + \text{c.c.}\right] \tag{5.175}$$

We interpret this result as follows. The field is equal to a carrier term multiplied by an envelope that propagates without distortion (this is a consequence of the first-order expansion of k_z, which neglects reshaping due to dispersion and diffraction). As sketched in Fig. 5.22, the spatial envelope propagates at an angle ρ with respect to the z axis. This is the direction in which the energy flows. The pulse fronts, given by constant $t - \partial k/\partial\omega z$, propagate with velocity $(\partial k/\partial\omega)^{-1}$ and remain perpendicular to the carrier propagation direction (z). Thus, the behavior of the pulse fronts in time is the same as described in Chapter 4 for the case of pulse propagation in isotropic media. A new feature, however, is that if the magnitude of the k vector (hence the refractive index) depends on $\mathbf{k}_\mathbf{T}$, the pulse fronts are no longer perpendicular to the direction of energy flow.

As discussed in Section 5.3.2, two waves can propagate in an anisotropic medium. For the o-wave the refractive index is independent of the propagation direction. Therefore, $\rho = 0$. For the e-wave the refractive index does depend on propagation angle. In the case of a birefringent medium, the e-wave refractive index is a function of θ, the propagation angle with respect to the optic axis. From eq. (5.175), we obtain[9]

$$\tan\rho = \frac{-1}{n(\theta)}\frac{\partial n}{\partial \theta} \tag{5.176}$$

For positive uniaxial media, for which the index becomes larger with increasing angle from the optic axis, the walk-off is in the direction closer to the optic axis.

By using eq. (5.48b), the formula for $n(\theta)$ for birefringent media, one can show that eq. (5.176) gives the same result for the walk-off angle as given earlier in eq. (5.51).

5.6.2 Velocity Matching in Broadband Noncollinear Three-Wave Mixing

The latter part of Section 5.5.3 gave a detailed discussion of velocity matching in broadband noncollinear three-wave mixing for the case of type I phase matching, where both signal

[9] To evaluate $\nabla_{\mathbf{k}_\mathbf{T}} n$, we consider $\mathbf{k}$ in spherical coordinates, as shown in Fig. 5.4. For variation in θ, the gradient formula gives $\nabla_{\mathbf{k}_\mathbf{T}} n = \hat{\theta}(1/|k|)\partial n/\partial\theta = \hat{\theta}(\partial n/\partial\theta)/(\omega_0 n(\theta)/c)$.

and idler are o-waves. The essential assumption was that the effective refractive index seen by the idler wave was independent of propagation angle. Here we extend the discussion to cover the case where idler is allowed to be an e-wave, or equivalently to the case where the idler sees an angle-dependent index. Our discussion will be valid for both type I and type II phase matching; signal and pump waves are both allowed to be either o- or e-waves, provided of course that phase matching is achieved.

We begin with the broadband phase-matching condition, eq. (5.164). Using $k_2 = \omega_2 n_2(\omega_2, \Omega_{12})/c$ and eq. (5.166), the broadband condition becomes

$$\frac{\partial k_1}{\partial \omega_1} \cos \Omega_{12} + \frac{\partial k_2}{\partial \omega_1} + \frac{1}{n_2} \frac{\partial n_2}{\partial \Omega_{12}} \sin \Omega_{12} \frac{\partial k_1}{\partial \omega_1} = 0 \tag{5.177}$$

Referring to Section 5.6.1 and eq. (5.176) to relate the angle derivative of n_2 to the idler walk-off angle ρ_2, we obtain

$$\frac{v_{g,2}}{\cos \rho_2} \cos(\Omega_{12} + \rho_2) = v_{g,1} \tag{5.178}$$

where as usual $v_{g,i} = (\partial k_i / \partial \omega_i)^{-1}$. If $\rho_2 = 0$, this reduces to our previous velocity matching result, eq. (5.167).

We can interpret this result in terms of the pulse fronts. First we must consider pulse tilts in the presence of both angular dispersion and spatial walk-off. Extending the analysis in Sections 4.2.2 and 5.6.1, we write

$$\mathbf{k_2} = \frac{\omega_2 n_2(\omega_2, \Omega_{12})}{c} [\hat{\mathbf{x}} \sin \Omega_{12} + \hat{\mathbf{z}} \cos \Omega_{12}] \tag{5.179}$$

Expanding to the first order in frequency, and assuming propagation nominally along $\hat{\mathbf{z}}$ (i.e., we define $\hat{\mathbf{z}}$ to denote the direction of $\mathbf{k_2}$ at zero frequency offset), we obtain

$$\frac{\partial \mathbf{k_2}}{\partial \omega_2} = \frac{\hat{\mathbf{x}}}{c} \left(\omega_2 n_2 \frac{\partial \Omega_{12}}{\partial \omega_2} \right) + \frac{\hat{\mathbf{z}}}{c} \left(n_{g,2} + \omega_2 \frac{\partial n_2}{\partial \Omega_{12}} \frac{\partial \Omega_{12}}{\partial \omega_2} \right) \tag{5.180}$$

where

$$n_{g,2} = n_2 + \omega_2 \frac{\partial n_2}{\partial \omega_2} = \frac{c}{v_{g,2}}$$

Again using eq. (5.176), the pulse front tilt angle is given by

$$\tan \gamma = \frac{\omega_2 n_2 (\partial \Omega_{12} / \partial \omega_2)}{n_{g,2} + \omega_2 (\partial n_2 / \partial \Omega_{12})(\partial \Omega_{12} / \partial \omega_2)} = \frac{\omega_2 n_2 (\partial \Omega_{12} / \partial \omega_2)}{n_{g,2} - \omega_2 n_2 \tan \rho_2 (\partial \Omega_{12} / \partial \omega_2)} \tag{5.181}$$

For e-waves, angular dispersion still gives rise to a tilt in the pulse fronts, but the magnitude of the tilt angle is affected by the spatial walk-off angle. Now, by using eq. (5.166) for the angular dispersion, and by applying the broadband phase-matching condition, eq. (5.178), one can show that

$$\tan \gamma = -\tan \Omega_{12} \tag{5.182}$$

This is the same as we obtained for the case when the idler was an o-wave. Thus, for both o-waves and e-waves, when the broadband phase-matching condition is satisfied, the pulse fronts of the signal and idler are parallel and propagate with the same velocity.

PROBLEMS

5.1. Discuss the differences in the approximations that are used in deriving eqs. (5.17) and (5.25), which are frequency- and time-domain formulations for the nonlinear propagation equation, respectively.

5.2. Starting from eq. (5.29), with d_{eff} real, show that total power is conserved. (*Hint*: Show that $\partial I_{\omega}/\partial z + \partial I_{2\omega}/\partial z = 0$.)

5.3. For perfect phase matching and continuous-wave SHG, verify that the time average of $\tilde{\mathbf{E}}_{2\omega} \cdot \partial \mathbf{P}_{\mathrm{NL}}/\partial t$ in eq. (5.36) is negative for the $2\omega_0$ wave. Conversely, verify for the ω_0 wave that the time average of $\tilde{\mathbf{E}}_{\omega} \cdot \partial \mathbf{P}_{\mathrm{NL}}/\partial t$ is positive.

5.4. Starting from the wave equation and assuming monochromatic, plane-wave fields in complex form, derive eq. (5.43). (*Hint*: use appropriate vector identities, which you may find in books on vector calculus.)

5.5. Solve eq. (5.45) for the case of a uniaxial crystal by setting the determinant to zero. You may assume that $\mathbf{k}$ is in the $y - z$ plane, for which $k_x = 0$, $k_y = k \sin\theta$, and $k_z = k\cos\theta$. Your solution should give expressions for the magnitude of the propagation constant (k) for both ordinary and extraordinary waves.

5.6. Verify that eq. (5.176) together with eq. (5.48b) yields the same spatial walk-off angle as eq. (5.51).

5.7. Use the time-domain formulation of the forced wave equation, eq. (5.25), to derive a formula for the second-harmonic field in the case of a pulsed input, assuming plane waves, perfect phase matching, and a nondepleted pump. Show that your result is equivalent to eq. (5.97), which was derived based on the frequency-domain formulation.

5.8. Compare the width of the experimental second-harmonic spectrum in Fig. 5.7b for the case of relatively weak focusing ($L/b \approx 2$) to the theoretical prediction of eq. (5.87). Speculate on reasons for any differences.

5.9. In writing eq. (5.115), we assumed that the same mth harmonic of the QPM grating dominates over the entire optical bandwidth. Consider a bandwidth-limited input pulse with FWHM duration Δt. Derive the relation that Δt must satisfy for this assumption to be justified. Evaluate your relation numerically for the specific case of periodically poled lithium niobate, first-order QPM, and 1.56-μm input pulses, for which l_c and $\Delta(v_g^{-1})$ are approximately 19 μm and 0.3 ps/mm, respectively. Repeat your evaluation for a hypothetical material for which l_c and $\Delta(v_g^{-1})$ are, respectively, 50 μm and 1 ps/mm.

5.10. **(a)** A bandwidth-limited Gaussian pulse with 200-fs FWHM pulse duration and 1560-nm center wavelength is passed through a normally dispersive material, which broadens the pulse by a factor of 10. Subsequently, the pulse is frequency doubled in an aperiodically poled lithium niobate nonlinear crystal. Design an aperiodic QPM grating (i.e., specify the locations of the transitions in the poling pattern) that will result in a second-harmonic pulse that is compressed to the bandwidth limit. Assume first-order QPM, a 2-cm crystal length, a coherence length $l_c = 19\ \mu\text{m}$, and group velocity mismatch $\Delta(v_g^{-1}) = 0.3$ ps/mm. Numerically, compute the effective filter function (spectral amplitude and phase) and the compressed pulse intensity profile.

(b) Repeat part (a) assuming that the 200-fs input pulse is broadened by a factor of 20 in the normally dispersive material.

5.11. In this problem we consider ultrabroadband phase matching in noncollinear optical parametric amplification in beta-barium borate (BBO). Assume a narrowband pump at 400-nm wavelength and type I ($o + o \to e$) phase matching in a crystal of length 2 mm. Note that BBO is a uniaxial crystal with index of refraction values that may be calculated from the following formulas (in which λ is wavelength in micrometers) [220]:

$$n_0^2 = 2.7359 + \frac{0.01878}{\lambda^2 - 0.01822} - 0.01354\lambda^2$$

$$n_e^2 = 2.3753 + \frac{0.01224}{\lambda^2 - 0.01667} - 0.01516\lambda^2$$

(a) Calculate the phase-matching angle and the angle (ψ_{13}) between pump and signal fields for which phase matching and signal–idler group velocity matching are simultaneously achieved at 600-nm signal wavelength.

(b) Using the settings calculated in part (a), plot the magnitude of the phase mismatch as a function of signal wavelength in the range 450 to 800 nm. (*Hint:* For each signal wavelength you should vary the signal–idler angle Ω_{12} for minimum phase mismatch.) Over what bandwidth is phase matching maintained (i.e., $|\Delta kL| \leq \pi$)?

5.12. Consider ultrabroadband parametric amplification of visible wavelength pulses, as in Figs. 5.18 and 5.19. Why is pumping at the second harmonic (400 nm) of the Ti:S laser preferred to pumping at the fundamental laser wavelength (800 nm)?

6

ULTRAFAST NONLINEAR OPTICS: THIRD ORDER

6.1 PROPAGATION EQUATION FOR NONLINEAR REFRACTIVE INDEX MEDIA

We now consider the nonlinear optics of ultrashort pulses in media with an intensity-dependent index of refraction. Some of this material was introduced briefly in Section 2.4, due to its importance in mode-locking of solid-state lasers. Note that because ultrafast nonlinearities are best incorporated into the wave equation in the time domain, whereas dispersive effects are most easily included in the frequency domain, derivation of the nonlinear propagation equations becomes rather intricate. Several authors have treated this problem using rigorous perturbation theory [266,267]. In this section we give a simple derivation of the basic nonlinear propagation equation without going into all the details of the full mathematical formalism. Related derivations may be found in [9,37]. Later in this chapter (Section 6.7) we describe a generalized nonlinear propagation equation, including higher-order terms that become important for sufficiently short pulses.

We begin with the wave equation for an isotropic, nonmagnetic, and source-free medium. The medium is allowed to be nonlinear through the nonlinear refractive index and may be nonuniform. Similar to Sections 1.2 and 5.2, we have the vector wave equation

$$\nabla \times \nabla \times \mathbf{E} = \nabla(\nabla \cdot \mathbf{E}) - \nabla^2 \mathbf{E} = -\mu_0 \frac{\partial^2 \mathbf{D}}{\partial t^2} \tag{6.1}$$

As in eq. (5.9), we take

$$\mathbf{D} = \varepsilon_0 \mathbf{E} + \mathbf{P}_{(1)} + \mathbf{P}_{\mathbf{NL}} = \mathbf{D}_{(1)} + \mathbf{P}_{\mathbf{NL}}$$

Ultrafast Optics, By Andrew M. Weiner

where $\mathbf{P}_{\mathbf{NL}}$ is the nonlinear polarization, $\mathbf{P}_{(1)}$ is the first-order polarization (linear in $\mathbf{E}$), and $\mathbf{D}_{(1)}$ is the first-order portion of the $\mathbf{D}$ field. For an isotropic medium, we write $\mathbf{D}_{(1)} = \varepsilon_{(1)}\mathbf{E} = n_{(1)}^2\varepsilon_0\mathbf{E}$, where $n_{(1)}$ is the linear portion of the refractive index. Using $\nabla \cdot \mathbf{D} = 0$, we obtain

$$\varepsilon_{(1)}\nabla \cdot \mathbf{E} = -\mathbf{E} \cdot \nabla\varepsilon_{(1)} - \nabla \cdot \mathbf{P}_{\mathbf{NL}} \tag{6.2}$$

Even if the nonlinear polarization is weak, as in Section 5.2, $\nabla \cdot \mathbf{E}$ can still be nonzero in a nonuniform medium. This means that the solutions of eq. (6.1) may depend on the orientation of the electric field [268]. In some waveguides, for example, the propagation constant may depend on the polarization of the electric field. When the $\mathbf{E} \cdot \nabla\varepsilon_{(1)}$ term can be neglected, the dependence on the polarization is removed, and eq. (6.1) can be replaced with the simpler scalar wave equation

$$\nabla^2 E - \mu_0\varepsilon_{(1)}\frac{\partial^2 E}{\partial t^2} = \mu_0\frac{\partial^2 P_{\mathrm{NL}}}{\partial t^2} \tag{6.3}$$

where the field variables are now taken as scalar quantities. We will use the scalar wave equation as the point of departure for our treatment of nonlinear propagation.

One obvious case where this is valid is for plane waves in a uniform, unbounded medium. We treat this case in Section 6.1.1. Another case is for bound modes in weakly guiding waveguides, where weakly guiding is defined to mean that the fractional change in the refractive index in the waveguiding structure is small [268]. This case is especially important because of the richness of nonlinear optical behavior that has been investigated in single-mode optical fibers. In a scalar treatment, we write

$$D = \varepsilon_0 n^2 E \qquad \text{where} \qquad n = n_0 + \delta n_{(1)} + \delta n_{\mathrm{NL}} \tag{6.4}$$

Here n_0 is the low-field value of the refractive index at some center frequency ω_0, $\delta n_{(1)}$ represents linear perturbations to n_0 due to variations in frequency or spatial index profile, and δn_{NL} are perturbations due to the nonlinearity. If the perturbations are taken to be sufficiently small that the second-order terms can be omitted, we have

$$D \approx \varepsilon_0\left(n_0^2 + 2n_0\delta n_{(1)} + 2n_0\delta n_{\mathrm{NL}}\right)E \approx \left(\varepsilon_{(1)} + 2n_0\varepsilon_0\delta n_{\mathrm{NL}}\right)E \tag{6.5}$$

We now make the identification

$$P_{\mathrm{NL}} \approx 2n_0\varepsilon_0\delta n_{\mathrm{NL}}E \tag{6.6}$$

Usually, the nonlinear refractive index δn_{NL} is taken as proportional simply to the instantaneous optical intensity. However, the case where the nonlinear index may be time delayed with respect to the optical intensity is also important in ultrafast optics and is considered later in the chapter. The relation between nonlinear index and nonlinear polarization given by eq. (6.6) is independent of the temporal form of the nonlinear index.[1]

[1] In an isotropic nonlinear index material, the scalar treatment is valid for either a linearly or circularly polarized input field, for which the nonlinear polarization is parallel to the optical field. For general elliptical input polarizations, a tensor relationship between the nonlinear polarization must be considered, and a vector wave equation results.

In the following sections we first derive the nonlinear propagation equation for plane waves in a uniform medium and then for weakly guiding waveguides.

6.1.1 Plane Waves in Uniform Media

In this section we assume plane waves in a uniform medium, so that the linear perturbations to the refractive index $\delta n_{(1)}$ depend only on frequency. The derivation is similar to that in Section 5.2.2. In our standard way, we express the field and nonlinear polarization in terms of slowly varying amplitude functions:

$$E = \mathrm{Re}\left\{a(z,t)e^{j(\omega_0 t - \beta_0 z)}\right\} \qquad \text{and} \qquad P_{\mathrm{NL}} = \mathrm{Re}\left\{\tilde{P}_{\mathrm{NL}}(z,t)e^{j\omega_0 t}\right\} \tag{6.7}$$

where $\beta_0 = \omega_0 n_0/c$ is the propagation constant at the carrier frequency ω_0. The Fourier transform of $a(z,t)$ is written

$$A(z,\tilde{\omega}) = \int a(z,t)e^{-j\tilde{\omega}t}\,dt \tag{6.8}$$

where $\tilde{\omega} = \omega - \omega_0$ is the offset from the carrier frequency. Substituting into eq. (6.3) and invoking the slowly varying envelope approximation in both space and time gives

$$\int \frac{d\tilde{\omega}}{2\pi}\left\{\frac{\partial A}{\partial z} + \frac{j}{2\beta_0}\left[\beta^2(\tilde{\omega}) - \beta_0^2\right]A\right\}e^{j\tilde{\omega}t} = \frac{-j\mu_0\omega_0 c}{2n_0}\tilde{P}_{\mathrm{NL}}\,e^{j\beta_0 z} \tag{6.9}$$

where $\beta^2(\omega) = \omega^2\mu_0\varepsilon_{(1)}$ is the square of the propagation constant. Using the approximation $\beta + \beta_0 \approx 2\beta_0$, we obtain

$$\int \frac{d\tilde{\omega}}{2\pi}\left\{\frac{\partial A}{\partial z} + j\left[\beta(\tilde{\omega}) - \beta_0\right]A\right\}e^{j\tilde{\omega}t} = \frac{-j\mu_0\omega_0 c}{2n_0}\tilde{P}_{\mathrm{NL}}\,e^{j\beta_0 z} \tag{6.10}$$

We now express β as a Taylor series expansion up to second order in the frequency offset, including also a linear, frequency-independent loss term:

$$\beta(\omega) = \beta_0 + \beta_1\tilde{\omega} + \frac{1}{2}\beta_2\tilde{\omega}^2 - \frac{j\alpha}{2} \tag{6.11}$$

where α is the loss coefficient. Using this expression for β, we carry out the inverse Fourier transform in eq. (6.10), with the result

$$\frac{\partial a}{\partial z} + \beta_1\frac{\partial a}{\partial t} - \frac{j\beta_2}{2}\frac{\partial^2 a}{\partial t^2} + \frac{\alpha}{2}a + \frac{j\mu_0\omega_0 c}{2n_0}\tilde{P}_{\mathrm{NL}}\,e^{j\beta_0 z} = 0 \tag{6.12}$$

Now assuming an instantaneous and frequency-independent Kerr nonlinearity ($\delta n_{\mathrm{NL}} = n_2 |a|^2$) and using eq. (6.6), we arrive at the basic propagation equation:

$$\frac{\partial a}{\partial z} + \beta_1 \frac{\partial a}{\partial t} - \frac{j\beta_2}{2} \frac{\partial^2 a}{\partial t^2} + \frac{j\omega_0 n_2}{c} |a|^2 a + \frac{\alpha}{2} a = 0 \tag{6.13}$$

Note that $|a|^2$ is normalized to give the optical intensity. This equation, together with its guided-wave cousin discussed below, serves as the point of departure for much of our discussion of nonlinear pulse propagation. Effects resulting from higher-order terms that have been ignored in our derivation will be discussed later in the chapter.

We mention that in practice, it is difficult to strictly satisfy the plane-wave approximation, which allowed us to eliminate any transverse spatial derivatives in the wave equation. For example, the usual Gaussian spatial profile of a propagating beam leads to a Gaussian nonlinear refractive index profile, which acts as a nonlinear lens. This leads to self-focusing, as discussed briefly in Section 2.4.4. For ultrashort pulses, the time dependence of this nonlinear lens couples the temporal and spatial coordinates. Furthermore, even for an initial plane wave, for powers above a certain critical power, the input beam is unstable and tends to breakup into self-focused filaments [69]. Such transverse instabilities may be avoided by using a guided-wave geometry, described below.

6.1.2 Nonlinear Propagation in Waveguides

Assuming weak guiding, we write the scalar wave equation for a guided wave geometry as

$$\nabla_T^2 E + \frac{\partial^2 E}{\partial z^2} - \mu_0 \varepsilon_{(1)} \frac{\partial^2 E}{\partial t^2} = \mu_0 \frac{\partial^2 P_{\mathrm{NL}}}{\partial t^2} \tag{6.14}$$

where $\nabla_T^2 = \partial^2/\partial x^2 + \partial^2/\partial x^2$, and $\varepsilon_{(1)}$, E, and δn_{NL} now include a transverse spatial dependence. In our treatment we consider the nonlinear term on the right-hand side of eq. (6.14) as a perturbation, which we initially set to zero. Rewriting the wave equation in the frequency domain, with $\tilde{E}$ as the temporal Fourier transform of E, gives

$$\nabla_T^2 \tilde{E} + \frac{\partial^2 \tilde{E}}{\partial z^2} + \omega^2 \mu_0 \varepsilon_{(1)}(x, y, \omega) \tilde{E} = 0 \tag{6.15}$$

Further rewriting the frequency-domain field as the product of a transverse spatial profile and a longitudinal propagation term

$$\tilde{E}(\omega) = u(x, y, \omega) e^{-j\beta(\omega) z} \tag{6.16}$$

yields

$$\nabla_T^2 u + (\omega^2 \mu_0 \varepsilon_{(1)} - \beta^2) u = 0 \tag{6.17}$$

Solving this eigenvalue equation yields expressions for the transverse mode profile u and the corresponding propagation constant β as a function of ω. Bounded solutions are found for discrete values of β. (For a general discussion of such solutions, see, for example, [268].)

Waveguide structures can be designed to yield only a single bound (or guided) transverse mode solution at a particular optical frequency. This provides a "built-in" spatial field profile, which mitigates the transverse instabilities associated with nonlinear propagation in free space.

To gain insight into the relation between the propagation constant, the index variation, and the mode profile, we adopt a transverse integration procedure [9,268]. Equation (6.17) is first multiplied by u^* and then integrated over the transverse spatial coordinates:

$$\iint u^* \nabla_T^2 u \, dx \, dy + \iint \left(\omega^2 \mu_0 \varepsilon_{(1)} - \beta^2\right) |u|^2 \, dx \, dy = 0 \tag{6.18}$$

This procedure is valid for bound modes, for which the field decays rapidly to zero outside the waveguide core, and hence for which the integrals produce finite values. The first integral in this expression can be integrated by parts to yield[2]

$$\iint u^* \nabla_T^2 u \, dx \, dy = \iint \nabla_T \cdot \left(u^* \nabla_T u\right) \, dx \, dy - \iint |\nabla_T u|^2 \, dx \, dy \tag{6.19}$$

The first integral on the right-hand side of eq. (6.19) can be converted into a surface integral by using the divergence theorem, eq. (1.27),

$$\iint \nabla_T \cdot \left(u^* \nabla_T u\right) \, dx \, dy = \int u^* \nabla_T u \cdot \hat{\mathbf{n}} \, dS \tag{6.20}$$

where the integral is over a contour at ∞. Since u vanishes at ∞ for bound modes, the value of the integral is zero. As a result, eq. (6.18) can be arranged to give the following formula for the propagation constant:

$$\beta^2(\omega) = \frac{\iint \left(\omega^2 \mu_0 \varepsilon_{(1)}(x, y, \omega) \, |u(x, y, \omega)|^2 - |\nabla_T u(x, y, \omega)|^2\right) \, dx \, dy}{\iint |u(x, y, \omega)|^2 \, dx \, dy} \tag{6.21}$$

This relation shows that β^2 is given by the spatial average of $\omega^2 \mu_0 \varepsilon_{(1)}$, which would give β^2 in a plane-wave situation, reduced by the average squared transverse spatial derivative of the mode. The latter term may be regarded as arising from the transverse wave vector content needed to construct the guided mode, which reduces the magnitude of the wave vector available for longitudinal propagation along the guide. Since the integral over $|\nabla_T u|^2$ is positive, the value of $\beta^2 - \omega^2 \mu_0 \varepsilon_{(1)}$ must be negative over some region of the transverse spatial profile. In a step-index fiber, therefore, the propagation constant satisfies

$$\omega^2 \mu_0 \varepsilon_{(1)}^{(\text{cladding})} < \beta^2 < \omega^2 \mu_0 \varepsilon_{(1)}^{(\text{core})} \tag{6.22}$$

where the left-hand inequality is needed to ensure exponential decay away from the central guiding core region.

We now return to the nonlinear problem, eq. (6.14). We follow a procedure similar to that in Section 6.1.1, but modified to account for the spatial mode profile. The electric field

[2] Here we use the identity $\nabla_T \cdot (u^* \nabla_T u) = \nabla_T u^* \cdot \nabla_T u + u^* \nabla_T^2 u$.

and nonlinear polarization are written

$$E = \mathrm{Re}\{a(z,t)e^{j(\omega_0 t - \beta_0 z)}u(x,y)\} \quad \text{and} \quad P_{\mathrm{NL}} = \mathrm{Re}\left\{\tilde{P}_{\mathrm{NL}}(x,y,z,t)e^{j\omega_0 t}\right\} \tag{6.23}$$

where β_0 and u now represent the propagation vector and spatial mode profile, as obtained by solving the unperturbed eigenvalue equation (6.17) for a particular fixed frequency ω_0.[3] Using $A(z, \tilde{\omega})$ as the Fourier transform of $a(z, t)$, as in eq. (6.8), eq. (6.14) can be rewritten

$$\int \frac{d\tilde{\omega}}{2\pi}\left\{\nabla_T^2 - \beta_0^2 - 2j\beta_0\frac{\partial}{\partial z} + \frac{\partial^2}{\partial z^2} + \omega^2\mu_0\varepsilon_{(1)}\right\}Aue^{j(\omega_0+\tilde{\omega})t} = \mu_0\frac{\partial^2 P_{\mathrm{NL}}}{\partial t^2}e^{j\beta_0 z} \tag{6.24}$$

We simplify this expression by invoking the slowly varying envelope approximation in space ($\partial^2 A/\partial z^2 \to 0$) and time ($\partial^2/\partial t^2 \to -\omega_0^2$) and by making use of eq. (6.17). With the additional approximation $\beta(\omega) + \beta_0 \approx 2\beta_0$, as in our treatment of the plane-wave case, we obtain

$$\int \frac{d\tilde{\omega}}{2\pi}\left\{\frac{\partial A}{\partial z} + j\left[\beta(\omega) - \beta_0\right]A\right\}ue^{j\tilde{\omega}t} = \frac{-j\mu_0\omega_0^2}{2\beta_0}\tilde{P}_{\mathrm{NL}}e^{j\beta_0 z} \tag{6.25}$$

We express β as a second-order Taylor series expansion in the frequency offset $\omega - \omega_0$, including a frequency-independent loss term, following eq. (6.11), and carry out the Fourier integral to obtain

$$\left\{\frac{\partial a}{\partial z} + \beta_1\frac{\partial a}{\partial t} - \frac{j\beta_2}{2}\frac{\partial^2 a}{\partial t^2} + \frac{\alpha}{2}a\right\}u = \frac{-j\mu_0\omega_0^2}{2\beta_0}\tilde{P}_{\mathrm{NL}}e^{j\beta_0 z} \tag{6.26}$$

We now assume an instantaneous nonlinear index ($\delta n_{\mathrm{NL}} = n_2\,|a|^2\,|u|^2$) and use eq. (6.6) for the nonlinear polarization. The dependence on the transverse coordinate is then eliminated by multiplying both sides of the equation by u^* and integrating over x and y. This yields the nonlinear propagation equation for the slowly varying envelope:

$$\frac{\partial a}{\partial z} + \beta_1\frac{\partial a}{\partial t} - \frac{j\beta_2}{2}\frac{\partial^2 a}{\partial t^2} + j\kappa|a|^2 a + \frac{\alpha}{2}a = 0 \tag{6.27a}$$

where

$$\kappa = \frac{\omega_0^2}{\beta_0 c^2}\left\{\frac{\iint n_0 n_2 |u|^4\,dx\,dy}{\iint |u|^2\,dx\,dy}\right\} \tag{6.27b}$$

[3] In eq. (6.23) the spatial profile is taken to be independent of optical frequency. The validity of this formulation within the scope of our perturbation treatment is connected with the fact that perturbations to the refractive index, δn, result in changes to the propagation constant, $\delta\beta$, that are, to first order, independent of changes induced in the mode profile δu [9].

Above $|a|^2$ is normalized such that $|a(t)|^2|u(x, y)|^2$ gives the instantaneous intensity. In a guided-wave geometry, it is often convenient to consider the instantaneous power carried by the guided mode. For this purpose we can introduce a new field variable a_p, such that

$$|a_p|^2 = |a|^2 \iint |u|^2\, dx\, dy \tag{6.28}$$

Clearly, $|a_p|^2$ gives power. Propagation equation (6.27a) can be rewritten in terms of a_p as follows:

$$\frac{\partial a_p}{\partial z} + \beta_1 \frac{\partial a_p}{\partial t} - \frac{j\beta_2}{2}\frac{\partial^2 a_p}{\partial t^2} + j\gamma |a_p|^2 a_p + \frac{\alpha}{2} a_p = 0 \tag{6.29a}$$

where

$$\gamma = \frac{\omega_0^2 n_0 n_2}{\beta_0 c^2 A_{\text{eff}}} \tag{6.29b}$$

and A_{eff} is an effective area, given by

$$A_{\text{eff}} = \frac{\left\{\iint |u|^2\, dx\, dy\right\}^2}{\iint |u|^4\, dx\, dy} \tag{6.29c}$$

In the expression for A_{eff} we have assumed that the spatial variation of the linear and nonlinear refractive indices is sufficiently small that n_0 and n_2 can be pulled outside the integral. The value of the effective area depends on the fiber parameters and wavelength. For step-index fibers A_{eff} is typically on the order of 1.5 to 2.5 times the actual core area [269].

6.1.3 Optical Fiber Types

Fiber geometries have been very heavily employed for research on guided-wave nonlinear optics. Three major classes of fibers are of primary interest for ultrafast optics studies.

Standard step-index glass optical fibers consist of a central circular core region, on the order of several micrometers in diameter, surrounded by a lower-index cladding region. The core refractive index is typically only a few parts in a thousand higher than the cladding. This makes step-index fibers an example of a weakly guiding system. Nevertheless, the small refractive index difference is sufficient to yield bound modes whose fields decay exponentially outside the core region while still satisfying the scalar wave equation. Furthermore, these fibers can be designed to support only a single bound mode within a specified wavelength band. Thus, the spatial profile of the propagating beam is set by the waveguide structure. As a result, the single-mode fiber (and other single-mode waveguide devices) avoids issues related to nonuniform beam profile, diffraction, and self-focusing that complicate nonlinear pulse propagation experiments performed in bulk media. Additional attractive features of single-mode fibers include low loss (as low as 0.2 dB/km at 1.55 μm) and enhancement of the nonlinear effects through confinement of the field into a small area (hence high intensity), which can be maintained over long propagation distances. Step-index fiber technology has been developed extensively due to its crucial role in the lightwave communications industry. With variations in waveguide design, fibers have been realized with a selection of dispersion and mode size characteristics. Concerning the former, step-index fibers exhibit normal (anomalous) dispersion at sufficiently short (long) wavelengths, with a crossover (zero-dispersion) wavelength which may be designed to occur between approximately 1.3 and 1.55 μ m. The importance of optical transmission for high-speed communications provides further stimulus for the study of nonlinear pulse propagation, since nonlinear optical effects often play a crucial role in constraining the performance of state-of-the-art lightwave systems.

A number of comprehensive references are available that can provide the interested reader with more detail on the characteristics, theory, and system applications of optical fibers (e.g, [270]). A text focusing on nonlinear optics in fibers, primarily standard step-index fibers, is also available [37].

Microstructured fibers consist of a series of air holes running parallel to the fiber axis [271,272]. Strong index contrast at the air–glass interfaces, coupled with rich scope for varied geometrical designs, enables realization of a wide range of fiber characteristics not possible using standard, weakly guided fiber technology. For ultrafast optics applications, one key advantage is the ability to shift the zero-dispersion point to the near-visible and visible operating wavelengths of popular mode-locked lasers. At the same time, the mode size may be made very small, leading to high nonlinearity. These properties have made it possible to use mode-locked oscillators to study and exploit important nonlinear optical processes, such as supercontinuum generation (see Section 6.8), which were previously accessible only with amplified femtosecond systems.

Hollow-core fibers consist of a hollow capillary tube, which may have an inner diameter on the order of 100 μ m, filled by a gaseous nonlinear medium such as argon or krypton. Linear propagation in hollow dielectric waveguides was investigated in [273]. Such structures support many modes rather than a single mode, and modes are leaky (inherently lossy) rather than bound. Propagation may be thought of in terms of a series of reflections from the gas–dielectric interface, where for angles sufficiently close to grazing incidence, the reflectivity becomes close to (but always less than) 1. For moderate lengths, typically about 1 m or below, the loss experienced by the fundamental spatial mode is manageable. Furthermore, higher-order spatial modes suffer increased loss, which helps to preserve a regular spatial profile. Gas-filled hollow-core fibers have been shown to possess favorable properties for nonlinear optics with high-power amplified femtosecond pulses [274–276]. The strength of the nonlinearity may be adjusted by tuning the pressure as well by selecting different gases, while the combination of relatively large core diameter, low nonlinearity, and high damage threshold leads to excellent power-handling capability. Pure third-order nonlinearities dominate over a large range of intensities, eventually limited by multiphoton ionization. For noble gases such as krypton, femtosecond multiphoton ionizaton occurs only at intensities beyond 10^{13} W/cm^2 [274], which corresponds to pulse energies of hundreds of microjoules for 100-fs pulses.

It is worth noting that of the fiber types discussed above, only standard step-index fibers fully adhere to the assumptions we used in deriving the nonlinear propagation equation. However, although their linear propagation characteristics may vary substantially, many key aspects of nonlinear pulse propagation, as discussed in the remainder of the chapter, are found to behave similarly across different fiber classes.

6.2 THE NONLINEAR SCHRÖDINGER EQUATION

Let us now discuss the nonlinear propagation equation. To be specific we consider eq. (6.29a); however, it is clear that the same discussion applies equally to eqs. (6.13) and (6.27a) with obvious minor modifications. Furthermore, for simplicity of notation we drop the subscript and use a instead of a_p. It should be remembered, however, that in this form of the equation, the field squared gives instantaneous power.

We now consider some special cases.

Pulse Propagation Setting the dispersion, nonlinearity, and loss all to zero, we retain only the first two terms in eq. (6.29a), which gives

$$\frac{\partial a}{\partial z} + \beta_1 \frac{\partial a}{\partial t} = 0 \tag{6.30}$$

The solution takes the general form $a(t - z/v_g)$, where $v_g = \beta_1^{-1}$ is the group velocity. An input pulse will propagate at the group velocity without distortion.

Propagation with Loss Considering again all the terms in eq. (6.29a), we make the substitution

$$a(z, t) = a'(z, t)e^{-\alpha z/2} \tag{6.31}$$

This leads to the new propagation equation

$$\frac{\partial a'}{\partial z} + \beta_1 \frac{\partial a'}{\partial t} - \frac{j\beta_2}{2}\frac{\partial^2 a'}{\partial t^2} + j\gamma e^{-\alpha z}\left|a'\right|^2 a' = 0 \tag{6.32}$$

In the case of pure linear propagation (nonlinear term set to zero), the effect of dispersion is independent of loss. The loss simply adds a multiplicative attenuation factor. When the nonlinearity is important, loss has the effect of reducing the strength of the nonlinearity term with increasing propagation in the fiber.

Lossless Propagation: The Nonlinear Schrödinger Equation We now assume that loss can be ignored, and change to a reference frame moving with the pulse. The new coordinates are

$$\begin{aligned} z' &= z \\ \tau &= \frac{t - z/v_g}{T_0} \end{aligned} \tag{6.33}$$

τ represents a dimensionless retarded time; T_0 is an arbitrary characteristic time and may be chosen to take on any value convenient for a particular problem. Substituting into eq. (6.29a) yields[4]

$$\frac{\partial a}{\partial z'} - \frac{j\beta_2}{2T_0^2}\frac{\partial^2 a}{\partial \tau^2} + j\gamma|a|^2 a = 0 \tag{6.34}$$

This equation is known as the *nonlinear Schrödinger equation* (NLSE); its form is the same as for the conventional time-dependent Schrödinger equation in quantum mechanics, but with a nonlinear potential term.

We can proceed further by introducing the characteristic length scale L_D and normalized distance variable ξ, defined by

$$L_D = \frac{T_0^2}{|\beta_2|} \quad \text{and} \quad \xi = \frac{z'}{L_D} \tag{6.35}$$

L_D is the characteristic dispersion length scale; it gives the distance at which dispersion will significantly broaden a bandwidth-limited input pulse of duration T_0 in the absence of nonlinearity. The propagation equation becomes

$$j\frac{\partial a}{\partial \xi} = -\frac{\operatorname{sgn}(\beta_2)}{2}\frac{\partial^2 a}{\partial \tau^2} + \gamma L_D\,|a|^2\,a \tag{6.36}$$

where sgn$(\cdots)$ returns the sign of its argument. Finally, we renormalize the field amplitude according to

$$a(z,\tau) = \sqrt{P_c}\,a(z,\tau) \tag{6.37}$$

with P_c chosen such that

$$\gamma P_c L_D = 1 \quad \text{and} \quad P_c = \frac{|\beta_2|}{\gamma T_0^2} \tag{6.38}$$

P_c represents a normalizing power which is chosen such that the nonlinear phase shift would equal 1 radian for a pulse propagating under the influence of nonlinearity alone over a distance L_D. Thus, for an input pulse with peak power P_c (in real units), the influence of dispersion and of nonlinearity are approximately equal. We arrive now at a simple, dimensionless form of the NLSE:

$$j\frac{\partial a}{\partial \xi} = \frac{-\operatorname{sgn}(\beta_2)}{2}\frac{\partial^2 a}{\partial \tau^2} + |a|^2\,a \tag{6.39}$$

This equation is particularly useful for discussion of soliton propagation.

[4] Carrying out the coordinate transformation is accomplished using the chain rule for derivatives, which gives $\partial/\partial z = \partial/\partial z' - (1/v_g T_0)(\partial/\partial\tau)$, $\partial/\partial t = (1/T_0)(\partial/\partial\tau)$.

Characteristic Propagation Regimes We can classify the nonlinear propagation problem into distinct propagation regimes based on the relationship between the pulse duration, the peak power, and the length of the nonlinear medium (L). Letting P_{peak} denote the peak power of the input pulse, we introduce the characteristic nonlinearity length scale L_{NL} [37],

$$L_{\text{NL}} = \frac{1}{\gamma P_{\text{peak}}} \tag{6.40}$$

L_{NL} gives the distance at which the nonlinearity will significantly broaden the spectrum of a bandwidth-limited input pulse in the absence of dispersion, as we discuss in Section 6.3. Assuming that T_0 has been chosen to represent the actual pulse duration, the propagation regimes are as follows:

- $L \ll L_D, L \ll L_{\text{NL}}$. Neither dispersion nor the nonlinearity have an appreciable effect. The pulse propagates undistorted in both time and frequency.
- $L_D \ll L \ll L_{\text{NL}}$. Dispersion significantly affects the temporal profile of the pulse. However, the nonlinearity does not play a role, and the spectrum remains unchanged.
- $L_{\text{NL}} \ll L \ll L_D$. Pulse propagation is dominated by the nonlinearity, with dispersion playing little role, at least initially.
- $L_D \ll L, L_{\text{NL}} \ll L$. Provided that L_{NL} is also shorter than the loss length (α^{-1}), both dispersion and the nonlinearity play an important role; their interaction strongly affects pulse propagation.

Another, related way to classify the nonlinear propagation problem is based on the peak input power. In particular, we define a dimensionless parameter N, such that

$$N^2 = \frac{L_{\text{D}}}{L_{\text{NL}}} = \frac{P_{\text{peak}}}{P_c} \tag{6.41}$$

N characterizes the relative strengths of the nonlinearity and dispersion: For $N^2 \gg 1$, the nonlinearity dominates the initial propagation, whereas for $N^2 \ll 1$, dispersion dominates. As long as N remains constant, the relative balance between nonlinearity and dispersion remains unchanged, even when the pulse width, power, dispersion, and so on, are individually varied. This leads to a very useful NLSE scaling law:

> If $\tilde{a}(\xi, \tau)$ is a solution to the normalized NLSE, a scaled version of this function, $\eta\tilde{a}(\eta^2\xi, \eta\tau)$, is also a solution.

Computational Approach Although in the following we encounter several cases where exact solution of the NLSE is possible, in general the nonlinear propagation must be analyzed numerically. The most popular computational approach is termed the *split-step Fourier method*. In this method linear and nonlinear contributions to the propagation problem are handled separately, with nonlinear terms computed in the time domain and linear terms calculated in the Fourier domain. Because the fast Fourier transform algorithm is used to convert between time and frequency domains, this method is computationally efficient. Here we describe only the most simple formulation of the split-step Fourier method. A more nuanced discussion and a list of references to the literature may be found in [37].

As a specific example, consider the propagation equation

$$\frac{\partial a}{\partial z} = \frac{j\beta_2}{2}\frac{\partial^2 a}{\partial t^2} - j\gamma\, |a|^2\, a \qquad (6.42)$$

This is similar to eq. (6.29a) but with z and t defined in a reference frame moving with the pulse, with a written in place of a_p, and without loss. Equation (6.42) may be solved numerically as follows:

1. Begin with an initial field $a(z = 0, t)$. Here $z = 0$ is taken as the starting point for the computation.
2. Fourier-transform into the frequency domain:

$$A(0, \tilde{\omega}) = \int dt\, a(0, t) \exp(-j\tilde{\omega}t)$$

3. Account for dispersion corresponding to propagation over a distance Δz, according to

$$A_D(\Delta z, \tilde{\omega}) = A(0, \tilde{\omega}) \exp\left(-j\beta_2\tilde{\omega}^2\Delta z/2\right)$$

4. Inverse-Fourier-transform back into the time domain:

$$a_D(\Delta z, t) = \frac{1}{2\pi}\int d\tilde{\omega} A_D(\Delta z, \tilde{\omega}) \exp(j\tilde{\omega}t)$$

5. Now account for the nonlinearity over the same distance Δz, according to

$$a(\Delta z, t) = a_D(\Delta z, t) \exp\left(-j\gamma|a_D(\Delta z, t)|^2\Delta z\right)$$

6. This completes the computation for propagation over a single spatial increment Δz. To continue the simulation, the result is now inserted in place of the initial field in step 1, and the process repeats.

As in any numerical method, some care is necessary to obtain accurate results. Because the split-step method ignores the noncommuting nature of dispersion and nonlinearity operators, some error is introduced. To make this error acceptably small, a sufficiently small step size (Δz) must be selected. As a rough guide, Δz should be made small compared to both L_D and L_{NL}. Comparison of simulation results performed with step sizes Δz and $\Delta z/2$ can be used to check the reliability of the computations. Another point is that both time and frequency vectors must be selected so that the field is sampled adequately. This is equivalent to stipulating that the pulse and its spectrum must fit completely within the time and frequency vectors established for the computation. Conversely, numerical instabilities may arise if the pulse (spectrum) extends all the way to edge of the time (frequency) vector. Such instabilities are linked to the periodicity introduced when any signal is modeled using discrete Fourier transforms. For example, any portion of a pulse extending beyond the late time extreme of the time vector wraps around and reappears at early times.

The split-step Fourier method may also be used to handle pulse propagation involving additional effects beyond those included in eq. (6.42). Higher-order dispersion terms and linear loss, if present, may be included in step 3 by modifying the exponential multiplier in an obvious way. More general forms of nonlinearity may be handled by modifying the formula in step 5. Interestingly, a fully analogous split-step method has been used extensively in simulating linear as well as nonlinear propagation of single-frequency spatial beams [277,278]. In spatial problems the algorithm is often termed the *beam propagation method*. The applicability of the split-step method to both temporal and spatial problems arises because of a duality in the mathematics describing spatial and temporal propagation (see Section 8.3.1). Finally, the split-step method may be employed for analysis of problems with propagation in space and time simultaneously, such as the nonlinear envelope equation introduced later in the chapter (see Section 6.7.1).

6.3 SELF-PHASE MODULATION

We now consider the phenomenon of *self-phase modulation* (SPM), in which an intense pulse transiently modifies the refractive index through the optical Kerr effect, which in turn affects the temporal phase of that pulse. We start with the nonlinear propagation equation in the form

$$\frac{\partial a}{\partial z} = \frac{j\beta_2}{2}\frac{\partial^2 a}{\partial t^2} - j\gamma|a|^2 a \tag{6.43}$$

where $a(z, t)$ is the field amplitude, z is the propagation distance into the medium, and t now denotes retarded time. Loss is ignored in this equation. In the following we consider two distinct cases: the first where dispersion is neglected, and the second where SPM occurs simultaneously with normal dispersion. The case where SPM occurs together with anomalous dispersion is considered beginning in Section 6.5.

6.3.1 Dispersionless Self-Phase Modulation

When L_D sufficiently exceeds the other length scales under consideration, dispersion may be neglected. The propagation equation then simplifies to

$$\frac{\partial a}{\partial z} = -j\gamma|a|^2 a \tag{6.44}$$

This equation is easily integrated to give

$$a(z, t) = a(0, t)e^{-j\gamma|a(0,t)|^2 z} \tag{6.45}$$

where $a(0, t)$ is the field temporal profile at the input of the nonlinear medium. In this pure SPM case, a time-dependent nonlinear phase shift $\Delta\phi(z, t)$ accumulates during propagation, leaving the temporal intensity profile unaffected. The temporal profile of $\Delta\phi(z, t)$ is directly proportional to that of the intensity. This time-dependent nonlinear phase shift leads to dynamic shifts in the instantaneous frequency, given by

$$\omega_{\text{inst}}(z,t) = \omega_0 + \Delta\omega(z,t) \tag{6.46a}$$

where

$$\Delta\omega(z,t) = \frac{\partial}{\partial t}(\Delta\phi) = -\frac{\partial}{\partial t}\left(\gamma|a|^2 z\right) \tag{6.46b}$$

Thus, the instantaneous frequency is modulated according to the time derivative of the intensity. The magnitude of the frequency modulation scales with the magnitude of the peak nonlinear phase shift. The left-hand column of Fig. 6.1 shows the calculated instantaneous frequency and power spectrum for a pulse with an input sech(t) amplitude profile with a peak nonlinear phase shift of about 6π. For the normal case of media with positive nonlinear refractive index ($n_2 > 0$), the frequency modulation assumes the form of an approximately linear up-chirp in the central, most intense region of the pulse. The spectral broadening connected with this chirp can be exploited successfully for optical pulse compression (Section 6.4). In the lower-intensity regions away from the pulse center, the chirp first becomes nonlinear and then reverses sign.

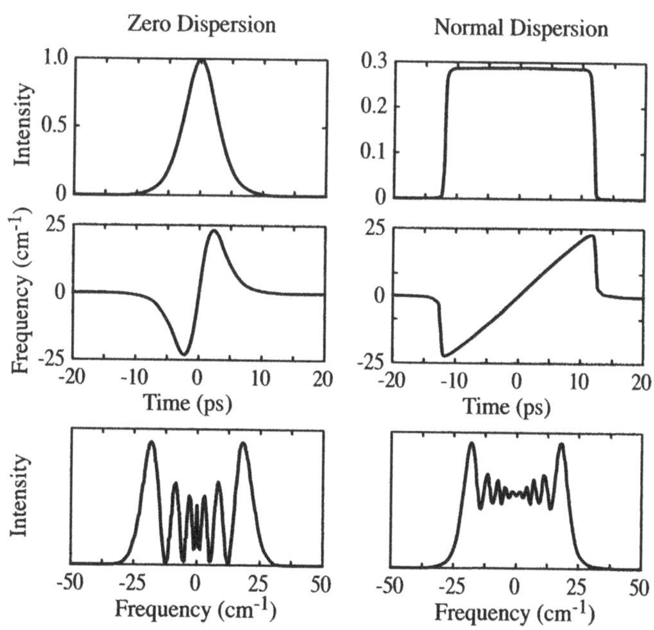

Figure 6.1 Relation between (top) the time-dependent output intensity profile $|a(z,t)|^2$ and (middle) instantaneous frequency shift $\Delta\omega(z,t)$ in SPM, as well as (bottom) resultant power spectrum. A medium with $n_2 > 0$ is assumed, and 1 cm^{-1} = 30 GHz. The left-hand side shows the case of pure (dispersionless) SPM with a peak nonlinear phase shift of about 6π. In this case the output intensity profile is identical to the input intensity profile. The right-hand column shows a calculation for SPM acting in concert with normal dispersion, with the peak intensity adjusted to yield the same spectral broadening. Adapted from [279], with permission. Copyright © 1982, American Institute of Physics.

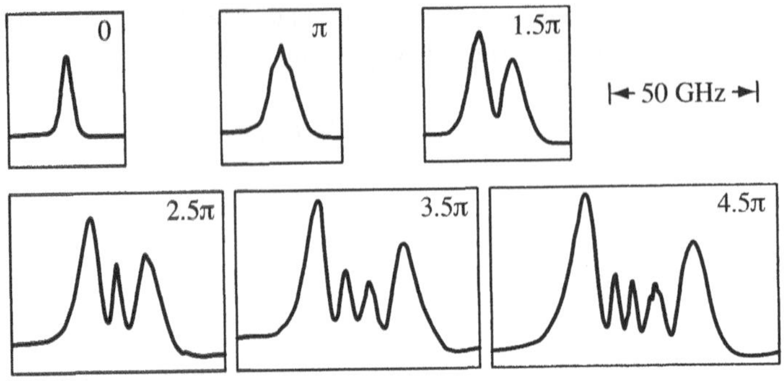

Figure 6.2 Experimental power spectra resulting from self-phase modulation. The curves are labeled by the peak nonlinear phase shift. Adapted from [38], with permission. Copyright © 1978, American Physical Society.

Data showing experimental power spectra resulting from SPM in a single-mode optical fiber are shown in Fig. 6.2 for different values of intensity (hence, peak nonlinear phase shift). The data are in good agreement with theoretical spectra (not shown) obtained by Fourier-transforming pulse solutions of the form eq. (6.45). The power spectra shown in Figs. 6.1 and 6.2 illustrate several important characteristics of SPM spectra. Under the assumption of a large nonlinear phase shift ($\Delta\phi_{\max} \gg 1$), these characteristics are described as follows:

1. The optical bandwidth subsequent to SPM is determined by the maximum excursion of the instantaneous frequency curve. The extrema in the instantaneous frequency can be found by setting the derivative of eq. (6.46b) equal to zero. For symmetric input pulses, the maximum frequency shift is given by

$$\Delta\omega_{\max} = \frac{\pm f_1 \phi_{\max}}{\Delta t} \tag{6.47}$$

where Δt is the input pulse width and f_1 is a pulse-shape-dependent numerical factor. The width (FWHM) of the input spectrum can be written $\Delta\omega_{\text{in}} = f_2/\Delta t$, where f_2 is another numerical factor. The spectral broadening factor is then given by

$$\frac{\Delta\omega_{\text{SPM}}}{\Delta\omega_{\text{in}}} = 2\frac{\Delta\omega_{\max}}{\Delta\omega_{\text{in}}} = \frac{2f_1}{f_2}\phi_{\max} \tag{6.48}$$

where $\Delta\omega_{\text{SPM}}$ is the full width of the output spectrum. In the case of a Gaussian input pulse, the result is

$$\frac{\Delta\omega_{\text{SPM}}}{\Delta\omega_{\text{in}}} \approx 1.03\,\phi_{\max} \tag{6.49}$$

Thus, the spectral broadening factor is approximately equal to the peak nonlinear phase shift.

2. The power spectra exhibit peaks at the frequency extrema. This behavior can be understood based on the instantaneous frequency picture. The amplitude of the power

spectrum at a particular instantaneous frequency offset $\Delta\tilde{\omega}$ is determined by the intensity at the time $\Delta\tilde{\omega}$ occurs and by the rate at which $\Delta\omega(t)$ passes through $\Delta\tilde{\omega}$. At the frequency extrema, $d(\Delta\omega)/dt = 0$, so a relatively broad range of times contributes to the same instantaneous frequency, leading to the peaks observed.

3. The power spectra exhibit a deep modulation. This can also be understood based on the instantaneous frequency picture. According to Fig. 6.1, there are actually two different time instants t_1' and t_2' which contribute to any particular $\Delta\tilde{\omega}$ (three different times for the special case $\Delta\tilde{\omega} = 0$). The amplitude of the power spectrum depends on the interference, hence the relative phase, between the contributions. Consider, for example, the case of a peak nonlinear phase shift $\Delta\phi_{\max} = 2m\pi$, with m an integer. The power spectrum exhibits a peak at $\Delta\omega = 0$, since the three contributions from $t = 0$ and from the far leading and trailing wings of the pulse are all in phase ($2m\pi$ phase difference). As $\Delta\tilde{\omega}$ is increased, the corresponding times t_1' and t_2' move closer together, and $\Delta\phi(t_1') - \Delta\phi(t_2')$, the difference between the corresponding nonlinear phases, is reduced. When the phase difference is reduced by π, a local minimum appears in the power spectrum; when the difference is decreased by a total of 2π, a second peak occurs, and so on. The process continues until the frequency extremum is reached, which is always a peak (corresponding to zero phase difference). Following this counting procedure for both halves of the power spectrum yields the following relations:

$$\text{number of peaks} \approx \frac{\Delta\phi_{\max}}{\pi} + 1 \tag{6.50a}$$

$$\text{number of minima} \approx \frac{\Delta\phi_{\max}}{\pi} \tag{6.50b}$$

The same result is also obtained for $\Delta\phi_{\max}$ equal to an odd multiple of π.

6.3.2 Dispersionless Self-Phase Modulation with Loss

We note that in the case of dispersionless SPM, the effect of loss can easily be included. Writing $a = a' \exp(-\alpha z/2)$, as in eq. (6.31), eq. (6.44) becomes

$$\frac{\partial a'}{\partial z} = -j\gamma|a'|^2 a' e^{-\alpha z} \tag{6.51}$$

Since $|a'|$ remains unchanged during propagation, this equation is easily integrated to give

$$a'(L,t) = a'(0,t)e^{-j\gamma|a'(0,t)|^2 L_{\text{eff}}} \qquad \text{where} \qquad L_{\text{eff}} = \frac{1}{\alpha}\left[1 - e^{-\alpha L}\right] \tag{6.52}$$

There are two simple limiting cases. When $\alpha L \ll 1$, the effect of loss is small and $L_{\text{eff}} = L$, while for $\alpha L \gg 1$, the effective length becomes $L_{\text{eff}} = \alpha^{-1}$. In the latter case the accumulated nonlinear phase shift is loss-limited.

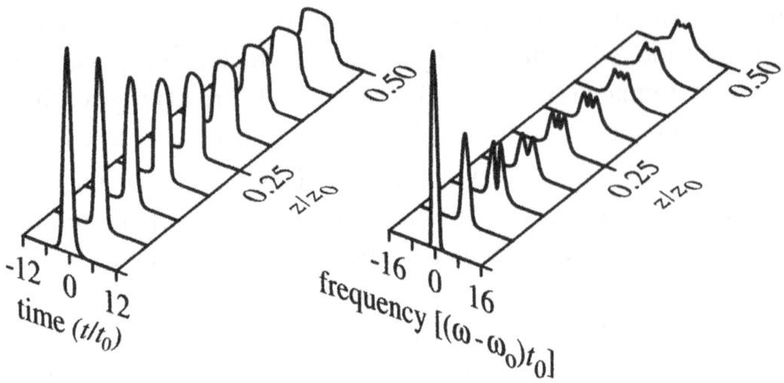

Figure 6.3 Evolution of temporal intensity profiles and power spectra for sech(t/t_0) input pulse under SPM with normal dispersion, for $N = 5$. Distance is normalized to $z_0 = (\pi/2)L_D$. From [280].

6.3.3 Self-Phase Modulation with Normal Dispersion

As we have seen, either SPM or normal group velocity dispersion acting alone on an ultrashort pulse leads to a frequency up-chirp. When SPM and normal dispersion act in concert, their interaction leads to several new effects, with important consequences.

These effects are illustrated in Figs. 6.1 and 6.3 for two cases satisfying $N > 1$. The effects of loss are ignored once more. Figure 6.3 shows the evolution of the temporal and spectral profiles of a sech(t/t_0) input pulse during nonlinear propagation in a short length of fiber calculated for simultaneous SPM and GVD [280]. The second column of Fig. 6.1 shows the output intensity profile, instantaneous frequency, and power spectrum for another case [279]. These output profiles can be contrasted with those in the first column of Fig. 6.1, in which the dispersion is set to zero but the intensity is chosen to yield the same spectral broadening as in the second column. Key features include the following:

1. In the initial section of fiber, SPM dominates, and the spectrum begins to broaden. However, as the spectrum broadens, dispersion plays an increasingly strong role. Consequently, temporal broadening becomes evident for fiber lengths well below the characteristic dispersion length L_D. As the pulse broadens, its intensity is reduced, and therefore, further spectral broadening through SPM proceeds more and more slowly. At sufficiently large propagation distances, dispersion becomes the stronger effect. These behaviors are quite distinct from the pure-SPM case, where there is no broadening in time and where the spectral broadening is essentially proportional to propagation length.
2. Nonlinear propagation eventually produces a nearly flat-topped intensity profile. At the same time, the shape of the chirp becomes remarkably linear, as evident from the instantaneous frequency plotted in Fig. 6.1. Furthermore, a very large fraction of the energy in the pulse falls within the linear chirp region, behavior that is again quite distinct from the zero-dispersion case. As discussed in greater detail in Section 6.4, these effects occur at a propagation distance which scales as $\sqrt{L_{NL} L_D}$.
3. Again in contrast to the pure SPM results, the power spectra are substantially smoother, and most of the deep modulation is washed out. This can be explained by the observation that although there are still two different time instants contributing to any particular $\Delta\tilde{\omega}$, the relative strengths of the contributions from the two time instants

are now very unbalanced. This reduces the modulation depth resulting from the interference between these two contributions.

We mention briefly an additional effect that occurs upon further propagation after a flat-topped intensity profile is obtained. Consider the instantaneous frequency profile shown in the right-hand column of Fig. 6.1. Under normal dispersion the center up-chirp region will continue to spread. However, the fast down-chirp at the leading edge of the pulse will continue to sharpen, since the slightly earlier, less red-shifted frequencies travel slower than the slightly later, more red-shifted frequencies. Eventually, two disparate groups of frequency components will overlap at the same time in the leading edge of the pulse. The interference between these overlapping frequencies produces a fast temporal modulation which has been termed *optical wave breaking* [281]. Similar behavior will occur in the down-chirped region at the trailing edge of the pulse. Further SPM acting on the rapidly modulated leading and trailing edges of the pulse leads to the creation of spectral sidelobes, which are (just barely) evident in the wings of the spectra in Fig. 6.3 for the largest propagation lengths. For further details, see [37,281].

6.3.4 Cross-Phase Modulation

Phase modulation effects may also be important when the input field consists of two or more distinct input waves that are distinguishable based on different optical frequencies, polarization, or (especially in the case of bulk media) propagation direction. In multimode waveguides, waves may also be distinguished based on different spatial modes. In the following we discuss the case of two waves at different frequencies, co-polarized and propagating collinearly in a single-mode guide.

We take the field envelope to be $a(z,t) = a_1(z,t) + a_2(z,t)$, where a_1 and a_2 are implicitly associated with frequency offset factors $\exp(j\tilde{\omega}_1 t)$ and $\exp(j\tilde{\omega}_2 t)$. Equation (6.43) becomes

$$\frac{\partial(a_1+a_2)}{\partial z} = j\frac{\beta_2}{2}\frac{\partial^2(a_1+a_2)}{\partial t^2} - j\gamma(a_1+a_2)\left(a_1^*+a_2^*\right)(a_1+a_2) \tag{6.53}$$

Multiplying out the nonlinear terms leads to contributions at frequency offsets $\tilde{\omega}_1, \tilde{\omega}_2, 2\tilde{\omega}_1 - \tilde{\omega}_2$, and $2\tilde{\omega}_2 - \tilde{\omega}_1$. Separating out the terms at different frequencies yields the following coupled equations for a_1 and a_2:

$$\begin{aligned} \tilde{\omega}_1 \text{ equation:} \quad & \frac{\partial a_1}{\partial z} = j\frac{\beta_2}{2}\frac{\partial^2 a_1}{\partial t^2} - j\gamma\left(|a_1|^2 + 2|a_2|^2\right)a_1 \\ \tilde{\omega}_2 \text{ equation:} \quad & \frac{\partial a_2}{\partial z} = j\frac{\beta_2}{2}\frac{\partial^2 a_2}{\partial t^2} - j\gamma\left(2|a_1|^2 + |a_2|^2\right)a_2 \end{aligned} \tag{6.54}$$

The new feature is a *cross-phase modulation* (XPM) interaction, where one wave modulates the phase of another. The XPM term includes an extra factor of 2 that does not appear in the SPM term. This is a general feature of XPM. In addition, the terms arising from eq. (6.53) proportional to $-j\gamma a_1^2 a_2^*$ and $-j\gamma a_2^2 a_1^*$ drive new fields at frequencies $2\tilde{\omega}_1 - \tilde{\omega}_2$ and $2\tilde{\omega}_2 - \tilde{\omega}_1$, respectively, which are not represented in eqs. (6.54). The latter interactions leading to the creation of new frequencies are termed *four-wave mixing*.

In the above we have implicitly assumed that the nonlinear refractive index is identical for SPM and XPM. This is a good approximation for co-polarized fields: We are considering the nonlinear index arising from an electronic nonlinearity, which is essentially instantaneous compared to most time scales of interest. Consequently, γ is at least approximately frequency independent. On the other hand, for cross-polarized fields the tensor nature of the nonlinear susceptibility becomes important. For an electronic nonlinearity in an isotropic medium, the nonlinear index for cross-polarized fields is three times less than for co-polarized fields [131]. In this case the factor of 2 multiplying the XPM terms in eq. (6.54) must be replaced by $\frac{2}{3}$.

6.4 PULSE COMPRESSION

The spectral broadening accompanying self-phase modulation (SPM) leads to the possibility of optical pulse compression, since the increased bandwidth can potentially support a shorter optical pulse. A simple block diagram of the process is shown in Fig. 6.4. The input pulse first propagates through a medium with a nonlinear refractive index. For sufficiently intense input pulses, SPM leads to a spectral broadening in the form of an up-chirp. If dispersion is significant, the temporal profile is also modified, as discussed earlier. Pulse compression is accomplished by using a dispersive delay line, which compensates the chirp due to SPM and brings different optical frequency components into synchronism. In many of the original experiments, the dispersive delay line was implemented using a pair of diffraction gratings, in either a single- or a double-pass geometry. Grating pairs provide the anomalous dispersion needed to compensate the up-chirp resulting from the nonlinear pulse propagation. The output pulse is both shorter and more intense than the input, since the input energy is compressed into a shorter time duration, ideally without loss.[5]

Normally, a guided-wave geometry is used in obtaining the SPM necessary for pulse compression. This prevents self-focusing and ensures that the nonlinear phase shift and spectral broadening are uniform across the transverse spatial beam profile. Most experiments employ fiber waveguides, which are categorized briefly in Section 6.1.3. Step-index optical fibers, for example, provide tight optical confinement and low loss, which are suitable for self-phase modulation and pulse compression at nanojoule pulse energies for fiber lengths ranging from millimeters to kilometers, depending on the input pulse duration [186,190,282–292]. Hollow-core fibers, which offer larger core diameter, low nonlinearity, and high damage threshold, may be used for compression of amplified pulses in the microjoule and millijoule range [274–276]. In principle, pulse compression

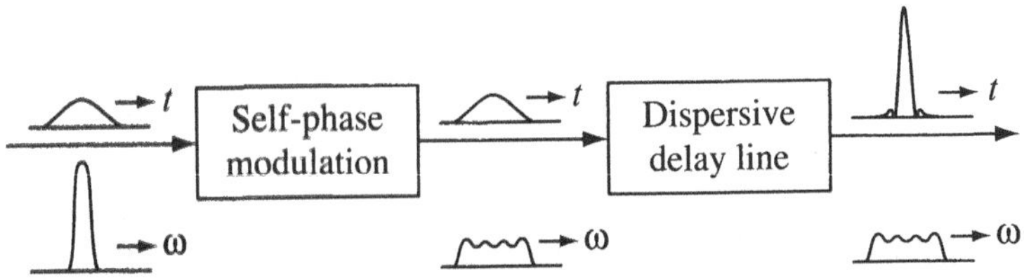

Figure 6.4 Block diagram view of optical pulse compression.

[5] In practice, the dispersive delay line may contribute unwanted insertion loss.

may also be performed using bulk nonlinear media, and a few results have been reported [293,294]. However, susceptibility to self-focusing limits such experiments to modest spectral broadening and pulse compression ratios. Furthermore, since the spectral broadening is spatially nonuniform, high-quality compression can only be achieved by selecting a portion of the spatial profile, which leads to high loss.

We can describe the pulse compression process mathematically as follows. We denote the field envelope after spectral broadening by $a_{\mathrm{SPM}}(t)$, which is expressed in the spectral domain as

$$a_{\mathrm{SPM}}(t) = \frac{1}{2\pi} \int d\omega \, |A_{\mathrm{SPM}}(\omega)| \, e^{j\psi_{\mathrm{SPM}}(\omega)} \tag{6.55}$$

The dispersive delay line adds a spectral phase function $\psi_{\mathrm{DDL}}(\omega)$. The resulting output field envelope is given by

$$a_{\mathrm{out}}(t) = \frac{1}{2\pi} \int d\omega \, |A_{\mathrm{SPM}}(\omega)| \, e^{j[\psi_{\mathrm{SPM}}(\omega)+\psi_{\mathrm{DDL}}(\omega)]} \tag{6.56}$$

An *ideal compressor* would satisfy

$$\psi_{\mathrm{DDL}}(\omega) = -\psi_{\mathrm{SPM}}(\omega) \tag{6.57}$$

That is, the spectral phase terms would cancel completely. In this case the frequency-dependent delay $\tau(\omega)$ is set to a constant, and the pulse is compressed to the bandwidth limit. On the other hand, in a practical compressor, it may not be possible to fully compensate the complete spectral phase. For example, using a grating pair as the dispersive delay line, one can vary the grating separation to vary the quadratic spectral phase. However, this does not provide a means of independent control of higher-order spectral phase terms. As a result, the output pulse may be compressed imperfectly, either because $\psi_{\mathrm{SPM}}(\omega)$ contains higher-order terms or because $\psi_{\mathrm{DDL}}(\omega)$ imparts undesired cubic or higher-order phase terms onto the spectrum in addition to the desired quadratic term.

As discussed earlier, when GVD is positive and sufficiently large, SPM and GVD working in concert lead to a substantially linear chirp and hence a predominantly quadratic spectral phase. This facilitates high-quality pulse compression, since the spectral phase required is relatively easy to match using practical pulse compressors [279,280,282,283]. Data illustrating the earliest experimental demonstration are shown in Fig. 6.5. Input pulses of

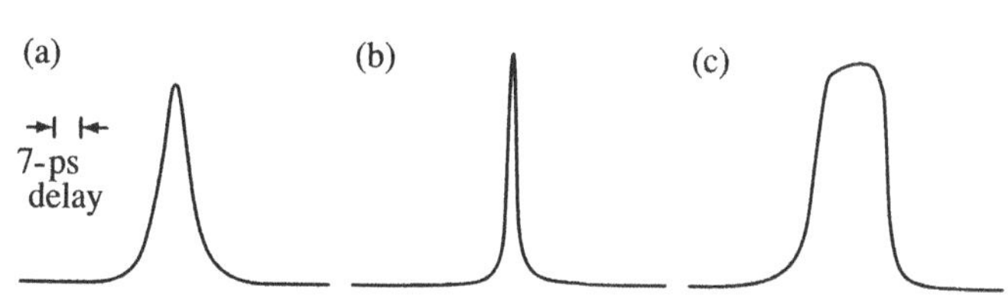

Figure 6.5 Pulse compression data resulting from combined SPM and GVD: (a) intensity autocorrelation of 5.5-ps input pulses; (b) autocorrelation of 1.5-ps compressed pulses; (c) intensity cross-correlation of 20-ps pulses directly after the optical fiber. The flat-topped intensity profile is characteristic of dispersive SPM in the normal dispersion regime. Adapted from [282], with permission. Copyright © 1981, American Physical Society.

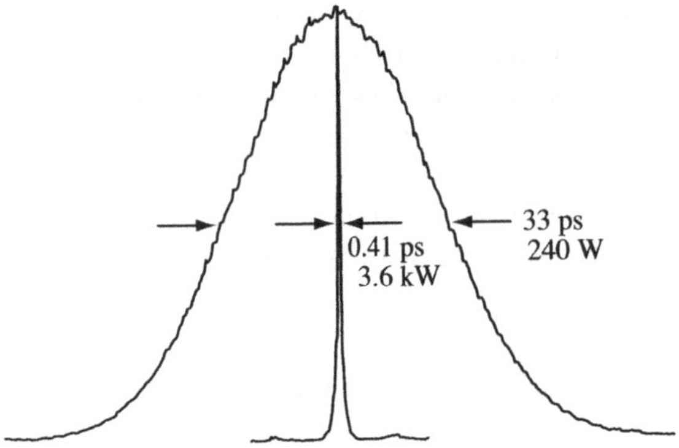

Figure 6.6 Eighty-fold pulse compression of frequency-doubled Nd:YAG laser pulses under dispersive SPM. The normalized autocorrelation traces of the input and compressed pulses are plotted on the same scale. From Fig. 3 of [295], with permission from Springer Science and Business Media.

5.5 ps at a wavelength of 587.4 nm were spectrally broadened in a 70-m length of single-mode fiber and then compressed using the anomalous dispersion from an atomic vapor cell. The output pulse was compressed to a duration of 1.5 ps and appears to have a clean shape with very little energy in the wings. The high quality of the compressed pulse can be attributed to the strong role of dispersion in the fiber. Direct evidence for the importance of dispersion in these experiments was obtained by performing intensity cross-correlation measurements of the pulses emerging from the fiber, using the compressed pulses as the reference. The pulses are broadened threefold to 20 ps and clearly exhibit the flat-topped intensity profile characteristic of dispersive SPM in the normal dispersion regime. Figure 6.6 shows another example, in which 532-nm pulses from a frequency-doubled Nd:YAG laser are compressed 80-fold from 33 ps to 0.41 ps [186,295]. The peak intensity is increased by a factor of 15, even accounting for losses in the double-pass grating compressor. The high quality of the compressed pulses again derives from the interaction of SPM and dispersion.

In contrast, pulse compression experiments must sometimes be performed in a regime where GVD is negligible. Compression of pulses from actively mode-locked Nd:YAG lasers at 1.06 μm constitutes an important example [290–292]. Here both the long input pulses (ca. 75 ps) and the reduced value of β_2 in single-mode fibers at the near-infrared laser wavelength lead to a situation where GVD plays a less important role. When SPM acts essentially alone, the instantaneous frequency follows the derivative of the temporal intensity profile, which results in strong nonlinear chirp components. A representative example is shown as the "unwindowed" trace in Fig. 6.7 [291,296,297]. Although the pulse duration is compressed approximately 90-fold to 0.8 ps, the wings of the pulse contain substantial energy, corresponding to portions of the spectrum with large nonlinear chirp and hence poor compression. Elimination of the nonlinearly chirped frequency components would lead to higher-quality compressed pulses. This has been accomplished by a technique called *spectral windowing* [161,291]. With reference to Fig. 6.8, we observe that most of the energy in the nonlinear chirp region occurs at large frequency offsets in the self-phase-modulated spectrum. These offending frequency components can be eliminated using a spectral window positioned as indicated in the figure. This was implemented experimentally by inserting a spatial window into the beam after the first pass through the double-pass grating pair, at a position where the optical frequency spectrum experiences a strong spatial separation. This spectral windowing

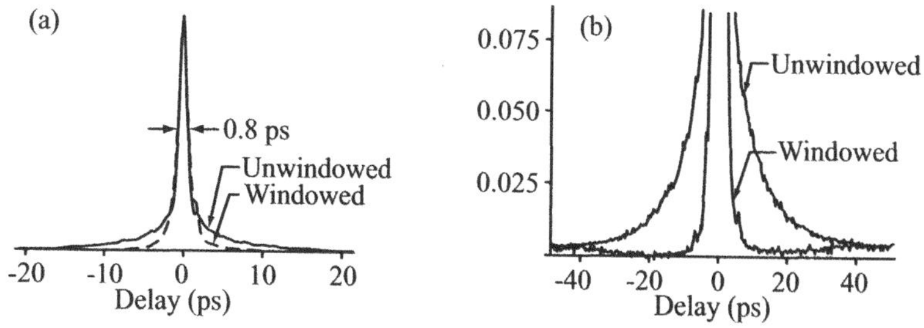

Figure 6.7 Autocorrelation measurements of windowed and unwindowed compressed pulses, performed under conditions of weak GVD: (a) full scale; (b) expanded vertical scale, dramatizing the degree of wing suppression. From [297].

resulted in a dramatic wing reduction, without appreciable pulse broadening (see Fig. 6.7). These results are emphasized by Fig. 6.7b, which shows autocorrelations of windowed and unwindowed pulses on an expanded scale. Note also the very weak sidelobes appearing at ±40 ps. These result from the down-chirp occurring in the low-intensity temporal wings of the self-phase-modulated pulse, which are not accessible to spectral windowing and which are broadened rather than compressed by subsequent anomalous dispersion.

For very short pulses, it becomes important to design compressors closely satisfying the ideal compressor condition. The cubic spectral phase term of grating pair pulse compressors, for example, is sufficient under typical conditions to degrade the compression of pulses in the sub-10-fs regime. This issue is especially acute for very short pulses, since in general the magnitude of the quadratic and linear frequency-dependent delays contributed by a dispersive delay line scale as

$$\left|\frac{\tau_{\text{quadratic}}}{\tau_{\text{linear}}}\right| \sim \left|\frac{\psi_3}{8\psi_2}\right| \Delta\omega \tag{6.58}$$

where ψ_2 and ψ_3 are the second- and third-order spectral phase coefficients and $\Delta\omega$ is the optical bandwidth. Data providing evidence of uncompensated cubic spectral phase are plotted in Fig. 6.9. Here 50-fs pulses from an amplified CPM dye laser were spectrally broadened in an optical fiber and compressed using a grating sequence alone. The upconversion signal obtained by mixing the compressed pulses with the original (unbroadened) pulses in a second-order nonlinear crystal was spectrally resolved and plotted as a

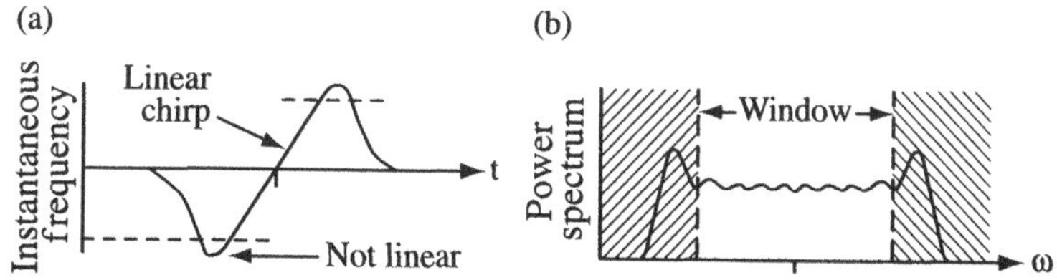

Figure 6.8 Spectral windowing concept (a spectral window blocks the frequency extrema of a spectrum broadened via self-phase modulation; this removes nonlinearly chirped frequency components and improves the quality of the compressed pulse): (a) instantaneous frequency; (b) power spectrum. From [297].

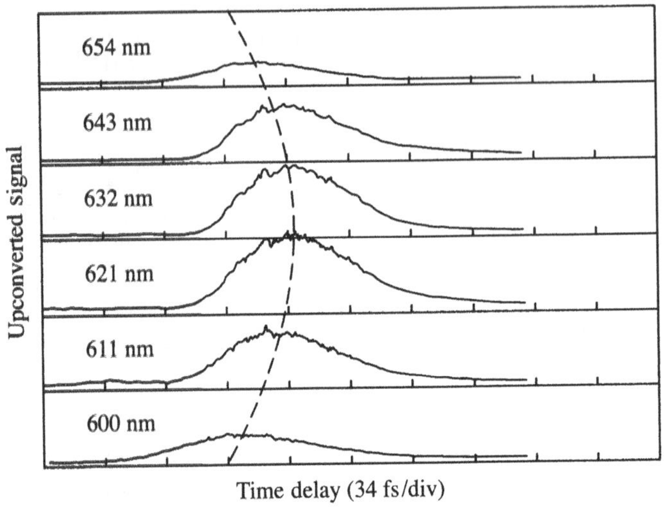

Figure 6.9 Experimental frequency-dependent delay measurement for dye laser pulses spectrally broadened in an optical fiber and then compressed using a grating sequence. The dashed line, connecting the peaks for different optical frequencies, clearly shows quadratic frequency-dependent delay. From [190].

function of delay. This provides data resembling a spectrogram which clearly show quadratic frequency-dependent delay. It was shown in [190,191] that the combination of a double-pass grating pair and a four-prism sequence allowed for simultaneous compensation of quadratic and cubic spectral phase terms. This led to compression of pulses from an amplified CPM dye laser to a record 6-fs duration [190], concluding a series of record-breaking pulse compression experiments in the early to mid-1980s [190,285–288], which had previously been limited, at least in part, by uncompensated cubic spectral phase. Later experiments performed on pulses generated from Ti:sapphire (Ti:S) laser systems led to even shorter, sub-5-fs pulses [275,276,289]. Compensation of higher-order spectral phase was again critical to the success of these experiments. In addition to grating-prism sequences, combinations of prism sequences with chirped mirrors have also been employed for compensation of up to third-order spectral phase [275,289], and compensation of terms up to and including quartic spectral phase has been reported by using a compressor consisting solely of specially designed chirped mirrors [276]. As optical bandwidths in pulse compression experiments continue to increase, compensation of spectral phase to even higher orders becomes necessary. For this reason, other compressor designs, including those based on programmable pulse shapers [298] (see Chapter 8), allowing nearly arbitrary control of spectral phase, have also become important [299,300].

Table 6.1 gives the characteristics of several representative pulse compression experiments. Clearly, there is a tremendous variation in the parameter range over which pulse compression experiments have been performed. Nevertheless, the NLSE scaling law (see Section 6.2) allows a relatively small number of calculations to span this large parameter space.

Exploiting these scaling relations, Tomlinson et al. [280] obtained results that predict the relationship between the input power, the optimum fiber length, and the achievable pulse compression ratio for a wide range of experimental parameters. Their approach was first

Table 6.1 Estimated Characteristics of Representative Pulse Compression Experiments[a]

Source laser					
Type	Nd:YAG	Frequency-doubled Nd:YAG	Synch.-pumped dye laser	Amplified CPM dye laser	Amplified Ti:S laser
Wavelength	1064 nm	532 nm	587 nm	620 nm	780 nm
Pulse duration	75 ps	33 ps	5.4 ps	50 fs	20 fs
Energy	7.5 nJ	5 nJ	3 nJ	10 nJ	80 μJ
Peak power	85 W	150 W	550 W	200 kW	4 GW
Nonlinear medium					
Fiber type	Single-mode	Single-mode	Single-mode	Single-mode	Argon-filled hollow core
Core diameter	7.3 μm	3.8 μm	4 μm	4 μm	160 μm
Length	400 m	105 m	30 m	9 mm	60 cm
Compressor type	Gratings	Gratings	Gratings	Gratings and prisms	Prisms and chirped mirrors
Characteristic parameters					
L_D	120 km	4.6 km	150 m	1.3 cm	320 cm
L_{NL}	3.5 m	20 cm	8 cm	0.2 mm	0.9 cm
N	185	150	43	7.6	19
z_{opt}	1.6 km	83 m	9 m	4.2 mm	43 cm
Compression results					
Output duration	~1 ps	410 fs	450 fs	6 fs	5 fs
Compression ratio	75	80	12	8	4
Calculated compression ratio[a]	115	94	27	5	12
References	[291,301]	[186]	[283]	[190]	[275]

[a] The calculated compression ratio is based on eq. (6.61a), which assumes a fiber of length z_{opt}.

to perform numerical simulations of pulse propagation with simultaneous SPM and GVD; then for each simulation the assumed compressor spectral phase was adjusted to yield the best pulse compression results. Figure 6.10 shows the calculated optimum compression ratio as a function of normalized fiber length for various normalized, input powers. The normalized input field is assumed to be of the form $\mathrm{sech}(t/t_0)$, and the normalized length and the input power are written in terms of

$$z_0 = \frac{\pi}{2} L_D = \frac{\pi t_0^2}{2|\beta_2|} \tag{6.59a}$$

and

$$N^2 = \gamma P_0 L_D \tag{6.59b}$$

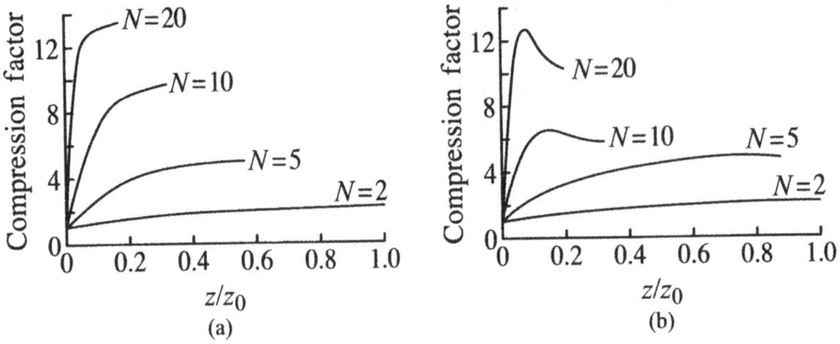

Figure 6.10 Calculated pulse compression ratio (ratio of input and output FWHM pulse durations) as a function of normalized fiber length for various normalized pulse amplitudes (N). Results are shown for (a) an ideal compressor and (b) an optimum quadratic compressor adjusted for maximum peak output power. Adapted from [280].

Results are given both for an ideal compressor and for the optimum quadratic compressor, that is, a compressor where

$$\psi_{\mathrm{DDL}}(\omega) = \frac{\psi_2}{2}(\omega - \omega_0)^2 \tag{6.60}$$

For the smaller values of N, the compression ratio increases slowly with length and is nearly the same for ideal and quadratic compressors. For higher-input powers, the compression ratio rises rapidly with length and then either saturates (ideal compressor) or reaches an optimum value beyond which it decreases (quadratic compressor). The rapid initial rise corresponds to the regime where GVD is relatively unimportant, while the saturation behavior corresponds to a regime where GVD dominates further pulse propagation. With increasing N the optimum compression (or the onset of saturation, in the optimum compressor case) occurs at shorter normalized lengths. Extrapolation of the quadratic compressor numerical results shown in Fig. 6.10 leads to approximate formulas, valid for $N \gg 1$, for the optimum compression ratio and fiber length (z_{opt}), given by

$$\frac{\Delta t_{\mathrm{compressed}}}{\Delta t_{in}} \approx \frac{1.6}{N} \tag{6.61a}$$

$$z_{\mathrm{opt}} \approx \frac{1.6 z_0}{N} \approx 2.5\sqrt{L_D L_{\mathrm{NL}}} \tag{6.61b}$$

where the pulse durations are defined in terms of the intensity FWHMs. These formulas are useful in describing experiments over a very wide parameter range. For example, the compression ratios calculated from eq. (6.61a) are in reasonable agreement with the actual compression data from Table 6.1, despite significant uncertainties in some of the parameters used for this comparison and even though some of these experiments were performed at fiber lengths differing significantly from z_{opt}.

The simulations also showed that the width of the uncompressed pulse leaving a fiber of length z_{opt} is roughly $3\Delta t_{\mathrm{in}}$, approximately independent of N. The quadratic spectral phase required for compression can therefore be estimated by setting the differential delay of the quadratic compressor ($\psi_2 \Delta\omega$) equal to $3\Delta t_{\text{in}}$, where $\Delta\omega$ is the bandwidth of the spectrally broadened pulse. Assuming that $\Delta\omega$ is inversely proportional to the compressed pulse duration together with eq. (6.61a) yields the scaling relation

$$\psi_2 \approx \frac{3.2 t_0^2}{N} \tag{6.62}$$

where the proportionality factor was evaluated from the numerical data. The required dispersion scales quadratically with pulse width. In the case of experiments with grating compressors, grating separations ranging from millimeters to many meters have been reported, depending on the duration of the input pulses, consistent with this result.

6.5 MODULATIONAL INSTABILITY

We now turn to nonlinear pulse propagation in the anomalous dispersion regime. In particular, we ask the question: Is continuous-wave (CW) light in a fiber stable? Or specifically, do small perturbations on an input CW beam propagating in the nonlinear regime remain bounded? We will see that the answer to this question reveals a fundamental difference between the normal and anomalous dispersion regimes [37,302].

Let us first assume an ideal CW input: $a(0, t) = \sqrt{P_0}$. Using a lossless nonlinear propagation equation in the form of eq. (6.43), one can easily verify that the solution is given simply by

$$a_0(z, t) = \sqrt{P_0} e^{-j\gamma P_0 z} \tag{6.63}$$

In effect, the nonlinearity increases the effective propagation constant from β_0 to $\beta_0 + \gamma P_0$.

Now we add a perturbation $\tilde{a}$, in general complex, which is taken to be much smaller than $\sqrt{P_0}$. The envelope function is written

$$a(z, t) = \left[\sqrt{P_0} + \tilde{a}(z, t)\right] e^{-j\gamma P_0 z} \tag{6.64}$$

which reduces to the unperturbed solution if $\tilde{a}$ is set to zero. Inserting this into eq. (6.43), canceling terms, and linearizing in $\tilde{a}$ (i.e., retaining only terms to the first power in $\tilde{a}$) yields an equation for the evolution of the perturbation:

$$\frac{\partial \tilde{a}}{\partial z} = \frac{j\beta_2}{2} \frac{\partial^2 \tilde{a}}{\partial t^2} - j\gamma P_0 (\tilde{a} + \tilde{a}^*) \tag{6.65}$$

Solutions can be written in the general form

$$\tilde{a}(z, t) = \alpha_1 \cos(Kz - \Omega t) + j\alpha_2 \sin(Kz - \Omega t) \tag{6.66}$$

where α_1 and α_2 are real, or as superpositions of such solutions. Substituting into eq. (6.65) yields

$$\begin{pmatrix} K & \dfrac{\beta_2\Omega^2}{2} \\ \dfrac{\beta_2\Omega^2}{2} + 2\gamma P_0 & K \end{pmatrix} \begin{pmatrix} \alpha_1 \\ \alpha_2 \end{pmatrix} = 0 \tag{6.67}$$

where the first and second rows of the matrices correspond to the $\sin(\cdots)$ and $\cos(\cdots)$ terms, respectively. Setting the determinant of the matrix to zero results in an important relation between K and γ, given by

$$K = \pm\frac{\beta_2\Omega}{2}\sqrt{\Omega^2 + \frac{4\gamma P_0}{\beta_2}} \tag{6.68}$$

In the normal dispersion regime for which $\beta_2 > 0$, K is always real. This means that the magnitude of a weak perturbation is maintained upon propagation but does not grow. For normal dispersion, therefore, CW light is stable. In the anomalous dispersion regime, $\beta_2 < 0$. For sufficiently large powers, K becomes imaginary, resulting in an exponentially growing perturbation solution. In this anomalous dispersion regime, the interplay between dispersion and nonlinearity can render the propagation of CW light unstable! This effect is known as the *modulational instability*. It arises in essence because SPM acts to create an up-chirp at peaks of the modulated intensity profile. Anomalous dispersion acts to compress the up-chirped regions, thereby reinforcing the initial intensity peaks. This leads to stronger SPM, which leads to further compression and intensity growth, and so on.

Figure 6.11 shows the calculated modulational instability gain spectrum assuming anomalous dispersion. The perturbation grows for all frequencies below a gain cutoff Ω_c,

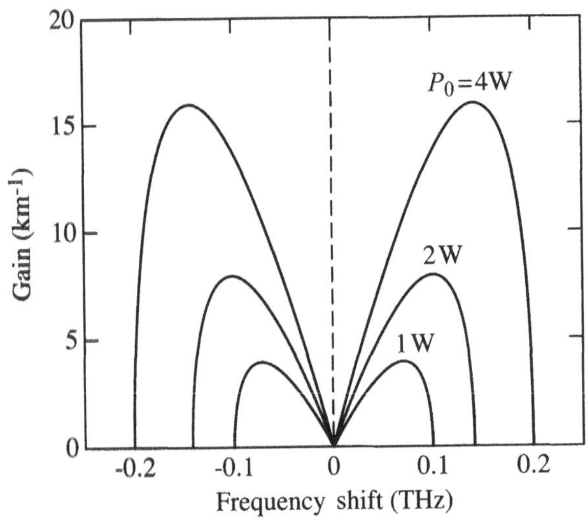

Figure 6.11 Modulational instability gain spectrum. Calculation is for $\beta_2 = -20$ ps^2 km^{-1} and $\gamma = 2$ W^{-1} km^{-1}. From [37]. Copyright © 1995, Academic Press, Inc.

given by

$$\Omega_c^2 = \frac{4\gamma P_0}{|\beta_2|} \tag{6.69}$$

The gain coefficient, expressed in terms of the perturbation amplitude, is given by $g = \text{Im}(K)$. The maximum gain is

$$g_{\max} = \frac{|\beta_2|\Omega_c^2}{4} = \gamma P_0 \tag{6.70}$$

which occurs at frequency

$$\Omega_{\max} = \pm\frac{\Omega_c}{\sqrt{2}} \tag{6.71}$$

Both the maximum attainable gain and the frequency where the maximum gain occurs grow with power. The gain bandwidth is sufficient to support growth of modulation features as short as $T_0 \sim \Omega_{\max}^{-1}$. Using the definitions of L_D and L_{NL} from Section 6.2, we find (using T_0 now as a characteristic pulse width) that $L_D/L_{\text{NL}} \sim 1/2$. This coincides with the range for which dispersion, nonlinearity, and their interaction are all expected to play an important role.

Our modulational instability analysis is valid only while the perturbation amplitude remains small. However, the growth of the perturbation eventually leads to a deep modulation and then to the generation of distinct steady-state pulses in which dispersive and nonlinear effects reach a balance. These pulses, called *solitons*, are discussed in Section 6.6.

The modulational instability may also be understood as a phase-matched four-wave mixing interaction (i.e., three input fields interacting nonlinearly to drive an output field). We consider the evolution of the field given by

$$a(z, t) = a_0(z) + a_1(z, t) + a_{-1}(z, t) \tag{6.72}$$

where a_0 is an unperturbed CW field, given by eq. (6.63), and a_1 and a_{-1} are weak harmonic perturbation fields, given by

$$a_{\pm 1}(z, t) = \tilde{a}_{\pm 1}(z)e^{-j(\beta_2\Omega^2 z/2)}e^{-2j\gamma P_0 z}e^{\pm j\Omega t} \tag{6.73}$$

The perturbation fields are written as the product of a spatially varying envelope $\tilde{a}_{\pm 1}(z)$ and exponential phase terms representing the effect of dispersion and of cross-phase modulation (XPM) from the CW field, respectively. As we shall see, writing these phase terms explicitly simplifies the ensuing math. We now insert the composite field (6.72) into the nonlinear propagation equation, eq. (6.43), linearize in the perturbation fields, and separate out terms corresponding to $e^{j\Omega t}$ and $e^{-j\Omega t}$. This yields a set of coupled equations for the perturbation field, which can be expressed by

$$\frac{\partial \tilde{a}_{\pm 1}}{\partial z} = -j\gamma P_0 e^{j\left(\beta_2\Omega^2 + 2\gamma P_0\right)z}\, \tilde{a}^*_{\mp 1} \tag{6.74}$$

The exponential term on the right-hand side of this equation is a phase mismatch term, arising in part from dispersion and in part from XPM. Efficient nonlinear optical interactions usually require zero phase mismatch, which occurs when

$$\beta_2\Omega^2 + 2\gamma P_0 = 0 \tag{6.75}$$

Ths relation can only be satisfied for $\beta_2 < 0$. In this anomalous dispersion regime, the nonlinear contribution to the propagation constant arising from XPM can balance the phase mismatch due to dispersion. This results in XPM-assisted phase matching, and the perturbation grows exponentially. The frequency for which eq. (6.75) is satisfied corresponds exactly to $\Omega_{\max}$ in eq. (6.71), the frequency for which the perturbation experiences the strongest growth. In the normal dispersion regime, phase matching is not possible, consistent with our previous results, which showed that for $\beta_2 > 0$, the perturbation does not grow.

6.6 SOLITONS

As we have already mentioned, nonlinear pulse propagation in the normal and anomalous dispersion regimes is qualitatively different. In the normal dispersion regime, SPM and group velocity dispersion reinforce each other, resulting in enhanced chirping and pulse-broadening effects. In the case of anomalous dispersion, SPM and dispersion act in opposite senses. This leads to the possibility of *solitons*,[6] pulses that can propagate without distortion under the joint effect of nonlinearity and dispersion. For detailed discussions of optical soliton propagation, see, for example, [37,308,309].

Our description begins with the dimensionless nonlinear Schrödinger equation, eq. (6.39). In the anomalous dispersion regime for which $\beta_2 < 0$, we have

$$j\frac{\partial a}{\partial \xi} = -\frac{1}{2}\frac{\partial^2 a}{\partial \tau^2} + |a|^2 a \tag{6.76}$$

This equation can be analyzed in detail using a mathematical approach known as the *inverse scattering method* [310]. For discussion of this method, see [37] and references therein. The most important solution, termed the *fundamental soliton* [82], is written

$$a(\xi, \tau) = \operatorname{sech}(\tau) e^{-j\xi/2} \tag{6.77}$$

[6] We confine our treatment to the anomalous dispersion regime, where isolated pulses propagate without distortion as *bright solitons*. This is the most important case in ultrafast optics. However, it is worth noting that soliton propagation is also possible in the normal dispersion regime in the form of *dark solitons*, where the soliton consists of a phase-modulated, pulselike reduction in the intensity of a continuous-wave background. For details, see [303–307].

The validity of this solution can be verified by direct substitution. In terms of real units, the soliton is written in a retarded time frame as

$$a(z,t) = \sqrt{P_c}\,\mathrm{sech}\left(\frac{t}{T_0}\right)\exp\left(\frac{-j|\beta_2|z}{2T_0^2}\right) \tag{6.78}$$

The fundamental soliton has several important properties:

- The pulse propagates without distortion under the combined influence of nonlinearity and dispersion.
- During propagation the pulse picks up a phase shift that is proportional to distance but constant across the pulse profile. This phase shift is the result of the nonlinear refractive index.
- The value of P_c was specified in our derivation of the NLSE [see eq. (6.38)]. Therefore, the fundamental soliton occurs only for special values of peak power, given in real units by

$$\frac{\gamma P_c T_0^2}{|\beta_2|} = 1 \tag{6.79}$$

  The soliton power and the pulse duration are such that nonlinearity and dispersion balance exactly.
- Using the NLSE scaling law, a family of equivalent fundamental soliton solutions can be written

$$a(\xi,\tau) = A\ \mathrm{sech}\,(A\tau)e^{-jA^2\xi/2} \tag{6.80}$$

  Although these solutions have different durations and peak powers, the relationship between power and duration in eq. (6.79) remains satisfied.
- The pulse area (i.e., the time integral of the field amplitude) is given by $\int a(\xi,\tau)\,d\tau = \pi$, independent of the value of A, while pulse energy is equal to $2A$. In real units, the area and the energy U are given, respectively, by

$$\text{area} = \pi\sqrt{\frac{|\beta_2|}{\gamma}} \qquad \text{and} \qquad U = \frac{2|\beta_2|}{\gamma T_0} \tag{6.81}$$

  The area is independent of pulse duration, while the energy depends inversely on pulse duration.

Soliton propagation has been found to be quite robust. As an important illustration of this robustness, we note that an input pulse whose area is close but not equal to the required soliton area will quickly evolve into an exact soliton upon propagation. Analysis based on the inverse scattering theory [311] shows that an input pulse corresponding to

$$a(\xi=0,\tau) = (1+\varepsilon)\mathrm{sech}(\tau) \qquad \text{where} \qquad |\varepsilon| < \frac{1}{2} \tag{6.82}$$

evolves into a soliton corresponding to the initial pulse shape

$$a(\xi = 0, \tau) = (1 + 2\varepsilon)\text{sech}((1 + 2\varepsilon)\tau) \tag{6.83}$$

Thus, the nonlinear propagation leads to formation of a fundamental soliton even for a factor-of-9 variation in input pulse energies. It is easy to show that the energy of the soliton in eq. (6.83), relative to that in the actual input pulse, is given by

$$\frac{U_{\text{soliton}}}{U_{\text{input}}} = \frac{1 + 2\varepsilon}{(1 + \varepsilon)^2} \tag{6.84}$$

The remaining energy evolves into a low-intensity dispersive wave, which spreads under the influence of group velocity dispersion. When the input pulse energy is close to the soliton energy, the energy that goes into the dispersive wave is small. It is also possible to form solitons from input pulses that are not secant hyperbolics. The fact that solitons propagate without distortion and are inherently robust and easy to form has led to extensive interest in the application of solitons to long-distance optical communications systems [309,312].

Pulses with sufficiently high input intensity can evolve into nonlinear waves comprised of sets of two or more fundamental solitons bound together through the nonlinear interaction. Such waves are known as *higher-order solitons*. A particularly well-known set of higher-order soliton solutions [37,310,311] comprised of N fundamental solitons occurs when the input pulse satisfies the initial condition

$$a(\xi = 0, \tau) = N\text{sech}(\tau) \tag{6.85}$$

where N is a positive integer. Unlike the fundamental $N = 1$ soliton, the temporal and spectral profiles of higher-order solitons both evolve during propagation. However, the pulse evolution is periodic, and the field returns to its original profile with period

$$z_0 = \frac{\pi}{2} L_D \tag{6.86}$$

Figure 6.12 shows the evolution of input pulses of the form eq. (6.85) over one soliton period, for $N = 2$ and $N = 3$, respectively. For $N = 2$ the behavior is still relatively simple; the pulse successively narrows and then broadens back to its original duration. For $N = 3$ (or higher), stronger pulse compression is possible, but the pulse evolution becomes more complex and entails both pulse compression and splitting. As the temporal profile is compressed or otherwise evolves, the optical spectra (not shown) broaden and develop structure. At a propagation distance z_0, the initial temporal and spectral profiles of the input pulse are restored. The compression evident at certain propagation distances in Fig. 6.12 is of potential interest for generation of shorter pulses. However, such higher-order soliton compression is accompanied by a broad pedestal containing substantial energy, which limits its practical application.

The first experimental demonstration of optical soliton behavior was reported by measuring the output pulses from a 700-m single-mode optical fiber excited by 7-ps input pulses at 1.55 μm [83]. The fiber parameters were estimated to be $\beta_2 \approx -20.4\,\text{ps}^2\,\text{km}^{-1}$ and $\gamma \approx 1.3$ $\text{W}^{-1}\,\text{km}^{-1}$ (the latter corresponding to an effective area of 100 μm^2 and $n_2 \sim 3 \times 10^{-16}$ $\text{cm}^2\,\text{W}^{-1}$). We have $T_0 = 4$ ps for a 7-ps FWHM secant hyperbolic input pulse. These

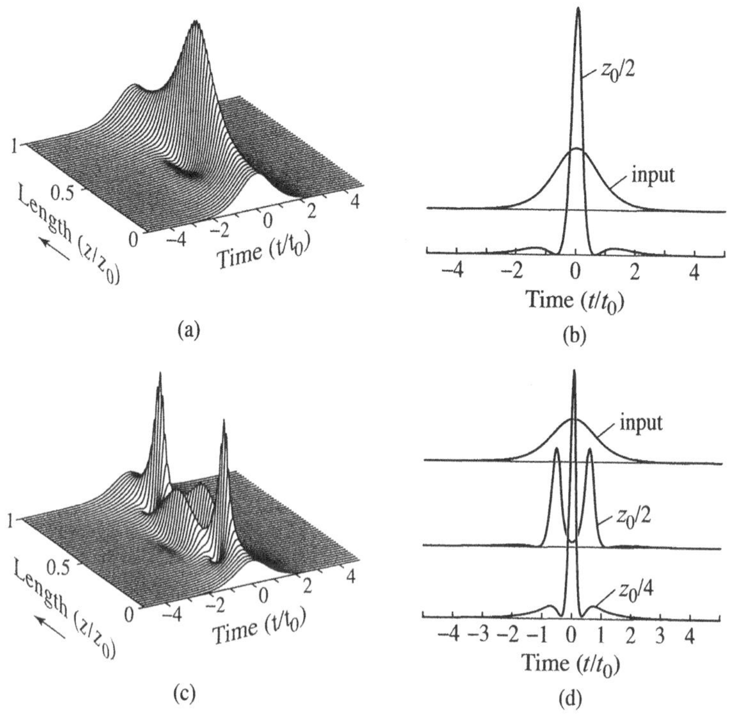

Figure 6.12 (a) Simulated intensity profile of an $N = 2$ soliton during propagation over one soliton period; (b) corresponding intensity profiles at input and distance $z_0/2$; (c) simulated intensity profile of an $N = 3$ soliton; (d) corresponding intensity profiles at input and distances $z_0/4$ and $z_0/2$. (*Note:* Intensities are normalized differently for $N = 2$ and $N = 3$ figures.)

numbers yield an estimated soliton period $z_0 = 1.26$ km and a peak power of roughly 1 W for the fundamental $N = 1$ soliton. Hence, the experiment is performed with a fiber equal to roughly one-half of the soliton period. Figure 6.13 shows intensity autocorrelation traces of the input and output pulses for various peak input powers. For a 0.3-W input, which is below the soliton power, dispersive pulse broadening is evident. At 1.2 W the output pulse duration is similar to that of the input, as expected near the soliton power. For the 5-W case, which should correspond to $N \approx 2$, the output pulse is strongly narrowed, which is

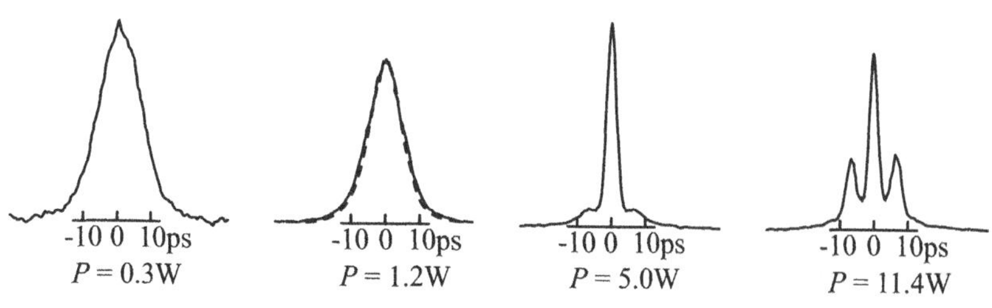

Figure 6.13 Autocorrelation traces of input pulse and output pulses from 700-m single-mode fiber for various peak input powers. The autocorrelation of the input pulse is shown as the dashed curve, overlaid with that of the output pulse at 1.2 W (corresponding to the $N = 1$ soliton). From [83], with permission. Copyright © 1980, American Physical Society.

expected for propagation of an $N = 2$ soliton over one-half-soliton period. The 11.4-W case, corresponding to a power roughly nine times higher than that of the fundamental soliton ($N \approx 3$), shows a triply peaked autocorrelation, which implies a doubly peaked intensity profile, again as expected for propagation over a one-half-soliton period. In view of experimental uncertainties, the agreement between experiment and theory is quite good.

The collision and interaction properties of solitons are also important. To discuss collisions, we need to consider soliton solutions with the more general form

$$a(\xi, \tau) = \text{sech}(\tau + \hat{\Omega}\xi)e^{j\hat{\Omega}\tau}e^{j(\hat{\Omega}^2-1)\xi/2} \tag{6.87}$$

This solution incorporates a normalized frequency shift $\hat{\Omega}$ away from the reference center frequency. Due to dispersion, this results in a change in the group velocity and in the wave vector. In real units the angular frequency shift is $\tilde{\omega} = \hat{\Omega}/T_0$, while the change in inverse group velocity and in the quadratic phase term corresponding to the change in wave vector are $\beta_2\tilde{\omega}$ and $\exp(-j\beta_2\tilde{\omega}^2 z/2)$, respectively. Equation (6.87) is a frequency-shifted version of a fundamental soliton. A similar generalization applies to any solution of eq. (6.76) [309]. That is:

> If $\tilde{a}(\xi, \tau)$ is any valid solution of the normalized NLSE, the frequency-shifted version $\tilde{a}(\xi, \tau + \hat{\Omega}\xi)\exp(j\hat{\Omega}\tau)\exp(j\hat{\Omega}^2\xi/2)$ is also a solution.

Solitons with different center frequencies may undergo collisions. Figure 6.14 shows a simulation of such a collision, taking as input a pair of well-separated fundamental soliton pulses with different center frequencies. Associated with the different frequency shifts are different group velocities. As a result, the trajectories of the two pulses in the ξ–τ plane have different slopes. When the trajectories meet and a collision occurs, there is a strong nonlinear interaction, leading to a complicated evolution of the pulse profile. Nevertheless, subsequent to the collision, the two original pulses emerge, with no change in intensity, pulse duration, or frequency. An important defining property of solitons is their robustness with respect to such collisions, with the output pulses unmodified except for phase and timing shifts.

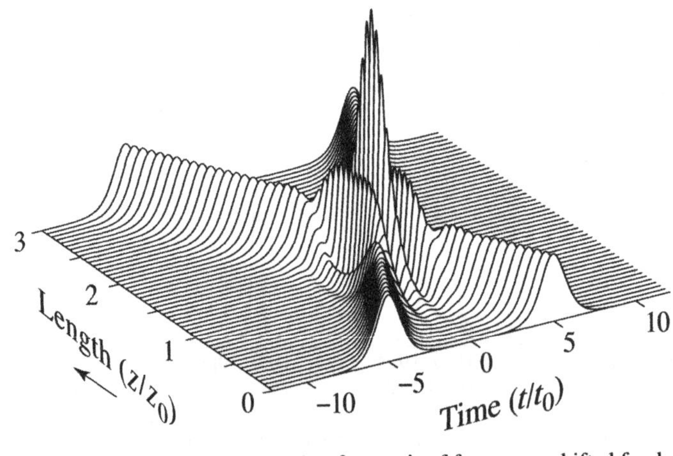

Figure 6.14 Nonlinear propagation simulation for a pair of frequency-shifted fundamental soliton input pulses, giving rise to a soliton collision. The input pulses are shifted by $10t_0$ in time and by $0.8t_0^{-1}$ in frequency (units: hertz).

Another manifestation of this robustness property relates to soliton propagation in the presence of slow variations in the propagation direction. Consider, for example, the case of a fiber with distributed gain. Studies have shown that if the gain over one soliton period is small, the soliton propagates adiabatically in the sense that the pulse duration decreases continuously as the soliton energy increases, with very little energy lost to dispersive waves [313]. In this adiabatic regime, the pulse duration after propagation is governed by eq. (6.81); the energy U and duration T after propagating distance z are related by

$$\frac{U(z)}{U(0)} = \mathcal{G} = \frac{T(0)}{T(z)} \tag{6.88}$$

The soliton is compressed with a compression factor equal to the energy gain $\mathcal{G}$. Note that although the gain per soliton period should be small, the overall gain and hence the compression factor can be large if the propagation distance is much larger than the soliton period. Similar adiabatic evolution of the soliton also occurs in the presence of slow variations of the dispersion β_2 and nonlinearity parameter γ [314,315]; in fact, it has been shown that the NLSE for a fiber with longitudinally decreasing dispersion can be mathematically transformed into a NLSE for a fiber with gain [316]. The case of fibers with slowly decreasing dispersion is of technological importance, since such fibers can be realized by tapering the fiber diameter during the fiber-pulling process [317].[7] Provided that the fractional change in dispersion per soliton period is small, the pulse evolves adiabatically, again with little energy lost to dispersive waves. The pulse duration is again governed by eq. (6.81), which gives [317,318]

$$\frac{T(z)}{T(0)} = \frac{\beta_2(z)}{\beta_2(0)} \tag{6.89}$$

Figure 6.15 shows simulation results for evolution of an input fundamental soliton during propagation in a fiber with slowly decreasing dispersion. A high-quality pulse is maintained during compression, with very little energy in the wings. This makes the dispersion-decreasing fiber technique a valuable pulse compression method experimentally [319–322], especially for applications in ultrafast communications.

6.7 HIGHER-ORDER PROPAGATION EFFECTS

We have seen that the nonlinear Schrödinger equation discussed in previous sections describes a wide range of nonlinear propagation phenomena. However, for sufficiently short pulses, approximations made in deriving the nonlinear Schrödinger equation break down. For example, in nonlinear fiber optics new effects not predicted by the nonlinear Schrödinger equation are often observed for pulse durations significantly below 1 ps. Therefore, in this section we derive a more accurate form of the nonlinear propagation equation and consider its consequences.

[7] Changing the fiber diameter modifies the waveguide contribution to the dispersion.

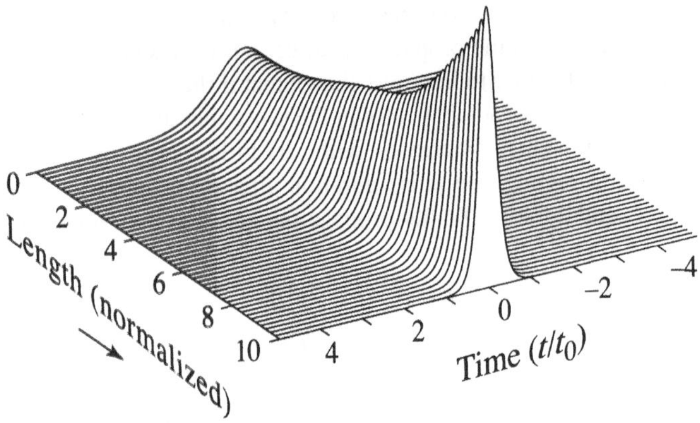

Figure 6.15 Nonlinear propagation simulation for an input fundamental soliton with $t_0 = 10$ ps in a fiber with slowly decreasing dispersion. The simulation assumes a total length of 78.5 km, a dispersion that is tapered linearly from $\beta_2 = -20$ ps^2 km^{-1} at the input to -5 ps^2 km^{-1} at the output, and a nonlinearity of $\gamma = 2.59$ km^{-1} W^{-1}. Loss is neglected. The output pulse is compressed by a factor of 4 compared to the input.

6.7.1 Nonlinear Envelope Equation in Uniform Media

We first consider propagation in a uniform bulk medium, retaining both temporal and spatial degrees of freedom. Our derivation follows the approach outlined in [131,323].

We express the electric field and the nonlinear polarization as follows:

$$\begin{aligned} E(x, y, z, t) &= \mathrm{Re}\left\{\frac{1}{2\pi}\int d\tilde{\omega}\, A(x, y, z, \tilde{\omega})\, e^{j(\omega_0+\tilde{\omega})t} e^{-j(\beta_0+\beta_1\tilde{\omega})z}\right\} \\ P_{\mathrm{NL}}(x, y, z, t) &= \mathrm{Re}\left\{\frac{1}{2\pi}\int d\tilde{\omega}\, Q(x, y, z, \tilde{\omega})\, e^{j(\omega_0+\tilde{\omega})t} e^{-j(\beta_0+\beta_1\tilde{\omega})z}\right\} \end{aligned} \tag{6.90}$$

Here Q is the Fourier transform of the nonlinear polarization envelope function. We assume that the medium is uniform and isotropic and insert these definitions into the scalar wave equation, eq. (6.3), which gives

$$\left[-(\beta_0 + \beta_1\tilde{\omega})^2 - 2j(\beta_0 + \beta_1\tilde{\omega})\frac{\partial}{\partial z} + \frac{\partial^2}{\partial z^2} + \beta^2(\tilde{\omega}) + \nabla_T^2\right] A = -\mu_0(\omega_0 + \tilde{\omega})^2 Q \tag{6.91}$$

where $\nabla_T^2 = \partial^2/\partial x^2 + \partial^2/\partial x^2$ and $\beta^2 = \mu_0\varepsilon_{(1)}(\omega_0 + \tilde{\omega})^2$. We rewrite the propagation constant as

$$\beta(\tilde{\omega}) = \beta_0 + \beta_1\tilde{\omega} + \tilde{D} \tag{6.92a}$$

where $\tilde{D}$ is a dispersion operator defined in the frequency domain by

$$\tilde{D} = \frac{\beta_2}{2}\tilde{\omega}^2 + \frac{\beta_3}{6}\tilde{\omega}^3 + \cdots \tag{6.92b}$$

Using these definitions, the propagation equation becomes

$$-2j\beta_0\left(1+\frac{\beta_1\tilde{\omega}}{\beta_0}\right)\left[\frac{\partial}{\partial z}+j\tilde{D}\right]A+\nabla_T^2 A+\mu_0\omega_0^2\left(1+\frac{\tilde{\omega}}{\omega_0}\right)^2 Q=-\left(\frac{\partial^2}{\partial z^2}+\tilde{D}^2\right)A \tag{6.93}$$

which can be rearranged to yield

$$\left(1+\frac{\tilde{\omega}}{\omega_0}\right)\left[\frac{\partial}{\partial z}+j\tilde{D}\right]A+\frac{j}{2\beta_0}\nabla_T^2 A+\frac{j\mu_0\omega_0 c}{2n_0}\left(1+\frac{\tilde{\omega}}{\omega_0}\right)^2 Q$$
$$=\frac{-j}{2\beta_0}\left(\frac{\partial^2}{\partial z^2}+\tilde{D}^2\right)A-\left(\frac{\omega_0\beta_1}{\beta_0}-1\right)\frac{\tilde{\omega}}{\omega_0}\left[\frac{\partial}{\partial z}+j\tilde{D}\right]A \tag{6.94}$$

To simplify, we now invoke the slowly varying envelope approximation *in space*. Specifically, we take

$$\frac{\partial A}{\partial z}\ll\beta_0 A \qquad \text{and} \qquad \tilde{D}A\ll\beta_0 A \tag{6.95}$$

Under these conditions the term in the first parentheses on the right-hand side of eq. (6.94) may be dropped. Physically, the requirement is that the effect of dispersion and the overall change in the envelope during propagation over a distance of one wavelength are small. The second term on the right-hand side of eq. (6.94) may be dropped when at least one of the following two conditions is met:

$$\tilde{\omega}A\ll\omega_0 A \tag{6.96a}$$

or

$$\left|\frac{\omega_0\beta_1}{\beta_0}-1\right|=\left|\frac{v_p}{v_g}-1\right|\ll 1 \tag{6.96b}$$

Here $v_p=\omega_0/\beta_0$ and $v_g=1/\beta_1$ are the phase velocity and group velocity, respectively. We discuss the physical meaning of these conditions shortly. Setting all of the right-hand side of eq. (6.94) to zero and dividing through by $(1+\tilde{\omega}/\omega_0)$ yields

$$\left[\frac{\partial}{\partial z}+j\tilde{D}\right]A+\frac{j}{2\beta_0}\frac{1}{(1+\tilde{\omega}/\omega_0)}\nabla_T^2 A=\frac{-j\mu_0\omega_0 c}{2n_0}\left(1+\frac{\tilde{\omega}}{\omega_0}\right)Q \tag{6.97}$$

To arrive at a time-domain propagation equation, we introduce

$$a(x,y,z,t')=\frac{1}{2\pi}\int d\tilde{\omega}\,A(x,y,z,\tilde{\omega})\,e^{j\tilde{\omega}t'}$$
$$q(x,y,z,t')=\frac{1}{2\pi}\int d\tilde{\omega}\,Q(x,y,z,\tilde{\omega})\,e^{j\tilde{\omega}t'} \tag{6.98}$$

where a and q represent the envelopes of the field and of the nonlinear polarization in terms of retarded time $t' = t - \beta_1 z$. The inverse Fourier transform of eq. (6.97) then gives

$$\frac{\partial a}{\partial z} + jDa + \frac{j}{2\beta_0}\left(1 - \frac{j}{\omega_0}\frac{\partial}{\partial t'}\right)^{-1} \nabla_T^2 a = \frac{-j\mu_0\omega_0 c}{2n_0}\left(1 - \frac{j}{\omega_0}\frac{\partial}{\partial t'}\right)q \tag{6.99a}$$

Here the $(1 - (j/\omega_0)(\partial/\partial t'))^{-1}$ factor operating on $\nabla_T^2 a$ can be evaluated in the frequency domain, and D, the dispersion operator in the time domain, is

$$D = \frac{-\beta_2}{2}\frac{\partial^2}{\partial t'^2} + \frac{j\beta_3}{6}\frac{\partial^3}{\partial t'^3} + \cdots \tag{6.99b}$$

Equation (6.99a) has been termed the *nonlinear envelope equation* (NEE) [323]. Its validity requires first the slowly varying envelope approximation in space, eq. (6.95), which means that the change in the field envelope A incurred during propagation over distances comparable to a wavelength is small. Furthermore, the different terms contributing to such changes must *individually* be weak on the wavelength scale. In the notation used earlier in this chapter, the requirement is that $L_D \gg \lambda$, $L_{NL} \gg \lambda$. Higher-order dispersion terms must be similarly weak. The validity of the NEE also depends on either condition (6.96a) or (6.96b) being satisfied. The first of these conditions is the slowly varying envelope approximation *in time*, which applies when the fractional bandwidth of the optical field is small. This means that pulses must be at least many optical cycles in duration. On the other hand, condition (6.96b) can be satisfied even for fields with large fractional bandwidth, provided that the group velocity and the phase velocity are approximately the same. This condition is found to be true for a wide range of transparent media far from resonances, provided that all the fields propagate in roughly the same direction. When (6.96b) is satisfied, the slowly varying envelope approximation in time need not be invoked; as a result the NEE can apply even for pulses approaching a single cycle.

To gain further insight, we rewrite the full electric field from eq. (6.90) as a function of retarded time in terms of the envelope function a. The result is

$$E(x, y, z, t') = \mathrm{Re}\left\{a(x, y, z, t')e^{j\omega_0 t'}e^{j(\omega_0\beta_1 - \beta_0)z}\right\} \tag{6.100}$$

When the phase velocity and group velocity are equal, the last exponential term above goes to unity; the phase shift normally associated with propagation drops out in the retarded time frame. However, when the velocities are unequal, this term gives a phase shift proportional to z. We observe that whenever (6.96b) is satisfied, the phase shift incurred during propagation over a one-wavelength distance remains small. Equivalently, the difference in phase delay and group delay for propagation over a wavelength must be much less than the optical period. Taken together, eqs. (6.95) and (6.96b) mean that the NEE will be valid as long as each of the individual terms contributing to changes in E, including the apparent phase slip due to difference in phase and group velocities, is small over one-wavelength propagation.

With appropriate additional approximations, the NEE can be simplified to yield the propagation equations considered earlier in this chapter. For example, if we assume plane waves and an instantaneous nonlinearity ($q = 2n_0\varepsilon_0 n_2|a|^2 a$), truncate the dispersion after the leading term, and make the slowly varying envelope approximation in time, the NEE then reduces to a form similar to that considered in Section 6.1.1. The correspondence is exact when eq. (6.13) is transformed into the retarded time frame and its loss set to zero.

The NEE includes several terms distinct from our earlier propagation equations. These are discussed in the following sections. Briefly, the important new features are:

- The form of the nonlinear polarization is not yet specified. This allows treatment of nonlinear index processes with finite response times. As we shall see, this gives rise to effects associated with Raman scattering (Section 6.7.3).
- Dispersion terms above β_2 can be retained as desired.
- The time derivative operating on the nonlinear polarization can give rise to an intensity-dependent group velocity and, in principle, to the formation of shock fronts. This shock term is discussed in Section 6.7.4.
- The ∇_T^2 term accounts for spatial propagation in bulk media; together with a nonlinear index, this can lead to self-focusing. Furthermore, as discussed in Section 6.7.5, the time derivative operator preceding the ∇_T^2 couples the spatial and temporal degrees of freedom.

6.7.2 Nonlinear Envelope Equation in Waveguides

We now consider the NEE in a waveguide. The derivation follows closely that in the preceding section, with modifications for the waveguide geometry as in Section 6.1.2. The electric field is taken as

$$E(x, y, z, t) = \mathrm{Re}\left\{\frac{1}{2\pi}\int d\tilde{\omega}\, A(z, \tilde{\omega})u(x, y)\, e^{j(\omega_0+\tilde{\omega})t}e^{-j(\beta_0+\beta_1\tilde{\omega})z}\right\} \tag{6.101}$$

u is the waveguide transverse mode profile (ignoring perturbations due to the nonlinearity), given by the solution of eigenvalue equation (6.17). We have simplified the problem at the outset by taking u to be frequency independent, which is a reasonable approximation for most situations encountered in nonlinear fiber optics. For the nonlinear polarization we use the same form as in eq. (6.90). Plugging into eq. (6.3) and using (6.17), we arrive at

$$-2j\beta_0\left(1+\frac{\beta_1\tilde{\omega}}{\beta_0}\right)\left[\frac{\partial}{\partial z}+j\tilde{D}\right]Au+\mu_0\omega_0^2\left(1+\frac{\tilde{\omega}}{\omega_0}\right)^2 Q=-\left(\frac{\partial^2}{\partial z^2}+\tilde{D}^2\right)Au \tag{6.102}$$

Using precisely the same approximations as in the preceding section, this equation simplifies to

$$\left[\frac{\partial}{\partial z}+j\tilde{D}\right]Au=\frac{-j\mu_0\omega_0^2}{2\beta_0}\left(1+\frac{\tilde{\omega}}{\omega_0}\right)Q \tag{6.103}$$

Using the same notation as in the preceding section, we perform the inverse Fourier transform to arrive at a propagation equation in the retarded time frame:

$$u\frac{\partial a}{\partial z}+jDau=\frac{-j\mu_0\omega_0^2}{2\beta_0}\left(1-\frac{j}{\omega_0}\frac{\partial}{\partial t'}\right)q \tag{6.104}$$

To proceed, we assume for the moment that the nonlinear polarization is associated with an instantaneous nonlinear index, so that $q = 2n_0\varepsilon_0 n_2|au|^2 au$. As in Section 6.1.2, we multiply both sides of the equation by u^* and integrate over the transverse coordinates, assuming that the spatial variation of n_0 and n_2 is sufficiently small that they can be pulled outside the integral. Following eq. (6.28), we also introduce a new field variable a_p which is normalized such that $|a_p|^2$ gives the power in the guided mode. The result is a nonlinear envelope equation for propagation within a single-mode waveguide:

$$\frac{\partial a_p}{\partial z} + jDa_p + j\gamma\left(1 - \frac{j}{\omega_0}\frac{\partial}{\partial t'}\right)|a_p|^2 a_p = 0 \tag{6.105}$$

where γ is defined exactly as in eq. (6.29b). This equation clearly resembles the nonlinear Schrödinger equation, but with the addition of a time derivative operating on the nonlinear polarization and with higher-order dispersion terms retained.

6.7.3 Delayed Nonlinear Response and the Raman Effect

Overview of Delayed Nonlinear Response To this point we have restricted our attention to nonlinear effects associated with electronic processes. Such nonlinearities occur when the optical field perturbs the electronic structure of the molecule (or crystal, etc.), which in turn modifies the polarizability and hence the refractive index of the material. For optical frequencies well below the electronic transitions (which typically fall in the ultraviolet for transparent media), the electronic contribution to the nonlinear index should have a response time on the order of the inverse of the typical transition frequencies. Because this is faster than one optical cycle, the electronic contribution to the nonlinear index is considered approximately instantaneous.

Although such electronic processes are very important, they are not the only processes that can contribute to the nonlinear refractive index. In particular, nonlinear index changes associated with optically induced motions of nuclei can also be important. Such nuclear nonlinearities arise because optically induced changes in electronic structure may exert forces on the associated nuclei whose consequent motions in turn modulate the electronic polarizability and the refractive index.

Since nuclei are considerably heavier than electrons, the vibrational motions of nuclei have resonant frequencies much lower than those of electronic resonances. Typical vibrational frequencies range from a few terahertz up to roughly 100 THz, corresponding to vibrational periods of tens to hundreds of femtoseconds. Since the vibrational period sets the time scale for the nonlinear response, experiments performed with pulses that are sufficiently short may sense a noninstantaneous or time-delayed vibrational contribution to the nonlinear refractive index. Note that in addition to vibrations, other nuclear motions, such as molecular rotations and laser-induced orientational ordering processes, can also provide a noninstantaneous contribution to the nonlinear index. Such effects can arise in a wide variety of media, including crystals, amorphous solids, liquids, and gases.

Nuclear contributions to the nonlinear index, and more generally to the third-order nonlinear susceptibility, are studied in detail in many texts on nonlinear optics, usually from a frequency-domain perspective [69,131,324,325]. Raman scattering is a particularly well-known effect arising from the nuclear contribution to the nonlinear index. In spontaneous Raman scattering, the interaction of a monochromatic laser beam with the material leads to generation of a small amount of light that is downshifted in frequency and scattered out of the incident beam. Spontaneous Raman is linear in the input field; scattering arises due to spontaneous time-dependent fluctuations in polarizability associated with nuclear motions. In each scattering event the energy associated with an incident photon is partitioned into a lower-energy scattered photon and a quantum of vibrational excitation (which is transferred to the medium). Peaks in the spontaneous Raman spectrum (as a function of frequency downshift) reveal the frequencies of vibrational resonances. Nonlinear Raman techniques based on interactions between two or more input fields with adjustable frequency offsets are also widely used; in these techniques the incident laser fields stimulate vibrational motion, which in turn yields a time-varying polarizability that acts back on the laser fields. Nonlinear Raman techniques typically produce output signals that are resonantly enhanced when the frequency offset between the laser fields matches a vibrational frequency and are therefore useful for vibrational spectroscopy.

In this section we adopt a time-domain perspective that ties in more directly with ultrafast optical effects. A detailed discussion of the third-order nonlinear susceptibility is given in [324], with special attention given to the relation between the time- and frequency-domain responses. In general, the polarization generated in response to an arbitrary pulsed electric field can be written

$$\begin{aligned} P_i(\mathbf{r},t) = {} & \varepsilon_0 \int dt' \tilde{\chi}_{ij}^{(1)}(t,t') E_j(\mathbf{r},t') \\ & + \varepsilon_0 \iint dt' dt'' \tilde{\chi}_{ijk}^{(2)}(t,t',t'') E_j(\mathbf{r},t') E_k(\mathbf{r},t'') \\ & + \varepsilon_0 \iiint dt' dt'' dt''' \tilde{\chi}_{ijkl}^{(3)}(t,t',t'',t''') E_j(\mathbf{r},t') E_k(\mathbf{r},t'') E_l(\mathbf{r},t'') \\ & + \cdots \end{aligned} \tag{6.106}$$

where $\tilde{\chi}_{ij}^{(1)}$, $\tilde{\chi}_{ijk}^{(2)}$, and $\tilde{\chi}_{ijkl}^{(3)}$ are time-dependent susceptibility tensors that give, respectively, the first-, second-, and third-order polarization response, and the notation $\tilde{\chi}^{(n)}$ is used to distinguish the time-dependent response function from the frequency-dependent response $\chi^{(n)}$ introduced in Chapter 5. Time invariance requires that the $\tilde{\chi}^{(n)}$ depend only on differences among the time arguments, and causality is also required. Otherwise, the form of eq. (6.106) is quite general, allowing for description of the full-time dependence of the nonlinear polarization. The first-order term, $\tilde{\chi}_{ij}^{(1)}$, is responsible for the linear polarization response. When the first-order susceptibility is not instantaneous [i.e., $\tilde{\chi}_{ij}^{(1)}(t,t')$ is not proportional to $\delta(t-t')$], the first-order polarization response in the frequency domain exhibits a frequency-dependent phase, which gives rise to the familiar dispersion effects of Chapter 4. Second-order nonlinearities were discussed extensively in Chapter 5; moreover, the second-order susceptibility tensor vanishes in centrosymmetric media, which are of primary concern in this chapter. Therefore, in the following we focus on the third-order nonlinear response.

The form of the third-order response simplifies considerably for the case of transparent, nonconducting media with electronic absorption resonances all well above the frequencies of the optical fields present [324]. In this situation the electrons respond nearly adiabatically both to the applied electric field and to the fluctuating electric fields associated with the nuclei. In a quantum mechanical calculation [324], this leads to the *Born–Oppenheimer approximation*, in which it is assumed that the electrons are always in the ground-state corresponding to the instantaneous configuration of the nuclei and fields. The resulting ground-state electron configuration is then used to calculate an effective potential acting on the nuclei. This procedure reveals a coupling between the electric polarization response of the medium and the nuclear coordinates. With the further restriction of purely "optical" nonlinear effects, in which both the fields and the polarization response occur only at frequencies well above the frequencies of the nuclear motions, the third-order polarization reduces to

$$P_i^{(3)}(t) = \varepsilon_0 \left\{ \sigma_{ijkl} E_j(t) E_k(t) E_l(t) + E_j(t) \int dt' d_{ijkl}(t-t') E_k(t') E_l(t') \right\} \qquad (6.107)$$

where the dependence on $\mathbf{r}$ has been suppressed. The first term represents the electronic nonlinearities with instantaneous response and is characterized by a tensor σ_{ijkl} which is independent of the time variables. The second term describes the nuclear contribution to the nonlinear polarization through the third-order response tensor d_{ijkl}. As a result of the approximations summarized above, the nonlinear response function depends on a single time difference only.

Further simplification is possible for the case of isotropic media (such as glass fibers), which are of great practical importance. The tensors representing the electronic and nuclear contribution to the nonlinear response, respectively, assume the forms

$$\sigma_{ijkl} = \frac{1}{6}\sigma \left(\delta_{ij}\delta_{kl} + \delta_{ik}\delta_{jl} + \delta_{il}\delta_{jk} \right) \qquad (6.108a)$$

$$d_{ijkl}(t) = d_a(t)\delta_{ij}\delta_{kl} + \frac{1}{2} d_b(t) \left(\delta_{il}\delta_{jk} + \delta_{ik}\delta_{jl} \right) \qquad (6.108b)$$

These tensor forms are useful for predicting nonlinear polarization effects in a variety of experimental geometries. By inserting these expressions into eq. (6.107), we obtain the following form for the nonlinear polarization vector in terms of the vector electric field $\mathbf{E}$:

$$\mathbf{P}^{(3)}(t) = \varepsilon_0 \left\{ \frac{\sigma}{2} \mathbf{E}(t) \left[\mathbf{E}(t) \cdot \mathbf{E}(t) \right] + \mathbf{E}(t) \int dt' d_a(t-t') \left[\mathbf{E}(t') \cdot \mathbf{E}(t') \right] \right.$$
$$\left. + \int dt' \mathbf{E}(t') d_b(t-t') \left[\mathbf{E}(t) \cdot \mathbf{E}(t') \right] \right\} \qquad (6.109)$$

The electronic contribution and the first part of the nuclear contribution (involving scalar response function d_a) are seen to have an isotropic character, in the sense that the nonlinear polarization vector is always aligned with the instantaneous applied electric field. On the other hand, the second part of the nuclear contribution (involving scalar response function d_b) has an anisotropic character, in the sense that nonlinear polarization can be oriented differently than the instantaneous field. The anisotropic (d_b) contribution to the nonlinearity leads to self-induced polarization changes of elliptically polarized fields [131,324]. The different forms of the σ_{ijkl} and d_{ijkl} tensors have been exploited for experimental determination of the relative strengths of electronic and nuclear contributions to the nonlinear index in various materials [324,326,327].

The Raman Response Function and Raman Gain In this chapter we are concerned primarily with self-action effects involving the nonlinear propagation of a single ultrashort pulse. We restrict our attention to linear polarization and isotropic media. We can then dispense with the tensor notation and write the nonlinear polarization simply as the summation of a scalar instantaneous electronic contribution and a scalar noninstantaneous nuclear contribution characterized by a single response function. The result is that the instantaneous nonlinear index, used in earlier derivations in this chapter ($n_2|a|^2$) is replaced as follows [328,329]:

$$n_2|a(t)|^2 \rightarrow n_{2e}|a(t)|^2 + n_{2R}\int dt'\, f(t-t')|a(t')|^2 \tag{6.110}$$

Here n_{2e} and n_{2R} denote the electronic and nuclear contributions to the nonlinear index, respectively, and $f(t)$ is the response function associated with the nuclear contribution. Since the nuclear contribution is closely connected with Raman scattering, $f(t)$ is commonly known as the *Raman response function*. In a similar vein, the nonlinear polarization envelope q of eq. (6.98) can be written

$$q = 2n_0\varepsilon_0 a(t)\left\{n_{2e}|a(t)|^2 + n_{2R}\int dt'\, f(t-t')|a(t')|^2\right\} \tag{6.111}$$

The Raman response function will be assumed to satisfy several important conditions:

- $f(t)$ is causal [i.e., $f(t) = 0$ for $t < 0$].
- $f(t)$ is real. This condition is equivalent to the assumption that there is no optical loss associated with the nuclear contribution to the nonlinear index other than coupling of energy into low-frequency nuclear motions. Nonlinear absorption processes are excluded.
- $f(t)$ will be normalized such that $\int dt f(t) = 1$.

With the latter condition it is useful to rewrite eq. (6.110) in the form

$$n_2|a(t)|^2 \rightarrow n_2\left\{(1-\alpha)|a(t)|^2 + \alpha\int dt'\, f(t-t')|a(t')|^2\right\} \tag{6.112}$$

In the case of long-pulse excitation (where the pulse duration is long compared to the time scale of the Raman response function) or continuous-wave light, the right-hand side of eq. (6.112) reduces back to $n_2|a(t)|^2$. We see then that α parameterizes the fraction of the long-pulse nonlinear index contributed by the nuclear response.

To gain a simple understanding of the effect of delayed nonlinear response, let us consider a highly simplified version of eq. (6.99a), in one dimension, with dispersion as well as the time derivative term set to zero. If in addition we retain only the delayed nonlinear index term ($\alpha = 1$), we obtain

$$\frac{\partial a}{\partial z} = \frac{-j\omega_0 n_2}{c}a(t)\int dt'\, f(t-t')|a(t')|^2 \tag{6.113}$$

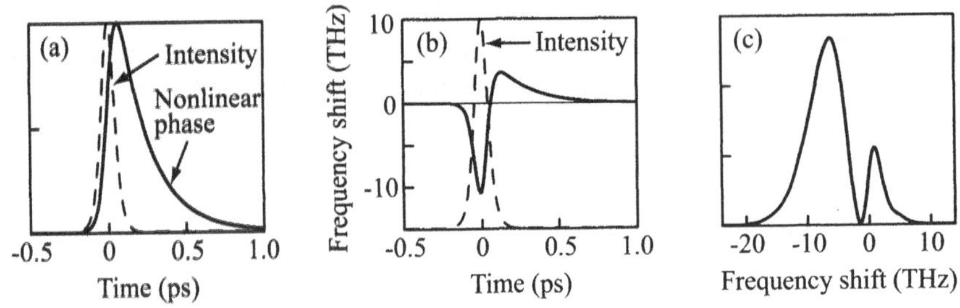

Figure 6.16 SPM with pure delayed nonlinear response: (a) input intensity profile and the negative of the nonlinear phase shift; (b) instantaneous frequency; (c) resulting power spectrum. The input pulse is a 100-fs FWHM secant hyperbolic, the Raman response function is a single-sided exponential with time constant 200 fs, and the peak nonlinear phase shift is 2π.

where t denotes retarded time. The only effect remaining is the self-phase modulation due to the delayed nonlinear response. This simplified propagation equation is easily integrated, with the result

$$a(z,t) = a(0,t)e^{j\Delta\phi} \qquad \text{where} \qquad \Delta\phi(z,t) = \frac{-j\omega_0 n_2 z}{c}\int dt'\, f(t-t')|a(0,t')|^2 \tag{6.114}$$

The effect of the delayed nonlinear response can be understood using the simple instantaneous frequency picture employed in Section 6.3.1, where time-varying instantaneous frequency shifts are given by the time derivative of the nonlinear phase shift $\Delta\phi$ [see eq. (6.46b)]. This is illustrated in Fig. 6.16, which shows the form of the time-dependent nonlinear phase and instantaneous frequency for an assumed single-sided exponential Raman response function. The key point is that the nonlinear phase is shifted to later time with respect to the input pulse. As a consequence, at a majority of instants within the input pulse, there is a shift to lower frequencies. This behavior turns out to be general: SPM with a delayed nonlinear response leads not only to spectral broadening but also to downshifts in the center of mass of the optical power spectrum.

We can also understand such downshifts from a frequency-domain perspective (e.g., [324,328]). Consider again a simplified version of eq. (6.99a), but now retaining dispersion and both instantaneous and delayed contributions to the nonlinear index:

$$\frac{\partial a}{\partial z} + jDa = \frac{-j\omega_0 n_2}{c} a(t)\left\{(1-\alpha)|a(t)|^2 + \alpha\int dt'\, f(t-t')|a(t')|^2\right\} \tag{6.115}$$

Let us assume that the total field is the superposition of two monochromatic contributions:

$$a(z,t) = a_p(z)e^{j\tilde{\omega}_p t} + a_s(z)e^{j\tilde{\omega}_s t} \tag{6.116}$$

where $\tilde{\omega}_p$ and $\tilde{\omega}_s$ represent the frequency offsets (with respect to ω_0) of a pump field (a_p) and a signal field (a_s), respectively. For the sake of discussion, let us assume that the signal field is at lower frequency than the pump field. We can keep track of the changes of the signal field, for example, by substituting into eq. (6.115) and retaining only the terms at frequency $\tilde{\omega}_s$. This yields

$$\frac{\partial a_s}{\partial z} + jDa_s = \frac{-j\omega_0 n_2}{c} a_s \left\{ |a_s|^2 + 2|a_p|^2 + \alpha \left[F(\tilde{\omega}_s - \tilde{\omega}_p) - 1 \right] |a_p|^2 \right\} \tag{6.117}$$

Here we have used the fact that the area under the $f(t)$ curve is unity and have introduced $F(\omega) = \int dt\, f(t) \exp(-j\omega t)$ as the Fourier transform of the delayed nonlinear response function. We can form an equation for $|a_s|^2$, the intensity of the signal beam, by multiplying eq. (6.117) by a_s^* and adding together the resulting equation and its complex conjugate. The result is

$$\frac{\partial |a_s|^2}{\partial z} = g|a_p|^2|a_s|^2 \qquad \text{where} \qquad g = \frac{4\pi n_2 \alpha}{\lambda} \text{Im} \left\{ F(\tilde{\omega}_s - \tilde{\omega}_p) \right\} \tag{6.118}$$

We see that the signal intensity experiences exponential gain (or loss) at a rate given by the product of the pump intensity $|a_p|^2$ and a Raman gain (loss) coefficient g that is proportional to the imaginary part of $F(\tilde{\omega}_s - \tilde{\omega}_p)$. Following an exactly analogous procedure, we can form an equation for the evolution of the pump intensity; the result is identical to eq. (6.118), but with subscripts p and s interchanged. For real $f(t)$ one has $F(u) = F^*(u)$. Also, if one assumes, for example, that the form of $f(t)$ is a simple single-sided exponential, one can easily show that Im $\{F(u)\}$ is less than zero for $u > 0$ and greater than zero for $u < 0$. A quantum mechanical calculation shows that this result holds for any Raman response function for a material in thermal equilibrium [324]. Using this information in eq. (6.118), we find that the lower-frequency signal field is amplified at the expense of the higher-frequency pump field, which is attenuated. This result is consistent with our discussion above based on SPM and an instantaneous frequency picture: The mean frequency shifts to the red. We can now also conclude that this result is independent of the dispersion.

This interaction forms the basis for Raman gain spectroscopy. In typical experiments a strong, fixed-frequency pump field and a weaker tunable signal field are directed into a Raman-active medium (i.e, a medium where the gain process analyzed above is nonzero), and the gain experienced by the signal field is recorded as a function of the pump-signal frequency offset. In suitable media, such measurements yield peaks at frequency offsets corresponding to resonant frequencies for motions within the medium (e.g., vibrations, rotations). From our analysis we see that the Raman gain is proportional to Im$\{F(\tilde{\omega}_s - \tilde{\omega}_p)\}$. It turns out that the Raman gain spectrum is closely related to, and can be calculated from, the spontaneous Raman spectrum (e.g., [69,131,324,325]). Therefore, spontaneous Raman measurements, which are easily performed, can be used to compute the Raman response function.

Raman gain data for fused silica, computed from spontaneous Raman spectra of silica optical fibers, are shown in Fig. 6.17a [329,330]. The most prominent feature is a broad peak centered at about 440 cm^{-1} (13.2 THz) and with peak gain given by $g\lambda = 9.9 \times 10^{-16}$ cm^2/W.[8] Inhomogeneous broadening arising from the variation in microscopic structure characteristic of a glass is responsible for the ca. 200 cm^{-1} width of this peak. Figure 6.17b shows the Raman response function calculated [329]. Given that $f(t)$ is

[8] A measure of frequency termed the *wave number* is defined by $\nu_{\text{cm}^{-1}} = 1/\lambda$ (units: cm^{-1}). To convert to frequency in hertz, use $\nu_{\text{Hz}} = c\nu_{\text{cm}^{-1}}$.

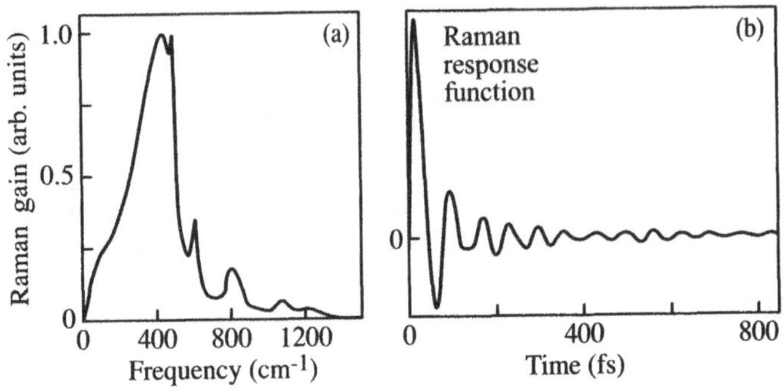

Figure 6.17 (a) Raman gain spectrum for fused silica, computed from spontaneous Raman spectra of silica optical fibers; (b) calculated Raman response function. Both Raman gain and the Raman response function are plotted in arbitrary units. Adapted from [329].

real and causal, it can be calculated directly from the Raman gain spectrum via

$$f(t) = \frac{-2}{\pi} \int_0^\infty \operatorname{Im}\left\{F(\tilde{\omega})\right\} \sin(\tilde{\omega}t)\, d\tilde{\omega} \qquad \text{for} \qquad t > 0 \tag{6.119}$$

where the notation $\tilde{\omega}$ is used to emphasize that F is a function of frequency *offset*. To a large extent, the response function is a rapidly damped sinusoid. The period is roughly 75 fs, consistent with the inverse of the peak frequency in the Raman gain spectrum. Measurements also reveal that the nuclear contribution to the nonlinear index is approximately 18% of the total ($\alpha = 0.18$) [327,329]. An interesting point is that $f(t) = 0$ at $t = 0$. We interpret this as follows. A very short impulse at $t = 0$ immediately imparts the nuclei with a vibrational velocity but does not lead to an instantaneous nuclear displacement. Therefore, there is no immediate change in polarizability and refractive index. The displacement develops fully in a quarter of a vibrational period, at which time the nonlinear index change assumes its maximum value. The subsequent nonlinear response reflects the oscillatory motion of the nuclei and is damped out due to the inhomogeneous distribution of vibrational frequencies.

For pulses that are long with respect to the Raman response function (equivalently for pulses with bandwidth small compared to the peak frequency of the gain spectrum), the delayed nonlinear response can be written in a simplified form. By using a first-order Taylor series expansion for $|a(t)|^2$, the term inside the braces in eq. (6.115) can be rewritten

$$\left\{ |a(t)|^2 - \alpha \frac{\partial |a(t)|^2}{\partial t} \int t'\, dt'\, f(t') \right\}$$

The integral, in turn, can be rewritten in terms of the Fourier transform of the Raman response function, as follows:

$$\int t'\, dt'\, f(t') = -\operatorname{Im}\left\{ \left. \frac{\partial F(\tilde{\omega})}{\partial \tilde{\omega}} \right|_{\tilde{\omega}=0} \right\}$$

Here we have used the fact that due to its even symmetry, the derivative of the real part of F vanishes at zero frequency. Since the derivative of the imaginary part of F is negative at zero frequency, we can now rewrite eq. (6.115) in the following approximate form:

$$\frac{\partial a}{\partial z} + jDa = \frac{-j\omega_0 n_2}{c}\left\{|a|^2 - T_R \frac{\partial |a(t)|^2}{\partial t}\right\} a \tag{6.120a}$$

where

$$T_R = \alpha \left| \mathrm{Im}\left\{ \left. \frac{\partial F(\tilde{\omega})}{\partial \tilde{\omega}} \right|_{\tilde{\omega}=0} \right\} \right| \tag{6.120b}$$

The parameter T_R is a characteristic time that can be used to estimate the strength of the Raman contribution to the nonlinear index relative to the instantaneous term for various pulse widths. It can be expressed in terms of the Raman gain spectrum as

$$T_R = \frac{\lambda}{4\pi n_2} \left. \frac{\partial g}{\partial \tilde{\omega}} \right|_{\tilde{\omega}=0} \tag{6.121}$$

There is significant uncertainty in the literature as to the exact value of T_R. In one reasonable rough approximation, the low-frequency Raman gain curve is taken as a simple straight line, with slope equal to the peak gain divided by the peak gain frequency [331], which yields $T_R \approx 3$ fs. Except for the uncertainty in the value of T_R, eq. (6.120a) is a reasonable model for propagation of pulses a few hundred femtoseconds in duration or longer.

The Raman response function of fused silica is especially important because of the numerous studies and practical applications of nonlinear pulse propagation in optical fibers. The delayed component of the nonlinear response can have a particularly noticeable effect in the case of solitons. Two important examples include the *soliton self-frequency shift* and *fission of higher-order solitons*, described below.

The Soliton Self-Frequency Shift We have seen that in the basic NLSE in the anomalous dispersion regime, fundamental ($N = 1$) solitons propagate without change in pulse shape or power spectrum. However, in experiments subpicosecond solitons experience a dramatic shift to longer wavelengths as they propagate [332]. This *soliton self-frequency shift* has been explained on the basis of Raman self-pumping or intrapulse Raman scattering, in which energy is transferred continuously from the high- to the low-frequency portion of the spectrum, and has been analyzed in [328].

To obtain a scaling law for the frequency shift, we first estimate the nonlinear phase shift associated with the Raman response. Referring to eq. (6.120a), we have approximately

$$\frac{\partial \phi_{\text{Raman}}}{\partial z} \approx -\frac{T_R}{|a(0)|^2} \frac{\partial |a(t)|^2}{\partial t} \frac{\partial \phi_{\text{elec}}}{\partial z} \tag{6.122}$$

where ϕ_{Raman} and ϕ_{elec} are the nonlinear phase shifts associated with the Raman response and the instantaneous electronic nonlinearity, respectively. From eq. (6.78), the phase shift for an unperturbed fundamental soliton is $-|\beta_2|z/2T_0^2$, which we take as an estimate for

ϕ_{elec}. We estimate the frequency shift associated with the Raman response by taking the time derivative of eq. (6.122) to get instantaneous frequency and then evaluating at the center of the pulse ($t = 0$). Taking $a(t) \sim \text{sech}(t/T_0)$, this procedure yields

$$\frac{\partial \omega_{\text{inst}}}{\partial z} \approx -\frac{T_R|\beta_2|}{T_0^4} \tag{6.123}$$

The frequency shift per unit length scales as the inverse fourth power of pulse width. For $T_R = 3$ fs and $|\beta_2| = 20\ \text{ps}^2\ \text{km}^{-1}$ (a typical value for standard single-mode fiber around 1.55 μm), the frequency shift $\nu = \omega_{\text{inst}}/2\pi$ is evaluated in terahertz as

$$\frac{\partial \nu}{\partial z} \approx -\frac{9.5 \times 10^{-3}}{T_0^4} \quad \text{THz km}^{-1} \tag{6.124}$$

with T_0 measured in picoseconds. A detailed calculation [328] predicts a soliton self-frequency shift that is roughly a factor of 2 smaller than eq. (6.123) but otherwise scales in the same way. Due to the strong pulse-width dependence, the soliton self-frequency shift can be a very important effect for pulse durations of a few hundred femtoseconds or below, but is typically weak or negligible for pulses with durations greater than several picoseconds.

Soliton Fission We have seen that the nonlinear propagation of higher-order solitons results in a periodic compression. The increased intensity and decreased duration of temporal features arising during this compression process can make higher-order solitons very sensitive to intrapulse Raman scattering. Delayed nonlinear response can lead to a decay or fission process in which a higher-order soliton breaks up into its constituent fundamental solitons [333–335]. Figure 6.18 shows simulation results performed using eq. (6.120a) for the case of a second-order ($N = 2$) soliton. For this calculation, $T_R = 0.01T_0$ ($T_0 = 300$ fs for $T_R = 3$ fs), and the shock term and third-order dispersion are set to zero to emphasize the effect of the Raman response. The intensity profiles, which are plotted for propagation over integer number of soliton periods, clearly show separation of the input pulse into a pair

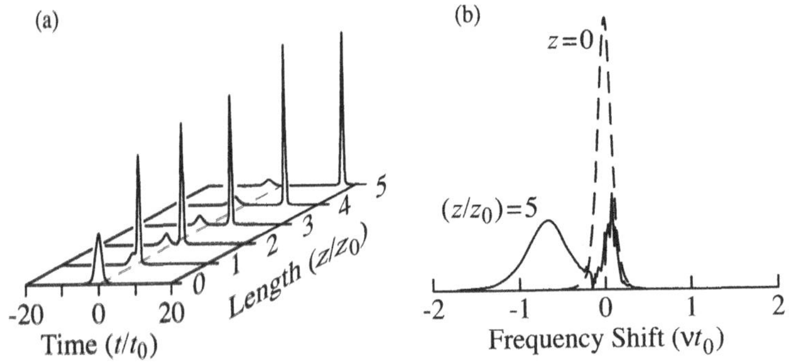

Figure 6.18 (a) Decay of a second-order ($N = 2$) soliton under the influence of delayed nonlinear response, calculated using eq. (6.120a) for $T_R = 0.01t_0$ ($t_0 = 300$ fs for $T_R = 3$ fs). The input pulse is of the form $a(t) \sim \text{sech}(t/t_0)$, and the shock term and third-order dispersion are set to zero. (b) Power spectra for input pulse and after propagation distance $5z_0$. Adapted from [37]. Copyright © 1995, Academic Press, Inc.

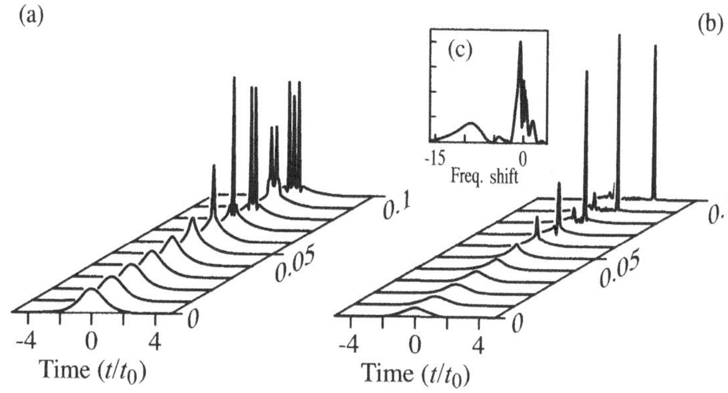

Figure 6.19 Simulation of tenth-order soliton propagation: (a) unperturbed nonlinear Schrödinger equation, $T_R = 0$, simulated up to normalized distance $\xi = 0.1$; (b) nonlinear Schrödinger equation simulated using eq. (6.120a) for $T_R = 0.01t_0$. The input pulse is of the form $a(t) \sim \text{sech}(t/t_0)$, and the shock term and third-order dispersion are set to zero. Compared to (a), the vertical scale (intensity) is compressed by a factor of 2.5. (c) Power spectrum at $\xi = 0.08$, plotted against normalized frequency shift νt_0. Adapted from [336].

of solitons. The pulses are clearly distinct after only two soliton periods. The shorter, more intense of these pulses experiences a strong soliton self-frequency shift. This is evident from Fig. 6.18b, which shows input and output spectra. Once the pulses separate, the frequency shift is expected to scale linearly with propagation length. For constant dispersion ($\beta_3 = 0$), this means that the intense pulse experiences a relative time delay proportional to propagation length squared.

Another set of simulations, for the case $N = 10$, is shown in Fig. 6.19. Without Raman scattering ($T_R = 0$, Fig. 6.19a), the pulse exhibits strong but symmetric pulse narrowing and splitting, with maximum compression at $\xi = 0.07$. With intrapulse Raman scattering ($T_R = 0.01$, Fig. 6.19b), the nonlinear pulse evolution is similar up to $\xi \approx 0.06$. However, around $\xi = 0.07 - 0.08$, the central compressed spike separates from the broad pedestal and moves toward later times; after separating cleanly from the central pedestal, the strong spike propagates essentially without distortion (except for soliton self-frequency shift). Figure 6.19c shows the power spectrum at $\xi = 0.08$. A broad spectral peak is present centered at a frequency shift of $-9/T_0$, a frequency shift equal to roughly 50 times the input spectral width. This red-shifted peak is associated with the intense soliton spike that has separated from the original pulse and provides clear evidence for the soliton self-frequency shift as a mechanism for the soliton fission process.

Soliton fission and formation of intense Raman-shifted solitons have been observed in a number of experiments [335,337–339]. The generation of a highly compressed pulse that is cleanly separated in frequency from the background is of interest from an applications perspective. In real experiments both the shock term and higher-order dispersion also have an influence; however, these simulations taking only the Raman response into account capture the most important effects. One additional interesting effect arises from higher-order dispersion: as the compressed pulse is downshifted in frequency, it sees a different (usually increased) value of β_2. Consequently, according to eq. (6.79), the pulse duration must broaden to preserve a balance between the dispersion and nonlinearity.

6.7.4 Self-Steepening

We consider now the time-derivative term (the shock term) in eq. (6.105). To concentrate on this term, we set dispersion equal to zero and assume an instantaneous nonlinear index. For simplicity we write a instead of a_p, although the field normalization is still such that the field squared gives instantaneous power. Our equation becomes

$$\frac{\partial a}{\partial z} + j\gamma\left(1 - \frac{j}{\omega_0}\frac{\partial}{\partial t'}\right)|a|^2 a = 0 \tag{6.125}$$

If we now write the field in terms of its amplitude and phase, $a = |a|\exp(j\phi)$, substitute into eq. (6.125), and look at the real part of the result, we obtain[9]

$$\frac{\partial |a|}{\partial z} + \frac{3\gamma|a|^2}{\omega_0}\frac{\partial |a|}{\partial t'} = 0 \tag{6.126}$$

This equation has a solution of the form $|a| = f(t' - z/u)$, where $f(t') = |a|(z = 0, t')$ is the input amplitude profile evaluated at $z = 0$ [340,341]. Substitution into eq. (6.126) yields

$$u = \frac{\omega_0}{3\gamma|a|^2} \tag{6.127a}$$

and

$$|a| = f\left(t' - \frac{3\gamma|a|^2 z}{\omega_0}\right) \tag{6.127b}$$

We see that the pulse amplitude profile is predicted to distort during propagation. Since the pulse profile remains unchanged during dispersionless self-phase modulation under the NLSE, the distortion can be attributed specifically to the additional time-derivative term in eq. (6.105). Physically, the distortion arises from an intensity-dependent group velocity; the expression for u written above gives the time-dependent change in the group velocity relative to the unperturbed value of group velocity. For positive values of the nonlinear index, the peak of the pulse is delayed relative to the wings. This leads to a progressive steepening of the trailing edge of the pulse. Figure 6.20 shows calculated values of the temporal intensity profile $|a|^2$ for various propagation distances which illustrate this *self-steepening effect.*

Our model predicts that the self-steepening effect should continue until the pulse develops a region where the temporal intensity becomes discontinuous (i.e., the derivative of the intensity, $\partial|a|^2/\partial t'$, becomes infinite). Such a feature is referred to as an *optical shock front*, in analogy with shock fronts that occur in acoustics, for example. A rough estimate of the propagation distance L_{shock} to generate such a shock is easily obtained: We require that the extra time delay experienced by the peak of the pulse is on the order of the input pulse width

[9] The imaginary part gives an equation for the phase evolution, which is coupled to the amplitude. The solution of the phase equation is more involved (see, e.g., [340]).

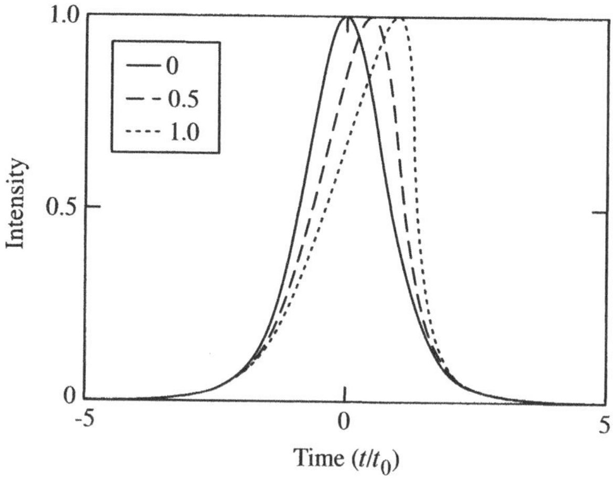

Figure 6.20 Nonlinear propagation simulation showing the self-steepening effect, assuming zero dispersion and an instantaneous nonlinear refractive index. The input pulse is of the form $a_0\,\mathrm{sech}(t/t_0)$, and the self-steepening parameter $3\gamma|a_0|^2 z/\omega_0 t_0$ takes on values 0, 0.5, and 1.

t_0, so that the trailing edge of the pulse catches up to the peak. We find then that

$$L_{\text{shock}} \approx \frac{\omega_0 t_0}{3\gamma P_0} \tag{6.128}$$

where P_0 represents the peak power of the input pulse. For a $\mathrm{sech}(t/t_0)$ input pulse, the shock actually forms at a distance 1.3 times that estimated in eq. (6.128). If we compare with the characteristic length for pure self-phase modulation, eq. (6.40), we obtain

$$\frac{L_{\text{shock}}}{L_{\text{NL}}} \approx \frac{\omega_0 t_0}{3} \tag{6.129}$$

The interesting result is that as long as the pulse duration is much greater than an optical cycle, the length for shock formation is much greater than the characteristic length for self-phase modulation. Equivalently, under the same condition, the self-steepening is weak compared to self-phase modulation. This is as expected, given that the slowly varying envelope approximation is built into the derivation of the NLSE in Section 6.1.

From an experimental perspective, observation of self-steepening in isolation is difficult since its effect is usually strongly intermixed with group velocity dispersion, delayed nonlinear response, and so on. However, numerical modeling suggests that the self-steepening process has an important influence on experiments in nonlinear fiber optics using very short pulses as well as on the phenomenon of white light continuum generation in bulk transparent media [342] discussed later in this chapter. Self-steepening can also affect solitons. As we have seen, for solitons self-phase modulation and dispersion essentially compensate each other. There is, however, an additional phase accumulation during soliton propagation, which may be viewed as a change in phase velocity due to the Kerr nonlinearity [see eq. (6.80)]. Analysis shows that due to the self-steepening term there is also a change in the group delay of the soliton [333], which may significantly exceed the nonlinear change in phase delay [343]. Finally, in the case of higher-order solitons, self-steepening can cause soliton fission [333] analogous to that caused by delayed nonlinear response.

6.7.5 Space–Time Focusing

We now discuss the ∇_T^2 term in the nonlinear envelope equation for bulk media. The ∇_T^2 operator, which involves spatial derivatives, is preceded by an operator involving either frequency offset, or equivalently, time derivatives, in eqs. (6.97) and (6.99a), respectively. This represents coupling between evolution of spatial and temporal field profiles.

A physical origin for this coupling, which has been termed *space–time focusing*, is described in [344]. Consider first the equation for linear propagation in both the slowly varying envelope and paraxial approximations:

$$\left[\frac{\partial}{\partial z} + j\tilde{D} + \frac{j}{2\beta_0}\nabla_T^2\right]A = 0 \tag{6.130}$$

This equation may be obtained from eq. (6.97) by setting both $\tilde{\omega}$ and nonlinear polarization (Q) to zero. Alternatively, this may be obtained by considering a superposition of plane waves with frequency $\tilde{\omega}$ propagating in various directions specified by $\mathbf{k} = k_x\hat{x} + k_y\hat{y} + k_z\hat{z}$, where the wave vector satisfies the dispersion relation $k_x^2 + k_y^2 + k_z^2 = \beta^2(\tilde{\omega})$. Taking the paraxial approximation, $k_T^2 = k_x^2 + k_y^2 \ll k_z^2$, this becomes

$$k_z - \beta(\tilde{\omega}) + \frac{k_T^2}{2\beta(\tilde{\omega})} \approx 0 \tag{6.131}$$

If we further set $\beta(\tilde{\omega}) \approx \beta_0$ in the denominator of the third term, we obtain

$$k_z - \beta(\tilde{\omega}) + \frac{k_T^2}{2\beta_0} \approx 0 \tag{6.132}$$

If we now convert to operators according to $k_z \leftrightarrow j\partial/\partial z$, $k_T^2 \leftrightarrow -\nabla_T^2$, and $\beta(\tilde{\omega}) \leftrightarrow \tilde{D}$, and use the result to operate on the field envelope A, as defined in eq. (6.90), we recover precisely eq. (6.130).

We can begin to go beyond the slowly varying envelope approximation by retaining $\beta(\tilde{\omega})$ in the denominator of the last term in eq. (6.131). Simplest is to expand β to first order in $\tilde{\omega}$, in which case we may write

$$k_z - \beta(\tilde{\omega}) + \frac{k_T^2}{2\beta_0}\left(1 - \frac{\tilde{\omega}\beta_1}{\beta_0}\right) \approx 0 \tag{6.133}$$

Compared to eq. (6.132), the difference is that the term proportional to $\tilde{\omega}$, namely $\beta_1\tilde{\omega}$, is replaced according to

$$\beta_1\tilde{\omega} \rightarrow \beta_1\tilde{\omega}\left(1 + \frac{k_T^2}{2\beta_0^2}\right) \tag{6.134}$$

Equivalently, the group velocity should apparently be modified according to

$$v_g \to v_g \left(1 - \frac{k_T^2}{2\beta_0^2}\right) \tag{6.135}$$

In the paraxial limit we may write $k_T/\beta_0 \approx \theta$, where θ represents propagation angle with respect to the z axis. Then the factor inside the parentheses is identified as the expansion of $\cos\theta$. We may therefore interpret the right-hand side of eq. (6.135) as the component of the group velocity projected along the propagation direction z. Field components propagating at large angles with respect to z progress more slowly with respect to the z reference direction than do field components propagating at small angles.

A similar result may be obtained from the frequency-domain version of the nonlinear envelope equation, eq. (6.97), if we set the right-hand side to zero and expand the prefactor multiplying ∇_T^2 to first order in $\tilde{\omega}$. This gives

$$\left[\frac{\partial}{\partial z} + j\tilde{D} + \frac{j}{2\beta_0}\left(1 - \frac{\tilde{\omega}}{\omega_0}\right)\nabla_T^2\right]A = 0 \tag{6.136}$$

If we use the same operator conversions as given below eq. (6.132), the third term inside the brackets (dropping a factor of $-j$) becomes

$$\frac{k_T^2}{2\beta_0}\left(1 - \frac{\tilde{\omega}}{\omega_0}\right)$$

This is approximately the same as the third term in eq. (6.133) provided that $\beta_1 \approx \beta_0/\omega_0$, which is equivalent to the approximation of eq. (6.96b). The conclusion is that the space–time focusing term in the nonlinear envelope equation includes increased group delay for field components traveling at increased angles with respect to the reference direction. An example of a nonlinear problem in which this is important is femtosecond pulse self-focusing. Here the self-focusing nonlinearity acts to increase the angular spectrum of the beam, which in turn leads to increased group delay for the new angular components. The net result, when averaged over the spatial degrees of freedom, is a nonlinear increase in group delay, similar in some respects to that considered in Section 6.7.4.

The full nonlinear envelope equation contains additional effects relevant to coupling between space and time (or frequency). It is well known, for example, that the wavelength λ is an important scale factor in diffraction problems: The spreading angle of an initially collimated beam is proportional to λ divided by the beam aperture, as is the diffraction-limited spot size of a beam focused by a lens. This effect is partially included in eq. (6.130) in the sense that for any fixed wavelength or frequency, the appropriate β_0 may be chosen corresponding to that wavelength. However, for pulses with broad frequency content, eq. (6.130) does not provide any means to handle the wavelength-dependent spatial propagation of the different frequency components contained within the pulse. In contrast, the frequency-domain formulation of the nonlinear envelope equation, eq. (6.97), contains a term that may be written without further approximation as

$$\frac{j\omega_0}{2\beta_0}\frac{1}{\omega}\nabla_T^2 A$$

The $1/\omega$ dependence of this term specifically accounts for the wavelength dependence of spatial propagation.

6.8 CONTINUUM GENERATION

When sufficiently intense pulses propagate through transparent matter, a dramatic spectral broadening is commonly observed. Such spectral broadening may significantly exceed that which is normally observed due to self-phase modulation acting alone and is then referred to as *continuum, supercontinuum,* or *white light generation.* Continuum generation has been observed in solid, liquid, and gas samples and in both bulk media and waveguides. Examples of spectra resulting from femtosecond pulse continuum generation are shown in Fig. 6.21. In Fig. 6.21a, 80-fs, 627-nm pulses from an amplified CPM dye laser system (250 μJ pulse energy, several gigawatts of peak power) were focused into a liquid jet of ethylene glycol [345]. The data show a narrow peak at the position of the input spectrum, merging into extremely broad wings that span the visible. The spectral broadening is asymmetric, favoring the blue. Figure 6.21b shows spectra generated when 140-fs, 796-nm pulses from an amplified Ti:S laser system (ca. microjoules of pulse energy, several megawatts to about 10 MW of peak power) were focused into liquid and solid media [346]. Extremely wide spectral tails are observed, with pronounced asymmetry extending farther to the blue. Figure 6.21c shows input and output spectra for an experiment in which 100-fs pulses from a Ti:S laser oscillator (800 pJ of pulse energy, ca. 8 kW of peak power) were transmitted through an air–silica microstructure fiber [271]. The resulting spectrum is relatively flat and extends over a two-octave span, from 390 to 1600 nm.

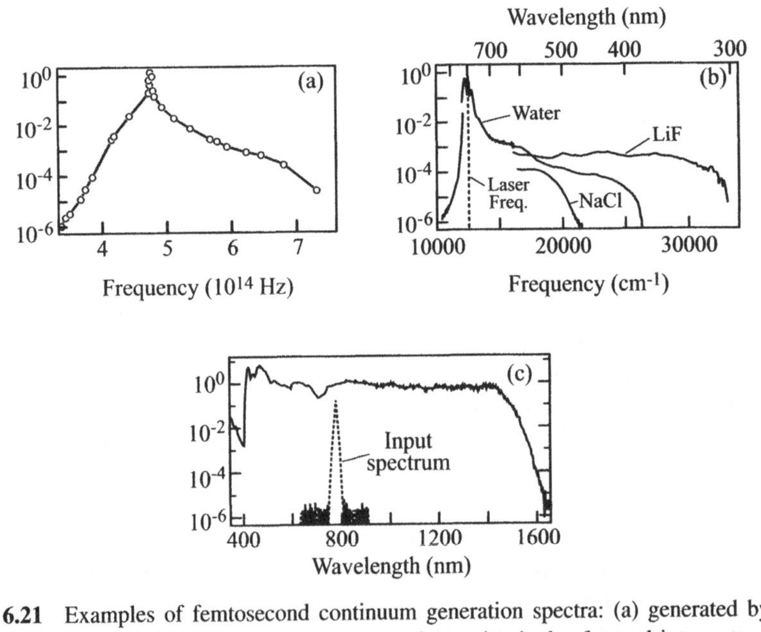

Figure 6.21 Examples of femtosecond continuum generation spectra: (a) generated by 250-μJ pulses in ethylene glycol; (b) generated by pulses of ca. microjoules focused into water, LiF, and NaCl samples; for clarity only $\lambda < 630$ nm is shown for LiF and NaCl; (c) generated by 800-pJ pulses in a 75-cm length of air–silica microstructure fiber. (a) Adapted from [345]. (b) Adapted from [346], with permission. Copyright © 1998, American Physical Society. (c) From [271].

Such supercontinuum sources are important for a number of uses. Applications in ultrafast science and technology include spectrally resolved ultrafast spectroscopy (Chapter 9), seeding of optical parametric amplifiers for optical frequency conversion (Section 5.5.3), and realization of octave-spanning spectra used in stabilization of femtosecond frequency combs (Section 7.5). Measurement and sensing schemes based on optical interference, such as optical coherence tomography (Section 3.3.3), benefit from the short coherence time associated with supercontinuum sources, which improves resolution. In addition, supercontinuum generation in the lightwave band around 1.55 μm has been used to realize a large number of wavelength-shifted replicas, starting from a single high-repetition-rate (e.g., 10-GHz) mode-locked pulse train, for experiments in wavelength-multiplexed optical communications.

Supercontinuum generation was originally observed in experiments with picosecond-duration light pulses [347,348]. Here we focus the discussion on experiments using femtosecond pulses. In general, the physics associated with continuum generation is complex and may vary depending on a number of factors, such as peak power, pulse duration, focusing, and material. For example, group velocity dispersion has a key influence on nonlinear propagation of femtosecond pulses but may be of only minor importance for picosecond pulse propagation in short samples. Modeling requires a treatment beyond the slowly varying envelope approximation and involves a mix of many of the processes discussed in this chapter (and sometimes others). Here we touch on only a few of the main concepts. For further information the reader is referred to detailed studies in the literature (e.g., [342,346,349–352]).

First, let us consider continuum generation in bulk media. Such experiments are usually performed in the normal group velocity dispersion regime. Some important observations are as follows:

- The threshold power for continuum generation is very close to the critical power for self-focusing. Experiments have shown that the very rapid spectral broadening characteristic of continuum generation occurs first as the beam self-focuses to a very small diameter. Accordingly, the sudden appearance of new spectral content is tied to an explosive rise in peak intensity accompanying beam collapse.
- Spectral broadening often occurs predominantly to the blue, at least at powers close to the continuum generation threshold. (At higher powers, additional broadening occurs to the red, making this asymmetry less visible.)
- Spectra may exhibit a strong and fairly regular modulation, especially when single-shot spectra are examined [350]. (Due to fluctuations, such modulation may be masked when many laser shots are averaged together.)

Simulations based on the nonlinear envelope equation provide insight into this behavior [342,352]. The proposed explanation goes roughly as follows:

- Initially, before group velocity dispersion plays a strong role, different temporal slices of the pulse self-focus independently. The peak of the pulse is the first to self-focus. When it does so, its intensity becomes very high, leading to strong spectral broadening. The spectrally broadened, self-focused pulse center now becomes susceptible to dispersion and spreads in time. When integrated over the spatial profile of the beam, this leads to a flow of energy away from the temporal center of the pulse and toward both earlier and later times. Thus, a strong interplay between self-phase modulation, self-focusing, and dispersion leads the initial pulse to split into two fragments. Such pulse-splitting behavior was first discussed in [353].

- The interaction of this pulse-splitting effect with self-steepening and space–time focusing pushes energy asymmetrically toward later times [344]. Thus, instead of two equal fragments, a substantially stronger fragment is produced toward the trailing edge of the initial pulse. This fragment is found to compress in time and form a shock on its back edge. The rapid decrease in intensity on this negative-going shock leads to a very strong blue shift.

Experimental data providing evidence for some of this behavior are shown in Fig. 6.22. These data were acquired by focusing pulses into a BK7 glass sample both above and below the critical power for self-focusing (P_{crit}), then employing a series of lenses and apertures to select the central spatial region of the transmitted beam for temporal and spectral characterization. Intensity autocorrelation and power spectrum data are shown for peak powers of 0.6, 3, and 4.8 MW, corresponding to 0.33, 1.7, and 2.7P_{crit}, respectively. At the lowest power, the autocorrelation and power spectrum were similar to those of the input pulses. At 3 MW peak power, the autocorrelation develops shoulders, which indicate the onset of pulse splitting. This splitting, resulting in a triply peaked autocorrelation, becomes very clear as the power is increased to 4.8 MW. The widths of the individual peaks are well below that of the input, suggesting that the pulse fragments are compressed in time. At this

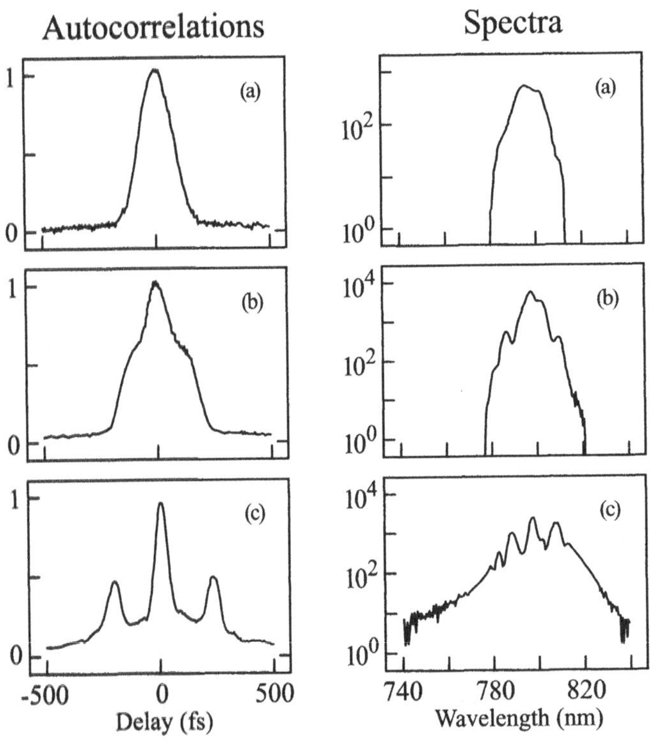

Figure 6.22 Experimental intensity autocorrelation and power spectrum data for pulses transmitted through a bulk sample of BK7 glass at input peak powers of (a) 0.6 MW, (b) 3 MW, and (c) 4.8 MW. From [354], with permission. Copyright © 1996, American Physical Society.

point the spectrum shows a clear modulation, and the asymmetric spectral tail extending to the blue is apparent. The period of the spectral modulation is approximately the inverse of the time separation between autocorrelation peaks, suggesting that the modulation arises due to spectral interference between pulse fragments. Slightly above the powers shown, the spectra were observed to broaden explosively. Thus, these data support the picture summarized above, in which dispersion strongly couples spatial and temporal degrees of freedom during self-focusing, as playing a key role in the onset of continuum generation.

Another process that is often discussed in the context of femtosecond supercontinuum generation is plasma generation through multiphoton ionization. Such multiphoton ionization is possible because of the high intensities, in the range of 10^{13} W/cm^2, that occur in the course of self-focusing. The resulting free electrons give a negative, steplike contribution to the refractive index, which acts to blue-shift the transmitted pulse. Multiphoton ionization and plasma generation are expected to oppose the self-focusing process, both through energy loss and because a negative nonlinear index contributes to defocusing. Researchers have differed, however, on the primacy of plasma generation in the continuum generation process (i.e., on whether multiphoton ionization and the resulting free electrons play a key role in initiating continuum generation or whether they merely act to limit self-focusing and spectral broadening after continuum generation has already occurred).

The physics of continuum generation in fibers differs from that of bulk media in several respects. First, because the spatial mode is defined by the waveguide, self-focusing is not a primary factor. Second, in fibers, waveguide dispersion may be designed to achieve zero dispersion or anomalous dispersion in the spectral region of interest. For example, the spectrum of Fig. 6.21c was achieved in air–silica microstructure fiber, where a very strong index contrast allows the zero-dispersion wavelength to be pushed to the (near-visible) spectral band where Ti:S lasers operate. In contrast, continuum generation in bulk materials usually occurs under conditions of strong normal group velocity dispersion. Third, lower intensities are required compared to bulk media, because the combination of low dispersion and guided-wave geometry allows nonlinear effects to accumulate over relatively long propagation distances. Therefore, electron generation is usually of less significance in continuum generation experiments in fibers. According to simulations based on the one-dimensional nonlinear envelope equation, the interplay between self-phase modulation and higher-order dispersion plays a principal role in the nonlinear propagation dynamics and spectral broadening [355].

Continuum generation in fibers is of significant practical interest because of a stable beam profile with high spatial coherence and because operation is possible at low pulse energies compatible with mode-locked laser oscillators. This permits pulse repetition rates much higher than with amplified systems.

PROBLEMS

6.1. Consider dispersionless self-phase modulation in the limit of large nonlinear phase shift.

(a) Derive the numerical factor in eq. (6.49) for Gaussian pulses giving the ratio of spectral broadening to input bandwidth in terms of the maximum nonlinear phase shift.

(b) Determine the corresponding numerical factor for secant hyperbolic input pulses.

6.2. The experimental power spectra reproduced in Fig. 6.2 exhibit an asymmetric tilt not present in the theoretical discussion of dispersionless self-phase modulation presented in the text. Can you suggest at least one plausible reason why this tilt may occur?

6.3. Calculate and plot intensity, nonlinear phase shift, instantaneous frequency, and power spectrum for a series of input pulses with varying degrees of asymmetry. Discuss the evolution of the power spectrum with increasing asymmetry in terms of the instantaneous frequency picture. For this problem assume dispersionless self-phase modulation and an instantaneous nonlinear refractive index. Choose parameters corresponding to a large nonlinear phase shift.

6.4. Consider fiber and grating pulse compression of pulses from a Nd:YAG laser. The input pulses are Gaussian with a 75-ps FWHM pulse duration, a 100-MHz repetition rate, a 1.06-μm center wavelength, and 1 W of average power (coupled into the fiber). The single-mode fiber is 250 m long and has normal dispersion with $\beta_2 = 23$ $\text{ps}^2\ \text{km}^{-1}$ and nonlinear coefficient $\gamma = 4\ \text{W}^{-1}\ \text{km}^{-1}$. The grating pair compressor is set up in a double-pass configuration with gratings ruled at 1200 lines/mm. Assume that pulses propagate in the fiber without loss and without dispersion.

(a) Compute and plot the power spectrum subsequent to fiber propagation. Compare your result (spectral broadening, number of peaks) to that expected theoretically based on the instantaneous frequency picture.

(b) Estimate the chirp (derivative of instantaneous frequency with time) at the center of the pulse, and use this information to predict the grating spacing required for optimum pulse compression. Assume that the gratings are used at the Littrow condition, $\theta_d = \theta_{\text{in}}$, and consider only the quadratic spectral phase of the grating pair (neglect higher-order spectral phase).

(c) Compute and plot the intensity profile obtained as a result of this quadratic compressor. Compare the pulse duration to that expected based on the bandwidth of the spectrally broadened pulse. Furthermore, select an appropriate gate function and then calculate and plot a spectrogram of the compressed pulse. Repeat for the case of an ideal compressor. Attempt to relate features present in your spectrograms to features observed in the compressed intensity profiles.

(d) Make a "back of the envelope" estimate of the tolerance with which the grating spacing must be set. That is, roughly how much can the grating spacing be changed before the compressed pulse width approximately doubles?

(e) In the case of a real (quadratic) compressor, the best compressed pulse (highest intensity, lowest sidelobes) may occur for a grating spacing slightly different than that calculated earlier. Compute intensity profiles and corresponding spectrograms of compressed pulses obtained as you slightly vary the grating spacing. Point out the highest intensity pulse and discuss in terms of its spectrogram.

6.5. Examine approximations made in Problem **6.4**. In particular:

(a) Comment on the assumption of no dispersion. Here it is useful to calculate L_D for both the input pulse and for a hypothetical bandwidth-limited pulse with spectral width equal to that obtained at the output of the fiber under dispersionless self-phase modulation.

(b) Estimate the actual nonzero third-order spectral phase (ψ_3) of the grating compressor. What is the approximate time spread in the compressed pulse due to ψ_3? How does this compare to the duration of the compressed pulse calculated in Problem **6.4**?

6.6. Write a code to solve the nonlinear Schrödinger equation based on the split-step Fourier method. Confirm your code by simulating pulse propagation for known test cases (e.g., dispersion only, self-phase modulation only, $N = 1$ solitons). Then apply your code for the following:

(a) Compute the temporal and spectral evolution of $N = 4$ and $N = 5$ solitons within one soliton period.

(b) Simulate the propagation of input pulses of the form eq. (6.82) for various values of ε in the range $0.5 \leq \varepsilon \leq 1.5$. Plot the temporal intensity profile after a propagation length sufficient that soliton and dispersive wave components can be distinguished visually (possibly using a logarithmic intensity scale). Compute the energies of soliton and dispersive wave components and compare to the prediction of eq. (6.84).

(c) Simulate the evolution of an $N = 1$ soliton with loss. The loss coefficient should be small enough that the soliton propagates adiabatically. Plot the variation in pulse duration with pulse energy (energy will decrease with propagation length) and compare to the prediction of eq. (6.81).

(d) Simulate propagation of an $N = 3$ soliton perturbed by nonzero cubic spectral phase. Adjust the parameters so that soliton fission is observed.

6.7. Discuss how to write a code based on the split-step Fourier method to simulate spatial–temporal propagation of a pulsed beam as described by the nonlinear envelope equation, eq. (6.99a). That is, although you should describe how each of the terms in the NEE will be handled in the computer algorithm, you are not required to actually write the code. For this problem you may assume an instantaneous nonlinear index effect.

6.8. Derive eq. (6.119), assuming that the Raman response function $f(t)$ is real and causal. Also derive a similar formula that gives $f(t)$ in terms of only $\text{Re}\{F(\Omega)\}$. [*Hint:* First express $f(t)$ as the sum of an even function and an odd function.]

6.9. Use eq. (6.119) to get a formula for the real part of the Raman spectrum, $\text{Re}\{F(\Omega)\}$, in terms of its imaginary part, $\text{Im}\{F(\Omega)\}$.

6.10. For a fixed propagation length, compare (approximately) the nonlinear group delay arising from self-steepening with the nonlinear phase delay arising from self-phase modulation. Assume an instantaneous nonlinear index and zero dispersion.

6.11. Determine the exact distance at which a shock first forms under the process of self-steepening for Gaussian and secant hyperbolic input pulses. Express your results in terms of FWHM pulse duration.

7

MODE-LOCKING: SELECTED ADVANCED TOPICS

In this chapter we continue our treatment of mode-locking by discussing selected additional advanced topics. This discussion makes use of the detailed understanding of ultrafast nonlinearities and group velocity dispersion developed in previous chapters. In Sections 7.1 to 7.3 we discuss lasers in which the effects of nonlinearity and dispersion per cavity round trip are often large. This is in contrast to the assumption of small effects per round trip used for the analysis of Chapter 2 and leads to new phenomena. Our discussion is divided into lasers operating in the anomalous dispersion regime (Sections 7.1 and 7.2), for which soliton effects play a key role, and lasers operating in the normal dispersion regime (Section 7.3). The treatment in these Sections focuses significantly but not exclusively on fiber lasers. In Section 7.4 we discuss issues of particular importance to mode-locked lasers operating in the few cycle regime. Finally, in Section 7.5 we revisit the basic frequency comb picture of mode-locking, first developed in Chapter 1, with a focus on the absolute frequencies of individual comb lines. This topic is of significant importance both for full specification of pulsed mode-locked fields and for application of mode-locked lasers for high-precision frequency metrology.

7.1 SOLITON FIBER LASERS: ARTIFICIAL FAST SATURABLE ABSORBERS

We have already seen in our previous study of mode-locked solid-state lasers (Chapter 2) that the interplay between nonlinear phase modulation and dispersion can lead to strong soliton-like behavior in laser operating characteristics. In particular, in the anomalous dispersion regime the relation between pulse width, energy, and dispersion in the mode-locked pulse solution was seen to satisfy that for a fundamental soliton. In this Section we discuss

Ultrafast Optics, By Andrew M. Weiner

passively mode-locked lasers based on artificial fast saturable absorbers, typically fiber lasers, where solitonlike phenomena play an important role not only in the pulse propagation (as in Section 2.4.5), but also in the self-amplitude modulation (SAM) process. Because of the large amount of material within the laser cavity (often, tens of meters of fiber), both dispersion and nonlinearity often make large individual contributions within a single round trip. Other mode-locked fiber laser configurations, which are of greater interest for practical applications as telecommunications signal sources, are considered in Section 7.2.1.

7.1.1 The Figure-Eight Laser

We focus our discussion on the *figure-eight laser* (F8L), which is one type of mode-locked fiber laser in which soliton effects play a prominent role [356–359]. A typical layout for a F8L is shown in Fig. 7.1. The laser consists of two fiber loops connected by a 2×2 coupler. The right-hand loop in the illustration is termed a *nonlinear amplifying loop mirror* (NALM) [360], discussed further below, and is responsible for both the laser gain and for the artificial fast saturable absorber action necessary for mode-locking. The left-hand loop, which contains an optical isolator to enforce unidirectional operation, is simply a feedback element that takes the output of the NALM and connects it back to the NALM input. The gain element is typically an erbium-doped fiber amplifier (EDFA), which is pumped at 980 nm and provides gain from about 1530 to 1570 nm in the telecommunications band. The polarization controllers are used to bias the polarization of the circulating light appropriately with respect to the small and essentially random birefringences present in normal circular-core single-mode fibers. Such polarization controllers are needed for practical implementations; however, their role is not fundamental to our discussion and is not detailed further. The laser output is taken through a 2×2 coupler, which is the fiber optic equivalent of the partially transmissive mirror generally used to extract output from a bulk optic laser.

In addition to the geometry shown in Fig. 7.1, the F8L can instead be configured with the EDFA gain element placed in the left-hand loop. In that case, the left-hand loop both ensures unidirectional operation and provides the laser's gain; the right-hand loop, now termed a *nonlinear optical loop mirror* (NOLM) [361,362], is responsible for pulse reshaping.

Before analyzing the nonlinear switching action of the loop mirror and its effect on laser operation, we note that the F8L is not the only passively mode-locked fiber laser

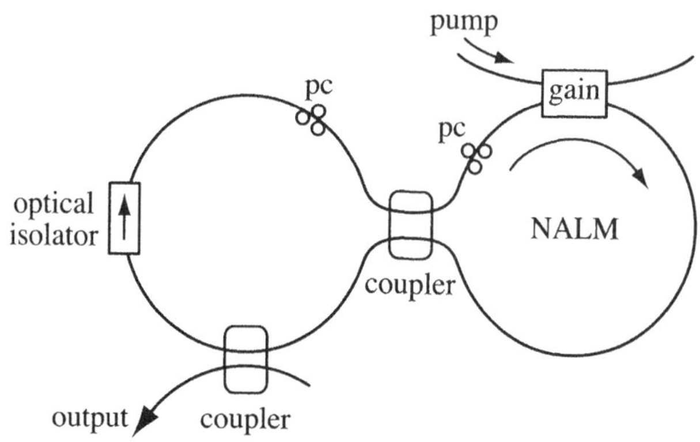

Figure 7.1 Typical layout for a figure-eight laser. pc, polarization controller. Adapted from [357].

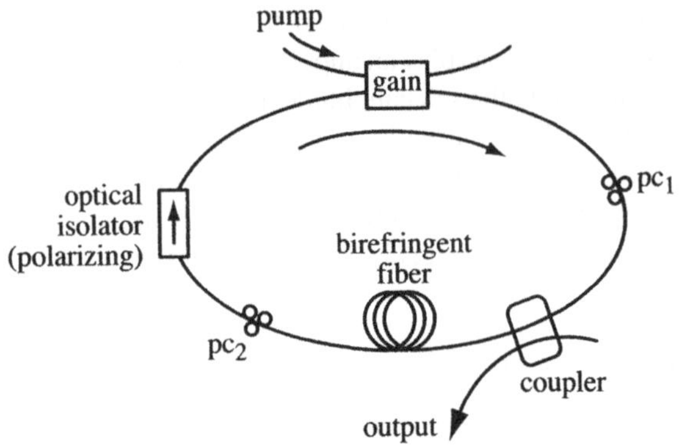

Figure 7.2 Layout for mode-locked fiber ring laser exploiting nonlinear polarization evolution. In this implementation the required polarizer function is included in the isolator. pc, polarization controller.

configuration for which soliton effects are important. Fiber lasers in which nonlinear polarization evolution forms the basis for artificial fast saturable absorber action have also been the subject of numerous studies (e.g., [363–365]). An example of such a laser [363], constructed in a ring configuration, is sketched in Fig. 7.2. The roles of the EDFA (gain), output coupler, and isolator (enforces unidirectional operation) should be clear. Other elements are incorporated to promote nonlinear polarization evolution; these include the two polarization controllers, a length of fiber (where the nonlinear changes in state of polarization take place), and a polarizer (in this case built into the isolator). The first polarization controller is adjusted to allow a large nonlinear polarization change, while the second polarization controller is adjusted such that the effect of the nonlinearity is to increase the alignment of the polarization with the polarizer, thereby increasing transmission through the cavity. This provides the self-amplitude modulation necessary for mode-locking. Nonlinear polarization evolution can occur either in birefringent or nonbirefringent fibers. In nonbirefringent (ideal circular-core) fibers, the nonlinear polarization evolution takes the form of ellipse rotation, an effect well known from bulk nonlinear optics studies [131,366]. As the name suggests, nonlinear propagation of elliptically polarized input light leads to a continuous rotation of the axes of the polarization ellipse. This effect can be understood by viewing elliptically polarized light as arising from a superposition of the two different circular polarization states with different intensities. The different intensities of the two circular polarization components, in conjunction with different SPM and XPM coefficients, causes the circularly polarized components to accumulate different nonlinear phases; this difference in nonlinear phases in turn leads to rotation of the polarization ellipse. Practically even circular-core single-mode fibers are usually at least weakly birefringent, as noted above. This complicates the analysis of the nonlinear polarization evolution, especially for the usual case of random birefringences that vary along the length of the fiber. Nevertheless, this does not alter the essential feature: Under nonlinear propagation the input polarization to a fiber can evolve in an intensity-dependent manner [367,368], and this forms the basis for self-amplitude modulation.

Compared to the F8L incorporating a NALM or NOLM, we see that the physical realization of the artificial fast saturable absorber appears quite different. Nevertheless, we shall see below that nonlinear operation of the NALM (or NOLM) is similar to nonlinear polarization evolution in that both cases exploit different nonlinear phase shifts incurred by two distinguishable contributions to the total field which add interferometrically. Accordingly, key aspects of laser operation that we discuss below for F8Ls also pertain to soliton fiber lasers based on nonlinear polarization evolution.

Switching Behavior of Nonlinear Loop Mirrors Returning now to the F8L, we analyze the nonlinear transmission of the loop mirror [361], which plays a key role in mode-locking. Figure 7.3 shows the geometry for our analysis. For the NALM the gain element is placed asymmetrically, close to one end of the loop. For the NOLM the gain element is absent. We can understand both the NALM and the NOLM by using the scattering matrix formalism introduced in Section 2.4.3. The fields injected into the loop (b_{1+} and b_{2+}) for input fields a_{1+} and a_{2+} incident onto the coupler may be written as

$$\begin{pmatrix} b_{1+} \\ b_{2+} \end{pmatrix} = \begin{pmatrix} \sqrt{\alpha} & \sqrt{1-\alpha} \\ \sqrt{1-\alpha} & -\sqrt{\alpha} \end{pmatrix} \begin{pmatrix} a_{1+} \\ a_{2+} \end{pmatrix} \tag{7.1}$$

where α is the power splitting ratio of the coupler.[1] Upon propagation around the loop, the fields are incident once again upon the coupler and are expressed by

$$\begin{pmatrix} b_{1-} \\ b_{2-} \end{pmatrix} = \begin{pmatrix} 0 & e^{g}e^{j\phi_{21}} \\ e^{g}e^{j\phi_{12}} & 0 \end{pmatrix} \begin{pmatrix} b_{1+} \\ b_{2+} \end{pmatrix} \tag{7.2}$$

where g is the gain coefficient (for the NOLM we set $g = 0$), and ϕ_{12} and ϕ_{21} are the phases accumulated in going around the loop from port 1 to port 2 or from port 2 to port 1, respectively. Finally, the output of the loop after a second passage through the coupler is given by

$$\begin{pmatrix} a_{1-} \\ a_{2-} \end{pmatrix} = \begin{pmatrix} \sqrt{\alpha} & \sqrt{1-\alpha} \\ \sqrt{1-\alpha} & -\sqrt{\alpha} \end{pmatrix} \begin{pmatrix} b_{1-} \\ b_{2-} \end{pmatrix} \tag{7.3}$$

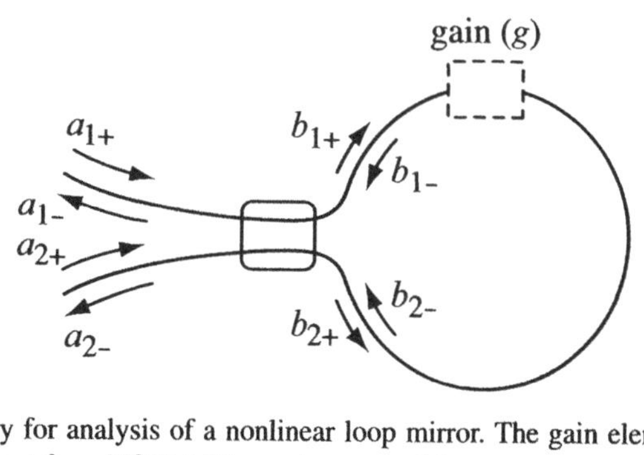

Figure 7.3 Geometry for analysis of a nonlinear loop mirror. The gain element is included for a NALM but is not present for a NOLM. The various a_i and b_i represent complex field amplitudes at indicated positions inside and outside the loop.

[1] In terms of eq. (2.84), we identify $\alpha = r^2$.

For the F8L we are interested in the case $a_{2+} = 0$ (enforced by the isolator). With this assumption we obtain the following expression for the output powers, normalized to the input power and the gain:

$$\left|\frac{a_{1-}}{a_{1+}}\right|^2 e^{-2g} = 4\alpha(1-\alpha)\cos^2\left(\frac{\Delta\phi}{2}\right)$$

$$\left|\frac{a_{2-}}{a_{1+}}\right|^2 e^{-2g} = 1 - 4\alpha(1-\alpha)\cos^2\left(\frac{\Delta\phi}{2}\right) \tag{7.4}$$

where $\Delta\phi = \phi_{12} - \phi_{21}$.

The transmission characteristics clearly depend on the phase difference $\Delta\phi$. In the absence of nonlinearity, the round-trip phases are identical ($\Delta\phi = 0$). When the coupler is exactly balanced as well (the splitting ratio is $\alpha = 0.5$), eq. (7.4) predicts 100% reflection (together with gain in the case of the NALM) and zero transmission! This explains the term *loop mirror*. When the splitting ratio is slightly detuned from 50%, some power is transmitted, but most of the power is still reflected. Nonlinear index effects can be used to imbalance the round-trip phases, resulting in a nonlinear transmission response. This is achieved slightly differently for the NOLM and NALM.

- For the NOLM the phase difference is written

$$\Delta\phi = -\gamma\left(|b_{1+}|^2 - |b_{1-}|^2\right)L = (1-2\alpha)\gamma|a_{1+}|^2 L \tag{7.5}$$

  where γ is the self-phase modulation coefficient from eq. (6.29b). When $|b_{1+}|^2 \neq |b_{1-}|^2$ (i.e., $\alpha \neq 0.5$), the counterpropagating fields have different intensities and thereby accumulate different phases under self-phase modulation (SPM). This leads to a nonlinear transmission effect, plotted in Fig. 7.4a. As the input power $|a_{1+}|^2$ is increased and the phase difference increases from zero, the transmission ($|a_{2-}|^2/|a_{1+}|^2$) also increases, reaching a first maximum of 100% at $\Delta\phi = \pi$. For even higher powers the transmission oscillates. We observe that nonlinear transmission occurs only for an imbalanced coupler. This leads to a design trade-off: (1) The switching power is proportional to $|1-2\alpha|^{-1}$ and is therefore reduced when the coupler becomes more imbalanced. (2) The on–off contrast of the nonlinear transmission function is given by $1-4\alpha(1-\alpha)$. Maximum extinction in transmission occurs for an exactly balanced coupler; the swing between on and off states decreases monotonically with increasing imbalance. Therefore, the design of a NOLM fundamentally requires a trade-off between reduced switching contrast and reduced switching power.
- For the NALM the important point is that the amplifier is placed very close to one end of the loop. In the limit one of the waves (in the figure the clockwise-propagating wave) is first amplified to high power and then undergoes SPM; the other wave undergoes SPM at low power and is only then amplified. The asymmetric amplifier location therefore imbalances the round-trip phases, so that a balanced coupler can be used.

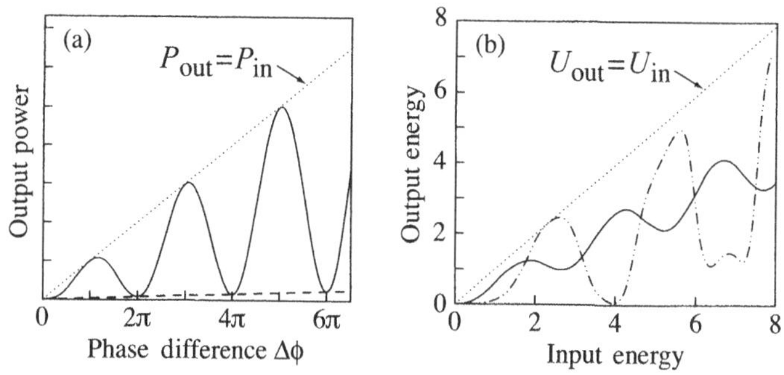

Figure 7.4 Calculated nonlinear transmission curves for a nonlinear optical loop mirror, with power-splitting ratio $\alpha = 0.4$. (a) Transmitted output power vs. nonlinear phase difference $\Delta\phi$ for continuous-wave light (or long square pulses without dispersion), from eq. (7.4). $\Delta\phi$ is proportional to input power through eq. (7.5). The upper and lower straight lines are guides to the eye for maximum transmission (100%, dotted line) and minimum transmission (dashed line), respectively. (b) Transmitted output energy vs. input energy for pulses. Dashed-dotted line; soliton pulses, for a loop length $L = 4z_0$; solid line; sech^2 pulses without dispersion, showing degraded switching contrast due to averaging over different intensities within the pulse. Energies are normalized to the energy of a fundamental soliton. (b) Adapted from [361].

Taking $\alpha = 0.5$, the phase difference becomes

$$\Delta\phi = -\gamma\frac{e^{2g}-1}{2}|a_{1+}|^2 L \tag{7.6}$$

An advantage of the NALM is that it allows both high on–off contrast and low switching power simultaneously.

In the simple discussion above, we considered SPM but neglected cross-phase modulation (XPM). For continuous-wave light, XPM in the nonlinear loop mirror is indeed an important effect. However, here we are interested in pulses. For reasonably short pulses the counter-propagating fields overlap only in a small region, and this justifies the neglect of XPM. Once we consider pulses, however, we should also consider dispersion and pulse reshaping. There are two main cases. First, for relatively long pulses, reshaping under dispersion may be unimportant. Therefore, the intensity profile remains constant during propagation around the loop. In this case the intensity $|a_{1+}|^2$ and the phase difference are both time dependent: $\Delta\phi(t) \sim |a_{1+}(t)|^2$. This has an interesting consequence: The nonlinear transmission itself varies as a function of time within the pulse. For mode-locking this constitutes a desirable self-amplitude modulation effect; the peak of the pulse experiences enhanced transmission compared to the wings (assuming, of course, that $\Delta\phi \leq \pi$). On the other hand, for applications in nonlinear optical self-switching, this time-dependent transmission function may be viewed as undesirable, since it degrades the energy contrast (integrated over pulse temporal profile) between maximum and minimum transmission states.

Second, for sufficiently short pulses, dispersion is important. A particularly interesting case relevant to mode-locked fiber lasers occurs when the pulses counterpropagating around the loop experience anomalous dispersion and behave as solitons. From Section 6.6 we know that solitons accumulate a nonlinear phase shift that is *uniform* across the pulse temporal profile, unlike the case of dispersionless SPM. Although this is strictly true only for exact soliton pulses, numerical studies reveal that this behavior remains approximately true over a significant range even for pulses with peak powers that deviate from the soliton condition. As a result, solitonlike pulses in a nonlinear loop mirror have a tendency not only to retain their pulse duration approximately upon propagation, but also to switch as a unit [361,362,369]. The effect that this has on the nonlinear transmission curve can be seen in Fig. 7.4b, which compares calculations of time-integrated transmission for solitonlike pulses (assuming a loop length of $4z_0$) and for pulses where dispersion is ignored. The results for pulses without dispersion show a relatively shallow modulation of the nonlinear transmission with increasing power; this results from averaging over different intensities within the pulse. On the other hand, the curve calculated for pulses propagating with anomalous dispersion (solitonlike pulses) exhibits a strong modulation in nonlinear transmission. The switching curve bears a close resemblance to the ideal single intensity curve (Fig. 7.4a), which is a consequence of the tendency of solitonlike pulses to switch as a unit. This behavior can be useful both for nonlinear optical switching applications and in mode-locking.

7.1.2 Energy Quantization

The F8L is capable of generating ultrashort pulses down to about 100 fs. Solitonlike propagation is important for both pulse shortening via nonlinear phase effects (interplay between SPM and GVD) and, as discussed in the preceding paragraph, for pulse shortening via self-amplitude modulation. The need to simultaneously optimize both processes leads to some interesting consequences for laser operation.

For simplicity of discussion, we consider the case of a NOLM (rather than a NALM) in a F8L.[2] Assume that the peak input power to the NOLM is $|a_{1+}|^2 = 2P_0$, where P_0 is the peak power of a fundamental soliton at the input pulse duration. Then for $\alpha = 0.5$, exact fundamental solitons would be launched into the NOLM in both directions. Now take $\alpha > 0.5$, but not by much (i.e., $|2\alpha - 1| \ll 1$). The launched pulses have different peak powers, $2\alpha P_0$ and $2(1-\alpha)P_0$, which are too high and too low, respectively, for a fundamental soliton. According to eqs. (6.82) and (6.83), both pulses nevertheless evolve into fundamental soliton pulses. Algebraic manipulation shows that to first order in $2\alpha - 1$, the peak powers of the resulting solitons are, respectively,

$$P_\pm \approx [1 \pm 2(2\alpha - 1)]\, P_0 \tag{7.7}$$

From eqs. (6.78) and (6.79), the solitons accumulate phase shifts equal to $-\gamma P_\pm L/2$ in propagating around the loop. The phase difference is then equal to

$$|\Delta\phi| \approx 2|2\alpha - 1|\gamma P_0 L \tag{7.8}$$

For maximum transmission through the NOLM, we require that $\Delta\phi = \pi$; hence,

$$P_0 \approx \frac{\pi}{2|2\alpha - 1|\gamma L} \tag{7.9}$$

[2] For the NALM we must also include the saturation of the optical amplifier within the loop mirror. This makes the analysis much more involved, since saturated gain, nonlinear switching operation of the loop mirror, and pulse parameters such as average and peak power are all coupled. For a discussion, see [370].

For a fixed power-splitting ratio and loop length, this fixes the power P_0 (since further increase in P_0 would reduce the nonlinear transmission and hence increase the loss in the laser). Then because of the relation between soliton power and duration, $T_0^2 = |\beta_2|/\gamma P_0$, the pulse duration is also determined, as is the pulse energy [see eqs. (6.79) and (6.81)]. Also using eq. (7.9), we find that

$$T_0 \sim \sqrt{|\beta_2| L} \tag{7.10}$$

That is, the minimum pulse width scales as the square root of the total cavity dispersion.

We have just seen that if the laser is to run at the optimum point on the nonlinear transmission curve (minimum loss), this fixes the peak power, pulse width, and pulse energy. Of course, our treatment is quite simplified, and in real operation at least some variation of these parameters is possible. Nevertheless, these considerations lead to behavior very different than what we are used to with conventional continuous-wave (CW) lasers. In particular, in CW lasers a continuous increase in the pump power leads to a continuous increase in the power circulating within the laser cavity, which in turn more strongly saturates the gain such that saturated gain always equals loss. In a soliton laser, as discussed above, the energy per pulse is essentially fixed or *quantized* and therefore incapable of increasing continuously with stronger pumping. Instead, additional pulses can appear within the laser cavity as the pump is increased. In this way the average power can grow as needed to saturate the gain appropriately, even with fixed energy per pulse. Note that there is nothing in the laser that dictates the relative spacings of the pulses when the laser is running with multiple pulses per round trip. Therefore, in this regime the laser output is typically a randomly spaced group of pulses, repeating with period equal to the cavity round-trip time [358,371,372].

Figure 7.5 shows data from a F8L which clearly demonstrate these points. Figure 7.5a shows a plot of output power vs. pump power. The threshold for CW oscillation is about 10 mW; when the pump power is increased further, to above about 40 mW, mode-locking self-starts. At this point, mode-locking is sustained even when the pump is turned down

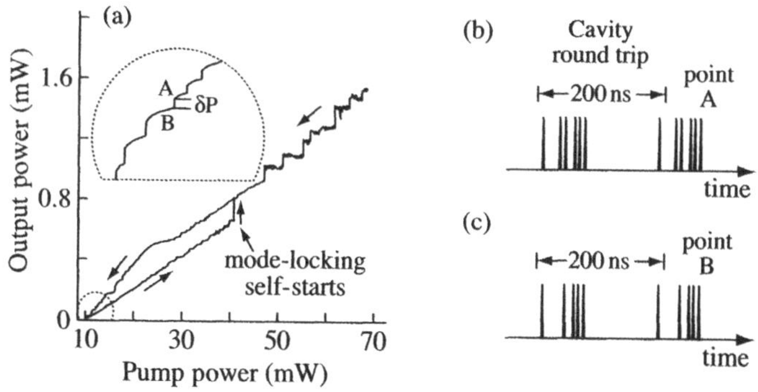

Figure 7.5 (a) Hysteretic output power vs. pump power curve for a figure-eight laser. The data from the lowest power regime, shown on an expanded scale in the inset, exhibit clear power jumps as the pump power is changed. (b,c) Oscilloscope measurements of intensity profiles, corresponding to points marked A and B in output vs. pump power curve. The power jump corresponds to the appearance or disappearance of one of the pulses circulating in the cavity. Adapted from [372]. Copyright ©1992, IEEE.

over a broad range, leading to a hysteretic output vs. input power curve. The data in the lowest pump power regime, which are expanded in the inset, clearly reveal abrupt changes in output power when the pump is reduced. Each power jump in Fig. 7.5a is associated with the appearance or disappearance of one circulating pulse from the laser cavity. This effect is illustrated in Fig. 7.5b and c, where oscilloscope traces of the output pulse trains are shown just above and below a power jump, respectively. The key point is that the number of recirculating pulses decreases by one when the power jumps down. Furthermore, when converted into a pulse energy, the power jump δP observed is consistent with that expected for a soliton of the pulse duration observed.

These observations clearly reveal soliton quantization behavior. Such soliton quantization phenomena have been widely observed in soliton laser studies for both NOLM/NALM-type fiber lasers and for fiber lasers mode-locked through nonlinear polarization evolution [363,373]. Although such behavior may be undesirable for many practical applications, these phenomena provide a striking example of the critical role that soliton nonlinear optics can sometimes play in defining laser operation.

7.1.3 Soliton Sidebands

Another interesting manifestation of soliton nonlinear optics arises because pulses propagating approximately as solitons within a laser cavity experience intensity variations due to the discrete or localized nature of the gain and loss. This is unlike the condition for true solitons, where the pulse duration and intensity remain constant during propagation, as discussed in Section 6.6. Since gain and loss must balance for steady-state laser operation, these intensity variations are periodic after a single round trip of the laser cavity. Because typical fiber lasers are constructed from different fiber types spliced together, the fiber dispersion and nonlinear coefficient may also vary within the laser cavity, often abruptly. Recirculating mode-locked pulses also see such dispersion and nonlinearity changes as periodic. As a result of such periodic variations, a series of sharp peaks can arise in the output spectrum from a soliton laser [374–376]. Examples of such spectral peaks observed in data from a figure-eight laser are shown in Fig. 7.6. In this Section we provide a simple theory for these striking spectral peaks and discuss their effect in limiting mode-locked pulse duration [376–379].

It is worth noting that the effect of periodic perturbations on solitonlike propagation was first encountered in long-distance soliton transmission for telecommunications [84,309,380–382]. In this case a series of lumped optical amplifiers may be placed periodically along the transmission link to compensate for propagation loss. Studies have shown that high-quality solitonlike propagation is still possible provided that the amplifier spacing (L) is much less than the characteristic dispersion length (L_D) [383]. This is because when the amplifier spacing is sufficiently small ($L \ll L_D$), neither the temporal profile nor the spectral profile of the pulse can change appreciably within a single amplifier period. In this case the soliton parameters are determined by the path-averaged value of intensity. Nonlinear phase and dispersion only need to cancel on a path-averaged basis, and the pulses are known as *path-averaged solitons* or *guiding center solitons*. On the other hand, when the amplification period is increased to L_D and beyond, solitons are increasingly perturbed, until the propagation goes unstable at $L = 8z_0$ (leading to large energy and pulse-shape fluctuations). Since an unperturbed soliton accumulates nonlinear phase $\pi/4$ for propagation over distance z_0, this instability condition corresponds to a nonlinear phase of 2π per amplifier period.

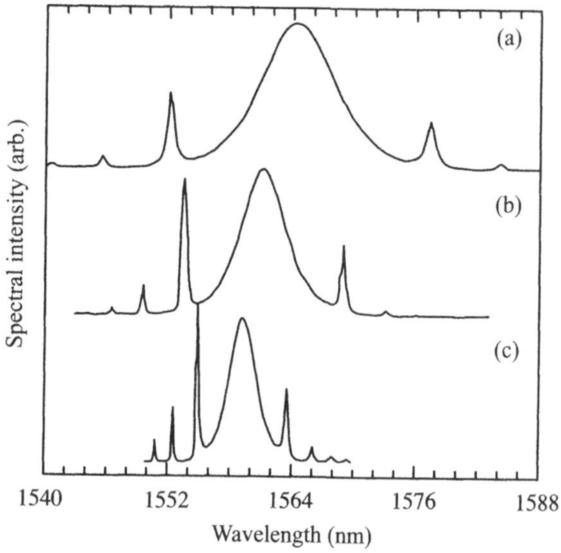

Figure 7.6 Spectra from three different figure-eight laser cavities, clearly illustrating the formation of soliton sidebands. The pulse duration and total cavity dispersion for each laser cavity are: (a) 300 fs, 78.8 fs/nm; (b) 470 fs, 216 fs/nm; (c) 1130 fs, 616 fs/nm. From [376], with permission. Copyright ©1993, American Institute of Physics.

The central point in our discussion of soliton sidebands is that abrupt changes in the soliton pulse parameters (e.g., due to lumped amplifiers) lead to creation of a dispersive wave. Dispersive waves, by definition, propagate according to the linear dispersion condition,

$$\beta_D(\omega) = \beta_0 + \beta_1(\omega - \omega_0) + \frac{\beta_2}{2}(\omega - \omega_0)^2 + \frac{\beta_3}{6}(\omega - \omega_0)^3 + \cdots \tag{7.11}$$

On the other hand, the soliton pulse itself propagates as a nonlinear wave according to

$$\beta_S(\omega) = \beta_0 + \beta_1(\omega - \omega_0) + \frac{\beta_2}{2}T_0^{-2} \tag{7.12}$$

where T_0 is the soliton pulse width parameter according to eq. (6.78). Since the mode-locked pulse is assumed to repeat periodically after every round trip through the laser cavity, new contributions to the dispersive wave are also generated periodically every trip through the cavity. The cumulative growth of the dispersive wave depends on the phase of the new contributions relative to those generated on previous round trips. In particular, the dispersive wave is generated most strongly when it is phase matched with the soliton wave (i.e., each new contribution is in phase with the dispersive wave already propagating). Mathematically, this condition is written

$$\beta_S(\tilde{\omega}_m)L = \beta_D(\tilde{\omega}_m)L + 2m\pi \tag{7.13}$$

where L is the round-trip cavity length, m is an integer, and $\tilde{\omega}_m$ is the offset of the mth sideband from the center frequency. Here β_S and β_D refer to the path-averaged values,

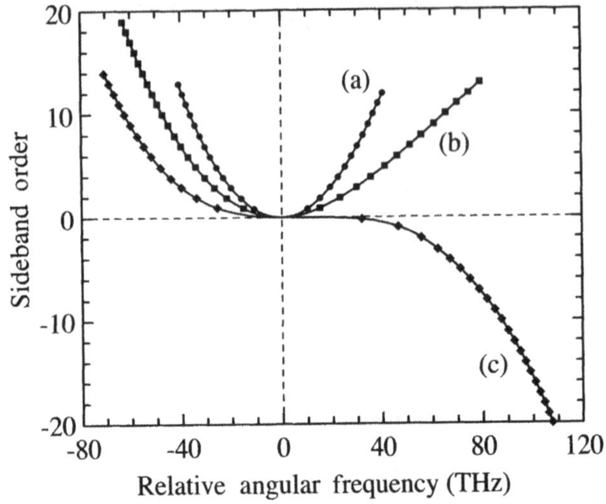

Figure 7.7 Soliton sideband order plotted as a function of frequency offset (from spectral peak) for a single laser cavity with different amounts of added dispersion-compensating fiber. Total cavity dispersions: (a) –0.093 ps^2; (b) –0.045 ps^2; (c) –0.015 ps^2. From [357].

and $\beta_D L$ is the total cavity dispersion. Combining these equations yields a formula for the frequencies at which phase matching occurs (i.e., the soliton sideband frequencies):

$$m = \frac{\beta_2 L}{4\pi}\left[T_0^{-2} - \tilde{\omega}_m^2\right] - \frac{\beta_3 L}{12\pi}\tilde{\omega}_m^3 \tag{7.14}$$

It is interesting to note that at the transition to instability ($L = 8z_0$), phase matching occurs for $\tilde{\omega}_m = 0$ (with $m = -1$). This corresponds to exact resonant coupling of the soliton center frequency to the dispersive wave [379].

Figure 7.7 shows a plot of the soliton sideband order plotted as a function of frequency offset (from the spectral peak) for a single figure-eight laser cavity into which different amounts of dispersion-compensating fiber (DCF) were added. The curves are well fit by cubic polynomials, in accord with our simple analysis. For the highest total cavity dispersion, the dispersion is dominated by β_2 and the plot is close to quadratic. As the average value of β_2 is reduced by adding increasing DCF, the β_3 term becomes more important, and the curves approach a cubic shape. The dependence of the soliton sideband resonances on the dispersion as described by eq. (7.14) can be used for characterization of the dispersion of fibers added into the laser cavity [379].

Transfer of energy into the soliton sidebands can act to limit the pulse width attained by a soliton laser [376,379]. Consider the case when the β_3 term may be neglected. Equation (7.14) is then rewritten

$$\tilde{\omega}_m = \pm\frac{1}{T_0}\sqrt{1 - \frac{4\pi m T_0^2}{\beta_2 L}} = \pm\frac{1}{T_0}\sqrt{1 + \frac{4\pi m L_D}{L}} \tag{7.15}$$

According to [384] the leading term determining the rate of energy transfer from the soliton to a phase-matched sideband is proportional to the soliton spectrum at the sideband frequency. Since the soliton spectrum is of the form $\mathrm{sech}^2(\pi\tilde{\omega}T_0/2)$, the loss rate Γ_m to the mth sideband is proportional to

$$\Gamma_m \sim \mathrm{sech}^2\left(\frac{\pi\tilde{\omega}_m T_0}{2}\right) \tag{7.16}$$

The highest energy loss will occur for the sideband closest to the soliton center frequency (i.e., $m = 1$). Assuming that $L \ll L_D$, the loss rate to the first sideband is described by

$$\Gamma_1 \sim \mathrm{sech}^2\left(\pi^{3/2}\sqrt{\frac{L_D}{L}}\right) = \mathrm{sech}^2\left(\pi^{3/2}\frac{T_0}{\sqrt{|\beta_2|L}}\right) \tag{7.17}$$

The loss is exponentially small when $L/L_D \ll 1$ but becomes large when L and L_D are comparable. As the pulse width decreases and the spectrum spreads within a fixed laser cavity, the loss to the sideband increases. Eventually, the increased loss overwhelms the mode-locking mechanisms, and the bandwidth and pulse duration are clamped. According to this picture, for the shortest pulses the argument of the $\mathrm{sech}^2(\cdots)$ in eq. (7.17) should be approximately constant. As a result, this model predicts that the shortest pulses in a soliton laser are again proportional to the square root of the dispersion:

$$T_0 \sim \sqrt{|\beta_2|L} \tag{7.18}$$

The dependence of minimum pulse width on dispersion is plotted in Fig. 7.8 for a variety of figure-eight soliton lasers. The data adhere to the square root of total cavity dispersion law predicted for pulse widths varying by more than an order of magnitude. Referring now to spectra shown in Fig. 7.6, we observe that the offset of the strongest sideband from the

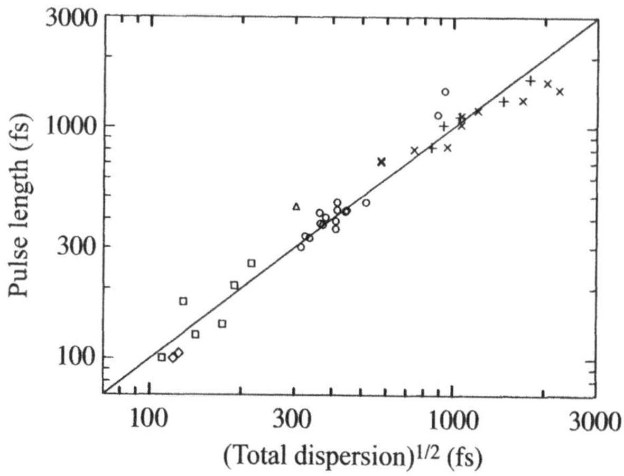

Figure 7.8 Variation of minimum pulse width with square root of total cavity dispersion for a variety of soliton lasers. From [379]. Copyright ©1994, IEEE.

peak of the spectrum scales roughly with the mode-locked bandwidth. The amplitude of the mode-locked spectrum at the strongest sideband (and hence the loss to the sidebands) remains roughly constant for these data, as expected from the model described above.

From a practical perspective, one finds that total cavity dispersion should be minimized in order to obtain the shortest pulses. In addition, high-quality pulses without solitonic sidebands can be obtained by inserting an optical bandpass filter into the laser cavity, to restrict the oscillating bandwidth to a range smaller than the frequency offset at which the first sideband would appear [385].

7.2 SOLITON MODE-LOCKING: ACTIVE MODULATION AND SLOW SATURABLE ABSORBERS

7.2.1 Harmonically Mode-Locked Soliton Fiber Lasers

Although Section 7.1 offers important insights into soliton phenomena affecting mode-locked pulse formation, the lasers discussed are nevertheless inadequate for applications such as high-rate telecommunications. In particular, such applications demand a high pulse repetition rate (e.g., 10 or 40 GHz) and high stability. Harmonic mode-locking offers a solution to these requirements.

Figure 7.9 shows one well-known harmonically mode-locked fiber laser geometry. The laser is constructed in a sigma configuration, in which a polarization maintaining (PM) loop section is connected to a non-PM straight section via a polarizing beamsplitter (PBS). An integrated optic intensity modulator, typically driven at about 10 GHz, harmonically mode-locks the laser. The idea is similar to active mode-locking (Section 2.2), except that the modulator is driven at a harmonic of the longitudinal mode spacing. The non-PM section includes a gain fiber as well as additional lengths of fiber wound around a piezoelectric

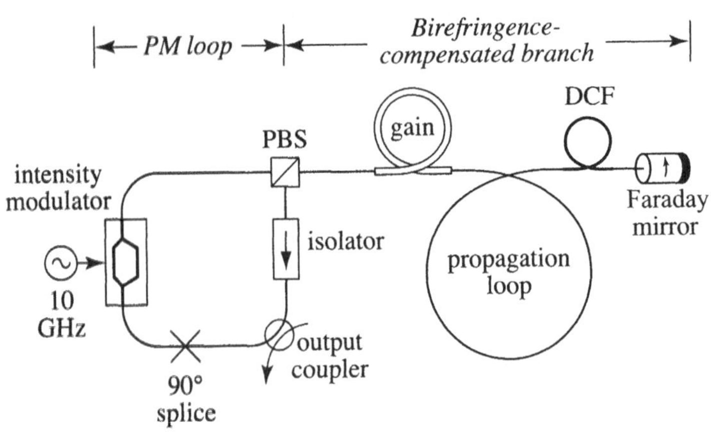

Figure 7.9 Harmonically mode-locked fiber soliton laser. The laser consists of a polarization-maintaining (PM) loop coupled to a non-PM but birefringence-compensated branch through a polarization beamsplitter (PBS). The non-PM loop consists of a fiber amplifier for gain, a propagation loop allowing for accumulation of nonlinear phase shift, and a length of dispersion-compensating fiber (DCF). Retroreflection through the non-PM loop via a Faraday mirror provides for birefringence compensation. Adapted from [20].

stretching element and dispersion compensation fiber (DCF). The fiber stretcher is used to adjust the overall cavity length such that the applied modulation frequency is an exact multiple of the cavity mode spacing. Such cavity length matching is critical for any actively mode-locked system and is accomplished under control of a feedback circuit that compares the timing of the output optical pulse train with the electrical drive signal applied to the modulator. The DCF, which has normal dispersion, reduces the average anomalous dispersion of the laser cavity to the regime where soliton propagation is possible at the available intracavity power. The Faraday mirror [386] consists of a 45° Faraday rotator[3] and end reflector. The double pass through the Faraday rotator transforms the input polarization into the orthogonal polarization state. Subsequent propagation back through the rest of the non-PM branch compensates any birefringence variations in the non-PM components and provides a high degree of immunity against fluctuations caused by time-varying birefringence changes. The polarization transformation in the non-PM branch, together with an isolator and a 90° fiber splice in the PM loop, provide for unidirectional operation in the PM loop, with light from the loop transmitted through the PBS into the non-PM branch, then reinjected from the non-PM branch back into the loop via reflection at the PBS. The laser of [20] had an effective cavity length of 192 m, which means that the fundamental cavity resonance frequency is of order 1 MHz. Therefore, for harmonic mode-locking at 10 GHz, roughly 10^4 pulses must circulate within the cavity. Such lasers have been made to produce high-repetition-rate trains of pulses of order 1-ps duration with high stability.

An advantage of the sigma configuration is that a single polarization laser, equivalent to a laser composed of all polarization-maintaining elements, can be realized even when some of the laser elements, such as special fibers, are unavailable in polarization-maintaining form. However, similar performance has been observed in harmonically mode-locked lasers in a simple loop configuration, similar to Fig. 7.2 with the addition of a modulator. Both PM and non-PM versions of such lasers have been reported (e.g., [387,388]). Generally, the operating principles of harmonically mode-locked lasers in sigma and loop configurations are similar.

There are several important points concerning laser operation:

- According to active mode-locking theory, eq. (2.20), the pulse width scales as the inverse square root of the modulation frequency applied. For mode-locking at high harmonic number, the pulse width predicted is substantially shorter than it would be for mode-locking at the fundamental cavity frequency. For example, in [20] the pulse width as predicted by active mode-locking theory is given as about 5 ps.
- The actual pulse width (e.g., ca. 1 ps in [20]) can be substantially below that predicted by active mode-locking theory. This effect is commonly observed in actively mode-locked lasers containing long lengths of anomalous dispersion fiber where both dispersion and self-phase modulation are strong (see, e.g., [389]). In such lasers soliton pulse-shaping effects dominate, and the dispersion, pulse energy, and duration follow the soliton law, eq. (6.81).
- A very important point is that in pure harmonic mode-locking in lasers with a long gain relaxation time, there is no guarantee that a pulse will be present in each transmission window of the modulator. This is because when the gain relaxation time is long

[3] A Faraday rotator is a nonreciprocal device that provides a certain amount of polarization rotation, here 45°, for both forward and backward propagation, independent of input polarization state.

compared to the pulse spacing, as it typically is in fiber or solid-state laser systems, the gain saturation depends only on the average power and therefore cannot stabilize the energies of individual pulses. Even if at some moment a pulse is present in each transmission window, the laser may be unstable against energy growth of some of the pulses at the expense of others which may decay and eventually drop out completely. Such problems are present in harmonically mode-locked lasers either with or without soliton effects. Therefore, other stabilizing mechanisms are needed. One possibility is to use nonlinear polarization evolution biased for limiting action [390]. Here the idea is to adjust the polarization states in the laser cavity such that intensity-dependent polarization changes degrade transmission through a polarizer within the laser. This negative self-amplitude modulation (SAM) effect discriminates against pulses whose intensity has grown higher than their neighbors and acts to equalize the intensities of all mode-locked pulses. Similarly, excessive nonlinear spectral broadening in conjunction with a bandwidth-limiting filter or a limited-gain bandwidth can discriminate against higher-intensity pulses, which see increased loss or decreased gain due to their increased bandwidth [391]. As a result of such limiting mechanisms, extremely stable harmonic mode-locking is possible, with pulses with equal energies and durations in every transmission window of the modulator [392]. Such sources have been tested in telecommunications transmission experiments and have yielded error rates below 10^{-12}, which provides clear evidence of extremely low probability of pulse dropouts [20,392].

It is interesting to note that limiting mechanisms as described above yield a negative SAM coefficient, which acting alone would inhibit mode-locking. Nevertheless, pulse formation is still possible since active mode-locking action dominates at relatively long pulse durations (recall that the strength of SAM effects scales inversely with pulse width). Therefore, in a properly designed laser of this type, the limiting effect due to negative SAM becomes important only after sufficiently short mode-locked pulses have been obtained.

7.2.2 The Net Gain Window in Soliton Mode-Locking

In our treatment of mode-locking in Chapter 2, we made extensive use of the concept of a net gain window, where net loss was required immediately on either side of a mode-locked pulse to prevent the growth of noise. An important point was that the net gain window was matched to the duration of the pulse, whether for active or passive mode-locking, and in the latter case for either slow or fast saturable absorbers. On the other hand, in our discussion of harmonically mode-locked fiber lasers (Section 7.2.1), we pointed out that due to soliton pulse shortening, pulses significantly shorter than those predicted by pure active mode-locking theory are observed. Such pulses essentially go to zero well within the region of positive net gain. Generation of pulses significantly shorter than the apparent gain window is also observed for solid-state lasers operating in the soliton regime, both with active mode-locking [393,394] and with passive mode-locking using slow absorbers [395,396]. The latter case is of significant technological importance, since semiconductor growth and fabrication technology allows realization of high-quality semiconductor saturable absorber mirrors, with significant flexibility for design of key absorber parameters (e.g, wavelength, saturable and nonsaturable loss, saturation energy, recovery time). Such semiconductor absorbers have been applied for successful mode-locking in the femtosecond regime for a variety of solid-state gain media, with self-starting operation and without the strong restrictions on cavity design associated with Kerr lens mode-locking [397,398].

In all such cases, it would seem that pulses much shorter than the net gain window should be unstable against growth in their wings. Fortunately, for solitonlike pulses a new stabilization process arises that permits stable operation at reduced pulse widths. In the following we analyze such stabilization for the specific case of active mode-locking, following the treatment of [393,394]. A theory for the passively mode-locked case which is similar in spirit has also been published [396], but because the mathematics are more involved we do not present this theory here. Our strategy for analysis of the actively mode-locked laser with soliton pulse shortening will be as follows:

1. Assume that soliton effects dominate. That is, steady-state pulses are sufficiently short that they are influenced primarily by the reactive effects (self-phase modulation and dispersion); pulse-reshaping effects arising from finite bandwidth and the active modulation are considered minimal.
2. Analyze the propagation of a weak but longer-duration background pulse postulated as a perturbation. The perturbation is assumed to experience a saturated gain determined by the soliton; this is justified by the detailed analysis given in [394]. Stability or instability of the soliton pulse solution is indicated, respectively, by attenuation or growth of the weak perturbation pulse.

We begin with the mode-locking equation in the presence of gain, spectral filtering, time-dependent loss modulation, self-phase modulation, and dispersion. Similar to Chapter 2, the master equation is written

$$\frac{\partial a(T,t)}{\partial T} = \frac{1}{T_R}\left\{\left(g - \ell_0 + D_g\frac{\partial^2}{\partial t^2} - \ell(t)\right) + j\left(D\frac{\partial^2}{\partial t^2} - \delta\,|a|^2\right)\right\} a(T,t)$$

$$\text{where}\quad D_g = \frac{g}{\Omega_g^2} \tag{7.19}$$

This formulation permits analysis of non-steady-state pulse dynamics but assumes that overall changes in the pulse per round trip are small. Here $a_{n+1}(t) - a_n(t)$ is approximated by $T_R[\partial a(T,t)/\partial T]$, where $a_n(t)$ and $a_{n+1}(t)$ represent the pulse after n and $n+1$ passes through the laser, respectively, T_R is the cavity round-trip time, T is a slow time variable that varies only on the scale of T_R (i.e., T/T_R gives the number of passes through the cavity), and t is the usual (fast) time variable that describes variations within individual pulses. Recall that all terms represent effects integrated over the full laser cavity (e.g., $D = \int \frac{1}{2}\beta_2\,dz$). For active mode-locking the time-dependent loss is written

$$\ell(t) = M(1 - \cos\omega_m t) \tag{7.20}$$

This mode-locking equation as written assumes that filtering arises from the finite bandwidth associated with saturated gain g, but if desired, this can easily be generalized to include a separate bandpass filter (by replacing D_g with $D_g + D_f$, where D_f represents the additional bandpass filter).

If the reactive terms on the right-hand side of eq. (7.19) dominate so that we neglect the real part, this reduces to the nonlinear Schrödinger equation [e.g., eq. (6.43)], with the mapping

$$\frac{T}{T_R} \mapsto z \qquad D \mapsto \frac{\beta_2}{2} \qquad \delta \mapsto \gamma \tag{7.21}$$

That is, the evolution in the slow time variable T/T_R in the laser equation maps onto evolution in distance z in the pulse propagation problem. When D is negative (anomalous dispersion), we have the fundamental soliton solution

$$a(T,t) = \sqrt{\frac{U}{2T_0}}\,\text{sech}\left(\frac{t}{T_0}\right)\exp\left(\frac{-j\Phi_0 T}{T_R}\right) \tag{7.22a}$$

where U and Φ_0 are the pulse energy and phase shift per round trip, given, respectively, by

$$U = \frac{4|D|}{\delta T_0} \qquad \text{and} \qquad \Phi_0 = \frac{|D|}{T_0^2} \tag{7.22b}$$

This equation is equivalent to the expression for a fundamental soliton given in Chapter 6, eq. (6.78). It is also consistent with the result found in Chapter 2 for mode-locking via artificial fast saturable absorbers in the case of large anomalous dispersion (when a chirp-free solitonlike solution was found).

It is interesting to consider the evolution of the pulse energy after passing through the cavity. The time derivative of energy U with respect to slow time variable T is written

$$\frac{dU}{dT} = \frac{d}{dT}\int_{-\infty}^{\infty} dt\, a(T,t)a^*(T,t) = \int_{-\infty}^{\infty} dt\left(a^*\frac{\partial a(T,t)}{\partial T} + a\frac{\partial a^*(T,t)}{\partial T}\right) \tag{7.23}$$

This expression can be evaluated by using eq. (7.19) for $\partial a(T,t)/\partial T$. The contributions from the reactive terms are easily shown to integrate to zero. On the other hand, the modulation, loss, and gain terms must be retained, as they do affect the energy. This yields

$$\frac{dU}{dT} = \left(\frac{1}{T_R}\right)\int_{-\infty}^{\infty} dt\left\{a^*(T,t)\left(g - \ell_0 + D_g\frac{\partial^2}{\partial t^2} - \ell(t)\right)a(T,t) + \text{c.c.}\right\} \tag{7.24}$$

In steady state the net change in soliton energy after a round trip through the laser cavity must be zero, which sets the value of the saturated gain. To perform the computation, we substitute in using eq. (7.22a) for a. We also assume, as in Chapter 2, that the mode-locked pulse is sufficiently short that the loss modulation function may be expanded near $t = 0$; that is,

$$\ell(t) \approx M_s t^2 \qquad \text{where} \qquad M_s = \frac{1}{2}M\omega_m^2 \tag{7.25}$$

By consulting an integral table [399], one obtains for the steady state:

$$g - \ell_0 = \frac{D_g}{3T_0^2} + \frac{\pi^2 M_s T_0^2}{12} \tag{7.26}$$

Now we consider the propagation of a weak, longer-duration background pulse which is assumed as a perturbation. Physically, such a background pulse might be expected to arise from continuum shed by solitonlike mode-locked pulses perturbed by the loss modulation and finite gain bandwidth. As long as the perturbation pulse is sufficiently weak, it will experience negligible SPM. Its evolution is then described by eq. (7.19), with $\delta = 0$ and saturated gain g determined by the soliton. In this limit, eq. (7.19) is the same as for active mode-locking, but with the addition of dispersion. If we now assume that the (perturbation) pulse $\tilde{a}(T, t)$ is localized near the peak of the loss modulation function, the solution is a chirped Gaussian,

$$\tilde{a}(T, t) = E_0 e^{-\Gamma t^2} e^{\lambda T/T_R} \qquad \text{where} \qquad \Gamma = \Gamma_r + j\Gamma_i \tag{7.27}$$

Here we have included the term $\exp(\lambda T/T_R)$ to account for the possibility of long-term growth or decay of the perturbation. Inserting eq. (7.27) into eq. (7.19) with $\delta = 0$ and requiring that the coefficients of the terms proportional to $\exp(-\Gamma t^2)$ and $t^2 \exp(-\Gamma t^2)$ independently sum to zero yields two complex equations:

$$\lambda = g - \ell_0 - 2D_g(1 + jD_n)\Gamma \tag{7.28a}$$

$$\Gamma^2 = \frac{1}{4T_A{}^4(1 + jD_n)} \tag{7.28b}$$

where

$$D_n = \frac{D}{D_g} \qquad \text{and} \qquad T_A = \left(\frac{D_g}{M_s}\right)^{1/4}$$

Here D_n describes the relative strength of pulse spreading due to dispersion compared to that due to bandwidth filtering; and consistent with Section 2.2, T_A gives the pulse width that occurs for pure active mode-locking ($D_n = 0$ and $\delta = 0$: no dispersion, no SPM). Equation (7.28b) for Γ determines both the duration and chirp of the Gaussian pulse. Now, to obtain an expression for λ, we substitute for Γ in eq. (7.28a), also using eq. (7.26) for g. The result is

$$\lambda = \frac{D_g}{3T_0^2} + \frac{\pi^2 M_s T_0^2}{12} - \frac{D_g}{T_A{}^2}\sqrt{1 + jD_n} \tag{7.29}$$

Stability of the short solitonlike pulse, eq. (7.22a), requires that $\mathrm{Re}\{\lambda\} \leq 0$. Using (7.29), the stability condition becomes

$$\frac{T_A{}^2}{T_0^2} + \frac{\pi^2 T_0^2}{4T_A{}^2} \leq 3\,\mathrm{Re}\left\{\sqrt{1 + jD_n}\right\} \tag{7.30}$$

We are most interested in the asymptotic case of large dispersion and short solitonlike pulse width, $T_0 \ll T_A$ and $|D_n| \gg 1$. In this limit the stability condition simplifies to

$$\frac{T_A}{T_0} \leq \sqrt{3}\left(\frac{|D_n|}{2}\right)^{1/4} \tag{7.31}$$

Or since we are considering different pulse shapes, we can rewrite in terms of intensity full width at half maximum pulse durations, which gives

$$\frac{T_{A\text{-FWHM}}}{T_{0\text{-FWHM}}} = \frac{1.66\, T_A}{1.76\, T_0} \leq 1.37|D_n|^{1/4} \tag{7.32}$$

These results show that stable mode-locked operation with solitonlike pulse durations well below that corresponding to pure active mode-locking is possible. The maximum pulse-width reduction factor increases as the one-fourth power of the normalized dispersion. For a given dispersion, increasing the pump level leads to a larger pulse energy, which in turn yields a shorter pulse duration [through eq. (7.22b)] until the minimum stable pulse duration is reached.

We can understand the pulse stabilization process as follows. As we know, our soliton-like pulses experience intensity variations as they propagate within the laser cavity, and this gives rise to low-intensity dispersive waves. If not attenuated, such dispersive waves would eventually form into the actively mode-locked pulse solution. However, since the dispersive waves spread in time, they see higher loss from the modulation function compared to the solitonlike pulses, which are held together with the aid of the nonlinearity and concentrated near the peak of the modulator transmission. This process is illustrated in Fig. 7.10a. With sufficient dispersion [i.e., eq. (7.30) is satisfied], the excess loss experienced by the weak

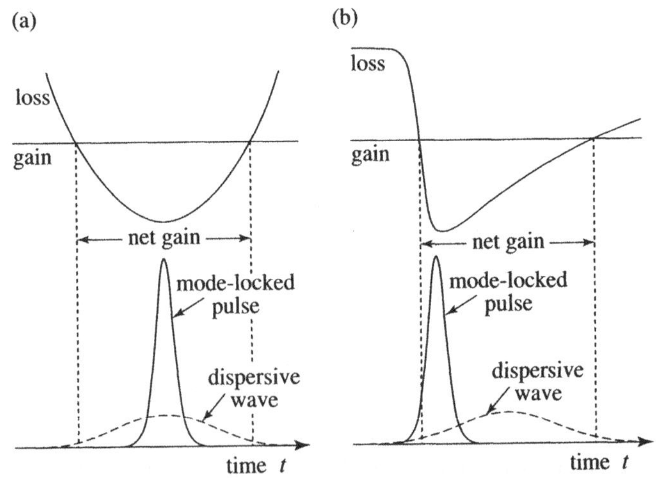

Figure 7.10 Net gain window for (a) an actively mode-locked laser with soliton pulse shortening and (b) a passively mode-locked laser with soliton pulse shortening and slow saturable absorber. In both cases mode-locked pulses shorter than the net gain window are stabilized due to the increased loss experienced by dispersive waves, which are spread in time.

background as it spreads in time is sufficient for stabilization (i.e., this extra loss suffered by the dispersive wave exceeds the additional loss incurred by the soliton due to its greater bandwidth and consequent greater susceptibility to spectral filtering). The amount of dispersion required for stabilization is increased for shorter pulses, since such shorter pulses are increasingly affected by bandwidth limitations.

A similar stabilization process occurs for soliton mode-locking with saturable absorbers [396,400], as depicted in Fig. 7.10b. Here the absorber saturates rapidly within the pulse duration but recovers more slowly. Accordingly, the leading edge of the pulse can see net loss, but for pulses much shorter than the recovery time, the trailing edge of the pulse sees net gain. However, as for the actively mode-locked case, perturbations to the soliton propagate under linear dispersion and hence spread into regions of time where there is increased loss. For sufficient dispersion this stabilizes short solitonlike pulse solutions.

A variety of experimental evidence is available to support the picture developed above. As one example [395] reports pulses as short as 300 fs from a solid-state (Ti:S) laser, mode-locked using a semiconductor saturable absorber mirror. Figure 7.11a shows a measurement of the time-resolved reflectivity of the absorber mirror after excitation with 100-fs pulses with fluence adjusted to equal that experienced within the laser. A rapid increase in reflectivity on a time scale comparable to the excitation pulses is observed due to absorption saturation in the mirror material, followed by exponential recovery with a > 10-ps time constant. Figure 7.11b shows a high-dynamic-range autocorrelation measurement of the 300-fs mode-locked pulses. The traces show no deviation from the exponential decay expected for an ideal sech^2 pulse shape and no evidence of a background pulse over the full six orders of magnitude measurement range. Since the laser cavity was configured such that Kerr lens mode-locking (KLM) did not play a role (this was verified experimentally), fast saturable absorber action cannot be invoked to explain these results. Furthermore, since the lifetime of the gain medium is on the order of microseconds, dynamic gain saturation also cannot play a role. Therefore, the observation of high-quality pulses, 35 times shorter than the absorber recovery time, is explained only via the soliton mode-locking model discussed above.

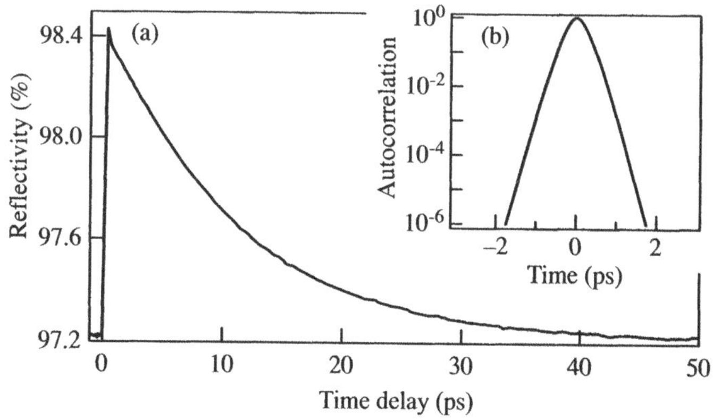

Figure 7.11 (a) Time-resolved reflectivity response of a semiconductor saturable absorber mirror used for mode-locking of a Ti:S laser, showing a recovery time of 10 to 20 ps; (b) high dynamic range autocorrelation of solitonlike pulses of about 300 fs. The pulses are much shorter than the absorber recovery time. From [395].

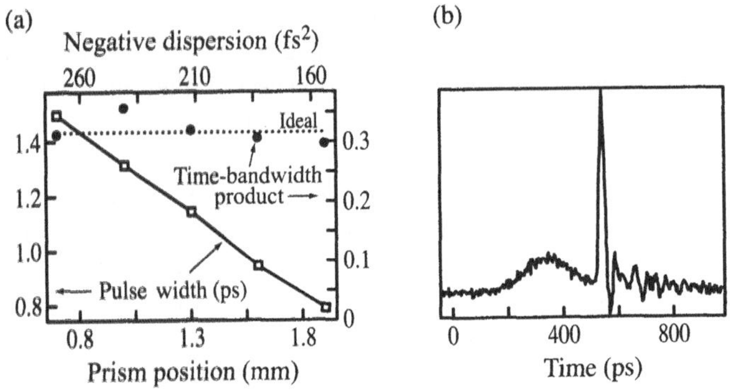

Figure 7.12 Data from a Nd:glass laser mode-locked via SPM, negative dispersion, and weak loss modulation: (a) pulse width and time–bandwidth product as negative intracavity dispersion is tuned (the behavior is as expected for sech^2 soliton pulses); (b) sampling oscilloscope measurement of laser output at the transition to instability. (a) Adapted from [393]; (b) adapted from [394].

References [393,394] report experimental study of an actively mode-locked solid-state (Nd:glass) laser employing an acousto-optic modulator for weak loss modulation and a prism pair to achieve adjustable anomalous dispersion. Although pulses of about 3 ps would be expected according to pure active mode-locking theory (no nonlinearity or dispersion), much shorter ca. 310 fs pulses whose autocorrelation closely fit that of an ideal sech^2 were observed. Since once again the laser cavity was configured such that KLM did not play a role, the very short pulse durations must be attributed to soliton shortening. Figure 7.12a shows data for a related laser configuration producing pulses on the order of 1 ps. Here the amount of negative intracavity dispersion was tuned by translating one of the prisms. Two points clearly indicate the solitonlike nature of the mode-locked pulses. First, the pulse width is clearly linear with the dispersion, as expected for a soliton [e.g., eq. (7.22b)].[4] Second, the time–bandwidth product remains very close to the ideal value of 0.315 expected for sech^2-shaped solitons. This is in clear contrast to the case of the Gaussian pulses that would be expected for active mode-locking without SPM. Such actively mode-locked Gaussians have a minimum time–bandwidth product of 0.44 (for $D_n = 0$), which increases with increasing dispersion. The experiments also provide evidence for our picture of dispersive waves acting as perturbations to the mode-locked soliton pulses. Figure 7.12b shows a measurement of the laser output when the laser is adjusted to operate at the transition to instability, recorded using a fast photodiode and sampling oscilloscope. The output consists of a short femtosecond pulse (too short to be resolved by the electrical detection system) accompanied by a weaker but much longer (ca. 100 ps) pulse representative of the instability. The background pulse is clearly broadened very substantially by the dispersion, as in our picture, even though in this case the dispersion is insufficient to achieve stabilization. Similar data showing the coexistence of a broad, low-intensity background with a short, mode-locked solitonlike pulse have also been reported in the case of insufficient negative dispersion (transition to instability) for solid-state lasers mode-locked via a slow saturable absorber [395] as well as for harmonically mode-locked fiber lasers [401].

Finally, note that according to eq. (7.31), pulse shortening can be achieved without limit if sufficient dispersion is available. However, the validity of the master equation of

[4] The pulse energy is assumed to remain constant, since neither the cavity loss nor the pump level are changed.

mode-locking upon which our treatment is based requires that changes to the pulse from each laser element must be small on a per-pass basis (in which case the discreteness of the individual laser elements and the order in which these elements are placed in the cavity can safely be ignored). From Section 7.1.3 we recall that solitons become unstable when subject to a periodic perturbation with period $8z_0$ or longer. This corresponds to a per-period nonlinear phase shift $\Phi_0 = 2\pi$ (which is no longer a small change). To avoid such instability and remain in the regime where the master equation is valid, the per-pass nonlinear phase shift should be limited (e.g., to $\Phi_0 \leq 0.1 \sim 1$) [394]. Under the condition that the nonlinear phase shift per round trip is limited to some value $\Phi_{\max}$, eq. (7.22b) implies that in addition to eq. (7.31), the pulse width is also subject to the inequality

$$T_0^2 \geq \frac{|D|}{\Phi_{\max}} \tag{7.33}$$

The value of $\Phi_{\max}$ depends on the strength of the loss modulation (active or passive) available to stabilize the pulse [402] as well as the degree of pulse-shape degradation (such as increased time–bandwidth product) that can be tolerated. Fiber lasers often permit larger values of $\Phi_{\max}$ compared to solid-state lasers, since in the former case SPM and dispersion may coexist in the same fiber, thereby presenting less discreteness.

7.3 STRETCHED PULSE MODE-LOCKING

7.3.1 Stretched Pulse Mode-Locked Fiber Laser

Our discussion in previous sections revealed important limitations governing minimum pulse durations and maximum peak powers and pulse energies generated by mode-locked soliton lasers. These limitations arose from the following:

- In lasers employing an artificial fast saturable absorber mechanism, the soliton plays a key role in the self-amplitude modulation process. The peak power of the soliton should not exceed the value required to reach the maximum of the nonlinear transmission curve.
- Soliton lasers experience instabilities unless the nonlinear phase shift per round trip remains small compared 2π. Equivalently, the soliton period z_0, which is proportional to the square of the pulse duration, should remain comparable to the cavity length or longer.

In both cases the relation between peak power, pulse duration, and pulse energy for fundamental solitons means that all three quantities are constrained. In this section we discuss a class of mode-locked lasers in which dispersion and nonlinear phase effects still play a key role, but which do not involve fundamental soliton propagation. In particular, we consider lasers in which large amounts of pulse stretching and compression occur on each pass through the laser. Compared to soliton lasers, such lasers are able to generate shorter pulses with higher energies.

Figure 7.13 shows the schematic layout of a laser that has been termed a *stretched pulse mode-locked fiber laser* [403–405]. Similar to Fig. 7.2, nonlinear polarization evolution results in artificial fast saturable absorber action. An isolator used to enforce unidirectional

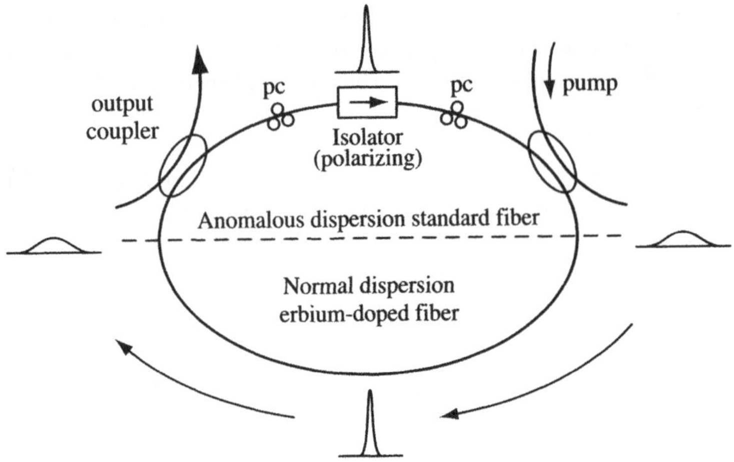

Figure 7.13 Layout for stretched-pulse fiber laser. pc, polarization controller. Adapted from [405]. Copyright ©1995, IEEE.

operation also acts as a polarizer, and polarization controllers before and after the isolator bias the ring for positive self-amplitude modulation (transmission through polarizer increases with increasing intensity). A new feature is that the laser is constructed out of separate positive (normal) and anomalous (negative) dispersion sections. Practically, the erbium-doped fiber section provides both the gain and positive dispersion (a typical number for the positive dispersion is about 0.07 ps^2 for the lasers of [403–405]), while the negative dispersion section may be constructed from standard fiber. The positive and negative dispersion sections are partially balanced such that the average cavity dispersion may be small relative to the dispersion in each section alone. Furthermore, unlike soliton lasers, the average cavity dispersion may be zero or positive.

Such lasers have been found to exhibit behavior quite different from that of soliton lasers. Figure 7.14 shows a comparison of data for two fiber ring lasers constructed according to Fig. 7.13. The only significant difference between the two lasers was the dispersion of the erbium-doped fiber, leading to estimated net cavity dispersions of −0.096 ps^2 and +0.004 ps^2, respectively. In the case of the negative dispersion, the power spectrum has a

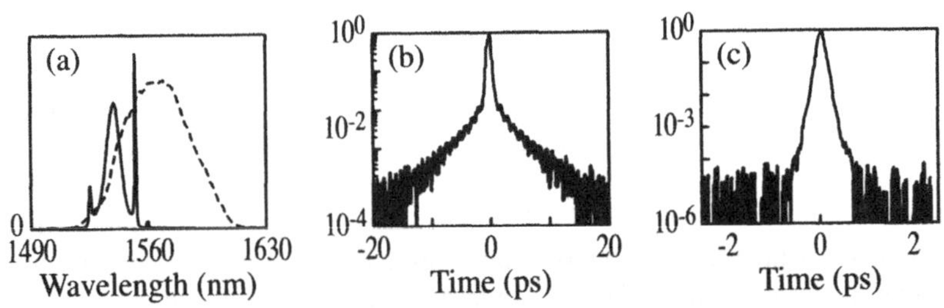

Figure 7.14 Performance comparison for fiber ring lasers in positive and negative dispersion regimes: (a) power spectra for negative dispersion (solid line) and positive dispersion (dashed line); (b) intensity autocorrelation for negative dispersion; (c) autocorrelation for positive dispersion. Adapted from [405]. Copyright ©1995, IEEE.

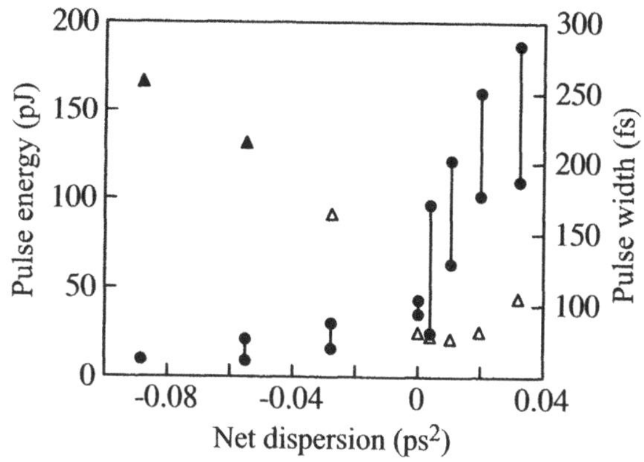

Figure 7.15 Pulse-width and pulse-energy data for fiber ring lasers as a function of net cavity dispersion. Joined circles, range of observed pulse energies; triangles, pulse widths estimated assuming either secant hyperbolic pulse shape (filled triangles) or Gaussian pulse shape (open triangles). From [404] with permission. Copyright ©1994, American Institute of Physics.

9-nm spectral width, and soliton sidebands indicative of soliton laser operation are clearly observed. The autocorrelation data imply a 450-fs pulse duration on top of a long exponentially decaying background representing the dispersive wave generated via periodic perturbations. The data for the positive dispersion case are strikingly different. The spectrum is 49 nm in width, substantially broader than the soliton laser case, while the autocorrelation implies a much shorter (90-fs) pulse duration. Both the spectrum and the autocorrelation are smooth and show no sign of spectral sidebands or a dispersive wave. Additionally, the output pulse energies are substantially different: 7 and 29 pJ for the negative dispersion (soliton laser) and positive dispersion cases, respectively. Pulses as short as 77 fs at energies of 90 pJ have been reported [403].

Figure 7.15 shows pulse width and pulse energy data for similar fiber lasers now plotted as a function of net cavity dispersion. Important observations concerning laser operation include the following:

- Pulse energies increase with decreasing negative dispersion, with a sharp increase when crossing into the positive dispersion regime. (The plot shows the range of pulse energies observed under stable mode-locked operation for different polarization controller settings, hence different self-amplitude modulation coefficients.)
- The minimum observed pulse duration decreases with decreasing negative dispersion, reaching its minimum value for small positive net dispersion. In the positive dispersion regime, the pulse duration is only weakly dependent on net dispersion.
- For positive net dispersion, pulses emerging from the laser are generally strongly chirped, with durations a factor of 10 or more above the Fourier transform limit (not shown in the figures). Short pulse durations are achieved only after compression via dispersive propagation in an appropriately selected length of external fiber. When the output coupler is placed as in Fig. 7.13, the external fiber lead must have negative dispersion in order for compression to occur. In contrast, for lasers with sufficient negative net dispersion, approximately transform-limited solitonlike pulses are obtained.

- Spectra are substantially broader under positive net dispersion and may exhibit significant power well outside the erbium gain region (roughly 1530 to 1565 nm). This provides evidence of strong spectral broadening via the phase nonlinearity.
- Soliton sideband generation is strongly suppressed for positive net dispersion. Furthermore, pulse shape and time–bandwidth product vary with net dispersion. For large negative dispersion, the pulse shape is close to secant hyperbolic, as customary for solitons, while for small negative dispersion, the pulse shape is closer to Gaussian. Increased time–bandwidth product is observed as the cavity is tuned into the positive dispersion regime.

7.3.2 Dispersion-Managed Solitons

The periodic stretching and compression that occurs within the stretched pulse laser bears a close resemblance to that which occurs with dispersion-managed solitons. The term *dispersion-managed soliton* refers to a type of stable nonlinear pulse propagation encountered in dispersion-managed optical fiber transmission links (i.e., in links constructed such that dispersion varies along the link). For example, as sketched in Fig. 7.16, the dispersion map may involve alternating but approximately balanced anomalous and normal dispersion sections. In this case, as in the stretched pulse laser, the overall path-averaged dispersion can be close to zero even though the local dispersion at any point in the system remains high. Compared to a system constructed from uniform low-dispersion fiber, but with the same average dispersion, unwanted nonlinearities due to four-wave mixing between different wavelength channels in wavelength-division-multiplexed (WDM) communications have been shown to be strongly reduced [406].

Strong dispersion management significantly modifies soliton propagation [407–409]. Figure 7.17 shows calculations that illustrate dispersion-managed soliton propagation. For these simulations the dispersion map is of the type depicted in Fig. 7.16, with 100-km fiber sections with dispersion alternating between −5.1 and +4.9 ps^2 km^{-1}. The nonlinear pulse propagation is plotted over one period of the dispersion map, using a log scale. A strong breathing of the pulse is clearly observed. The pulse reaches a minimum duration midway through each section. At these points it is unchirped and shows a Gaussian-like profile near its center, with wings that are exponential but modulated. At the boundaries between sections the pulses reach their maximum duration and are strongly chirped. The power spectrum (not shown) also evolves, with greater (lesser) bandwidth in the anomalous (normal) dispersion

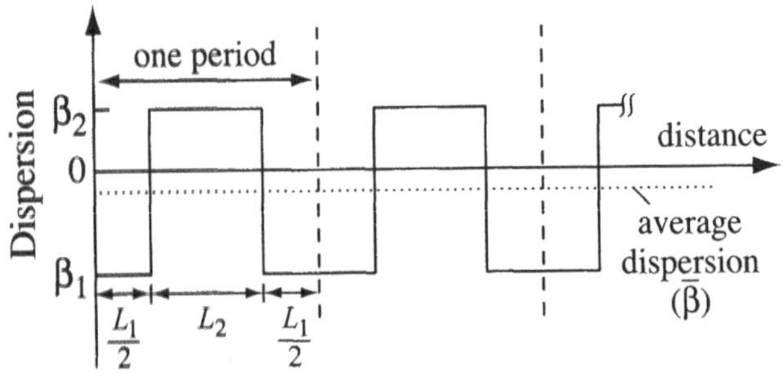

Figure 7.16 Representative periodic dispersion map. The average dispersion is significantly less than dispersion of individual fiber segments.

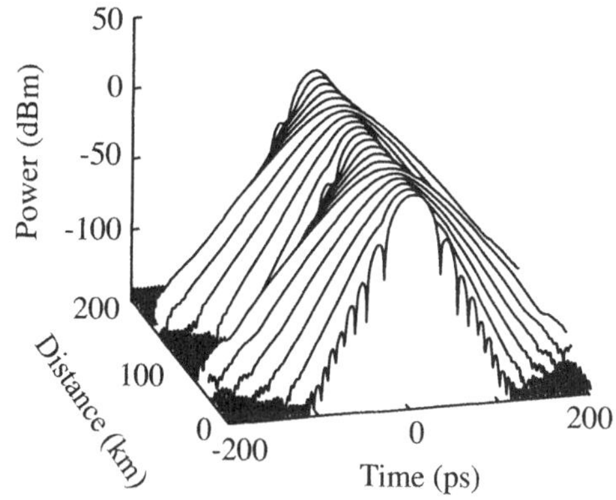

Figure 7.17 Simulated dispersion-managed soliton propagation through one period of a dispersion map consisting of alternating 100-km negative and positive dispersion segments. From [409]. Copyright ©1997, IEEE.

section. Simulations performed over hundreds of periods of the dispersion map demonstrate that this mode of propagation can be extremely stable.

Provided that appropriate input power is selected, such behavior appears to be rather general for the case of strong dispersion map; that is,

$$\frac{2|\beta_i - \bar{\beta}|L_i}{t_p^2} \geq 1$$

where β_i and L_i are the dispersion and length of the ith section and $\bar{\beta}$ is the path-average dispersion. A few key points that characterize dispersion-managed solitons are as follows:

- Within a single period of the dispersion map, propagation is dominated by strong linear dispersion. However, over longer distances nonlinearity allows us to compensate for the weak average dispersion and supports stable propagation.
- The pulse shapes are significantly different than the $\text{sech}(t/t_p)$ soliton pulses found for uniform anomalous dispersion fiber.
- The pulse energy required to achieve stable propagation is enhanced compared to a fundamental soliton of the same duration in a uniform fiber with the same average anomalous dispersion. This is of practical benefit for fiber communications, because higher pulse energies allow greater signal-to-noise ratio.
- Stable pulse propagation similar to that in Fig. 7.17 is not limited to the anomalous dispersion regime, but can also be obtained for zero or even small positive path-average dispersion [409]. It is this positive dispersion regime of dispersion-managed solitons that is most closely analogous to stretched-pulse mode-locking. The existence of solitons even for positive path-average dispersion is explained by invoking the variation in optical bandwidth within a single period of the dispersion map [410,411]. The bandwidth is increased in the anomalous dispersion section, which increases the effect of the anomalous dispersion, while the bandwidth and hence the effect of dispersion are decreased in the normal dispersion section. The net result of the bandwidth

variation is that the effective dispersion that acts on the dispersion-managed soliton is shifted toward the anomalous dispersion regime compared to the path average. When this effect is strong enough, self-phase modulation and group velocity dispersion can still self-compensate in an average sense, even for zero or positive net dispersion.

7.3.3 Theoretical Issues

Solid-state laser mode-locking theory, discussed in Section 2.4.5, revealed a strong asymmetry with respect to the sign of intracavity dispersion. Similar to the present discussion, solitonlike behavior was found for negative dispersion, while strongly chirped output pulses were predicted for positive dispersion. However, there are also strong contrasts. According to Figs. 2.19 and 2.20, greatest bandwidth (and hence shortest pulses, after compression of any chirp) is obtained near zero dispersion, while according to the data in Figs. 7.14 and 7.15, the greatest bandwidth and shortest pulses are clearly achieved for positive dispersion. The theory of section 2.4.5 also fails to account for the dramatic increase in pulse energy in the positive dispersion regime. These differences are explained by large pulse stretching and compression effects during each pass through the laser, as sketched in Fig. 7.13. Accordingly, this mode of laser operation has been termed *stretched-pulse mode-locking*. Such effects are not handled by the theory of Section 2.4.5, which explicitly requires that all the effects are small on a per-pass basis.

Early investigations of mode-locking under conditions of large changes in the pulse per pass were reported in [412]. Generally, when the per-pass effects are large, the order in which the elements are placed in the laser cavity becomes important, and it is difficult to arrive at an analytical theory. Reference [405] treats the case where large amounts of pulse stretching and compression occur *under linear dispersion* on each pass, but with the net cavity dispersion sufficiently well balanced that the *net* dispersive effects per pass are small. With the further approximation that nonlinearity and all other effects remain small per pass, an analytical formulation is obtained that successfully predicts the behavior observed in stretched-pulse mode-locking. The theory uses the model of a chirped Gaussian pulse with two minima in duration (occurring halfway through the positive and negative dispersion sections, respectively) per round trip. The self-phase modulation (and similarly, the self-amplitude modulation) nonlinearity is taken simply as an integral over a pulse that varies in width according to linear dispersive propagation, as follows:

$$\Delta\Phi(t) = 2\gamma|a_0|^2 \int_{-L/2}^{L/2} \frac{dz}{\sqrt{1+(z/b)^2}} \exp\left(\frac{-2t^2/t_p^2}{1+(z/b)^2}\right) \tag{7.34}$$

Here eq. (4.24) is used for the intensity of the dispersed pulse as a function of distance z, $b = t_p^2/2|\beta_2|$ denotes a characteristic dispersion distance for Gaussian pulses, and γ is the self-phase modulation coefficient from eq. (6.29b). The factor of 2 is included because of the two pulse minima per round trip, and L denotes the individual lengths of the positive and negative sections (which for simplicity are taken to be identical except for the sign of dispersion). To proceed further, the exponential is expanded up to order t^2. This is justified when intracavity pulse stretching and compression are large; in that case the nonlinear phase profile, which results from successive self-phase modulation of a pulse that varies in width, extends over a time interval long compared to the minimum pulse duration.

Introducing $u = z/b$, we have

$$\Delta\Phi(t) \approx 2\gamma|a_0|^2 b \int_{-L/2b}^{L/2b} du \left\{ \frac{1}{\left(1+u^2\right)^{1/2}} - \left(\frac{2t^2}{t_p^2}\right) \frac{1}{(1+u^2)^{3/2}} \right\} \tag{7.35}$$

Finally, using results from an integral table[5] and approximating in the limit $L/b \gg 1$ gives

$$\Delta\Phi(t) \approx \frac{4\gamma|a_0|^2 b}{\mu} \left\{ 1 - \frac{2\mu t^2}{t_p^2} \right\} \quad \text{where} \quad \mu = \frac{1}{\ln(L/b)} \tag{7.36}$$

Using this expression for the self-phase modulation term and a similar expression for self-amplitude modulation, a mode-locking equation can be written as follows:

$$\left\{ (g - \ell - j\psi) + \left(D_g + jD\right) \frac{\partial^2}{\partial t^2} + (\gamma_0 - j\delta_0)\,|a_0|^2 \left(1 - \frac{2\mu t^2}{t_p^2}\right) \right\} a(t) = 0 \tag{7.37}$$

where γ_0 and δ_0 denote the SAM and SPM coefficients, respectively. Compared to mode-locking equation (2.118a), the difference is that the $|a|^2$ in the SPM and SAM terms is replaced by $(1 - 2\mu t^2/t_p^2)|a_0|^2$. The t^2 dependence leads to Gaussian solutions, consistent with the assumed model. By comparing to a Gaussian intensity function expanded up to second order in time ($\sim 1 - 2t^2/t_p^2$), we can see that μ represents a factor by which the strengths of the SPM and SAM processes are reduced. This comes about because of strong pulse stretching within the laser cavity, which for fixed pulse energy reduces the average intensity available to drive the nonlinearities. Conversely, for a given *minimum* pulse width, higher pulse energy can be tolerated without overdriving the self-amplitude modulation beyond the maximum of the nonlinear transmission curve and without producing excessive nonlinear phase shift per pass. Assuming 77-fs Gaussian pulses and taking $\beta_2 L = 0.07$ ps^2 gives a nonlinearity suppression factor $\mu^{-1} = 3.5$. This factor is of the same order as the energy increases observed in stretched-pulse mode-locking.

It is worth noting that the simple theory outlined above does not consider cubic dispersion, which may be an important effect for the shortest pulse widths and net cavity dispersions near zero. It is also important to note that the μ factor in the mode-locking equation is itself an implicit function of the mode-locked pulse width. For long pulses there is little pulse stretching, which means that the full self-amplitude modulation is available to assist in initiation of mode-locking. However, for short pulses the stretching factor continues to increase with further decreases in pulse width, thereby more strongly limiting excess nonlinearity.

Summarizing, operation of stretched-pulse lasers under small net positive dispersion eliminates factors such as sideband generation that limit soliton lasers, while large pulse-stretching effects suppress limits in mode-locked performance due to excess nonlinearity.

[5] From [399] we have $\int \frac{du}{(1+u^2)^{1/2}} = \ln\left(u + \sqrt{1+u^2}\right)$ and $\int \frac{du}{(1+u^2)^{3/2}} = \frac{u}{\sqrt{1+u^2}}$.

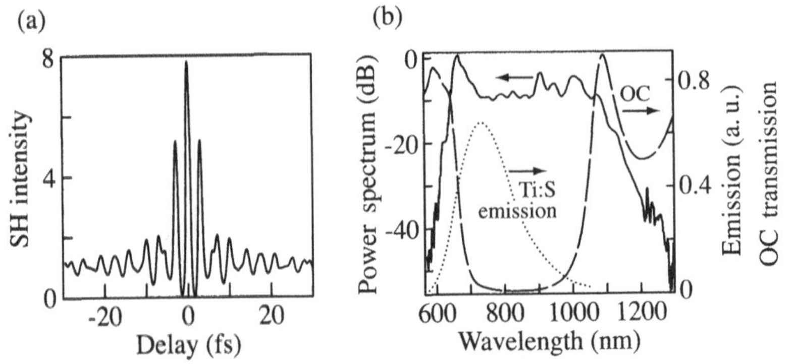

Figure 7.18 Representative measurements of few-cycle Ti:S lasers: (a) interferometric autocorrelation data indicating a pulse width of roughly 5 fs; (b) power spectrum (log plot) of laser optimized for a bandwidth greater than one octave. The transmission spectrum of the output coupler is also shown. The emission spectrum of Ti:S (from [415]) is also superimposed onto part (b). (a) Adapted from [413]; (b) adapted from [414].

7.4 MODE-LOCKED LASERS IN THE FEW-CYCLE REGIME

Mode-locked lasers based on broadband solid-state gain media have progressed to the point where pulses only a few optical cycles in duration can be generated directly. Especially spectacular results have been obtained with Ti:Sapphire (Ti:S) lasers. After the first demonstration in the early 1990s of Kerr lens mode-locking at the 100-fs time scale [52], over the next decade a number of groups contributed to the development of Ti:S lasers, ultimately producing pulses as short as approximately 5 fs. Figure 7.18a shows an interferometric autocorrelation measurement of a Ti:S laser generating pulses on this extremely short time scale, while Fig. 7.18b shows the optical power spectrum of a Ti:S laser optimized for extremely broad bandwidth. The laser spectrum contains discernible frequency content from 600 nm to beyond 1200 nm and is appreciably broader than the Ti:S gain spectrum (as given, e.g., in [415]). The greater than one octave (greater than a factor of 2) frequency span is important for applications discussed in Section 7.5.

Mode-locking of solid-state lasers based on artificial fast saturable absorbers, including Kerr lens mode-locking, was discussed in Section 2.4. Although many of these concepts remain valid, additional considerations become important for very short pulse durations and very broad bandwidths. A few of the important points are as follows:

- Careful control and minimization of dispersion over a very broad spectral range is imperative. One early series of papers demonstrated a clear correlation between reduced cubic spectral phase and shorter mode-locked pulses [56,57,416,417]. Initially, higher-order dispersion was decreased by using shorter Ti:S crystals (less overall dispersion to be compensated) in conjunction with dispersion-compensating prism sequences constructed of materials providing a low $|\psi_3/\psi_2|$ ratio. In at least one case, this resulted in a laser with sub-10-fs pulse duration that appeared to be limited by a fourth-order spectral phase [417]. Further progress was achieved through the introduction of chirped mirrors, especially double-chirped mirrors (Section 4.6.3), either replacing prisms entirely [58,59,414,418] or working in conjunction with prisms

[35,36,413,419]. For lasers operating over the broadest bandwidths, even the dispersion of the air in the laser cavity is important and should be taken into account in the dispersion compensation design. Furthermore, even the small ripple in the group delay vs. frequency characteristics of double-chirped mirrors poses a limitation. One solution is to utilize pairs of such mirrors, where the group delay ripple in one mirror is designed to be approximately equal and opposite that in the other (e.g. [420]).

- The high-reflectivity bandwidth of the laser cavity mirrors becomes a key issue. The bandwidth of conventional quarter-wave stack dielectric mirrors (Section 4.6.2) depends only on the refractive index ratio between high- and low-index layers. Bandwidths achieved using practical dielectric materials are insufficient to support extremely broad spectra such as that shown in Fig. 7.18b. Metal mirrors can provide large bandwidth but have relatively high loss. However, double-chirped mirrors that provide high reflectivity over an octave of bandwidth have been realized [420]. Such mirrors then play a dual role in the mode-locked laser cavity: They provide for both broad reflectivity bandwidth and dispersion control.
- The partially transmissive output coupler mirror can also play a special role. Available output couplers are based on quarter-wave stack designs and deliver lower-reflectivity bandwidth than state-of-the-art chirped mirrors. One example is shown in Fig. 7.18, where the transmission of the coupler increases from about 1% to 80% near the edges of the laser spectrum. Although this does introduce another filtering effect into the laser cavity, if the mode-locking is strong enough, it can also provide a benefit: The wings of the spectrum transmitted through the output coupler are enhanced and sometimes peaked. This can broaden the output spectrum compared to the intracavity field and aid in reducing the pulse duration in the full-width at half-maximum sense [35,36,414].
- Changes of the pulse as it propagates through different elements in the laser cavity become large, even on a single-pass basis. Such changes have been documented via spectral and pulse-duration measurements of light coupled out from different positions of a Ti:S laser cavity (e.g., [417]). Within one trip of the laser cavity, pulse stretching and compression due to linear dispersion are the dominant effects on the pulse propagation. In fact, in the steady state, the intracavity pulse propagation may closely resemble that of the dispersion-managed solitons of Section 7.3.2.

The latter point is illustrated by Fig. 7.19, which shows simulations of pulse propagation in a mode-locked laser cavity, with parameters representative of a Ti:S laser operating in the sub-10-fs regime. Here the simulations include linear dispersion, self-phase modulation, saturable gain with finite gain bandwidth, and KLM modeled as a fast saturable absorber. The dispersion map includes a positive dispersion section with nonlinearity (the Ti:S crystal) and a negative dispersion section with no nonlinearity (representing prisms or other dispersion-compensating elements). The simulations reveal significant pulse dynamics within a single round trip. The time-domain traces show pronounced pulse breathing, with minimum pulse durations precisely in the center of negative and positive cells of the dispersion map. The behavior is qualitatively similar to that shown in Fig. 7.17 for a pure dispersion-managed soliton (no gain, no filtering, no saturable absorber action, evenly distributed nonlinearity). This supports the picture that propagation is dominated by linear dispersion and phase nonlinearity, with other processes acting as perturbations that stabilize against growth of noise, smooth out pulse substructure, and so on. The spectral plots illustrate that the optical bandwidth also expands and contracts, with minima and maxima

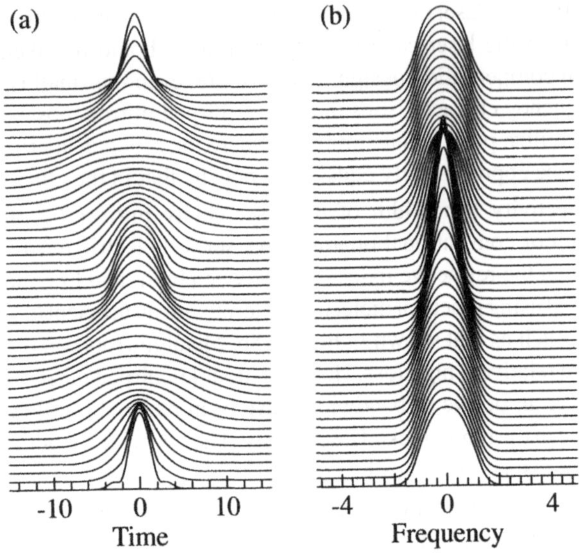

Figure 7.19 Simulation of pulse propagation in one round trip within a few-cycle mode-locked Ti:S laser, with a dispersion map like Fig. 7.16 with average dispersion $\bar{\beta} = 0$. The negative dispersion segment has no nonlinearity. (a) Time-domain profile (linear scale). (b) spectral profile (linear scale). Time and frequency are expressed in normalized units. From [421].

at the midpoints of the positive and negative dispersion cells, respectively. As noted in Section 7.3.2, it is the asymmetry in bandwidths between positive and negative dispersion sections that makes possible stable pulse propagation even into the regime of positive net dispersion.

Kerr lens mode-locked Ti:S lasers have been reported to generate pulses deep into the sub-10-fs regime for a number of different cavity variations. For purposes of discussion, one such cavity arrangement, corresponding to the laser generating the octave-spanning spectrum of Fig. 7.18, is sketched in Fig. 7.20. Dispersion compensation is achieved using double-chirped mirrors arranged in matched pairs yielding low group delay ripple. The

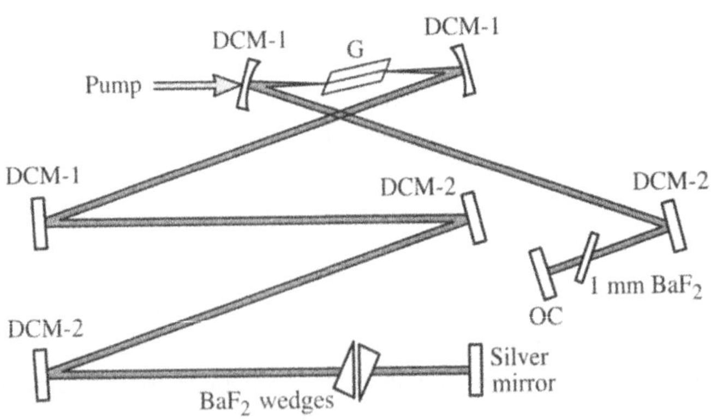

Figure 7.20 Cavity arrangement of an octave-spanning Ti:S laser. DCM-1 and DCM-2 are, respectively, two different mirror designs together comprising a double-chirped mirror pair optimized for low group delay ripple. Adapted from [414].

prismless design is advantageous because small misalignments of the resonator do not affect the intracavity dispersion; in comparison to prism-compensated lasers, this leads to improved long-term stability [58,59,414,418]. A thin plate of BaF_2 is placed on one side of the Ti:S crystal; a pair of wedges made of the same material are placed on the other side. Moving one of the wedges transversely in or out of the beam allows dispersion fine tuning and optimization of the generated spectrum without disturbing alignment. BaF_2 is used because it has the lowest known $|\psi_3/\psi_2|$ ratio over the operating wavelength band. Placing the BaF_2 dispersion elements on both sides of the laser crystal allows for a symmetric dispersion map, with equal negative dispersion on either side of the laser crystal. In this case one period of the dispersion map corresponds to a half round trip (rather than a full round trip) through the laser. In this configuration dispersion-managed soliton propagation yields an optimally compressed pulse (minimum duration, maximum intensity) at the center of the laser crystal for both directions of propagation. This is believed to increase the effective nonlinearity (compared to nonsymmetric dispersion maps that provide a compressed pulse in the laser crystal only once per round trip) and therefore to enhance mode-locking [35,36,413,421].

Finally, although short-pulse Ti:S lasers are seen to have many points in common with stretched-pulse mode-locked fiber lasers, it is worth remembering that there is also a fundamental difference: Pulse propagation is one-dimensional in fiber lasers but inherently three-dimensional in a Kerr lens mode-locked laser. This can give rise to space–time focusing effects in which the laser spatial profile within the cavity is coupled to the temporal as well as spectral degrees of freedom [422]. In this regard simulations suggest that the nonlinear laser dynamics favor an operating condition under which the point of spatial focus (minimum beam size) within the laser crystal coincides with the point of temporal focus (minimum pulse duration) [423]. Simulations also suggest that rapid pulse spreading after the temporal focus point due to dispersion prevents overdriving the nonlinearity (i.e., suppresses catastrophic self-focusing), even though the peak intracavity power can exceed the critical power for self-focusing [424].

7.5 MODE-LOCKED FREQUENCY COMBS

As we have seen, mode-locked laser technology now makes possible generation of optical pulses in the few-cycle regime. In this regime the time scale of the pulse envelope approaches that of the optical carrier, and it becomes worthwhile to reexamine the role of the optical phase. This leads to new insight into the relationship between time- and frequency-domain pictures of mode-locked pulse trains and to the use of femtosecond frequency combs from mode-locked lasers as a leading edge tool for high-precision frequency metrology (an accomplishment for which the Nobel Prize in Physics was awarded in 2005 [425]). Key concepts related to the development of mode-locked frequency combs and their applications are developed in [426–433], as well as a wealth of other papers.

7.5.1 Comb Basics

Let us begin by considering mode-locked pulses as the superposition of optical frequencies, as in Section 1.4:

$$e(z,t) = \mathrm{Re}\left\{\sum_m E_m e^{j(\omega_m t - \beta_m z + \phi_m)}\right\} \tag{7.38}$$

where z represents position within the cavity (for convenience a ring laser geometry is assumed, with the physical distance around the ring denoted L). However, unlike the earlier treatment which assumed propagation in vacuum, we write the propagation constant $\beta(\omega)$ as[6]

$$\beta(\omega) = \beta_0 + \beta_1\tilde{\omega} \tag{7.39}$$

Here the notation is that of Section 4.1: $\beta_0 = \omega_0/v_p$ and v_p are, respectively, the propagation constant and the phase velocity evaluated at a reference frequency ω_0, $\tilde{\omega}$ is the frequency offset from ω_0, and β_1 is equal to the inverse of the group velocity v_g. Combining these expressions, we can rewrite the mode-locked pulses as

$$e(z,t) = \mathrm{Re}\left\{ e^{j(\omega_0 t - \beta_0 z)} \sum_m E_m e^{j[\tilde{\omega}_m(t-\beta_1 z)+\phi_m]} \right\} \tag{7.40}$$

On the other hand, using eq. (7.39) together with the cavity resonance condition $\beta_m L = 2m\pi$, we find that

$$\omega_{m+p} - \omega_m = \frac{2\pi p v_g}{L} = p\Delta\omega \qquad \text{where} \qquad \Delta\omega = \frac{2\pi v_g}{L} \tag{7.41}$$

Thus the frequency spacing is inversely proportional to the *group* delay $T = L/v_g$. Let us now take the reference frequency ω_0 to represent one of the cavity resonances, with mode index m', and let the integer $p = m - m'$ denote the difference in mode indices between frequencies ω_m and ω_0. We also specialize to a single position within the laser cavity: for simplicity, $z = 0$. With these notations the mode-locked pulse train becomes

$$e(z=0,t) = \mathrm{Re}\left\{ e^{j\omega_0 t}\; e^{j\phi_{m'}} \sum_p E_{p+m'} e^{j(p\Delta\omega t + \tilde{\phi}_p)} \right\} \tag{7.42}$$

where the common phase term $\phi_{m'}$ is factored out and $\tilde{\phi}_p = \phi_{p+m'} - \phi_{m'}$. Here we can introduce the envelope function

$$a(t) = \sum_p E_{p+m'} e^{j(p\Delta\omega t + \tilde{\phi}_p)} \tag{7.43}$$

which repeats periodically with the group delay T. For $\tilde{\phi}_p$ equal to a constant, $a(t)$ represents a train of bandwidth-limited pulses. However, our discussion below is valid whether or not the pulses are bandwidth limited.

[6] Our reason for adopting a first-order expansion for β is as follows: (1) An evenly spaced comb of frequencies, which is necessary to obtain a periodic pulse train solution, is retained in this approximation; (2) Intuitively, in steady-state mode-locking, dispersive effects and nonlinear effects cancel each other out.

We now examine the carrier frequency ω_0. From the cavity resonance condition, we have

$$\frac{\omega_0 L}{v_p} = 2\pi m' \tag{7.44}$$

However, we can also write the carrier frequency in terms of the frequency spacing $\Delta\omega$:

$$\omega_0 = m''\Delta\omega + \delta\omega \tag{7.45}$$

where $\delta\omega$ is known as the comb offset frequency and m'' is an integer chosen such that $|\delta\omega| \leq \Delta\omega/2$. Combining these equations yields an expression for $\delta\omega$:

$$\delta\omega = \Delta\omega\left(\frac{v_p}{v_g}m' - m''\right) \tag{7.46}$$

When the phase and group velocities are equal, as in vacuum, the offset frequency is equal to zero, $\delta\omega = 0$. In this case when the comb of frequencies defined by $n\,\Delta\omega + \delta\omega$, with n an integer, is extended to low frequency, one of the comb frequencies falls precisely at zero. On the other hand, in real media the phase and group velocities generally differ, and the frequency comb is offset by $\delta\omega$ from an imaginary frequency grid with identical frequency spacing ($\Delta\omega$) but passing through zero. Such a frequency comb with offset is illustrated in Fig. 7.21a [429,432]. In general, the frequencies of the individual comb lines may be written (in rad/s) as

$$\omega_m = m\Delta\omega + \delta\omega \tag{7.47}$$

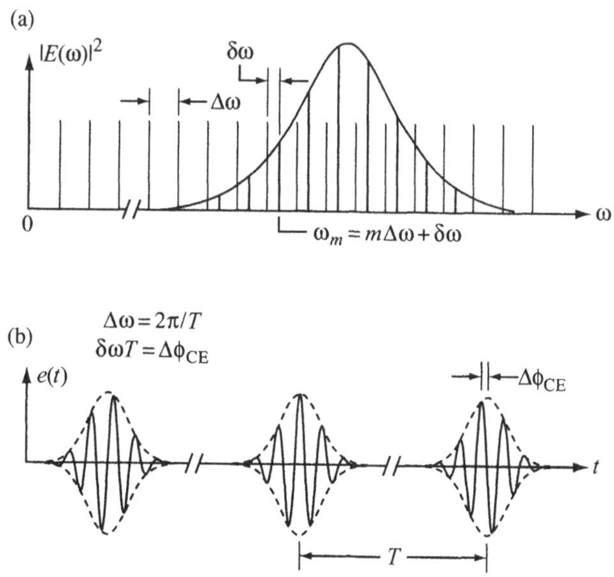

Figure 7.21 (a) Mode-locked frequency comb, illustrating the offset frequency $\delta\omega$; (b) mode-locked pulse train, illustrating pulse-to-pulse phase shift $\Delta\phi_{CE}$. Adapted from [432].

or alternatively, in hertz as

$$f_m = mf_{\text{rep}} + f_{\text{CEO}} \tag{7.48}$$

where $f_{\text{rep}} = 1/T$ is the pulse repetition frequency and $f_{\text{CEO}} = \delta\omega/2\pi$ is the carrier-offset frequency.

When $\delta\omega \neq 0$, the optical carrier (and hence the overall electric field) is *not* periodic in T. To see this, consider the field at time $t = qT + \tilde{t}$ [i.e., after an integer number (q) of round trips in the laser cavity]. Using eq. (7.45) for ω_0, we have after some simplification

$$e(z = 0, t = qT + \tilde{t}) = \text{Re}\left\{e^{j(\omega_0\tilde{t}+\phi_{m'}+q\delta\omega T)}\ a(\tilde{t})\right\} \tag{7.49}$$

The interpretation is that the field is equal to a periodic part multiplied by a term representing an additional phase shift accumulated on each pass through the laser. This is illustrated in Fig. 7.21b [429,432]. The additional phase shift per pass ($q = 1$) is denoted

$$\Delta\phi_{\text{CE}} = \delta\omega T \tag{7.50}$$

We conclude that the carrier-envelope phase ϕ_{CE} evolves in a way that is directly connected to the offset frequency. Such a pulse-to-pulse phase shift was introduced in our earlier treatment of mode-locking (e.g., Section 2.4.5); however, at that time we did not pursue its consequences.

We can gain insight by examining the expression for $\Delta\phi_{\text{CE}}$ in more detail. Using first eq. (7.46) and then eq. (7.44), we have

$$\Delta\phi_{\text{CE}} = \frac{\omega_0 L}{v_g} - 2\pi m'' = \frac{\omega_0 L}{v_g} - \left(\frac{\omega_0 L}{v_p} - 2\pi m'\right) - 2\pi m'' \equiv \omega_0\left(\frac{L}{v_g} - \frac{L}{v_p}\right) \tag{7.51}$$

In the last expression we have dropped the phase terms equal to integer multiples of 2π, which are not physically meaningful. As a result, we find that the per-pass change in the carrier-envelope phase is determined by the difference between the per-pass group and phase delays. Interestingly, we have already seen an essentially identical effect in our treatment of the nonlinear envelope equation in Section 6.7.1. In particular, according to eq. (6.100), when the electric field is written in a retarded time frame traveling with the envelope (i.e., traveling at the group velocity), a phase term identical to that in eq. (7.51) must be included to account for the different propagation velocities of the carrier and the envelope.

7.5.2 Measurement Techniques

Let us now turn our attention to measurement of the effects discussed above. The first measurement of $\Delta\phi_{\text{CE}}$ was performed using an interferometric cross-correlation method using pulses from a sub-10-fs laser oscillator [426]. The setup is similar to that for interferometric intensity autocorrelation (Section 3.4.1, Fig. 3.8), but with one arm of the interferometer incorporating an extra delay equal to one cavity round-trip time (T). This means that the apparatus looks at the intensity correlation between successive pulses emitted from the laser. The measurement principle (see Fig. 7.22) is that a difference in the phase and group delays through the cavity will show up as a shift of the interferometric fringes by

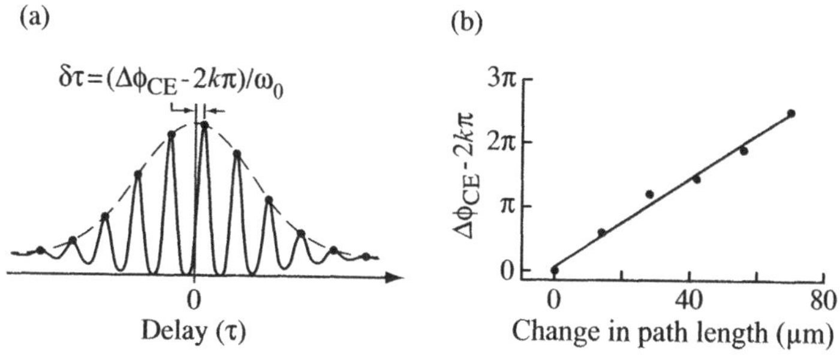

Figure 7.22 Measurement of pulse-to-pulse phase evolution ($\Delta\phi_{\mathrm{CE}}$) via interferometric intensity cross-correlation. (a) Schematic representation of interferometric cross-correlation between pulse n and pulse $n+1$. Mathematically, the quantity plotted is of the form $G_2(\tau) \sim \int dt\ |e_n(t)e_{n+1}(t-\tau)|^4$. (b) Data on pulse-to-pulse phase change as path length $\delta\ell_g$ though an intracavity glass wedge is varied. Adapted from [426]; modified for consistency with the sign conventions of this book.

an amount $\delta\tau$ with respect to the peak of the intensity correlation envelope $G_2(\tau)$. This shift provides a direct measure of $\Delta\phi_{\mathrm{CE}}$ to within an integer multiple of 2π through the relation

$$\omega_0\delta\tau = \Delta\phi_{\mathrm{CE}} - 2k\pi \tag{7.52}$$

where it is assumed that the extra delay T is added into the fixed arm of the interferometer. In contrast, in an interferometric intensity *autocorrelation* measurement, the interferometric fringes are always exactly centered at $\tau = 0$. Note that from a practical perspective, these effects are best observed for very short (few-cycle) input pulses. As the pulse duration increases, it becomes more difficult to determine the center (peak) of the correlation envelope precisely, and ultimately the measurement fails.

An example of data on the pulse-to-pulse phase shift, measured according to this cross-correlation technique, is shown in Fig. 7.22b. A thin fused-silica glass wedge was placed inside the laser such that the path length ℓ_g through the wedge could be tuned by translating it into the beam. Using eq. (7.51) together with (4.14), we obtain $\Delta\phi_{\mathrm{CE}} = -2\pi L\, dn/d\lambda$. The change in $\Delta\phi_{\mathrm{CE}}$ with path length through the wedge becomes $\Delta\phi_{\mathrm{CE}} = -2\pi(dn/d\lambda)\delta\ell_g$. Numerically, the value of $dn/d\lambda$ at 800 nm is approximately $-0.017\ \mu\mathrm{m}^{-1}$ [426], which means that $\Delta\phi_{\mathrm{CE}}$ should vary over a 2π range with only a 60-μm change in ℓ_g. The data of Fig. 7.22b confirm quantitatively the dependence predicted.

Experiments also confirm the relationship between pulse-to-pulse phase shift ($\Delta\phi_{\mathrm{CE}}$) and comb offset frequency ($\delta\omega$), as shown in Fig. 7.23. Here $\Delta\phi_{\mathrm{CE}}$ was measured as discussed above, but with the cross-correlator set for a delay equal to twice the cavity group delay [$q = 2$ in eq. (7.49)]. The offset frequency was measured using the nonlinear interferometer method described below. The data confirm the linear relationship, including the expected proportionality constant.

In principle, then, this cross-correlation method can be used to determine $\delta\omega$ through eq. (7.50). However, the limited delays that can practically be implemented (small q) limit the precision with which frequency can be measured. Furthermore, since cross-correlation is typically a relatively slow measurement, it cannot detect rapid variations in the offset frequency. Therefore, to measure the comb offset frequency directly, one would like to compare at least one of the comb frequencies against some reference. Although there are many possible approaches, a self-referencing scheme based on the f-to-$2f$ or

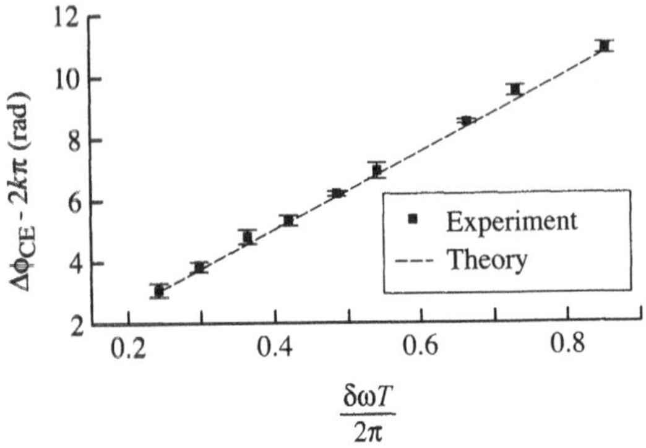

Figure 7.23 Measurement of pulse-to-pulse phase change $\Delta\phi_{CE} - 2k\pi$ as a function of comb offset frequency normalized to pulse repetition rate ($\delta\omega T/2\pi$). The offset frequency was measured via the f-to-$2f$ interferometer technique. The slope is close to the value of 2 expected for an interferometer delay set to $2T$. Adapted from [432].

nonlinear interferometer has proved to be particularly advantageous. As shown schematically in Fig. 7.24, this approach involves second-harmonic generation (SHG) of an input optical signal whose spectrum is required to span an octave (i.e., a factor of 2 in frequency). The octave-spanning spectrum can be obtained either directly from an appropriately optimized laser (Section 7.4) or through extreme nonlinear spectral broadening: for instance, in an appropriately optimized fiber (Section 6.8). The key idea is that for an octave-spanning spectrum, the low-frequency portion of the frequency-doubled spectrum will overlap with the high-frequency portion of the fundamental (undoubled) spectrum. The overlapping spectral region is selected by an optical filter and then directed onto a photodetector. This results in a heterodyne beat signal between the fundamental and frequency-doubled comb lines

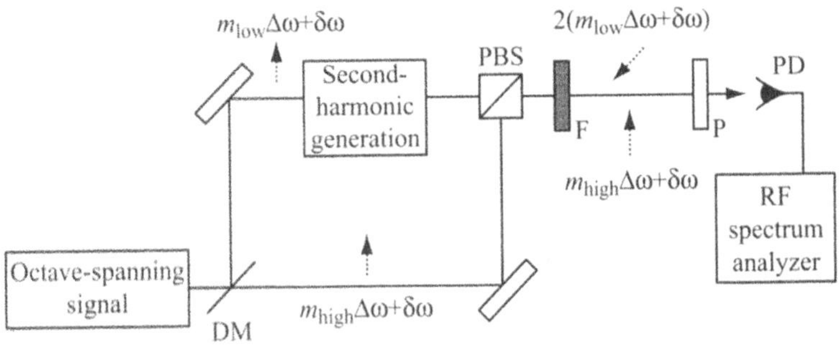

Figure 7.24 f-to-$2f$ or nonlinear interferometer for measuring the comb offset frequency. A dichroic mirror (DM) separates the high- and low-frequency portions of the input octave-spanning spectrum. After second-harmonic generation in the low-frequency arm, the fields are recombined in a polarization beamsplitter (PBS) and passed through an optical filter (F). The arrows depict the different polarization orientations of fundamental and second-harmonic fields. A polarizer (P) converts changes in relative phase between fundamental and second-harmonic fields into intensity changes observable by a photodetector (PD).

which can be measured via an RF spectrum analyzer.[7] The total field at the photodetector includes both the fundamental field with frequency components of the form

$$e^{j(m_{\text{high}}\Delta\omega+\delta\omega)t}$$

and the second-harmonic field with frequency components of the form

$$e^{2j(m_{\text{low}}\Delta\omega+\delta\omega)t}$$

where m_{high} and m_{low} are both large positive integers characteristic of higher- and lower-frequency regions of the comb, respectively, and m_{high} is of order $2m_{\text{low}}$. Because the photodetector responds to time-average intensity, its output includes difference frequency terms of the form

$$e^{\pm j[(2m_{\text{low}}-m_{\text{high}})\Delta\omega+\delta\omega]t}$$

$2m_{\text{low}} - m_{\text{high}}$ is an integer that can be positive, negative, or zero. In terms of positive frequencies, the detector output signal contains frequencies of the form $\tilde{m}\,\Delta\omega \pm \delta\omega$, where $\tilde{m}$ is a nonnegative integer. When $\tilde{m}$ is sufficiently small, the photodetector output is conveniently analyzed using standard RF techniques.

Figure 7.25 shows a typical experimental RF spectrum of the photodetector output. The strong lines at approximately 450 and 540 MHz are harmonics of the pulse repetition rate (ca. 90 MHz). The weaker lines correspond to the heterodyne beats due to the interference between fundamental and second-harmonic fields (f-to-$2f$ interference). According to our discussion, the difference between the f-to-$2f$ beat frequencies and the repetition rate frequencies yields the comb offset frequency. For example, in the data of Fig. 7.25, the offset frequency is approximately either 38 or 52 MHz. The ambiguity arises because the f-to-$2f$ interference (and heterodyne beating in general) depends only on the *absolute value* of the difference between the fundamental and second-harmonic comb frequencies.

Various variations of the f-to-$2f$ self-referencing technique are possible [427,434]. For example, one can form a beat signal between the third harmonic and the second harmonic of the low- and high-frequency regions of the comb, respectively. In this case, the photodiode output includes terms of the form

$$e^{\pm j[(3m_{\text{low}}-2m_{\text{high}})\Delta\omega+\delta\omega]t}$$

When m_{high} is of order $1.5m_{\text{low}}$, the beat frequencies are again easily analyzed via RF techniques to yield the offset frequency $\delta\omega$. This approach can be applied for optical signals with lower fractional bandwidth, at the cost of increased complexity (more nonlinear optical frequency conversion steps).

[7] Several practical details are worth mentioning. First, in addition to the optical filter, phase-matching considerations also typically limit the bandwidth of the second-harmonic field. Second, for SHG in the common type I phase-matching configuration (see Chapter 5), the second-harmonic (SH) field is polarized perpendicular to the input field. Therefore, a polarizing beam-splitter may be used advantageously to combine fundamental and SH fields. Changes in relative phase between fundamental and SH fields lead to polarization changes in the combined field, which are converted to intensity changes by a polarizer placed before the photodetector.

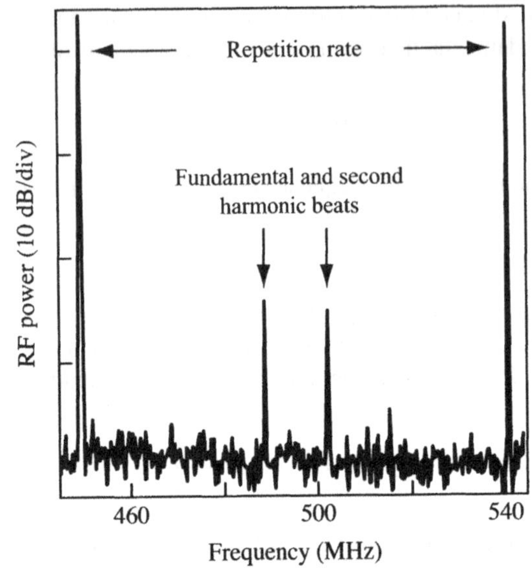

Figure 7.25 Measured RF spectrum of photocurrent detected at output of f-to-$2f$ interferometer. The strong lines are harmonics of the pulse repetition rate, while the two weak lines are heterodyne beats due to the f-to-$2f$ interference. From [432].

7.5.3 Stabilization of Frequency Combs

Measurement of the absolute comb frequencies raises the prospect of their stabilization. For a Ti:S laser, the difference between the group and phase delay is about 140 fs (280 fs) for a single (double) pass through a 2-mm-gain crystal [427]. Even for a ring (single pass) cavity, the pulse-to-pulse phase shift is $\Delta\phi_{CE} \approx 100\pi$, corresponding to an offset frequency about 50 times the pulse repetition frequency. These effects are sufficiently large that significant variations are expected in response to environmental fluctuations. According to eq. (7.47), the entire frequency comb can be stabilized if just two quantities are fixed: the offset frequency and the repetition rate. A schematic view of a feedback stabilization scheme is shown in Fig. 7.26. A portion of the mode-locked pulse train is directed onto a photodetector, producing an output signal that contains frequency components at the laser repetition frequency (and harmonics). This signal is compared with an RF reference oscillator (e.g., via a mixer). The difference frequency output of the mixer is connected through appropriate electronics back to the laser, to lock the laser repetition rate to the reference. Another portion of the laser output is directed to a nonlinear interferometer to measure the comb offset frequency. In cases where the laser spectrum is insufficiently broad to operate the nonlinear interferometer, a supercontinuum stage may be included to obtain the needed optical bandwidth. The output of the nonlinear interferometer is mixed with a second RF reference signal and fed back to the laser in order to lock the offset frequency.[8]

Several actuation mechanisms are available to convert the applied feedback control signals into changes in laser frequency. Simplest is control of the repetition rate, which can be achieved using a piezoelectric element to tune the laser cavity length (e.g., by translating

[8] When available, the use of an octave-spanning laser is preferred. This avoids the introduction of noise processes (e.g., amplitude noise-to-phase noise conversion) that may arise during supercontinuum generation.

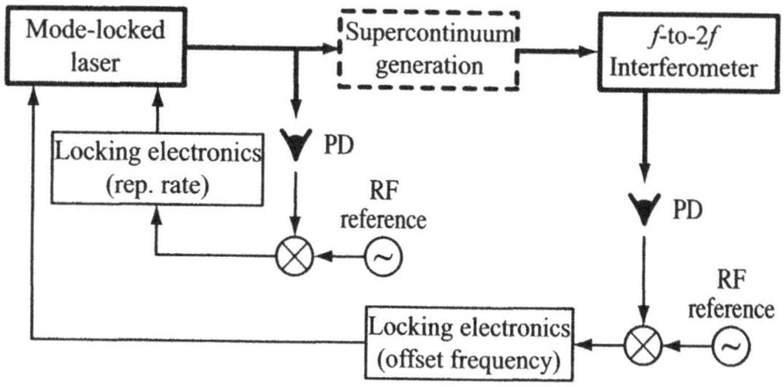

Figure 7.26 Mode-locked laser frequency stabilization scheme. The laser repetition rate is measured using a photodetector (PD) and locked to an RF reference, while the comb offset frequency is measured using a nonlinear interferometer and locked to a second RF reference. The supercontinuum generation setup is included to obtain an octave-spanning spectrum and is omitted when sufficient optical bandwidth is available directly from the mode-locked laser.

the position of an end mirror). This is similar to what is done in actively mode-locked lasers. For full control of the optical comb, an additional actuation mechanism providing different changes to the group and phase delays is needed. Possibilities include the following:

- *Modulation of the pump power to the laser crystal.* The resulting modulation in laser power leads to changes in both group and phase delays due to nonlinear index effects. Experimentally, the changes in group and phase delays are found to differ, as modulation of the pump power leads to clear changes in $\Delta\phi_{CE}$ [426]. Theoretically, this effect is explained by an analysis of perturbations to soliton propagation due to the self-steepening term in the nonlinear propagation equation, which predicts a nonlinear group delay equal to twice the nonlinear phase delay [343].
- *Modification of the intracavity material path length.* For example, the BaF_2 wedge in the laser of Fig. 7.20 can be translated to increase or decrease the total amount of material, which changes $\Delta\phi_{CE}$ though eq. (7.51). However, this changes intracavity group velocity dispersion simultaneously, which is undesirable. A solution is to fabricate a composite wedge composed of two different materials. By exploiting the different dispersive characteristics of the two materials, a composite structure can be designed whereby $\Delta\phi_{CE}$ may be modified with little change to group velocity dispersion [435].

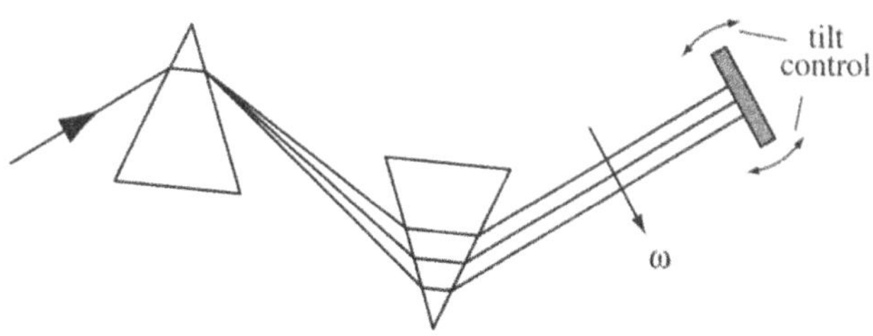

Figure 7.27 Mirror and prism pair apparatus for tuning group delay. Angle-tuning a reflecting mirror gives rise to a linear spatial phase, which becomes a linear spectral phase when coupled with spectral dispersion. This results in tunable group delay, decoupled from phase delay.

- *Manipulation of prism sequences* (restricted to lasers using intracavity prisms for dispersion control). Referring to Fig. 7.27, we note that different optical frequencies are shifted spatially (transverse to the propagation direction) after a single pass through an intracavity prism pair. Adjusting the angle of the turnaround mirror imparts a spatially varying phase shift onto the laser beam. Because frequency is also spatially varying, this results in imposition of a linear spectral phase $\psi(\omega) \sim \omega$. Since group delay is given by $-\partial\psi/\partial\omega$, angle-tuning the mirror provides adjustable group delay. On the other hand, for a frequency incident at the pivot point of the mirror, there is no change in phase delay, since the mirror–prism separation remains fixed at that point. This decouples the control of the phase and group delays, as desired. Tuning of group delay in this manner may be considered a simple example of Fourier transform optical pulse shaping, discussed in detail in Chapter 8. If desired, phase delay may still be adjusted in the conventional manner by translating the mirror parallel to the optical axis.

7.5.4 Applications

Optical Frequency Metrology The development of self-referenced mode-locked laser frequency combs has had an enormous impact on high-precision optical frequency metrology. Reviews of the basic principles may be found, for example, in [432,434]. A key point is that the second is currently defined in terms of the cesium-133 atomic transition at 9.192631770 GHz (i.e., a microwave frequency). Measurement of the absolute frequencies of optical transitions therefore requires a link to microwave frequencies. Previously, such links required highly complex and expensive harmonic frequency chains, which were unsuitable for widespread deployment. On the other hand, the frequency comb from a mode-locked laser can provide a stable set of frequency markers across the visible in a relatively simple apparatus. According to eq. (7.47), the positions of all the markers in a frequency comb are determined in terms of just two microwave frequencies: the repetition rate and the offset frequency. Therefore, a single mode-locked laser employing the comb stabilization procedure outlined in the preceding provides a direct link between microwave and optical frequencies needed for absolute frequency measurements.

Figure 7.28 depicts a schematic of the apparatus for high-precision frequency measurement using a mode-locked comb. The mode-locked laser is frequency stabilized as in Fig. 7.26, with the caveat that a high-stability microwave clock is used as the time base for the RF references to which the laser is locked. The output of a continuous-wave (CW)

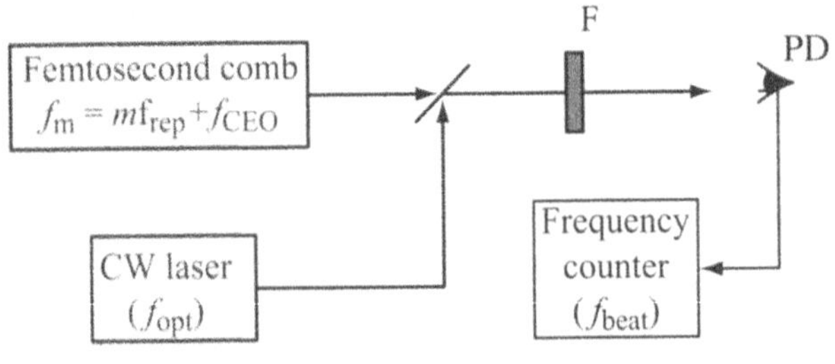

Figure 7.28 Setup for optical frequency metrology using a femtosecond comb. The stabilized comb and a narrow-linewidth CW laser locked to a transition of interest are combined, pass through an optical filter (F), and are detected by a photodetector (PD). The resulting beat note is directed to a frequency counter, which allows determination of the unknown input frequency f_{opt} with high precision.

laser that has been frequency-locked to the optical transition of interest (frequency f_{opt}) is combined with the mode-locked frequency comb. At this point an optical filter may be used to select a relatively narrow region of the comb centered approximately on the CW laser frequency. Optical-to-electrical conversion via a photodetector generates a series of RF beats corresponding to difference frequencies between the CW laser and the nearby comb lines. Typically, an electronic tracking oscillator (not shown) is locked to one of the beat notes following the photodetector, effectively filtering the RF beats to obtain a high-quality timing signal with improved signal-to-noise ratio and corresponding to a single beat note [436]. Finally, an RF frequency counter records the beat frequency f_{beat}. The unknown optical frequency is then given by

$$f_{opt} = mf_{rep} + f_{CEO} \pm f_{beat} \tag{7.53}$$

The $\pm$ sign arises because the beat frequency depends only on the *absolute value* of the difference between the CW laser frequency and the comb lines. (As already mentioned, a similar sign uncertainty arises in determination of the offset frequency.) However, f_{opt} can usually be determined by other means to within much better than half the repetition rate. This resolves the sign uncertainties, determines the correct value of integer m, and yields the desired absolute frequency measurement [429].

It is interesting that the concept of using the frequency comb from a mode-locked laser for high-precision spectroscopy was envisioned as early as the late 1970s [437]. However, this vision became a reality only more than two decades later, after development of reliable mode-locked solid-state lasers, means for obtaining octave spanning spectra, and the f-to-$2f$ self-referencing technique.

Optical Clocks Related to optical frequency metrology, one can also use mode-locked frequency combs for applications in low-noise frequency synthesis and optical clocks. Whereas the goal in metrology is to measure absolute optical frequencies tied to a frequency standard in the microwave region, the idea in the synthesis and clock applications is the converse (e.g., to generate a high-purity microwave tone starting from a frequency standard in the optical region).

Reviews of modern (atomic) clock technologies, including prospects for optical clocks, may be found in [434,438]. Briefly, any clock consists of two main components: an oscillator producing periodic events or clock ticks, and a counter accumulating and displaying each tick. In atomic clocks there is the added feature that the oscillator is locked to a quantum mechanical transition, which provides better long-term frequency stability than any physical artifact (e.g., quartz crystal for a crystal oscillator, optical cavity for a laser oscillator). The fractional frequency instability (smaller is better) of an atomic clock is fundamentally given by

$$\sigma \sim \frac{\Delta\nu}{\nu_0} \frac{1}{S/N} \tag{7.54}$$

where ν_0 and $\Delta\nu$ are the frequency and the linewidth of the clock transition, respectively, and S/N is the signal-to-noise ratio. This equation leads to two important observations. First, there is a clear advantage to higher transition frequencies (faster clock ticks) and narrower linewidths. The transition linewidth in so-called Cs fountain clocks is about 1 Hz, corresponding to a fractional linewidth of roughly 10^{-10}. Because optical frequencies are up to 10^5 higher than the frequency of a Cs clock, an improvement in fractional linewidth by a similar factor of 10^5 should be possible if transitions with comparable linewidths can be accessed. A second observation is that frequency instability is reduced for increased signal-to-noise ratio. Since S/N usually improves as the square root of averaging time (this applies when white or frequency-independent noise dominates), it is useful to compare the fractional frequency instability of different clocks at constant averaging time. At a 1-s averaging time, cesium fountain clocks provide $\sigma \sim 10^{-13}$, whereas optical frequency standards with fractional instability $\sigma \sim 10^{-15}$ or better at 1 s have been projected [434,439]. If attained, such optical clocks would yield a four-order-of-magnitude decrease in averaging time required to reach a specified frequency instability compared to Cs clock technology.

Despite their significant promise, one of the difficult challenges in realizing optical clocks is that the transition frequencies are too high for direct counting, unlike the microwave case. Self-referenced femtosecond combs now offer a workable approach for reading out the ticks of an optical frequency clock. The setup is similar to Fig. 7.28, but with a few important modifications:

- As in the metrology case, a CW laser locked to a transition of interest (frequency f_{opt}) beats with a nearby line of the femtosecond comb. However, unlike the metrology case for which f_{opt} is to be determined, here f_{opt} is a frequency standard representing the ticking of the clock.
- Also different from the metrology case, the photodetector beat signal is mixed with an RF reference and fed back as one of the control signals to stabilize the comb, thus locking the beat note to the RF reference. (The self-referencing scheme is used to provide the second feedback stabilization signal to the laser, as previously.)
- The output from a second photodetector connected directly after the stabilized mode-locked laser provides the clock signal, which lies in the microwave region (frequency f_{rep} or harmonic).

The output frequency may be expressed as

$$f_{\text{rep}} = \frac{f_{\text{beat}} + f_{\text{opt}} - f_{\text{CEO}}}{m} \tag{7.55}$$

where we assume here that f_{opt} is below the nearest comb frequency. An important point is that the integer m is typically of order 10^6, which means that the frequency fluctuations in both the optical standard as well as the RF references (f_{beat} and f_{CEO}) are strongly suppressed. The phase noise power spectral density (see Section 3.9) is suppressed by m^2, leading to prospects for extremely low-jitter microwave timing signals.

Absolute Phase and Extreme Nonlinear Optics The phase structure of ultrashort pulses is also important for time-domain applications (i.e., applications in which the temporal profile of an individual ultrashort pulse is of prime importance, as opposed to its long-term coherence over many pulse repetition periods). An example is the *extreme* nonlinear optics regime, in which the electric field strongly affects the state of atomic or other gas-phase media [221]. In contrast to perturbative nonlinear optics (Chapters 5 and 6), in which the nonlinear polarization may be understood in terms of envelope functions and intensity

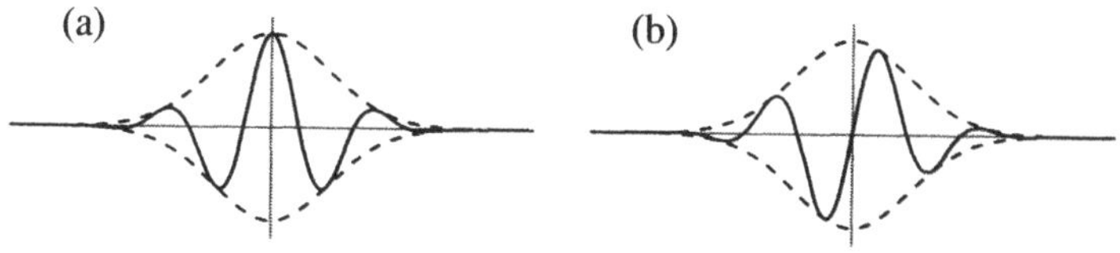

Figure 7.29 Electric field waveforms corresponding to identical Gaussian envelope functions with duration of approximately one and one-half cycles. In (a) $\phi_{CE} = 0$, and in (b) $\phi_{CE} = -\pi/2$. The pulse energy is held constant. The difference in carrier envelope phase accounts for about a 10% change in peak electric field amplitude.

averaged over multiple cycles, in the strong field regime the nonlinear polarization arising from motions of bound electrons responds directly to the instantaneous electric field with important nonperturbative dynamics on a time scale fast compared to an optical cycle. In one important process, within a single cycle the optical field first drives outer-shell electrons far away from the atomic core and then slams them back into the core, generating radiation at high harmonics (factors of tens or even hundreds) of the input frequency.

To look at the phase structure, we rewrite eq. (7.49) with $q = 0$ to represent the individual pulse centered near time $t = 0$ as

$$e(z = 0, t) = \mathrm{Re}\left\{ e^{j(\omega_0 t + \phi_{CE})} \, a(t) \right\} \tag{7.56}$$

This form differs from that used in earlier chapters in the inclusion of a carrier-envelope phase shift ϕ_{CE}. Physically, this term represents a temporal shift between the peak of the envelope function and the optical frequency oscillations. Figure 7.29 shows plots of the electric field for few-cycle pulses with identical Gaussian envelope function, but with different carrier-envelope phases (0 and $-\pi/2$). Two principal differences are observed:

- The $\phi_{CE} = 0$ case shows a unique peak in the field amplitude, which coincides with the peak of the envelope function, while $\phi_{CE} = -\pi/2$ leads to a pair of peaks which are of opposite polarity but equally strong.
- If pulse energy is kept fixed, the single strong peak for $\phi_{CE} = 0$ has larger instantaneous intensity compared to either of the main peaks for $\phi_{CE} = -\pi/2$.

For pulses that are many cycles long, these differences have little noticeable effect on the overall electric field waveform or its peak intensity. However, for pulses approaching the single-cycle regime, these differences can have a profound effect on strong-field processes such as high harmonic generation, due to their very strong intensity dependence.

It is finally important to point out that $\Delta\phi_{CE}$ and ϕ_{CE} are, in fact, distinct quantities. The former refers to the *change* in carrier-envelope phase from one pulse to the next, while the latter refers to the *absolute* carrier-envelope phase for a specific single pulse. $\Delta\phi_{CE}$ can be measured using the cross-correlation or self-referencing techniques described earlier in this section, while the value of ϕ_{CE} itself is not revealed thereby. Still, the comb stabilization methods described earlier have an important impact for strong-field experiments, as they allow such experiments to be performed with *repeatable* settings of ϕ_{CE}, even if the value of the settings are not known a priori.

PROBLEMS

7.1. Consider the self-amplitude modulation (SAM) coefficient of the nonlinear optical loop mirror in the small-signal limit. Assume that the phase difference $\Delta\phi$ is biased (e.g., using polarization effects) so that its value at zero intensity can be set to a desired value $\Delta\phi_0$ that may be nonzero. Give a formula for the maximum possible SAM coefficient and determine values of phase bias $\Delta\phi_0$ and coupling ratio α necessary to achieve this maximum.

7.2. From the data in Fig. 7.6 and 7.8, estimate the nonlinear phase shifts per pass for the solitons propagating in the various lasers represented. How do your results compare to a per-pass phase shift of 2π, for which instabilities are expected? [*Note:* The dispersion numbers in Fig. 7.8 represent $\beta_2 L$ (averaged over the laser cavity), while the dispersion numbers in Fig. 7.6 are related to $\beta_2 L$ through eq. (4.18).]

7.3. For active mode-locking with dispersion (but no SPM), eq. (7.29) tells us that in general $\text{Im}\{\lambda\} \neq 0$. Discuss what physical effect this represents.

7.4. For active mode-locking with dispersion (but no SPM), give the ratio of the pulse duration with normalized dispersion D_n relative to that with no dispersion. Also give the time–bandwidth product.

7.5. Consider a mode-locked laser limited by cubic spectral phase. Assuming that mode-locking action can compensate for 25% pulse broadening per pass, give a formula for the maximum allowable value of ψ_3 as a function of full-width half-maximum pulse width Δt. You may assume a Gaussian pulse shape. Repeat for a laser limited by a fourth-order spectral phase (ψ_4). Compare your results to parameters given in [417]: $\Delta t = 11$ fs for a laser limited by $|\psi_3| = 230\ \text{fs}^3$ per pass; 8.5 fs for a laser limited by $|\psi_4| = 1430\ \text{fs}^4$; 5 fs (simulation) for a laser limited by $|\psi_4| = 120\ \text{fs}^4$.

7.6. Consider propagation of a short pulse at 800-nm wavelength in air. Over what propagation distance does a carrier-envelope phase shift ($\Delta\phi_{\text{CE}}$) of 2π accumulate? Assuming a 5-fs pulse duration, how does this compare to the distance at which the pulse duration approximately doubles? Use the following formula [440] for the refractive index of dry air under standard conditions (where λ is wavelength in μm):

$$(n-1)\times 10^8 = 8091.37 + \frac{2,333,983}{130-\lambda^{-2}} + \frac{15,518}{38.9-\lambda^{-2}}$$

7.7. Referring to Section 4.1.2, obtain an expression for the ratio of the distance needed for a pulse to approximately double in duration under linear dispersion to the distance needed to accumulate a carrier-envelope phase shift of 2π. Express your result in terms of the number of cycles in the input pulse.

7.8. Figure 7.22a depicts an intensity cross-correlation with the peak near zero shifted to later delay for positive $\Delta\phi_{\text{CE}}$. Using the definitions of Section 7.5, construct a careful argument showing that the sign of shift in the cross-correlation is as depicted.

7.9. Consider pulses with a constant Gaussian temporal envelope function and constant energy. Work out the maximum fractional variation in peak electric field amplitude as the carrier-envelope phase ϕ_{CE} is varied from 0 to 2π. Your result will be a function of pulse duration (which you may express in terms of intensity FWHM of the envelope function) normalized to the optical oscillation period. Since very short pulses are of interest, be sure to compute energy by integrating the square of the instantaneous real electric field.

7.10. Work out a formalism to specify the form of the electric field, valid even for very short pulses approaching the single-cycle regime, subject to the following requirements: (1) you must be able to vary the carrier-envelope phase ϕ_{CE} freely without any change in the shape of the power spectrum or the pulse energy; (2) the power spectrum must go to zero at zero-frequency, corresponding to the fact that zero-frequency fields do not propagate. Use your formalism to plot out a family of electric field waveforms as ϕ_{CE} is varied, both for approximate single-cycle pulses and for pulses approximately five cycles in duration. Comment on your results.

8

MANIPULATION OF ULTRASHORT PULSES

As we have seen in previous chapters, mode-locked lasers are a powerful technology for generation of ultrafast optical impulses. In this chapter we consider methods for manipulating, processing, and amplifying such impulses. In Section 8.1 we discuss Fourier transform pulse shaping, in which mode-locked laser pulses are reshaped into new waveforms according to user specification by parallel manipulation of the optical frequency spectrum. Section 8.2 covers additional important pulse-shaping methods. Pulse-processing techniques that exploit large amounts of dispersion and chirp are treated in Section 8.3, including the possibility of temporal imaging systems based on clever combinations of dispersion and phase modulation. Finally, in Section 8.4 we discuss amplification methods that bring femtosecond laser pulses to unprecedented power levels, allowing access to new regimes of laser–matter interaction.

8.1 FOURIER TRANSFORM PULSE SHAPING

In this section we present an in-depth discussion of a Fourier transform pulse-shaping technique that allows controllable transformation of mode-locked pulses into complex ultrafast optical waveforms according to user specification. The ability to engineer ultrashort-pulse waveforms optimized for specific applications has had a significant impact in fields as diverse as ultrafast spectroscopy, nonlinear fiber optics and fiber communications, and nonlinear microscopy.

The femtosecond pulse-shaping methods we describe are based on the well-known concept of a linear, time-invariant filter. Linear filtering can be described in either the time or the frequency domain, as depicted in Fig. 8.1. In the time domain, the filter is characterized by an impulse response function $h(t)$. The output of the filter $e_{\text{out}}(t)$ in response

Ultrafast Optics, By Andrew M. Weiner

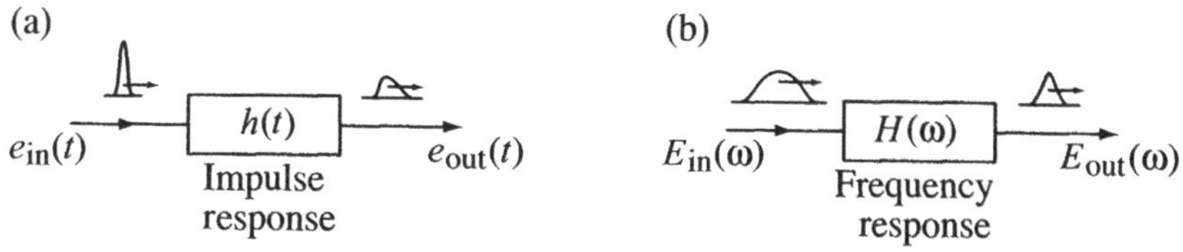

Figure 8.1 Pulse shaping by linear filtering: (a) time-domain view; (b) frequency-domain view.

to an input pulse $e_{in}(t)$ is given by the convolution of $e_{in}(t)$ and $h(t)$:

$$e_{out}(t) = e_{in}(t) * h(t) = \int dt'\, e_{in}(t')h(t - t') \tag{8.1}$$

If the input is a delta function, the output is simply $h(t)$. Therefore, for a sufficiently short input pulse, the problem of generating a specific output pulse shape is equivalent to the task of fabricating a linear filter with the desired impulse response. In the frequency domain, the filter is characterized by its frequency response $H(\omega)$. The output of the linear filter $E_{out}(\omega)$ is the product of the input signal $E_{in}(\omega)$ and the frequency response $H(\omega)$, that is,

$$E_{out}(\omega) = E_{in}(\omega)H(\omega) \tag{8.2}$$

where $E_{in}(\omega)$, $E_{out}(\omega)$, and $H(\omega)$ are the Fourier transforms of $e_{in}(t)$, $e_{out}(t)$, and $h(t)$, respectively. For a delta function input pulse, the input spectrum $E_{in}(\omega)$ is equal to unity, and the output spectrum is equal to the frequency response of the filter. Therefore, due to the Fourier transform relations, generation of a desired output waveform can be accomplished by implementing a filter with the required frequency response. Fourier transform optical pulse shaping is described most naturally by means of this frequency-domain point of view.

Figure 8.2 shows the basic pulse-shaping apparatus, which consists of a pair of diffraction gratings and lenses (see also Section 4.3) and a pulse-shaping mask. The individual

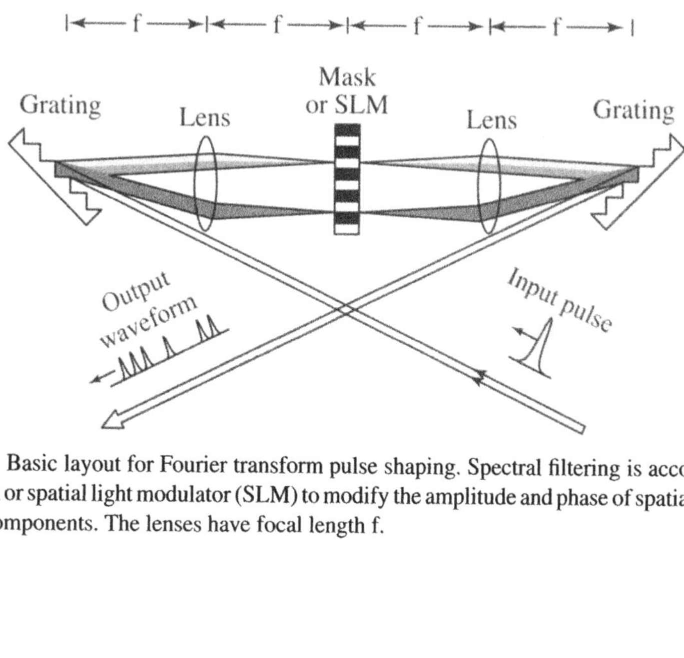

Figure 8.2 Basic layout for Fourier transform pulse shaping. Spectral filtering is accomplished by using a mask or spatial light modulator (SLM) to modify the amplitude and phase of spatially dispersed frequency components. The lenses have focal length f.

frequency components contained within the incident ultrashort pulse are angularly dispersed by the first diffraction grating, and then focused to small diffraction-limited spots at the back focal plane of the first lens, where the frequency components are spatially separated along one dimension. Essentially, the first lens performs a Fourier transform which converts the angular dispersion from the grating to a spatial separation at the back focal plane. Spatially patterned amplitude and phase masks or a programmable spatial light modulator (SLM) are placed in this plane to manipulate the spatially dispersed optical Fourier components. After a second lens and grating recombine all the frequencies into a single collimated beam, a shaped output pulse is obtained, with the output pulse shape given by the Fourier transform of the pattern transferred by the masks or SLM onto the spectrum. A key point is that because waveform synthesis is achieved by parallel modulation in the frequency domain, waveforms with serial modulation on the femtosecond time scale can be generated without requiring ultrafast modulators. This technique is quite general and has been applied with minor modifications for time scales from approximately a few femtoseconds to almost a nanosecond, a difference of more than five orders of magnitude.

For this technique to work as desired, one requires that in the absence of a pulse-shaping mask, the output pulse should be identical to the input pulse. Therefore, the grating and lens configuration must be truly free of dispersion. This can be guaranteed if the lenses are set up as a unit magnification telescope, with the gratings located at the outside focal planes of the telescope. A simple argument for the zero-dispersion behavior of this 4-f configuration was presented in Section 4.3; a mathematical treatment is given in Section 8.1.3. Note that this dispersion-free condition depends on several approximations (e.g., that the lenses are thin and free of aberrations, that chromatic dispersion in passing through the lenses or other elements that may be inserted into the pulse shaper is small, and that the gratings have a flat spectral response). For very short pulses (e.g., in the 20-fs range and below), increased care must be taken to satisfy these approximations. For example, both the chromatic aberration of the lenses in the pulse shaper and the dispersion experienced in passing through the lenses can become important effects. By using spherical mirrors instead of lenses, these problems can be avoided and dispersion-free operation obtained even at very short pulse durations [441].

Pulse-shaping methods have been discussed in a number of review articles. Early pulse-shaping studies that employed relatively broad, picosecond pulses are described in [187]. Later articles that review femtosecond and programmable pulse shaping may be found in [155,298]. Related experiments, in which picosecond pulse shaping is implemented in the context of fiber and grating pulse compression, are discussed in [297,442]. The latter experiments may be considered an extension of the spectral windowing technique described in Section 6.4.

8.1.1 Examples of Pulse Shaping Using Fixed Masks

We now discuss several examples of pulse shaping using fixed spatial masks, obtained from a pulse shaper excited by input pulses of about 100 fs [443]. Further examples using programmable spatial light modulators are discussed in Section 8.1.2. Figure 8.3 shows waveforms generated by placing an opaque mask with two isolated slits into the pulse shaper, resulting in a pair of distinct and isolated spectral peaks. Waveforms were measured via intensity cross-correlation, using a reference pulse directly from the laser (because the

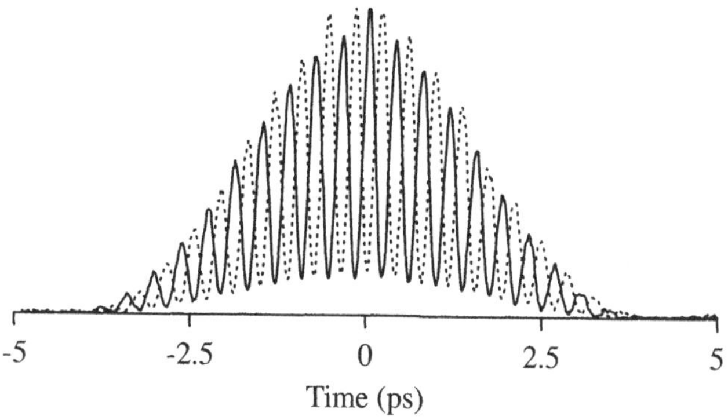

Figure 8.3 Intensity cross-correlation traces of optical tone bursts resulting from a pair of isolated optical frequency components. Solid line, optical frequencies in phase; dotted line, optical frequencies shifted by π. Adapted from [443].

reference is substantially shorter than the shaped pulse, this yields a good measurement of the shaped intensity profile). The two frequencies selected by the pulse shaper interfere in the time domain, producing an intensity profile sinusoidally modulated at 2.6 THz. In one experiment an additional phase mask was used to impose a relative phase shift of π between the two spectral components. The resulting waveform is shown in Fig. 8.3 as the dotted line. The two distinct frequencies still interfere to produce a modulated burst. However, the π phase shift leads to an interchange in the positions of the peaks and nulls of the time-domain waveform. This shift in the time-domain interference features reflecting spectral phase variations forms the basis of a characterization method for measuring spectral phase profiles of unknown ultrashort pulses discussed in Section 3.7.2. The data in Fig. 8.3 represent a time-domain analog of the well-known double-slit interference experiment of classical optics.

As a second example, we consider generation of ultrafast square pulses. The spectrum of a square pulse of duration T is shaped as a *sinc* function, given by

$$E(\tilde{\omega}) = E_0 T \frac{\sin(\tilde{\omega}T/2)}{\tilde{\omega}T/2} \tag{8.3}$$

The corresponding mask is specified by

$$M(x) = \frac{\sin(\pi x/x_0)}{\pi x/x_0} \tag{8.4}$$

where $M(x)$ is the spatial masking function, $x_0 = (2\pi/T)(\partial x/\partial\omega)$, and $\partial x/\partial\omega$ is the spatial dispersion at the masking plane. To implement the desired filtering function, both phase and amplitude masks were fabricated on fused-silica substrates using microlithographic patterning techniques. The phase mask is used to impart the required alternating sign to the filter and is realized by etching a relief pattern into the substrate. The amplitude mask, fabricated via spatial patterning of a metal film, features a transmission function that varies continuously with position. Due to the combination of fast and slow temporal features

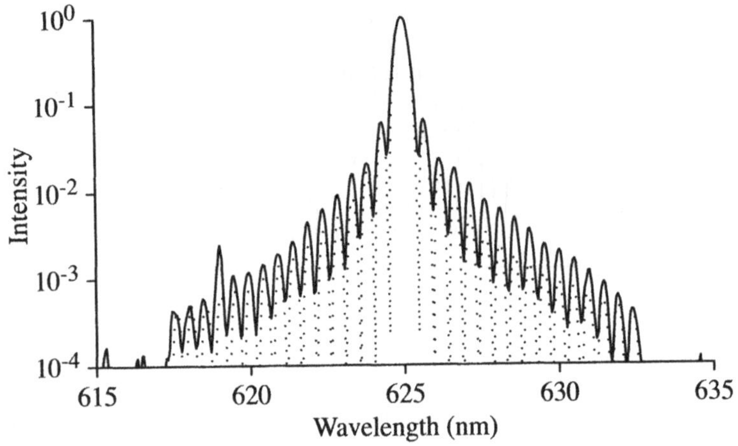

Figure 8.4 Semilog plot of power spectrum of an optical square pulse. From [443].

(<100-fs rise and fall times, pulse duration in the picosecond range), the amplitude mask is required to produce a series of sidelobes over a large dynamic range. Figure 8.4 shows a semilog plot of a power spectrum after filtering in the pulse shaper. The data show 15 sidelobes on either side of the central peak and a dynamic range approaching 10^4:1. The dotted line, which is an actual sinc function squared, is in good agreement with the data, although the zeros in the data are washed out due to the finite spectral resolution of the measurement device. Figure 8.5a shows an intensity cross-correlation measurement of a 2-ps square pulse produced by using masks truncated after five sidelobes on either side of the main spectral peak. The rise and fall times of the square pulse are found to be on the order of 100 fs, limited by the duration of the input pulse. The ripple arises because of the truncation of the spectrum and is in good qualitative agreement with the theoretical intensity profile (not shown). Square pulses with reduced ripple have also been obtained by avoiding truncation of the spectrum and instead using a more gentle apodization or tapering of the spectrum. An experimental example of such a smooth square pulse is plotted in Fig. 8.5b.

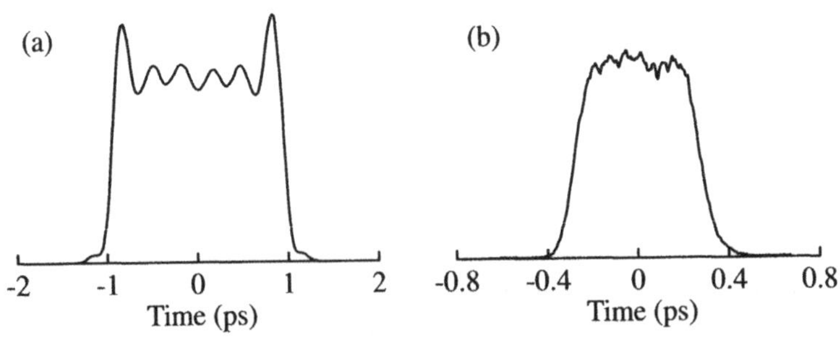

Figure 8.5 Optical square pulses: (a) measurement of a 2-ps square pulse; (b) measurement of a femtosecond square pulse with reduced ripple. (a) From [443]. (b) From [444]. Copyright © 1989, IEEE.

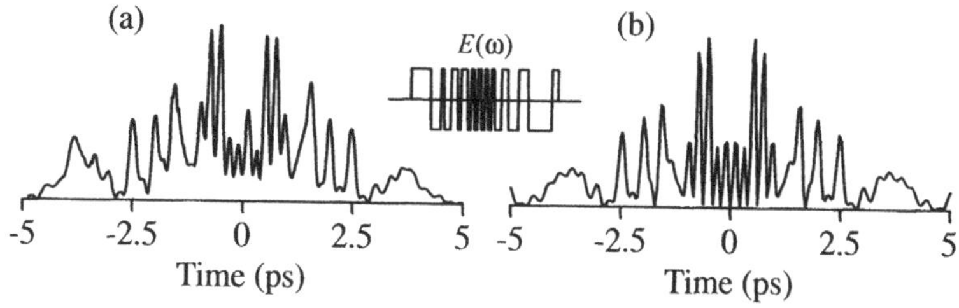

Figure 8.6 Ultrafast pseudonoise bursts generated by encoding with a pseudorandom phase filter (shown as inset): (a) intensity cross-correlation trace; (b) theoretical intensity profile. From [445].

At this point we also discuss pulse shaping using phase-only filters. Phase-only filters have the advantage, in situations where they are adequate, of no inherent loss. One interesting example (Fig. 8.6) involves encoding of femtosecond pulses by utilizing pseudorandom phase patterns to scramble (encode) the spectral phase [445]. Here the clear aperture of the mask is divided into 44 equal pixels, each of which corresponds to a phase shift of either zero or π. Spectral encoding spreads the incident femtosecond pulse into a complicated pseudonoise burst within an $\sim$ 8-ps temporal envelope. The peak intensity is reduced by more than an order of magnitude compared to that of an uncoded pulse of the same optical bandwidth. The theoretical intensity profile, which is simply the magnitude squared of the Fourier transform of the spectral phase mask, is in excellent agreement with the experimental data. An important feature is that because the phase-only filtering is ideally lossless, the spectral phase modulation can be undone by using a second pulse shaper with a conjugate phase mask, with the result that the pseudonoise burst is decoded (restored) back to the original ultrashort pulse duration. Figure 8.7 shows intensity cross-correlation traces of pulses that (a) are not coded, (b) are coded and subsequently decoded, and (c) are coded and then incorrectly decoded (i.e., by using the wrong pseudorandom mask). Here the pseudorandom phase patterns are length-31 maximal-length sequences, which are well known and widely applied in fields such as spread-spectrum communications, precision ranging, and radar [447,448]. The decoding experiments may be considered as an example of a generalized dispersion-compensation operation. Furthermore, the ability to perform optical encoding and decoding forms the basis of a novel concept for ultrashort-pulse fiber optic telecommunications [446,449]. In this optical code-division multiple-access (O-CDMA) scheme, multiple independent data channels are assigned different spectral phase codes, which allow them to share a common fiber optic transmission channel. The key is that the O-CDMA receiver is able to recognize the desired data in the presence of interference from other channels by exploiting the strong intensity contrast between correctly and incorrectly decoded ultrashort pulses, as illustrated in Fig. 8.7.

In some cases only the temporal intensity profile of an output pulse is of interest, and this greatly increases the degrees of freedom available for filter design. In particular, phase-only filters can be designed to yield the desired temporal intensity profile. An interesting example is the use of periodic phase-only spectral filters to produce high-quality pulse trains [450]. This concept is illustrated in Fig. 8.8. We assume that the input pulse has bandwidth B corresponding to duration $\delta t = 0.44B^{-1}$ (for a Gaussian pulse). First, in the case of amplitude filtering, all but a periodically spaced set of frequency components are blocked. The resulting pulse train consists of a series of pulses, each of duration δt, with a pulse spacing Δf^{-1} equal to the inverse of the frequency periodicity. The duration of the

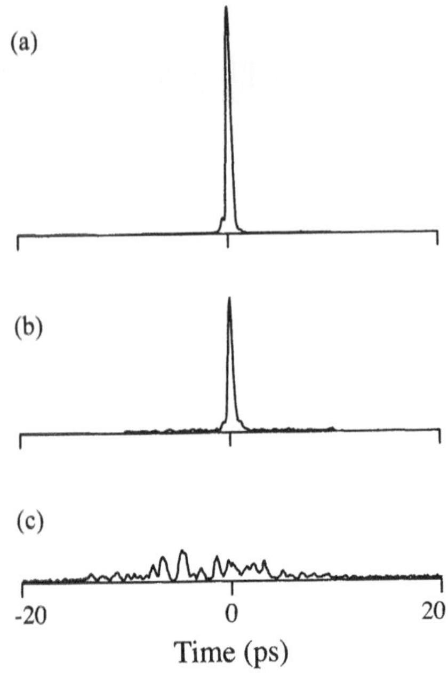

Figure 8.7 Encoding and decoding of ultrashort pulses at 1.55 μm, performed using a pair of fiber-pigtailed pulse shapers and length-31 maximal length sequence phase codes: (a) cross-correlation of pulse that is not coded; (b) pulse that is coded and then correctly decoded; (c) pulse that is coded but then decoded incorrectly. The vertical scales in (b) and (c) are normalized to the peak intensity in part (a). Note the strong intensity contrast between pulses decoded correctly and incorrectly. Adapted from [446]. Copyright © 1998, IEEE.

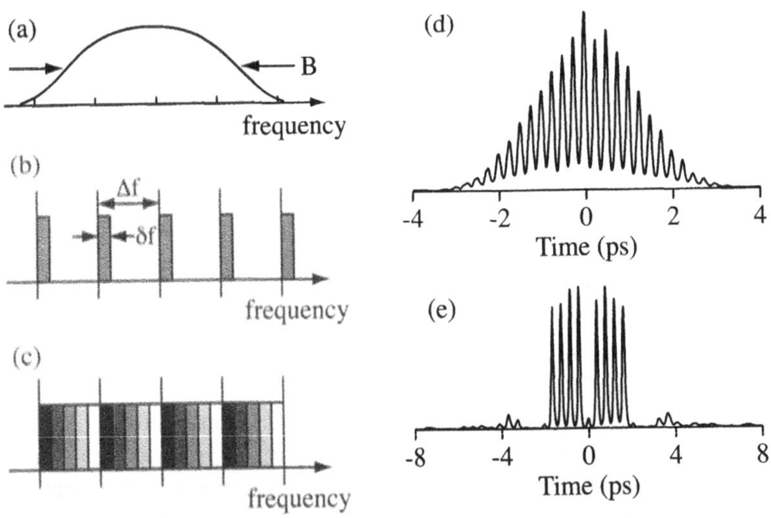

Figure 8.8 Generation of high-repetition-rate pulse trains by periodic spectral filtering: (a–c) schematic view of (a) input spectrum, (b) periodic amplitude filter, and (c) periodic phase filter; (d,e) examples of pulse trains generated by periodic phase-only filtering. Adapted from [450].

pulse train is about δf^{-1}, and the number of pulses in the train is of order $\Delta f/\delta f$, where δf is the spectral width of the individual passbands of the filter. Since the total transmission is only $(\Delta f/\delta f)^{-1}$, any amplitude filter designed to produce a large number of pulses will be inherently inefficient.

High-quality pulse trains can be generated without loss by phase-only filtering. The phase response of the filter should vary periodically with frequency, and the amplitude response should be constant. The repetition rate of the resulting pulse train remains equal to Δf, but the envelope of the pulse train depends on the structure of the phase response within a single period of the phase filter. The output intensity profile $I(t)$ is related to the spectral amplitude $E(\omega)$ according to

$$I(t) = \left(\frac{1}{2\pi}\right)^2 \int d\Omega\, e^{j\Omega t} \int d\omega\, E^*(\omega)E(\omega+\Omega) \tag{8.5}$$

The intensity is equal to the inverse Fourier transform of the autocorrelation of the filtered spectrum. To obtain a burst of pulses under a smooth envelope, one must choose a phase response with an autocorrelation that consists of a series of evenly spaced, isolated spectral peaks. This can be achieved by using pseudorandom phase sequences with sharp autocorrelation peaks as the building blocks of the periodic phase filter. Again, suitable pseudorandom phase sequences are well known in fields such as spread-spectrum communications. Figure 8.8d shows a measurement of a pulse train generated by using a spectral phase filter constructed out of periodic repetitions of a length-15 maximal-length sequence code (i.e., $\Delta f/\delta f = 15$). The pulse train is clean, and the pulses are well separated. As in the case of amplitude filtering, the number of pulses is expected to be of order $\Delta f/\delta f$, which is confirmed by the data. Note that the optical phase is constant from pulse to pulse in trains produced by amplitude filtering, unlike the phase filtering case, where the optical phase varies.

Pulse trains with different envelopes can be generated by varying the details of the phase response within a single period of the periodic phase filter. Figure 8.8e shows a pulse train generated by using a mask corresponding to a binary-phase computer-generated hologram that was designed to split a single-input beam spatially into eight replica beams with the central beam missing. The time-domain result consists of eight pulses with a missing central pulse. These time-domain results, achieved by using a phase filter originally designed for spatial beam-forming applications, point to a close analogy between time-domain Fourier pulse shaping and classical spatial-domain Fourier processing [168].

8.1.2 Programmable Pulse Shaping

An important attribute of pulse shaping is its programmability, allowing waveform definition under computer control. Programmable pulse shaping is accomplished by using a spatial light modulator (SLM) at the masking plane in Fig. 8.2. A number of technologies, including liquid-crystal arrays, acousto-optics, and deformable mirrors, have been utilized as the SLM for programmable pulse shaping. A detailed discussion may be found in [298] and references therein. Here we limit our discussion to the case of liquid-crystal modulator arrays (LCMs).

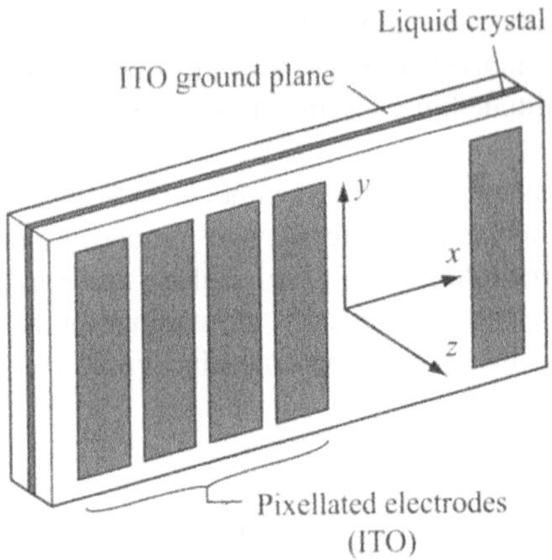

Figure 8.9 Electronically addressed liquid-crystal spatial light modulator.

Liquid-Crystal Modulator Arrays LCM arrays have been configured primarily for either phase-only or phase-and-amplitude operation in pulse-shaping applications. Figure 8.9 shows a schematic of an electronically addressed, phase-only liquid-crystal SLM [451]. A thin layer of a nematic liquid crystal is sandwiched between two pieces of glass. The nematic liquid crystal consists of long, thin, rodlike molecules, which in the absence of an electric field are aligned with their long axes along the y direction. When an electric field is applied (in the z direction), the liquid crystal molecules tilt along z, causing a refractive index change for y-polarized light. (The refractive index change for x-polarized light is usually minimal.) A maximum phase change of at least 2π is required for complete phase control. To apply the required electric field, the inside surface of each piece of glass is coated with a thin, transparent, electrically conducting film of indium tin oxide (ITO). One piece is patterned into a number of separate electrodes (or pixels) with the corresponding fan-out for electrical connections. Typical commercially available LCMs comprise 128 to 640 pixels with 100-μm center-to-center pixel spacing and gaps between pixels on the order of 2 μm. The modulator array is controlled by a special drive circuit that generates separate, variable-amplitude signals to achieve independent, gray-level phase control of each of the modulator elements.

Figure 8.10 shows an example of real-time pulse-shaping data measured using an early 32-element LCM array and 75-fs input pulses. In this experiment half of the pixels were connected to a constant-amplitude drive signal, while the other half were switched at a 20-Hz rate between the different drive levels. In one state, the phases are all the same, and the output is a single pulse similar to the input pulse. In the other state, half of the pixels experience a relative phase shift of π. With the particular set of connections chosen in this experiment, this results in a pulse doublet. The pulse intensity profiles (top trace) were obtained by using a real-time cross-correlation setup which was driven in synchronism with the time-varying drive signal (bottom trace). The data demonstrate complete switching from a single pulse to a pulse doublet within 25 ms. Under appropriate drive conditions, such LCM arrays are capable of switching in about 10 ms. It is important to point out, however, that switch-on and switch-off times are usually not the same and depend on the bias voltage levels.

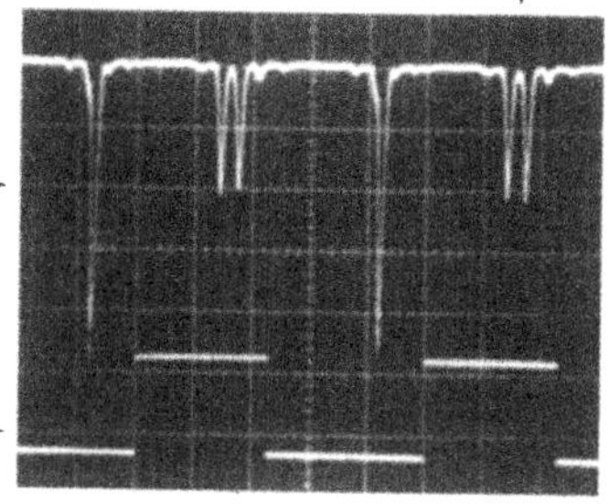

Figure 8.10 Real-time pulse-shaping data. Top trace: Pulse intensity profiles measured using a real-time cross-correlation setup, showing switching between two distinct waveforms within 25 ms. In one case the phases were all set to zero; in the other case the spectral phases were programmed according to the sequence $0000\pi\pi\pi\pi0000\pi\pi\pi\pi\ldots$. Bottom trace: Time-varying drive signal to the liquid-crystal array. Adapted from [452].

Mathematically, we can describe the transformation of light associated with passage through a birefringent element by a matrix formalism. Assuming the liquid-crystal molecules are aligned along y at zero applied field, we write

$$\begin{pmatrix} E_x^{(\text{out})} \\ E_y^{(\text{out})} \end{pmatrix} = \begin{pmatrix} e^{j\phi_0} & 0 \\ 0 & e^{j[\phi_0+\Delta\phi(V)]} \end{pmatrix} \begin{pmatrix} E_x^{(\text{in})} \\ E_y^{(\text{in})} \end{pmatrix} \tag{8.6}$$

Here ϕ_0 is the approximately voltage-independent phase shift seen by x-polarized light, and $\Delta\phi(V)$ is the voltage-dependent birefringence. In the limit of sufficiently high voltage, the liquid-crystal molecules tilt to align parallel to z, and $\Delta\phi$ goes to zero. For the case of pulse shaping with an LCM array, eq. (8.6) describes the operation of any single pixel. Different pixels are described by similar equations, but with independent values of V for different pixels.

For input light polarized along y, the liquid-crystal cell operates as a pure phase modulator, with no polarization rotation. This is the mode of operation most useful for pulse shaping. However, to use the LCM for gray-level (continuous) phase control, a careful phase vs. voltage calibration is required. This can be accomplished by using the array as an amplitude modulator. Now the laser is linearly polarized, with its polarization rotated 45° relative to the alignment direction of the liquid crystals. The fractional power transmission through a subsequent crossed polarizer is easily shown to be

$$\frac{|E^{(\text{out})}|^2}{|E^{(\text{in})}|^2} = \sin^2\left(\frac{\Delta\phi(V)}{2}\right) \tag{8.7}$$

Thus, the phase calibration can be extracted from the voltage-dependent transmission through the crossed polarizer. Alternatively, LCM arrays can also be calibrated *in situ* within the pulse shaper by using pulse characterization techniques such as spectral interferometry or FROG to measure the spectral phase changes imposed onto an incident pulse.

For general pulse-shaping operation, independent control of spectral phase and amplitude is desired. As discussed above, a single LCM array together with a polarizer can be used as an amplitude modulator; however, this also leads to a phase modulation which depends on the amplitude modulation level. Therefore, for independent phase and amplitude control the use of two liquid-crystal arrays is required. In such a two-layer LCM array, two individual arrays of the type depicted in Fig. 8.9 are attached permanently, one on top of the other, with their pixels aligned and angular orientations fixed. In the most common configuration, the long axes of the liquid-crystal molecules in the first and second layers are aligned at 45° and −45° with respect to the x axis, respectively [453]. When a voltage is applied to a pixel in one of the SLMs, the liquid-crystal molecules in that pixel are rotated toward z, resulting in a phase modulation for the component of light parallel to the liquid-crystal axis in that SLM. The operation of the two-layer LCM is conveniently analyzed by expressing the input field in terms of its components along ±45° directions [i.e., by working in the $\hat{\mathbf{x}}' = (\hat{\mathbf{x}} + \hat{\mathbf{y}})/\sqrt{2}$ and $\hat{\mathbf{y}}' = (\hat{\mathbf{y}} - \hat{\mathbf{x}})/\sqrt{2}$ basis set]. In this basis the field resulting after passing through both liquid-crystal layers is written

$$\begin{pmatrix} E_{x'}^{(\text{out})} \\ E_{y'}^{(\text{out})} \end{pmatrix} = \frac{1}{\sqrt{2}} \begin{pmatrix} e^{j\Delta\phi_1(V_1)} & 0 \\ 0 & 1 \end{pmatrix} \begin{pmatrix} 1 & 0 \\ 0 & e^{j\Delta\phi_2(V_2)} \end{pmatrix} \begin{pmatrix} E_x^{(\text{in})} + E_y^{(\text{in})} \\ E_y^{(\text{in})} - E_x^{(\text{in})} \end{pmatrix} \tag{8.8}$$

where subscripts refer to the first and second liquid-crystal layers, respectively, and the voltage-independent phase term $\exp[j(\phi_1 + \phi_2)]$ is omitted. Multiplying through and converting back to the original x, y basis set yields

$$\begin{pmatrix} E_x^{(\text{out})} \\ E_y^{(\text{out})} \end{pmatrix} = e^{j\overline{\Delta\phi}(V_1,V_2)} \begin{pmatrix} \cos(\Delta\Phi) & j\sin(\Delta\Phi) \\ j\sin(\Delta\Phi) & \cos(\Delta\Phi) \end{pmatrix} \begin{pmatrix} E_x^{(\text{in})} \\ E_y^{(\text{in})} \end{pmatrix} \tag{8.9}$$

where $\overline{\Delta\phi} = \frac{1}{2}[\Delta\phi_1(V_1) + \Delta\phi_2(V_2)]$ is the average of the voltage-dependent birefringences of the two layers, and $\Delta\Phi = \frac{1}{2}[\Delta\phi_1(V_1) - \Delta\phi_2(V_2)]$ depends on their difference.

Equation (8.9) describes several modes of operation:

- To obtain independent amplitude and phase control, the input field is linearly polarized along either $\hat{\mathbf{x}}$ or $\hat{\mathbf{y}}$, and the output field passes through a polarizer, again along either $\hat{\mathbf{x}}$ or $\hat{\mathbf{y}}$. Here let us consider the case where the input field and the polarizer are both along $\hat{\mathbf{x}}$. Equation (8.9) now reduces to

$$\mathbf{E}^{(\text{out})} = E^{(\text{in})} e^{j\overline{\Delta\phi}(V_1,V_2)} \cos[\Delta\Phi(V_1, V_2)]\hat{\mathbf{x}} \tag{8.10}$$

  The output phase depends on $\overline{\Delta\phi}$ and the output amplitude depends on $\Delta\Phi$. Since these quantities may be controlled independently, independent control of phase and amplitude is possible.

- Consider now the case where the input field is again along $\hat{\mathbf{x}}$ but there is no output polarizer. The field is now written

$$\mathbf{E}^{(\mathrm{out})} = E^{(\mathrm{in})} e^{j\overline{\Delta\phi}} \{\cos(\Delta\Phi)\hat{\mathbf{x}} + j\sin(\Delta\Phi)\hat{\mathbf{y}}\} \tag{8.11}$$

The phase depends on $\overline{\Delta\phi}$ as before, while control of $\Delta\Phi$ allows transformation of the output state of polarization [453,454].[1] (Such polarization control in pulse shaping requires the use of a spectral disperser without polarization-dependent loss. This requirement can be a significant challenge for practical diffraction gratings.)

- For applications involving optical fibers, the polarization of the input light is usually scrambled due to small random birefringences in the fiber. In this case it may be desirable to be able to perform pulse shaping independent of the input state of polarization [456]. For phase-only shaping this may be achieved by setting $\Delta\Phi = 0$ (i.e., the two liquid-crystal layers are programmed for identical phase shifts). With $\Delta\phi_1 = \Delta\phi_2$, eq. (8.9) reduces to

$$\begin{pmatrix} E_x^{(\mathrm{out})} \\ E_y^{(\mathrm{out})} \end{pmatrix} = e^{j\Delta\phi_1(V_1)} \begin{pmatrix} 1 & 0 \\ 0 & 1 \end{pmatrix} \begin{pmatrix} E_x^{(\mathrm{in})} \\ E_y^{(\mathrm{in})} \end{pmatrix} \tag{8.12}$$

The phase shift is independent of the input state of polarization, the amplitude response is unity, and the output state of polarization is unchanged compared to the input. This yields a polarization-independent, phase-only modulation functionality.

Gray-Level Phase Control In addition to reconfigurability, one of the advantages of programmable pulse-shaping modulators is the ability to implement continuous or gray-level phase control. Such gray-level control can be used to accomplish a number of interesting pulse-shaping tasks. Here we discuss two examples, both of which were performed using liquid-crystal arrays. Related experiments have also been performed with other types of spatial light modulators.

In the first example, the modulator array imparts a linear phase sweep onto the spectrum, which corresponds in the time domain to a pulse position modulation. This process relies on the fact that if $f(t)$ and $F(\omega)$ are a Fourier transform pair, the delayed signal $f(t-\tau)$ has Fourier transform $F(\omega)\exp(-j\omega\tau)$. This is a special case of the general relation between spectral phase ψ and frequency-dependent delay, given by

$$\tau(\omega) = -\frac{\partial\psi(\omega)}{\partial\omega} \tag{8.13}$$

Thus, a pulse can be retarded (or advanced) by imposing a linear phase sweep onto its spectrum. The delay τ is given by the relation

$$\tau = -\frac{\delta\phi}{2\pi\delta f} \tag{8.14}$$

[1] More complete control of the output polarization state is possible using a modified geometry in which the orientations of the liquid-crystal molecules in the two layers differ by 45° rather than 90° [455].

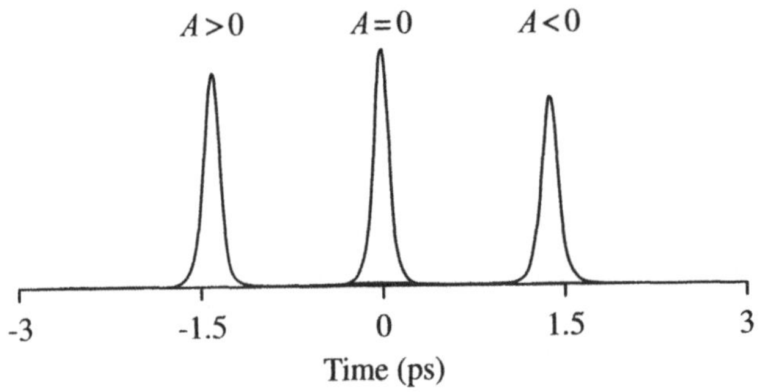

Figure 8.11 Measurements of temporally shifted pulses achieved by placing linear phase ramps, $\psi(\tilde{\omega}) = A\tilde{\omega}$, onto the spectrum. Data are shown for phase changes per pixel of 0 and $\pm\pi/4$, respectively. Adapted from [451]. Copyright © 1992, IEEE.

where $\delta\phi$ and δf are, respectively, the imposed phase change per pixel and the change in optical frequency from one pixel to the next. Figure 8.11 shows cross-correlation measurements of temporally shifted pulses achieved by applying a linear spectral phase. For these experiments the array consisted of 128 pixels, δf was approximately 0.092 THz, and in different measurements $\delta\phi$ was set to 0 and to $\pm\pi/4$, respectively. This results in output pulses at 0 and at ± 1.38 ps, in close agreement with eq. (8.14). These data clearly demonstrate the ability to shift the pulse position by many pulse widths by using spectral phase manipulation.

In these experiments the modulator was set to provide the required phase sweep modulo 2π, so that for each pixel the phase is in the range 0 to 2π, even though the total equivalent phase sweep was 32π. Because pixellated arrays can be programmed to provide the desired phase function modulo 2π, large phase sweeps can be achieved using modulators with limited maximum phase change per pixel. However, in any case of a smoothly varying target phase function, as here, the phase change from one pixel to the next should remain small enough that the staircase phase pattern that accompanies a modulator array with discrete pixels is a sufficiently good approximation to the desired phase function. This point is discussed further in Section 8.1.3.

Gray-level phase control can also be used to achieve programmable compression of chirped optical pulses. As one example, we discuss experiments in which a programmable pulse shaper was used to complement fiber dispersion-compensation techniques in propagating sub-500-fs pulses over an optical fiber link [457,458]. The link consisted of 50 km of standard single-mode fiber (SMF) concatenated to a matching length of dispersion-compensating fiber (DCF). Since SMF and DCF have dispersion with opposite signs at the operating wavelength of 1542 nm, the fiber lengths can be adjusted to cancel all of the lowest-order dispersion (i.e., phase varying quadratically with frequency). In addition, since the derivatives of the dispersion with respect to frequency also have opposite signs, the dispersion slope (i.e., cubic spectral phase variations) can be partially compensated. In the experiments in [458], input pulses of about 460 fs are first broadened about 10^4 times to about 5 ns in propagating through the SMF, then recompressed by the DCF to within a factor of 2 of the input pulse duration. Intensity cross-correlation measurements of the input and output pulses after 50-km fiber propagation are shown in Fig. 8.12. The remaining broadening and distortion of the output pulse are characteristic of cubic spectral phase variation.

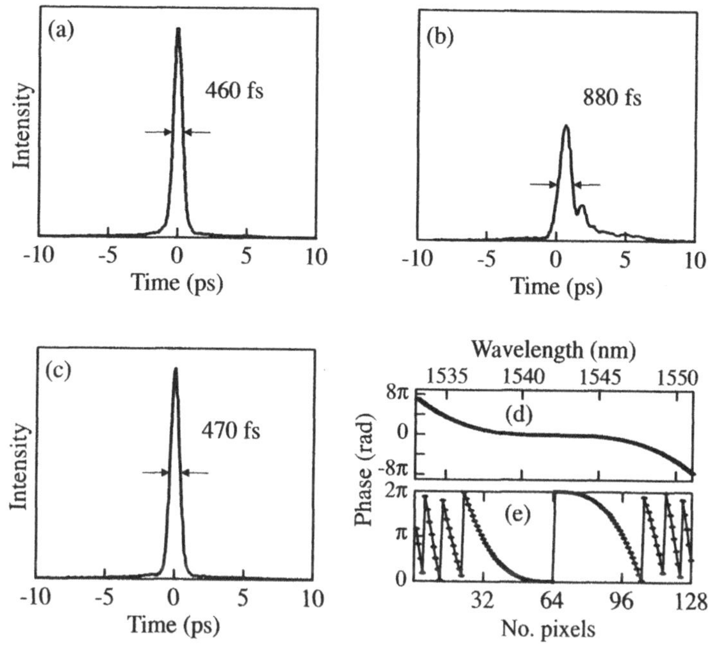

Figure 8.12 Distortionless transmission over 50-km fiber via spectral phase equalization: (a–c) intensity correlation measurements of (a) input pulse, (b) distorted pulse after propagation through 50-km fiber and partial compensation by dispersion-compensating fiber, (c) restored pulse after spectral phase equalization in a pulse shaper; (d) unwrapped and (e) actual spectral phase profile programmed onto the liquid-crystal modulator. Adapted from [458].

At this point the pulse shaper is used to apply a spectral phase function which is equal and opposite to the estimated residual phase variation. Both the actual applied phase function, which is a cubic function modulo 2π, as well as the unwrapped version showing a continuous cubic phase, are plotted for purposes of illustration. Application of the inverse spectral phase function results in a completely recompressed pulse with essentially the original pulse duration (ca. 470 fs) and no observable distortion. Thus, in these experiments all-fiber techniques are used for coarse dispersion compensation, while a programmable pulse shaper is used as a spectral phase equalizer to fine-tune away any remaining dispersion. In a completely analogous fashion, programmable pulse shapers are also used extensively in compensation of residual dispersion in chirped pulse amplifier systems and in few-cycle pulse generation.

Control Strategies for Programmable Pulse Shaping Pulse shapers incorporating spatial light modulators may be operated either under open loop control or in a feedback (adaptive control) configuration (Fig. 8.13). In the open-loop configuration, the output waveform desired is specified by the user, and reasonable knowledge of the input pulse is also usually available. Therefore, the desired transfer function is known, and one simply programs the pulse-shaping SLM to provide this transfer function. If additional linear distortion is present between the input and output (e.g., phase aberration in a femtosecond amplifier, not shown), the pulse shaper can be programmed so that its transfer function also includes precompensation for such distortion. This open-loop strategy requires precise calibration

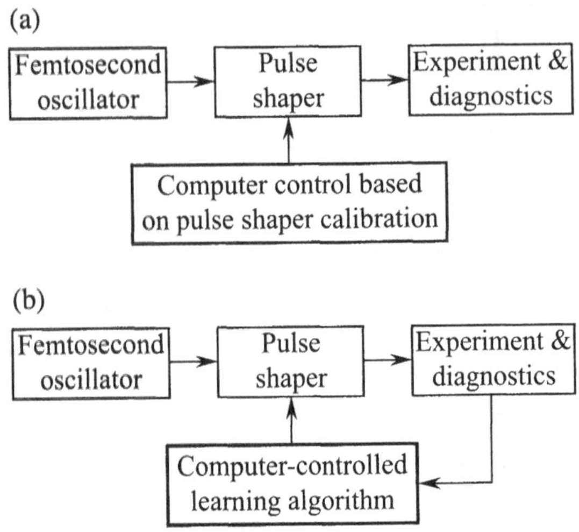

Figure 8.13 Control strategies for programmable pulse shaping: (a) open-loop control; (b) feedback or adaptive control.

of the SLM. However, precise characterization of the input pulse is often unnecessary, provided that the input pulse is known to be shorter than the shortest features desired in the shaped output waveform. All of the pulse-shaping data given earlier in this section are examples of open-loop control.

The ability to program a pulse shaper under computer control also leads to an alternative adaptive control strategy (Fig. 8.13b) [459–461]. In these experiments one usually starts with a random spectral pattern programmed into the pulse shaper, which is updated iteratively according to a stochastic optimization algorithm based on the difference between a desired and a measured experimental output. In this scheme there is no need to explicitly program the pulse shaper. Experimental applications of pulse shaping under the adaptive control scheme include compensation of residual chirps in chirped-pulse amplifiers and in few-cycle pulse compression, as well as optimization of strongly nonlinear processes. The latter includes research in a field termed *coherent control*, wherein shaped laser waveforms can be a tool for controlling quantum mechanical motions, with application, for example, to laser-controlled chemistry. Often, the quantum mechanics of even relatively simple molecules is not known with sufficient accuracy to predict the appropriate laser waveform. In such cases the adaptive control approach can be used to search for the laser waveform that gives the best experimental result, such as optimizing the yield of a particular photochemical product [462,463].

8.1.3 Pulse-Shaping Theory

It is important to analyze the output field from the pulse shaper quantitatively. This will reveal fundamental limits governing pulse-shaping operation. Our treatment draws on publications analyzing grating pair compressors [464] as well as pulse shapers [301,465].

In our analysis we calculate the propagation of the input signal to different planes of interest within the pulse-shaping apparatus, as indicated in Fig. 8.14. For reasons that will become clear later, we also calculate focusing of the field by an additional lens subsequent to the second diffraction grating.

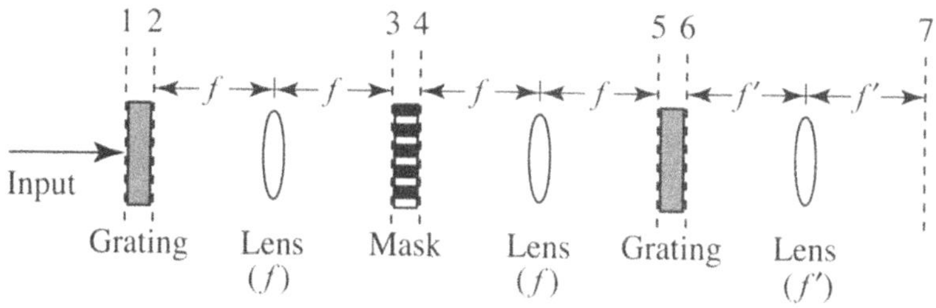

Figure 8.14 Reference diagram of a pulse shaper used in theoretical analysis. For simplicity the gratings are depicted in transmission mode, and their tilt is not shown. The dashed lines refer to different planes at which the field is calculated. The calculation includes the possibility of focusing into a single-mode fiber at the output (plane 7).

We start by writing the input field in plane 1, immediately before the first diffraction grating, as

$$e_{\text{in}}(x,t) = \text{Re}\left\{\hat{e}_{\text{in}}(x,t)e^{j\omega_0 t}\right\} = \text{Re}\left\{a(t)s(x)e^{j\omega_0 t}\right\} \tag{8.15}$$

For simplicity later we have introduced the notation $f(x,t) = \text{Re}\{\hat{f}(x,t)e^{j\omega_0 t}\}$. Typically, we will take the input spatial profile as Gaussian, that is,

$$s(x) = e^{-x^2/w_{\text{in}}^2} \tag{8.16}$$

For now, however, we keep the analysis general. As derived in Section 8.5.2, the field immediately after the grating may be written

$$e_2(x,t) = \sqrt{\beta_a}\,\text{Re}\left\{\int \frac{d\tilde{\omega}}{2\pi} A(\tilde{\omega})s(\beta_a x)e^{-j\gamma\tilde{\omega}x}e^{j(\omega_0+\tilde{\omega})t}\right\} \tag{8.17a}$$

where

$$\beta_a = \frac{\cos\theta_i^{(0)}}{\cos\theta_D^{(0)}} \quad \text{and} \quad \gamma = \frac{2\pi}{\omega_0 d\cos\theta_D^{(0)}} \tag{8.17b}$$

Here the $e^{-j\gamma\tilde{\omega}x}$ factor imparts the variation in diffraction angle with frequency (assuming diffraction into the $m = -1$ order); and the beam size is scaled by the inverse of an astigmatism factor β_a, which results due to the difference in input and output angles.

Our analysis also exploits the Fourier transform property of a lens, as derived in Section 8.5.1. Specifically, we consider propagation of a scalar, monochromatic, one-dimensional field (x variation only) in the Fresnel or paraxial regime, characterized by spatial variations slow compared to the wavelength scale. For an input field $s_{\text{in}}(x)$ at a plane a distance f in front of a thin lens with focal length f, the resulting field at an output plane a distance f

behind the lens is given by

$$s_{\text{out}}(x) = \sqrt{\frac{j}{\lambda f}} \int dx' \, s_{\text{in}}(x') e^{jkxx'/f} \sim S_{\text{in}}\left(\frac{kx}{f}\right) \tag{8.18}$$

where as usual $k = \omega/c = 2\pi/\lambda$, and $S_{\text{in}}(k_x)$ refers to the spatial Fourier transform of the input spatial profile $s_{\text{in}}(x)$, as defined in eq. (8.109).

Using this Fourier transform property in conjunction with eq. (8.17a) for the field just after the grating, we immediately obtain the field in the back focal plane (Fourier plane) of the first lens (plane 3):

$$\hat{e}_3(x,t) = \sqrt{\frac{j}{\beta_a \lambda f}} \int \frac{d\tilde{\omega}}{2\pi} A(\tilde{\omega}) S\left(\frac{k}{\beta_a f}(x - \alpha\tilde{\omega})\right) e^{j\tilde{\omega}t} \tag{8.19a}$$

where

$$\alpha = \frac{f\gamma}{k} = \frac{\lambda^2 f}{2\pi c d \cos\theta_D^{(0)}} \tag{8.19b}$$

is a spatial dispersion parameter, which describes the proportionality between spatial displacement and optical frequency. In the case of a Gaussian input beam, eq. (8.16), we have

$$S(k_x) \sim e^{-w_{\text{in}}^2 k_x^2/4} \tag{8.20}$$

Therefore, the field at the Fourier plane is written

$$\hat{e}_3(x,t) \sim \sqrt{j} \int \frac{d\tilde{\omega}}{2\pi} A(\tilde{\omega}) e^{-(x-\alpha\tilde{\omega})^2/w_0^2} e^{j\tilde{\omega}t} \tag{8.21a}$$

where

$$w_0 = \frac{f\lambda\beta_a}{\pi w_{\text{in}}} \tag{8.21b}$$

is the radius of the focused beam at the Fourier plane (for any single-frequency component).

These equations describe the basic workings of a spectrometer, in which the spectral dispersion of the grating causes a separation of different optical frequencies at the Fourier plane of a focusing element (lens). This system becomes a tunable filter if a slit is placed at the Fourier plane. Consider the case of a delta-function slit (i.e., a slit much narrower than the spot size of an individual frequency), centered at transverse position x'. For a Gaussian input beam, the power spectral density of the light transmitted through the slit, normalized to the input power spectral density, is proportional to

$$\frac{|E_{\text{out}}(\tilde{\omega})|^2}{|A(\tilde{\omega})|^2} \sim e^{-2\alpha^2(\tilde{\omega}-\omega')^2/w_0^2} \sim \exp\left[-\frac{1}{2}\left(\frac{\lambda w_{\text{in}}}{cd\cos\theta_i^{(0)}}\right)^2 (\tilde{\omega} - \tilde{\omega}')^2\right] \tag{8.22}$$

The center frequency of the passband, $\tilde{\omega}' = x'/\alpha$, is determined by the transverse position of the slit. The spectral resolution of the spectrometer depends on the spot size of the individual frequencies at the Fourier plane (w_0) and the rate of spatial dispersion with frequency (determined by α). From eq. (8.22) the frequency resolution is inversely proportional to the quantity $\lambda w_{\text{in}}/cd \cos\theta_i^{(0)}$. This same quantity appeared in our discussion of the dispersion associated with grating pairs, Section 4.3, where it was identified as the physical delay variation across the output-diffracted beam. Hence, we find that the spectral resolution of a grating spectrometer is inversely proportional to the physical delay variation associated with the grating. Fundamentally, this relationship is a manifestation of the linkage between angular dispersion and tilted intensity fronts, as explained in Section 4.2.2. Consequently, a similar relationship holds for any spectrometer based on angular dispersion, independent of whether the disperser is a grating, a prism, or some other element.

Now we add the spatial mask, with a complex transmission that we denote as $M(x)$, which is the key to pulse-shaping action. The field in plane 4 directly after the mask is simply

$$\hat{e}_4(x, t) = M(x)\hat{e}_3(x, t) \tag{8.23}$$

We consider two cases. First we consider the case where the spectral resolution is very high. Then we return to examine the consequences of a finite spectral resolution.

In a pulse shaper, specifying very high spectral resolution is equivalent to specifying that the spot size of any individual frequency at the Fourier plane is small compared to the spatial scale over which $M(x)$ varies. In this limit the spatial mask modifies the amplitude and phase of individual frequency components but does not modify their spatial profiles. Mathematically, we take $S(\sim)$ in eq. (8.19a) as a delta function and replace $M(x)$ with $M(\alpha\tilde{\omega})$. This gives

$$\hat{e}_4(x, t) = \sqrt{\frac{j}{\beta_a \lambda f}} \int \frac{d\tilde{\omega}}{2\pi} M(\alpha\tilde{\omega}) A(\tilde{\omega}) S\left(\frac{k}{\beta_a f}(x - \alpha\tilde{\omega})\right) e^{j\tilde{\omega}t} \tag{8.24}$$

To get to the plane prior to the second grating (plane 5), we simply perform another spatial Fourier transform:

$$\hat{e}_5(x, t) = \frac{j}{\sqrt{\beta_a}\,\lambda f} \int \frac{d\tilde{\omega}}{2\pi} M(\alpha\tilde{\omega}) A(\tilde{\omega}) e^{j\tilde{\omega}t} \iint dx'\, dx''\, s(x') e^{jkx'(x''-\alpha\tilde{\omega})/\beta_a f} e^{jkxx''/f} \tag{8.25}$$

Performing the integral over x'' first and using the fact that the Fourier transform of $e^{jk_x x}$ is $2\pi\delta(k_x)$, we arrive at

$$\hat{e}_5(x, t) = j\sqrt{\beta_a}\, s(-\beta_a x) \int \frac{d\tilde{\omega}}{2\pi} M(\alpha\tilde{\omega}) A(\tilde{\omega}) e^{j\gamma\tilde{\omega}x} e^{j\tilde{\omega}t} \tag{8.26}$$

Finally, we take into account the diffraction off the second grating. With the antiparallel grating arrangement, the incident and diffracted angles onto the second grating are equal to the diffracted and incident angles onto the first grating, respectively. This means that the effect of this second grating may be expressed as

$$\hat{e}_6(x,t) = \frac{1}{\sqrt{\beta_a}}\hat{e}_5\left(\frac{x}{\beta_a},t\right)e^{-j\gamma\tilde{\omega}x/\beta_a} \tag{8.27}$$

and

$$e_6(x,t) = \mathrm{Re}\left\{js(-x)\int\frac{d\tilde{\omega}}{2\pi}M(\alpha\tilde{\omega})A(\tilde{\omega})e^{j(\omega_0+\tilde{\omega})t}\right\} \tag{8.28}$$

The diffraction off the second grating removes the angular dispersion, resulting in a spatial profile identical to that at the input (except for a flip in x due to passage through a telescopic lens pair). The frequency filter resulting from the mask is simply a scaled version of the mask complex transmission profile. In the time domain, the shaped field is written

$$e_6(x,t) = \mathrm{Re}\left\{js(-x)\left[a(t) * h^{(0)}(t)\right]e^{j\omega_0 t}\right\} \tag{8.29a}$$

where

$$h^{(0)}(t) = \frac{1}{2\pi}\int d\tilde{\omega}\, M(\alpha\tilde{\omega})e^{j\tilde{\omega}t} \tag{8.29b}$$

The shaped field in time is given by the convolution of the input pulse and the infinite resolution impulse response function $h^{(0)}(t)$, which is equal to a scaled version of the inverse transform of the spatial masking function. In this high spectral resolution limit, the pulse-shaping response reflects the spatial mask in a very simple way. This picture provides an excellent description of most of the data described in connection with Section 8.1.1. A further point is that the only pulse reshaping is dictated by the mask; there is no dispersion. This is consistent with the simple argument concerning the dispersion properties of the 4-f grating and lens apparatus presented in Section 4.3.

Turning now to the case of finite spectral resolution, we notice that in general the electric field subsequent to the spatial mask is a nonseparable function of both space and frequency. This occurs because the spatial profiles of the focused spectral components may be altered by the mask (e.g., some spectral components may impinge on abrupt amplitude or phase steps on the mask, whereas others may not). This leads to different diffraction effects for different spectral components and results in an output field that couples space and time beyond the simple and reversible effects of spectral dispersion.

On the other hand, from an applications perspective, one is usually interested in generating a spatially uniform output beam with a single prescribed temporal profile. To obtain an output field that is a function of frequency (or time) only, one must perform an appropriate spatial filtering operation. In the following we analyze the case where such spatial filtering is implemented by focusing into a single-mode optical fiber placed in a Fourier plane of the second diffraction grating. This situation is of practical interest for applications related to optical communications. A similar mode selection operation could also be performed by coupling into a regenerative amplifier for high-power applications. Approximately, such spatial filtering can be performed simply by placing an iris after the pulse-shaping setup.

For our analysis we are interested in taking the masked field from plane 4 and propagating it to the input face of the fiber (plane 7). The portion of the field at plane 7 that corresponds to the single guided spatial mode of the fiber is transmitted; any remaining portion of the field is not guided and is therefore eliminated. Denoting the spatial mode of the fiber as u_F and the field at plane 7 as $\hat{e}_7$, the coupled field is given by

$$\hat{e}_{\text{out}}(x,t) = \frac{\int dx\, \hat{e}_7(x,t) u_F^*(x)}{\int dx\, u_F(x) u_F^*(x)}\, u_F(x) \tag{8.30}$$

Here the first factor gives the complex amplitude of the coupled field, and the second is the spatial mode.

Although we now have the machinery to propagate the masked field given by eqs. (8.19a) and (8.23) to plane 7, it is simpler to propagate the fiber mode backward to the masking plane (plane 4) and perform the overlap integral there. The backward propagation involves a spatial Fourier transform via a lens with focal length f', diffraction off the grating, and a second spatial Fourier transform via a lens of focal length f.[2] After some algebra the fiber mode after back-propagation, $u_4(x)$, is obtained as

$$u_4(x) \sim u_F\left(\frac{f'}{f\beta_a}(\alpha\tilde{\omega} - x)\right) \tag{8.31}$$

Similar to eq. (8.30), the amplitude of the field that will eventually couple into the fiber is proportional to

$$H(\tilde{\omega}) \sim \frac{\int dx\, M(x) S\left[(k/\beta_a f)(x - \alpha\tilde{\omega})\right] u_4^*(x)}{\int dx\, u_4(x) u_4^*(x)} \tag{8.32}$$

where $H(\tilde{\omega})$ is defined via eq. (8.2). The fiber mode is usually approximated as a Gaussian, $u_F(x) = \exp(-x^2/w_F^2)$. Taking the input to the pulse shaper now as Gaussian as well, eq. (8.16), we have

$$H(\tilde{\omega}) \sim \frac{\int dx\, M(x) e^{-(x-\alpha\tilde{\omega})^2/w_0^2} e^{-f'^2(x-\alpha\tilde{\omega})^2/(f\beta_a w_F)^2}}{\int dx\, e^{-2f'^2(x-\alpha\tilde{\omega})^2/(f\beta_a w_F)^2}} \tag{8.33}$$

The most interesting case is when $f\beta_a w_F/f' = w_0$. This is the case when the input field as transformed by the pulse shaper and subsequent lens is perfectly mode-matched to the fiber. In the absence of masking, the entire input field is coupled successfully into the fiber without loss. With this choice of w_0, eq. (8.33) reduces to

$$H(\tilde{\omega}) = \left(\frac{2}{\pi w_0^2}\right)^{0.5} \int dx\, M(x) e^{-2(x-\alpha\tilde{\omega})^2/w_0^2} \tag{8.34}$$

[2] For backward propagation, j is replaced by $-j$ in the formula for spatial Fourier transform by a lens, eq. (8.118), and in the spectral dispersion term introduced by the grating, eq. (8.125).

and the field coupled into the fiber is given by

$$\hat{e}_{\text{out}}(x,t) = \left(\frac{2}{\pi w_0^2}\right)^{0.5} e^{-x^2/w_0^2} \int \frac{d\tilde{\omega}}{2\pi} A(\tilde{\omega}) e^{j\tilde{\omega}t} \int dx\, M(x) e^{-2(x-\alpha\tilde{\omega})^2/w_0^2} \tag{8.35}$$

This result shows that the effective filter in the frequency domain is the mask function $M(x)$ convolved with the spatial *intensity* profile of the beam. The main effect of this convolution is to limit the spectral resolution of the pulse shaper to $\delta\omega = (\ln 2)^{0.5} w_0/\alpha$, in terms of the full-width at half-maximum of the power spectrum. Physical features on the mask smaller than about w_0 are smeared out by the convolution, and this limits the finest features that can be transferred onto the spectrum. By comparing with eq. (8.22), we see that the resolution of a pulse shaper is a factor of $\sqrt{2}$ better than that of a spectrometer built out of the same components. This is a result of the diffraction effects treated above, which did not enter into our discussion of the spectrometer. A related point is that the precise resolution of a spectrometer, pulse shaper, or similar device depends on the details of the measurement performed to determine the resolution.

One consequence of eq. (8.34) is that wavelength components impinging on mask features which vary too fast for the available spectral resolution are in part diffracted out of the main beam and eliminated by the spatial filter. This can lead to phase-to-amplitude conversion in the pulse-shaping process [301]. Conversely, in the limit $w_0 \to 0$, the apparatus provides perfect spectral resolution, and the effective filter is just a scaled version of the mask, following eq. (8.28).

The effect of finite spectral resolution and phase-to-amplitude conversion is illustrated by Fig. 8.15a. The data show the power spectrum measured for a pulse spectrally encoded using a pseudorandom binary phase code in the context of a fiber optic code-division multiple-access communications experiment, as discussed in Section 8.1.1. As for Fig. 8.7, the encoding process impresses a pseudorandom sequence of 31 phases, each either zero or π, onto the spectrum. Even though the code itself is phase-only, a series of dips is present

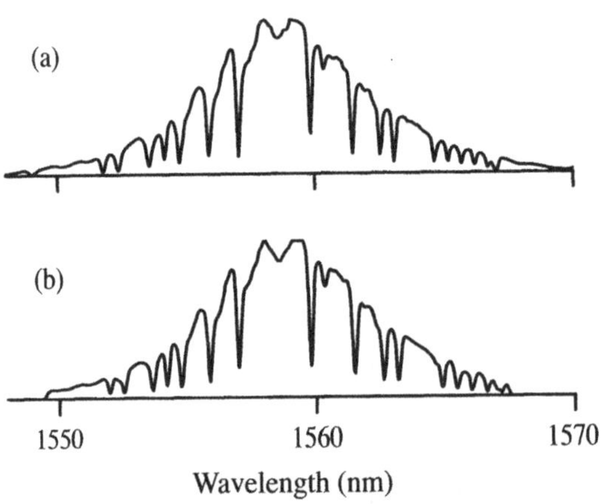

Figure 8.15 (a) Experimental and (b) theoretical power spectra resulting from spectral encoding using a pseudorandom binary-phase mask. For frequency components impinging on phase steps on the mask, diffraction takes power out of the main beam, leading to the spectral dips observed. Adapted from [446]. Copyright © 1998, IEEE.

in the encoded spectrum. Each dip corresponds to a π phase jump in the spectral code and results from power diffracted out of the main beam which is not refocused into the output fiber. Figure 8.15b shows the theoretical spectrum, simulated on the basis of eq. (8.34), the known phase code, and the appropriate values for α and w_0. The agreement between experiment and simulation is excellent. Furthermore, the spectral dips evident in these data fully account for the approximately 40% decrease in peak intensity observed in the decoded pulse of Fig. 8.7b. These experiments, and others like them, show that eq. (8.34) provides an appropriate theoretical description, including the effect of diffraction losses due to mask features, provided that a suitable spatial filter is employed following the pulse-shaping apparatus.

The effect of finite spectral resolution can be understood in the time domain by examining the impulse response $h(t)$. From the inverse Fourier transform of eq. (8.34), the impulse response function can be written as follows:

$$h(t) = h^{(0)}(t)g(t) \tag{8.36a}$$

where

$$g(t) = \exp\left(-\frac{w_0^2 t^2}{8\alpha^2}\right) = \exp\left(-\frac{\beta_a^2 t^2}{2\gamma^2 w_{\text{in}}^2}\right) \tag{8.36b}$$

Thus, the impulse response is the product of the infinite-resolution impulse response and a time aperture function $g(t)$. The time aperture function restricts the time window in which the tailored output pulse can accurately reproduce the response of the infinite-resolution mask. The full-width half-maximum duration of this time window (in terms of intensity) is given by

$$T = \frac{4(\ln 2)^{0.5}\alpha}{w_0} = \frac{2(\ln 2)^{0.5}\lambda w_{\text{in}}}{cd\cos\theta_i^{(0)}} \tag{8.37}$$

The time window is proportional to the number of grating lines illuminated by the input beam multiplied by the period of an optical cycle. For fixed grating parameters, a larger time window can only be obtained by expanding the input beam diameter. The shortest feature in the output-shaped pulse is of course governed by the available optical bandwidth.

These results impose limits on the complexity of shaped pulses, which can be understood in terms of schematic frequency and time-domain plots in Fig. 8.16. The shortest temporal feature that can be realized, δt, is inversely related to the total bandwidth B by $B\delta t \approx 0.44$,

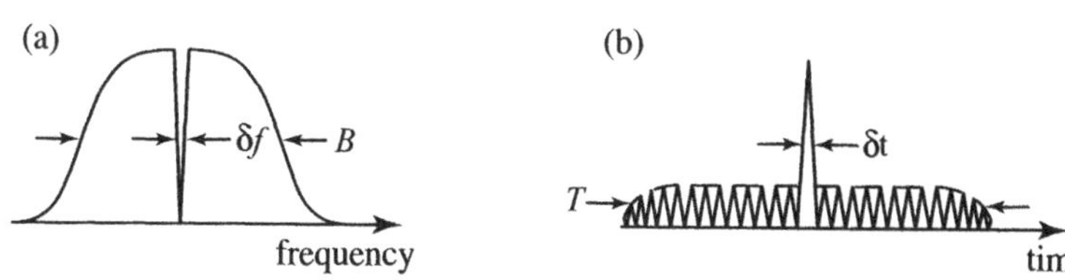

Figure 8.16 (a) Frequency- and (b) time-domain plots used to illustrate limits on the complexity of shaped pulses. From [298], with permission. Copyright © 2000, American Institute of Physics.

and the maximum temporal window T is inversely related to the finest achievable spectral feature δf by $\delta f T \approx 0.44$. Here T, δt, B, and δf all refer to full widths at half maximum. We define η as a measure of the potential complexity of the shaped pulse, as follows:

$$\eta = \frac{B}{\delta f} = \frac{T}{\delta t} \tag{8.38}$$

Thus, η describes the number of distinct spectral features that may be placed into the available bandwidth, or equivalently, the number of independent temporal features that may be synthesized into a waveform. This measure is related to the maximum time–bandwidth product: $BT = 0.44\eta$. An expression for the complexity η is given in terms of the grating and input beam parameters as follows:

$$\eta = \frac{\Delta\lambda}{\lambda} \frac{\pi w_{\text{in}}}{(\ln 2)^{0.5} d \cos\theta_i^{(0)}} \tag{8.39}$$

where $\Delta\lambda$ is the bandwidth (in units of wavelength). For 100-fs input pulses, input beam diameters of a few millimeters, and typical grating parameters, time windows of tens of picoseconds and complexities of 100 to a few hundred may be obtained. Larger time windows are possible using larger input beams; higher complexities are possible using either larger input beams or shorter pulse widths.

Note that eqs. (8.37) and (8.39) are upper bounds on the time window and complexity as set by fundamental optics. Achieving these limits depends on having a sufficient number of spatial features available in the pulse-shaping mask or SLM.

An additional limit arises in the particular case of pixellated SLMs, such as electronically addressable liquid crystal spatial light modulators [451]. In particular, a pixellated SLM can only produce a staircase approximation to a smooth spectral profile. We denote the desired spatial mask corresponding to a desired infinite-resolution impulse response function $h^{(0)}(t)$ as $M^{(0)}(x)$. For a pixellated SLM, the actual mask that can be programmed onto the SLM is written

$$M(x) = \left\{ M^{(0)}(x) \sum_n \delta(x - n\,\delta x) \right\} * \mathrm{sq}\left(\frac{x}{\delta x}\right) \tag{8.40}$$

where δx is the pixel spacing and sq(x) is the unit square function equal to 1 for $|x| \leq \frac{1}{2}$ and equal to 0 otherwise. The requirement that the actual spectral modulation should approximate a smooth function, despite the fixed finite size of the individual modulator elements, limits the temporal range over which pulse shaping can be achieved successfully. Essentially, this is a sampling limitation: The spectrum must vary sufficiently slowly that it is sampled adequately by the fixed modulator elements. For the case of pulse position modulation, for example, the requirement that the phase change per pixel satisfies $\delta\phi \ll \pi$ can be inserted into eq. (8.14) to yield $|\tau| \ll 1/2\delta f$, where $\delta\phi$ and $\delta f = \delta x/2\pi\alpha$ are the phase change and optical frequency span per pixel, respectively.

The impulse response function of the pulse shaper can be obtained by performing an inverse Fourier transform, using eq. (8.40) for the spatial masking function. The overall impulse response function, including both the effects of pixellation and finite resolution, becomes

$$h(t) = \left\{ \sum_n h^{(0)}(t - n\delta f^{-1}) \right\} \operatorname{sinc}(\pi \delta f t) \cdot g(t) \tag{8.41}$$

where $\operatorname{sinc}(u) = \sin(u)/u$. Assuming for the moment that $g(t)$ is much wider than the sinc function in eq. (8.41), the result of pixellation is to produce an output pulse which is the convolution of the input pulse not only with the desired impulse response function $h^{(0)}(t)$, but also with a series of replica impulse response function occurring at times $t = n\delta f^{-1}$. The entire result is weighted by a temporal window function, $\operatorname{sinc}(\pi\delta f t)$, which has its first zeros at $t = \pm\delta f^{-1}$. If $h^{(0)}(t)$ is nonzero only during a time interval $|t| \ll \delta f^{-1}$, the actual shaped pulse will closely approximate the shaped pulse desired, except for a few low-amplitude replicas. On the other hand, if $h^{(0)}(t)$ takes on a significant amplitude for times extending out to $|t| \sim (2\delta f)^{-1}$, then a sort of temporal aliasing, in which the replica waveforms blend into the desired central waveform, will occur.

It is worth mentioning that in actual experiments, if desired the focused spot size at the masking plane can be set approximately equal to the pixel size. In this case the overall spectral filter function from eq. (8.34) is a smoothed version of the spatial profile corresponding to the pixellated SLM. The corresponding Gaussian time window function significantly reduces the replica pulses predicted by eq. (8.41). This allows the experimentalist to give up some spectral resolution for the purpose of smoothing of pixellation effects and elimination of replica pulses.

Returning now to a general discussion, one may also use the framework developed above but without invoking spatial filtering to analyze space–time coupling effects that arise due to diffraction from spatial variations in the pulse-shaping mask. A simple analytical expression that describes this space–time coupling is obtained in [465]. The procedure is to take the masked field represented by eqs. (8.19a) and (8.23), perform the spatial Fourier transform to get to the second diffraction grating, then use eq. (8.27) to obtain the output field in plane 6 after the final grating. After significant manipulation, one obtains

$$\hat{e}_6(x, t) = a(t) * \left[h^{(0)}(t)\, s\left(\frac{\beta_a t}{\gamma} - x \right) \right] \tag{8.42}$$

It is interesting that this formula does not explicitly include any time window. However, the impulse response function is shifted transversely in space in proportion to time. This gives insight into the origin of the time delays that can be obtained in pulse shaping: Components of the masked beam are deflected to slightly different beam paths, with shorter paths corresponding to advances in time and longer paths to delays. In both cases, delay is specified relative to the delay of undeflected field components. An important point is that the deflected field components separate from the undeflected ($t = 0$) beam for time shifts of order T, as defined by eq. (8.37) for a Gaussian beam. Spatial filtering therefore leads to the time window discussed above; for a spatial filter that selects the matched Gaussian mode, we recover exactly the results derived previously. We can also identify an alternative point of view for the origin of the time aperture effect. Because field components corresponding to larger time offsets in the impulse response function suffer larger transverse shifts, they also impinge on the output fiber with increased incidence angles. However, optimum coupling to the fiber mode occurs at normal incidence, with an angular acceptance range inversely proportional to the mode size. For sufficiently large temporal and spatial shifts, incidence angles are increased beyond the angular acceptance range and are suppressed.

8.2 OTHER PULSE-SHAPING TECHNIQUES

8.2.1 Direct Space-to-Time Pulse Shaping

We now consider another pulse-shaping configuration, which we refer to as the direct space-to-time (DST) pulse shaper [305,466–470]. For the DST shaper the impulse response is a scaled version of a spatial mask; this is in contrast to the Fourier transform (FT) shaper, where the impulse response is related to the inverse Fourier transform of the spatial mask. From a practical point of view, the DST pulse shaper is of interest for high-speed parallel-to-serial conversion, where direct mapping between parallel (spatial) input data and serial (time-domain) output data is usually desired. From a fundamental perspective, investigation of the DST shaper leads us to a deeper understanding of the concept of a spectrometer.

A schematic representation of the DST pulse shaper is shown in Fig. 8.17. Similar to an FT shaper, an input short pulse is directed onto a grating and lens setup. However, unlike the FT shaper, there is only a single grating and lens, the spatial mask is placed at the grating, and a narrow slit is placed at the Fourier plane. The optical arrangement becomes identical to that of a spectrometer, with the addition of a spatially patterned input beam.

We can quickly analyze the response of the DST shaper by using results derived in Section 8.1.3. For an input field given by eq. (8.15), with temporal and spatial profiles just prior to the grating of $a(t)$ and $s(x)$, respectively, the field at the back focal plane of the lens is given by eq. (8.19a). If we now assume that a very narrow (delta-function) slit is placed at transverse position $x = 0$, the filtered spectrum immediately after the slit is proportional to

$$\hat{e}_{\text{out}}(t) \sim \int \frac{d\tilde{\omega}}{2\pi} A(\tilde{\omega}) S\left(-\frac{\gamma\tilde{\omega}}{\beta_a}\right) e^{j\tilde{\omega}t} = a(t) * s\left(\frac{\beta_a t}{\gamma}\right) \tag{8.43}$$

In words, the output temporal profile is determined by the input field convolved with a scaled representation of the input spatial profile. The space-to-time conversion constant (seconds per meter) is given by

$$\frac{\gamma}{\beta_a} = \frac{\lambda}{cd\cos\theta_i^{(0)}} \tag{8.44}$$

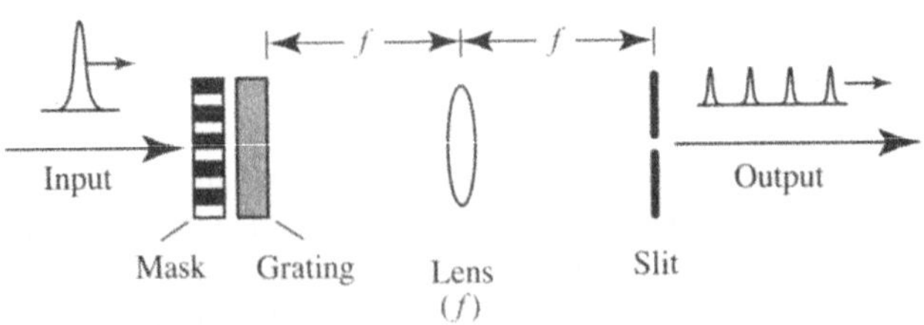

Figure 8.17 Direct space-to-time pulse shaper. For simplicity the grating is depicted in transmission mode, and its tilt is not shown.

The spatial profile just before the diffraction grating is the product of the input beam profile with the masking function $m(x)$. For a Gaussian input beam, we write

$$s(x) = m(x)\exp\left(-\frac{x^2}{w_{\text{in}}^2}\right)\exp\left(-\frac{jkx^2}{2R}\right) \tag{8.45}$$

For the time being, we assume flat phase fronts ($R = \infty$) (i.e., the beam has a waist at the grating). In this case the output temporal field becomes

$$\hat{e}_{\text{out}}(t) \sim a(t) * \left\{ m\left(\frac{\beta_a t}{\gamma}\right)\exp\left(-\frac{\beta_a^2 t^2}{\gamma^2 w_{\text{in}}^2}\right)\right\} \tag{8.46}$$

The impulse response function consists of a scaled version of the mask multiplied by a Gaussian temporal window function corresponding to the input spatial beam profile. Note that the window function is the same as for the Fourier transform pulse shaper, eq. (8.36b), except for a factor of $\sqrt{2}$ difference in the time aperture.

Figure 8.18 shows experimental data confirming these predictions. For Fig. 8.18a the mask consisted of a gold-on-glass film which was completely opaque except for 20 evenly spaced transparent rectangular openings arranged along a line. The cross-correlation measurement of the output waveform shows a train of 20 pulses, one per opening. Figure 8.18b shows data for a similar gold film, but with some of the openings blocked. For each opening that remains (marked "1" on the figure), a pulse appears in the output

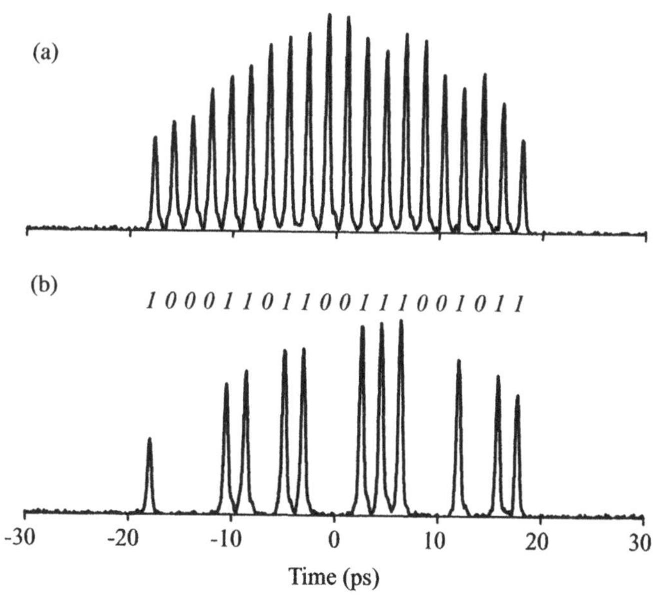

Figure 8.18 Pulse trains from a DST pulse shaper: (a) 20 evenly spaced pulses, resulting from a mask with 20 evenly spaced apertures; (b) ultrafast data packet, resulting from similar mask but with some of the apertures blocked. The pattern of open and closed apertures is denoted with "1"s and "0"s. Adapted from [468].

waveform, whereas for each opening that is blocked (marked "0"), the pulse disappears completely. In both cases the pulse spacing is in agreement with that predicted based on the experimental parameters and the time-to-space conversion constant of eq. (8.44). Furthermore, the output pulse trains are observed to roll off with a roughly Gaussian window. This roll-off arises from the Gaussian beam profile incident on the mask, as predicted by eq. (8.46).

To gain intuitive understanding, consider the case without a spatial mask [i.e., $m(x) = 1$]. This is simply a spectrometer with a Gaussian input beam. We know from Section 8.1.3 that for any single frequency, the spot size at the Fourier plane and hence the bandwidth transmitted through a narrow slit is inversely proportional to the input spot size. Since the temporal spread in the output waveform is inversely proportional to the transmitted bandwidth, we find immediately that the output pulse duration is proportional to the input spot size. Thus, the impulse response function of a conventional spectrometer already yields a simple proportionality between the spatial description of the input beam and the temporal description of the output. Our discussion above shows that this proportionality generalizes to essentially arbitrary input beam profiles.

It is also helpful to recall that the intensity front of a pulse is always tilted when subject to angular dispersion. This tilt means that the delay varies across the spatial aperture of a beam diffracted by a grating. One may then understand the operation of the DST shaper as follows: A spatial mask that blocks certain spatial locations of the input field is also blocking the corresponding temporal instants of the output field. Using the pulse tilt analysis of Section 4.2.2, one may easily show that this picture yields a space-to-time conversion constant identical to that of eq. (8.44).

Let us now generalize to the case where the narrow slit is centered at transverse location $x = x'$ rather than $x = 0$ as before. For the conventional spectrometer, such translation of the output slit is used to tune the center frequency of the optical passband. Again using eq. (8.19a), we have

$$\tilde{e}_{\text{out}}(t) \sim \int \frac{d\tilde{\omega}}{2\pi} A(\tilde{\omega}) S\left[\frac{\gamma}{\beta_a}\left(\frac{x'}{\alpha} - \tilde{\omega}\right)\right] e^{j\tilde{\omega}t} \tag{8.47}$$

The quantity $S\left[(\gamma/\beta_a)\left((x'/\alpha) - \tilde{\omega}\right)\right]$ represents the spectral response function. Translation of the slit simply shifts the spectral response function to a new center frequency position $\tilde{\omega}' = x'/\alpha$. The ability to shift the spectral response function of the DST pulse shaper via slit position is completely analogous to the behavior of the conventional spectrometer, but generalized to essentially arbitrary passband profiles.

Data demonstrating this property are shown in Fig. 8.19. For these measurements the input mask to the DST pulse shaper was the same 20-opening gold-on-glass mask as for Fig. 8.18a, and the output power spectrum in response to a short pulse input was measured for various transverse positions of the output slit. Since as shown previously, this mask generates a train of regularly spaced pulses in the time domain, the output power spectrum is expected to consist of a regularly spaced series of spectral peaks, consistent with the data. Furthermore, the entire pattern of spectral peaks shifts uniformly in frequency, without distortion, in proportion to the shift in slit position. The variation in the *amplitudes* of the different power spectra is simply a reflection of the bandwidth of the input pulse, expressed through the $A(\tilde{\omega})$ factor.

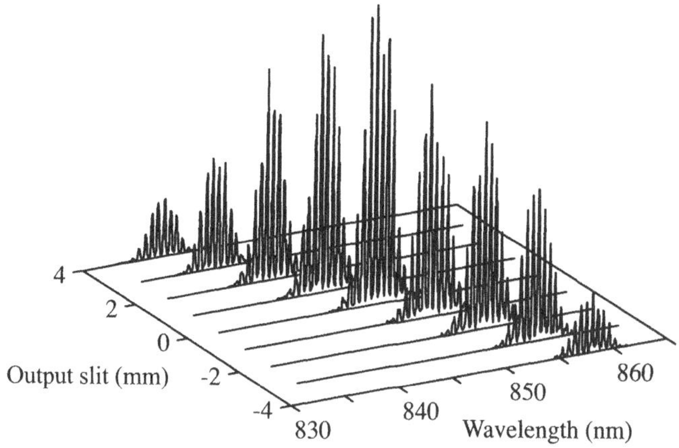

Figure 8.19 Spectra from the DST pulse shaper as a function of transverse position of the output slit. The spectra translate along the frequency axis according to the translation of the slit. Adapted from [470]. Copyright © 2001, IEEE.

By evaluating the integral in eq. (8.47), we obtain the response of the DST shaper to an input pulse. This gives

$$\hat{e}_{\text{out}}(t) = a(t) * \left\{ s\left(\frac{\beta_a t}{\gamma}\right) \exp\left(\frac{jx't}{\alpha}\right) \right\} \tag{8.48}$$

This result predicts that the pulse response of the DST shaper is independent of slit position x', except for a frequency shift. That is, the temporal intensity profile does not vary with slit position. This property has also been confirmed by experiment [470].

Finally, let us consider the case where the input beam has a curved phase front [i.e., R is finite in eq. (8.45)]. For a narrow slit at $x = 0$, the output temporal profile becomes

$$\hat{e}_{\text{out}}(t) \sim a(t) * \left\{ m\left(\frac{\beta_a t}{\gamma}\right) \exp\left(-\frac{\beta_a^2 t^2}{\gamma^2 w_{\text{in}}^2}\right) \exp\left(-\frac{jk\beta_a^2 t^2}{2\gamma^2 R}\right) \right\} \tag{8.49}$$

An interesting point is that for a sufficiently short input pulse, the output intensity profile is independent of R. However, the quadratic spatial phase at the input is converted into a quadratic temporal phase at the output. Thus, the output pulse picks up a frequency modulation, which broadens its spectrum. The case without an input mask, $m(x) = 1$, is particularly informative. This represents a conventional spectrometer, but with a converging or diverging beam incident on the grating. Physically, we expect the spectral resolution of the spectrometer to be degraded, since the field no longer comes to a focus in the plane of the slit. This is indeed true, but what we learn in addition from eq. (8.49) is that the spectral broadening arises from the frequency modulation that is induced in the impulse response function. Further analysis [470] shows that when the slit is moved to the plane of the new focus, the intensity profile continues to remain invariant; however, the frequency modulation disappears and the spectral resolution is restored to its original value. Similar behavior occurs for a collimated input beam ($R = \infty$) when the slit is moved out of the focal plane of the lens. Again, the spectral broadening that occurs because the field does not focus at the plane of the slit is seen to arise from a frequency modulation on the impulse response function [470].

8.2.2 Acousto-optic Dispersive Filters

In this section we consider a different type of pulse shaper, based on acousto-optic interactions in a birefringent photoelastic medium. For pulse-shaping applications such devices are generally restricted to systems with limited pulse repetition rates (e.g., amplified femtosecond systems).

Figure 8.20 shows the basic geometry.[3] The input optical field is launched with polarization along one of the principal axes of a birefringent crystal. In the absence of any interaction, light is extinguished at the output by a crossed polarizer. An acoustic wave launched collinearly with the optical wave produces an additional, periodic stress-induced birefringence that couples light from the input polarization into the orthogonal polarization state (which is transmitted by the polarizer). When the acoustic wave is driven by a single electrical tone, the result is an optical bandpass filter, with center frequency that can be tuned by varying the electrical (hence acoustic) frequency. In this mode of operation, the device is known as an acousto-optic tunable filter [473,474]. For pulse-shaping applications the optical input is a pulse, and the acoustic wave is time-varying, corresponding to a superposition of different acoustic frequencies [471,475,476]. As we shall see, in this mode the optical pulse is shaped under control of the acoustic waveform, which in turn is conveniently programmed by connecting a radio-frequency arbitrary waveform generator to the acoustic transducer. In this pulse-shaping mode of operation, the device has been termed an *acousto-optic programmable dispersive filter* (AOPDF) [471,476].

The interaction between two optical fields and one acoustic wave in the AOPDF bears similarity to the three-wave optical interactions studied in Section 5.5. As in the three-wave optical interactions, phase matching plays a key role. Consider an input optical field with frequency ω_1 and polarization corresponding to refractive index n_1, which is coupled to

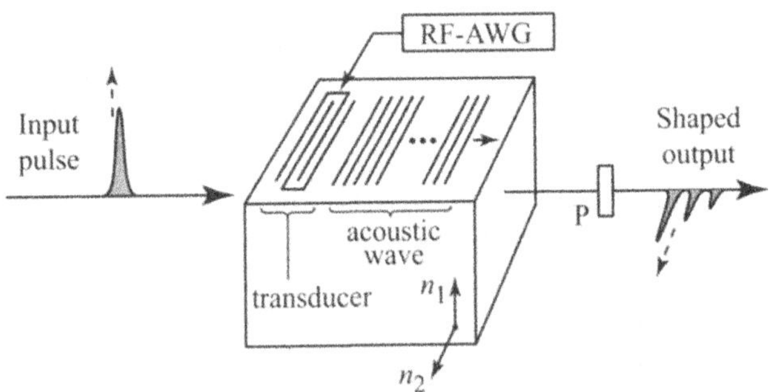

Figure 8.20 Geometry of an acousto-optic programmable dispersive filter. Pulse shaping is achieved via interaction of co-propagating acoustic and optical waves in a birefringent photoelastic medium. The acoustic wave is controlled by means of a radio-frequency arbitrary waveform generator (RF-AWG), and the optical output is taken through a polarizer (P).

[3] This simple collinear geometry is sufficient to explain the essentials of pulse-shaping operation. Considerations related to the quasi-collinear as opposed to truly collinear beam geometries of real devices are discussed in [471,472].

a second optical field with frequency ω_2 and index n_2 by an acoustic wave at frequency Ω (may be positive or negative) and propagation velocity v. Conservation of energy then requires

$$\omega_2 = \omega_1 + \Omega \tag{8.50}$$

while conservation of momentum, or phase matching, dictates that

$$\frac{\omega_2 n_2}{c} = \frac{\omega_1 n_1}{c} + \frac{\Omega}{v} \tag{8.51}$$

Combining these equations and making use of the fact that the acoustic velocity is orders of magnitude smaller than the speed of light, we obtain

$$\omega_1 \approx \frac{\Omega}{\alpha} \quad \text{and} \quad \omega_2 \approx \omega_1 \tag{8.52a}$$

where

$$\alpha = (n_2 - n_1)\frac{v}{c} \tag{8.52b}$$

These expressions, which are valid to lowest order in v/c, already show that phase matching, hence optical transmission, occurs at an optical frequency selectable through the choice of acoustic frequency.

Consider now time-varying optical fields with time profiles $e_1(t)$ and $e_2(t)$ with Fourier transforms $E_1(\omega)$ and $E_2(\omega)$. We take e_1 as the input pulse and e_2 as the shaped pulse that will be generated. If the acoustic waveform is given by $s_{ac}(t)$ with Fourier transform $S_{ac}(\Omega)$, the phase-matching conditions suggest that under weak conversion conditions, the field generated is given in the Fourier domain by

$$E_2(\omega) \sim E_1(\omega) S_{ac}(\alpha\omega) \tag{8.53}$$

The time-domain waveform is obtained by taking the inverse Fourier transform, with the result

$$e_2(t) \sim e_1(t) * s_{ac}\left(\frac{t}{\alpha}\right) \tag{8.54}$$

The impulse response of the AOPDF is a directly scaled version of the applied acoustic waveform. The scaling factor α can be either positive or negative, depending on whether the input polarization is aligned with the fast or slow birefringent axis. For the case where α is negative, the impulse response is time-reversed compared to the applied acoustic wave.

The simple results derived above are valid in the limit of high spectral resolution (i.e., sufficiently narrow phase-matching bandwidth). To quantify effects due to finite spectral resolution, we can apply the results derived for sum frequency generation in Section 5.5.1.

In particular, using eq. (5.148), the response to an optical impulse $a_1(t) \sim \delta(t)$ is given by

$$a_2(t_2') \sim \mathrm{sq}\left(\frac{t_2'}{T} - \frac{1}{2}\right) \tilde{s}_{\mathrm{ac}}\left(\frac{t_2'}{\alpha}\right) \tag{8.55}$$

where L is the interaction length, $T = (n_1 - n_2)L/c$, $t_2' = t - n_2 L/c$ is the retarded time corresponding to a reference frame moving with the generated shaped pulse, $\mathrm{sq}(u)$ is a unit square function centered at $u = 0$, and a_1, a_2, and $\tilde{s}_{\mathrm{ac}}$ refer to complex envelope functions in notation similar to that of eq. (5.135). The implied carrier frequencies ω_1, ω_2, and Ω are assumed to correspond to a phase-matched triplet.

We interpret this result as follows. The transit time τ for light to pass through the AOPDF is bounded by $n_f L/c < \tau < n_s L/c$, where n_f and n_s are the fast and slow refractive indices, respectively. Consider the case where the input pulse is launched along the fast axis, $n_1 = n_f$. From a time-domain perspective, reshaping occurs when the acoustic wave induces portions of the light to travel along the slow axis through parts of the device. The relative transit time, referred to that of the unshaped pulse, is increased by an amount that falls in the range zero to $|T|$. This leads to a rectangular time aperture function for the pulse-shaping operation. On the other hand, the acoustic wave propagates so slowly that it appears essentially frozen within an optical transit time. Hence, only those portions of the acoustic wave launched in the time interval $-T_{\mathrm{ac}} < t < 0$, where $T_{\mathrm{ac}} = L/v$ is the acoustic transit time, contribute to the pulse-shaping operation.

The frequency response of the AOPDF is obtained by Fourier transforming the impulse response. In the case of a continuous, single-frequency acoustic wave ($\tilde{s}_{\mathrm{ac}} = 1$), one obtains a passband given by

$$A_2(\tilde{\omega}) \sim \mathrm{sinc}\left(\frac{\tilde{\omega} T}{2}\right) A_1(\tilde{\omega}) \tag{8.56}$$

The FWHM frequency resolution is $\delta f = 0.886 T^{-1}$.

We are now in a position to discuss the complexity or time–bandwidth product that may be attained in AOPDF pulse shaping. In the absence of factors constraining the optical bandwidth (see below), the optical and acoustic bandwidths are related simply through $B_{\mathrm{opt}} = B_{\mathrm{ac}}/|\alpha|$. We may therefore write the optical time–bandwidth product as

$$B_{\mathrm{opt}} T = \left| \frac{B_{\mathrm{ac}}}{\alpha} \frac{(n_2 - n_1)L}{c} \right| = B_{\mathrm{ac}} \frac{L}{v} = B_{\mathrm{ac}} T_{\mathrm{ac}} \tag{8.57}$$

Fundamentally, the optical time–bandwidth product can be as large as that of the acoustic wave. Practically, this translates into time–bandwidth products and pulse-shaping complexities of several hundred.

It is worth considering typical numbers. The AOPDF of [476] was based on a TeO_2 crystal characterized by a length of 2.5 cm, an index difference $|n_2 - n_1| \approx 0.04$, and an acoustic velocity $v \approx 10^5$ cm/s. These parameters yield $\alpha \approx 1.3 \times 10^{-7}$ (i.e. a scaling factor of roughly 10^7 between phase-matched acoustic and optical frequencies). The acoustic transducer allowed for an operating bandwidth of 20 MHz around a 52.5-MHz acoustic center frequency, corresponding to an optical bandwidth of 150 THz around a 375-THz

(800-nm) center frequency. The time aperture for pulse shaping is $T \approx 3.3$ ps. The acoustic transit time is $T_{ac} = 25$ μs, which yields a time–bandwidth product of 500.

Based on such numbers, we may draw several conclusions:

- High-complexity pulse-shaping operations may be achieved.
- Time apertures are limited to the several picosecond range. Therefore, the full potential for pulse-shaping complexity is obtained only in systems involving very short pulses (e.g., a few tens of femtoseconds and below).
- The maximum rate at which the pulse-shaping function may be reprogrammed is limited by the transit time of the acoustic wave through the device. Therefore, application of AOPDFs is generally restricted to amplified femtosecond systems where the time between pulses exceeds the acoustic transit time. This allows independent selection of the pulse shape for each laser shot. As an important special case, each laser pulse may be chosen to have the same waveform. In contrast, for higher-repetition-rate systems, the traveling-wave nature of the acoustic wave means that successive laser pulses see different acoustic waveforms. Successive pulses are shaped differently and cannot be controlled independently.

In addition, the combination of short pulse durations and significant device length means that material dispersion is important. Such dispersion must be compensated to obtain desired pulse-shaping operation. Although dispersion compensation may be achieved by suitable programming of the AOPDF itself, the finite time aperture for pulse shaping limits the amount of dispersive spreading that may be compensated. Assuming that second-order spectral phase dominates, this leads to an upper bound on the optical bandwidth over which self-compensation is possible:

$$B_{opt} \leq \frac{|T|}{2\pi\psi_2} \tag{8.58}$$

For the 2.5-cm TeO_2 AOPDF discussed above, the second-order spectral phase is given as $\psi_2 = 12{,}500\,\text{fs}^2$. This restricts the optical bandwidth to approximately 40 THz and the optical time–bandwidth product to roughly 140.

For these reasons the AOPDF is often employed in amplified femtosecond systems in configurations such as that of Fig. 8.21. (For a discussion of the chirped pulse amplifier scheme assumed in the figure, see Section 8.4.2). In such a scheme the dispersive pulse stretcher and compressor may be designed such that the second-order spectral phase of the entire system, including amplifier, stretcher, AOPDF, and compressor, is in balance. In this case the AOPDF does not need to be programmed to self-compensate its own dispersion, and the restriction of eq. (8.58) is lifted. In such configurations the AOPDF is attractive

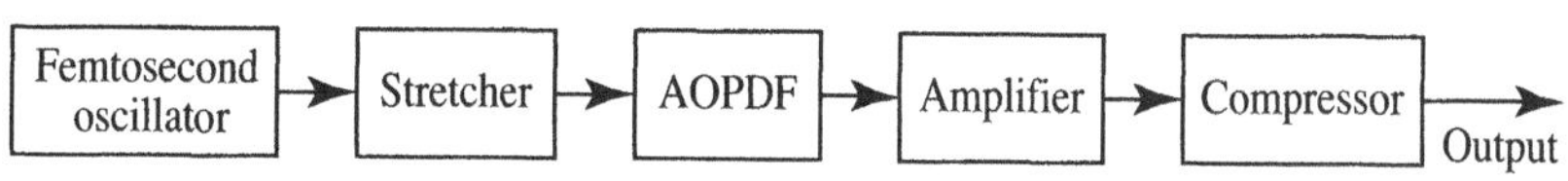

Figure 8.21 Typical configuration for using an acousto-optic programmable dispersive filter in an amplified femtosecond system.

due to its compact size and robust physical nature. Useful functionalities of the AOPDF in amplified configurations include the following:

- Phase filtering to compensate for residual higher-order spectral phase from the overall system. Here the stretcher and compressor provide large and approximately balanced dispersions of opposite sign, while the AOPDF provides a smaller but fully programmable dispersion. This provides an important dispersion fine-tuning capability, which is challenging to achieve over very large bandwidths with fixed components. In this respect the function is similar to that depicted in Fig. 8.12, where a programmable Fourier transform shaper provided programmable dispersion fine-tuning capability in a system involving long lengths of fibers with dispersions of opposite signs.
- Amplitude filtering to mitigate spectral narrowing arising due to the finite gain bandwidth of the amplifying medium. Here the AOPDF is programmed to provide a transmission response shaped according to the inverse of the gain spectrum. This increases the optical bandwidth that can be obtained at the output of the amplifier. Together with dispersion fine-tuning, this leads to shorter output pulses.
- Arbitrary pulse shaping according to experimental needs. Such arbitrary shaping may be overlaid with the other functions cited above. That is, the AOPDF is programmed to achieve desired output waveforms while simultaneously compensating for residual dispersion and gain narrowing.

We close this section by pointing out the similarity of the pulse-shaping mechanism with that involved in second-harmonic generation using tailored quasi-phase-matched (QPM) gratings (Section 5.4.3) and in sum and difference frequency generation (Sections 5.5.1 and 5.5.2). Each of these cases involves coupling between distinct waves with different velocities; pulse shaping is achieved by controlling the longitudinal position at which conversion from one wave to another occurs. The differences lie in what mechanism is used to control the coupling: fixed QPM gratings for second-harmonic generation, temporal walk-off between two input waves both traveling at light velocities in sum and difference frequency generation, and slowly propagating acoustic waves for the AOPDF.

8.3 CHIRP PROCESSING AND TIME LENSES

In this section we describe methods of pulse processing in which large amounts of dispersion are used to advantage. We begin by discussing a close analogy that exists between diffraction of spatial beams and dispersive propagation of pulses. Appealing to the familiar case of spatial optics provides insight into some of the temporal processing systems we consider.

8.3.1 Space–Time Duality

We considered short pulse propagation in detail in Chapter 6. In a retarded time frame moving with the pulse, propagation in a lossless, linear medium is described by

$$\frac{\partial a}{\partial z} = \frac{j\beta_2}{2}\frac{\partial^2 a}{\partial t^2} \tag{8.59}$$

where only the lowest-order dispersion term is retained. Here $a(t)$ is a slowly varying envelope function in time, and either plane waves or a single-mode waveguide are assumed (so that spatial variations do not enter). By transforming to the frequency domain, the solution may easily be written as

$$a(t,z) = \frac{1}{2\pi}\iint d\tilde{\omega}\, dt'\, a_{\text{in}}(t')\, e^{-j\beta_2\tilde{\omega}^2 z/2}\, e^{j\tilde{\omega}(t-t')} \tag{8.60}$$

The multiplication by a quadratic phase function in the frequency domain is equivalent to convolution with a quadratic phase function in the time domain. By performing the integral over $\tilde{\omega}$, we obtain

$$a(t,z) = \sqrt{\frac{-j}{2\pi\beta_2 z}}\int dt'\, a_{\text{in}}(t')\, e^{j(t-t')^2/2\beta_2 z} \tag{8.61}$$

As derived in Section 1.3.3, paraxial propagation of scalar, monochromatic fields along the z direction is given by

$$\frac{\partial u}{\partial z} = \frac{-j}{2k}\left(\frac{\partial^2 u}{\partial x^2} + \frac{\partial^2 u}{\partial y^2}\right) \tag{8.62}$$

Here u refers to a slowly varying envelope function in space. For the case of one-dimensional fields (no y variation), the solution is again written formally by transforming to the Fourier domain:

$$u(x,z) = \frac{1}{2\pi}\iint dk_x\, dx'\, u_{\text{in}}(x')\, e^{jk_x^2 z/2k}\, e^{jk_x(x-x')} \tag{8.63}$$

By performing the integral over k_x, we obtain

$$u(x,z) = \sqrt{\frac{j}{\lambda z}}\int dx'\, u_{\text{in}}(x') e^{-jk(x-x')^2/2z} \tag{8.64}$$

This is essentially the same as the Fresnel diffraction integral derived in Section 8.5.1. However, here we are considering only the envelope function, so the e^{-jkz} factor in eq. (8.113) does not appear.

By comparing these equations, we see that plane-wave pulse propagation and paraxial propagation of one-dimensional monochromatic fields are described by the same mathematics [70,477,478]. One may transform between the equations for plane-wave pulse propagation and one-dimensional spatial propagation by making the following identifications:

$$a \leftrightarrow u \qquad \beta_2 \leftrightarrow -\frac{1}{k} \qquad t \leftrightarrow x \tag{8.65}$$

One interesting difference physically is that k is a positive quantity, whereas the dispersion β_2 may be either positive or negative.

Of particular interest to us will be the fact that propagation of both temporal and spatial fields may be described as the response of a quadratic phase filter [479]. Let us consider two limiting cases. First, taking the temporal example, we consider the low dispersion limit, $\beta_2 z \to 0$. In this case the coefficient of the quadratic phase term in eq. (8.61) is very large, $1/2\beta_2 z \to \infty$, and only the small time interval for which $t - t' \approx 0$ contributes to the integral. Therefore, in this limit we may write

$$e^{j(t-t')^2/2\beta_2 z} \longrightarrow \sqrt{2j\pi\beta_2 z}\,\delta(t-t') \tag{8.66}$$

where we used the result that $\int dt\, e^{-\Gamma t^2} = (\pi/\Gamma)^{1/2}$. Substituting now into eq. (8.61) yields the reassuring result that for sufficiently small dispersion, the input pulse is unchanged. Analogously, in the spatial domain we find that the input field is unchanged for $2z/k \to 0$. This regime is known as the *near field*.

The second limiting case is that of large dispersion. Assume that the input pulse is time limited [i.e., $a(t) = 0$ for $|t| > T/2$].[4] Then if $|\beta_2 z| \gg T^2/8$, the t'^2 term in eq. (8.61) may be neglected. In the terminology of Chapter 6, this corresponds to the regime $z \gg L_D$. The output pulse becomes

$$a(t,z) \approx \sqrt{\frac{-j}{2\pi\beta_2 z}}\, e^{jt^2/2\beta_2 z} \int dt'\, a_{\text{in}}(t')\, e^{-jtt'/\beta_2 z} \tag{8.67a}$$

$$a(t,z) \approx \sqrt{\frac{-j}{2\pi\beta_2 z}}\, e^{jt^2/2\beta_2 z}\, A\left(\frac{t}{\beta_2 z}\right) \tag{8.67b}$$

where A is the Fourier transform of a. Thus, we have the important result that in the limit of large dispersion, the output pulse is a scaled version of the input spectrum, multiplied by a quadratic temporal phase function. The output pulse is strongly chirped, and optical frequency is mapped onto time, with mapping factors

$$\frac{\partial\omega}{\partial t} = \frac{1}{\beta_2 z} \quad \text{and} \quad \frac{\partial f}{\partial t} = \frac{1}{2\pi\beta_2 z} \tag{8.68}$$

The spatial analog follows immediately. If the input field is limited in space, $u(x) = 0$ for $|x| > X/2$, the output field for $|z/k| \gg X^2/8$ is given by

$$u(x,z) \approx \sqrt{\frac{j}{\lambda z}}\, e^{-jkx^2/2z}\, U\left(\frac{-kx}{z}\right) \tag{8.69}$$

where U is the Fourier transform of u. The output field is equal to a quadratic spatial phase function, multiplied by a scaled version of the spatial Fourier transform of the input field. This limit is known as the *far-field* or *Fraunhofer diffraction regime* [9,168]. Note that requiring $|z/k| \gg X^2/8$ is equivalent to stipulating that the spatial extent of the diffracted field is much larger than that of the input field.

[4] In practice, our results will also hold for pulses that are not strictly time limited as long as $a \to 0$ sufficiently quickly for $|t| > T/2$.

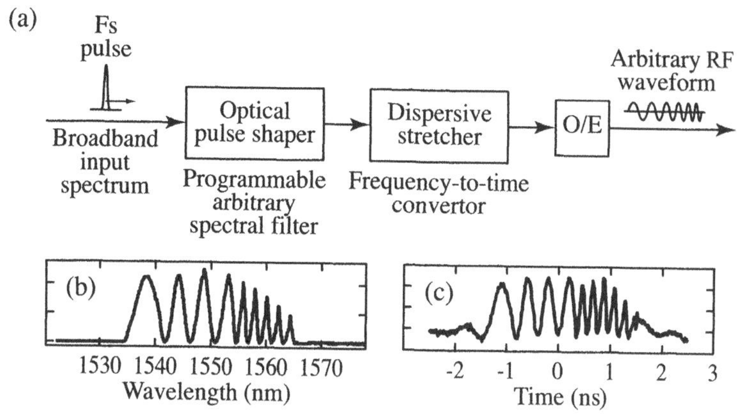

Figure 8.22 RF waveform generation via frequency-to-time mapping: (a) block diagram; (b,c) experimental data showing (b) shaped optical spectrum and (c) temporal profile of resulting frequency-stepped RF waveform. (b,c) Adapted from [481]. Copyright © 2005, IEEE.

8.3.2 Chirp Processing

The Fourier transform inherent in pulse propagation in the large dispersion limit is useful for pulse manipulation and measurement. Consider, for example, the setup of Fig. 8.22a, which exploits frequency-to-time mapping in a dispersive medium for radio-frequency (RF) electrical arbitrary waveform generation at frequencies beyond the capabilities of conventional electronic technologies [480,481]. Here short optical pulses from a mode-locked laser are first passed through a Fourier transform pulse shaper, which is used to impress an arbitrary filter function onto the optical spectrum. The shaped pulses are then dispersed in a long length of single-mode fiber. The chromatic dispersion of the fiber results in a time-domain optical waveform exhibiting the shape of the optical filter function applied in the pulse shaper, following eq. (8.67b). Thus, after optical-to-electrical (O/E) conversion using a fast photodetector (which produces an output electrical signal proportional to the optical intensity), the RF waveform exhibits the shape of the optical power spectrum.

Figure 8.22 also shows an example of experimental results. Here the laser bandwidth was approximately 30 nm centered around 1550 nm, and the fiber length was 5.5 km. This yields a frequency-to-time conversion constant of about 90 ps/nm. In this example the optical power spectrum has been shaped to exhibit a frequency-stepped sinusoidal intensity modulation. After stretching, this spectrum gives rise to an RF waveform with abrupt frequency modulation, exhibiting one half cycle at 1.25 GHz, two and a half cycles at 2.5 GHz, and four cycles at 5 GHz. The temporal profile of the RF waveform generated closely resembles a scaled version of the optical power spectrum, as expected. The 2.7-ns time aperture of the generated waveform is consistent with the optical bandwidth and the frequency-to-time conversion constant, while the sense of the RF waveform (longer wavelength spectral features arrive later in time) is consistent with the anomalous dispersion of the fiber. Generation of waveforms as shown in Fig. 8.22, with multi-gigahertz instantaneous bandwidth and abrupt frequency changes on a cycle-by-cycle basis, is not possible using current electronic techniques.

It is important to consider limits governing waveform generation via this method. Fundamentally, frequency-to-time mapping occurs only when $|\beta_2 z| \gg T^2/8$. Since the duration of the input waveform T and the width of its finest spectral feature δf are related through $\delta f T \sim 1$, we find that there is a minimum spectral feature size consistent with the large dispersion requirement:

$$|\beta_2 z|\delta f^2 \gg \frac{1}{8} \tag{8.70}$$

Now assuming that the conditions for frequency-to-time mapping are satisfied, spectral feature δf is chirped to duration

$$\delta t_{\text{chirp}} = 2\pi\delta f|\beta_2 z| \tag{8.71}$$

Combining these equations, we have

$$\delta t_{\text{chirp}} \gg \frac{\pi}{\sqrt{2}}\sqrt{|\beta_2 z|} \tag{8.72a}$$

and

$$\delta t_{\text{chirp}} \gg \frac{\pi}{4\delta f} \tag{8.72b}$$

The latter result has a simple interpretation: The minimum temporal feature that may be generated via frequency-to-time mapping must exceed the bandwidth-limited pulse duration of the corresponding spectral feature. As a simple measure of the minimum time feature that can be generated, we define $\delta t_{\text{chirp-min}}$ as equal to the right side of eq. (8.72a):

$$\delta t_{\text{chirp-min}} = \frac{\pi}{\sqrt{2}}\sqrt{|\beta_2 z|} \tag{8.73}$$

The minimum temporal feature scales as the square root of the total dispersion. Note that in this limit, signs of distortion should already be visible in the finest features arising from the frequency-to-time mapping process.

It is also interesting to note that the bandwidth limit for the shortest features in the input pulse is given by $\delta t \sim B^{-1}$, whereas the duration of the chirped pulse is given by $T_{\text{chirp}} = 2\pi B|\beta_2 z|$. Here B is the optical bandwidth. Multiplying these expressions and using eq. (8.73), we obtain

$$\delta t^2_{\text{chirp-min}} = \frac{\pi}{4} T_{\text{chirp}}\delta t \tag{8.74}$$

The minimum temporal feature in frequency-to-time mapping scales as the geometric mean of the bandwidth-limited duration of the input pulse and the time aperture of the chirped output.

Finally, we provide a bound on waveform complexity η. Defining complexity according to eq. (8.38) and using eq. (8.70), we have

$$\eta = \frac{B}{\delta f} \sim B\sqrt{|\beta_2 z|} \tag{8.75}$$

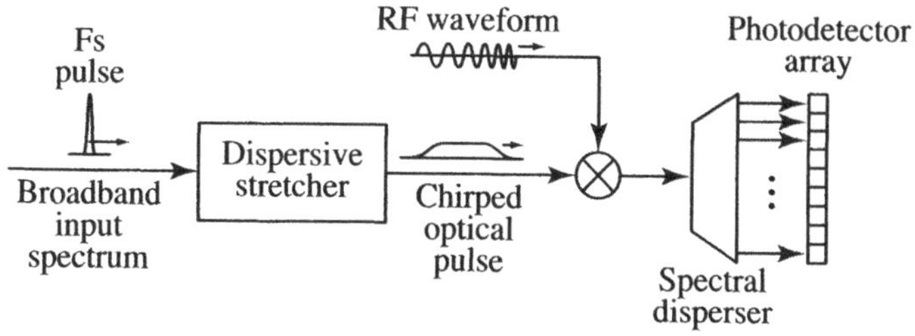

Figure 8.23 RF waveform measurement via time-to-frequency mapping.

The complexity fundamentally is limited by the optical bandwidth and the square root of the dispersion.

Converse to the frequency-to-time mapping described above for RF waveform generation, time-to-frequency mapping may be used for measurement of fast RF waveforms. The block diagram is shown in Fig. 8.23. A modulator is used to impress a fast electrical signal onto an optical pulse. Prior to the modulator, the optical pulse is chirped in a dispersive medium to a duration exceeding that of the electrical signal of interest. Because of the dispersion, once again there is linear mapping between time and frequency. Therefore, we may think of different temporal samples of the RF waveform being modulated onto different frequency components of the optical beam. Subsequent to the modulator, the optical frequencies are spatial dispersed, using a grating for example, and directed onto a photodetector array. Different detector elements therefore respond to different temporal samples of the RF waveform. This time-to-frequency mapping process allows parallel detection of electrical signals too fast to be resolved by an individual photodetector and oscilloscope. This approach has been applied for measurement of freely propagating electromagnetic transients in the terahertz regime [482,483] as well as for parallelization of lower-frequency RF signals in photonically aided analog-to-digital conversion systems [484].

Again it is important to consider temporal resolution limits. Assume that an RF temporal feature of duration δt_{RF} is modulated onto the optical signal. Such a feature creates new frequency components in a bandwidth approximately equal to $\delta t_{\mathrm{RF}}^{-1}$. Unless this bandwidth remains less than the original optical frequency sweep present within the identical temporal span δt_{RF}, the time-to-frequency mapping process will be disturbed. Using eq. (8.68) for the frequency sweep of the chirped input, we obtain

$$\delta t_{\mathrm{RF}}^2 \gg 2\pi |\beta_2 z| \tag{8.76}$$

Neglecting numerical factors of order unity, this limit is the same as eq. (8.72a) for frequency-to-time mapping. A similar result was obtained by [485] for the special case of measuring single-cycle terahertz transients. The temporal resolution again scales as the square root of the total dispersion.

8.3.3 Time Lens Processing

In the preceding section we explored forms of pulse processing exploiting large dispersion (i.e., large quadratic spectral phase variation). Here we discuss additional processing possibilities that arise when phase modulation (i.e., quadratic temporal phase variation) is

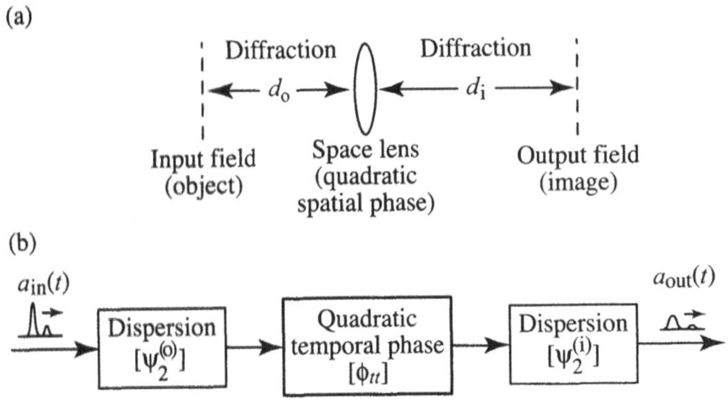

Figure 8.24 Analogy between (a) spatial and (b) temporal imaging systems.

also available. The ability to impose both temporal and spectral phase variation leads to temporal imaging systems in complete analogy with well-known spatial imaging systems [478,486]. Temporal imaging systems can potentially provide a number of functionalities for ultrafast optical waveforms, such as expansion and compression (magnification and demagnification) in the time domain, time reversal, reduction of timing jitter, and waveform measurement.

The space–time imaging analogy is illustrated in Fig. 8.24. A conventional spatial imaging system entails propagation (diffraction) from the input plane to the lens, imposition of a quadratic spatial phase by the lens, and propagation (diffraction) from the lens to the output plane. The analogous temporal imaging system involves a first dispersive delay line (the temporal equivalent of diffraction), imposition of a quadratic temporal phase, and then a second dispersive delay line. The element that imparts quadratic temporal phase has been termed a *time lens*. Practically time lenses have been implemented using either electro-optic phase modulators [486,487] or nonlinear mixing of the input signal beam with a highly chirped control beam in a second-order nonlinear crystal [488,489].

The conditions for spatial imaging are quickly reviewed in a simple ray optics picture. Applying the propagation matrices of Section 1.3.3, a ray optics description of the spatial imaging system of Fig. 8.24 is given by

$$\begin{pmatrix} x_{\text{out}} \\ x'_{\text{out}} \end{pmatrix} = \begin{pmatrix} 1 & d_i \\ 0 & 1 \end{pmatrix} \begin{pmatrix} 1 & 0 \\ -1/f & 1 \end{pmatrix} \begin{pmatrix} 1 & d_o \\ 0 & 1 \end{pmatrix} \begin{pmatrix} x_{\text{in}} \\ x'_{\text{in}} \end{pmatrix}$$
$$= \begin{pmatrix} 1 - d_i/f & d_o + d_i - d_o d_i/f \\ -1/f & 1 - d_o/f \end{pmatrix} \begin{pmatrix} x_{\text{in}} \\ x'_{\text{in}} \end{pmatrix} \tag{8.77}$$

Here x and x' designate the spatial position of a ray and its slope, respectively. The latter is equivalent to the normalized Fourier transform variable k_x/k in a wave optics picture. The input (or *object*) field is a distance d_o in front of the lens, while the output (or *image*) field is a distance d_i behind the lens. Note that the notation "o" and "i" refer to "object" and "image," respectively, *not* to "output" and "input." The imaging condition is $d_o + d_i - d_o d_i/f = 0$, which is equivalent to

$$\frac{1}{d_o} + \frac{1}{d_i} = \frac{1}{f} \tag{8.78}$$

When this imaging condition is satisfied, we may write

$$x_{\text{out}} = \left(1 - \frac{d_i}{f}\right) x_{\text{in}} = \left(-\frac{d_i}{d_o}\right) x_{\text{in}} \tag{8.79}$$

Each spatial position at the object (input) plane is mapped to a unique spatial position at the image (output) plane with magnification $M = -d_i/d_o$.

The temporal imaging problem may be understood in an equivalent manner by introducing the concept of a temporal ray and temporal ray propagation matrices [489,490]. A temporal ray is specified by its time coordinate t and normalized frequency offset $\tilde{\omega}/\omega_0$. The matrix representing temporal propagation in a dispersive medium is analogous to spatial propagation in free space and is written

$$\begin{pmatrix} t_{\text{out}} \\ \tilde{\omega}_{\text{out}}/\omega_0 \end{pmatrix} = \begin{pmatrix} 1 & -\omega_0\psi_2 \\ 0 & 1 \end{pmatrix} \begin{pmatrix} t_{\text{in}} \\ \tilde{\omega}_{\text{in}}/\omega_0 \end{pmatrix} \tag{8.80}$$

where $\psi_2 = -\beta_2 z$ is the accumulated second-order spectral phase.

The propagation matrix for a quadratic phase modulator, or time lens, is determined by analogy with the spatial case. A spatial lens imparts a quadratic phase term of the form $\exp(jkx^2/2f) = \exp(j\phi_{xx}x^2/2)$, where ϕ_{xx} is the second spatial derivative of the imposed phase. Hence, the spatial focal length satisfies $f = k/\phi_{xx}$. We also note that in traversing a spatial lens, a spatial ray bends by an amount $\Delta k_x/k = -\phi_x/k = -\phi_{xx}x/k$ (here ϕ_x is the first spatial derivative of the phase). These points are consistent with the form of the propagation matrix for a lens. In the temporal case, we write the quadratic phase term imposed by the time lens as $\exp(j\phi_{tt}t^2/2)$, where ϕ_{tt} is the second derivative of the phase with time. The frequency shift imposed by the time lens is $\Delta\tilde{\omega}/\omega_0 = \phi_{tt}t/\omega_0$. By analogy, then, the focal length of a time lens is given by

$$f_T = -\frac{\omega_0}{\phi_{tt}} \tag{8.81}$$

and its propagation matrix is written

$$\begin{pmatrix} t_{\text{out}} \\ \tilde{\omega}_{\text{out}}/\omega_0 \end{pmatrix} = \begin{pmatrix} 1 & 0 \\ -1/f_T & 1 \end{pmatrix} \begin{pmatrix} t_{\text{in}} \\ \tilde{\omega}_{\text{in}}/\omega_0 \end{pmatrix} \tag{8.82}$$

Using these temporal matrices, the temporal imaging system of Fig. 8.24 is described by eq. (8.77), but with the substitutions

$$t \leftrightarrow x \qquad \frac{\tilde{\omega}}{\omega_0} \leftrightarrow x' \qquad -\omega_0\psi_2^{(o)} \leftrightarrow d_o \qquad -\omega_0\psi_2^{(i)} \leftrightarrow d_i \qquad f_T \leftrightarrow f$$

Here $\psi_2^{(o)}$ and $\psi_2^{(i)}$ are the accumulated second-order phase from the object (input) plane to the time lens and from the time lens to the image (output) plane, respectively. The temporal imaging condition is written

$$\frac{1}{\omega_0 \psi_2^{(o)}} + \frac{1}{\omega_0 \psi_2^{(i)}} = -\frac{1}{f_T} \quad \text{or} \quad \frac{1}{\psi_2^{(o)}} + \frac{1}{\psi_2^{(i)}} = \phi_{tt} \tag{8.83}$$

When this imaging condition is satisfied, we may write

$$t_{\text{out}} = \left(-\frac{\psi_2^{(i)}}{\psi_2^{(o)}} \right) t_{\text{in}} = M_T t_{\text{in}} \tag{8.84}$$

and

$$\tilde{\omega}_{\text{out}} = \phi_{tt} t_{\text{in}} - \frac{\psi_2^{(o)}}{\psi_2^{(i)}} \tilde{\omega}_{\text{in}} = \frac{\phi_{tt} t_{\text{out}}}{M_T} - \frac{\psi_2^{(o)}}{\psi_2^{(i)}} \tilde{\omega}_{\text{in}} \tag{8.85}$$

Each temporal sample at the object (input) plane is mapped to a unique temporal sample at the image (output) plane with magnification M_T given above. Furthermore, at the image plane the frequency offset of a ray is both scaled by the inverse magnification (M_T^{-1}) and shifted in proportion to temporal position. This time-dependent frequency shift means that the temporal image is chirped. Such frequency shifts are easily understood using the spatial analogy: Rays passing through a spatial lens are bent (acquire a spatial frequency shift) in proportion to the displacement of the ray from the lens axis. This corresponds to the converging or diverging nature of spatial images.

We may gain further insight by analyzing the imaging problem from a wave optics perspective. The spatial imaging analysis is well known in classical optics texts and may be formulated using the Fresnel diffraction integral, similar to our derivation of the Fourier transform property of a lens in Section 8.5.1. Here we focus on the analysis for temporal imaging.

The field at the image plane may be written

$$a_{\text{out}}(t) \sim \iint dt'\,dt''\, a_{\text{in}}(t') \exp\left(\frac{-j(t'-t'')^2}{2\psi_2^{(o)}} + \frac{j\phi_{tt}(t'')^2}{2} - \frac{j(t-t'')^2}{2\psi_2^{(i)}} \right) \tag{8.86}$$

Here eq. (8.61) has been applied twice to handle propagation from the object plane to the time lens and from the time lens to the image plane, and the quadratic phase function corresponding to the time lens has been applied after the first propagation. If we apply the temporal imaging condition, eq. (8.83), all the terms quadratic in t'' cancel out. This leaves

$$a_{\text{out}}(t) \sim e^{-jt^2/2\psi_2^{(i)}} \iint dt'\,dt''\, a_{\text{in}}(t') e^{-jt'^2/2\psi_2^{(o)}} e^{jt''t'/\psi_2^{(o)}} e^{jt''t/\psi_2^{(i)}} \tag{8.87}$$

Carrying out the t'' integral yields the delta-function term $\delta\left(t' + \psi_2^{(o)} t/\psi_2^{(i)}\right)$. This allows the t' integral to be performed easily, with the result

$$a_{\text{out}}(t) \sim a_{\text{in}}\left[\left(-\frac{\psi_2^{(o)}}{\psi_2^{(i)}}\right)t\right]\exp\left(\frac{-j(\psi_2^{(i)} + \psi_2^{(o)})t^2}{2(\psi_2^{(i)})^2}\right) \tag{8.88}$$

The output field is proportional to a scaled version of $a_{\text{in}}(t)$, with magnification factor precisely as given by the ray optics result, eq. (8.84). The output field additionally includes a quadratic phase factor, corresponding to a linear frequency modulation also predicted by the ray optics picture. In particular, if we set $\tilde{\omega}_{\text{in}} = 0$ (bandwidth-limited input pulse), eqs. (8.85) and (8.88) predict an identical frequency modulation.

Going beyond the predictions of the ray optics theory, the wave optics formulation enables discussion of the resolution of temporal imaging systems. The principal issue limiting resolution is the fact that practical time lenses are unable to provide perfect quadratic phase modulation over infinite time aperture. Denote the useful time aperture over which quadratic phase modulation is available as T_L. Then to avoid distortions in the imaging process, the input field after propagating to the position of the time lens must have duration less than T_L. If the input waveform has duration T_{in} and inverse bandwidth $\delta t_{\text{in}} \sim B_{\text{in}}^{-1}$, this requirement becomes

$$T_{\text{in}} < T_L \tag{8.89a}$$

and

$$2\pi|\psi_2^{(o)}|\delta t_{\text{in}}^{-1} < T_L \tag{8.89b}$$

The former condition is trivial. Let us therefore focus on the latter condition. We specifically consider the large magnification case (i.e., a time microscope). For $|M_T| \gg 1$, we have from eqs. (8.83) and (8.84) that $|\psi_2^{(i)}| \gg |\psi_2^{(o)}|$, $\psi_2^{(o)} \approx \phi_{tt}^{-1}$, and $\omega_0|\psi_2^{(o)}| \approx |f_T|$. Substituting into eq. (8.89b), we obtain

$$\delta t_{\text{in}} > \frac{2\pi}{\omega_0}\frac{|f_T|}{T_L} \qquad \text{or} \qquad \delta t_{\text{in}} > \frac{2\pi}{\omega_0} f_T^{\#} \tag{8.90}$$

where the temporal f-number $f_T^{\#} = |f_T|/T_L$ is defined as the ratio of the focal length and aperture of the time lens. The minimum temporal feature size that can be imaged successfully is the temporal f-number times the optical period. This is in close analogy with spatial imaging systems, where the spatial resolution is given by the (spatial) f-number times the optical wavelength. The minimum temporal feature size may also be expressed as

$$\delta t_{\text{in}} > \frac{2\pi}{|\phi_{tt}|T_L} \qquad \text{or} \qquad \delta t_{\text{in}} > \frac{2\pi}{B_L} \tag{8.91}$$

This result shows that the temporal resolution is limited to the inverse of the maximum range of frequency shifts, or bandwidth ($B_L = |\phi_{tt}|T_L$), that the time lens is able to impose.

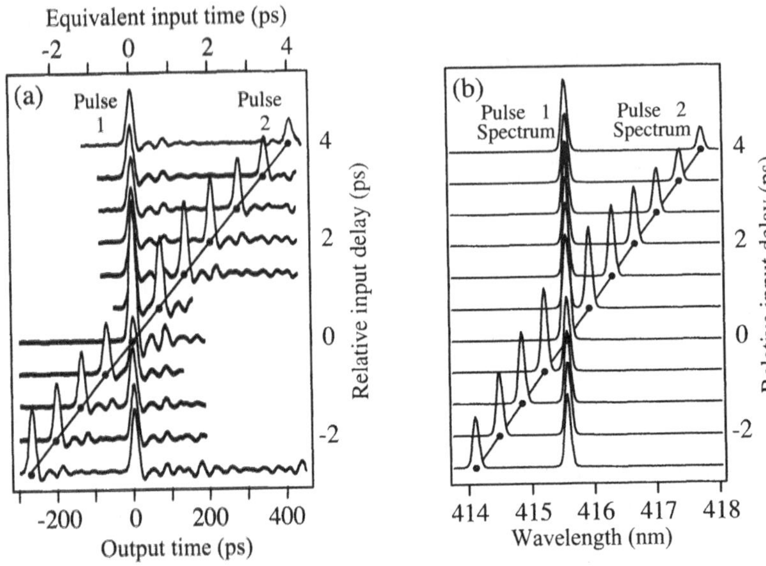

Figure 8.25 (a) Temporal images of pulse pairs with various relative delays, measured via a time microscope with magnification $M_T = 103$; (b) corresponding output spectra. Adapted from [488].

These temporal resolution limits are not restricted to the case of large magnification. Consider, for example, the case of large demagnification, in which the imaging system is used to compress an input waveform to much shorter duration. Here the shortest temporal features (δt_{out}) are found at the output plane. By using an argument similar to that given above, one may show that for $|M_T| \ll 1$, the resolution is still governed by eqs. (8.90) and (8.91), but with δt_{out} replacing δt_{in}.

We close this section with an experimental example of time microscope action [488]. Here the input waveforms consist of a pair of pulses of about 100 fs derived from an 830-nm Ti:S laser, with relative delay that varies in steps of 667 fs in different experimental trials. Input and output dispersions $\psi_2^{(o)} = -0.176\,\text{ps}^2$ and $\psi_2^{(i)} = 17.6\,\text{ps}^2$ were implemented using grating-and-lens systems. The time lens with $\phi_{tt} = -5.623\,\text{ps}^{-2}$ and $f_T^{\#} \approx 80$ was implemented by noncollinear sum frequency mixing of the signal waveform with a highly chirped pump pulse (also derived by passing a pulse from the Ti:S laser through a grating pair).[5] This sum frequency mixing implementation results in significantly better temporal resolution than is generally possible using electro-optic phase modulators (due to limits in technologically feasible modulation indices). Figure 8.25a shows data at the output of the time microscope. For maximum input separations of just a few picoseconds, the output pulse pair separations range up to several hundred picoseconds. The data correspond to a magnification $M_T = 103$, in good agreement with the value calculated (100). As a result, the ultrafast input waveforms are expanded to a time scale for which direct measurement via a conventional photodetector-oscilloscope pair is possible. This leads to interesting possibilities for single-shot waveform recording. Figure 8.25b shows output spectra corresponding to the series of temporal images shown in Fig. 8.25a. Clearly, the output spectra are scaled

[5] The experimental dispersion and phase modulation parameters have been converted to be consistent with the notation in this book.

versions of the temporal images. This is a manifestation of the frequency modulation term in eq. (8.88). Because the frequency modulation is sufficiently strong that the output waveforms are in the large chirp limit, frequency and time are mapped on to each other, as in Section 8.3.2.

8.4 ULTRASHORT-PULSE AMPLIFICATION

Mode-locked oscillators generate pulses with energies typically on the order of nJ. For 100-fs pulses at a 100-MHz repetition rate, this corresponds to peak powers of order 10 kW and maximum intensities of order 10^{12} W/cm^2 (even assuming extremely tight focusing to the limit of a square wavelength). Amplification is needed to reach the substantially higher powers needed for many experiments. Femtosecond pulses amplifiers providing energies in a range approximately 1 mJ to 1 J are now available (in a few cases substantially higher energies have been obtained), representing amplification factors of about 10^6 to 10^9. With such amplification new regimes of laser–matter interaction are accessible, as reviewed in [221]. Briefly, these regimes may be described as follows:

- Up to approximately 10^{13} W/cm^2, the applied laser electric field remains weak compared to static atomic Coulomb fields. Assuming that there are no direct absorptions, bound electrons remain bound and in the ground state. In this regime, nonlinear optical interactions are adequately described via a perturbative approach, as discussed in Chapters 5 and 6.
- In the intensity range between approximately 10^{13} and 10^{15} W/cm^2, multi-photon ionization by simultaneous absorption of N or more photons becomes important (here N is the minimum number of photons required for ionization). However, in this intermediate regime, the nonlinear polarizations contributed by bound and ionized (free) electrons remain comparable. The ionized (free) electrons give rise to a self-defocusing nonlinearity that may act to balance the self-focusing nonlinearity associated with bound electrons. An interesting result is the possibility of self-channeling of intense pulses over distances of hundreds of meters in air.
- Above approximately 10^{15} W/cm^2 we enter the strong-field regime, where applied laser fields are much larger than atomic fields (at least for the most weakly bound, outer electrons). Here weakly bound electrons are ionized within a single laser cycle, and effects associated with such free electrons now dominate. One important new effect arises when the electric field first drives ionized electrons far away from the parent ion, then slams some of them back into the parent ion later during the optical cycle. This results in high harmonic generation, yielding coherent radiation in the difficult-to-reach extreme ultraviolet and soft x-ray regions of the spectrum, including under some conditions, attosecond bursts of short-wavelength radiation.
- Finally, for visible and near-infrared radiation at intensities above approximately 10^{18} W/cm^2, free electrons are accelerated up to velocities approaching the speed of light and to kinetic energies approaching their rest energy. In short, laser–electron interactions and electron dynamics become relativistic. This relativistic nonlinear optics regime is reviewed in [491] and leads to new phenomena and new concepts, such as laser-driven accelerators.

In the following we discuss the basic principles of femtosecond pulse amplification. We begin with an analysis of energy extraction in saturated amplifiers and a description of amplifier configurations used to achieve high gain together with good energy extraction. At this level the discussion is general and applies to narrowband as well as broadband (femtosecond) amplifiers. We then describe issues specially relevant for femtosecond pulse amplification, focusing on the powerful chirped pulse amplification concept. Our treatment draws on a general treatment of solid-state amplifiers [492] as well as review articles targeting specific issues associated with ultrashort-pulse amplification [491,493,494].

In addition to amplifiers based on solid-state laser materials, parametric amplifiers based on nonlinear optical materials are the subject of increased investigation. Some aspects of optical parametric amplification (OPA) were discussed in Section 5.5.3, including the potential for extremely large amplification bandwidth. In this treatment we focus on gain media based on solid-state laser materials.

8.4.1 Amplification Basics

Most broad-bandwidth gain media used in femtosecond pulse amplification may be modeled as four-level systems (see Fig. 1.5). Following Section 2.3.1, we express the amplification of the field according to

$$e_{\text{out}}(t) = \exp\left(\frac{1}{2}\sigma_G NL\right)e_{\text{in}}(t) = \left(e^{g}\right)e_{\text{in}}(t) \tag{8.92}$$

Here σ_G is the gain cross section, N is the population density of the upper laser level, L is the length of the amplifying medium, and $g = \ln(e_{\text{out}}/e_{\text{in}})$ is the gain coefficient for the field. In this very simple formula we have neglected the frequency dependence of the amplifying transition. For energetic pulses we must keep track of the gain saturation. When the pulse duration is short compared to the gain relaxation time and the pumping rate, a homogeneously broadened gain medium saturates according to eq. (2.48a):

$$N(t) = N^{(i)}e^{-U(t)/U_G} \quad \text{and} \quad g(t) = g^{(i)}e^{-U(t)/U_G} \tag{8.93a}$$

where $N^{(i)}$ and $g^{(i)}$ are, respectively, the small-signal population density and gain coefficient (no saturation), $U(t)$ is the pulse energy integrated up to time t, and

$$U_G = \frac{\hbar\omega_0 A_G}{\sigma_G} \tag{8.93b}$$

is the saturation energy, which is proportional to the beam cross section A_G and the photon energy ω_0. The population density of the laser transition and consequently the gain coefficient saturate exponentially with integrated pulse energy. In addition, it is useful to keep track of the energy U_{sto} stored in the gain medium, which is written simply as

$$U_{\text{sto}} = N^{(i)}\hbar\omega_0 A_G L \tag{8.94}$$

This energy must be supplied externally, which imposes requirements on the pump source for the amplifier. The stored energy is related to the saturation energy and gain according to

$$g^{(i)} = \frac{1}{2}\left(\frac{U_{\text{sto}}}{U_G}\right) \tag{8.95}$$

where the factor of 1/2 arises because our gain coefficient is defined in terms of the fields.

The description above, in which pulse energy is taken implicitly as independent of longitudinal position, is appropriate for laser oscillators where typically the gain is small, but needs to be modified in the case of high-gain amplifiers. In particular, the evolution of pulse energy within the amplifier should be considered. Again assuming that the bandwidth of the gain medium is much larger than the bandwidth of the pulse, we may formulate this problem in terms of the instantaneous power P [495]:

$$\frac{\partial P(z,t)}{\partial z} = \sigma_G N(z,t)P(z,t) \tag{8.96}$$

This expression ignores any linear loss in the gain medium. Assuming that the medium is pumped uniformly, so that $N^{(i)}$ is independent of z, this becomes

$$\frac{\partial P(z,t)}{\partial z} = \sigma_G N^{(i)} e^{-U(z,t)/U_G} P(z,t)$$

$$\text{with} \qquad U(z,t) = \int_{-\infty}^{t} dt'\, P(z,t') \tag{8.97}$$

We are interested in the total integrated pulse energy, $U(z) = \int_{-\infty}^{\infty} dt\, U(z,t)$. Integrating eq. (8.97) over time gives

$$\frac{dU(z)}{dz} = \sigma_G N^{(i)} \int_{-\infty}^{\infty} dt\, e^{-U(z,t)/U_G} P(z,t) = \sigma_G N^{(i)} U_G \left(1 - e^{-U(z)/U_G}\right) \tag{8.98}$$

Here in performing the integral, we have made the identification $dU(z,t) = P(z,t)\, dt$. The resulting differential equation for the spatial buildup of the total energy may be integrated directly. For an amplifier of length L, the output energy $U(L)$ is related to the input pulse energy $U(0)$ by

$$\ln\left(\frac{e^{U(L)/U_G} - 1}{e^{U(0)/U_G} - 1}\right) = \sigma_G N^{(i)} L \tag{8.99}$$

or

$$U(L) = U_G \ln\left(e^{\sigma_g N^{(i)} L}[e^{U(0)/U_G} - 1] + 1\right) \tag{8.100}$$

Within the assumptions enumerated above (homogeneous broadening, no loss, neglect of frequency-dependent effects), this equation is valid over the full range of amplifier parameters. Two limiting cases are of particular interest:

- For sufficiently small input energy, such that $e^{2g^{(i)}}(U(0)/U_G) \ll 1$, we have

$$U(L) = U(0)e^{\sigma_G N^{(i)} L} = U(0)e^{2g^{(i)}}$$

  The amplifier remains in the small-signal regime; the pulse energy is increased exponentially according to the small-signal gain factor.
- For input energy such that $U(0)/U_G \gg 1$, we have

$$U(L) = U(0) + N^{(i)}\hbar\omega_0 A_G L = U(0) + U_{\rm sto}$$

  This is the large-signal limit, in which the input pulse is already at the saturation level. The output energy grows linearly with amplifier length, and ideally all of the energy stored in the medium is extracted.

These limiting cases point to a fundamental trade-off in amplifier design. The small-signal regime is most favorable for achieving high gain but is associated with low energy extraction. The large-signal (saturated) regime allows efficient energy extraction but leads to low gain.

Materials for Femtosecond Amplifiers Similar to gain materials for femtosecond mode-locked lasers, amplifying media for femtosecond pulses must possess broad optical bandwidth. For this reason laser dyes and vibronic solid-state materials have been widely utilized in ultrashort-pulse amplifiers. Dye amplifiers were heavily investigated during the 1980s, in conjunction with mode-locked dye laser oscillators (e.g., [496]). Broadband solid-state materials, particularly Ti:S, came to dominate femtosecond amplifier technology beginning in the 1990s, around the same time that solid-state gain media became the material of choice for mode-locked oscillators.

Our discussion above allows us to elucidate key reasons why solid-state gain materials are preferred. Relevant physical parameters for laser dyes and for the broadest bandwidth and most widespread solid-state amplifying medium, Ti:S, are given in Table 8.1. In addition to obvious practical differences associated with liquid vs. solid-state media, perhaps the most striking difference is the three-order-of-magnitude-higher saturation fluence associated with Ti:S. On the one hand, if we assume simply that amplified fluences will be on the order of the saturation fluence, this means that the cross-sectional diameter of a dye amplifier must be roughly 30 times higher than in a Ti:S amplifier of comparable output energy. For example, for a 10-mJ system this corresponds to apertures of about 1 mm for Ti:S vs. 3 cm for dyes. Equation (8.95) tells us that at constant small-signal gain, one can store

Table 8.1 Approximate Physical Parameters Relevant for Amplification: Laser Dyes and Ti:Sapphire

Material	Bandwidth	Saturation Fluence, $h\nu/\sigma_G$	Gain Lifetime, τ_G
Dye	2.5×10^{13} Hz	1 mJ cm^{-2}	~1 ns
Ti:S	10^{14} Hz	1 J cm^{-2}	3 μs

approximately three-order-of-magnitude-higher energy density in Ti:S compared to dyes. For high-power systems, the lower beam aperture possible in solid-state amplifiers generally implies a significant advantage in practicality and cost. On the other hand, eq. (8.95) also tells us that if we try to keep both the aperture and output energy of an amplifier constant, a dye amplifier will have three-order-of-magnitude-higher small-signal gain. Having too high a small-signal gain can be a serious problem, as this leads to strong amplification of spontaneous emission. The resulting amplified spontaneous emission can both deplete the population inversion (wasting energy) and act as an unwanted background that limits the on–off contrast of the amplified pulse. Generally, such amplified spontaneous emission is much more serious in dye amplifiers than in amplifiers constructed from high-energy-storage solid-state laser materials.

Amplifier Configurations Since amplifier design involves a trade-off between gain and energy extraction, high-energy amplifiers are often configured as shown in Fig. 8.26. The input pulse is first sent to a preamplifier that is optimized to provide most of the overall gain. However, because the energy is not yet very high, the beam cross section can be kept small. This reduces the fraction of the pump power that is needed for the preamplifier stage. In femtosecond systems the input to the preamplifier is typically on the order of 1 nJ, and may be amplified to the range 1 to 10 mJ (representing gain up to about 10^7). To achieve higher energies, the output of the preamplifier is directed to one or more low-gain power amplifier stages. Here the amplifier operates under saturation and is optimized for efficient energy extraction. In the case of multiple power amplifier stages, the beam cross section and the pump power are typically increased from stage to stage to accommodate increasing pulse energy. Different materials may be used for later power amplifier stages, where higher saturation fluence and higher energy storage are preferred (corresponding to lower gain cross section). Using such schemes, amplifiers that produce output pulses in the petawatt regime (1 PW = 10^{15} W) have been demonstrated! The first reported petawatt system generated a pulse energy of 660 J at a duration of 440 fs, corresponding to peak power of 1.5 PW, which could be focused to an intensity of $7 \times 10^{20}\ \mathrm{W \cdot cm^{-2}}$ [497].

Since solid-state media typically provide low gain per pass, to achieve high overall gain it is common (especially in preamplifiers) to arrange for multiple passes through the amplifying medium. Such multiple-pass designs also help to address simultaneously the conflicting demands of high gain and efficient energy extraction. Two basic schemes are commonly used, described below.

In the multipass scheme (Fig. 8.27), a series of mirrors is used to pass the beam through the amplifier crystal sequentially at different angles. Thus, the different passes are separated geometrically. Geometric complexity typically limits the number of passes to eight or so in femtosecond amplifiers. Consequently, for an overall gain of 10^7, gains per pass on the order of 10 are needed.

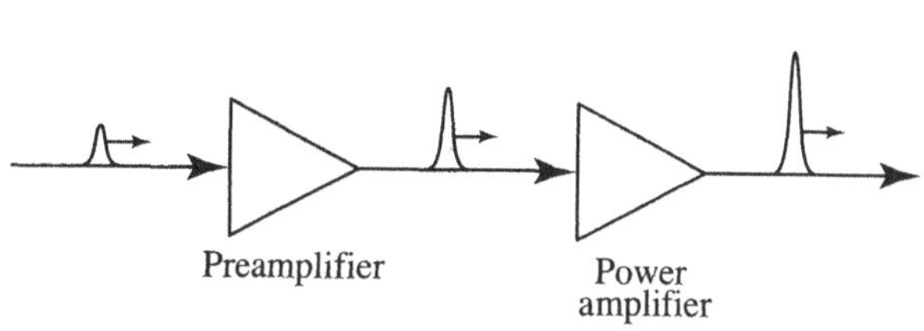

Figure 8.26 Multiple-stage design for high-energy amplification.

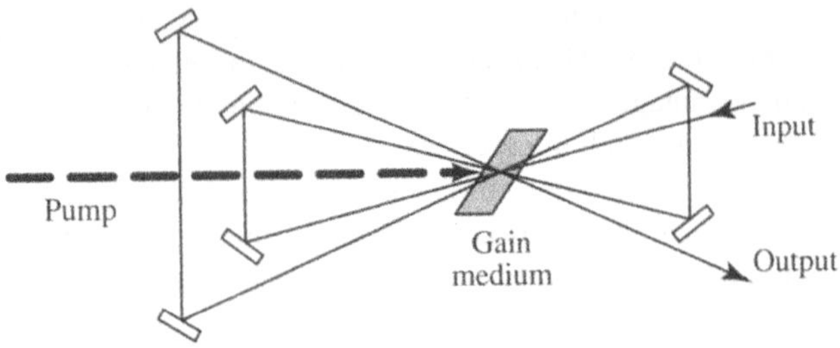

Figure 8.27 Generic multipass amplifier. Although not shown, curved mirrors are often used to refocus the beam for each pass through the amplifier.

In the regenerative amplifier scheme the pulse to be amplified is trapped inside a laser resonator using a polarization gating approach. In the configuration illustrated in Fig. 8.28, light in the resonator passes repeatedly through a thin-film polarizer, a quarter-wave plate, and an electro-optic Pockels cell. Operation of a regenerative amplifier involves three distinct phases.

1. During the pump phase, energy from the pump laser is accumulated in the gain crystal. During this phase any light emitted into the resonator by the laser crystal (spontaneous emission) is polarized by passing through the thin-film polarizer, rotated by 90° in double passing through the wave plate, then reflected out of the cavity by the polarizer. (The Pockels cell is set for zero retardation during this phase, hence no polarization rotation.) This acts to suppress buildup of amplified spontaneous emission.
2. At the start of the amplification phase, the pulse to be amplified is directed onto the polarizer, with polarization such that it reflects into the cavity. The return light reflected by the cavity end mirror is polarization rotated upon double passing through the wave plate and is now transmitted by the polarizer toward the gain medium. Before the pulse returns to the polarization optics section of the cavity (right-hand side of the figure), a voltage step is applied to the Pockels cell so that it also act as a quarter-wave plate. Together the wave plate and activated Pockels cell act as a half-wave plate, which in double pass does not affect the polarization. Hence, the pulse remains polarized in the direction transmitted by the polarizer and is trapped in the resonator, allowing repeated passes through the gain crystal.
3. Finally, in the switch-out phase the voltage applied to the Pockels cell is deactivated rapidly. The next double pass through the wave plate (and now deactivated Pockels

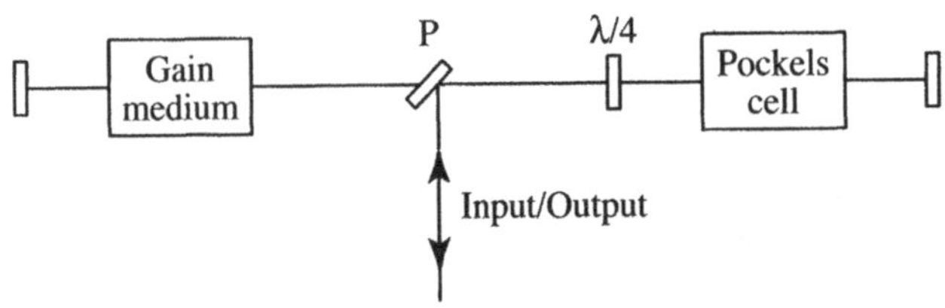

Figure 8.28 Generic regenerative amplifier. P, thin-film polarizer; $\lambda/4$, quarter-wave plate. An additional Faraday rotator and polarizer (not shown) are typically used to separate the input and output beams.

cell) rotates the polarization by 90°, so that upon its next encounter with the polarizer the pulse is reflected out of the cavity. The timing of the switch-out can be adjusted to tune the total number of passes through the gain medium and hence the overall amplification factor.

Fundamentally, multipass and regenerative amplifier schemes are similar, with the main difference being the method of separating the beams (different angles vs. time gating of the polarization). Practically, regenerative amplifiers tend to operate at larger number of passes through the gain medium at lower gain per pass and deliver higher efficiency (easier to obtain good overlap of pump and signal beams). A disadvantage of the regenerative amplifier scheme, at least for femtosecond applications, is larger material path lengths. This leads to larger material dispersion, which makes it more difficult to achieve complete pulse compression in chirped pulse amplification systems (see the following section).

8.4.2 Special Issues in Femtosecond Amplifiers

Our discussion above centers on behavior that is generally applicable to laser amplifiers, ultrashort pulse or otherwise. Now we discuss features that are specific to femtosecond pulse amplification.

Nonlinear Beam Propagation Arguably the most profound new feature of ultrashort-pulse amplification concerns methods to overcome nonlinear index phenomena. As we know from our discussion of Kerr lens mode-locking (Section 2.4.4), propagation of an intense beam in a nonlinear index material induces a nonlinear lens and self-focusing. Such nonlinear lensing is beneficial in KLM lasers, as it provides the mode-locking mechanism. In a high-power amplifier, however, nonlinear lensing resulting in a tightly confined self-focused beam can be catastrophic, as the increased intensities associated with the self-focused beam can easily exceed the damage threshold of amplifier materials. One way to estimate the importance of self-focusing effects is by comparing peak power with the critical power above which self-focusing is expected. A formula for the critical power is given in eq. (2.113); for Ti:S the numerical value is $P_{crit} \approx 2.6$ MW [73]. Consider as an example a 100-fs pulse amplified to an energy of 1 mJ. Such a pulse has a peak power of 10 GW, so self-focusing is expected to be a very serious problem.

Another way to characterize the importance of self-focusing effects is in terms of the B *integral* (beam breakup integral), written

$$B = \frac{2\pi}{\lambda} \int n_2 I(z)\, dz \tag{8.101}$$

where I is the intensity and n_2 is the nonlinear index. The B integral gives the total nonlinear phase shift accumulated over the length of the amplifier. The significance of this expression derives from a theory that treats self-focusing as an instability phenomenon, where even an infinitesimally small modulation in the spatial intensity profile is amplified during nonlinear propagation, eventually to very large scale (see e.g., [492]). The case of a continuous-wave input beam with a small, one-dimensional spatial modulation is exactly analogous to the case of a plane-wave input beam with a small modulation in time propagating in a medium with anomalous dispersion. We analyzed the latter (time-domain) case in Section 6.5 and found that the initial modulation grows exponentially. For both space- and time-domain problems, the peak exponential gain factor is found to be equal to the accrued nonlinear phase shift—hence to the B integral. Typically in amplifier design, one desires $B \leq 2$. Let us consider a numerical example: A 100-fs pulse at 800-nm wavelength passes through a 1-cm-long Ti:S amplifier, with fluence of 1 J cm^{-2} (equal to the saturation fluence). Taking $n_2 = 3.2 \times 10^{-16}\,cm^2W^{-1}$ [492], we obtain $B \sim 87.5\pi$. Although this is a very rough calculation (it does not take into account multiple amplifier passes or changes of pulse energy with distance, for example), it indicates again that self-focusing and nonlinear beam breakup are very serious issues.

Chirped Pulse Amplification Thus, the peak power that may safely propagate within a laser amplifier is seriously limited by nonlinear index effects. Optical damage issues also constrain attainable peak powers. An approach known as *chirped pulse amplification* (CPA) overcomes these limitations [498,499]. The idea, sketched in Fig. 8.29, is as follows:

1. The input pulse is passed through a first dispersive system, leading to a highly chirped stretched pulse. The stretched pulse preserves the bandwidth of the input pulse but lowers the peak power by the stretching factor.
2. The stretched pulse is amplified. Either single- or multiple-stage amplifier designs may be employed. Because of the peak power reduction associated with stretching, self-focusing and optical damage effects are suppressed significantly.
3. The amplified stretched pulse is sent into a second dispersive system, where the sign of the dispersion is opposite that in the stretching stage. As a result, pulse compression is achieved. With proper dispersion balance the amplified pulse is compressed to the bandwidth limit. In this way only the compressor stage is subject to the full peak power of the final amplified pulse.

It is interesting that ideas analogous to CPA were first developed in the context of radar during World War II [500]. Radar systems are often subject to peak power limitations at the transmitter. Therefore, long pulses are desired to increase the energy transmitted, which increases the range over which the radar system can operate. However, using long transmit pulses will degrade range resolution if the pulses are bandwidth limited. Chirped radar solves this problem by transmitting long chirped pulses. After scattering off a target, the low-energy return signal is compressed to the bandwidth limit, thereby providing high temporal (hence range) resolution.

In our laser amplifier application, both self-focusing and optical damage issues are addressed by CPA. An estimate for the desired stretching factor needed to mitigate damage is given in [501]. Take the damage threshold of the mirrors used in the amplifier chain as 5 GW/cm^2. To operate at the saturation fluence of the gain medium while remaining below the mirror damage threshold, the pulse duration must remain above a minimum value. For a

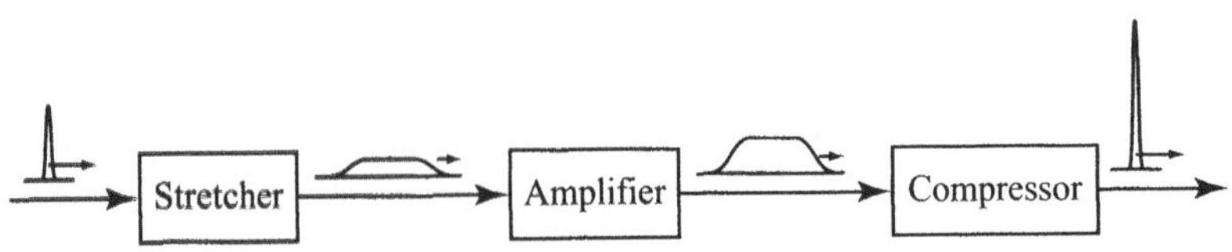

Figure 8.29 Chirped pulse amplification concept.

1-J/cm^2 saturation fluence (Ti:S), the duration of the stretched pulse should exceed 200 ps. In the case of even higher energy storage materials that may be employed in power amplifier stages in larger femtosecond amplifier systems (e.g., Nd:glass with a 5-J/cm^2 saturation fluence), even longer stretched pulses are needed.

The stretcher and compressor needed for CPA are usually implemented using diffraction gratings, as discussed in Section 4.3. In particular, the stretcher is typically formed using the concept of Fig. 4.6. Here the grating–lens separation is set to be less than the focal length, which makes the effective grating separation negative and hence yields normal dispersion. The compressor is typically configured as a Treacy grating pair (Fig. 4.5), which provides anomalous dispersion. The use of the Treacy pair configuration in the compressor stage is favored because it avoids passing the amplified output through material (i.e., lenses), thereby reducing associated nonlinearities and optical damage issues. From a practical perspective both the stretcher and compressor are operated in a double-pass geometry, in which a flat mirror is used to reflect the rays emerging after a single pass through the grating pair back through the apparatus. This reassembles the various optical frequencies, which are spatially separated after a single pass through the grating pair, back into a single overlapped beam. The double-pass compressor geometry is sketched in Fig. 8.30.

Because very large pulse stretching and compression ratios are commonly used in CPA, it is important that the dispersive properties of stretcher and compressor be carefully matched, not only for the lowest order of dispersion (ψ_2) but for all orders. Ideally, the dispersion of the Treacy pair compressor and the grating–lens stretcher are in fact matched to all orders, provided that identical gratings are used at identical angles and with effective grating separations equal and opposite. That is,

$$\psi_2^{\text{(stretch)}}(\omega) + \psi_2^{\text{(compress)}}(\omega) = 0 \tag{8.102}$$

where the notation $\psi_2(\omega)$ includes quadratic, cubic, and all higher-order-spectral phase terms. Practically, however, achieving high-quality compression requires considerable care. A few of the key issues are as follows:

- In typical femtosecond amplifier systems, pulses experience a considerable amount of material dispersion due to passage through the gain medium and other optical

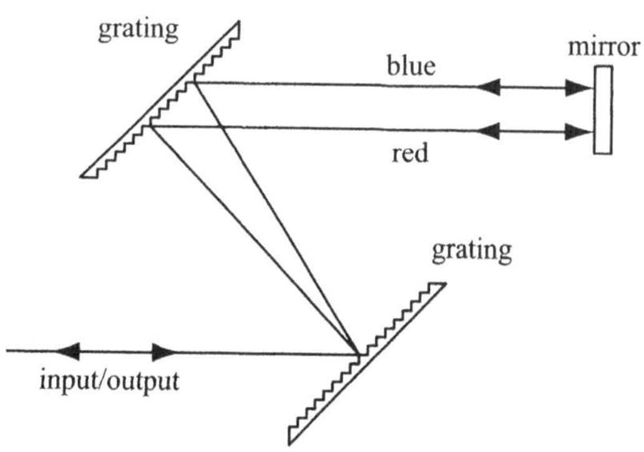

Figure 8.30 Double-pass grating pair pulse compressor geometry. The turnaround mirror may be tilted slightly in the vertical direction (orthogonal to spectral-dispersion direction) to separate the input and output beams.

components. Such material dispersion must be included in the overall dispersion balance. Mathematically, the requirement is

$$\psi_2^{(\text{stretch})}(\omega) + \psi_2^{(\text{material})}(\omega) + \psi_2^{(\text{compress})}(\omega) = 0 \tag{8.103}$$

Now the effective grating separations of stretcher and compressor must be imbalanced to compensate the lowest-order material dispersion, and higher orders of dispersion become imbalanced as well. Two degrees of freedom that are easily accessed to improve the dispersion balance are the grating angles (which may be set differently in stretcher and compressor). Nevertheless, equalizing the frequency-dependent delay to much less than the pulse duration in view of stretching ratios on the order of 10^4 is a difficult challenge, especially for pulses below a few tens of femtoseconds. For this reason, programmable devices such as pulse shapers and acousto-optic dispersive filters (Sections 8.1 and 8.2.2, respectively) are used in some systems as higher-order spectral phase equalizers, in a manner analogous to that illustrated in Fig. 8.12.

- Spectral clipping must be minimized. Because optical frequencies separate during passage through stretcher and compressor, spatial clipping at the edges of gratings or other optical elements introduces a spectral windowing effect. The resulting sharp cutoff, even in the low-intensity wings of the optical spectrum, causes wings and ringing in the compressed time-domain waveform, which may degrade unacceptably the pulse on-off contrast for high-field experiments. We know from Section 4.3 that the required grating size scales in proportion to the time aperture over which the pulse is stretched. To minimize spectral clipping, the grating size (which may already be quite large and hence expensive in some high-power systems) must be scaled up by an additional safety factor.
- Aberrations in the focusing optics in stretchers may also introduce phase errors. Again because optical frequencies separate within the stretcher, spatial phase errors in the lenses translate into spectral phase errors, similar to the operation of a pulse shaper. For this reason the lenses in the stretcher may be replaced by spherical mirrors (avoids chromatic aberrations), and special reflective optics configurations are used to reduce other aberrations [502,503].

Gain Narrowing Another issue arises from the finite bandwidth of the amplifying medium. In the small-signal regime, we may take the frequency response of the amplifier into account by writing

$$E_{\text{out}}(\omega) = \exp\left(\frac{1}{2}\sigma_G(\omega)NL\right)E_{\text{in}}(\omega) = \left(e^{g(\omega)}\right)E_{\text{in}}(\omega) \tag{8.104}$$

Here E_{out} and E_{in} are the Fourier transforms of their time-domain counterparts from eq. (8.92). Because the frequency-dependent gain cross section appears as the argument of an exponential, the overall amplification bandwidth can be significantly narrower than the bandwidth of the gain cross section itself. This gain narrowing effect is most pronounced in the case of large amplification.

Denote the gain coefficient and gain cross section at the peak of the gain curve $g(\omega_0)$ and $\sigma_G(\omega_0)$, respectively. Then the frequency-dependent power gain is given by

$$\frac{|E_{\text{out}}(\omega)|^2}{|E_{\text{in}}(\omega)|^2} = e^{2g(\omega)} = \exp\left(2g(\omega_0)\frac{\sigma_G(\omega)}{\sigma_G(\omega_0)}\right) \tag{8.105}$$

Compared to power gain at the line center, the power gain may be written

$$\frac{e^{2g(\omega)}}{e^{2g(\omega_0)}} = \exp\left(2g(\omega_0)\left[\frac{\sigma_G(\omega)}{\sigma_G(\omega_0)} - 1\right]\right) \tag{8.106}$$

As an example, suppose that an amplifier provides a factor of 10^9 small-signal (power) gain at the line center, so that $2g(\omega_0) = \ln(10^9) = 20.7$. Then from eq. (8.106) the power gain drops by a factor of 2 when $\sigma_G(\omega) = 0.967\sigma_G(\omega_0)$. That is, a 3.3% decrease in the gain cross section corresponds to a 50% change in the overall gain. For a Lorentzian line shape, eq. (2.3), the FWHM bandwidth of the overall power gain is only 19% as large as the FWHM of the gain cross section. In addition to spectral narrowing, gain narrowing may also lead to shifts in center frequency for input spectra offset from the gain peak.

This gain-narrowing effect hinders amplification of the shortest pulses produced from mode-locked lasers. In a laser oscillator the spectrum of a mode-locked pulse may actually exceed the gain bandwidth of the laser medium, thanks to the action of strong phase nonlinearities. For a high-gain amplifier, however, the effective bandwidth is less than that of the gain medium.

Gain saturation in conjunction with chirp may also lead to spectral distortion. For pulse energies sufficient to induce significant saturation, the leading edge of the pulse experiences the highest gain, whereas later portions of the pulse see gain that is increasingly saturated (reduced). Since pulses in CPA systems are strongly chirped (usually from low to high frequency), the red portion of the pulse is preferentially amplified. Under some conditions this results in a significant shift of the amplified pulses to the red.

Figure 8.31 illustrates spectral distortion arising from the amplification process. Figure 8.31a shows the calculated output spectrum from a Ti:S amplifier with gain equal

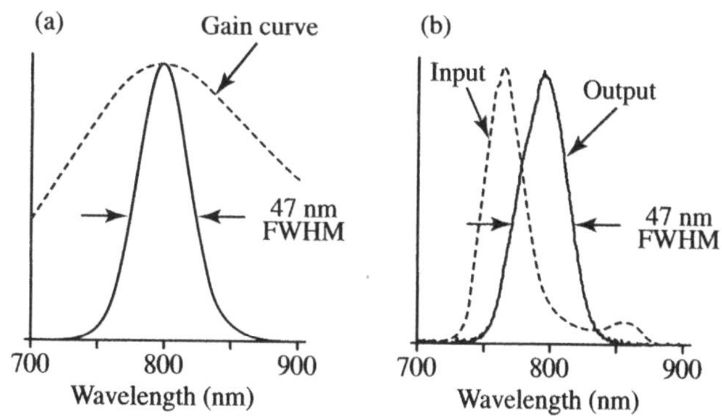

Figure 8.31 Gain narrowing in Ti:S: (a) calculated output spectrum assuming flat input spectrum and gain of 10^7 (the gain cross section is also shown); (b) experimental amplified spectrum resulting with optimally preshaped input spectrum. From [493], with permission. Copyright © 1998, American Institute of Physics.

to 10^7, assuming an infinitely broad and flat input spectrum and neglecting saturation. The frequency dependence of the gain cross section $\sigma_G(\omega)$ is also depicted. Even with an infinitely broad input spectrum, the output is narrowed to 47 nm, sufficient only to support a pulse on the order of 20 fs. For a real input pulse with finite input spectrum centered at the gain peak, even narrower output bandwidth is expected.

On the other hand, by appropriately preshaping the input spectrum, spectral narrowing effects may be mitigated, at least in part. Figure 8.31b shows an experimental preshaped input spectrum and the resulting output spectrum after amplification. The output spectrum shows an obvious red shift, as discussed above. More important, the amplification process actually leads to spectral broadening, reaching the same 47-nm width as before. This desirable result is achieved by suppressing the spectral content of the input at frequencies corresponding to the peak of the gain curve while increasing the weight of the input spectrum at frequencies off gain center. Referring to eq. (8.104), the point is to shape the input signal $E_{\text{in}}(\omega)$ so that its curvature is opposite that at the peak of the gain function $e^{g(\omega)}$. The product of these factors, which yields the output spectrum (neglecting nonlinear effects), is then flattened compared to either factor alone. Gain narrowing can also be alleviated by placing a weak intracavity filter inside a regenerative amplifier. In this case also the filter is selected to have reduced transmission at the gain peak, so that the intracavity net gain factor inside the regenerative amplifier is flattened. Using this regenerative gain filtering approach, an amplified bandwidth of about 100 nm has been demonstrated at the 5-mJ level [504].

8.5 APPENDIX

8.5.1 Fresnel Diffraction and Fourier Transform Property of a Lens

We consider paraxial propagation of a scalar, monochromatic, one-dimensional field (x variation only) along the z direction. The results we shall derive are one-dimensional versions of standard formulas found in textbooks such as [9,168]. We write the field as

$$e(x, z, t) = \text{Re}\left\{s(x, z)e^{j\omega t}\right\} \tag{8.107}$$

where $s(x, z)$ is a z-dependent complex spatial amplitude function. The transverse spatial variation of s is taken to be slow relative to the wavelength scale, so that scalar paraxial propagation is valid. The input field, taken as $s(x, 0) = s_{\text{in}}(x)$, can be expressed in terms of its spatial Fourier transform according to

$$s_{\text{in}}(x) = \frac{1}{2\pi}\int dk_x\, S_{\text{in}}(k_x)e^{-jk_x x} \tag{8.108}$$

where the spatial Fourier transform pair is taken as

$$S(k) = \int dx\, s(x)e^{jkx} \qquad \text{and} \qquad s(x) = \frac{1}{2\pi}\int dk\, S(k)e^{-jkx} \tag{8.109}$$

In eq. (8.108) the input field is decomposed into a superposition of plane waves. Since each individual plane wave propagates very simply according to eq. (1.25), the propagation of the overall field is easily formulated as follows:

$$s(x, z) = \frac{1}{2\pi} \int dk_x \, S_{\text{in}}(k_x) e^{-jk_x x} e^{-jk_z z} \tag{8.110a}$$

where

$$k_x^2 + k_z^2 = k^2 = \frac{\omega^2}{c^2} \tag{8.110b}$$

We have taken the refractive index to be equal to 1 (free-space propagation).

The paraxial regime is characterized by small propagation angles (i.e., $|k_x| \ll k$). Therefore, we expand k_z to lowest order in k_x,

$$k_z \approx k - \frac{k_x^2}{2k} \tag{8.111}$$

Using this approximation in eq. (8.110a) leads to

$$s(x, z) = \frac{1}{2\pi} e^{-jkz} \iint dx' dk_x \, s_{\text{in}}(x') e^{jk_x(x'-x)} e^{jk_x^2 z/2k} \tag{8.112}$$

By performing the integral over k_x and using the identity $\int_{-\infty}^{\infty} du \, e^{-u^2} = \sqrt{\pi}$, we arrive at

$$s(x, z) = \sqrt{\frac{j}{\lambda z}} \, e^{-jkz} \int dx' \, s_{\text{in}}(x') e^{-jk(x-x')^2/2z} \tag{8.113}$$

This result, which is known as the *Fresnel diffraction integral*, expresses the field after propagation as the convolution of the input field with a quadratic phase function.

Using the Fresnel diffraction integral, we now go on to calculate propagation of the field from the front focal plane of a lens to its back focal plane. That is, we start at the plane a distance f in front of the lens and propagate until the plane a distance f behind the lens, where f is the focal length. The field $s_1(x)$ immediately before the lens is given by eq. (8.113) as

$$s_1(x) = \sqrt{\frac{j}{\lambda f}} \, e^{-jkf} \int dx' \, s_{\text{in}}(x') e^{-jk(x-x')^2/2f} \tag{8.114}$$

Passage through a thin lens imparts a simple quadratic phase function $\exp(jkx^2/2f)$, such that the field after the lens s_2 is

$$s_2(x) = s_1(x) e^{jkx^2/2f} \tag{8.115}$$

Applying the Fresnel integral once more to obtain the field $s_{\text{out}}(x)$ at the back focal plane gives

$$s_{\text{out}}(x) = \frac{j}{\lambda f} \iint dx' dx'' \, s_{\text{in}}(x') \exp\left(\frac{-jk[x'^2 + (x - x'')^2 - 2x'x'']}{2f} \right) \tag{8.116}$$

where we have now dropped the constant phase term. Rearranging the exponential terms and completing the square gives

$$s_{\text{out}}(x) = \frac{j}{\lambda f} \int dx' \, s_{\text{in}}(x') e^{jkxx'/f} \int dx'' \exp\left(\frac{-jk[x + x' - x'']^2}{2f} \right) \tag{8.117}$$

Finally, after evaluating the x'' integral, we obtain

$$s_{\text{out}}(x) = \sqrt{\frac{j}{\lambda f}} \int dx' \, s_{\text{in}}(x') e^{jkxx'/f} = \sqrt{\frac{j}{\lambda f}} \int dx' \, s_{\text{in}}(x') e^{2j\pi xx'/\lambda f} \tag{8.118}$$

We see that propagation between the focal planes of a lens results in a spatial Fourier transform of the input field profile, with scale factor $k_x = 2\pi x/\lambda f$. We use this Fourier transform property in analyzing grating and lens optical pulse shapers.

8.5.2 Wave Optics Model of a Grating

Let us now consider the effect of a diffraction grating on a pulsed field in space and time, following a treatment from [464]. Our findings will be useful for analysis of grating pulse shapers.

We express the input field in the plane immediately before the diffraction grating as

$$e_{\text{in}}(x_1, t) = \text{Re}\{a(t)s(x_1)e^{j\omega_0 t}\} \tag{8.119}$$

Referring to Fig. 8.32, we consider a coordinate system aligned with a reference ray incident on the grating at angle $\theta_i^{(0)}$; x_1 refers to the transverse coordinate in the plane of incidence perpendicular to the reference ray. We assume paraxial propagation and ignore the second transverse coordinate (y). Rewriting the input field in terms of its temporal and spatial Fourier amplitudes, we have

$$e_{\text{in}}(x_1, t) = \text{Re}\left\{ \frac{1}{(2\pi)^2} \iint d\tilde{\omega} \, dk_x \, A(\tilde{\omega}) S(k_x) e^{j(\omega_0 + \tilde{\omega})t} e^{-jk_x x} \right\} \tag{8.120}$$

We identify the field as a superposition of monochromatic plane-wave components, with different transverse wave vectors (k_x) and frequencies ($\omega = \omega_0 + \tilde{\omega}$). For each component k_x is related to the incident angle onto the grating via $\sin\tilde{\theta}_i = k_x/k$, where $\tilde{\theta}_i$ represents differences of the incident angle relative to the reference ray.

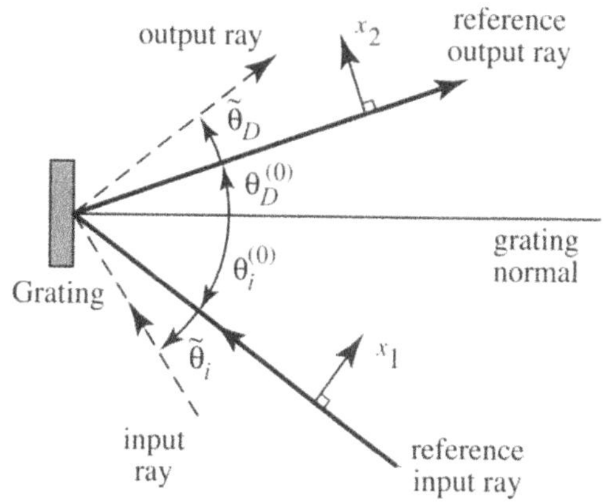

Figure 8.32 Reference diagram for wave optics analysis of a grating.

Now let the input beam encounter the grating. We consider each individual $(\tilde{\omega}, \tilde{\theta}_i)$ pair as an independent plane wave, for which the grating equation (4.59) can be written

$$\sin\left[\theta_D^{(0)} + \tilde{\theta}_D\right] = \sin\left[\theta_i^{(0)} + \tilde{\theta}_i\right] + \frac{2\pi mc}{(\omega_0 + \tilde{\omega})d} \tag{8.121}$$

where $\theta_D^{(0)}$ and $\tilde{\theta}_D$ are, respectively, the reference ray diffraction angle and deviations in the diffraction angle caused by changes in frequency or input angle, and m is the diffraction order. Expanding to first order yields

$$\tilde{\theta}_D = \frac{\cos\theta_i^{(0)}}{\cos\theta_D^{(0)}}\tilde{\theta}_i - \frac{2\pi mc}{\omega_0^2 d\cos\theta_D^{(0)}}\tilde{\omega} \tag{8.122}$$

Taking into account the change in output angle, the transverse wave vector k_x' perpendicular to the reference diffracted ray (along the x_2 direction in Fig. 8.32) is obtained to first order as

$$k_x' = \beta_a k_x + \gamma\tilde{\omega} \tag{8.123a}$$

where

$$\beta_a = \frac{\cos\theta_i^{(0)}}{\cos\theta_D^{(0)}} \tag{8.123b}$$

and

$$\gamma = \frac{-2\pi m}{\omega_0 d\cos\theta_D^{(0)}} \tag{8.123c}$$

Usually, the grating is used in the -1 order ($m = -1$), for which $\gamma = 2\pi/\omega_0 d \cos\theta_D^{(0)}$. Using the expression for k_x' in place of k_x in eq. (8.120), we arrive at

$$e_{\text{out}}(x_2, t) = c_1 \text{Re}\left\{ \frac{1}{(2\pi)^2} \iint d\tilde{\omega}\, dk_x\, A(\tilde{\omega}) S(k_x) e^{j(\omega_0+\tilde{\omega})t} e^{-j\beta_a k_x x_2} e^{-j\gamma\tilde{\omega}x_2} \right\} \tag{8.124}$$

where c_1 is a constant. Now performing the inverse spatial transform, we have

$$e_{\text{out}}(x_2, t) = \sqrt{\beta_a} \text{Re}\left\{ \frac{1}{2\pi} \int d\tilde{\omega}\, A(\tilde{\omega}) s(\beta_a x_2) e^{-j\gamma\tilde{\omega}x_2} e^{j(\omega_0+\tilde{\omega})t} \right\} \tag{8.125}$$

where the value of c_1 is selected to conserve energy. The effect of the grating is two-fold: The beam size is scaled by the inverse of an astigmatism factor β_a, and spectral dispersion is introduced through the factor $e^{-j\gamma\tilde{\omega}x_2}$.

PROBLEMS

8.1. Consider pulse shaping with a single-layer liquid-crystal device. Assuming that the input polarization is aligned for amplitude modulator operation, according to eq. (8.7), work out an expression for the complex field amplitude subsequent to the final polarizer. In particular, express the phase response as a function of the programmed amplitude response.

8.2. For a Gaussian input beam, use the Fourier transform property of a lens, eq. (8.18), to calculate the spot size at the focal plane. Show that your answer is in agreement with the spot size calculated using the Gaussian beam formalism of Section 1.3.3.

8.3. The pulse-shaping theory of Section 8.1.3 assumes that the output field from the shaper is coupled into a single-mode optical fiber, which performs a Gaussian mode selection operation. The fiber is taken to be perfectly mode-matched to the input field as transformed by the pulse shaper and subsequent lens. Consider what happens if the fiber is not mode-matched (i.e., the radius of the fiber mode w_F is not equal to the optimum value). How do the spectral resolution and insertion loss of the pulse shaper depend on w_F?

8.4. Consider a pulse shaper equipped with a pixellated spatial light modulator with N uniformly spaced pixels and frequency span δf per pixel. You may assume that the intrinsic resolution of the pulse shaper is infinite (i.e., the beam radius w_0 at the Fourier plane is much less than the pixel size).

(a) Work out an expression for the maximum quadratic spectral phase ψ_2 that may be obtained without significant distortion. What about the maximum cubic spectral phase ψ_3? Evaluate your expressions for $N = 256$ and $\delta f = 0.1$ THz.

(b) The pulse shaper is programmed for linear spectral phase, which results in pulse delay. Express the relative intensities of the main time-shifted peak, as well as that of the largest replica pulse, as a function of the applied delay. Evaluate your expressions for $N = 128$ and $\delta f = 0.025$ THz.

8.5. Derive eq. (8.42), which describes space–time coupling effects that occur in pulse shaping in the absence of external spatial filtering. You may make use of the other theoretical expressions developed in the text.

8.6. Consider an acousto-optic programmable dispersive filter that is to be programmed simultaneously for self-dispersion compensation and for programmable delay. That is, the pulse-shaping operation is simply to delay the input pulse by an amount between zero and the pulse-shaping time aperture T. Discuss effects that arise if the fast and slow optical axes have different dispersion (ψ_2). How does the required acoustic waveform change as a function of delay?

8.7. Consider pulse propagation as in eq. (8.61). Referring to a pulse as sketched in Fig. 8.16, how small must the propagation distance z be if even the finest features of the input pulse are to remain essentially unchanged upon propagation?

8.8. Show that when eq. (8.67b) holds, the output pulse duration is much larger than that of the input pulse.

8.9. Consider the data of Fig. 8.22. Using the numbers supplied in the text, discuss the extent to which the experimental example satisfies the constraints enumerated in Section 8.3.2.

8.10. Consider a time lens in the limit of large demagnification. Show that the resolution (minimum output temporal feature size) is governed by eqs. (8.90) and (8.91), with δt_{out} replacing δt_{in}.

8.11. Consider the data of Fig. 8.25. Using the numbers supplied in the text, estimate the aperture T_L of the time lens and compare the frequency modulation in the temporal image waveforms with that expected from theory. Also discuss the bandwidths of the individual pulses in the output from the time lens.

8.12. Consider a perfectly balanced grating pulse stretcher and compressor pair in which a 50-fs FWHM Gaussian input pulse at 1 μm wavelength is first stretched to 500 ps, then compressed. Estimate how large a grating is needed to minimize spectral clipping such that the resulting wings in the compressed pulse are at a level below 10^6 of peak intensity. Assume that gratings with 1200 lines/mm are used near the Littrow angle.

8.13. A regenerative amplifier is constructed using an amplifier material with a Lorentzian gain profile with small-signal bandwidth of 5×10^{13} Hz centered at 800 nm. The power gain per pass is equal to 3, and an overall power gain of 10^7 is desired at the line center. A passive intracavity filter with a sinusoidal spectral power transmission profile is used to alleviate gain narrowing via the regenerative gain filtering approach. Neglecting saturation effects, specify the form of the filter (modulation depth and period) if the overall amplification bandwidth is to be maintained at (**a**) 2.5×10^{13} Hz or (**b**) 5×10^{13} Hz. How many additional passes through the amplifying material are required in each case to reach the power gain specified?

9

ULTRAFAST TIME-RESOLVED SPECTROSCOPY

9.1 INTRODUCTION TO ULTRAFAST SPECTROSCOPY

A major application of femtosecond pulses is for time-resolved studies of ultrafast dynamical processes in chemical, solid-state, and biological materials. Just as high-speed (microsecond) electronic flashes have been used for many decades to make stop-action photographs of rapid (microseconds and slower) mechanical motions of macroscopic objects [2], ultrashort laser pulses are a unique tool for excitation and stop-action measurements of ultrafast microscopic and quantum mechanical processes within materials. Examples include scattering of photoexcited charge carriers in a solid-state crystal lattice, dynamics of solvated chemical species interacting with their fluctuating liquid-state environment, oscillatory wave packet motions arising due to the interference between coherently excited quantum mechanical eigenstates, and many more. Detailed treatments of time-resolved spectroscopy of ultrafast physical processes in chemical and semiconductor systems may be found in [505,506] and [249], respectively.

A large number of different ultrafast spectroscopy schemes may be understood within the framework of the pump–probe approach (e.g., [113]), illustrated in Fig. 9.1. A short pump pulse first impinges upon the sample under investigation and excites it in some way. This excitation induces a change in some material property $\mathcal{P}$, which in most cases of interest to us is an optical property. We may write

$$\mathcal{P}(t) \rightarrow \mathcal{P}_0 + \Delta\mathcal{P}(t - t_0) \tag{9.1}$$

where $\mathcal{P}_0$ and $\Delta\mathcal{P}$ are the initial value of property $\mathcal{P}$ and the change in $\mathcal{P}$ induced by a pump pulse arriving at time t_0, respectively. A second pulse, termed the *probe*, arrives at the sample with time delay τ relative to the pump pulse. Often, the intensity of the probe

Ultrafast Optics, By Andrew M. Weiner

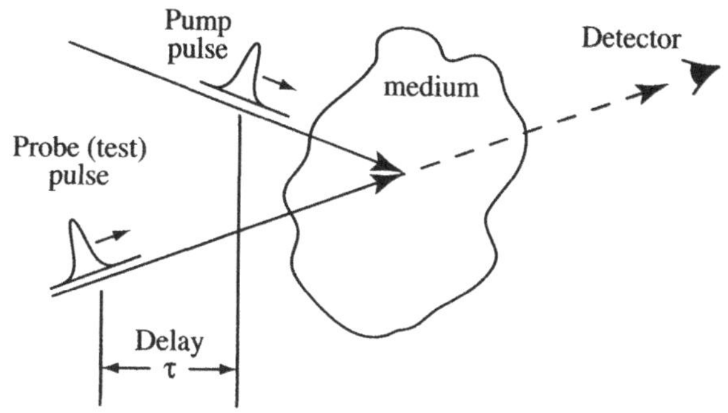

Figure 9.1 Pump–probe approach for time-resolved spectroscopy.

pulse is kept small compared to that of the pump pulse. The idea is that by detecting either the probe pulse itself, subsequent to its interaction with the sample, or by detecting some other effect induced by the interaction of the pump and probes pulse with the sample, one monitors $\Delta\mathcal{P}$ at time $t - t_0 = \tau$. By performing a series of measurements in which delay τ is varied, one maps out the full time dependence of the material response function $\Delta\mathcal{P}(\tau)$. Essentially, this is a correlation function approach, analogous to that common in pulse measurement as discussed in Chapter 3, but adapted for spectroscopic purposes. By relating the measured material response function to a microscopic model of the physical processes under investigation, one attempts to gain insight into the dynamics of these processes.

There are many different material properties that may be probed, depending, for example, on what information is desired, as well many different probing methods. A few examples of material properties and probing methods follow.

- *Transmission measurements* yield data on changes in time-resolved absorption and are often used to probe phenomena related to relaxation out of photoexcited energy states. *Reflectivity measurements* may provide similar information but are generally sensitive to changes in refractive index as well as changes in absorption.
- *Fluorescence yield* out of a state of interest may also be used to monitor the occupation of that state. In some studies the fluorescence intensity is time resolved by detecting it subsequent to a nonlinear interaction such as sum frequency generation gated by a short probe pulse.
- *Refractive index* changes may be sensed not only through reflectivity changes, but also via interferometric detection methods or via changes in the optical spectrum induced by time-varying phase. Index changes may also be observed by exploiting effects related to their spatial profile. Index changes induced by a single pump pulse typically vary radially, which may lead to focusing or deflection of a probe beam (detectable, respectively, as changes in size or position of the output probe beam). Index changes induced by a noncollinear pump pulse pair may form a spatially modulated transient grating, from which a delayed probe beam may be diffracted.
- *Dichroism* and *birefringence* refer, respectively, to anisotropic absorption and anisotropic refractive index (i.e., absorption and refractive index that depend on the polarization of the electric field). Changes in absorption (index) induced by a polarized pump pulse are often anisotropic, resulting in induced dichroism (birefringence). Measurements of time-resolved dichroism and birefringence provide information on reorientation dynamics of anisotropic molecules, for example.

- *Coherent vibrations* either of isolated molecules or of crystalline lattices may be observed via the third-order nonlinear polarization response associated with such vibrations (see Section 6.7.3). Pump–probe experiments may detect either a frequency-shifted (Raman shifted) output field or may directly resolve time-varying refractive index changes associated with the vibrations.
- *Structural information* concerning atomic-level order in crystalline samples may be probed by several different methods. For example, the transition from ordered to disordered structure in femtosecond laser-induced melting may be probed by techniques such as surface second-harmonic generation that are sensitive to order. Changes in lattice constant induced by a pump pulse may be measured by observing diffraction patterns of laser-generated x-ray or electron pulses that Bragg diffract from the crystal lattice.

Pump–probe measurements come in many flavors related to modes of detection and choice of pump and probe beams. In terms of detection, most pump–probe measurements utilize a detector with a slow response time compared to that of the dynamics of interest. In such time-integrated detection, time resolution comes about due to the ability to scan the pump–probe delay precisely. Furthermore, pump and probe pulses may be either degenerate (pulses are essentially identical, are derived directly from the same laser, and occupy the same spectral band) or nondegenerate. A representative experimental geometry for degenerate pump–probe transmission measurements is shown in Fig. 9.2. The pump and probe pulses are focused into an absorbing sample noncollinearly with controllable relative delay τ. The average power in the probe beam is measured using a time-integrating detector as a function of τ. Similar to autocorrelation measurements with repetitive pulse systems (Section 3.4.2), one commonly uses a chopper to modulate the pump beam and a lock-in amplifier to detect the component of the probe average power modulated at that frequency. In this way only the change in probe power induced by the pump beam is recorded.

In nondegenerate experiments pump and probe beams are spectrally nonoverlapping. Here the probe beam is often derived from the same laser as the pump, but frequency-

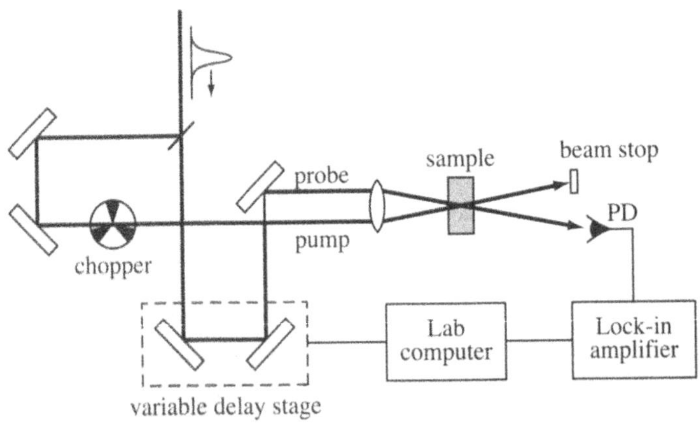

Figure 9.2 Representative experimental setup for pump–probe measurements in transmission geometry. The chopper and lock-in amplifier pull out the change in the power of the transmitted probe induced by the presence of the pump. PD, photodetector.

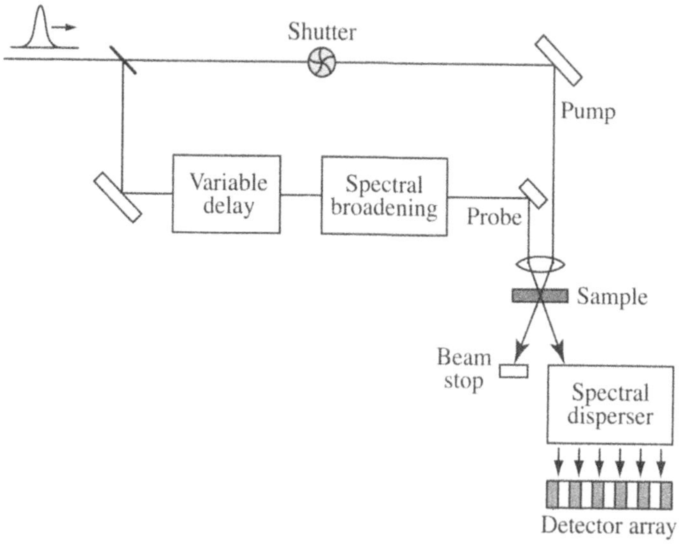

Figure 9.3 Experimental setup for spectrally resolved pump–probe measurements. To obtain differential spectra, data are collected with the shutter alternately passing and blocking the pump beam.

conversion processes are used to shift the frequency of the pump, the probe, or both. For example, the pump may lie in the visible spectrum, while the probe beam may be elsewhere in the visible, in the ultraviolet or infrared, or even in more exotic frequency regions such as terahertz or x-ray. Furthermore, the process of continuum generation (see Section 6.8) provides access to probe pulses with extremely broad frequency content. This makes possible spectrally resolved pump–probe experiments, depicted schematically in Fig. 9.3. The arrangement is similar to that for degenerate pump–probe except that prior to the sample, the probe is spectrally broadened to cover the frequency range of interest; and after passing through the sample, the probe is spectrally dispersed and then measured in parallel via a (time-integrating) photo-detector array. This gives the transmitted probe spectrum. Typically, spectra are recorded both with the pump on and with the pump blocked; subtracting these yields the differential transmission spectrum. Such differential spectra are then recorded as a function of probe delay. The resulting time- and frequency-resolved pump–probe data often provide much richer information than is possible in degenerate measurements (which provide only time resolution).

In contrast to the examples given above, some measurement schemes directly resolve the temporal profile of the probe pulse transmitted or of some other optically induced effect. Figure 9.4 shows the basic setup for experiments in which the time dependence of emitted fluorescence is measured [248]. Here a pump pulse photoexcites an absorbing sample, which then radiates fluorescence into a wide solid angle as a result of its photorelaxation processes. In conventional fluorescence techniques, the light emitted is collected and analyzed as a function of frequency to yield a (time-integrated) fluorescence spectrum. Figure 9.4 adds an ultrafast time gate activated by a delayed probe pulse to time resolve the evolution of the fluorescence spectrum. The time gating operation is typically implemented through sum frequency generation in a thin nonlinear crystal (see Section 5.5.1). Time-resolved fluorescence spectroscopy is especially well suited for tracing time-dependent energy distributions of molecules or charge carriers within an excited electronic energy manifold or band.

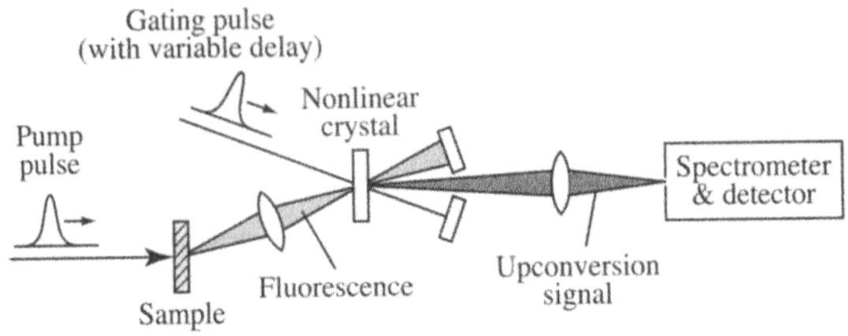

Figure 9.4 Experimental setup for time-resolved fluorescence spectroscopy.

Many other measurement schemes based on the basic concept of pump and probe are possible. A few additional examples are described later in the chapter.

Many of the topics we consider in this chapter involve resonant laser–material interactions (i.e., light is absorbed). This is in contrast to the nonresonant nonlinear interactions discussed in Chapters 5 and 6. In the following we discuss several important methods of ultrafast spectroscopy, all based on the pump–probe concept. In the course of this treatment, we sometimes illustrate measurement techniques by referring to ultrafast processes occurring in representative physical systems. However, we do not attempt to broadly survey the diverse range of physical systems that have been studied via the tools of ultrafast time-resolved spectroscopy. For detailed discussion of ultrafast processes in specific physical systems, the reader is directed to monographs such as [249,505,506] or to the proceedings of specialty conferences such as the International Conference on Ultrafast Phenomena (e.g, [507]).

The remainder of this chapter is organized as follows. In Section 9.2 we give a detailed analysis of the degenerate pump–probe measurement technique, including both induced grating and orientational relaxation phenomena. Spectrally resolved pump–probe is illustrated in Section 9.3 by referring to experiments revealing energy relaxation in both organic molecules and semiconductors. In Section 9.4 we review the quantum mechanics formalism needed for a discussion of coherent ultrashort-pulse phenomena and spectroscopy. Coherent wave packets excited in matter systems via resonant ultrashort-pulse excitation are discussed in Section 9.5; examples involving semiconductor nanostructures as well as molecules are presented. Measurement of phase relaxation phenomena is covered in Section 9.6, and coherent vibrations generated in response to nonresonant ultrashort pulse excitation are discussed in Section 9.7.

9.2 DEGENERATE PUMP–PROBE TRANSMISSION MEASUREMENTS

In this section we analyze degenerate pump–probe measurements in a transmission configuration, as sketched in Fig. 9.2. We begin with a scalar treatment useful for co-polarized pump and probe pulses.

9.2.1 Co-polarized Fields: Scalar Treatment

Denote the absorption coefficient of the sample as $\alpha(t) = \alpha_0 + \Delta\alpha(t)$, where α_0 and $\Delta\alpha(t)$ represent the static absorption and the time-dependent absorption change induced by the field, respectively. We assume that the absorption change may be expressed as the convolution of the optical intensity $I(t)$ and a causal impulse response function $h^{(\alpha)}(t)$,

$$\Delta\alpha(t) = -\int dt'\, I(t')h^{(\alpha)}(t-t') \tag{9.2}$$

The minus sign is inserted in this equation so that a response function with positive sign corresponds to the usual case in which the intensity acts to reduce or saturate the absorption. For a sample of thickness L, we express the effect of the absorption on the field according to

$$\frac{\partial E}{\partial z} = -\frac{\alpha}{2}E \qquad \text{and} \qquad E_{\text{out}} = E_{\text{in}}e^{-\alpha L/2} \approx E_{\text{in}}[1-(\alpha_0+\Delta\alpha)L/2] \tag{9.3}$$

We have already made several assumptions in this simple formulation:

- The total absorption is small; hence, the exponential attenuation factor may be expanded to first order in the absorption. In this limit the optical intensity that drives absorption changes may be taken as independent of z, as in eq. (9.2).
- Pump and probe intensities are sufficiently small that changes in the sample response may be considered as perturbations that scale linearly with applied intensity.
- The absorption is homogeneously broadened to a bandwidth large compared to the bandwidth of pump and probe pulses. Consequently, the absorber is modeled as frequency independent, and dephasing (a concept introduced later in this chapter) is taken to be instantaneous.
- Dispersion may be neglected. Pulse shapes are unchanged in propagating through the sample, except as affected by the time-varying absorption change.

We now analyze the pump–probe experiment with two further simplifying assumptions:

- Pump and probe pulses are degenerate and are identical except for time delay, intensity, and propagation direction.
- Pump and probe pulses have identical linear polarization and therefore may be analyzed as scalars.

We write the input pump field E_p and input probe (or test) field E_t in the following form:

$$E_p(\mathbf{r},t) = \text{Re}\{\hat{a}_p(\mathbf{r},t)e^{j\omega_0 t}\} = \text{Re}\{a_p(t)e^{j(\omega_0 t-\mathbf{k_p}\cdot\mathbf{r})}\} \tag{9.4a}$$

$$E_t(\mathbf{r},t) = \text{Re}\{\hat{a}_t(\mathbf{r},t)e^{j\omega_0 t}\} = \text{Re}\{a_t(t)e^{j(\omega_0 t-\mathbf{k_t}\cdot\mathbf{r})}\} \tag{9.4b}$$

where

$$a_t(t) = a_p(t-\tau) \tag{9.4c}$$

In general, pump and probe pulses will have different amplitudes. However, for simplicity we set the amplitudes equal, which will not affect our results in a significant way. We proceed by inserting the total field, taken as the superposition of pump and probe fields ($E = E_p + E_t$), into eqs. (9.2) and (9.3). Using $I = |\hat{a}_p + \hat{a}_t|^2$, we get

$$E_{\text{out}} = \left\{1 - \frac{1}{2}\left[\alpha_0 + \Delta\alpha(t)\right] L\right\} \text{Re}\left\{a_p(t)e^{j(\omega_0 t - \mathbf{k_p}\cdot\mathbf{r})} + a_t(t)e^{j(\omega_0 t - \mathbf{k_t}\cdot\mathbf{r})}\right\} \tag{9.5a}$$

where

$$\Delta\alpha(t) = -\int dt'\, h^{(\alpha)}(t - t')\left\{|a_p(t')|^2 + |a_t(t')|^2 + \left(a_p(t')a_t^*(t')e^{-j(\mathbf{k_p}-\mathbf{k_t})\cdot\mathbf{r}} + \text{c.c.}\right)\right\} \tag{9.5b}$$

For the pump–probe measurement we are interested only in the field component emerging in the $\mathbf{k_t}$ direction, given by

$$\begin{aligned}\text{Re}&\left\{a_t(t)\left[1 - \frac{\alpha_0 L}{2} + \frac{L}{2}\int dt'\, h^{(\alpha)}(t - t')\left(|a_p(t')|^2 + |a_t(t')|^2\right)\right]e^{j(\omega_0 t - \mathbf{k_t}\cdot\mathbf{r})}\right\}\\ &+\frac{L}{2}\,\text{Re}\left\{a_p(t)\int dt'\, h^{(\alpha)}(t - t')a_p^*(t')a_t(t')e^{j(\omega_0 t - \mathbf{k_t}\cdot\mathbf{r})}\right\}\end{aligned} \tag{9.6}$$

For the usual case of a slow detector, the measurement records time-integrated probe intensity (i.e., probe energy transmitted). We assume also that the experiment isolates that portion of the probe energy modulated by the pump. If the absorption and change in absorption are also small, so that only lowest-order terms need be retained, the change in probe energy U_t due to the pump beam is

$$\Delta U_t(\tau) = \left(1 - \frac{\alpha_0 L}{2}\right)(L)\left[\gamma(\tau) + \beta(\tau)\right] \tag{9.7a}$$

where

$$\gamma(\tau) = \int dt\, |a_t(t)|^2 \int dt'\, h^{(\alpha)}(t - t')|a_p(t')|^2 \tag{9.7b}$$

$$\beta(\tau) = \frac{1}{2}\int dt\, \left\{a_t(t)a_p^*(t)\int dt'\, h^{(\alpha)}(t - t')a_p(t')a_t^*(t') + \text{c.c.}\right\} \tag{9.7c}$$

and a_p and a_t are related through eq. (9.4c).

There are two terms [113]. The first term $\gamma(\tau)$ is expected: The pump intensity saturates the absorption, whose dynamics are probed by the delayed probe. In the usual case where absorption is decreased ($\Delta\alpha < 0$), the probe energy transmitted is increased. Now introducing the change of variables $t - t' = \tau - \tau'$, we obtain

$$\gamma(\tau) = \int d\tau'\, h^{(\alpha)}(\tau - \tau')\int dt\, |a_p(t)|^2|a_p(t - \tau')|^2 \tag{9.8}$$

The second integral (over t) is proportional to the intensity autocorrelation of the laser pulse, $G_2(\tau')$ in the notation of Chapter 3. The temporal resolution of the measurement is limited by the width of the intensity autocorrelation, and the overall form of the $\gamma(\tau)$ saturation term is given by the absorption change impulse response function convolved with the intensity autocorrelation. Response function dynamics slow compared to the autocorrelation width are clearly resolved, while dynamics fast compared to the autocorrelation are not resolved at all. For dynamics occurring on a time scale comparable to the autocorrelation width, one may attempt to deconvolve the data to extract the response function, since the intensity autocorrelation may be determined independently using the methods of Section 3.4. However, such deconvolution efforts are complicated by the additional coherent coupling process described below.

The second term denoted $\beta(\tau)$ is unexpected and results from coherent coupling of the pump into the probe direction. This process, illustrated in Fig. 9.5, arises when the pump and probe overlap in time and therefore interfere to produce a spatially modulated intensity pattern. Since the intensity is spatially modulated, the induced absorption change is also spatially modulated. In other words, an absorption grating is induced. Such a spatial modulation term is clearly evident from eq. (9.5b). The induced grating scatters some of the pump light precisely in the direction of the transmitted probe; as a result, the scattered pump and transmitted probe add coherently on the detector. This process introduces an unwanted coherent coupling artifact to the recorded pump–probe signal, which masks features of the absorption response function near zero delay.

This $\beta(\tau)$ coherent coupling term has the following properties:

- $\beta(\tau)$ is symmetric with respect to τ and is nonzero only for small delays on the order of a pulse width or less (i.e., pump and probe pulses must have temporal overlap).
- Coherent coupling arises only in degenerate pump–probe measurements (i.e., pump and probe must have spectral overlap). In nondegenerate experiments, pump and probe do not have spectral overlap and their time-average interference vanishes. Hence, the coherent coupling term vanishes as well.

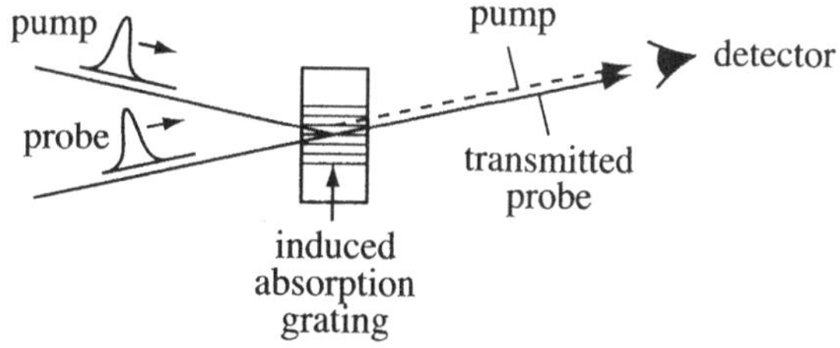

Figure 9.5 When pump and probe are temporally overlapped in pump–probe measurements, they interfere to produce a spatially modulated intensity pattern. This leads to an induced absorption grating, which scatters some of the pump light into the direction of the transmitted probe. This process introduces a coherent coupling contribution to the recorded pump–probe signal.

- When the absorption dynamics are very fast compared to the pulse width, the response function may be taken as a delta function. The coherent coupling term becomes

$$\beta(\tau) = \int dt\, |a_p(t)|^2 |a_p(t-\tau)|^2 \tag{9.9}$$

In this fast response limit, $\beta(\tau)$ takes the form of the intensity autocorrelation function, as does the saturation term $\gamma(\tau)$. The pump–probe signal reflects the pulse duration, not the material dynamics.

- When the absorption response function may be considered as a unit step function (i.e., the photo-induced absorption response has very fast rise time but recovers very slowly compared to the pulse duration), the coherent coupling term becomes

$$\beta(\tau) = \frac{1}{2}\left|\int dt\, a_p(t) a_p^*(t-\tau)\right|^2 \tag{9.10}$$

Here $\beta(\tau)$ assumes the form of the electric field autocorrelation envelope squared. This occurs because the absorption grating responsible for coherent coupling is the result of time-integrated interference between pump and probe fields.

- Independent of the form of $h^{(\alpha)}(t)$, saturation and coherent coupling terms always give equal contributions at zero delay [i.e., $\beta(0) = \gamma(0)$].

It is important to note that transmitted pump and probe energies are simultaneously increased as a result of this coherent coupling process [508]. Such enhanced transmission as a result of optically induced gratings played an important role in the operation of colliding pulse mode-locked dye lasers, as discussed in Section 2.3.2.

Figure 9.6 shows a plot of the full pump–probe signal as well as individual contributions from saturation and coherent coupling terms, computed for the case of very slow absorption recovery [$h^{(\alpha)}(t)$ is taken as a step function] and a transform-limited Gaussian pulse shape. The saturation term is simply the integral of the intensity autocorrelation function and therefore reaches half of its final value at $\tau = 0$. However, due to coherent coupling, the overall signal at $\tau = 0$ is exactly equal to the signal at large τ. This feature is independent of the pulse shape assumed and results from the step function material response. There is a small overshoot of about 15% near $\tau = 0.8t_p$ (both of these values are dependent on pulse shape). An important point is that this overshoot and its fast recovery come about due to the coherent coupling term and *do not reflect a fast material response*.

An experimental example in which pulses of about 7-ps duration were used for transmission pump–probe measurements of cresyl violet dye in solution is shown in Fig. 9.7. On the time scale of the pulses, the absorption response function is well modeled as a step function (essentially instantaneous rise time with very slow recovery). The data using bandwidth-limited pulses (Fig. 9.7a) are similar to those calculated in Fig. 9.6. Figure 9.7b shows data in which the laser spectrum was broadened approximately tenfold (without appreciable increase in pulse duration) via nonlinear propagation in an optical fiber. The saturation component of the pump–probe signal continues to follow the integrated intensity autocorrelation function and is largely unchanged. However, the coherent coupling term, which has a width related to the laser coherence time, is now shortened to a narrow spike

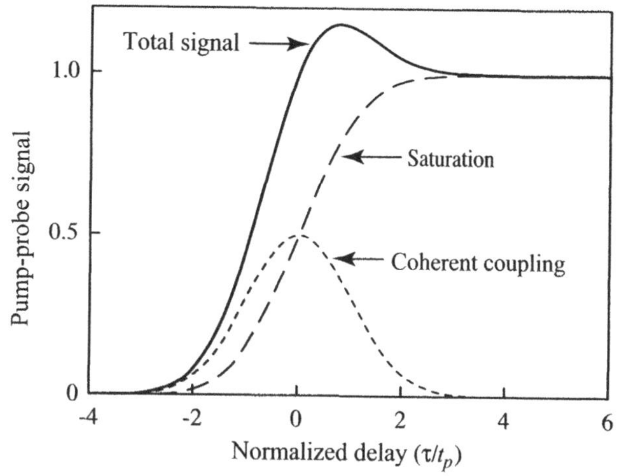

Figure 9.6 Calculated degenerate pump-probe signal for a Gaussian pulse shape, $a_p(t) \sim \exp\left(-t^2/t_p^2\right)$, and for slow absorption recovery $h^{(\alpha)}(t) \sim u_{-1}(t)$, where $u_{-1}(t)$ represents the unit step function. The signal is normalized to its value at large delay. Both the overall pump–probe signal and individual saturation and coherent coupling contributions are shown.

clearly distinguishable near $\tau = 0$. This dependence on laser coherence is sometimes useful, as here, to clearly identify the coherent coupling contribution. We emphasize that the narrow coherence spike is merely an indicator of reduced laser coherence and *does not reflect a fast material response*. Because of such effects, one must exercise great caution in degenerate pump–probe experiments before assigning fast peaks near zero delay to sample dynamics, unless such peaks exhibit decay clearly slower than the pulse durations.

9.2.2 Vector Fields and Orientational Effects

In this section we discuss pump–probe transmission experiments in which pump and probe fields are no longer co-polarized. Such experiments lead to new effects for samples made up of anisotropic building blocks, which may reorient on an ultrafast time scale.

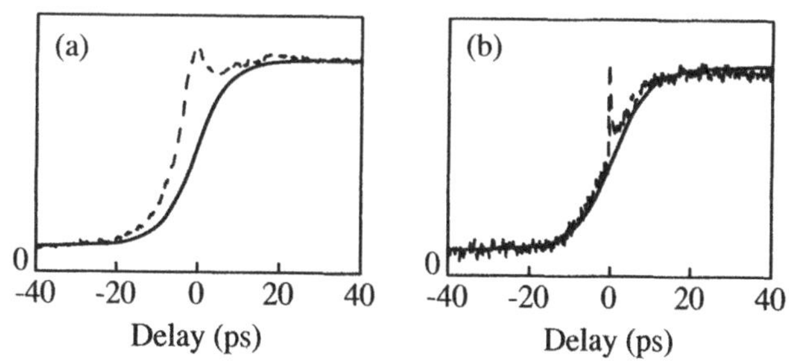

Figure 9.7 Pump–probe data for cresyl violet dye solution with pulses of about 7 ps: (a) pulses are nearly bandwidth limited; (b) pulses are highly chirped. Solid lines are integrals of measured intensity autocorrelation functions. The rise of the saturation signal is related to the pulse duration, while the width of the coherent coupling peak depends on the laser coherence time. Adapted from [508].

To analyze experiments in which pump and probe may have different polarizations, it is helpful to utilize a tensor response function formalism [508–511], as introduced briefly in Section 6.7.3. To lowest order the pump–probe signal comes about through the third-order nonlinear polarization. In the regime of interest here, we write

$$\hat{P}_i^{(3)}(\mathbf{r},t) = \varepsilon_0 \hat{a}_j(\mathbf{r},t) \int dt' \tilde{\chi}_{ijkl}^{(3)}(t-t') \left[\frac{1}{2}\hat{a}_k^*(\mathbf{r},t')\hat{a}_l(\mathbf{r},t') + \text{c.c.}\right] \tag{9.11}$$

$\tilde{\chi}_{ijkl}^{(3)}$ is the third-order polarization response function tensor (which is, of course, causal). As usual, the subscripts take on values 1, 2, 3 (equivalently, x, y, z), denoting an orthogonal basis set in Cartesian coordinates. $\hat{P}_i^{(3)}$ and the various $\hat{a}_j$ represent vector components of the nonlinear polarization and the electric field, respectively, in a complex envelope function representation similar to eq. (9.4a) (i.e., $e^{j\omega_0 t}$ time dependence factored out but spatial variations remain). The specific form of the response function results from the assumption of a resonant, essentially instantaneous, and very broadband absorption process. That is, the absorption responds directly to a generalized optical intensity, represented in eq. (9.11) as the $\hat{a}_k^*\hat{a}_l + \text{c.c.}$ term.

The symmetry of the response function tensor will play an important role in our discussion later in this section. We specialize to the case of isotropic media [131,509], for which the only nonzero elements of the response function are of the form $\tilde{\chi}_{iiii}^{(3)}$ (all four subscripts the same) or $\tilde{\chi}_{iijj}^{(3)}$, $\tilde{\chi}_{ijij}^{(3)}$, or $\tilde{\chi}_{ijji}^{(3)}$ (subscripts take on two different values, in pairs). For isotropic media the response tensor also satisfies

$$\tilde{\chi}_{iiii}^{(3)}(t) = \tilde{\chi}_{iijj}^{(3)}(t) + \tilde{\chi}_{ijij}^{(3)}(t) + \tilde{\chi}_{ijji}^{(3)}(t) \qquad \text{with} \qquad i \neq j \tag{9.12}$$

Also, because eq. (9.11) is unchanged upon interchange of indices k and l in the driving fields, we have the additional permutation symmetry [509,511]

$$\tilde{\chi}_{ijij}^{(3)}(t) = \tilde{\chi}_{ijji}^{(3)}(t) \tag{9.13}$$

From Poynting's theorem [see eq. (5.36)] the product $\mathbf{E} \cdot \partial \mathbf{P}_{\mathrm{NL}}/\partial t$ represents power transfer from the field into the medium (i.e., absorption). In our notation the time-average intensity change due to absorption is proportional to

$$\Delta I \sim \frac{\omega_0 L}{2} \operatorname{Im}\left\{\hat{a}_i^*(t)\hat{P}_i^{(3)}(t)\right\} \tag{9.14}$$

Directly on resonance we expect the material response to be dominated by absorption, in which case the nonlinear response function should be pure imaginary. We may then write

$$h_{ijkl}^{(\alpha)}(t) = \operatorname{Im}\left\{\tilde{\chi}_{ijkl}^{(3)}(t)\right\} \tag{9.15}$$

where $h_{ijkl}^{(\alpha)}(t)$ is the tensor generalization of the absorption impulse response function introduced earlier.

As in our scalar treatment, we are interested only in the change in energy of the beam emerging from the sample in the probe direction. The change in probe energy due to the pump beam is given by

$$\Delta U(\tau) \sim \gamma(\tau) + \beta(\tau) \tag{9.16a}$$

where

$$\gamma(\tau) = \int dt\, \mathrm{Im}\left\{ a_{t,i}^*(t) a_{t,j}(t) \int dt'\, \tilde{\chi}_{ijkl}^{(3)}(t-t') a_{p,k}^*(t') a_{p,l}(t') \right\} \tag{9.16b}$$

and

$$\beta(\tau) = \int dt\, \mathrm{Im}\left\{ a_{t,i}^*(t) a_{p,j}(t) \int dt'\, \tilde{\chi}_{ijkl}^{(3)}(t-t') a_{p,k}^*(t') a_{t,l}(t') \right\} \tag{9.16c}$$

Here $a_{p,i}$ and $a_{t,i}$ are vector components of pump and probe pulses, respectively, as in eqs. (9.4a) and (9.4b).

Let us first consider these expressions for parallel, linear pump and probe polarizations. The saturation term immediately yields the same expression for $\gamma(\tau)$ as our earlier scalar treatment, eq. (9.7b), but with $h_{1111}^{(\alpha)}$ in place of $h^{(\alpha)}$. The expression for $\beta(\tau)$ reduces to eq. (9.7c) only when $\tilde{\chi}_{1111}^{(3)}$ is purely imaginary. However, the nonlinear polarization may also have a real part, corresponding to an optically induced change in refractive index; this may arise, for example, when the laser frequency is tuned off center of the absorption resonance. When $\tilde{\chi}_{1111}^{(3)}$ has nonzero real part, a new coherent coupling term that is antisymmetric in delay occurs in addition to the term revealed in our earlier analysis [512]. The finding that $\gamma(\tau) = \beta(\tau)$ at $\tau = 0$ remains valid. For the remainder of this section we assume a purely imaginary nonlinear response.

We now consider the case of pump and probe pulses with orthogonal linear polarizations, with pump and probe fields specified as

$$\mathbf{E_p}(\mathbf{r}, t) = \hat{\mathbf{z}}\, \mathrm{Re}\left\{ a_p(t) e^{j(\omega_0 t - \mathbf{k_p}\cdot\mathbf{r})} \right\} \tag{9.17a}$$

$$\mathbf{E_t}(\mathbf{r}, t) = \hat{\mathbf{x}}\, \mathrm{Re}\left\{ a_t(t) e^{j(\omega_0 t - \mathbf{k_t}\cdot\mathbf{r})} \right\} \tag{9.17b}$$

where $a_t(t) = a_p(t-\tau)$, as previously. The contributions to the pump–probe signal now become

$$\gamma(\tau) = \int dt\, |a_p(t-\tau)|^2 \int dt'\, h_{xxzz}^{(\alpha)}(t-t') |a_p(t')|^2 \tag{9.18a}$$

$$\beta(\tau) = \frac{1}{2} \int dt\, \left\{ a_p^*(t-\tau) a_p(t) \int dt'\, h_{xzzx}^{(\alpha)}(t-t') a_p^*(t') a_p(t'-\tau) + \text{c.c.} \right\} \tag{9.18b}$$

Both saturation and coherent coupling terms retain a form similar to that obtained for the parallel polarization case. However, saturation and coherent coupling arise from different tensor components of the nonlinear response function and may therefore have different relative amplitudes. An important point is that, in general, the coherent coupling term remains important even for pump–probe with perpendicular polarizations. At first this may seem surprising, since in our scalar treatment we attributed coherent coupling to the spatially modulated intensity of the combined pump and probe fields. In the perpendicular polarization case, there is no spatial modulation of the intensity. There is, however, a spatial modulation in the polarization state of the combined pump and probe fields. The mechanism whereby this leads to an optically induced grating is elucidated later in the section.

Let us now consider a specific physical system: liquid solutions of dye molecules. This example illustrates interesting photophysics and at the same time lends itself to direct calculation of the nonlinear response tensor elements. As sketched in Fig. 9.8, individual dye molecules may often be represented as linear, while the orientations of different molecules are randomly distributed. That is, individual molecules are highly anisotropic, while the ensemble of molecules on a macroscale remains isotropic. The statement that an individual molecule is linear means more precisely that the electronic wave function in the portion of the molecule responsible for optical absorption (chromophore) is delocalized mainly along a single direction. In such a case, light polarized parallel to the chromophore is absorbed strongly, whereas light polarized in the perpendicular direction is absorbed only weakly or not at all. In terms of absorption cross section, we set $\sigma_\perp = 0$; only the cross section associated with light polarized along the chromophore ($\sigma_\parallel$) plays a role.

Let us calculate the saturation component of the pump–probe transmission signal assuming a volume density of N_0 dye molecules that are initially in the ground electronic state with isotropic orientation. The pump pulse is linearly polarized along $\hat{z}$, and the orientation of any specific molecule is specified by angles θ, ϕ in standard spherical coordinates. We assume that the molecules may be modeled as four-level systems and that pump and probe pulses are resonant with the $1 \leftrightarrow 2$ transition as sketched in Fig. 1.5. We further assume that the $2 \rightarrow 3$ relaxation is very fast, whereas the $3 \rightarrow 4$ relaxation is very slow compared to the pulses. With these assumptions, molecules respond according to the slow saturable absorber model, eq. (2.44), with the modification that only the component of the field along

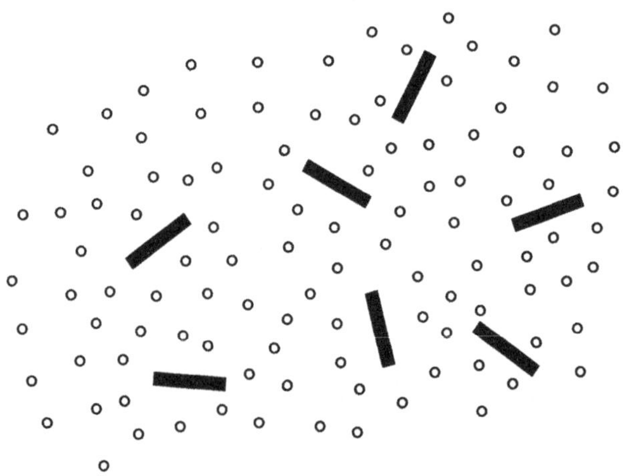

Figure 9.8 Notional view of linear dye molecules oriented isotropically in solution. Filled rectangles, dye molecules; open circles, solvent molecules.

the long axis of the molecule influences absorption and its saturation. Immediately subsequent to a weak pump pulse of energy U_p, the density of molecules remaining in the ground state, hence contributing to further absorption is

$$N(\theta, \phi) = \frac{N_0}{4\pi}\left(1 - \frac{U_p \cos^2\theta}{U_A}\right) \tag{9.19}$$

where $\sigma_\|$ is used in the expression for saturation energy U_A. The pulse energy responsible for saturation is multiplied by $\cos^2\theta$, which represents the effect of projecting the pump field along the axis of a dye molecule oriented at polar angle θ. Therefore, the absorption saturates anisotropically. The absorption seen by a probe pulse polarized along the i direction and arriving shortly after the pump is given by

$$\alpha_i = \frac{\sigma_\| N_0}{4\pi}\iint d\phi\, d\theta \sin\theta \cos^2 \mu_{i,\theta,\phi}\left(1 - \frac{U_p \cos^2\theta}{U_A}\right) \tag{9.20}$$

This expression is obtained by integrating over all molecular orientations, with absorption occurring only according to the projection of the probe onto the various molecules and with $\mu_{i,\theta,\phi}$ giving the angle between the probe polarization along i and a dye molecule oriented along θ, ϕ. For a probe pulse polarized along $\hat{z}$, parallel to the pump, we have $\mu_{i,\theta,\phi} = \theta$ and the absorption becomes

$$\begin{aligned}\alpha_z &= \frac{\sigma_\| N_0}{4\pi}\iint d\phi\, d\theta \sin\theta \cos^2\theta\left(1 - \frac{U_p \cos^2\theta}{U_A}\right)\\ &= \sigma_\| N_0\left(\frac{1}{3} - \frac{U_p}{5U_A}\right)\end{aligned} \tag{9.21}$$

For a probe polarized along $\hat{x}$, we have $\cos\mu_{i,\theta,\phi} = \sin\theta\cos\phi$, and the absorption immediately after the pump pulse becomes

$$\begin{aligned}\alpha_x &= \frac{\sigma_\| N_0}{4\pi}\iint d\phi\, d\theta \sin^3\theta \cos^2\phi\left(1 - \frac{U_p \cos^2\theta}{U_A}\right)\\ &= \sigma_\| N_0\left(\frac{1}{3} - \frac{U_p}{15U_A}\right)\end{aligned} \tag{9.22}$$

The same result is obtained for a probe polarized along $\hat{y}$ or for any other probe polarization orthogonal to the pump polarization.

Already a few key points emerge from our analysis:

- The small-signal absorption coefficient is independent of probe polarization. This is as expected: The ensemble of molecules is isotropic, and therefore the linear optical properties of the ensemble are isotropic as well.
- The value of the small-signal absorption coefficient, $\sigma_\| N_0/3$, is a factor of 3 smaller than would be obtained for a similar ensemble of spherical molecules (individual

molecules are isotropic in their optical response, with the same cross section $\sigma_{\parallel}$ in all directions).

- A linearly polarized pump pulse leads to a saturation (reduction) of the absorption, observable using a probe pulse with any polarization. However, the magnitude of the absorption change observed is three times larger for a parallel-polarized probe than for a perpendicular-polarized probe. Thus, the pump pulse induces dichroism. Even though the unperturbed molecular ensemble is isotropic, the anisotropy of individual molecules in the ensemble allows for anisotropic third-order nonlinear effects. No dichroism would be induced in an ensemble of spherical molecules.

Our discussion so far assumes that the orientations of individual molecules are frozen in place. However, in the liquid state both solvated dye molecules and surrounding solvent molecules experience motion, orientationally as well as translationally. After sufficient elapsed time, individual molecules reorient, and memory of their initial orientations is lost. The dynamics of such orientational relaxation provides information about interactions between solvated and solvent molecules in the liquid state. When orientational relaxation is complete, the sample no longer "remembers" the pump polarization direction; the dichroism disappears and the ensemble is once again isotropic. Absorption saturation may persist in this long time limit, but depends only on the total number of molecules excited out of the ground state. Assuming that recovery of the ground-state population is still negligible, the change in ground-state population is computed by integrating eq. (9.19) over 4π solid angle, with the result

$$\Delta N = \left(\frac{-U_p}{3U_A}\right) N_0 \tag{9.23}$$

The absorption coefficient after complete orientational relaxation (α_∞) is

$$\begin{aligned}\alpha_\infty &= \frac{\sigma_{\parallel}(N_0 + \Delta N)}{4\pi} \iint d\phi\, d\theta \sin\theta \cos^2\theta \\ &= \sigma_{\parallel} N_0 \left(\frac{1}{3} - \frac{U_p}{9U_A}\right)\end{aligned} \tag{9.24}$$

This expression has been calculated assuming a z-polarized probe, but an identical result is obtained independent of probe polarization.

If we assume exponential relaxation processes, our results may be written in the following empirical form:

$$\Delta\alpha_z(t) = -\frac{\sigma_{\parallel} N_0 U_p}{U_A}\left(\frac{1}{9} + \frac{4}{45} e^{-t/\tau_{\text{or}}}\right) e^{-t/\tau_{\text{pop}}} \sim -h_{zzzz}^{(\alpha)}(t) \tag{9.25}$$

$$\Delta\alpha_x(t) = -\frac{\sigma_{\parallel} N_0 U_p}{U_A}\left(\frac{1}{9} - \frac{2}{45} e^{-t/\tau_{\text{or}}}\right) e^{-t/\tau_{\text{pop}}} \sim -h_{xxzz}^{(\alpha)}(t) \tag{9.26}$$

Here τ_{or} is the orientational relaxation time. We have also included exponential recovery of the ground-state population with time constant τ_{pop}. Orientational relaxation leads to a decrease of the absorption saturation seen by a parallel-polarized probe and to an increase of the absorption saturation seen by a perpendicular-polarized probe, while population relaxation decreases the saturation observed at an equal rate for either probe polarization. We also see that we have arrived at forms for two independent components of the nonlinear response function.

It is also interesting to consider the case of a probe pulse polarized at an arbitrary angle β with respect to the pump (which is still polarized along $\hat{z}$). For the case of polarization in the x–z plane, the input probe pulse may be written

$$\mathbf{E_t}(\mathbf{r}, t) = \mathrm{Re}\left\{(\hat{z}\cos\beta + \hat{x}\sin\beta)a_t(t)e^{j(\omega_0 t - \mathbf{k_t}\cdot\mathbf{r})}\right\} \tag{9.27}$$

If we assume the pulse durations are short compared to the time scale of the relaxation, the nonlinear polarization induced by a probe pulse arriving subsequent to the pump is

$$\hat{\mathbf{P}}^{(3)}(\mathbf{r}, t) \sim jU_p a_t(t)\left[\hat{\mathbf{z}}\cos\beta\, h_{zzzz}^{(\alpha)}(\tau) + \hat{\mathbf{x}}\sin\beta\, h_{xxzz}^{(\alpha)}(\tau)\right] \tag{9.28}$$

The nonlinear polarization and associated dichroism results in a delay-dependent rotation of the probe, which may be detected by measuring the transmission of the probe through a crossed polarizer [513]. Alternatively, one may simply measure the transmitted probe energy without a polarizer. In the latter case the change in probe energy is obtained by applying eq. (9.14) and integrating over time. The result using eqs. (9.25) and (9.26) also is given by

$$\Delta U_t \sim U_p U_t \left(\frac{1}{9} + \frac{2}{45}(2\cos^2\beta - \sin^2\beta)\, e^{-\tau/\tau_{\mathrm{or}}}\right) e^{-\tau/\tau_{\mathrm{pop}}} \tag{9.29}$$

Figure 9.9 shows an example of data on the orientational relaxation of Nile blue dye in solution for three different probe polarizations. The population relaxation is sufficiently long (nanosecond scale) that it does not show up during the limited time aperture of the plotted data. Each trace shows an initial rapid rise in transmission due to the saturation induced by

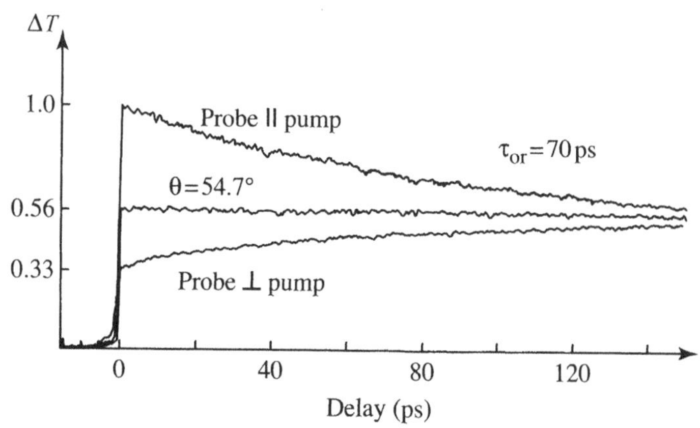

Figure 9.9 Orientational relaxation of Nile blue dye in solution observed with different probe polarizations. From [514].

the pump. As predicted above, the initial transmission step is three times larger for a parallel probe than for a perpendicular probe. Parallel and perpendicular probe traces decay to a common level (equal to $\frac{5}{9}$ of the initial parallel probe value) with an orientational relaxation time of 70 ps. For significantly larger delays the system recovers to an isotropic state, so the output signal is independent of probe polarization. The data for intermediate probe angles lie between those for parallel and perpendicular polarization cases. The data for $\beta = 54.7°$, corresponding to $\cos^2\beta = 1/3$, is particularly interesting. For this "magic angle" the initial step in transmission is exactly equal to that which remains after orientational relaxation; the orientational relaxation term drops out completely, and only population decay plays a role.

So far in this section we have analyzed the induced saturation at times beginning shortly after the pump pulse. That is, temporal overlap of pump and probe pulses has been excluded, which allows us to neglect coherent coupling effects. To analyze the coherent coupling, we need to know the $h^{(\alpha)}_{xzzx}$ component of the nonlinear response. This is obtained by using eqs. (9.12) and (9.13) together with expressions for $h^{(\alpha)}_{zzzz}$ and $h^{(\alpha)}_{xxzz}$ given above. The result is

$$h^{(\alpha)}_{xzzx}(t) = h^{(\alpha)}_{xzxz}(t) \sim \left(\frac{1}{15}\right)\frac{\sigma_{\parallel} N_0 U_p}{U_A} e^{-t/\tau_{\mathrm{or}}}\, e^{-t/\tau_{\mathrm{pop}}} \tag{9.30}$$

For times short compared to orientational relaxation, we have $h^{(\alpha)}_{xzzx}(t) = h^{(\alpha)}_{xxzz}(t)$. Consequently, for pulses short compared to τ_{or}, coherent coupling and saturation terms retain equal amplitude at zero delay, even for perpendicular polarization experiments. On the other hand, for $t \gg \tau_{\mathrm{or}}$, $h^{(\alpha)}_{xzzx}(t) \to 0$. Therefore, for pulses long compared to τ_{or}, the coherent coupling term vanishes in the perpendicular polarization geometry.

Let us now discuss the physics of coherent coupling in pump–probe experiments with perpendicular polarization. Near zero delay when pump and probe pulses are overlapped temporally, the polarization state of the combined pump and probe fields depends on their relative phase. Assuming equal pump and probe amplitudes, the overall polarization changes from 45° linear to circular to −45° linear to circular and then back to 45° linear as the relative phase progresses from 0 to $\pi/2$, π, $3\pi/2$, and then to 2π. The effect is as depicted in Fig. 9.5, except that it is the polarization state of the light rather than its intensity that is spatially modulated. This spatial modulation in polarization state implies a corresponding spatial modulation in the angular orientations for which molecules are most strongly excited to the upper state. The result is a dichroism grating in which the strength and orientation of the induced dichroism is spatially modulated. This dichroism grating can scatter the pump pulse into the direction and polarization state of the probe, which accounts for the coherent coupling. Induced gratings such as those described here, termed *orientational gratings* [510], are distinct from concentration gratings that arise in parallel polarization experiments (for which the overall concentration of molecules excited to the upper state is spatially modulated). For large times such that orientational relaxation has occurred, the system is everywhere restored to an isotropic condition. No further orientational information remains, and the induced grating has disappeared.

Experiments confirming the predicted relationships between coherent coupling and saturation terms were reported in [511]. Figure 9.10a shows pump–probe data for a solution of cresyl violet dye dissolved in ethylene glycol, for parallel as well as perpendicular polarizations. On a pulse duration time scale of about 1 ps, neither population nor

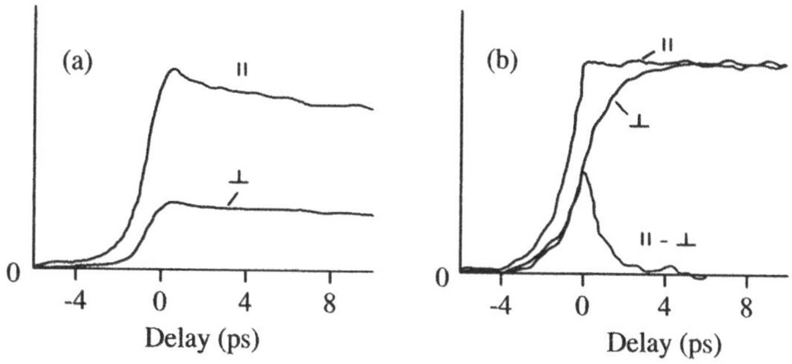

Figure 9.10 Comparison of co-polarized (||) and perpendicular-polarized (⊥) pump–probe transmission data with fast and slow orientational relaxation. (a) Cresyl violet dye solution, slow orientational relaxation. The difference (not shown) of the || curve and three times the ⊥ curve is essentially zero, showing that saturation and coherent coupling terms are reduced identically for perpendicular polarization. (b) Amorphous silicon, fast orientational relaxation. The difference of the || and the ⊥ curves yields only a peak centered at zero delay, showing that coherent coupling is strongly reduced for perpendicular polarization, while saturation is unaffected. Adapted from [511].

orientational relaxation plays a role. Therefore, from our discussion above, we have $h^{(\alpha)}_{xxzz} \approx h^{(\alpha)}_{xzzx} \approx \frac{1}{3}h^{(\alpha)}_{zzzz}$. The pump–probe signal for parallel polarization should be three times as large as that for perpendicular polarization, but because saturation and coherent coupling terms retain equal weighting, the shape of the pump–probe curves should be the same. These predictions are confirmed by the data. Figure 9.10b shows pump–probe data for amorphous silicon using a comparable pulse duration. For such semiconductor samples, orientational effects arise from anisotropic distributions of photoexcited charge carriers in momentum space, a rather different physical effect than that responsible for orientational effects in dye solutions. Nevertheless, our analysis of the pump–probe signal in terms of response functions is still valid. The key point for our discussion here is that for this semiconductor sample, orientational relaxation occurs very rapidly relative to the pulse duration. Hence, we have $h^{(\alpha)}_{xxzz} \approx h^{(\alpha)}_{zzzz}$ and $h^{(\alpha)}_{xzzx} \approx 0$. As a result, the amplitude of the saturation term for parallel and perpendicular polarization is the same. However, for perpendicular polarization the coherent coupling term vanishes, which leads to a clear difference in the overall pump-probe signal. Subtracting the perpendicular polarization data from the parallel polarization data yields the coherent coupling contribution to the parallel polarization case. As expected, it is centered at $\tau = 0$, has duration comparable to the pulse width, and has amplitude approximately equal to one-half the total signal at $\tau = 0$. Additional measurements (not shown) on the same amorphous silicon sample but performed with non-transform-limited pulses confirm that the difference between parallel and perpendicular polarization data in Fig. 9.10b is associated with the polarization dependence of the coherent coupling.

9.3 NONDEGENERATE AND SPECTRALLY RESOLVED PUMP–PROBE: CASE STUDIES

In Section 9.2 we analyzed the degenerate pump–probe technique with a focus on the detailed form of the signals obtained. Such degenerate measurements have the practical advantage that only a single laser is required. On the other hand, ultrafast systems that provide synchronized short pulses at different optical frequencies have become common. From the analysis perspective, the use of nondegenerate pump and probe pulses gets rid of the coherent coupling artifact that was discussed in depth in Section 9.2. More important, nondegenerate measurements (using continuum probe pulses, for example) allow study of time- and frequency-resolved pump–probe data, resulting in much richer information than possible in degenerate measurements alone. In the following we adopt a case study approach in which we survey studies of ultrafast energy relaxation dynamics, first in large organic molecules in solution, specifically dye molecules, and then in the direct bandgap semiconductor GaAs. Both of these serve as model systems that were heavily investigated early in the history of femtosecond spectroscopy. Our intent is to illustrate new information that becomes available with nondegenerate and spectrally resolved measurement techniques, and in the course of this discussion, to touch upon examples of interesting ultrafast photophysics.

9.3.1 Femtosecond Pump–Probe Studies of Dye Molecules

An energy diagram relevant to the study of dye molecules is sketched in Fig. 9.11 [515]. This diagram shows the energies of the electronic ground-state and first electronic excited-state manifolds (denoted S_0 and S_1, respectively) as a function of generalized molecular coordinates Q. Q is sketched in one dimension for convenience but actually represents a multidimensional space defining the positions of the nuclei that compose the molecule. In a Born–Oppenheimer approximation picture, described mathematically in Section 9.5.2, electronic and nuclear motions are taken to be separable; for a given electronic state, energies are calculated as a function of nuclear positions, which are assumed to be clamped during the calculation. For different electronic states, minimum energy occurs at different nuclear configurations. We may understand this by recalling that it is the attraction between negatively charged electrons and positively charged nuclei that overcomes repulsive

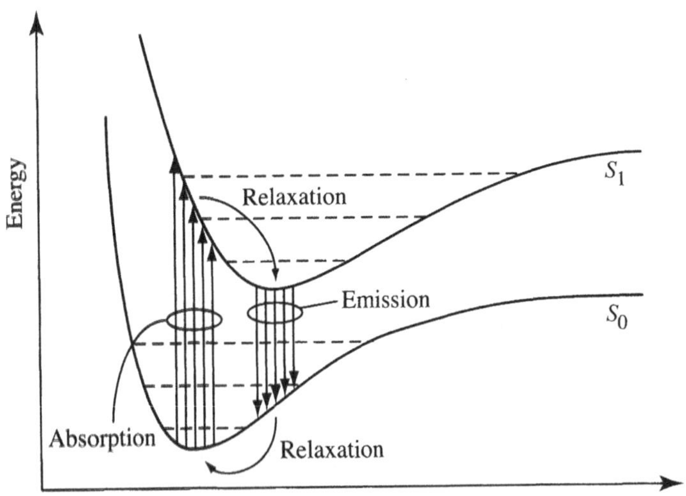

Figure 9.11 Potential energy surfaces relevant to photoexcitation and relaxation dynamics in molecules. Vibrational levels bound within the potential surfaces are depicted as dashed lines.

nuclear–nuclear forces to bind the molecule together. The result is an effective potential acting on the nuclei, which depends on the spatial distribution of the electrons relative to the nuclei and hence on the electronic wave function.

According to the Franck–Condon principle, electronic transitions occur vertically (i.e., without displacement of the nuclei). In other words, the molecular coordinates Q immediately after an electronic transition are identical to those immediately prior to the transition. This approximation, like the Born–Oppenheimer approximation of the preceding paragraph, is motivated by the much heavier mass of nuclei compared to that of electrons, which results in vibrational (nuclear) motions much slower than characteristic time scales for electron motions. Following excitation the molecule rapidly relaxes out of the initially photoexcited levels, transferring excess energy into molecular vibrations [516–518]. As a result of this vibrational relaxation process, the molecule rearranges itself to reach the lowest-energy configuration within the S_1 manifold, typically within a picosecond. On a slower time scale, other processes, such as orientational relaxation (tens to hundreds of picoseconds; see Section 9.2.2) and radiative relaxation (hundreds of picoseconds to several nanoseconds), may be observed. In the following we focus on some of the fast initial relaxation processes.

Note that for descriptions of laser operation, the dye molecule is often modeled as a four level system (see Fig. 1.5). Pumping occurs from low-energy states within the S_0 manifold (level 1) to excited states in S_1 (level 2); stimulated emission occurs from low-energy states near the bottom of S_1 (level 3) to excited states within S_0 (level 4). As discussed above, fast vibrational relaxation is responsible for taking the initially photoexcited molecule from level 2 to level 3, where it contributes to a population inversion, and then from level 3 to level 4 (subsequent to stimulated or spontaneous emission). This picture correctly predicts that the molecule's fluorescence and gain spectra are red-shifted with respect to its absorption spectrum.

Figure 9.12 shows degenerate pump–probe data for Nile blue dye in methanol, performed with pulses 70 fs in duration. These data are of interest not only because they reveal rapid

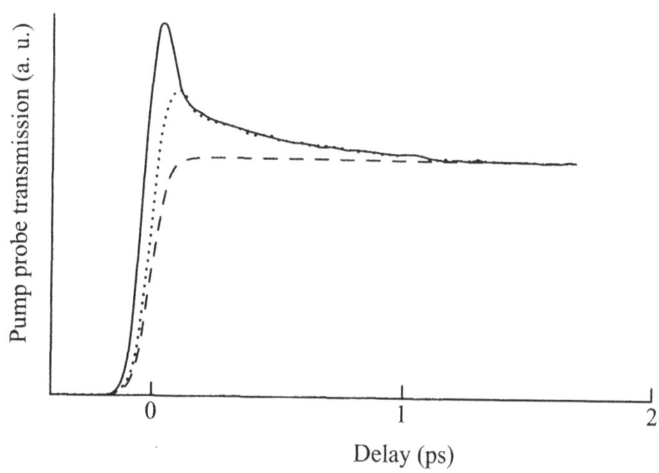

Figure 9.12 Transmission pump–probe data for Nile blue dye in methanol. Plots include the unprocessed data (solid line), data after approximate removal of coherent coupling term (dotted line), and the integral of the intensity autocorrelation (dashed line). From [516].

dynamics immediately following photoexcitation but also because they illustrate limits to the information that may be extracted from such degenerate experiments. The data show a fast rise, with strong peaking near zero delay, followed by a subpicosecond decay to a constant level. As explained previously, such degenerate pump–probe data contain contributions both from the desired absorption dynamics as well as from the coherent coupling artifact. Accordingly, the data are processed with the intent of removing the coherent coupling term, which can be accomplished at least approximately by independently measuring and then subtracting the squared electric field autocorrelation function (scaled to one-half of the $\tau = 0$ signal level). Recalling that the likelihood of absorption is proportional to the population difference $N_1 - N_2$ between initial and excited states, we may attribute the remaining bleaching signal to two mechanisms:

1. *Depopulation of the electronic ground state* S_0 (decrease in population N_1 of level 1). This effect is long-lived and recovers slowly via radiative relaxation. Over the short time window shown in Fig. 9.12, ground-state depopulation may be modeled as a step function contribution to the absorption response function. This yields a contribution proportional to the integral of the intensity autocorrelation function to the pump–probe saturation term.
2. *Introduction of population into the initially photoexcited levels in* S_1 (increase in population N_2 of level 2). This population of molecules gives rise to stimulated emission at the probe frequency, which further decreases the net absorption. This effect manifests itself as an additional signal beyond the integrated intensity autocorrelation and exhibits a rapid decay indicative of the time for molecules to relax out of the initially photoexcited states. These data may be satisfactorily fit with a biexponential absorption response, with a slower time constant of order 400 fs and a faster time constant on the order of the pulse duration.

These degenerate pump–probe data unequivocally reveal fast molecular dynamics following photoexcitation. Similar effects are seen in other dye molecules. However, such measurements are not able to reveal the dynamics after the decay out of the initially photoexcited states. In particular, neither the time required for molecules to relax to the lowest-energy states in S_1, nor the details of the relaxation process within the excited-state manifold, are revealed.

Spectrally resolved data for cresyl violet dye in solution are plotted in Fig. 9.13. In these experiments, configured as illustrated in Fig. 9.3, the sample was excited by 60-fs pump pulses centered at 618 nm, and a fiber and grating pulse compressor (see Section 6.4) was used to obtain probe pulses that were both spectrally broadened and compressed in time to 10 fs. Figure 9.13a shows a comparison of absorption spectra taken for negative delay (prior to arrival of the pump pulse) and for delay near zero (pump and probe coincide). Consistent with our earlier discussion, a prompt absorption decrease is seen in the frequency region near the pump. However, the spectrally resolved data immediately provide new information as well: The absorption change is spectrally nonuniform, comprising spectral holes both at the pump wavelength and at positions symmetrically shifted up and down in frequency from the pump. Similar effects are observed with other dye molecules, including Nile blue.

To understand these data, we must introduce the concept of vibronic transitions [515,519]. The energy surfaces introduced in connection with Fig. 9.11 refer only to an effective *potential energy* without consideration of nuclear motions (*kinetic energy*). However, nuclear

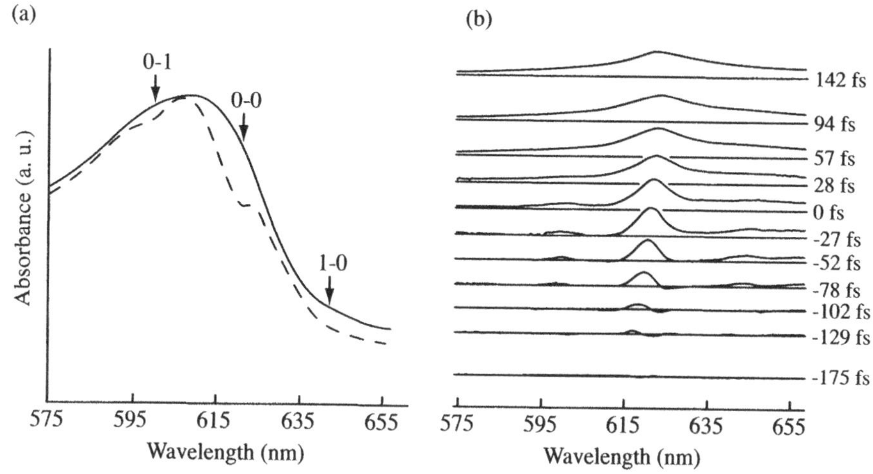

Figure 9.13 Spectrally resolved pump–probe data for cresyl violet dye in solution: (a) absorbance spectrum near zero time delay before (solid line) and after (dashed line) excitation with 60-fs pump pulse; (b) differential transmittance spectra at various relative time delays. Adapted from [517].

motions are also possible and give rise to discrete vibrational levels bound within the effective potential. Representative energy levels of such vibrational states within S_0 and S_1 are shown as dashed lines in Fig. 9.11. In a vibronic transition, absorption of a photon results in simultaneous transitions between both electronic and vibrational states. The absorption spectrum of a large dye molecule in equilibrium is strongly affected by vibronic transitions and may be written in the form [519]

$$\alpha(\omega) \sim \sum_{i,f} P_i \chi_{if} \mathcal{L}_{if}(\omega - \omega_{if}) \tag{9.31}$$

where ω_{if} is the transition frequency between the initial vibrational level in the electronic ground state and the final vibrational level in the electronic excited state; χ_{if} is the Franck–Condon factor, which quantifies the degree of spatial overlap of initial and final vibrational wave functions; P_i gives the thermal equilibrium occupation probability of the initial state; and the $\mathcal{L}_{if}$ are the line shapes of the various transitions.

The pump pulse in these experiments is resonant with the transition between the lowest vibrational levels in the S_0 and S_1 electronic states (termed the 0–0 *transition*). Thus, the prompt appearance of the central hole designated 0–0 in Fig. 9.13a results from the direct absorption process, with both depopulation of the initial state and population of the final state contributing to the bleaching. As explained in [517], the prompt replica holes are manifestations of vibronic transitions associated with a vibrational mode at about 600 cm^{-1} known from Raman scattering. In particular, the 0–1 hole is associated with the transition between the partially depopulated ground vibrational state in S_0 to the first excited vibrational level in S_1, while the 1–0 hole is associated with the transition from the thermally populated first excited vibrational level in S_0 to the partially populated ground vibrational level in S_1. Although a dye molecule has a large number of degrees of freedom and hence a large number of vibrational modes, the observation of distinct replica holes indicates that only a few modes are optically active, in the sense that their occupation numbers are altered significantly during the optical transition. Hence, these data show that pumping with a sufficiently short pulse induces a nonequilibrium distribution of population among the vibronic levels.

The observation of spectral holes also indicates that over short time scales, dye molecules appear to be inhomogeneously broadened. The concepts of homogeneous and inhomogeneous broadening are treated in some detail later in the chapter. Briefly, by *inhomogeneous broadening* we mean that instantaneous vibronic spectra are shifted (or distorted) in a different way for different individual molecules. The widths of the spectral holes reflect the $\mathcal{L}_{if}$ line shape functions in eq. (9.31), convolved with the pump pulse spectrum. The spectral shifts and distortions responsible for inhomogeneous broadening may be explained on the basis of coupling between the few optically active modes and the large number of modes that are optically inactive. In a real space picture, these effects are related to differences in instantaneous molecular configurations in the multidimensional nuclear coordinate space. The contributions of individual molecules are not distinguished in the linear absorption spectrum, which is simply the ensemble average over the full inhomogeneous distribution, but are brought out in the spectrally resolved pump–probe.

Figure 9.13b shows a progression of differential transmission spectra recorded at various pump–probe delays. The spectral holes are observed to broaden and then merge into a smooth continuum on a time scale on the order of 100 fs. These data show that for such large molecules in solution, inhomogeneous broadening effects are short lived; individual molecules quickly lose memory of their specific initial absorption spectrum. This fast relaxation of the induced nonthermal population distribution occurs on a time scale similar to that of the fast initial decay observed in degenerate pump–probe data such as those of Fig. 9.12.

Additional interesting relaxation processes continue to occur on a variety of time scales. Temporally and spectrally resolved fluorescence measurements [518], for example, suggest that full equilibration of vibrational energy among the modes of the molecule may continue for several hundred femtoseconds. Even after the vibrational population reaches a thermal distribution, the temperature characterizing the vibrational distribution may exceed the solvent temperature significantly. Vibrational cooling in which excess vibrational energy diffuses into the solvent appears to occur on a time scale of several picoseconds [519].

9.3.2 Femtosecond Pump–Probe Studies of GaAs

Direct bandgap semiconductors such as gallium arsenide (GaAs) are of significant technological importance both in optoelectronics (diode lasers, photodetectors, etc.) and in high-speed electronics. These materials have been studied extensively via the tools of time-resolved spectroscopy, both because of fundamental interest in nonequilibrium charge carrier physics and because of practical implications for understanding and design of high-speed devices. Like large molecules, semiconductors exhibit a number of ultrafast physical processes on a variety of time scales. Reference [249] classifies the relaxation of a semiconductor excited out of thermal equilibrium by an ultrashort pulse into four temporally overlapping stages. These are summarized briefly as follows:

- *Coherent regime* ($\leq$ 200 fs). In this regime excitations within the semiconductor, such as optical polarization, remain in phase with each other and with the electric field of the pulse giving rise to such excitations. Processes such as carrier–carrier scattering and momentum relaxation rapidly act to disrupt (or dephase) the coherence, with characteristic time scale on the order of 200 fs or significantly faster. Ultrafast spectroscopy methods that provide information on coherence (and loss of coherence) are discussed later in the chapter.

- *Nonthermal regime* (≤ 2 ps). Excitations remaining immediately after loss of coherence (e.g., electron–hole pairs) are generally not yet thermalized; that is, the distribution of such excitations (e.g., as a function of energy) cannot yet be characterized by a temperature. Processes causing exchange of energy among carriers as well processes involving interactions between the carriers and lattice act to bring the carriers into a thermal distribution. Time scales depend on carrier densities but are typically a few hundred femtoseconds for either electrons or holes to thermalize among themselves, while a few picoseconds may be required for electrons and holes to reach a common temperature.
- *Hot carrier regime* (ca. 1–100 ps). Even after the charge carriers reach thermal distributions, they are not yet in thermal equilibrium with the crystal lattice. That is, the carrier distributions may be characterized by temperatures much higher than that of the lattice itself. Cooling of the hot carrier population, in which the carriers lose energy to the lattice, is mediated by various phonon modes of the semiconductor. The time required for carriers to cool to the lattice temperature is typically on the order of picoseconds to about 100 ps (depending, for example, on the temperature at which experiments are performed).
- *Isothermal regime* (≥ 100 ps). After the carrier populations and the lattice reach equilibrium with each other (at the lattice temperature), the carrier densities are still large compared to thermal equilibrium. Finally, carriers decay via radiative and nonradiative recombination processes to restore the semiconductor to its initial state, with typical time scales ranging from a few hundred picoseconds to a few nanoseconds.

In this section we discuss a few examples of experiments that provide information on early relaxation of photoexcited carriers, primarily in the nonthermal and hot carrier regimes.

Due to the periodic crystal lattice and associated periodic potential felt by electrons moving in a semiconductor lattice, electron wave functions consist of a periodic part multiplied by a plane-wave term with wave vector $\mathbf{k}$. The energy states of the electrons are specified as a function of the wave vector or equivalently of the crystal momentum $\hbar\mathbf{k}$ [520,521]. Energy band diagrams, in which allowed electron energies are plotted as a function of wave vector, are key to understanding optical spectroscopy studies in semiconductors. A simplified schematic view of the energy band structure for GaAs is shown in Fig. 9.14 (see also [249]). The conduction band has minimum energy at $\mathbf{k} = 0$, denoted the Γ point. There are additional energy minima at symmetry points termed L and X for wave vector magnitudes at the maximum of the allowed zone; these additional minima influence both ultrafast carrier relaxation dynamics and high-field transport. There are three valence bands: the *heavy hole, light hole*, and *split-off bands*. Heavy and light hole bands have identical energy extrema at $\mathbf{k} = 0$, while the split-off band has its energy extremum also at $\mathbf{k} = 0$ but at lower energy. In undoped samples at low temperature, the valence bands are full of electrons, whereas the conduction band is empty. Therefore, photons with energy exceeding the bandgap may be absorbed, leading to generation of electrons and holes, respectively, in conduction and valence bands. The minimum bandgap for transitions from heavy and light hole bands to the conduction band is located at $\mathbf{k} = 0$ and is approximately 1.42 eV at room temperature. Note that because visible photons have wavelength on the scale of a micrometer, the wave vector (momentum) associated with a photon is small relative to the

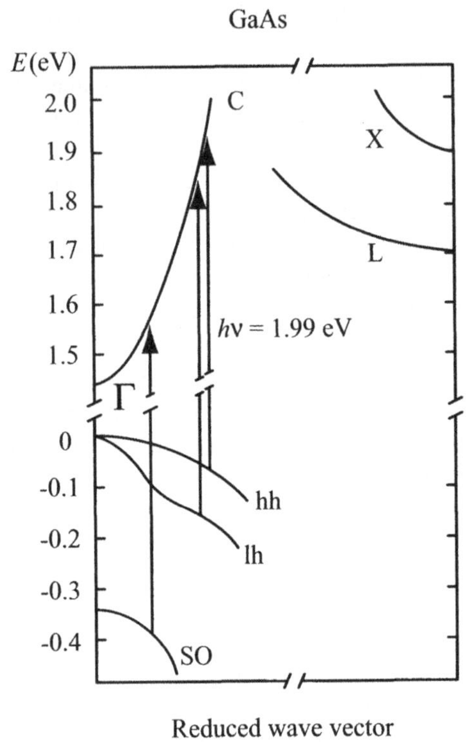

Figure 9.14 Simplified view of the energy band structure for the direct bandgap semiconductor GaAs. Adapted from [522], with permission. Copyright © 1987, American Institute of Physics.

scale of electron wave vectors (momenta) displayed on the energy band diagram (for which wave vectors range up to the inverse of the lattice periodicity, of order angstroms). Consequently, optical absorption events are approximately vertical, in the sense that the momentum of an electron is approximately unchanged in the transition from valence to conduction bands.

Data from spectrally resolved pump–probe measurements on GaAs at room temperature are shown in Fig. 9.15 [522,523]. In these experiments, performed according to the basic geometry of Fig. 9.3, the pump pulses (60 fs in duration, centered at 625 nm wavelength or 2.0 eV photon energy) were derived from an amplified mode-locked dye laser, while the probe pulses were spectrally broadened via continuum generation. As indicated from the GaAs energy band diagram, absorption of the ultrashort pump pulse initially populates distinct distributions of conduction band states about 500 and 150 meV above the minimum of the Γ valley, due to transitions from heavy and light hole bands and from the split-off band, respectively. Due to the larger curvature of the conduction band relative to the valence bands, most of the photon energy in excess of the bandgap is deposited into the electrons; photoexcited holes have relatively little excess energy. Furthermore, due to the larger density of states in the valence bands, photoexcited electrons are more effective at saturating the absorption than are photoexcited holes. For these reasons the absorption pump–probe measurements of [522,523] are primarily (but not exclusively) sensitive to the electron distributions within the conduction band.

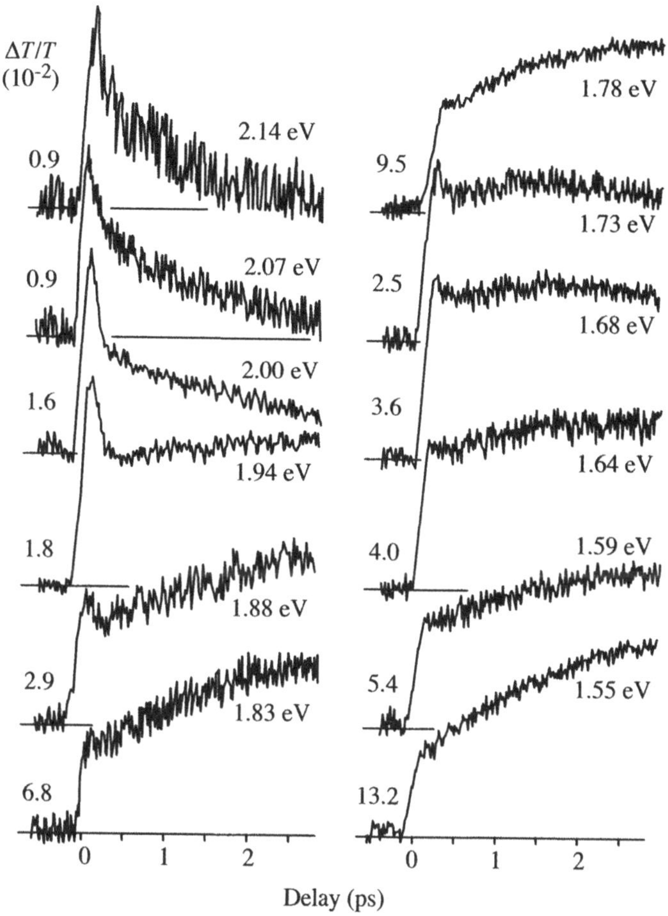

Figure 9.15 Spectrally resolved pump–probe data for GaAs at room temperature. The time dependence of the differential transmission is plotted for various probe frequencies. For each trace the magnitude of the differential transmission is also given. From [522], with permission. Copyright © 1987, American Institute of Physics.

Consistent with our earlier discussion, several processes are expected to influence the pump–probe data. At the earliest times the photoexcitation is expected to produce carrier distributions peaked at the specific energies where absorption is allowed to occur. Carrier–carrier and carrier–lattice interactions act rapidly to scatter carriers out of the initially photoexcited states and to produce broad and eventually thermal carrier distributions. Transitions from heavy and light hole bands give rise to electrons whose initial excess energy places them at or above the energy minima of the L and X valleys within the conduction band. Therefore, scattering out of the Γ valley and into L and X valleys is allowed. Such intervalley scattering is asymmetric; due to lower curvature and hence larger density of states in the L and X valleys compared to the Γ valley, intervalley scattering out of the Γ valley is expected to be much more rapid than scattering from L and X back into the Γ valley. Finally, interpretation of the data is expected to be rendered more difficult, due to the presence of carrier distributions initially excited with different excess energy.

With these considerations in mind, the data shown in Fig. 9.15 have been interpreted as follows:

- For probe energies in the range approximately 1.88 to 2.07 eV, the differential transmission exhibits a fast peak that recovers on the time scale of the pulse width. These peaks that appear for probe energies near the pump energy are evidence of nonthermal carrier distributions centered around the initially photoexcited states. The rapid decay of these peaks indicates that at the experimental carrier density of 10^{18} cm^{-3}, scattering processes very quickly spread the carriers into a broad distribution. The ca. 200-meV range over which these peaks occur is significantly broader than the bandwidth of the pump pulse; this may be attributed both to the superposition of "spectral holes" excited from three different valence bands as well as due to the rapid scattering out of initially populated states. Note that fast initial peaks indicative of a nonthermal carrier distribution are also observed at 1.68 and 1.73 eV. These are attributed to carriers created by a transition from the split-off band but probed via transitions involving the heavy and light hole bands.
- All probe energies show very rapid rise of at least part of the absorption change. This provides further evidence of rapid scattering out of initially populated states, resulting in a broad distribution of carriers.
- For the highest available probe energies (2 to 2.14 eV), the absorption change exhibits a slow decay component (few-picosecond time scale). This is probably a manifestation of carrier cooling, which reduces the carrier concentration at high excess energies. For the lowest probe energies (1.55 and 1.59 eV), the absorption change exhibits a slow rise (also with a few-picosecond time scale) subsequent to its rapid initial rise. This indicates an increasing concentration of carriers near the conduction band minimum. This increase may arise both from carrier cooling and from slow return of carriers from the L and X valleys via intervalley scattering processes.
- Data for intermediate probe energies in the range 1.78 to 1.94 eV exhibit an apparently anomalous slow rise in differential transmission, which is explained by the contribution of the split-off band to the probe absorption. In particular, for these probe energies, transitions from the split-off band to near the bottom of the conduction band are possible. Thus, the slow rise at these intermediate probe energies arises from accumulation of carriers near the bottom of the conduction band as probed by transitions from the split-off band. This effect is not seen for probe energies below 1.78 eV, for which transitions from the split-off band are not allowed.

This experiment successfully illustrates signatures of several ultrafast processes following short-pulse photoexcitation of GaAs; nevertheless, the presence of different processes occurring at the same time poses challenges for full interpretation. For quantitative interpretation, comparison to detailed modeling studies is helpful [524,525]. Other measurements providing complementary information are also desirable.

Time-resolved fluorescence spectroscopy, as depicted in Fig. 9.4, is one such complementary technique. An experiment in which the fluorescence from room-temperature GaAs, excited by 500-fs pump pulses at 2.04 eV, was gated with similar 500-fs time resolution and then analyzed spectrally was reported in [526]. Figure 9.16a shows the resulting fluorescence spectra for delays of 1, 2, and 10 ps after the excitation pulse. Two points are of importance:

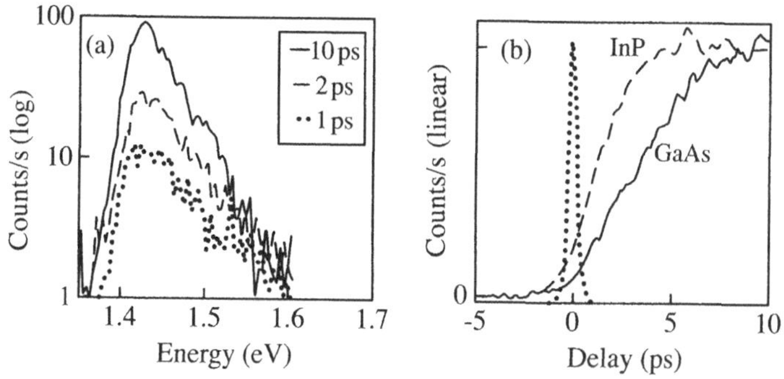

Figure 9.16 (a) Fluorescence spectra of GaAs plotted at three time delays after excitation. (b) Near band-edge fluorescence intensity as a function of time delay after excitation. Solid curve, GaAs at 1.44 eV; dashed curve, InP at 1.35 eV; dotted curve, cross-correlation of scattered laser light, showing both the time zero and the temporal response of the measurement system. Both sets of data are for samples held at room temperature. (a) Adapted from [526], with permission. Copyright © 1987, American Physical Society. (b) Adapted from [248]. Copyright © 1988, IEEE.

1. In each case the fluorescence spectra peak near the bandgap energy (1.42 eV) and exhibit exponential decay for energies sufficiently far above the bandgap. These observations point to carrier distributions that are thermalized well within the time resolution of these experiments. The shape of the high-energy spectral tail is dominated by the Boltzmann factor

$$p(E) \sim \exp\left(\frac{-E}{kT}\right) \tag{9.32}$$

   where $p(E)$ is the probability that a state an energy E above the band minimum is occupied, T is temperature, and $k \approx 8.62 \times 10^{-2}$ meV/K is the Boltzmann constant. The high-energy tail of the fluorescence is seen to decay more rapidly with increasing time, corresponding to reduced temperature (carrier cooling).
2. The overall intensity of the fluorescence signal increases dramatically over time. This is true both for the near-bandgap fluorescence and for the spectrally integrated signal. The fluorescence intensity in these experiments depends principally on the electron density within the Γ valley of the conduction band. The slow rise in fluorescence intensity is then indicative of a slow rise in electron density. This is attributed to rapid intervalley scattering (into X or L) immediately after photoexcitation, followed by much slower return of carriers back into the Γ valley.

Additional evidence supporting this interpretation is obtained by comparing data for different semiconductors. Figure 9.16b shows the rise of near-bandgap fluorescence for GaAs and for indium phosphide (InP). Although these semiconductors generally have similar characteristics, and in particular have similar band structures and only slightly different bandgaps, the rise time for the near-bandgap fluorescence is dramatically faster in InP. This is explained based on one significant difference in the band structures: The energy difference between the minima of the L and Γ valleys is about 300 meV higher in InP than for GaAs. As a result, intervalley scattering in InP is forbidden under the conditions of this experiment, and carriers must remain in the Γ valley. Thus, the InP data are influenced only by carrier cooling, while the GaAs data reflect both carrier cooling and intervalley transfer dynamics. Detailed comparison with Monte Carlo simulations gave an average Γ–L transfer time of about 100 fs for electrons with an initial 500 meV of excess energy in the GaAs experiments, but a much slower 2-ps L–Γ return time for electrons at the bottom of the L valley. The simulations also revealed that these returning electrons act to heat the electrons in the Γ valley, thereby slowing the overall carrier cooling process. Experimentally, the important point is that the spectral shape of the fluorescence reflects the distribution function of the carriers, while the integrated fluorescence provides complementary information about carrier density.

As a final example of this type of data, Fig. 9.17 shows time-resolved fluorescence spectra for an InGaAs semiconductor film at a temperature of 10 K. The bandgap of InGaAs is about 800 meV, so photoexcited carriers have an initial kinetic energy of about 1 eV. These spectra were acquired with 10-ps time resolution. With increasing time delay, the fluorescence spectra become narrower, the high-energy exponential tails grow steeper, and the intensity of the near-bandgap fluorescence grows. These characteristics are again consistent with carrier cooling. Compared to the GaAs time-resolved fluorescence experiments performed at room temperature, here cooling is observed over a much longer time scale, since the lattice is held at a much lower temperature. On this longer time scale, the intervalley scattering processes discussed earlier are less important, and carrier cooling is the dominant effect.

An interesting technical observation is that the signal-to-noise ratio in Fig. 9.17 is much better than that in Fig. 9.16. This may be explained by recalling that the time resolution in an upconversion gating experiment is limited by group velocity walk-off between the two input fields (see Section 5.5.1). Therefore, higher time resolution necessitates a shorter

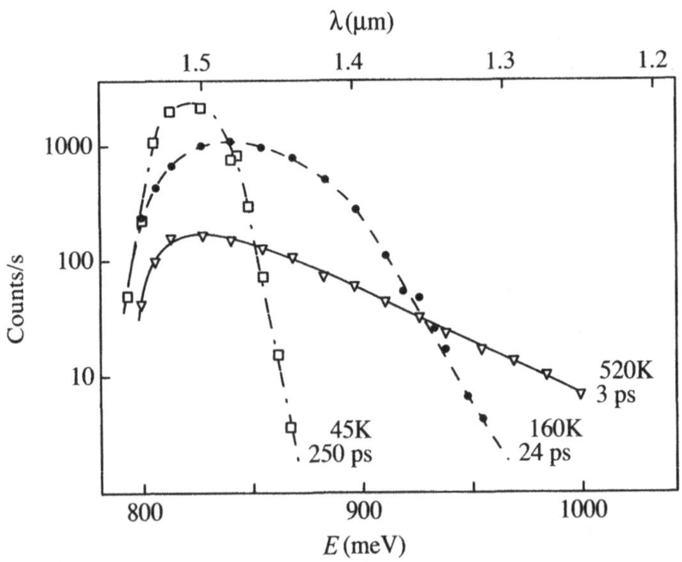

Figure 9.17 Fluorescence spectra of InGaAs plotted at three time delays after excitation, for a 10-K lattice temperature. As a result of carrier cooling, the decay rate of the high-energy tail increases with time. Adapted from [527], with permission. Copyright © 1984, American Institute of Physics.

nonlinear crystal, which results in lower signal levels. The difference in signal-to-noise ratio in these different experiments is then understood as a consequence of the trade-off between time resolution and signal intensity in upconversion gating.

9.4 BASIC QUANTUM MECHANICS FOR COHERENT SHORT-PULSE SPECTROSCOPIES

Earlier sections in this chapter deal with incoherent light–matter interactions, in which absorption of energy from the laser pulse and dynamics of resulting excited-state populations were of primary interest. Phase did not play an important role. For these topics we were able to use an essentially classical description of the field–material interactions. Nevertheless, many of the processes discussed are clearly quantum mechanical in nature; and a few quantum mechanical concepts, such as transitions between energy levels and wave function overlap, were introduced implicitly when convenient. In the following sections we discuss coherent processes in which phase (of the electric field, of material excitations, or both) is of key importance. Here a quantum mechanical language becomes much more natural. In the following we provide a brief overview of basic quantum mechanics relevant to a description of selected areas of ultrashort-pulse spectroscopy in which coherence is important. Our treatment is necessarily sketchy, and the reader without prior background is referred to any of a wide range of quantum mechanics texts. Examples of texts providing such background in a quantum electronics and nonlinear optics context, upon which we draw in the discussion below, include [7,69,131,219,528,529].

9.4.1 Some Basic Quantum Mechanics

Our starting point is Schrödinger's equation,

$$H\,|\psi\rangle = i\hbar\frac{\partial\,|\psi\rangle}{\partial t} \tag{9.33}$$

H is the Hamiltonian operator, corresponding to the total energy of the system, $\hbar \approx 1.045\ \mathrm{J\cdot s}$ is Planck's constant, and $|\psi\rangle$ represents the time-dependent state function, or wave function, of the system. Quantum mechanics holds that all observables of the physical system may be expressed in terms of the wave function. For our purposes it will suffice to write $|\psi(t)\rangle$ as a function of spatial coordinates [i.e, equivalent to $\psi(\mathbf{r}, t)$]; however, when convenient the state function may be written as a function of other coordinates, such as wave vector $\mathbf{k}$. The quantity $|\psi|^2$ is associated with the probability density of finding a particle in the vicinity of position $\mathbf{r}$ and is normalized such that

$$\int |\psi|^2 dV = 1 \tag{9.34}$$

We shall write the Hamiltonian as

$$H = H_0 + H_1 \tag{9.35}$$

Here H_0 is the Hamiltonian for the matter system itself, considered to be time independent. If we focus on a single electron within a matter system, H_0 may be written as

$$H_0 = \frac{-\hbar^2\nabla^2}{2m} + V(\mathbf{r}) \tag{9.36}$$

where the first term represents the kinetic energy (in quantum mechanics $\mathbf{p} = -i\hbar\nabla$ is the momentum operator) and the second term is a spatially dependent potential function representing the interaction of the electron with the rest of the matter. H_1 represents the interaction with a weak time-dependent applied electromagnetic field, which will be considered as a perturbation. In the electric dipole approximation, we have

$$H_1 = -\mu \cdot \mathbf{E} \tag{9.37}$$

where $\mu = -e\mathbf{r}$ is the electric dipole moment operator and $-e$ is the charge of the electron. Both the dipole moment and the field are vectors, and the light–matter interaction depends on the projection of one vector onto the other. However, for simplicity of notation, in the following we write μ and E as scalars.

If we first consider the case where no external field is applied, the Hamiltonian is simply H_0 and is time independent. Schrödinger's equation then possesses solutions of the form

$$|\psi(\mathbf{r}, t)\rangle = e^{-j\omega_k t}\, |\phi_k(\mathbf{r})\rangle \tag{9.38}$$

where the $|\phi_k\rangle$ are functions only of space that satisfy the eigenvalue equation

$$H_0\, |\phi_k(\mathbf{r})\rangle = E_k\, |\phi_k(\mathbf{r})\rangle \qquad \text{where} \qquad E_k = \hbar\omega_k \tag{9.39}$$

Here E_k is the energy of the kth eigenfunction. The eigenfunctions may be chosen to form a complete, orthonormal set of basis functions and may be normalized such that

$$\langle\phi_k|\, \phi_l\rangle = \delta_{kl} \tag{9.40}$$

In this Dirac notation, the product of a *bra* $\langle u|$ and a *ket* $|v\rangle$ is a shorthand for the integral of u^*v over space [i.e., $\langle u|\, v\rangle = \int dV\, u^*(\mathbf{r})v(\mathbf{r})$]. In general, the total wave function may be written as a superposition over the eigenstates,

$$|\psi(\mathbf{r}, t)\rangle = \sum_k c_k e^{-j\omega_k t}\, |\phi_k(\mathbf{r})\rangle \tag{9.41}$$

Since H_0 is independent of time, the expansion coefficients c_k are constants. The probability that a measurement finds the system in the kth eigenstate is $|c_k|^2$. Using eq. (9.34), it is easily shown that $\sum_k |c_k|^2 = 1$. For systems consisting of an ensemble of identical quantum systems, the $|c_k|^2$ are associated with the population densities in the various eigenstates.

Quantum mechanics tells us that the expectation value $\langle \mathcal{O} \rangle$ of an observable associated with an operator $\mathcal{O}$ is given by

$$\langle \mathcal{O} \rangle = \langle \psi | \, \mathcal{O} \, | \psi \rangle = \int dV \, \psi^* \mathcal{O} \psi \tag{9.42}$$

For our purposes we will be particularly interested in the macroscopic polarization in a matter system. We have seen in Chapters 5 and 6 that the polarization appears as a source term in the wave equation for the electric field. In many ultrafast spectroscopy experiments, excitations photoinduced within the matter system are observed by measuring how a probe pulse is affected by the polarization associated with such excitations. Within the framework of eq. (9.41), the expectation value for the electric polarization associated with a single quantum system is written

$$\langle \mu \rangle = \langle \psi | \, \mu \, | \psi \rangle = \sum_{k,l} c_k^* c_l \mu_{kl} e^{j\omega_{kl} t} \tag{9.43a}$$

where

$$\mu_{kl} = \langle \phi_k | \, \mu \, | \phi_l \rangle = \int dV \, \phi_k^* \, \mu \, \phi_l \tag{9.43b}$$

is called a dipole moment matrix element and $\omega_{kl} = \omega_k - \omega_l$ is proportional to the energy difference between eigenfunctions k and l. As we shall see, the strength of a single photon absorption transition depends on the dipole moment matrix element between the corresponding states. Note that from eq. (9.43b), we have $\mu_{lk} = \mu_{kl}^*$. A similar relation holds for any operator corresponding to a physical observable, since observables may only take on real number values. Such operators possessing the property $\mathcal{O}_{lk} = \mathcal{O}_{kl}^*$ are called *Hermitian*.

Simplification occurs when the matter system exhibits inversion symmetry. In particular, the eigenstates possess definite parity; any specific eigenstate must be either an even or odd function of $\mathbf{r}$. Since the electric dipole moment operator has odd parity, its matrix elements μ_{kl} are nonzero only when $|\phi_k\rangle$ and $|\phi_l\rangle$ possess opposite parity. In particular, $\mu_{kk} = 0$ for any k. These considerations lead to an important selection rule in systems with inversion symmetry: single-photon absorption transitions are possible only between states of opposite parity. Further simplification occurs for problems in which the frequency of the electromagnetic field applied is tuned sufficiently close to a particular ω_{kl} energy difference that only the corresponding two eigenstates need be retained. For this case of a *two-level atom*, the expectation value of the dipole moment becomes

$$\langle \mu \rangle = c_1 c_2^* \mu_{21} e^{j\omega_{21} t} + \text{c.c.} \tag{9.44}$$

For an ensemble of systems that are identical and identically prepared, the macroscopic polarization density becomes

$$P = N \langle \mu \rangle \tag{9.45}$$

where N is the volume density. These results describe an oscillatory polarization that varies sinusoidally in time. To obtain such polarization, the system must be prepared into a superposition of eigenstates.

We now consider the response of a two-level atom with an applied electromagnetic field. For definiteness we take levels 1 and 2 as the lower and higher energy eigenstates, respectively. Schrödinger's equation becomes

$$(H_0 + H_1)|\psi\rangle = i\hbar\frac{\partial|\psi\rangle}{\partial t} \tag{9.46}$$

where H_1 is given by eq. (9.37) and the electric field is written in our standard form as

$$E(t) = \text{Re}\left\{a(t)e^{j\omega_0 t}\right\} \tag{9.47}$$

For a two-level atom the wave function becomes

$$|\psi(\mathbf{r},t)\rangle = c_1(t)e^{-j\omega_1 t}|\phi_1(\mathbf{r})\rangle + c_2(t)e^{-j\omega_2 t}|\phi_2(\mathbf{r})\rangle \tag{9.48}$$

where the expansion coefficients c_1 and c_2 are now time dependent as a result of the applied field. Substituting into the Schrödinger equation and using eq. (9.39) yields

$$H_1|\psi\rangle = i\hbar\left(\dot{c}_1(t)e^{-j\omega_1 t}|\phi_1(\mathbf{r})\rangle + \dot{c}_2(t)e^{-j\omega_2 t}|\phi_2(\mathbf{r})\rangle\right) \tag{9.49}$$

To make progress, we multiply both sides of this equation by $\phi_k^*(\mathbf{r})$ and integrate over space. In the Dirac notation this is equivalent to inserting a bra term $\langle\phi_k|$ at the left of each side of eq. (9.49). Using the orthogonality of the eigenfunctions, we have the result

$$\langle\phi_k|H_1|\psi\rangle = i\hbar\dot{c}_k e^{-j\omega_k t} \tag{9.50}$$

Evaluating now for $k = 1$ and $k = 2$, we obtain

$$\begin{aligned}\dot{c}_1 &= \frac{i\mu_{12}}{2\hbar}c_2(t)a(t)e^{j(\omega_0-\omega_{21})t}\\ \dot{c}_2 &= \frac{i\mu_{21}}{2\hbar}c_1(t)a^*(t)e^{-j(\omega_0-\omega_{21})t}\end{aligned} \tag{9.51}$$

Here we have retained exponential terms involving the frequency difference $\omega_0 - \omega_{21}$ and neglected exponential terms involving the frequency sum $\omega_0 + \omega_{21}$ that are far from resonance. This is known as the *rotating wave approximation* and is justified provided that the laser frequency is nearly equal to the transition frequency, $\omega_0 \approx \omega_{21}$, and the time variation of envelope function $a(t)$ is slow compared with the oscillation period of the light field.

When the electric field is sufficiently weak, equations (9.51) may be analyzed via a perturbation theory approach. Here the c_k are expanded as

$$c_k(t) = \sum_{n=0}^{\infty} c_k^{(n)}(t) = c_k^{(o)} + c_k^{(1)}(t) + c_k^{(2)}(t) + \cdots \tag{9.52}$$

where the perturbation appears as a factor n times in terms written with a superscript $^{(n)}$. For example, the electric field (a or a^*) does not appear in the $c^{(0)}$, appears once in the $c^{(1)}$, twice in the $c^{(2)}$, and so on. The philosophy is that perturbation terms become successively weaker as the perturbation order (n) increases. Within this framework the problem can be solved recursively, according to

$$\begin{aligned} \dot{c}_1^{\,(n)} &= \frac{i\mu_{12}}{2\hbar} c_2^{(n-1)}(t)\, a(t) e^{j(\omega_0-\omega_{21})t} \\ \dot{c}_2^{\,(n)} &= \frac{i\mu_{21}}{2\hbar} c_1^{(n-1)}(t)\, a^*(t) e^{-j(\omega_0-\omega_{21})t} \end{aligned} \tag{9.53}$$

and

$$\begin{aligned} c_1^{(n)}(t) &= \frac{i\mu_{12}}{2\hbar} \int^t dt'\, c_2^{(n-1)}(t')\, a(t') e^{j(\omega_0-\omega_{21})t'} \\ c_2^{(n)}(t) &= \frac{i\mu_{21}}{2\hbar} \int^t dt'\, c_1^{(n-1)}(t')\, a^*(t') e^{-j(\omega_0-\omega_{21})t'} \end{aligned} \tag{9.54}$$

These results are valid for $n \geq 1$. The expansion coefficients for $n = 0$ are regarded as initial conditions, and the perturbation response is assumed to be zero prior to the application of the electric field. One common special case arises when the system starts in the ground state [i.e., $c_1^{(0)} = 1$, $c_2^{(0)} = 0$]. The first-order perturbation response becomes simply

$$\begin{aligned} c_1^{(1)}(t) &= 0 \\ c_2^{(1)}(t) &= \frac{i\mu_{21}}{2\hbar} \int^t dt'\, a^*(t') e^{-j(\omega_0-\omega_{21})t'} \end{aligned} \tag{9.55}$$

Using eqs. (9.44) and (9.45), the lowest-order polarization response is

$$P^{(1)}(t) = \frac{-Ni\,|\mu_{21}|^2}{2\hbar} e^{j\omega_{21}t} \int^t dt'\, a(t') e^{j(\omega_0-\omega_{21})t'} + \text{c.c.} \tag{9.56}$$

The lowest-order polarization is first order in the electric field. Note, however, that the polarization is not instantaneous in the field; rather, it is proportional to the Fourier component of the field at the transition frequency. The quadrature phase relationship between the polarization and the field is as required for an absorption process [compare, e.g., eq. (9.14)]. The leading term in the upper-level population change, given by $N|c_2^{(1)}|^2$, is second order in the field, as expected, and may be obtained from eq. (9.55). The lower-level population change is equal to $N(1 - |c_2^{(1)}|^2)$.

Although these considerations make sense, there are serious problems with this formulation. For example, the polarization excited by a weak delta-function input pulse continues to oscillate at frequency ω_{21} indefinitely. This defies our expectation that a finite excitation should result in a time-limited response. The observation that the excited polarization depends only on the Fourier amplitude of the field exactly at the transition frequency contradicts our physical experience that absorption transitions have finite linewidth. These problems arise due to neglect of relaxation processes. Relaxation terms are most conveniently introduced via the density matrix formalism, which is explained in the following.

9.4.2 The Density Matrix

In Section 9.4.1 we considered Schrödinger's equation for a single particle. Full knowledge of the Hamiltonian was assumed, and the probability amplitudes c_k were calculated explicitly. Most optical spectroscopy experiments, however, are performed on large ensembles of particles, in which individual particles cannot be prepared identically. As an example, consider a thermal population of molecules, in which different molecules are initially in different states, distributed according to a Boltzmann probability function. Furthermore, in complicated systems involving large numbers of degrees of freedom, the Hamiltonian is usually not known precisely, and the number of degrees of freedom quickly becomes too large for explicit calculations, except in a statistical sense. This is where the density matrix comes in. Rather than requiring precise knowledge of the initial state of the system, for density matrix analyses it is sufficient to know the probability that the system is in different initial states. Treatment of complicated systems may be simplified by confining the analysis to a small number of optically active modes, while treating the remaining large number of modes that are not directly optically active as a reservoir. Examples include large molecules in solution and photoexcited charge carriers in semiconductors. In the former case a few chromophore modes of the molecule may dominate the absorption process, leaving a large number of degrees of freedom for translational and rotational motions, both of the molecule itself and of the surrounding solvent, that do not themselves participate strongly in absorption events but do couple to the active modes. In the latter case both vibrations of the crystal lattice and electrons far removed in energy from the optical transition may be considered as a reservoir. Because the number of modes in the reservoir is usually taken to very large compared to the number of explicitly treated modes, the interaction between the optically active modes and the reservoir is assumed not to affect the properties of the reservoir. The optically active modes, however, are affected by such coupling. This is handled via introduction of relaxation terms into the density matrix equations, which represent the statistical effect of the reservoir on the optically active modes.

Following [528], we consider an ensemble of quantum particles, each of which may be in any of a number of different possible state functions $|\psi_n\rangle$ with probability p_n. Such a situation is termed a *mixed state*. Similar to eq. (9.41), each of the possible state functions may be expanded in terms of a complete set of orthonormal basis functions $|\phi_k\rangle$ according to

$$|\psi_n\rangle = \sum_k c_{nk} e^{-j\omega_k t} |\phi_k\rangle \tag{9.57}$$

In a mixed state the expectation value of an operator $\mathcal{O}$ becomes

$$\langle \mathcal{O} \rangle = \sum_n p_n \langle \psi_n | \mathcal{O} | \psi_n \rangle \tag{9.58}$$

This may be manipulated to obtain

$$\begin{aligned} \langle \mathcal{O} \rangle &= \sum_{nk} p_n \langle \psi_n | \mathcal{O} | \phi_k \rangle \langle \phi_k | \psi_n \rangle \\ &= \sum_{nk} p_n \langle \phi_k | \psi_n \rangle \langle \psi_n | \mathcal{O} | \phi_k \rangle \\ &= \sum_k \langle \phi_k | \rho \mathcal{O} | \phi_k \rangle \end{aligned} \tag{9.59}$$

Here we have used the fact that

$$\sum_k |\phi_k\rangle \langle\phi_k| = \mathcal{I} \tag{9.60}$$

where $\mathcal{I}$ is the identity operator, and we have defined the density operator as

$$\rho = \sum_n p_n |\psi_n\rangle \langle\psi_n| \tag{9.61}$$

Note that the expectation value may be written compactly as

$$\langle\mathcal{O}\rangle = \mathrm{Tr}(\rho\mathcal{O}) \tag{9.62}$$

where $\mathrm{Tr}(\cdots)$ represents the sum of the diagonal elements of a matrix, known as its *trace*. Since the expectation value of any observable $\mathcal{O}$ may be computed from eq. (9.62) if ρ is known, the density operator contains all the physically observable information about a system.

By making use of 9.57, we may rewrite the density operator as

$$\rho = \sum_{nkl} p_n c_{nk} c_{nl}^* e^{j(\omega_l - \omega_k)t} |\phi_k\rangle \langle\phi_l| \tag{9.63}$$

The matrix elements of ρ are then obtained as

$$\rho_{mq} = \langle\phi_m| \rho \left|\phi_q\right\rangle = \sum_n p_n c_{nm} c_{nq}^* e^{j(\omega_q - \omega_m)t} \tag{9.64}$$

All the terms contributing to the matrix elements contain products of two c_{nk} expansion coefficients (with one complex conjugated). The diagonal elements of the density matrix take on the form

$$\rho_{kk} = \sum_n p_n |c_{nk}|^2 \tag{9.65}$$

This gives the total probability that the system is in the kth eigenstate. Since probability is normalized such that $\sum_n p_n = 1$, it follows that $\mathrm{Tr}(\rho) = 1$. The off-diagonal elements represent coherent superpositions of eigenstates [i.e., for $m \neq q$, we may have $\rho_{mq} \neq 0$ if and only if both $|\phi_m\rangle$ and $\left|\phi_q\right\rangle$ are occupied with nonzero probability and there is an ensemble average phase relationship between c_{nm} and c_{nq}]. We saw earlier that such superpositions are important for generating optical frequency polarizations. Assuming again eigenstates with definite parity, the macroscopic polarization of a system of two-level atoms is written in density matrix terms as

$$P = N\langle\mu\rangle = N(\rho_{12}\mu_{21} + \rho_{21}\mu_{12}) = N\rho_{12}\mu_{21} + \mathrm{c.c.} \tag{9.66}$$

Here we have used $\rho_{mj} = \rho_{jm}^*$, which follows from eq. (9.64).

To analyze problems with time-dependent perturbations, it is sufficient to determine the time dependence of ρ. Taking the time derivative of ρ as defined in eq. (9.61) and requiring that all of the $|\psi_n\rangle$ satisfy Schrödinger's equation, we have

$$i\hbar\dot{\rho} = \sum_n p_n \left(H\left|\psi_n\right\rangle\left\langle\psi_n\right| - \left|\psi_n\right\rangle\left\langle\psi_n\right| H\right) = [H, \rho] \tag{9.67}$$

where the notation $[H, \rho] = H\rho - \rho H$ is called the *commutator*. We now take $H = H_0 + H_1 + H_r$, where H_1 is an applied perturbation such as an applied laser field and H_r incorporates the relaxation processes and is considered to be time independent [69]. As discussed earlier, this may represent coupling to the degrees of freedom not considered in the formulation of H_0 (i.e., to a thermal reservoir). Using the eigenfunctions of H_0 as basis functions, eq. (9.67) yields

$$i\hbar\dot{\rho}_{mj} = i\hbar\left\langle\phi_m\right|\dot{\rho}\left|\phi_j\right\rangle = (E_m - E_j)\rho_{mj} + [H_1, \rho]_{mj} + [H_r, \rho]_{mj} \tag{9.68}$$

Commonly, the relaxation processes are approximated as exponential. In this case eq. (9.68) is modified to read

$$\dot{\rho}_{mj} = -i\omega_{mj}\rho_{mj} - \frac{i}{\hbar}[H_1, \rho]_{mj} - \frac{\rho_{mj} - \rho_{mj}^{(0)}}{\tau_{mj}} \tag{9.69}$$

Here $\rho_{mj}^{(0)}$ represents the equilibrium value of ρ_{mj}, τ_{mj} is real and positive, and since $\rho_{mj} = \rho_{jm}^*$, $\tau_{mj} = \tau_{jm}$. For diagonal elements of the density matrix, the equation of motion simplifies to

$$\dot{\rho}_{jj} = -\frac{i}{\hbar}[H_1, \rho]_{jj} - \frac{\rho_{jj} - \rho_{jj}^{(0)}}{\tau_{jj}} \tag{9.70}$$

The τ_{jj} represent the time constants for the populations of the various states to relax to their equilibrium values. For thermal equilibrium the values of $\rho_{jj}^{(0)}$ may be determined via the Boltzmann distribution. For sufficiently low temperatures this simplifies to $\rho_{jj}^{(0)} = 1$ for the lowest-energy state ($j = 1$) and $\rho_{jj}^{(0)} = 0$ for states at higher energy ($j > 1$). For off-diagonal elements of the density matrix, $m \neq j$, we take $\rho_{mj}^{(0)} = 0$. Physically, this means that in equilibrium the system has no net coherence; such coherences are induced only through the action of an external time-dependent perturbation (H_1). For off-diagonal elements we may therefore write

$$\dot{\rho}_{mj} = -i\omega_{mj}\rho_{mj} - \frac{i}{\hbar}[H_1, \rho]_{mj} - \frac{\rho_{mj}}{\tau_{mj}} \qquad (m \neq j) \tag{9.71}$$

Off-diagonal elements decay both as a result of population relaxation out of the relevant levels and additionally as a result of elastic interactions, such as collisions in a gas, that randomize the phase but leave the population intact. Accordingly, relaxation of the off-diagonal terms is commonly called *dephasing*. The conditions under which dephasing assumes the exponential form assumed above are discussed further in Section 9.6.

Let us now apply the density matrix formalism to the problem of a two-level atom driven by an applied electromagnetic field. As before, we follow a perturbation theory approach and make the rotating wave approximation. This yields the following equations of motion:

$$\begin{aligned}
\dot{\rho}_{12}^{(n)} &= \frac{i\mu_{12}}{2\hbar} a(t) e^{j\omega_0 t} \left(\rho_{22}^{(n-1)} - \rho_{11}^{(n-1)} \right) + i\omega_{21} \rho_{12}^{(n)} - \frac{\rho_{12}^{(n)}}{T_2} \\
\dot{\rho}_{21}^{(n)} &= \frac{i\mu_{21}}{2\hbar} a^*(t) e^{-j\omega_0 t} \left(\rho_{11}^{(n-1)} - \rho_{22}^{(n-1)} \right) - i\omega_{21} \rho_{21}^{(n)} - \frac{\rho_{21}^{(n)}}{T_2} \\
\dot{\rho}_{22}^{(n)} &= \frac{i\mu_{21}}{2\hbar} a^*(t) e^{-j\omega_0 t} \rho_{12}^{(n-1)} - \frac{i\mu_{12}}{2\hbar} a(t) e^{j\omega_0 t} \rho_{21}^{(n-1)} - \frac{\rho_{22}^{(n)}}{T_1} \\
\dot{\rho}_{11}^{(n)} &= \frac{i\mu_{12}}{2\hbar} a(t) e^{j\omega_0 t} \rho_{21}^{(n-1)} - \frac{i\mu_{21}}{2\hbar} a^*(t) e^{-j\omega_0 t} \rho_{12}^{(n-1)} - \frac{\rho_{11}^{(n)}}{T_1}
\end{aligned} \tag{9.72}$$

Here we have assumed that the number of two-level atoms is conserved, and hence we set $\tau_{11} = \tau_{22}$. Following established terminology we denote this quantity T_1, which commonly is referred to as the *population, energy*, or *longitudinal relaxation time*. Similarly, we denote the quantity $\tau_{12} = \tau_{21}$ as T_2, which is commonly referred to as the *phase* or *transverse relaxation time*, or simply the *dephasing time*. The terms *longitudinal* and *transverse relaxation* derive from the field of magnetic resonance.

To proceed we adopt the initial conditions $\rho_{11}^{(0)} = 1$, and $\rho_{22}^{(0)} = \rho_{12}^{(0)} = \rho_{21}^{(0)} = 0$. We observe that the off-diagonal elements of ρ are driven by the product of the field and the population difference, whereas the population terms are driven by products of the field and the off-diagonal elements. Therefore, the off-diagonal terms will arise at odd orders of the perturbation expansion, and the diagonal elements will appear at even orders. We observe further that the first two equations above are complex conjugates of each other, consistent with the requirement that $\rho_{21} = \rho_{12}^*$, while the last two equations above imply that $\rho_{11}^{(n)} = -\rho_{22}^{(n)}$ for $n \geq 1$, or $\dot{\rho}_{11} = -\dot{\rho}_{22}$. As a result, it is sufficient to solve for ρ_{12} and ρ_{22}. The solutions with $n \geq 1$ are as follows:

$$\begin{aligned}
\rho_{12}^{(n)}(t) &= \frac{i\mu_{12}}{2\hbar} e^{j\omega_0 t} \int^t dt'\, a(t') \left\{ \rho_{22}^{(n-1)}(t') - \rho_{11}^{(n-1)}(t') \right\} e^{-(j\delta\omega + T_2^{-1})(t-t')} \\
\rho_{22}^{(n)}(t) &= \int^t dt'\, e^{-(t-t')/T_1} \left\{ \frac{i\mu_{21}}{2\hbar} a^*(t') e^{-j\omega_0 t'} \rho_{12}^{(n-1)}(t') + \text{c.c.} \right\}
\end{aligned} \tag{9.73}$$

Here $\delta\omega = \omega_0 - \omega_{21}$ is the detuning of the laser center frequency from the transition frequency. We will be interested in terms up to third order in the field. The first-order term is written

$$\rho_{12}^{(1)}(t) = \frac{-i\mu_{12}}{2\hbar} e^{j\omega_0 t} \int^t dt'\, a(t') e^{-(j\delta\omega + T_2^{-1})(t-t')} \tag{9.74}$$

The second-order term, which represents the lowest-order change in the population, is given by

$$\rho_{22}^{(2)}(t) = \frac{|\mu_{12}|^2}{4\hbar^2} \int^t \int^{t'} dt'dt''\, e^{-(t-t')/T_1} \left\{ a^*(t')a(t'')e^{-(j\delta\omega+T_2^{-1})(t'-t'')} + \text{c.c.} \right\} \quad (9.75)$$

The third-order term will be useful in time-resolved four-wave mixing techniques for measuring ultrafast dephasing, described in Section 9.6. The expression is

$$\rho_{12}^{(3)}(t) = \frac{i\mu_{12}|\mu_{12}|^2}{4\hbar^3} e^{j\omega_0 t} \int^t \int^{t'} \int^{t''} dt'dt''dt'''\, e^{-(j\delta\omega+T_2^{-1})(t-t')} e^{-(t'-t'')/T_1}$$
$$\times\, a(t') \left\{ a^*(t'')a(t''')e^{-(j\delta\omega+T_2^{-1})(t''-t''')} + \text{c.c.} \right\} \quad (9.76)$$

Please note that our treatment above assumes homogeneous broadening (i.e., all particles are treated as identical). In Section 9.6 the discussion is extended to include inhomogeneous broadening.

9.5 WAVE PACKETS

In Section 9.3 we considered examples of ultrafast energy relaxation in large molecules and semiconductors. The time evolution of the excited matter system was viewed to be incoherent, in that phase effects were not important. Here we discuss examples in which markedly different, coherent (often oscillatory) time evolution effects may be observed. These effects, which are inherently quantum mechanical in nature, have been observed in a wide range of materials, including atoms [530,531], molecules in solution and in gas phase [532–536], and semiconductors [537].

To begin our discussion, consider the simple example of a three-level system consisting of a single electronic ground state $|g\rangle$ and a pair of closely spaced eigenstates $|e_1\rangle$ and $|e_2\rangle$ both within an electronic excited state. If the system is excited at $t = 0$ with an approximately delta function input pulse, the first-order perturbation response yields a contribution to the wave function of the form

$$\left|\psi^{(1)}\right\rangle = c_1 |e_1\rangle e^{-j\omega_1 t} + c_2 |e_2\rangle e^{-j\omega_2 t} \quad (9.77)$$

The coefficients c_i depend on the dipole moment matrix elements between ground and excited states as well the frequency content of the excitation pulse at the ground-to-excited state transition frequencies. Provided that both c_1 and c_2 are nonzero, eq. (9.77) represents a superposition state that evolves periodically in time with period $T = 2\pi/\omega_{21}$. This oscillatory time evolution can have interesting physical consequences. To generate such a superposition state, the bandwidth of the input pulse must exceed the ω_{21} frequency difference, so that both $|e_1\rangle$ and $|e_2\rangle$ may be excited simultaneously. Equivalently, if bandwidth limited, the duration of the input pulse should be short compared to the period T.

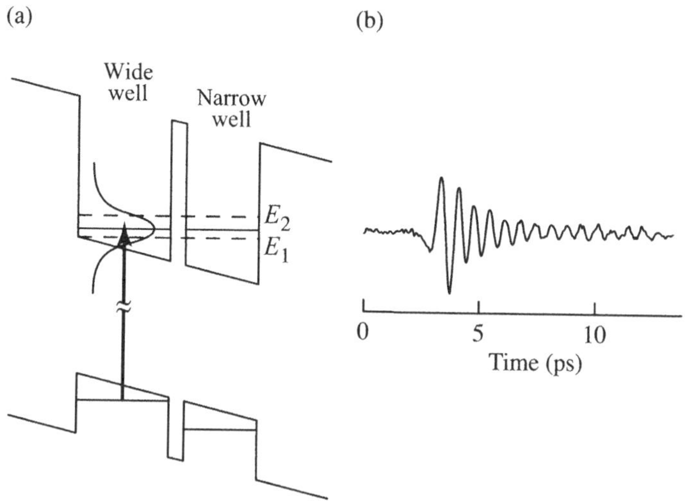

Figure 9.18 (a) Energy diagram of asymmetric coupled quantum well structure. The bias field is adjusted so that conduction band energies of isolated narrow and wide wells are degenerate (solid line), leading to wave functions delocalized across the two wells with energy-level splitting (dashed lines). The center frequency and bandwidth of the laser pulse are selected to excite a wave packet localized initially in the wide well. (b) Measured time-dependent terahertz field radiated from coupled quantum well structure at 6.7-kV/cm bias. The sustained ringing is evidence of oscillatory wave packet motion. Adapted from [537], with permission. Copyright © 1992, American Physical Society.

9.5.1 Example: Semiconductor Quantum Wells

Figure 9.18a shows an example of a physical system whose behavior may be understood via this model. A series of thin layers of GaAs and AlGaAs are epitaxially grown to form an asymmetric double-quantum-well sample. Because of the bandgap difference between the materials, the GaAs layers form low-energy quantum wells surrounded by high-energy AlGaAs barriers. Electrons and holes are largely confined to the quantum well regions. The band-edge absorption displays a prominent exciton peak associated with the Coulomb attraction between an electron photoexcited into the conduction band and the positively charged hole left behind in the valence band. The energies of these exciton peaks may be manipulated both via quantum confinement effects, in which the squeezing of a charge carrier within a quantum well leads to an increase in energy, and via application of a static electric field. The double-quantum-well sample of [537] consisted of a closely spaced narrow well and wide well pair, with different confinement energies in each well. By tuning the strength of the applied static field, the relevant electron states in the narrow and wide wells were brought into degeneracy (equal energy). The energies of the relevant hole states, for which quantum confinement effects are smaller, remained quite different. Therefore, it was possible to choose a laser pulse with center frequency and bandwidth such that transitions were allowed to either of the electron states but from only one of the hole states. Hence, the three-level model is appropriate.

Let us use $|n\rangle$ and $|w\rangle$ to denote the wave functions for electrons confined, respectively, in the narrow and wide wells. Now if we assume a small coupling between $|n\rangle$ and $|w\rangle$, these states are no longer eigenstates. Such coupling arises due to the exponential wave function tails that pass through the thin barrier and penetrate slightly into the adjacent well.

As a result, the $|n\rangle$ and $|w\rangle$ states mix to form new eigenstates that are split in energy. For our purposes we write the new eigenfunctions as

$$|e_2\rangle = \frac{1}{\sqrt{2}}|w\rangle + \frac{1}{\sqrt{2}}|n\rangle$$

$$|e_1\rangle = \frac{1}{\sqrt{2}}|w\rangle - \frac{1}{\sqrt{2}}|n\rangle \tag{9.78}$$

Equation (9.78) tells us that the excited-state wave functions entering the three-level model of eq. (9.77) are themselves (symmetric and antisymmetric) superpositions of electrons, respectively, in the narrow and wide wells.

Let us now consider excitation by a short laser pulse at $t = 0$ from the valence band of the wide well up to the conduction band states. For a vertical transition the conduction band electrons must be created in the wide well, which corresponds to a superposition of the antisymmetric and symmetric $|e_1\rangle$ and $|e_2\rangle$ states. The overall time evolution is described by eq. (9.77) with $c_1 = c_2 = 1/\sqrt{2}$. Using this result together with eq. (9.78), it is easy to show that the probabilities for finding a conduction band electron in the wide well or narrow well are, respectively,

$$\left|\left\langle w|\psi^{(1)}\right\rangle\right|^2 = \cos^2(\omega_{21}t)$$

$$\left|\left\langle n|\psi^{(1)}\right\rangle\right|^2 = \sin^2(\omega_{21}t) \tag{9.79}$$

Here $\hbar\omega_{21}$ is the energy splitting between the symmetric and antisymmetric superposition states. The photoexcited conduction band electrons oscillate in real space between the wide and narrow well. The coherent superposition state resulting in such motions via the time-varying interference of the constituent wave functions is known as a *wave packet*.

Since the hole in these experiments remains localized in its original (wide) well, the electron wave packet motion leads to a time-varying macroscopic polarization. In turn, this time-varying polarization acts as a source term in the wave equation (see Section 5.2), and therefore the sample radiates an electric field at the oscillation frequency. Figure 9.18b shows a measurement of such a radiated field, which is in the terahertz range in these experiments (see Chapter 10 for discussion of measurement techniques for terahertz radiation). The radiated field persists for nearly 10 ps and consists of more than 10 discernible cycles. This means that dephasing of the coherence induced between $|e_1\rangle$ and $|e_2\rangle$ ($\rho_{e_1e_2}$ in the density matrix picture) is relatively slow. In further experiments (not shown) the static bias field was adjusted such that $|w\rangle$ and $|n\rangle$ states have substantially different energies. As expected, the long oscillatory tail disappears, since away from degeneracy the electrons remain localized, and wave packet motion is no longer possible.

9.5.2 Molecules

Wave packet effects have also been observed in molecules. As discussed in Section 9.3.1, these systems are usually discussed within the Born–Oppenheimer approximation in which electronic and nuclear motions are separated. Here we develop this picture mathematically in order to see implications for wave packet generation. We then discuss some experimental examples.

Theory Following [515], we write the Hamiltonian for a molecule approximately (spins for example are not considered) as

$$H = -\frac{\hbar^2}{2m}\sum_i \nabla_i^2 - \sum_A \frac{\hbar^2}{2M_A}\nabla_A^2 - \sum_{A,i}\frac{Z_A e^2}{r_{A,i}} + \sum_{A>B}\frac{Z_A Z_B e^2}{R_{AB}} + \sum_{i>j}\frac{e^2}{r_{ij}} \tag{9.80}$$

These terms represent, in order: kinetic energy of the electrons, kinetic energy of the nuclei, Coulomb attraction between electrons and nuclei, Coulomb repulsion between nuclei, and Coulomb repulsion between electrons, respectively. According to the Born–Oppenheimer theory, the total molecular wave function is separated into an electron part $|\psi^e\rangle$ and a nuclear part $|\chi\rangle$,

$$|\Psi(\mathbf{r}, \mathbf{R}, t)\rangle = \left|\psi^e(\mathbf{r}; \mathbf{R}, t)\right\rangle |\chi(\mathbf{R}, t)\rangle \tag{9.81}$$

where $\mathbf{r}$ and $\mathbf{R}$ represent the set of electron and nuclear coordinates, respectively.[1] At this point the wave functions are still functions of time. Consistent with the notion that nuclear motions are slow compared to characteristic time scales for electron motion, the electron wave function is taken to vary adiabatically with $\mathbf{R}$. Using eq. (9.81) in the Schrödinger equation, one obtains separate equations for the electronic and nuclear wave functions, written in time-independent form as

$$\left\{-\frac{\hbar^2}{2m}\sum_i \nabla_i^2 - \sum_{A,i}\frac{Z_A e^2}{r_{A,i}} + \sum_{i>j}\frac{e^2}{r_{ij}}\right\}\left|\psi^e(\mathbf{r}; \mathbf{R})\right\rangle = E_e(\mathbf{R})\left|\psi^e(\mathbf{r}; \mathbf{R})\right\rangle$$
$$\left\{-\sum_A \frac{\hbar^2}{2M_A}\nabla_A^2 + E_e(\mathbf{R}) + \sum_{A>B}\frac{Z_A Z_B e^2}{R_{AB}}\right\}|\chi(\mathbf{R})\rangle = E_{\text{total}}\,|\chi(\mathbf{R})\rangle \tag{9.82}$$

Because the electronic wave function usually varies only slowly with $\mathbf{R}$, terms involving derivatives of $|\psi^e\rangle$ with respect to the nuclear coordinates have been neglected. Furthermore, in writing the equation for the electronic wave function, the nuclear–electron attraction term is taken as approximately independent of time (slow nuclear motions), so that the time-independent form of Schrödinger's equation is allowed. The result of these approximations is that the electronic wave function is taken to vary adiabatically with the quasistatic nuclear coordinates. Furthermore, the electron energy as a function of nuclear coordinates, together with the nuclear–nuclear repulsion terms, forms a potential energy surface governing the nuclear motions. Such potential energy surfaces are depicted in Fig. 9.11, as are representative vibrational levels bound within the potential energy surfaces.

[1] $\mathbf{R}$ is used similarly to the Q of Section 9.3.1.

In this picture we can write individual eigenstates in the form

$$\left|\Psi_{m,N}(\mathbf{r},\mathbf{R},t)\right\rangle = \left|\psi_m^e(\mathbf{r};\mathbf{R})\right\rangle \left|\chi_{m,N}(\mathbf{R})\right\rangle e^{-j\omega_{m,N}t} \tag{9.83}$$

Here $\left|\psi_m^e\right\rangle$ is the mth electronic state eigenfunction and is a function of both the electron coordinates $\mathbf{r}$ and the nuclear coordinates $\mathbf{R}$; and $\left|\chi_{m,N}\right\rangle$, which is a function of the nuclear coordinates only, is the Nth vibrational eigenstate within the potential energy surface of the mth electronic state manifold. The total eigenstate $\left|\Psi_{m,N}\right\rangle$ has energy $\hbar\omega_{m,N}$. For our discussion of wave packets, we consider transitions from the ground electronic and vibrational state ($m = 0$ for the S_0 electronic manifold, $N = 0$) to the full set of vibrational states in the S_1 electronic manifold ($m = 1$, various N). The initial state then is given by

$$\left|\Psi^{(0)}(\mathbf{r},\mathbf{R},t)\right\rangle = \left|\psi_0^e(\mathbf{r};\mathbf{R})\right\rangle \left|\chi_{0,0}(\mathbf{R})\right\rangle e^{-j\omega_{0,0}t} \tag{9.84}$$

and the excited state, which appears in the first-order perturbation theory, is written

$$\left|\Psi^{(1)}(\mathbf{r},\mathbf{R},t)\right\rangle = \left|\psi_1^e(\mathbf{r};\mathbf{R})\right\rangle \sum_N c_N^{(1)} \left|\chi_{1,N}(\mathbf{R})\right\rangle e^{-j\omega_{1,N}t} \tag{9.85}$$

Let us assume now that we weakly excite the system with a very short pulse at time $t = 0$. Inserting a delta function for the field in eq. (9.55), the resulting c_N coefficients are proportional to the dipole moment matrix element connecting the ground state with the Nth vibrational state in S_1 [i.e., $c_N^{(1)} \sim i\mu_{(1,N)(0,0)}$]. To evaluate the matrix elements, we note that $\mu = \mu^e + \mu^A$; the total dipole moment is the sum of the dipole associated with electron positions (μ^e) and that associated with nuclear positions (μ^A). The matrix elements are written

$$\begin{aligned}\mu_{(1,N)(0,0)} &= \left\langle\chi_{1,N}(\mathbf{R})|\chi_{0,0}(\mathbf{R})\right\rangle \left\langle\psi_1^e(\mathbf{r};\mathbf{R})\right| \mu^e \left|\psi_0^e(\mathbf{r};\mathbf{R})\right\rangle \\ &\quad + \left\langle\chi_{1,N}(\mathbf{R})\left|\mu^A\right|\chi_{0,0}(\mathbf{R})\right\rangle \left\langle\psi_1^e(\mathbf{r};\mathbf{R})|\psi_0^e(\mathbf{r};\mathbf{R})\right\rangle\end{aligned} \tag{9.86}$$

Due to the orthogonality of the electronic wave functions, the second term vanishes. The matrix elements become

$$\mu_{(1,N)(0,0)} = \mu_{10}^e(\mathbf{R}) \left\langle\chi_{1,N}(\mathbf{R})|\chi_{0,0}(\mathbf{R})\right\rangle \tag{9.87}$$

where $\mu_{10}^e(\mathbf{R})$ is the electronic matrix element between S_1 and S_0. Although in general μ_{10}^e varies with the nuclear coordinates $\mathbf{R}$, this variation is usually slow. Therefore, following common practice, we neglect this variation in the following and treat μ_{10}^e as a constant. The nuclear wave functions come in through the Franck–Condon factors, which are the projections of the S_1 vibrational eigenstates onto the initial S_0 ground vibrational eigenstate.

Substituting into eq. (9.85) yields an excited-state wave function given by

$$\left|\Psi^{(1)}\right\rangle \sim i\mu_{10}^{e}\left|\psi_1^e(\mathbf{r};\mathbf{R})\right\rangle \sum_N \left|\chi_{1,N}(\mathbf{R})\right\rangle \left\langle \chi_{1,N} \middle| \chi_{0,0}\right\rangle e^{-j\omega_{1,N}t} \tag{9.88}$$

The initial excited-state wave function at $t = 0$, immediately after the very short excitation pulse, becomes

$$\left|\Psi^{(1)}(\mathbf{r},\mathbf{R},t=0)\right\rangle \sim i\mu_{10}^{e}\left|\psi_1^e(\mathbf{r};\mathbf{R})\right\rangle \left|\chi_{0,0}(\mathbf{R})\right\rangle \tag{9.89}$$

where we have used eq. (9.60). The interesting point is that the initial nuclear wave function prepared in S_1 following the electronic transition is simply the initial nuclear wave function in S_0. We see then the Born–Oppenheimer approximations used in writing eqs. (9.81) and (9.82) imply the concept of vertical transitions as a direct consequence.

Although multiple vibrational eigenfunctions are excited, they interfere at $t = 0$ to reproduce the initial vibrational eigenfunction in S_0. However, for $t > 0$ the time-dependent phase factors in eq. (9.88) lead to time-varying interference effects. Physically, such interference is associated with motions of the nuclei, which we describe as wave packets moving on the S_1 potential energy surface. As described by [538], the linear absorption spectra of molecules may be formulated in terms of the Fourier transform of the time-dependent overlap between the initial state and the time-dependent excited-state wave packet.

We may gain some insight into wave packet motions by considering the simplest possible case of a one-dimensional harmonic S_1 potential energy surface. This is relevant to an optical transition that is strongly coupled to a single vibrational mode of the molecule. As pointed out in our treatment of active mode-locking, the eigenstates of Schrödinger's equation for this potential are Hermite Gaussians, as given by eq. (2.30a), and the energy levels are evenly spaced. The excited-state wave function may be written

$$\left|\Psi^{(1)}\right\rangle \sim i\left|\psi_1^e(\mathbf{r};X)\right\rangle \sum_N c_N^{(1)}\left|\chi_{1,N}(X)\right\rangle e^{-j(N+\frac{1}{2})\Delta\omega t} \tag{9.90}$$

where X refers to the displacement from the minimum energy position of the S_1 potential energy surface in the assumed one-dimensional coordinate system for nuclear motion, the $\left|\chi_{1,N}(X)\right\rangle$ are the Hermite Gaussian nuclear eigenfunctions in S_1, and $\hbar\Delta\omega$ is the energy spacing between levels. For $N = 0, 2, 4, \ldots$ ($N = 1, 3, 5, \ldots$) the nuclear eigenfunctions are even (odd) functions of X. Even with minimal detailed analysis, we may quickly discern general features of the wave packet motion.

- As per discussion above, the initial wave packet within the electronic excited state is centered at the position of the S_0 energy minimum and is displaced from the S_1 energy minimum. We shall call the amount of initial displacement X_0.
- Except for an unimportant overall phase, the motion is periodic; the nuclear positions repeat at times equal to multiples of $T = 2\pi/\Delta\omega$.
- At times $(2p + 1)T/2$ (with p an integer), the phase factor for the Nth eigenstate takes on $(-1)^N$ dependence; all the odd eigenstates are multiplied by -1. The overall wave function in the nuclear coordinates is the same as that created at $t = 0$, except that the

wave packet is now centered at $X = -X_0$ (i.e., it has moved to the opposite side of the S_1 energy minimum).

- In general, the mean position of the wave packet may be written

$$\langle X \rangle = \sum_{N\,\text{even}} \sum_{N'\,\text{odd}} \left(c_N^{(1)}\right)^* c_{N'}^{(1)} X_{NN'} e^{j(N-N')\Delta\omega t} + \text{c.c.} \tag{9.91}$$

where $X_{NN'} = \langle \chi_{1,N} | X | \chi_{1,N'} \rangle$, which is nonzero only when N and N' refer to states of opposite parity. Note that all the quantities appearing in the expression for $\langle X \rangle$ are real except for the exponential phase factor. Therefore, when the exponential is purely imaginary, which occurs at times $t = (2p+1)T/4$, we have $\langle X \rangle = 0$; the wave packet is centered at the S_1 minimum energy position. Furthermore, using the properties of Hermite Gaussians (e.g., [9]), one may easily show that $X_{NN'} = 0$ unless $N' = N-1$ or $N' = N+1$. Using this in eq. (9.91) yields

$$\langle X \rangle = X_0 \cos(\Delta\omega t) \sim \sum_N c_N^{(1)} c_{N+1}^{(1)} X_{N,N+1} \cos(\Delta\omega t) \tag{9.92}$$

Summarizing these points, the wave packet oscillates periodically from one side of the S_1 minimum energy position to the other. Just as with a classical harmonic oscillator, the mean position of the wave packet varies as a cosinusoidal function of time.

The harmonic oscillator is of course a limiting case, and a range of different behaviors may be obtained, depending on the potential energy surface. For example, if the harmonic potential considered above becomes weakly anharmonic, the energy spacing between levels changes slowly with energy. If the anharmonicity is small enough, oscillatory behavior will still be observed. However, the contributions to the wave packet motion corresponding to different N get out of phase with time, and the wave packet spreads. Such wave packet spreading is analogous to spreading of light pulses in dispersive media. Because the phases are still known, this effect is distinct from irreversible dephasing processes characterized by a T_2 time in Section 9.4.2, which also act to dampen wave packet motion. In other cases, the evolution may be quite different, such as cases where the wave packet disperses rapidly or moves rapidly away from its original position on a multidimensional potential energy surface, never to return. Additionally, for finite-duration input pulses, details of wave packet motion may often be manipulated through choice of bandwidth, center frequency, chirp, or excitation pulse shape.

Experimental Examples As a first example, we discuss excitation of Nile blue dye molecules in solution with pump pulses that are short compared to a single vibrational cycle [534]. The bandwidth of the sub-10-fs laser pulses were large enough essentially to span the entire S_0 to S_1 absorption band. Hence, the entire manifold of Franck–Condon states are excited simultaneously and impulsively with a single pulse. Such coherent excitation of a superposition of vibrational levels results in time-dependent nuclear wave packet motion, which may be observed through its effect on the absorption spectrum. For example, at any given time the stimulated emission spectrum is expected to peak at an energy corresponding to the vertical distance between S_1 and S_0 potential energy surfaces at the position of the S_1 nuclear wave packet at that time. Figure 9.19a shows the spectrally resolved pump–probe differential transmittance data. The spectra exhibit quite

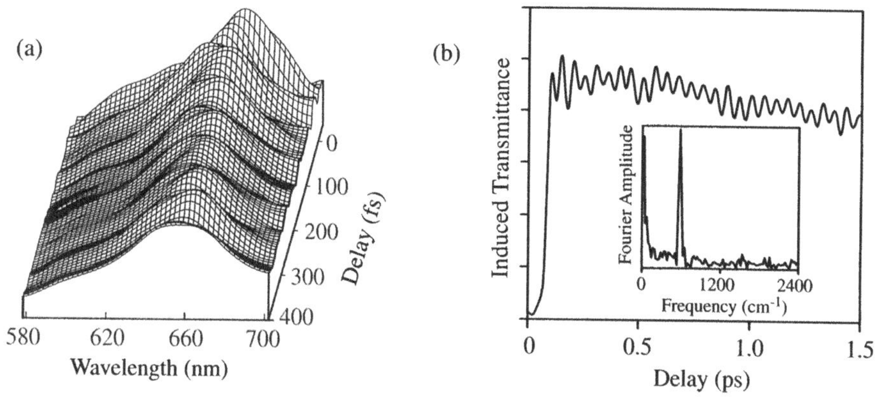

Figure 9.19 (a) Spectrally resolved pump–probe transmission data for impulsively excited Nile blue dye molecules; (b) pump–probe signal at selected probe wavelength (666 nm) and its Fourier transform. Adapted from [534].

visible oscillatory features, with both the amplitude and position of the peak absorption varying quasi-periodically with a period of about 60 fs. Figure 9.19b displays a cut through the time-resolved transmittance spectra for wavelengths near 660 nm. The oscillation is clearly observed for the entire time record of 1.5 ps, corresponding to more than 25 cycles. The frequency of the oscillation is very close to a 590-cm^{-1} vibrational mode known from Raman scattering.

These experiments have been modeled successfully using a time-dependent wave packet theory assuming displaced one-dimensional harmonic potential energy surfaces [539]. The choice of harmonic energy surfaces reflects the long lifetime of the oscillations observed, which would rapidly get out of phase if the vibrational energy-level spacing was not constant. The analysis reveals two different contributions to the absorption oscillations observed. One contribution comes from the nuclear motions on the excited-state (S_1) potential energy surface, as discussed above. A second contribution comes from nuclear motions in the ground (S_0) electronic state. Such motions may arise due to interactions in which the molecule is first excited from S_0 to S_1, then driven back down to S_0 after a short delay (allowing some initial displacement of the S_1 wave packet). Hence, the wave function reinjected into the electronic ground state is displaced compared to the S_0 energy minimum, thereby initiating nuclear motions within S_0. Pulses approximately one-quarter vibrational period in duration were found to be optimum for excitation of ground-state wave packets via this mechanism. Such pulse duration is long enough to allow displacement of the S_1 wave packet but still short enough to excite a vibrational coherence. Another mechanism for exciting ground-state wave packets, more commonly observed in nonresonant laser–matter interactions, is discussed in Section 9.7.

It is worth contrasting this experiment with the time and spectrally resolved absorption measurement [517] discussed in connection with Fig. 9.13 for a similar dye molecule. The main difference is the duration of the pump pulse: 60 fs for Fig. 9.13 vs. $\leq$10 fs for Fig. 9.19. In both cases the probe is short ($\leq$10 fs). The former case results in dynamic spectral hole burning, without oscillations. This is because the duration of the pump pulse is approximately equal to the vibrational period. The pump pulse is too long (and its bandwidth too narrow) to excite more than a single vibrational state; hence, a wave packet is not generated. In the case considered in this section, the pump pulse is much shorter and may

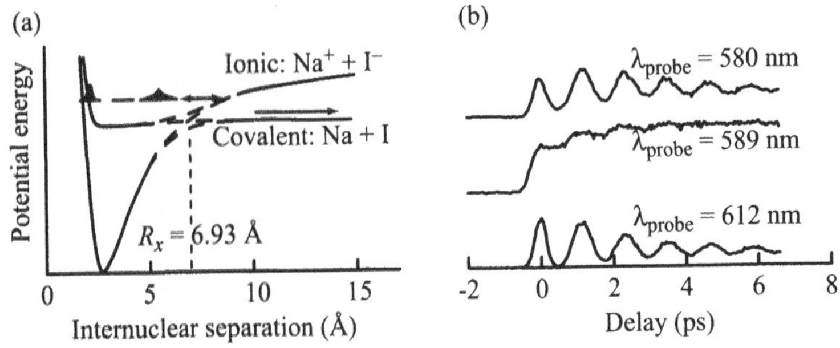

Figure 9.20 Laser-induced fluorescence measurements of wave packet dynamics in the NaI dissociation reaction; (a) potential energy surfaces, including an avoided crossing (dashed line) at internuclear separation R_x; (b) time-integrated fluorescence probed both on resonance with neutral sodium atoms (589 nm) and off resonance (580 nm, 612 nm). Adapted from [536].

therefore coherently excite a superposition of vibrational states, resulting in wave packet action.

As a second example, we consider femtosecond dissociation reaction dynamics of a small molecule, NaI, in the gas phase [535,536,540]. The relevant potential energy surfaces are sketched in Fig. 9.20a as a function of the reaction coordinate X (the interatomic distance). In the ground electronic state NaI has an ionic bond, while in the excited state the bond is covalent. At a relatively large interatomic distance ($X \sim 7$ Å), there is an avoided crossing between the two potential energy curves. Hence, the lower curve is ionic when X is small but becomes covalent when X is sufficiently large. The opposite is true for the upper energy curve. Absorption of a short pulse creates a wave packet that propagates on the upper potential energy surface, as in the preceding example. However, when the wave packet reaches the avoided crossing region, it has two choices. One possibility is that it remains on the upper energy surface. In this case, initially, X continues to increase, but the motion then stops and reverses direction, resulting in oscillatory wave packet dynamics bound within the upper potential energy well. The second possibility is that the wave packet "jumps" to the lower energy surface. In this case the atoms continue to separate, culminating in dissociation into neutral Na and I.

The dynamics of NaI photoexcited with a pump pulse of about 100 fs at a wavelength of 310 nm have been studied using a pump–probe laser-induced fluorescence technique. A variable-wavelength probe pulse, derived by spectral filtering after a continuum generation cell, was used to further excite the already photo-excited NaI to a higher energy state (not shown in the figure). The resulting fluorescence was collected and measured using a time-integrating detector. Figure 9.20b shows the time-integrated fluorescence signal as a function of probe delay for three probe wavelengths. When the probe is tuned to 589 nm, resonant with the D line of neutral free Na atoms, the induced fluorescence corresponds to sodium atoms sufficiently far removed from the iodine that the Na and I atoms are essentially noninteracting; the molecule is dissociated. The fluorescence signal rises with probe delay, indicating growing probability for dissociation. However, rather than a smooth rise such as an exponential, the growth occurs in a series of steps. The other traces are obtained when the probe is tuned away from 589 nm, so that is not resonant with the free sodium absorption resonance. In this case, fluorescence arises from perturbed sodium that is part of a $Na^+ + I^-$ complex still within the upper electronic energy state. The wavelength at which the perturbed sodium absorbs depends on position X along the reaction coordinate. Hence, different interatomic distances may be probed via choice of probe wavelength. The pronounced oscillations evident in these nonresonant probe cases correspond to the upper-state wave packet passing through the region of interatomic distance sensitive to the specific probe wavelength. Importantly, the oscillation period remains the same as the period between steps in the 589-nm probe trace. This allows a very interesting interpretation: During each period of wave packet oscillation, there is a finite probability of crossing to the lower, dissociative energy surface. The remaining molecules stay on the upper energy surface, continue to oscillate, and have the opportunity to dissociate each subsequent time that they approach the avoided crossing region. Under this picture dissociation occurs in bursts. Data such as that in Fig. 9.20b provide information on the shapes of the potential energy curves, the probability for crossing to the dissociated state, the coupling between ionic and covalent potential energy surfaces, and the time for the reaction to occur. As an additional point, the broadening of the peaks associated with later wave packet oscillation cycles provides evidence for wave packet dispersion associated with nonuniform energy spacings of vibrational states within the upper state well.

Femtosecond studies of the NaI photodissociation reaction, and related studies of other small molecules in the gas phase, resulted in the first direct observation of the atomic-level transition dynamics from reactants to products in a chemical reaction. The development of such femtosecond transition-state spectroscopy was recognized with the 1999 Nobel Prize for Chemistry [541].

9.6 DEPHASING PHENOMENA

In this section we return to the subject of dephasing. We begin by considering how dephasing shows up in linear optical spectroscopies, in both the frequency and time domains, and provide a simple discussion of dephasing in the homogeneous broadening limit. We also introduce other simple models of dephasing, inhomogeneous broadening and multilevel absorbers, that are physically relevant in common matter systems. We then proceed to a more detailed analysis of dephasing from the perspective of a stochastic oscillator. Finally, we discuss nonlinear spectroscopies based on two- and three-pulse four-wave mixing that allow us to distinguish between homogeneous and inhomogeneous dephasing, including techniques termed two-dimensional spectroscopies, which provide further insight into dephasing and line-broadening mechanisms.

9.6.1 Linear Spectroscopies

Linear spectroscopies probe the response of the matter system to first order in the laser field applied. Such response may be analyzed with the help of the first-order perturbation term of the density matrix, eq. (9.74). For homogeneous broadening (all particles are treated as identical), the general expression for the first-order polarization for a pulsed field is written

$$P^{(1)}(t) = \frac{1}{2}\left\{\tilde{P}^{(1)}(t)e^{j\omega_0 t} + \text{c.c.}\right\} \tag{9.93a}$$

with

$$\tilde{P}^{(1)}(t) = -\left(\frac{Ni|\mu_{12}|^2}{\hbar}\right)\int^{t} dt'\, a(t')e^{-j\delta\omega(t-t')}e^{-(t-t')/T_2} \tag{9.93b}$$

Here we have also used eq. (9.66), and $\delta\omega = \omega_0 - \omega_{21}$ is defined as the detuning of the laser carrier frequency from the transition frequency. We now consider two limiting cases: a continuous-wave input field and a delta-function input.

For a continuous-wave (single-frequency) input field, we take $a_{\text{in}}(t) = a_0$, a constant. At this point we assume an optically thin medium (the effect of the medium on the incident field is sufficiently small that we may assume the same laser field at all distances into the sample). We may then perform the integral indicated in eq. (9.93b) to obtain the induced polarization as

$$\hat{P}^{(1)}(\delta\omega) = -\frac{N|\mu_{12}|^2T_2}{\hbar}\frac{j + \delta\omega T_2}{1 + (\delta\omega T_2)^2}a_0 \tag{9.94}$$

where the notation $\hat{P}^{(1)}(\delta\omega)$ is used here to denote the complex spectral amplitude function of the first-order polarization. The effect of the polarization may be calculated by inserting it as a source term in Maxwell's equations, as in Section 5.2. The propagation of a multifrequency field with complex spectral amplitude $A(\omega)$ is described by

$$\frac{\partial A(\omega)}{\partial z} = \frac{-j\omega\mu_0 c}{2n}\hat{P}^{(1)}(\omega) = -\left(\frac{\alpha(\omega)}{2} + j\beta_r(\omega)\right)A(\omega) \tag{9.95a}$$

$\alpha(\omega)$ and $\beta_r(\omega)$ are the absorption coefficient and propagation constant, respectively, given by

$$\alpha(\omega) = \frac{\alpha_0}{1 + (\delta\omega T_2)^2} \qquad \text{where} \qquad \alpha_0 = \frac{\omega_0\mu_0 c N|\mu_{12}|^2T_2}{n\hbar} \tag{9.95b}$$

and

$$\beta_r(\omega) = \frac{-\alpha_0}{2}\frac{\delta\omega T_2}{1 + (\delta\omega T_2)^2} \tag{9.95c}$$

In Section 2.2 we introduced a Lorentzian, plotted in Fig. 2.3, to model the line shape of a gain medium. Comparison shows that the current result for the electric susceptibility, and for the absorption and propagation constants, are of exactly the same form, except for an overall factor of -1. This factor arises because the population is inverted for a gain medium, while our current result applies to a noninverted absorbing medium. The absorption has a peak centered at ω_{21} with a FWHM linewidth in hertz units of $1/\pi T_2$. Accordingly, linear absorption spectroscopy, in which the frequency of a tunable continuous-wave laser is swept through the resonance, provides a simple measurement of dephasing: The dephasing time may be inferred from the inverse width of the absorption spectrum.

For a pulse much shorter than T_2, with spectrum centered at the transition frequency ($\omega_0 = \omega_{21}$), the induced polarization becomes

$$p^{(1)}(t) = \frac{1}{2}\left\{\left[\left(\frac{-Ni|\mu_{12}|^2\mathcal{A}_{\text{in}}}{\hbar}\right)e^{-t/T_2}\right]e^{j\omega_{21}t} + \text{c.c.}\right\} \qquad (t \geq 0) \qquad (9.96a)$$

where

$$\mathcal{A}_{\text{in}} = \int_{-\infty}^{\infty} dt'\, a_{\text{in}}(t') \qquad (9.96b)$$

This formula is valid for times subsequent to the finish of the applied pulse. $\mathcal{A}_{\text{in}}$, called the *pulse area*,[2] gives the spectral amplitude of the field at the transition frequency ω_{21}. The polarization oscillates at ω_{21} and persists after the end of the exciting field. The lifetime of this transient polarization is equal to the dephasing time T_2. Thus, the absorption linewidth is inversely related to the polarization lifetime. Accordingly, our assumption of a pulse duration much less than T_2 implies an absorption linewidth narrow compared to the bandwidth of the exciting pulse. Such transient polarization effects are seen only for absorption that is narrowband compared to the applied field. For broadband absorption the familiar notion of an induced polarization that is instantaneous in the applied field may be retained.

The effect of the polarization on the field may be calculated by inserting it as a source term in Maxwell's equations, according to eq. (5.25). For a medium of length L, this gives an output field of the form[3]

$$a_{\text{out}}(t) \approx a_{\text{in}}(t) - \frac{\mu_0\omega_{21}c|\mu_{12}|^2NL}{2n\hbar}\mathcal{A}_{\text{in}}\, e^{-t/T_2}\, u_{-1}(t) \qquad (9.97)$$

where $u_{-1}(t)$ is the unit step function. In the time domain, the narrowband absorber leads to formation of a long exponential tail following the initial pulse. The polarity is opposite that of the incident pulse (i.e., there is a π phase shift). The result is a decrease in the pulse area corresponding to a reduction in the spectral amplitude of the pulse at the transition frequency, as expected for a resonant absorber.

The coherent polarization tail induced by short-pulse excitation, and consequent tail induced on the pulse itself, have been termed *free induction decay*. These effects are observable in coherent optics experiments on a variety of time scales, where the characteristic time scale depends on the linewidth of the absorber [69,542,543]. Here we discuss an example in the femtosecond regime, in which midinfrared pulses probe the intersubband absorption of a GaAs–AlGaAs quantum well sample [544]. In these experiments short pulses from a Ti:S laser are incident on a difference frequency generation crystal, phase-matched to produce radiation centered at 12.5 μm (see Section 5.5.2). The resulting midinfrared pulses are a few hundred femtoseconds in duration. When held at low temperature, the quantum well

[2] It is more common to define pulse area according to $(\mu_{12}/\hbar)\int dt'\, a_{\text{in}}(t')$, which is dimensionless. For our discussion, which is confined to linear effects, definition (9.96b) is sufficient.

[3] Compared to eq. (5.25), we have moved to a retarded time frame moving at the group velocity and have taken into account differences in the definitions of the slowly varying amplitude functions in eqs. (5.18) and (9.47).

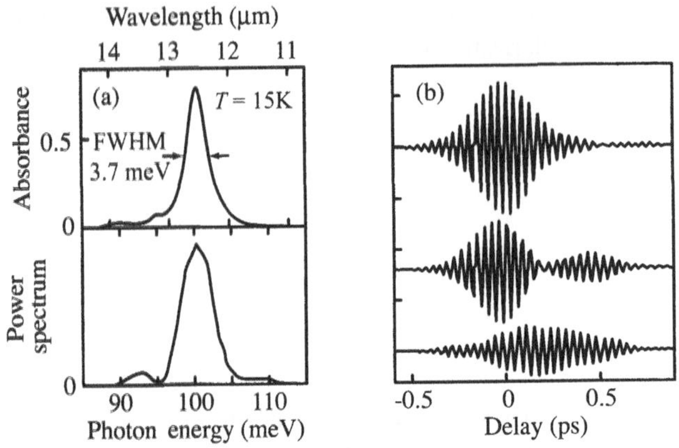

Figure 9.21 Data for quantum well intersubband absorber probed by midinfrared femtosecond pulses: (a) power spectrum of midinfrared excitation pulses and absorption spectrum of intersubband transition; (b) electro-optic sampling measurements of excitation field (top) and field after absorbing sample (middle), and their difference (bottom). Adapted from [544], with permission. Copyright © 2001, American Institute of Physics.

absorber has a linewidth of 3.7 meV (ca. 0.9 THz), as shown in Fig. 9.21a, and a dephasing time of about 300 fs. In this case the line width and the T_2 time of the absorber are comparable to the bandwidth and duration of the input pulse, respectively. Figure 9.21b shows the midinfrared pulses without (top) and with (middle) the intersubband absorption. These data were obtained via electro-optic sampling (see Chapter 10), which measures the actual electric field, including oscillations at the carrier frequency. After passing through the absorber, the trailing edge of the original pulse is reshaped such that it first decays to zero, then recovers to form a second lobe several hundred femtoseconds in duration. The zero crossing is indicative of a π phase shift in the field. The difference between the top and middle traces is also plotted. This trace shows that the main (early) part of the pulses with and without absorber are in phase (since their difference tends to cancel) but out of phase on the trailing edge (since the difference operation causes the fields to add constructively). These data, especially the π phase shift in the field, are as expected for free induction decay.

It is also interesting to consider the effect of propagation in optically thick media, where absorption is large and pulse reshaping becomes important. We retain the assumption of weak fields, for which the first-order perturbation theory results remain valid. By using eq. (9.95a), the field at position z into the sample may be written down immediately as

$$a(z,t) = \frac{1}{2\pi} \int d\tilde{\omega}\, e^{j\tilde{\omega}t} A_{\text{in}}(\tilde{\omega}) e^{-\alpha(\tilde{\omega})z/2} e^{-j\beta_r(\tilde{\omega})z} \tag{9.98}$$

where $A_{\text{in}}(\tilde{\omega})$ is the Fourier transform of $a_{\text{in}}(t)$. This frequency-domain or linear dispersion theory result is computationally efficient compared to the direct application of eq. (9.93b), for which the propagation region must be broken down into a series of incremental distances, with the time-dependent polarization computed at the beginning of an increment used to calculate a modified electric field waveform, which in turn is used to compute a new time-dependent polarization at the beginning of the next increment, and so on. Here we are most interested in the case of short laser pulses with bandwidth much greater than the absorption width. Based on eq. (9.98), we may make the following general comments:

- For $\alpha_0 L \gg 1$, the laser spectrum will be greatly attenuated in the vicinity of the absorption resonance. However, because the laser spectrum is wide compared to the absorption width, much of the spectrum propagates without significant attenuation. Consequently, the fractional energy loss in the pulse remains small.
- On the other hand, frequency-dependent phase effects associated with $\beta_r(\omega)$ persist much farther away from line center than do absorption effects. Therefore, dispersion is expected to play a very significant role in the pulse propagation.

The pulse area, defined similarly to eq. (9.96b) as the integral of field amplitude over time, turns out to be a useful metric for characterizing pulse propagation. We may relate the area $\mathcal{A}$ to the spectral amplitude at $\tilde{\omega} = 0$:

$$\mathcal{A}(z) = \left| \int_{-\infty}^{\infty} dt\, a(z,t) \right| = |A(z, \tilde{\omega} = 0)| \tag{9.99}$$

Rewriting eq. (9.95a) as

$$\frac{\partial \left[A(z,\tilde{\omega}) e^{j\beta_r(\tilde{\omega})z} \right]}{\partial z} = \frac{-\alpha(\tilde{\omega})}{2} \left[A(z,\tilde{\omega}) e^{j\beta_r(\tilde{\omega})z} \right] \tag{9.100}$$

we are able to show that

$$\frac{\partial \mathcal{A}(z)}{\partial z} = -\frac{\alpha_0}{2} \mathcal{A}(z) \tag{9.101}$$

Thus, the pulse area decays exponentially upon propagation.

This result [545] is the weak field limit of a more general pulse area theorem [69,542] (applies also at high fields) which describes pulse evolution in coherent laser–matter interactions. Interestingly, although the integral of the field (pulse area) decays to zero in an optically thick absorber, the integral of power (pulse energy) does not. This implies strong pulse reshaping. Such reshaping effects were observed in experiments in which weak 7-ps pulses were tuned to the 589-nm absorption resonance in sodium and passed through a sodium vapor cell [545]. The cell was adjusted for different optical densities up to $\alpha_0 L = 400$, and the dephasing time was estimated at 100 ps. Output pulses were measured via intensity cross-correlation using 0.5-ps reference pulses generated via fiber and grating pulse compression. Figure 9.22 shows data for various values of $\alpha_0 L$. The output develops a pronounced trailing edge with a series of subpulses developing as the peak absorption increases. We may interpret the first subpulse as the free induction decay tail expected from eq. (9.97). Upon continued propagation this free induction decay tail itself excites a second free induction decay tail, which becomes the second subpulse, and so on. From our discussion of free induction decay, each subpulse is expected to be π out of phase with the previous portion of the field. In this way the area of the pulse tends to zero since the contributions of different subpulses with alternating polarities tend to cancel. On the other hand, the pulse energy does not change significantly: The reduction in pulse energy as a

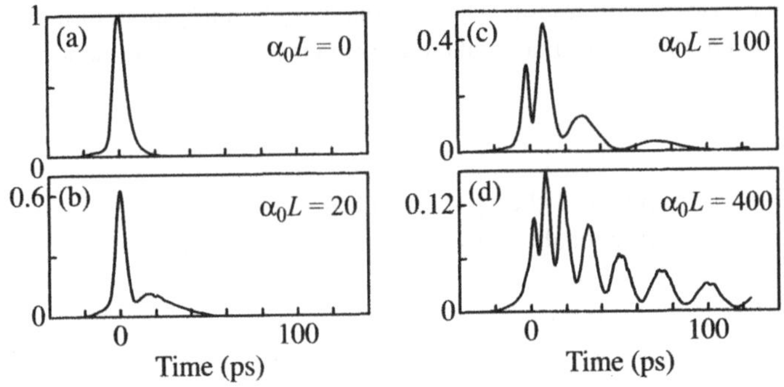

Figure 9.22 Intensity cross-correlation traces of 7-ps input pulses after propagation through a resonant sodium vapor absorption cell with (a) $\alpha_0 L = 0$, (b) $\alpha_0 L = 20$, (c) $\alpha_0 L = 100$, and (d) $\alpha_0 L = 400$. Adapted from [545], with permission. Copyright © 1984, American Physical Society.

result of propagation through the vapor cell was only 7% (11%) for $\alpha_0 L = 5$ ($\alpha_0 L = 400$). Hence, dispersive effects associated with the frequency dependence of $\beta_r(\omega)$ represent the primary mechanism for continued pulse reshaping.

Let us now return briefly to the relationship between time-domain pulse propagation and the polarization spectrum. Consideration of eq. (9.94) and following equations shows that absorption and propagation (dispersion) line-shape functions follow the Fourier transform of the assumed exponential transverse relaxation (dephasing) function. This relationship is general. Consider eq. (9.93b) with $\delta\omega$ set to zero (i.e., we choose the carrier frequency ω_0, which we may select at our convenience, to coincide with the transition frequency ω_{21}). Instead of exponential dephasing, we assume a causal but otherwise arbitrary polarization relaxation function $h_T(t)$. $h_T(t)$, also known as the *transverse relaxation function*, acts as an impulse response function for the first-order polarization, similar to the first-order time-dependent susceptibility tensor of eq. (6.106). The time-dependent first-order polarization amplitude function becomes

$$\tilde{P}^{(1)}(t) = -\left(\frac{Ni|\mu_{12}|^2}{\hbar}\right)\int^t dt'\, a(t')h_T(t-t') \tag{9.102}$$

The complex spectral amplitude is written

$$\hat{P}^{(1)}(\tilde{\omega}) = \int dt\, \tilde{P}^{(1)}(t)e^{-j\tilde{\omega}t} = -\frac{Ni|\mu_{12}|^2}{\hbar}A(\tilde{\omega})H_T(\tilde{\omega}) \tag{9.103a}$$

where

$$H_T(\tilde{\omega}) = \int dt\, h_T(t)e^{-j\tilde{\omega}t} \tag{9.103b}$$

Written in this form, the absorption spectrum is proportional to the real part of $H_T(\tilde{\omega})$:

$$\alpha(\tilde{\omega}) \sim \mathrm{Re}\{H_T(\tilde{\omega})\} \tag{9.104}$$

The conclusion is that first-order time- and frequency-domain spectroscopies are equivalent: The linear absorption spectrum may be obtained from the Fourier transform of the polarization relaxation function. The converse is also true: The polarization relaxation function may be obtained from the absorption spectrum. Measurement of $\beta_r(\omega)$ is not necessary, since the causality of the polarization relaxation function ensures that $\beta_r(\omega)$ may be determined from $\alpha(\omega)$, or vice versa.

9.6.2 Models of Dephasing

So far we have assumed homogeneously broadened media, primarily with exponential dephasing. We now discuss dephasing processes in more detail. First we introduce additional simple models, involving inhomogeneous broadening and multilevel absorbers, that are physically relevant to line broadening and dephasing in common systems (see Fig. 9.23). We then outline a more sophisticated treatment of dephasing based on stochastic oscillator theory.

Inhomogeneous Broadening In inhomogeneous broadening we relax the assumption that all absorbers are identical. In particular, we allow for the possibility that different absorbing particles have shifted resonance (transition) frequencies. Such frequency shifts may arise from a variety of mechanisms (e.g., Doppler shifts in atomic vapors or differences in local environment for impurities doped into a solid host). To some level of approximation, we may also consider frequency shifts associated with liquid solutions of absorbing molecules interacting with different local solvent configurations as a form of inhomogeneous broadening. However, such solvent configurations vary relatively rapidly, leading to time variations in the individual frequency shifts. It is simplest to consider *static* inhomogeneous broadening, in which the frequency shifts experienced by individual absorbers are taken to be independent of time (or at least to vary slowly compared to time scales of interest). A schematic view of an inhomogeneously broadened spectrum is shown in Fig. 1.4a. When the range of inhomogeneous frequency shifts is large compared to the linewidth of an individual absorber, the absorption spectrum is determined mainly by the distribution of resonance frequencies.

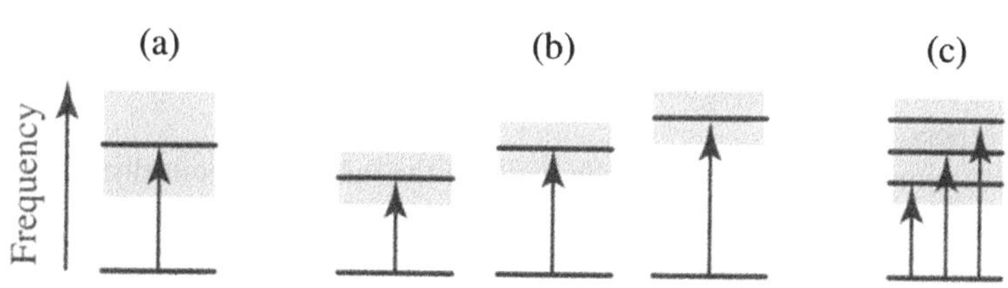

Figure 9.23 Comparison of three simple models of line broadening, sketched such that the three different systems have approximately the same spectrum when probed in linear absorption: (a) homogenously broadened two-level system, for which all the absorbing particles have identical resonance frequency; (b) inhomogeneously broadened two-level absorber, for which the resonance frequency is shifted differently for different absorbing particles; (c) multilevel system, for which even an individual absorber has multiple resonant transitions (which may be smeared together) within the laser bandwidth. Note that although a single isolated lower level is shown, transitions connecting multiple lower levels to multiple upper levels are also possible.

In the case of static inhomogeneous broadening, the first-order polarization response to a pulsed laser field may be written

$$\tilde{p}^{(1)}(t) = -\frac{Ni|\mu_{12}|^2}{\hbar} \int d\omega_{21}\, G(\omega_{21} - \Omega_0) \int dt'\, a(t') e^{-j(\omega_0 - \omega_{21})(t-t')} h_T(t - t') \tag{9.105}$$

This result follows from eq. (9.93b), but assuming a general form for the transverse relaxation function and integrating over a distribution $G(\omega_{21} - \Omega_0)$ of resonant frequencies. Ω_0 represents the center of the frequency distribution, and G is normalized such that its integral over frequency is equal to 1. The matrix elements and transverse relaxation functions are assumed to be the same for all absorbers, independent of the frequency shifts. For convenience taking the laser carrier frequency to coincide with the center of the inhomogeneous distribution, $\omega_0 = \Omega_0$, we may perform the frequency integral to obtain

$$\tilde{p}^{(1)}(t) = -\frac{Ni|\mu_{12}|^2}{\hbar} \int dt'\, a(t') \left\{2\pi g(t - t') h_T(t - t')\right\} \tag{9.106a}$$

where

$$g(t) = \frac{1}{2\pi} \int d\tilde{\omega}\, G(\tilde{\omega}) e^{j\tilde{\omega}t} \tag{9.106b}$$

$g(t)$, which is the inverse Fourier transform of the inhomogeneous frequency distribution function, assumes the role of a polarization relaxation function. The overall polarization relaxation function is the product of that associated with individual absorbers (homogeneous dephasing) with that arising due to the inhomogeneous broadening. By Fourier transforming with respect to t, we may also obtain the frequency response of the polarization, with the result

$$\hat{p}^{(1)}(\tilde{\omega}) = -\frac{Ni|\mu_{12}|^2}{\hbar} A(\tilde{\omega}) \left\{ \int d\tilde{\omega}'\, G(\tilde{\omega}') H_T(\tilde{\omega} - \tilde{\omega}') \right\} \tag{9.107}$$

The interpretation is that the overall polarization response spectrum, and hence the overall absorption spectrum, is the weighted summation of the spectra of the individual frequency-shifted absorbers.

Let us examine two limits. First, if the width of the inhomogeneous distribution is very narrow compared to the homogenous linewidth, we may approximate $G(\omega)$ as a delta function. In this case we recover the formulas of Section 9.6.1, and the homogenous broadening model applies. Conversely, if the width of the inhomogeneous distribution is very wide compared to the homogenous linewidth, the overall absorption spectrum reflects the inhomogeneous frequency distribution, as stated above, and the decay of the coherent polarization response in time is dictated simply by the inverse of the inhomogeneously broadened linewidth. This is the inhomogeneous broadening limit. Interestingly, it is sometimes possible to tune a given absorber system between homogeneous and inhomogeneous broadening limits: for example, by adjusting the pressure of an atomic vapor or the temperature of an impurity-doped solid host.

Multilevel Systems In condensed matter systems, energy levels are often closely spaced. For example, the room-temperature absorption spectrum of a dye molecule is composed of a manifold of unresolved vibrational and rotational levels; in crystalline semiconductors the levels merge into energy bands. Unlike atomic vapors, in such samples it is typically not possible to excite just a single transition. To gain insight into the behavior of such multilevel systems, we consider a simplified model in which the ground state is an isolated level and the excited state is a continuum characterized by a density-of-states function $D(\omega_{21} - \Omega_0)$ and a dipole moment $\mu(\omega_{21} - \Omega_0)$, which is also allowed to depend on the transition frequency. In contrast to our discussion of inhomogeneous broadening, each individual molecule has a series of transitions, but each transition is assumed to be homogeneously broadened (i.e., all molecules are taken to be identical). Again taking $\omega_0 = \Omega_0$, the first-order polarization in response to a pulsed field becomes

$$\tilde{P}^{(1)}(t) = -\frac{Ni}{\hbar} \int d\tilde{\omega}\, D(\tilde{\omega}) |\mu(\tilde{\omega})|^2 \int dt'\, a(t') e^{j\tilde{\omega}(t-t')} h_T(t-t') \tag{9.108}$$

After performing the integration over frequency, we obtain

$$\tilde{P}^{(1)}(t) = -\frac{Ni}{\hbar} \int dt'\, a(t') \left\{ 2\pi g'(t-t') h_T(t-t') \right\} \tag{9.109a}$$

where

$$g'(t) = \frac{1}{2\pi} \int d\tilde{\omega}\, D(\tilde{\omega}) |\mu(\tilde{\omega})|^2 e^{j\tilde{\omega}t} \tag{9.109b}$$

When several transitions are excited simultaneously, interferences between the various coherences leads to dephasing faster than that due to one line individually. Mathematically, this is expressed by the $g'(t)$ function. As in the case of inhomogeneous broadening, we may Fourier-transform eq. (9.109a) to obtain the frequency response of the polarization, with a result similar to eq. (9.107).

An important point is that in first-order spectroscopy, the polarization impulse response function and the complex absorption spectrum are always a Fourier transform pair. Hence, observing the free induction decay of the coherent polarization in the time domain does not provide additional information compared to measurement of the absorption spectrum. Despite the different physics, it is therefore difficult to determine via linear spectroscopy whether a sample with a smooth absorption spectrum is a homogeneously broadened two-level system, an inhomogeneously broadened two-level system, or a homogeneously (or inhomogeneously) broadened multilevel system.

Stochastic Oscillator Treatment A more general treatment of dephasing is possible by considering a stochastic oscillator model in which an ensemble of two-level systems is assumed to experience random fluctuations of the resonant frequency. Such frequency fluctuations may arise due to random interaction forces between the two-level systems and their fluctuating environment. The stochastic oscillator has been discussed by Kubo [546,547], whose analysis naturally draws extensively upon the mathematics of random processes. Here we discuss some of the key implications of Kubo's analysis while delving only lightly into the relevant mathematics.

We consider an oscillator whose coordinate x is described by

$$\frac{dx(t)}{dt} = j\omega(t)x \tag{9.110a}$$

where

$$\omega(t) = \omega_0 + \delta\omega(t) \tag{9.110b}$$

$\delta\omega$ is taken to be a real zero-mean random process, which is further assumed to be both stationary in time and ergodic (time averages and ensemble averages are equivalent). In terms of two-level systems, x may represent an off-diagonal element of the density matrix, as in eq. (9.71), but with both the applied perturbation (e.g., the applied laser field) and the phenomenological damping term turned off. Equation (9.110a) may be integrated to obtain

$$x(t) = x(0)e^{j\omega_0 t} \exp\left(j\int_0^t dt'\, \delta\omega(t')\right) \tag{9.111}$$

The ensemble average of x is then written as

$$\langle x(t)\rangle = x(0)e^{j\omega_0 t} \left\langle \exp\left(j\int_0^t dt'\, \delta\omega(t')\right)\right\rangle \tag{9.112}$$

Finally, we form the time correlation function describing the oscillator motion, with the result

$$\left\langle x(t)x^*(0)\right\rangle = |x(0)|^2 e^{j\omega_0 t}\Phi(t) \tag{9.113a}$$

where

$$\Phi(t) = \left\langle \exp\left(j\int_0^t dt'\, \delta\omega(t')\right)\right\rangle \tag{9.113b}$$

is called the *relaxation function of the oscillator.* $\Phi(t)$ plays a role similar to that of the polarization relaxation function $h_T(t)$, which we introduced previously. Physically, the frequency fluctuations integrate to form phase fluctuations, the variance of which grows with integration time. When the phase fluctuations become so large that their distribution is essentially uniform over a range of 2π, the ensemble-averaged oscillator motion drops to zero. At this point we say that the ensemble of oscillators has dephased.

Further analysis of the relaxation function requires information on the statistics of the frequency fluctuations. Here we shall assume that $\delta\omega(t)$ is a Gaussian random process. This assumption is physically reasonable when the frequency fluctuations are allowed to take on continuous values and arise from interactions with a reservoir with very large number of degrees of freedom. Expanding the exponential in eq. (9.113b), we obtain

$$\Phi(t) = 1 + \sum_{n=1}^{\infty} \frac{j^{2n}}{(2n)!} \int_0^t \cdots \int_0^t dt_1 \cdots dt_{2n}\, \langle \delta\omega(t_1)\delta\omega(t_2)\ldots\delta\omega(t_{2n})\rangle \tag{9.114}$$

Here the odd terms in the exponential expansion have been set to zero for zero-mean Gaussian frequency fluctuations. The nth term in the summation depends on the ensemble average of the product of the fluctuations at $2n$ times. A moment theorem for real Gaussian random processes [548] allows this expression to be simplified. In particular, for a real, zero-mean, Gaussian random process y, the joint moments satisfy

$$\langle y(t_1)y(t_2)\cdots y(t_{2n})\rangle = \sum_{\text{pairs}} \left\langle y(t_{\mu_1})y(t_{\mu_2})\right\rangle \ldots \left\langle y(t_{\mu_{2n-1}})y(t_{\mu_{2n}})\right\rangle \tag{9.115}$$

The ensemble average of the product of $2n$ fluctuation terms is expressed as the sum of products of n ensemble averages, where each ensemble average now involves only the product of a pair of fluctuation terms. The summation is over all of the $(2n)!/(2^n n!)$ distinguishable combinations of $2n$ different time samples taken in pairs; ordering is not important. Using this expression in eq. (9.114) yields

$$\begin{aligned}\Phi(t) &= 1 + \sum_{n=1}^{\infty} \left(\frac{(-1)^n}{(2n)!}\right)\left(\frac{(2n)!}{2^n n!}\right)\left[\int_0^t \int_0^t dt_1\, dt_2\, \langle \delta\omega(t_1)\delta\omega(t_2)\rangle\right]^n \\ &= \exp\left(\frac{-1}{2}\int_0^t \int_0^t dt_1\, dt_2\, \langle \delta\omega(t_1)\delta\omega(t_2)\rangle\right)\end{aligned} \tag{9.116}$$

Furthermore, because $\delta\omega(t)$ is stationary, the ensemble averages depend only on time differences. We therefore introduce the two-time frequency fluctuation correlation function $\Psi(t_1, t_2) = \Psi(t_1 - t_2)$, given by

$$\Psi(\tau) = \frac{1}{\Delta^2}\langle \delta\omega(t)\delta\omega(t-\tau)\rangle \qquad \text{where} \qquad \Delta^2 = \left\langle (\delta\omega)^2\right\rangle \tag{9.117}$$

This allows us to rewrite the relaxation function as

$$\Phi(t) = \exp\left(\frac{-\Delta^2}{2}\int_0^t \int_0^t dt_1 dt_2\, \Psi(t_1 - t_2)\right) \tag{9.118}$$

We can simplify by performing the double integration over coordinates $\tau = t_1 - t_2$ and $\eta = t_1 + t_2$. This gives

$$\begin{aligned}\Phi(t) &= \exp\left(\frac{-\Delta^2}{2}\int_{-t}^{t} \frac{d\tau}{\sqrt{2}} \int_{|\tau|}^{2t-|\tau|} \frac{d\eta}{\sqrt{2}}\Psi(\tau)\right) \\ &= \exp\left(\frac{-\Delta^2}{2}\int_{-t}^{t} d\tau\,(t - |\tau|)\Psi(\tau)\right) \\ &= \exp\left(-\Delta^2 \int_0^t d\tau\,(t-\tau)\,\Psi(\tau)\right)\end{aligned} \tag{9.119}$$

The factors of $1/\sqrt{2}$ above are needed to give the proper length increments along the τ and η directions. With our assumption of a Gaussian random process, the relaxation function depends only on the two-time frequency correlation function.

Let us now consider the simple case in which the correlation function decreases monotonically with delay τ. We characterize the frequency fluctuations by their root-mean-square amplitude Δ and by their correlation time, which we take here as

$$\tau_c = \int_0^{\infty} d\tau \, \Psi(\tau) \tag{9.120}$$

τ_c represents the time over which the system has memory of its frequency fluctuations. We may now consider the behavior of the relaxation function at times much shorter, or much longer, than the correlation time:

- For $t \ll \tau_c$, we may use $\Psi(\tau) \approx 1$ in eq. (9.119). This yields

$$\Phi(t) = \exp\left(\frac{-\Delta^2 t^2}{2}\right) \qquad (t \ll \tau_c) \tag{9.121}$$

  At short times the relaxation function shows Gaussian behavior.

- For $t \gg \tau_c$, we may set the upper limit of the integral in eq. (9.119) to infinity. This yields

$$\Phi(t) \sim \exp\left(-\Delta^2 \tau_c t\right) \qquad (t \gg \tau_c) \tag{9.122}$$

  At long times the relaxation function is exponential.

These two expressions cross at $t = 2\tau_c$, for which $\Phi(t) = \exp(-2\Delta^2\tau_c^2)$. Again we may distinguish two limiting cases, now based on the value of $\tau_c\Delta$:

- For $\tau_c\Delta \gg 1$, the relaxation function decays to a very small value even at a time short compared to τ_c. Hence, the short time limit of the relaxation function, eq. (9.121), dominates. Both the dephasing and its Fourier transform, the complex absorption spectrum, are approximately Gaussian. Essentially static inhomogeneous broadening dictates the dephasing behavior. The Gaussian line-shape function reflects the assumed Gaussian distribution of the frequency fluctuations.
- For $\tau_c\Delta \ll 1$, the relaxation function shows almost no decay at times out to τ_c. Accordingly, the dephasing is dominated by the long time limit of the relaxation function, eq. (9.122). The relaxation function may now be rewritten

$$\Phi(t) = \exp\left(\frac{-t}{T_2}\right) \qquad \text{where} \qquad T_2 = \frac{1}{\Delta^2 \tau_c} \tag{9.123}$$

  This is the limit of pure homogeneous broadening. We see now that this limit applies only (a) for times long compared to τ_c, and (b) when the phase change integrated over each independent frequency fluctuation (or "collision") is small. Accordingly, many collisions are needed to randomize the phase. In this limit the absorption linewidth is much narrower than the frequency fluctuation distribution function.

To illustrate this discussion, let us consider the specific case where the two-time correlation function is exponential:

$$\Psi(\tau) = \exp\left(\frac{-|\tau|}{\tau_c}\right) \tag{9.124}$$

Evaluation of eq. (9.119) yields a relaxation function of the form [506]

$$\Phi(t) = \exp\left\{-\Delta^2\left[\tau_c t - \tau_c^2\left(1 - e^{-t/\tau_c}\right)\right]\right\} \tag{9.125}$$

As expected, predominantly Gaussian relaxation, associated with an inhomogeneous broadening character, is seen for $\tau_c\Delta \gg 1$, while predominantly exponential dephasing, associated with homogeneous broadening behavior, is observed for $\tau_c\Delta \ll 1$. In all cases nonexponential behavior is observed when t is very small ($t \ll \tau_c$). However, in many practical experiments, the time scales that are probed are sufficiently long that τ_c may be approximated as zero. Within the ultrafast optics community, the terminology *Markovian dynamics* is commonly used to refer to this limit. Conversely, experiments at time scales sufficiently short that the finite value of τ_c plays a role are said to probe *non-Markovian dynamics*.

Finally, it is interesting to note that our earlier discussion of static inhomogeneous broadening, as in eqs. (9.106a) and (9.107), does not follow our results from the stochastic oscillator theory. Assuming a Gaussian distribution of static frequency shifts, and taking $h_T(t)$ as exponential, eq. (9.106a) predicts exponential behavior at short times and Gaussian behavior at long times. This is opposite to the behavior expected from the stochastic oscillator model. The resolution to this paradox is that the static inhomogeneous limit must be understood as resulting from frequency fluctuations governed by two independent random processes. For example, for impurity absorbers in a low-temperature solid host, spatially varying strain fields give rise to essentially static local frequency shifts, for which $\tau_c\Delta \gg 1$ is satisfied and for which τ_c may often be taken as much longer than any time scales of interest. Independently, vibrations of the host material give rise to rapid frequency fluctuations, for which $\tau_c\Delta \ll 1$ may be satisfied. We see then that the validity of the static inhomogeneous broadening model depends on the ability to make a clear separation between processes responsible for the inhomogeneous and homogeneous dephasing.

9.6.3 Measurement of Dephasing Using Transient Gratings

We have seen that linear spectroscopies are unable to distinguish between homogenous and inhomogeneous dephasing mechanisms. Further insight is available from nonlinear spectroscopies. In the frequency domain, saturation of an inhomogeneously broadened absorption line by a spectrally narrow laser field leads to spectral hole burning (see Fig. 1.4b); measurement of the hole written within the absorption spectrum gives the homogeneous line shape. Observation of transient spectral hole burning in spectrally resolved pump–probe measurements of dye molecules in solution was discussed in Section 9.3.1. Here we concentrate on measurement techniques in which the relaxation of coherent polarization transients are probed directly in the time domain.

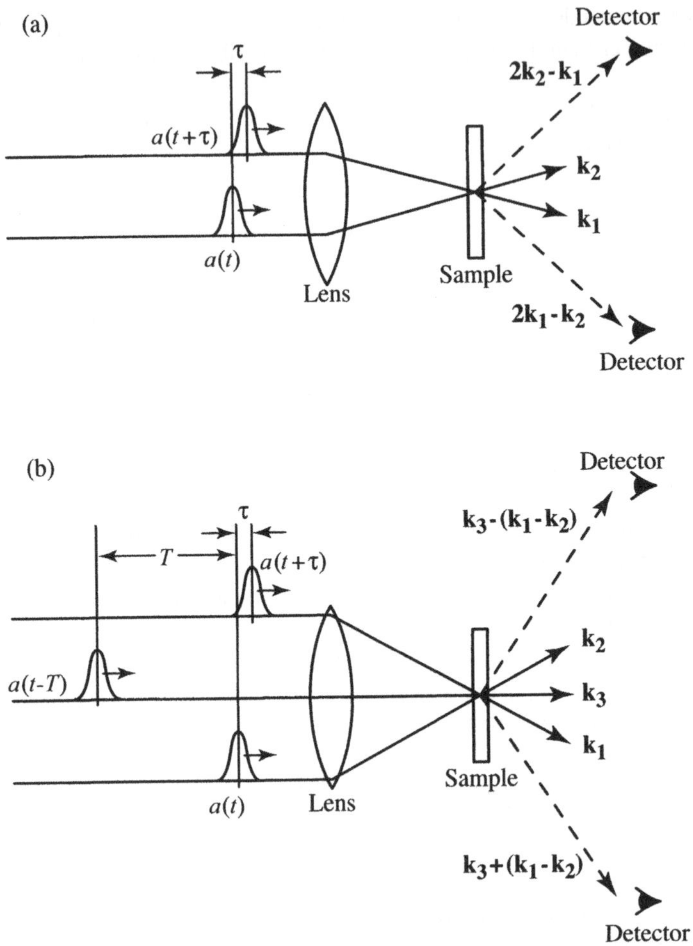

Figure 9.24 Geometry for (a) two-pulse and (b) three-pulse transient four-wave mixing experiments for measurement of dephasing.

Figure 9.24 shows the geometry of transient grating or transient four-wave mixing experiments that are useful for dephasing studies. A pair of pulses is incident noncollinearly on a resonant sample with wave vectors $\mathbf{k}_1$ and $\mathbf{k}_2$, respectively. Due to interference effects the population change induced through absorption of the pulse pair exhibits a transverse spatial modulation. This constitutes an optically induced grating. In three pulse transient grating experiments, an additional delayed probe pulse with wave vector $\mathbf{k}_3$ scatters off the induced grating into directions $\mathbf{k}_3 \pm (\mathbf{k}_1 - \mathbf{k}_2)$ [549]. In the two-pulse version, the same two pulses that write the grating simultaneously act as probe pulses, leading to scattering in directions $2\mathbf{k}_2 - \mathbf{k}_1$ and $2\mathbf{k}_1 - \mathbf{k}_2$ [550].

Three pulse-scattering experiments have often been configured to measure grating lifetimes [510,551]. In that case the delay between the two pulses writing the grating is set to zero, and the scattered energy is measured as a function of the delay of the third (probe) pulse. This configuration is applicable both to samples with finite dephasing times and to those with essentially instantaneous dephasing. Such transient grating experiments provide information complementary to pump–probe measurements of transient absorption. For example, the transient grating technique is equally sensitive to induced absorption gratings and induced refractive index gratings, while pump–probe transmission senses only changes in absorption. Furthermore, additional physical channels, such as diffusion processes, that do not usually affect transmission measurements may contribute to the decay of transient gratings.

The basis for obtaining dephasing information from transient grating studies is as follows. Normally, we think of induced gratings occurring when the writing pulses are overlapped in time, so that their electric fields interfere directly. However, even when the writing pulses are temporally separated, grating formation can still occur provided that the delay is less than the lifetime of the coherent polarization. In this case the electric field of the second pulse may be said to interfere with the coherent polarization induced by the first. By measuring the grating amplitude as a function of the delay between writing pulses, one measures the dephasing. This is accomplished by recording the energy in either or both of the scattered output directions as a function of the pulse delays.

Let us now analyze transient grating dephasing measurements, using the simple model of a two-level atom with static inhomogeneous broadening. From eq. (9.75) the population grating written by an input field of the form

$$e(\mathbf{r}, t) = \mathrm{Re}\left\{\left[a(t+\tau)e^{-j\mathbf{k}_1\cdot\mathbf{r}} + a(t)e^{-j\mathbf{k}_2\cdot\mathbf{r}}\right]e^{j\omega_0 t}\right\} \tag{9.126}$$

is given by

$$\gamma_{\tilde{\omega}_{21}}(\mathbf{r}, t) \sim \int^t\int^{t'} dt'\,dt''\,e^{-(t-t')/T_1}\left\{e^{(j\tilde{\omega}_{21}-T_2^{-1})(t'-t'')}\right.$$
$$\left.\times\left[a^*(t'+\tau)a(t'')e^{-j(\mathbf{k}_2-\mathbf{k}_1)\cdot\mathbf{r}} + a^*(t')a(t''+\tau)e^{-j(\mathbf{k}_1-\mathbf{k}_2)\cdot\mathbf{r}}\right]\right\} + \text{c.c.} \tag{9.127}$$

Here we have taken the laser frequency to be coincident with the center of the inhomogeneous distribution, $\omega_0 = \Omega_0$, and introduced the detuning $\tilde{\omega}_{21} = \omega_{21} - \Omega_0$. $\gamma_{\tilde{\omega}_{21}}$ is the grating written into the set of absorbers characterized by resonant frequency $\tilde{\omega}_{21}$, and only the spatially modulated terms in the population change are retained. In the two-pulse scattering experiment, the third-order polarization terms responsible for the scattering are obtained from the quantity

$$\tilde{p}^{(3)}(\mathbf{r}, t) \sim \int d\tilde{\omega}_{21}\,G(\tilde{\omega}_{21})\int^t dt'\,\gamma_{\tilde{\omega}_{21}}(\mathbf{r}, t')$$
$$\times\left\{a(t'+\tau)e^{-j\mathbf{k}_1\cdot\mathbf{r}} + a(t')e^{-j\mathbf{k}_2\cdot\mathbf{r}}\right\}e^{(j\tilde{\omega}_{21}-T_2^{-1})(t-t')} \tag{9.128}$$

In three-pulse scattering experiments, the induced grating is interrogated by a third pulse with envelope function $a(t-T)\exp(-j\mathbf{k}_3\cdot\mathbf{r})$. Often, the delay T is made sufficiently large that we may attribute the writing of the grating solely to the pulses specified in eq. (9.126). In that case the third-order polarization terms responsible for scattering in directions

$\mathbf{k_3} \pm (\mathbf{k_1} - \mathbf{k_2})$ are contained within the quantity

$$\tilde{P}^{(3)}(\mathbf{r}, t) \sim \int d\tilde{\omega}_{21}\, G(\tilde{\omega}_{21}) \int^{t} dt'\, a(t' - T) e^{-j\mathbf{k_3}\cdot\mathbf{r}} \times \gamma_{\tilde{\omega}_{21}}(\mathbf{r}, t') e^{(j\tilde{\omega}_{21} - T_2^{-1})(t-t')} \tag{9.129}$$

For experiments with a time-integrating detector, we measure the output energy U in one or more of the scattering directions. Since we are considering directions that are background-free and since the scattered field is directly proportional to the corresponding third-order polarization source term, the output is given by

$$U_{\mathbf{k}_{\text{out}}} \sim \int_{-\infty}^{\infty} dt \left| \tilde{P}^{(3)}_{\mathbf{k}_{\text{out}}}(t) \right|^2 \tag{9.130}$$

where $\tilde{P}^{(3)}_{\mathbf{k}_{\text{out}}}(t)$ refers to the polarization terms with wave vector in the $\mathbf{k}_{\text{out}}$ direction.

Two-Pulse Scattering Now let us analyze two-pulse scattering [550] assuming that the pulses are delta functions (i.e., the pulses are much shorter than the inverse bandwidth of the inhomogeneously broadened absorption line). Using eq. (9.128), the relevant nonlinear polarization terms are written

$$\begin{aligned} P_{2\mathbf{k_2}-\mathbf{k_1}}(t) &\sim g(t-\tau)\, e^{-(t+\tau)/T_2} \qquad (t \geq 0\,, \tau \geq 0) \\ P_{2\mathbf{k_1}-\mathbf{k_2}}(t) &\sim g(t - 2|\tau|)\, e^{-t/T_2} \qquad (t \geq |\tau|\,, \tau \leq 0) \end{aligned} \tag{9.131}$$

Here $g(t)$ is the inverse Fourier transform of $G(\tilde{\omega}_{21})$ and the polarizations are nonzero only under the conditions listed. These polarization terms are sketched in Fig. 9.25 in both homogeneous and inhomogeneous broadening limits. For scattering in the $2\mathbf{k_2} - \mathbf{k_1}$ direction, the $\mathbf{k_1}$ pulse induces the initial (first-order) coherent polarization; the delayed $\mathbf{k_2}$ pulse both interferes with the first-order polarization to write a population grating and then interacts with this grating to form the coherent third-order polarization responsible for generation of the scattered field. For delta-function input pulses, this time ordering is

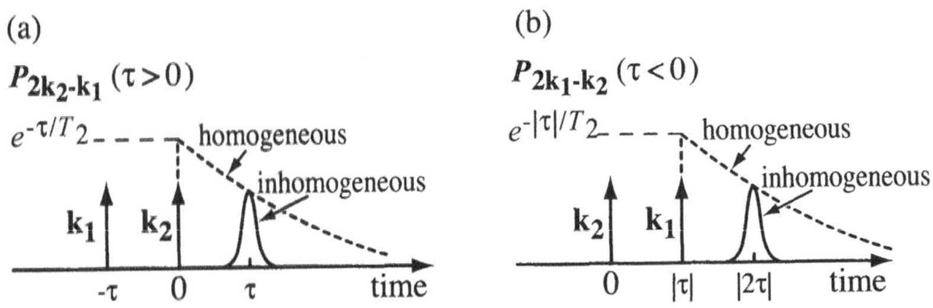

Figure 9.25 Time dependence of coherent polarization excited in two-pulse transient grating experiments with delta-function input pulses (shown as vertical arrows), for both homogeneous and inhomogeneous broadening: (a) polarization in $2\mathbf{k_2} - \mathbf{k_1}$ direction with $\tau > 0$; (b) polarization in $2\mathbf{k_1} - \mathbf{k_2}$ direction with $\tau < 0$. The polarizations are zero for the opposite choice of delay.

strictly enforced, which means that $P_{2\mathbf{k}_2-\mathbf{k}_1}$ is nonzero only for $\tau \geq 0$. In the homogeneous broadening case, the curve shows the familiar single-sided exponential decay. The initial amplitude of the polarization is $\exp(-\tau/T_2)$, which reflects the first-order polarization decay between the time of the first and second pulses. Since for delta-function pulses, the grating is written and read out with no time delay, population decay does not enter into our result. In the strong inhomogeneous broadening case (inhomogeneous linewidth $\gg T_2^{-1}$), the third-order polarization is a narrow pulse, delayed by τ with respect to the readout ($\mathbf{k}_2$) pulse and with duration determined by the width of the inhomogeneous distribution. The delay in the appearance of this burst of coherent polarization introduces a second factor of $\exp(-\tau/T_2)$ into its peak amplitude. $P_{2\mathbf{k}_1-\mathbf{k}_2}$ displays similar behavior except that it is nonzero only for $\tau \leq 0$ and that the curves are shifted to later times by $|\tau|$ to coincide with the time position of the $\mathbf{k}_1$ pulse, which reads out the grating for this scattering direction.

To understand the delay in the onset of the third-order polarization in the inhomogeneous broadening case, consider eq. (9.127) with delta-function pulses. The grating in the population of absorbers centered at frequency $\tilde{\omega}_{21}$, immediately after the second pulse to arrive, is

$$\gamma_{\tilde{\omega}_{21}} \sim e^{-|\tau|/T_2} \cos\{(\mathbf{k}_1 - \mathbf{k}_2) \cdot \mathbf{r} - \tilde{\omega}_{21}\tau\} \tag{9.132}$$

Thus, the population gratings corresponding to different subgroup frequencies are spatially shifted with respect to each other. The total population, integrated over all the absorbers, becomes spatially uniform when $|\tau|$ exceeds the inverse of the inhomogeneous width; the total grating amplitude approaches zero due to the destructive interference of the various grating components, as does the initial third-order polarization (again integrated over all the absorbers). However, the third-order polarization contains an $\exp(j\tilde{\omega}_{21}t)$ phase term, which evolves at different rates for different $\tilde{\omega}_{21}$. This phase evolution can cause interference which is initially destructive to transform into constructive interference at a specific time. For example, for $\tau > 0$ the third-order polarization contributions from different population subgroups come into phase at $t = \tau$ to form a phased array for scattering into the $2\mathbf{k}_2 - \mathbf{k}_1$ direction. Similarly, for $\tau < 0$ the contributions to the polarization rephase at $t = 2|\tau|$ for scattering into the $2\mathbf{k}_1 - \mathbf{k}_2$ direction. This delayed rephasing behavior that occurs for inhomogeneous broadening is equivalent to the weak field limit of a phenomenon known as the *photon echo* [69,542,543,552].

Unlike traditional photon echo experiments performed on relatively narrowband inhomogeneous absorbers, for which the delayed output (the echo) can be resolved directly, in ultrafast optics one usually works with time-integrated quantities. Working from eq. (9.131), we can easily work out expressions for the scattered energy measured by a time-integrating detector. For the $2\mathbf{k}_2 - \mathbf{k}_1$ direction the results in the pure homogeneous broadening and strong inhomogeneous broadening limits are respectively

$$\begin{aligned} U_{2\mathbf{k}_2-\mathbf{k}_1}(\tau) &\sim u_{-1}(\tau)\, e^{-2\tau/T_2} \qquad \text{(homogeneous)} \\ &\sim u_{-1}(\tau)\, e^{-4\tau/T_2} \qquad \text{(inhomogeneous)} \end{aligned} \tag{9.133}$$

Here $u_{-1}(\cdot)$ is the unit step function. These results are valid for delta-function pulses, and in the inhomogeneous broadening case, for delays greater than the inverse inhomogeneous linewidth. Identical results are obtained for scattering into the $2\mathbf{k}_1 - \mathbf{k}_2$ direction, except that the sign of τ is reversed. In the case of pulses with finite duration (but remaining much shorter than T_2), the only change is that the rise of the energy vs. delay curve for homogeneous broadening occurs over a delay scale comparable to the pulse duration. For inhomogeneous broadening the rising edge occurs over a delay scale determined by the longer of the pulse duration and the inverse inhomogeneous linewidth.

Assuming that the dephasing is well resolved, eq. (9.133) shows that the scattered energy vs. delay curves decay with a time constant proportional to the homogeneous dephasing time T_2, for both pure homogeneous broadening (time constant $T_2/2$) and wide inhomogeneous broadening (time constant $T_2/4$). Comparing the dephasing behavior obtained from the scattered energy curve with that predicted from the inverse Fourier transform of the absorption spectrum provides a diagnostic for inhomogeneous broadening. In particular, in the case of wide inhomogeneous broadening, the scattered energy decays with a time constant much longer than the inverse absorption width.

For pulse widths that are not much shorter than T_2, it becomes tricky to distinguish between homogeneous and inhomogeneous broadening. One finds that scattering curves peak at a delay greater than zero (referring now to the $2\mathbf{k_2} - \mathbf{k_1}$ direction) and are asymmetric, even for T_2 much less than the pulse duration, and for both homogeneous and inhomogeneous broadening. This behavior arises because the second pulse in two-pulse scattering has two functions: It both participates in writing the grating and scatters off the grating. The details of the scattering curve depend on the population relaxation time (the peak shifts toward zero delay for smaller T_1), on the pulse shape (which in general is not known unless special care is taken to measure it), and on the ratio of homogeneous and inhomogeneous broadening. Because phenomena such as rapid vibrational relaxation are often present in condensed matter samples of interest for ultrafast dephasing studies, the mixing of T_1 and T_2 effects in two pulse scattering complicates interpretation of the data.

Figure 9.26 shows two-pulse scattering data for a thin film of GaAs excited by pulses of about 6 fs at room temperature. The experiments were performed for different laser intensities, and consequently, different photoexcited carrier densities. The data show exponential decay for delays greater than approximately the pulse duration, with the decay rate increasing with carrier density. These experiments were interpreted using a simple inhomogeneous broadening model, in which the mechanism for inhomogeneous broadening is simply the distribution of carriers in a band made up of states with different wave vectors (**k**) and with different interband transition frequencies. According to eq. (9.133), in this model T_2 is equal to four times the decay constant in the data; the estimated T_2 times range from 44 fs at the lowest carrier density tested to 14 fs at the highest. These data point to carrier–carrier scattering as the dominant dephasing mechanism in GaAs under these conditions [553].

Figure 9.27 shows another example of two-pulse scattering data, in this case for a dense potassium vapor sample. In a dense atomic vapor, interactions (collisions) between atoms provide the main dephasing mechanism, in contrast to Doppler broadening, which dominates at low atomic densities. A virtue of experiments in such atomic systems is a clear separation of time scales, typically not present in condensed matter systems, which enables detailed comparison with the stochastic oscillator model. In particular, in the experiments of [554], the dephasing time (many picoseconds), the time between collisions (ca. 2 ps), the collision duration (ca. 0.5 ps), and the pulse duration (70 fs) were all well separated. The broad spectral content of the femtosecond excitation pulses excites both the potassium D_1 and D_2 lines. These lines are well separated and result in a pronounced periodic beating in the scattering signal. The dephasing dynamics are obtained from the envelope (maxima) of this beat signal and may be interpreted in terms of the stochastic oscillator model, suitably

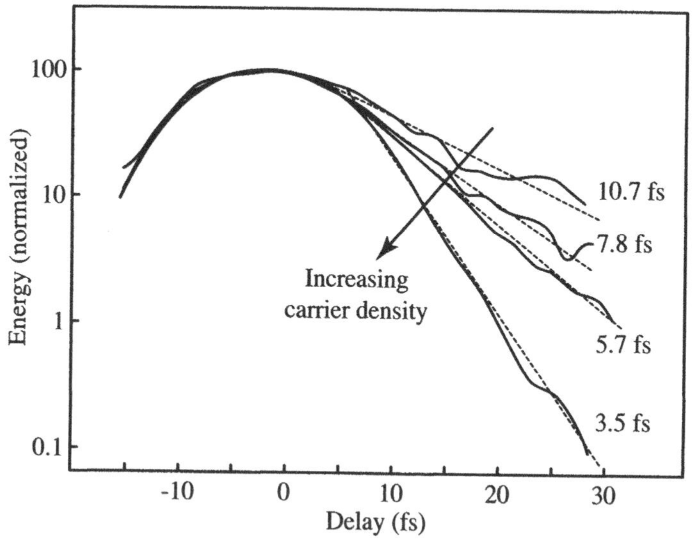

Figure 9.26 Scattered energy in the $\mathbf{2k_2} - \mathbf{k_1}$ direction for two-pulse transient grating experiments performed on room-temperature GaAs, at carrier densities of 2×10^{17}, 3.5×10^{17}, 8×10^{17}, and $7 \times 10^{18}\ \mathrm{cm}^{-3}$, respectively. The data are plotted logarithmically, and their decay is fit by straight lines indicative of exponential decay. The decay constants, equal to $T_2/4$ in an inhomogeneous broadening model, are given. Adapted from [553], with permission. Copyright © 1988, American Physical Society.

extended to handle nonlinear spectroscopy (e.g., [506,555]). At delays greater than a few picoseconds, longer than the characteristic time for collisions, the decay is exponential and may be described by Markovian dynamics. For delays below approximately 2 ps, the scattered energy vs. delay curve flattens out. This is a signature of non-Markovian dynamics,

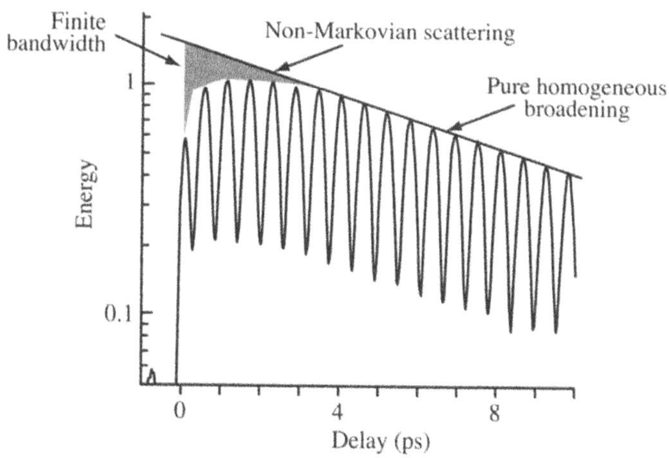

Figure 9.27 Scattered energy for two-pulse transient grating experiments performed on dense potassium vapor. Different physical regimes are labeled. Adapted from [554], with permission. Copyright © 2005, American Physical Society.

which become important when probing time scales less than or equal to the correlation time τ_c of the frequency fluctuations. In this regime the dephasing shows primarily inhomogeneous behavior, which arises due to a distribution of interatomic distances and interaction potentials (which are not yet averaged out by atomic motion). At very short times the scattering signal shows a finite rise time, limited by the pulse duration and the inverse of the inhomogeneous width.

Three-Pulse Scattering We now consider the three-pulse scattering scheme, again using a static inhomogeneous broadening model [549]. The delay T of the third pulse is assumed to meet the following criteria:

- T is much greater than both the pulse duration and the T_2 dephasing time. Pulse 3 is chosen to arrive several pulse widths after pulses 1 and 2, when the formation of the grating is complete.
- The population does not change appreciably within the duration of pulse 3. This is valid when pulse 3 arrives after all fast population relaxation processes, such as fast vibrational relaxation within the electronic excited state, are complete.
- $T \ll T_1$: pulse 3 arrives before interband relaxation erases the grating.

This picture is appropriate within the framework of a three-level system, composed, for example, of an electronic excited state coupled to a reservoir and subject to rapid intraband relaxation and a slowly recovering ground state. This implies a separation of time scales for relaxation processes. When this separation of time scales applies, and the delay T is chosen as specified, the three pulse-scattering curves assume a simple form that facilitates interpretation of experimental data.

Let us first consider the case of delta-function pulses. From eq. (9.129), the third-order nonlinear polarization becomes

$$
\begin{aligned}
P_{\mathbf{k_3}+\mathbf{k_2}-\mathbf{k_1}} &\sim g(t-T-\tau)\, e^{-T/T_1}\, e^{-(|\tau|+t-T)/T_2} \qquad (t \geq T) \\
P_{\mathbf{k_3}+\mathbf{k_1}-\mathbf{k_2}} &\sim g(t-T+\tau)\, e^{-T/T_1}\, e^{-(|\tau|+t-T)/T_2}
\end{aligned}
\tag{9.134}
$$

Here we have simplified some of the population relaxation terms by setting $\exp(-\tau/T_1) \approx 1$; this approximation is retained throughout our following analysis. For homogeneous broadening the third-order polarization is symmetric in τ for both scattering directions; this is unlike two-pulse scattering. For wide inhomogeneous broadening, the third-order polarization appears as an echo, similar to two-pulse scattering but now peaking at $t = T + |\tau|$. For $\tau > 0$ rephasing occurs only for scattering in the $\mathbf{k_3} + \mathbf{k_2} - \mathbf{k_1}$ direction; rephasing for the $\mathbf{k_3} + \mathbf{k_1} - \mathbf{k_2}$ direction occurs only for $\tau < 0$. For inhomogeneously broadened media the delayed rephasing behavior that occurs in three-pulse scattering is formally equivalent to the weak field limit of an effect known as the *stimulated photon echo* [556].

The scattered energy is obtained by using eq. (9.130). For the $\mathbf{k_3} + \mathbf{k_2} - \mathbf{k_1}$ direction, the results for homogeneous broadening and for strong inhomogeneous broadening are,

respectively,

$$U_{\mathbf{k_3}+\mathbf{k_2}-\mathbf{k_1}}(\tau) \sim e^{-2T/T_1}\, e^{-2|\tau|/T_2} \qquad \text{(homogeneous)}$$
$$\sim e^{-2T/T_1}\, e^{-4|\tau|/T_2}\, u_{-1}(\tau) \qquad \text{(inhomogeneous)} \tag{9.135}$$

Identical results are obtained for the $\mathbf{k_3}+\mathbf{k_1}-\mathbf{k_2}$ direction but with the sign of τ reversed. The inhomogeneous broadening results are valid for $|\tau|$ greater than the inverse inhomogeneous linewidth. The important point is that the scattering curves are symmetric in τ for homogeneous broadening and identical in both scattering directions. With inhomogeneous broadening the scattering curves become asymmetric, with opposite sense of asymmetry in the two scattering directions. These differences constitute a clear signature for inhomogeneous broadening that is not available in two-pulse scattering.

It is also worth analyzing the scattering curves with both arbitrary input pulses and transverse relaxation function. We introduce the notation

$$\gamma_{\tilde{\omega}_{21}} = \frac{1}{2}\left\{\tilde{\gamma}_{\tilde{\omega}_{21}} e^{-j(\mathbf{k_1}-\mathbf{k_2})\cdot\mathbf{r}} + \tilde{\gamma}^*_{\tilde{\omega}_{21}} e^{-j(\mathbf{k_2}-\mathbf{k_1})\cdot\mathbf{r}}\right\} \tag{9.136}$$

and apply eq. (9.127) to obtain the following expression for $\tilde{\gamma}_{\tilde{\omega}_{21}}$:

$$\tilde{\gamma}_{\tilde{\omega}_{21}} \sim \iint dt'\, dt''\, \left\{h_{\tilde{\omega}_{21}}(t'-t'')\, a^*(t')\, a(t''+\tau) + h^*_{\tilde{\omega}_{21}}(t'-t'')\, a(t'+\tau)\, a^*(t'')\right\} \tag{9.137}$$

where

$$h_{\tilde{\omega}_{21}}(t) = h_T(t) e^{j\tilde{\omega}_{21} t} \tag{9.138}$$

is the transverse relaxation function multiplied by a time oscillation term corresponding to frequency $\tilde{\omega}_{21}$. Here $\tilde{\gamma}_{\tilde{\omega}_{21}}$ is evaluated for times at which grating formation is complete, and for simplicity we have now omitted the $\exp(-T/T_1)$ population relaxation factor. By changing variables, we may rearrange eq. (9.137) to read

$$\tilde{\gamma}_{\tilde{\omega}_{21}}(\tau) \sim \int d\tau'\, \left\{h_{\tilde{\omega}_{21}}(\tau'+\tau)\, \Gamma^*_a(\tau') + h^*_{\tilde{\omega}_{21}}(\tau'-\tau)\, \Gamma_a(\tau')\right\} \tag{9.139}$$

where $\Gamma_a(\tau)$ is the electric field amplitude correlation function, as given in eq. (3.15).

For homogeneous broadening the scattered energy is proportional simply to $|\tilde{\gamma}_{\tilde{\omega}_{21}}(\tau)|^2$. Since it is easily seen that $\tilde{\gamma}_{\tilde{\omega}_{21}}(\tau) = \tilde{\gamma}^*_{\tilde{\omega}_{21}}(-\tau)$, the scattering remains a symmetric function of τ for arbitrary pulses. Identical scattering curves are obtained in both scattering directions. Furthermore, by using the connection between the field correlation function and the power spectrum, eq. (3.21), we may write

$$\tilde{\gamma}_{\tilde{\omega}_{21}}(\tau) \sim \int \frac{d\tilde{\omega}}{2\pi} |A(\tilde{\omega})|^2 \left[H_T(\tilde{\omega} - \tilde{\omega}_{21}) + H_T^*(\tilde{\omega} - \tilde{\omega}_{21})\right] e^{j\tilde{\omega}\tau}$$
$$\sim \int \frac{d\tilde{\omega}}{2\pi} |A(\tilde{\omega})|^2 \alpha(\tilde{\omega} - \tilde{\omega}_{21})\, e^{j\tilde{\omega}\tau} \qquad (9.140)$$

In the latter expression we have used eq. (9.104). The grating amplitude is equal to the inverse Fourier transform of the product of the absorption spectrum and laser power spectrum. Hence for homogeneous broadening the scattered energy as a function of delay may be predicted on the basis of these two easily measurable quantities. The detailed form of the homogeneous broadening scattering curves has been verified experimentally [549].

By extending this analysis, one may derive analogous expressions with inhomogeneous broadening [549]. It turns out that the electric field autocorrelation still provides all the pulse-shape information needed to specify the scattered energy curves. However, the expressions are rather cumbersome. Instead, we gain insight from numerical results. Figure 9.28a shows curves calculated for the scattering of Gaussian pulses into direction $\mathbf{k_3} + \mathbf{k_2} - \mathbf{k_1}$, for wide inhomogeneous broadening. Identical curves are obtained for the $\mathbf{k_3} + \mathbf{k_1} - \mathbf{k_2}$ direction, but with the delay axis reversed. The characteristic exponential decay is evident only for $T_2 > 2t_p$ or so. Even for small dephasing times, however, the rising edge and peak of the scattering curves exhibit a clear shift toward positive delay. Figure 9.28b shows the peak shift, defined as the total separation between the peaks of the scattering curves for the two output directions, as a function of T_2. For small dephasing times the peak shift is proportional to T_2. Accordingly, the peak shift can serve as a sensitive indicator of inhomogeneous broadening even for T_2 times below the pulse width.

Figure 9.29 shows three-pulse scattering data for cresyl violet dye in a polymer film that demonstrate this peak shift behavior. These experiments were performed with 70-fs laser pulses at a variety of sample temperatures, and data were recorded for both scattering directions. The scattering curves taken at low temperature (15 K) exhibit clear asymmetry, illustrated by the 60-fs peak shift; this shows that inhomogeneous broadening is present. With increasing temperature the asymmetry and peak shift decrease. The peak shift is

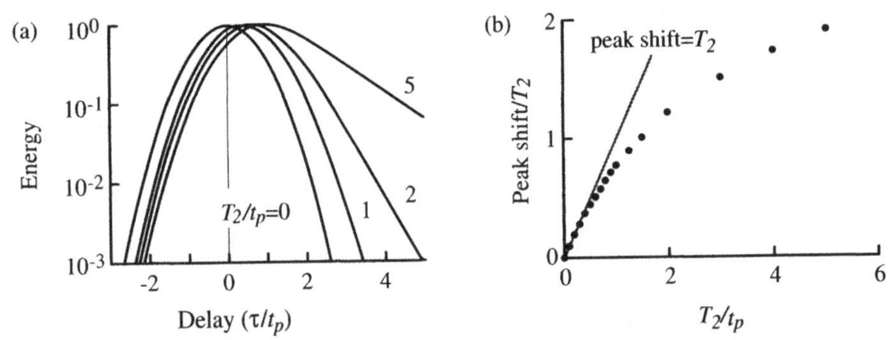

Figure 9.28 (a) Energy in three-pulse scattering experiments, as a function of delay, calculated for Gaussian pulses, $a(t) \sim \exp(-t^2/t_p^2)$, and for different dephasing times, $T_2/t_p = 0$, 1, 2, and 5. Wide inhomogeneous broadening is assumed. (b) Shift between the delay peaks for the two scattering directions. (a) Adapted from [559]; (b) adapted from [549].

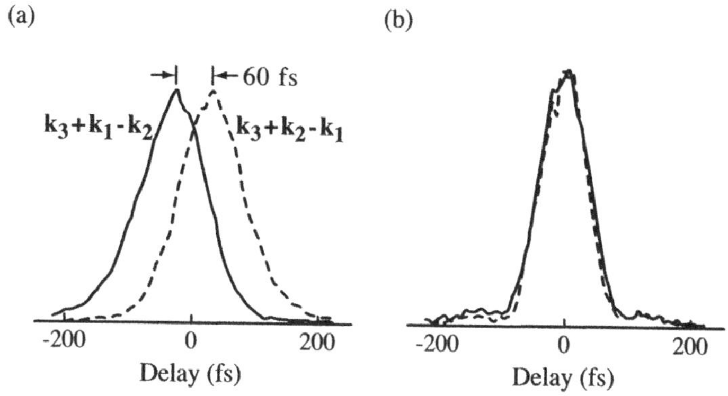

Figure 9.29 Three-pulse scattering data for cresyl violet dye in a polymer host at temperatures of (a) 15 K and (b) 290 K. Adapted from [557].

no longer evident in the room-temperature data, for which homogeneous broadening apparently dominates. The transition to homogeneous broadening was attributed in a multilevel picture to smearing of the vibrational structure within the spectra of individual dye molecules [557].

As an example of more complex physics, we now consider the possibility of *spectral cross-relaxation*, also known as *spectral diffusion*, which occurs when the resonant frequencies of individual absorbers are not fixed but can migrate within an inhomogeneous distribution. Velocity-changing collisions provide such a mechanism in gases [543]; for molecules in solution, the mechanism could involve changes in either the surrounding solvent configuration or in intramolecular bath modes that perturb the optically active modes. Spectral cross-relaxation is also possible for mobile absorbers doped into solid hosts. With reference to the static inhomogeneous broadening model, we now allow the correlation time of the slow frequency fluctuations to be finite rather than infinite, while remaining much longer than the homogeneous dephasing time T_2.

In the presence of spectral cross-relaxation, the three-pulse scattering curves become sensitive to the delay of the third pulse. For delays T longer than the characteristic spectral cross-relaxation time T_3, the rephasing behavior discussed previously in connection with inhomogeneous broadening can no longer occur. This is because the individual gratings, considered for example in eq. (9.132), are no longer distinct. There exists then only a grating in the total population, the amplitude of which is equal to the sum of the complex amplitudes of the individual gratings. The total grating is obtained by integrating eq. (9.140) over the inhomogeneous distribution, with the result

$$\begin{aligned}\tilde{\gamma}_{\text{TOTAL}}(\tau) &= \int d\tilde{\omega}_{21}\, G(\tilde{\omega}_{21})\tilde{\gamma}_{\tilde{\omega}_{21}}(\tau) \\ &\sim \int \frac{d\tilde{\omega}}{2\pi} |A(\tilde{\omega})|^2 \left\{ \int d\tilde{\omega}_{21}\, G(\tilde{\omega}_{21})\alpha_{\text{HOM}}(\tilde{\omega}-\tilde{\omega}_{21}) \right\} e^{j\tilde{\omega}\tau} \qquad (9.141)\end{aligned}$$

Here α_{HOM} refers to the homogeneous absorption line shape corresponding to a particular absorber, and the quantity within the braces is the full (inhomogeneous) absorption spectrum. Equation (9.141) is of the same form as eq. (9.140) for homogeneous broadening. Consequently, for $T \gg T_3$, the scattered energy is again a symmetric function of τ and equal in both scattering directions. However, the width of the scattering curves reflect the inverse of the full absorption width rather than the homogeneous dephasing time T_2.

In general, the scattering curves vary continuously from their inhomogeneous form at $T \ll T_3$ to the homogeneous form, eq. (9.141), at $T \gg T_3$. This evolution can be analyzed if we specify a model for the cross-relaxation. Following [558], the probability per unit time that a dipole initially resonant at $\tilde{\omega}'_{21}$ jumps to a new resonance frequency $\tilde{\omega}_{21}$ is assumed to have the following simple form:

$$P(\tilde{\omega}'_{21} \to \tilde{\omega}_{21}) = \frac{G(\tilde{\omega}_{21})}{T_3} \tag{9.142}$$

This formula takes the resonance frequency of an absorber to be completely randomized after just a single "collision," which occurs in characteristic time T_3. After its creation by pulses 1 and 2, the amplitude $\tilde{\gamma}_{\tilde{\omega}_{21}}$ of the grating at the frequency $\tilde{\omega}_{21}$ evolves according to

$$\frac{\partial \tilde{\gamma}_{\tilde{\omega}_{21}}}{\partial t} = -\left(\frac{1}{T_1} + \frac{1}{T_3}\right)\tilde{\gamma}_{\tilde{\omega}_{21}} + \frac{G(\tilde{\omega}_{21})}{T_3}\int d\tilde{\omega}'_{21}\,\tilde{\gamma}_{\tilde{\omega}'_{21}} \tag{9.143}$$

This differential-integral equation can be solved analytically [559] for delta-function input pulses and a Gaussian inhomogeneous distribution,

$$G(\tilde{\omega}_{21}) = \frac{1}{\delta\omega\sqrt{\pi}}\exp\left(-\frac{\tilde{\omega}_{21}^2}{\delta\omega^2}\right)$$

Here $\delta\omega$ sets the width of the inhomogeneous distribution. The resulting scattered energy curves are plotted in Fig. 9.30 for several values of T/T_3, for normalized inhomogeneous

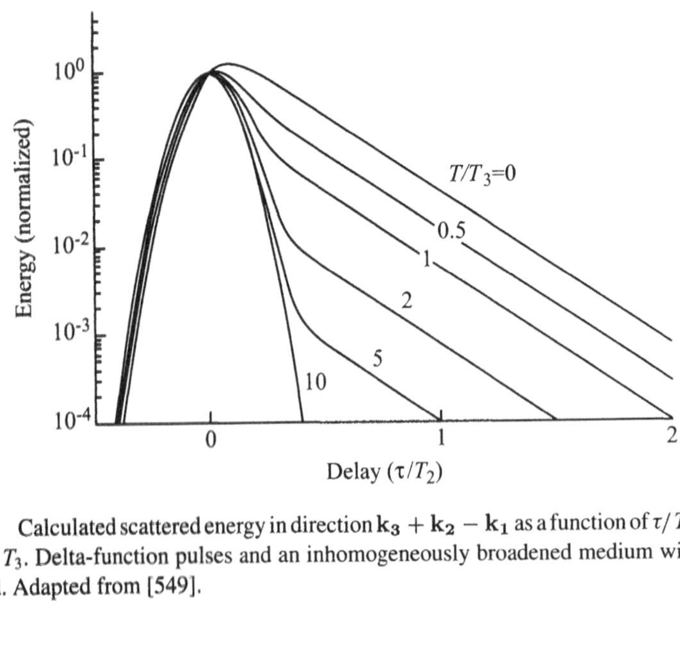

Figure 9.30 Calculated scattered energy in direction $\mathbf{k_3} + \mathbf{k_2} - \mathbf{k_1}$ as a function of τ/T_2 for different values of T/T_3. Delta-function pulses and an inhomogeneously broadened medium with $\delta\omega T_2 = 10$ are assumed. Adapted from [549].

broadening parameter $\delta\omega T_2 = 10$. The population decay time T_1 is assumed to be much longer than the time scale of interest. As the delay of the third pulse is increased, the scattering curves become narrower and more symmetric. The scattering efficiency for $\tau = 0$ is insensitive to spectral cross-relaxation since all gratings are created in phase. The conclusion is that three-pulse scattering can yield information about spectral cross-relaxation as well as dephasing, by acquiring data for several values of T and looking for a transition from curves that are asymmetric to symmetric in τ.

Such effects have been observed in three-pulse experiments performed with Nile blue dye in solution [560]. The pulse durations were in the range 6 to 8 fs, short enough that the pulse bandwidth (160 nm centered at 620 nm) spans the absorption spectrum of the dye. Figure 9.31 shows scattering data for three different values of delay T. Two principal effects are evident. First, the curves are markedly asymmetric at small T, providing clear evidence for inhomogeneous broadening behavior at short times. Second, the curves become more symmetric as T increases, which provides evidence for spectral cross-relaxation and evolution toward homogeneous broadening. Reference [560] analyzes these experiments in terms of the stochastic oscillator model, extended to the regime of nonlinear spectroscopy. In this language the data give information on the two-time correlation function of the frequency fluctuations and provide evidence of a finite correlation time (i.e., non-Markovian dynamics). The ability to map the frequency correlation function by comparing data obtained at different T has been further explored in [561–563].

In another case, three-pulse scattering was used to study dephasing dynamics in CdSe nanocrystals [564]. Quantum confinement in such nanocrystals reduces the continuous band of energy states found in bulk samples to a series of discrete electronic transitions. These transitions are substantially inhomogeneously broadened due to variations in nanocrystal size and the effect of such variations on the quantum confinement. Furthermore, the electronic transitions couple strongly to the longitudinal optical (LO) phonon mode of the

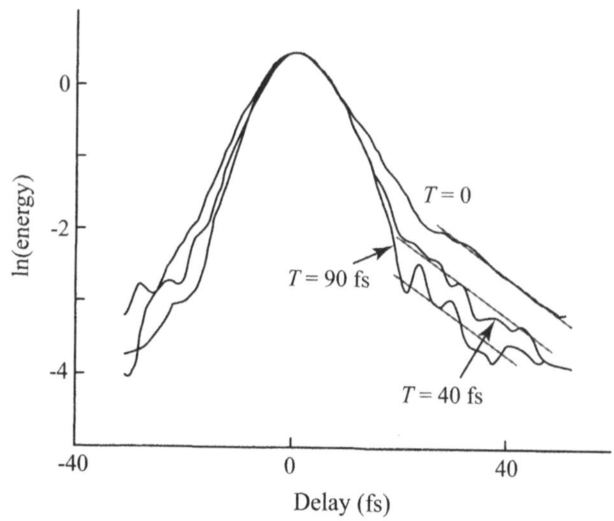

Figure 9.31 Scattered energy in a three-pulse scattering experiment for Nile blue dye molecules in solution, for delays $T = 0$, $T = 40$ fs, and $T = 90$ fs. Adapted from [560], with permission. Copyright © 1991, American Physical Society.

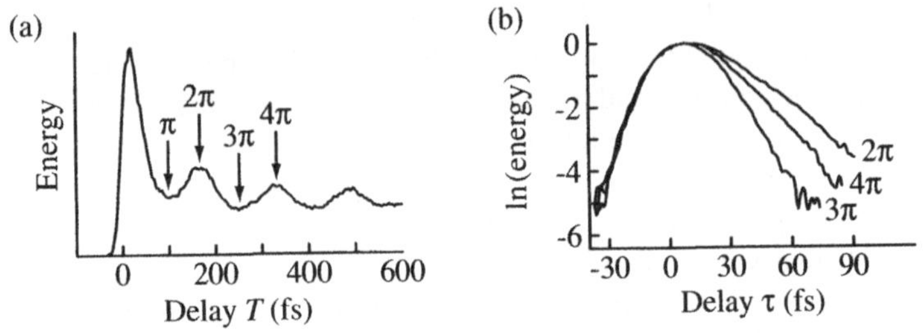

Figure 9.32 Three-pulse scattering data for CdSe nanocrystals: (a) scattered energy as a function of pulse 3 delay (T) for fixed delay $\tau = 33$ fs between pulses 1 and 2 (values of T for which $\omega_{LO}T = n\pi$ are indicated for $n = 1 \ldots 4$); (b) scattered energy vs. τ for $\omega_{LO}T = 2\pi$, 3π, and 4π. Adapted from [564], with permission. Copyright © 1993, American Physical Society.

lattice. This electronic–vibrational coupling gives rise to wave packets which introduce quantum beat behavior into the dephasing dynamics. That is, both stochastic dynamical interactions with a continuum of reservoir modes and coupling to a specific periodic vibrational mode contribute to dephasing of the electronic transition. These effects are illustrated by data obtained using pulses of about 15 fs in a blue-green spectral region overlapping with the excitonic absorption peak. Figure 9.32a shows scattered energy data as a function of delay between the first and third pulses (T), for fixed 33-fs delay between pulses 1 and 2 (τ). The data show clear modulation with a period of 163 fs, equal to the period of the 205-cm^{-1} LO phonon mode, damped on a subpicosecond time scale associated with vibrational dephasing. The explanation is that at times equal to odd multiples of half the phonon period, $\omega_{LO}T = (2m + 1)\pi$, nuclei are maximally displaced, and the coherent electronic polarization is diminished. On the other hand, at even multiples of the phonon period, $\omega_{LO}T = 2m\pi$, the nuclei are back at their equilibrium positions, and the influence of the phonon motion on the electronic polarization is suppressed. Figure 9.32b shows scattered energy as a function of delay between pulses 1 and 2, for various fixed values of T. The asymmetric decay associated with inhomogeneous broadening is clearly observed. At $\omega_{LO}T = 2\pi$, the exponential decay constant is 20 fs, corresponding to a T_2 of approximately 80 fs. At $\omega_{LO}T = 3\pi$, corresponding to maximum nuclear displacement, the decay constant is much faster (9 fs); evidently, the electronic polarization is strongly modulated by the nuclear wave packet motion. At $\omega_{LO}T = 4\pi$, the phonon contribution is suppressed significantly and the echo decay becomes longer again. Full suppression is not achieved due to vibrational damping. It is interesting that interpretation of data such as that in Fig. 9.32 depends on two types of memory effects. One type of memory is the electronic phase memory, associated with the two-time frequency correlation function of eq. (9.117), which we have discussed throughout this section. The second is memory within the reservoir itself, here manifested through the coherent phonon motion, which is not considered in the stochastic oscillator model.

9.6.4 Two-Dimensional Spectroscopy

We have seen that transient four-wave mixing, using time-integrating detectors, may be used to distinguish between homogeneous and inhomogeneous broadening and to examine the correlation function of the frequency fluctuations that give rise to such broadening. However, these methods still provide only limited information in the case of congested multilevel systems. New spectroscopy methods that acquire nonlinear spectroscopy data for display in two (or more) dimensions help to unravel the contributions of individual transitions in multilevel systems and in particular the couplings between various transitions. By measuring the strength of such couplings and their time evolution, it becomes possible to glean information on molecular structure, energy transport within complex structures, chemical dynamics, and many-body interactions. Two-dimensional optical spectroscopies may be considered the optical analog of related techniques in nuclear magnetic resonance (NMR), where sophisticated sequences of radio-frequency pulses have for some time been employed to extract key information in structural biology [565]. Optical two-dimensional spectroscopies are in use both at infrared frequencies, where primarily vibrational excitations and their couplings are investigated, and in the visible, where electronic excitations are accessible. The theory of two-dimensional optical spectroscopy is treated in [506]. Here our discussion takes a more qualitative flavor, drawing on reviews in [566,567].

Although various experimental arrangements for two-dimensional (2D) spectroscopy are possible, the most general is based on three-pulse transient four-wave mixing, with the important addition that the scattered pulse is resolved in time. As sketched in Fig. 9.33, this is accomplished through interference with a reference pulse, which is combined collinearly with the scattered field. Either electric field cross-correlation or spectral interferometry techniques may be used to obtain the scattered field profile. As discussed in Section 3.3, in electric field cross-correlation fringes in the energy of combined reference and scattered fields are recorded as the reference pulse delay is varied, while in spectral interferometry fringes in the power spectrum of combined reference and scattered fields are recorded for fixed relative delay. The quantities of interest are various representations of the scattered field, related through

$$s(\tau, t; T) \xrightarrow{\mathcal{F}(t)} S(\tau, \omega_t; T) \xrightarrow{\mathcal{F}(\tau)} S_{2D}(\omega_\tau, \omega_t; T) \tag{9.144}$$

Here $s(\tau, t; T)$ denotes the time-domain interferogram, which is a function of time t within the scattered field (determined experimentally through the delay of the reference pulse) and of the delay between writing pulses (τ). The delay T of the readout pulse (pulse 3) is taken as an additional parameter; T is also known as the waiting time or mixing time. The spectral representation of the field $S(\tau, \omega_t; T)$ is related to the time-domain interferogram by Fourier transform, denoted $\mathcal{F}(t)$, with respect to the t variable. The 2D spectrum $S_{2D}(\omega_\tau, \omega_t; T)$ is obtained by a second Fourier transform, denoted $\mathcal{F}(\tau)$, in this case with respect to variable τ. When the experiment is configured to acquire the time-domain interferogram through

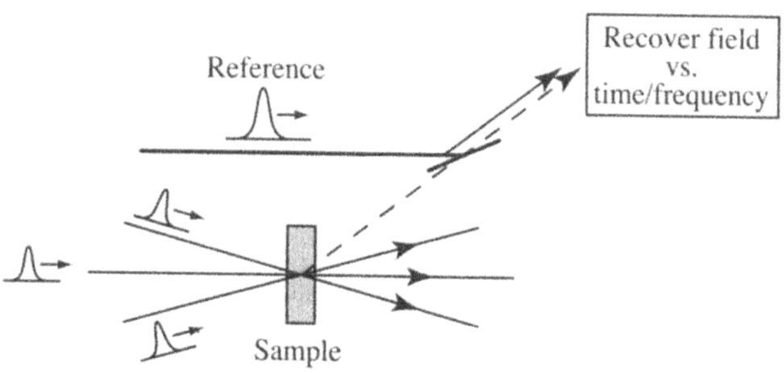

Figure 9.33 Setup for two-dimensional spectroscopy experiments.

electric field cross-correlation, both Fourier transforms are performed in software. In contrast, when the field is extracted from a spectral interferometry measurement, the data are represented as a function of ω_t directly; only the transform with respect to τ remains to be computed.

The 2D spectrum may be interpreted, approximately, by associating the Fourier transform with respect to τ with the absorption spectrum sensed by the writing pulses and the Fourier transform with respect to t with the coherent emission spectrum of the four-wave mixing signal. The latter is related to the differential absorption spectrum sensed by the readout pulse at time T, which is affected by the various relaxation phenomena that may occur during the waiting period. Since S_{2D} is a representation of the full complex field, it has both real and imaginary parts, associated, respectively, with absorption and dispersive (refractive index) effects. Experimentally, significant care must be taken to determine the phase zero of the reference pulse, which is needed in order to correctly separate the real and imaginary parts of the 2D spectrum. (As in any interferometry technique, the experimentalist must also ensure that the phase remains sufficiently stable during the course of the measurements.)

Figure 9.34 shows a highly schematic view of the real, or absorptive, part of a 2D spectrum for fixed waiting time T. Important points are as follows:

- Individual transitions, as would be seen in a linear absorption spectrum, appear as features along the $\omega_\tau = \omega_t$ diagonal. Such features may have both different magnitude and different sign, which indicate, respectively, the strength of the effect and its character (induced bleaching or induced absorption). In the figure each of the diagonal features (labeled 1, 2, and 3) are shaded in the same way to indicate that their sign is the same. (Typically, along the diagonal we expect to see induced bleaching.)
- Shapes of individual features provide information about homogeneous and inhomogeneous broadening. In the figure feature 1 is drawn as a circle, which signifies homogeneous broadening. Features 2 and 3 are depicted as ellipses elongated along

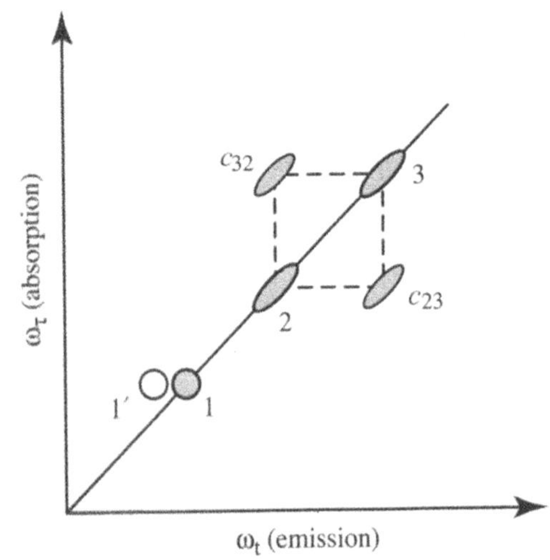

Figure 9.34 Two-dimensional absorption spectrum. Shading (features 1, 2, 3, c_{23}, and c_{32}) represents a positive sign, corresponding to induced bleaching. The unshaded feature (feature 1′) corresponds to a negative sign (i.e., induced absorption). Dashed lines connecting cross-peaks c_{23} and c_{32} to diagonal features 2 and 3 are a guide to the eye to emphasize the coupling.

the diagonal. Such elongation indicates inhomogeneous broadening. Major and minor axes of the ellipses give inhomogeneous and homogeneous contributions to the linewidth, respectively.

- If similar 2D spectra were portrayed for increasing values of T (not shown), we might expect to see elliptical features 2 and 3 spread in the direction perpendicular to the diagonal. This would be a signature of spectral cross-relaxation. With complete spectral cross-relaxation, features 2 and 3 would become circles, signifying complete loss of memory of the original inhomogeneous distribution with sufficiently large waiting time.
- Features that do not fall along the diagonal provide new information unavailable from time-integrated four-wave mixing. Feature 1′ is drawn without shading to signify that its sign is opposite that of feature 1. This could represent an excited-state absorption. The position of feature 1′ (i.e., $\omega_t < \omega_\tau$) corresponds to an absorption change induced at a frequency lower than the frequency of the radiation inducing the change. Such behavior might arise in infrared spectroscopy, where as a result of anharmonicity, transition frequencies decrease as one climbs a vibrational ladder.
- The feature labeled c_{23} is positioned at the same absorption frequency as feature 2 and at the same emission frequency as feature 3. The converse is true for feature c_{32}. These *cross-peaks* signify coupling, represented in the figure by dashed lines, between diagonal transitions 2 and 3. Such cross-peaks constitute a key aspect of 2D spectroscopy that helps to extract microscopic information from the nonlinear response. In infrared 2D spectroscopy, for example, features 2 and 3 may represent vibrational modes within a protein molecule, which are anharmonically coupled to yield the cross-peaks. Because such coupling may depend strongly on the separation and relative orientation of the vibrational modes, information on molecular structure is revealed through the strengths of the cross-peaks.

Another interesting point of two-dimensional spectroscopy is its generality: many common forms of nonlinear spectroscopy can be extracted from the 2D spectrum. One obvious example is time-integrated three-pulse four-wave mixing, which is given by

$$U(\tau; T) = \int dt\, |s(\tau, t; T)|^2 \tag{9.145}$$

where $s(\tau, t; T)$ is related to the 2D spectrum $S_{2D}(\omega_\tau, \omega_t; T)$ through eq. (9.144). Time-integrated two-pulse four-wave mixing is obtained similarly by setting $T = 0$. Time-resolved hole burning and spectrally resolved pump–probe signals, such as those shown in Figs. 9.13 and 9.19, respectively, may also be computed from the 2D spectrum. For example, assuming an impulsive pump pulse, the results of a spectrally resolved pump–probe experiment should be given by [566]

$$S_{srpp}(\omega_t, T) \sim \omega_t \,\mathrm{Re}\left\{ E_{pr}(\omega_t) \int d\omega_\tau\, S_{2D}(\omega_\tau, \omega_t; T) \right\} \tag{9.146}$$

Here $E_{pr}(\omega_t)$ represents the spectral amplitude of the probe field, which may be omitted if the probe is sufficiently broadband and has flat phase. In the case of measurements over a very broad spectral range, the ω_t prefactor is needed to correctly convert the polarization response of the medium to absorption; compare eq. (5.17). Since spectrally resolved pump–probe is insensitive to the phases of pump and probe signals, eq. (9.146) is of practical use in phasing of 2D spectra. That is, one may perform independent spectrally resolved pump–probe and 2D spectroscopy experiments and then adjust the reference phase during the software processing of the 2D data until eq. (9.146) is satisfied.

Figure 9.35 shows an example of experimental infrared 2D spectra, obtained for deuterated phenol dissolved in a mixed benzene–CCl_4 solvent. The linear vibrational absorption spectrum (Fig. 9.35a) of the OD stretch of phenol in the mixed solvent shows two distinct peaks: one at lower frequency associated with a bound phenol–benzene complex, the second at higher frequency associated with free (unbound) phenol. For comparison, spectra of phenol in pure benzene (dominated by phenol–benzene complexes) and in pure CCl_4 (dominated by free phenol) are also shown. The 2D spectra provide insight into the dynamics of the complexation process, which plays an important role in many aspects of chemistry. Figures 9.35b and c show the 2D spectra, specifically $\mathrm{Re}\{S_{2D}(\omega_\tau, \omega_t; T)\}$, for the mixed-solvent case. For a short waiting time, $T = 200$ fs, the spectrum shows a pair of inhomogeneously broadened diagonal peaks, representing bleaching of the $0 \rightarrow 1$ transition of free and complexed phenol, respectively. A similar set of peaks appear below the diagonal, but with opposite sign, representing induced absorption on the $1 \rightarrow 2$ free and complexed transitions. (Note that the roles of the horizontal and vertical axes in Fig. 9.35 are switched compared to those in Fig. 9.34.) No cross-peaks are evident; on this fast time scale there is no significant coupling or exchange between free and complexed species. The

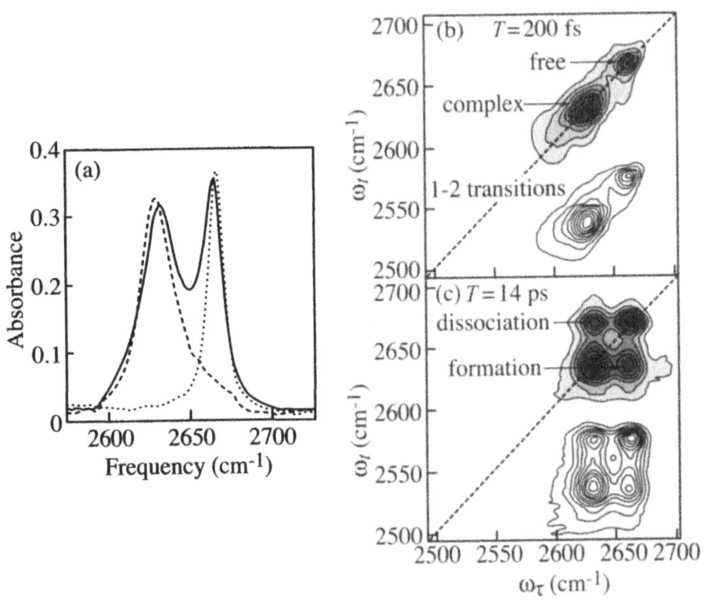

Figure 9.35 Two-dimensional spectroscopy of solute–solvent complexation: (a) linear infrared absorption spectra for phenol in mixed benzene–CCl_4 solvent (solid line), phenol–benzene complex (dashed line), and free phenol (dotted line); (b) real part of the 2D spectra for phenol in mixed benzene–CCl_4 for two different waiting times, $T = 200$ fs (upper) and $T = 14$ ps (lower). In the 2D spectra, shading represents induced bleaching; unshaded contours represent induced absorption. From [568].

situation is quite different for longer waiting time, $T = 14$ ps. Now both the $0 \to 1$ transition and its $1 \to 2$ replica show four distinct peaks. The original peaks have become round, indicating essentially complete spectral cross-relaxation. Most important, fully developed cross-peaks are now apparent. One such peak may be attributed to initial absorption (ω_τ) by the phenol–benzene complex; the complex subsequently dissociates to yield vibrationally excited free phenol, hence an upshifted ω_t bleaching signal. Conversely, the other cross-peak corresponds to initial absorption by free phenol, which subsequently binds to benzene to form a complex; this yields an ω_t bleaching signal at lower frequency than the initial absorption. These data demonstrate that exchange between free and complexed phenol in this mixed solvent occurs significantly faster than 14 ps. Additional studies performed for intermediate values of T indicate an 8-ps characteristic time scale for the phenol–benzene bond formation and bond-breaking dynamics [568].

9.7 IMPULSIVE STIMULATED RAMAN SCATTERING

In previous sections we treated coherent, resonant light–matter interactions, in which real absorption takes place. Here we discuss an interesting coherent but nonresonant effect in short-pulse spectroscopy, in which refractive index dynamics play the key role. The case of a delayed nonlinear index response was covered in Section 6.7.3. In this earlier section we adopted a nonlinear response function formalism and emphasized the effect of noninstantaneous index on the short pulse and the relation to frequency domain Raman spectroscopy. Here we take a perspective focusing more on coherent excitations, such as phonons and molecular vibrations and rotations, that may be induced in the matter by a short pulse. We will be particularly interested in the case of a coherent matter excitation that has a relatively sharp linewidth, corresponding to a well-defined oscillation frequency, interacting with a light pulse with duration shorter than the oscillation period of the coherent matter excitation. This class of interaction is commonly termed *impulsive stimulated Raman scattering* (ISRS).

Raman scattering may be treated classically in terms of the nonlinear polarization associated with a mode with normal coordinate Q [131]. As a specific case we consider vibrations of a collection of noninteracting molecules, for which Q is the deviation of the molecular coordinate (e.g., internuclear distance) from its equilibrium value. However, the treatment is general and may be applied to optical phonon modes in crystals, molecular rotations, and other coherent excitations. In a simple scalar picture, the polarization is taken as $P = N\alpha E$, where N is the volume density of molecules and α is the polarizability of an individual molecule. For Raman scattering to occur, the polarizability must vary with the molecular motion, that is,

$$\alpha = \alpha_0 + \left(\frac{\partial \alpha}{\partial Q}\right) Q \tag{9.147}$$

Here α_0 is the polarizability at the equilibrium molecular position and $\partial\alpha/\partial Q$ is the molecular differential polarizability. The equation of motion for the molecular vibrations is

$$\frac{d^2 Q}{dt^2} + 2\gamma \frac{dQ}{dt} + \Omega^2 Q = \frac{1}{2M}\left(\frac{\partial \alpha}{\partial Q}\right)\langle E^2 \rangle \tag{9.148}$$

Here M is the reduced mass, Ω is the frequency of the vibrational mode, γ represents phenomenological damping (e.g., vibrational dephasing), and the angular brackets denote time averaging over the optical period. The right-hand side of this equation represents a force term by which the field drives the vibrational motion and is obtained from the relation

$$F = \frac{\partial W_{\text{pol}}}{\partial Q} = \frac{\partial}{\partial Q}\left\{\frac{1}{2}\left[\alpha_0 + \left(\frac{\partial \alpha}{\partial Q}\right) Q\right]\left\langle E^2\right\rangle\right\} \tag{9.149}$$

That is, the force F exerted is equal to the derivative of the energy W_{pol} associated with the polarization of a molecule with respect to its normal coordinate Q. Practically, this means that the force driving molecular vibrations is proportional to optical intensity, $F \sim I(t)$.

ISRS may be analyzed by considering eq. (9.148) for the case of an input pulse short compared to the vibrational period [569,570]. In particular, for delta function input, the first term on the left-hand side of eq. (9.148) dominates, and the equation is easily integrated to yield the vibrational velocity immediately after the driving pulse:

$$\left.\frac{dQ}{dt}\right|_{t=0^+} = \frac{1}{2M}\frac{\partial \alpha}{\partial Q}\int dt \left\langle E^2\right\rangle \tag{9.150}$$

Subsequent to the excitation pulse, the vibrational motion exhibits damped free oscillations, with eq. (9.150) as the initial condition. Working in a retarded time frame that moves with the group velocity of the driving pulse, we have

$$Q \approx \frac{1}{2M\Omega}\frac{\partial \alpha}{\partial Q}\left(\int dt \left\langle E^2\right\rangle\right)\sin(\tilde{\Omega}t)e^{-\gamma t} \tag{9.151}$$

Here $\tilde{\Omega}^2 = \Omega^2 - \gamma^2$, which gives $\tilde{\Omega} \approx \Omega$ for damping slow compared to the vibrational period.

Equation (9.151) describes coherent vibrational motion initiated through the action of a sufficiently short optical excitation pulse. The motion is coherent in the sense that there is a macroscopic vibrational amplitude across the sample and the oscillations possess a well-defined phase with respect to the excitation pulse. In particular, immediately after a delta-function input pulse, the vibrational amplitude is zero (molecules are still in their equilibrium positions), but the vibrational velocity is at its maximum. This is in contrast to wave packets excited in molecules through absorption of a short pulse (Section 9.5), for which the vibrational amplitude (velocity) is maximum (zero) immediately subsequent to the laser pulse. In ISRS the vibrational amplitude is proportional to the excitation pulse energy and (again unlike interactions involving resonant absorption) is not sensitive to optical phase. The assumption of an input pulse short compared to the vibrational period implies that the optical spectrum has bandwidth large compared to vibrational frequency. This provides a connection to the traditional frequency-domain view of stimulated Raman scattering, in which two lasers tuned for a frequency offset of Ω mix to drive the vibration. In ISRS, pairs of frequencies with frequency difference Ω may be found within the bandwidth of a single short pulse.

It should be noted that ISRS is effective only when the pulse duration is short compared to the vibrational period. Considering eq. (9.148) in the frequency domain, we find that the induced vibrational amplitude should be proportional to the Fourier transform of the forcing term, $\tilde{I}(\omega) = \int dt\, I(t)e^{-j\omega t}$, evaluated at $\omega = \tilde{\Omega}$. Note that $\tilde{I}(\omega)$ is a baseband function. As the pulse duration approaches the vibrational period, the bandwidth of $I(t)$ becomes less than the vibrational frequency and the driving term $\tilde{I}(\Omega)$ becomes small.

Several techniques are available to probe the coherent motion. One possibility is to use the transient grating geometry of Fig. 9.24b. Here pulses 1 and 2, set for zero relative delay ($\tau = 0$), excite the coherent motion, and pulse 3 reads it out as a function of delay T. In the transient grating geometry, the amplitude of the coherent motion is spatially modulated according to the spatially modulated intensity of the excitation pulse pair:

$$Q(\mathbf{r}, t) \sim \frac{\partial \alpha}{\partial Q}(1 + \cos[(\mathbf{k_1} - \mathbf{k_2}) \cdot \mathbf{r}]) \sin(\tilde{\Omega} t) e^{-\gamma t} \tag{9.152}$$

Since the change in refractive index induced by the coherent motion is proportional to $N(\partial\alpha/\partial Q)Q$, eq. (9.152) represents a dynamic refractive index grating that will diffract an appropriately timed probe pulse. Recording the diffracted energy in either of output directions $\mathbf{k_{out}} = \mathbf{k_3} \pm (\mathbf{k_2} - \mathbf{k_1})$ provides a measure of $Q(t)$. Mathematically, the interaction of the probe pulse with the index grating produces an optical polarization proportional to

$$P_{\mathbf{k}_{\text{out}}}(t) \sim N \frac{\partial \alpha}{\partial Q} Q(t) a(t - T) \tag{9.153}$$

According to eq. (9.130), the scattered intensity is proportional to the squared polarization magnitude, which is then integrated over time for a slow detector. For a probe pulse much shorter than the vibrational period, we may evaluate $Q(t)$ at the time of the probe pulse; the scattered energy is approximately proportional to $|Q(T)|^2$.

Figure 9.36a shows ISRS data for α-perylene molecular crystal detected in the transient grating geometry [571–573]. The signal consists of a sharp peak at $t = 0$, arising from the essentially instantaneous nonlinearity associated with distortion of the electronic wave function, followed by a long tail corresponding to optical phonons excited through ISRS. A

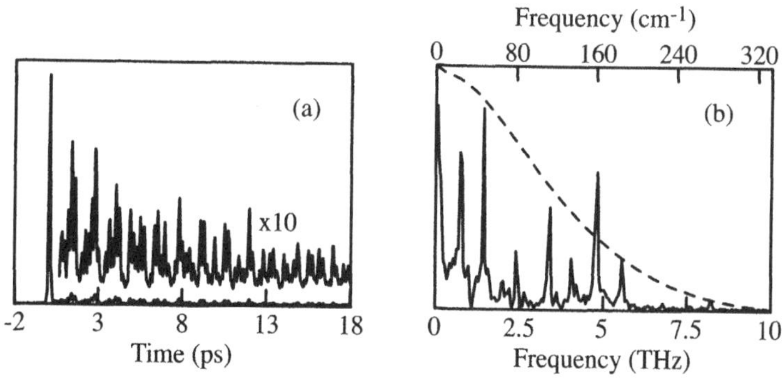

Figure 9.36 (a) Single-pulse ISRS scattering data for α-perylene molecular crystal at low temperature. The femtosecond excitation drives several phonon modes whose combined response yields the beating pattern observed following the strong initial electronic response. (b) Fourier transform of the scattering data with the electronic scattering peak suppressed (solid line) and of the electronic scattering peak (dashed line). From [573].

few points are of note. First, the intensity of the ISRS signal is relatively weak, approximately 4% of that arising from the electronic scattering peak. Second, the ISRS diffraction intensity presents an irregular pattern that is caused by beating among several simultaneously excited phonon modes. A Fourier transform of the scattering data (with the electronic peak removed) is shown in Fig. 9.36b. Because the scattering signal is proportional to the square of the time-dependent refractive index change induced by the phonons, the Fourier transform consists of a complicated series of lines at frequencies corresponding to sum and difference frequencies of perylene modes at 33, 56, 80, and 104 cm^{-1} (which are known from spontaneous Raman scattering). Lines are also visible at some of the fundamental phonon frequencies because of mixing with a nonoscillatory component of the signal arising from electronic excited states generated through two-photon absorption. As a comparison, the dashed curve in Fig. 9.36b shows the Fourier transform of the electronic scattering peak. This curve represents the instrumental response arising from the 75-fs duration of the excitation and probe pulses. For this pulse duration the ability to excite and monitor coherent vibrational motion rolls off for frequencies above about 3 THz (100 cm^{-1}), although well-resolved features are evident out to 5.5 THz.

Although such data clearly demonstrate the ability to excite and monitor coherent vibrational motion through ISRS, they also reveal some limitations:

- Small vibrational response, at least in comparison with the electronic scattering peak, which provides an internal intensity calibration
- Lack of mode selectivity (a short pulse excites any Raman-active mode whose vibrational frequency falls within the instrumental response)
- Difficulty in determining the individual excitation amplitudes of different phonons, individual phonon decay times, and so on

These issues may be addressed by driving the system with a train of pulses [572,573]. When the repetition rate of the pulse train matches the vibrational frequency of a particular mode, that mode is selectively amplified. This process is closely analogous to the way a child on a swing may be pushed repetitively to reach an increased amplitude of oscillatory motion. Because the forcing function for a pulse train becomes spectrally narrow, only vibrational frequencies matched to the pulse repetition rate (or its harmonics) are effectively excited. This forms the basis for mode selectivity.

These points are demonstrated by the data of Fig. 9.37 [572,573]. Figure 9.37a shows the intensity profile of a 2.39-THz (79.6-cm^{-1}) excitation-pulse train generated via an optical pulse shaper and timed to match the period of the 80-cm^{-1} vibrational mode. Figure 9.37b shows the resulting ISRS data, read out as usual using a single probe pulse. Initially, the signal is dominated by the crystal's electronic response to each of the pulses in the sequence. However, the vibrational signal builds up as the pulse train progresses. By the end of the pulse train, the ISRS signal is approximately as intense as the strongest electronic scattering peak. Furthermore, in contrast to the single-pulse experiments, the ISRS signal after the pulse train has ended appears to be oscillatory at a single frequency (equal to twice the vibrational frequency), and the dephasing of the excited mode is now clearly discernible. These results constitute a simple example of the use of shaped femtosecond waveforms for control over molecular motion.

The coherent motion excited through ISRS may also be sensed via the frequency shifts it imparts onto a probe beam [569]. We have seen that the quantity $N(\partial\alpha/\partial Q)Q(t)$ represents

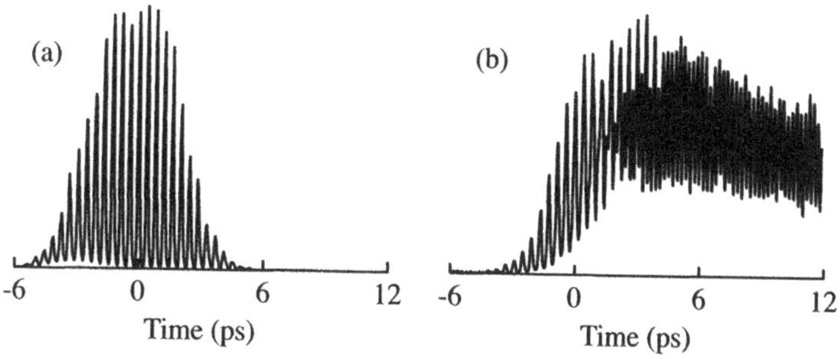

Figure 9.37 Multiple-pulse ISRS scattering data for α-perylene molecular crystal at low temperature: (a) pulse train used for repetitive excitation; (b) scattered energy as a function of probe pulse delay T, demonstrating selective amplification of the 80-cm^{-1} mode. From [573].

a time-varying contribution to the refractive index, which may therefore act as a source of phase modulation. One can show that the self-action effect of this phase modulation on the same single short pulse initiating the coherent motion is always a spectral red shift. This result is independent of the sign of the molecular differential polarizability, since the phase modulation is proportional to $(\partial\alpha/\partial Q)^2$; compare eqs. (9.151) and (9.153). In contrast, in experiments using a distinct probe pulse with duration short compared to the vibrational period, the probe pulse may be spectrally shifted either to the red or to the blue, or may remain unshifted, depending on its delay.

The experiments described so far were performed in the weak excitation regime, corresponding to small vibrational amplitudes and small effects of the coherent motion on the light pulses. However, by using sufficiently intense amplified femtosecond systems, together with gas-phase samples where optical damage is not a primary concern, strong excitation is possible. In this regime coherent vibrational or rotational motions may reach large amplitude, leading to large swings in the refractive index. This has led to interest in *molecular modulators*, which provide refractive index modulation at very rapid rates, corresponding to the frequency of the excited vibrational or rotational mode [570,574,575]. Such molecular modulators can produce dramatic effects on a probe pulse.

In one example, 30-fs, 130-μJ pulses at 800-nm wavelength excited the 775-cm^{-1} symmetric stretch vibrational mode in gaseous SF_6 (44-fs vibrational period) [570]. This highly symmetric molecule was chosen because it exhibits no rotational Raman spectrum and only a single strong vibrational line. Thus, a clean single-frequency coherent motion is excited despite the lack of mode selectivity in ISRS. Weaker probe pulses at 400 nm were generated via second-harmonic generation. Due to group velocity mismatch in the SHG crystal (see Section 5.4), the probe pulses were broadened to several hundred femtoseconds. Data showing the output probe spectrum is presented in Fig. 9.38. Several points are of interest:

- The spectral broadening takes the form of discrete sidebands. Because the probe pulse is now longer than the vibrational period, the refractive index associated with the molecular coherence runs through several modulation cycles within the duration of the probe. Hence, the generation of spectral sidebands reflects the resulting quasi-periodic refractive index (hence, phase) modulation experienced by the probe field.

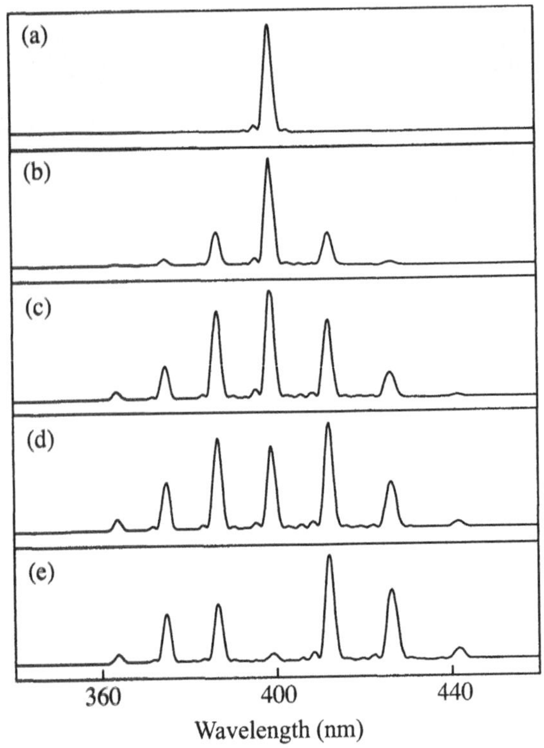

Figure 9.38 Measured output spectra of a 400-nm probe pulse spectrally broadened via interaction with SF_6 gas that has been vibrationally excited through ISRS. The broadening depends on the gas pressure: (a) $p = 0$; (b) $p = 346$ mbar; (c) $p = 395$ mbar; (d) $p = 410$ mbar; (e) $p = 470$ mbar. From [570], with permission. Copyright © 1999, American Physical Society.

- The spectral broadening is approximately symmetric, with sidebands generated equally to the red and to the blue. This is in contrast to most traditional Raman scattering experiments, where spectra are shifted preferentially to the red. In this experiment the quasi-periodic index variation associated with the molecular modulator shifts the spectrum alternately to the red and to the blue, leading to frequency generation on both sides of the original spectrum.
- The effects are large. At the highest pressure six new sidebands are visible, and the energy in the original probe spectrum is strongly depleted. This indicates that the phase modulation reaches an amplitude of several π. Furthermore, there is a notable increase in bandwidth, even beyond the bandwidth of the pump pulse.

The bandwidth enhancement produced via ISRS excitation of such a molecular modulator, followed by means for appropriate spectral phase control, provides the possibility of pulse compression. Several experiments have demonstrated that such pulse compression may be realized [574,576,577], with compression down to the few-cycle range [577]. Advantages of this approach to pulse compression include both high power-handling capability and the ability to work in spectral regions, such as the ultraviolet, that are difficult to access via other techniques. It is worth noting that standard two-color Raman excitation schemes have also been used extensively for driving such molecular modulators. A sufficiently large number of frequency sidebands has been generated to realize field transients as fast as a single optical cycle [575].

PROBLEMS

9.1. Consider pump–probe transmission experiments in which induced absorption changes remain small, $|\Delta\alpha L| \ll 1$, but where the static absorption $\alpha_0 L$ may not be considered small. How should eqs. (9.2) and (9.3) be modified? Derive an expression, accurate to first order in $\Delta\alpha$, that gives the output field from a sample of length L in terms of the input field. Can you express your results in terms of an effective length L_{eff}?

9.2. Show that the coherent coupling term $\beta(\tau)$ arising in degenerate pump–probe measurements, eq. (9.7c), is symmetric with respect to τ. Assume the nonlinear polarization to be purely imaginary (pure absorption response).

9.3. Show that when the absorption response function is a unit step function, the coherent coupling term arising in degenerate pump–probe measurements is proportional to the squared envelope of the electric field autocorrelation function. That is, derive eq. (9.10).

9.4. Analyze the coherent coupling term as given by eq. (9.16c). Assume parallel linear polarizations for pump and probe pulses, but allow both imaginary and real parts of the nonlinear response function $\tilde{\chi}^{(3)}(t)$ to be nonzero.

(a) Show that the contribution to the coherent coupling arising from the real part of the nonlinear response is an odd function of delay.

(b) Generate plots of the full pump–probe signal vs. delay, both for bandwidth-limited Gaussian pulses and for Gaussian pulses with appreciable chirp, assuming that both real and imaginary parts of the nonlinear response are step functions (but with various relative amplitudes). Comment on your results. Orientational relaxation may be neglected.

9.5. Analyze the pump–probe absorption saturation signal for the case of linear dye molecules with orientational relaxation, as in Section 9.2.2.

(a) The pump is linearly polarized along $\hat{z}$, while the probe is circularly polarized in the x–z plane.

(b) Both pump and probe have identical right-hand-circular polarization.

9.6. Consider a solution of ideal planar absorbing molecules (light polarized anywhere in the plane of the molecule is absorbed equally strongly, but light polarized perpendicular to the plane is not absorbed at all). As in Section 9.2.2, work out the various components of the nonlinear response tensor, both with full orientational memory and with full orientational relaxation. For what "magic angle" does the effect of orientational relaxation drop out of the pump–probe data?

9.7. From the time-resolved fluorescence spectra of Fig. 9.16a, estimate the electron temperatures in the experiment of [526] at time delays of 1, 2, and 10 ps.

9.8. In the perturbation theory analysis of the two-level atom, the upper-level population change to second order in the field is obtained as $|c_2^{(1)}|^2$, with $|c_2^{(1)}|$ evaluated using eq. (9.55). The lower-level population change may be obtained either from $1 - |c_2^{(1)}|^2$ or by taking the perturbation theory calculation to second order to get $c_1^{(2)}$ directly. Show that these two methods of calculating the lower-level population change yield the same result.

9.9. Simplify eq. (9.75) for second-order population change assuming that the dephasing time T_2 is much less than the pulse width t_p. Consider two cases: **(a)** $T_1 \ll t_p$ and **(b)** $T_1 \gg t_p$. Discuss the meaning of your results physically.

9.10. Consider static inhomogeneous broadening, where the transverse relaxation function $h_T(t)$ and the matrix element μ_{21} are both allowed to be functions of the frequency shift (ω_{21}). By performing an analysis similar to that in Section 9.6.2, work out the relationship between the overall polarization relaxation function and the overall absorption spectrum.

9.11. Consider the expansion of the relaxation function in the stochastic oscillator model, as in eqs. (9.113b) and (9.114), but now without the assumption that the frequency fluctuations $\delta\omega$ are Gaussian. The assumption that the fluctuations are zero mean is retained. Are the odd terms in the expansion of the exponential still zero? Justify your answer. If the answer is "no," construct an example of a random process for which the odd terms do not vanish.

9.12. Analyze the strength of the impulsive stimulated Raman scattering (ISRS) signal in the transient grating geometry as a function of pulse width. Assume Gaussian excitation pulses. Consider two cases: **(a)** the probe pulse is very short; **(b)** the probe pulse is identical to the excitation pulses.

9.13. Show that when a single ultrashort pulse induces coherent vibrational motion in a Raman-active medium through ISRS, the pulse acquires a spectral red shift due to the resulting phase modulation.

9.14. Consider a long probe pulse that has been spectrally broadened due to the influence of coherent vibrational motion in a gas sample.

(a) Assuming that the vibrational motion is harmonic, estimate the peak nonlinear phase shifts needed to yield the various spectra shown in Fig. 9.38.

(b) Assuming that the spectral phases are fully compensated, work out or estimate the resulting compressed probe waveforms.

10

TERAHERTZ TIME-DOMAIN ELECTROMAGNETICS

Mode-locked lasers are capable of generating pulses orders of magnitude shorter than the time scales accessible by electronic circuit technologies. By using ultrafast optics in conjunction with appropriate electronic materials, it becomes possible to generate and characterize electrical transients with durations in the deep subpicosecond range and with terahertz bandwidths. This is well beyond what is possible using direct electronic solutions. Accordingly, application of ultrafast optical tools to problems in electronics and electromagnetics has attracted considerable research and practical attention and has become an active field in its own right. A useful early review is given in [578]. Here we refer to this field as *terahertz time-domain electromagnetics*.

In this chapter we provide a brief survey of selected topics in terahertz time-domain electromagnetics, with an emphasis on basic techniques. First we discuss optical generation and measurement of ultrafast electrical signals on transmission lines. These represent simple circuit structures. We then cover excitation and detection of freely propagating terahertz radiation (terahertz beams), including application to terahertz time-domain spectroscopy.

10.1 ULTRAFAST ELECTROMAGNETICS: TRANSMISSION LINES

10.1.1 Photoconductive Generation and Sampling

Optical generation of ultrafast transients in electrical circuits has relied principally on photoconductors. Usually, the photoconductor is incorporated into a gap in a high-speed microwave transmission line to form an optically actuated switch. In general, the advantage of a transmission-line geometry is that the capacitances and inductances associated with the electrodes are distributed and couple to support traveling-wave voltages and currents,

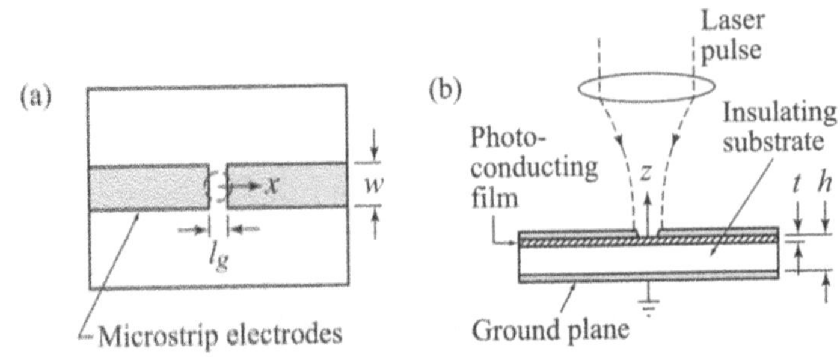

Figure 10.1 Photoconductive switch in microstrip transmission-line geometry: (a) top view; (b) side view. Adapted from [579]. Copyright © 1983, IEEE.

with no intrinsic limit to response time. Detailed treatments of transmission lines may be found in standard microwave texts (e.g., [5]).

A basic photoconductive switch geometry compatible with integrated-circuit fabrication techniques is shown in Fig. 10.1 [579,580]. A thin photoconductive film is grown on an insulating substrate and then metalized to form a microstrip transmission line. The microstrip structure consists of a thin metal line and a conducting ground plane, which are fabricated on opposite surfaces of the substrate. Let us assume that the transmission line is biased at one end with a constant-voltage source V_b. In the absence of optical excitation, the bias voltage drops primarily across the gap, provided that the dark conductivity of the photoconductor is sufficiently low. At $t = 0$ a short optical pulse illuminates the photoconductor, resulting in generation of mobile charge carriers and suddenly increased conductivity. In effect, this switches a portion of the bias voltage from the gap onto the initially unbiased section of transmission line. This results in voltage and current transients that travel down the transmission line away from the photoconductive gap at a velocity equal to the speed of light divided by the effective refractive index. The rise time and duration of the electrical transient depends on the optical pulse duration, the photoconductor recovery time, and circuit effects associated with the capacitance of the photoconducting gap. For photoconductors with long recovery times, such as high-quality semi-insulating semiconductors with typical nano-second carrier lifetimes, the photoconductive switch generates steplike electrical transients. On the other hand, shorter carrier lifetimes, which may reach the subpicosecond range, may be obtained by ion implantation or by special growth conditions in order to introduce carrier trapping centers into the semiconductor [581–583]. With sufficiently fast recovery times, the photoconductive switch generates impulsive electrical transients.

An analysis of such photoconductive switches is given in [579]. The framework, shown in Fig. 10.2, is that of a small photoconductive gap inserted into an otherwise ideal transmission

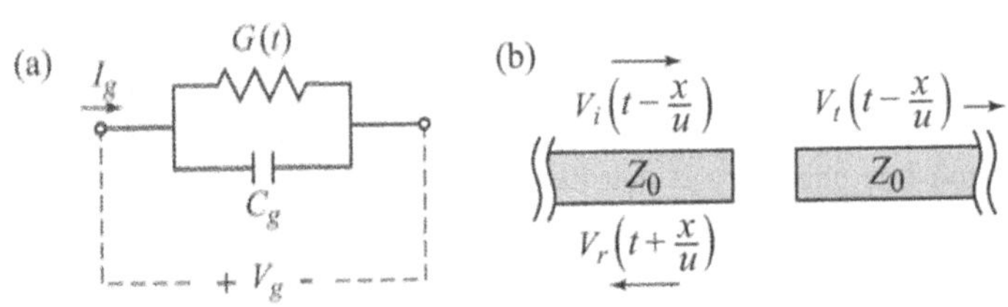

Figure 10.2 Analysis framework for photoconductive switch: (a) lumped-circuit-element model of photoconducting gap; (b) transmission-line model with embedded gap.

line at position $x = 0$. The gap is modeled as a lumped circuit element consisting of a resistor with conductance $G(t)$ (conductance is the inverse of resistance) in parallel with a capacitance C_g. The capacitance accounts for charges that pile up on the electrodes at either end of the gap and is typically small, on the order of tens of femtofarads. The lumped-element treatment assumes a gap sufficiently small that retardation effects associated with finite propagation time across the gap may be neglected. The current I_g into the gap and the voltage V_g across the gap are then related by

$$I_g(t) = C_g \frac{dV_g}{dt} + G(t)V_g(t) \tag{10.1}$$

For our purposes the conductance may be approximated simply as

$$G(t) = \frac{q[\mu_e n_e(t) + \mu_h n_h(t)]wt}{l_g} \tag{10.2}$$

where μ_e, μ_h, $n_e(t)$, and $n_h(t)$ are the mobility and volume density of photogenerated electrons and holes, respectively, and q is the charge of an electron, taken as positive. The length and width of the photoconducting gap are l_g and w, respectively, and the thickness of the photoconducting film is t. This formula assumes that the carrier densities and the electric field are spatially uniform and that the field is not altered by carrier injection. Furthermore, ultrafast carrier dynamics such as momentum relaxation that might lead to time-dependent mobilities are not considered. A more general expression for the conductance, which applies when these simplifying conditions do not hold, is given in [579].

Voltage and current transients on the transmission lines are modeled as waves traveling at velocity u. The voltage and current signals on the left and right sections of transmission line are written

$$V(x < 0, t) = V_i\left(t - \frac{x}{u}\right) + V_r\left(t + \frac{x}{u}\right) \tag{10.3}$$

$$V(x > 0, t) = V_t\left(t - \frac{x}{u}\right) \tag{10.4}$$

$$I(x < 0, t) = I_i\left(t - \frac{x}{u}\right) + I_r\left(t + \frac{x}{u}\right) \tag{10.5}$$

$$= \left\{V_i\left(t - \frac{x}{u}\right) - V_r\left(t + \frac{x}{u}\right)\right\} \Big/ Z_0 \tag{10.6}$$

$$I(x > 0, t) = I_t\left(t - \frac{x}{u}\right) = \frac{V_t\left(t - \frac{x}{u}\right)}{Z_0} \tag{10.7}$$

Here V_i, V_r, and V_t are voltage waves excited by a source at the left and incident onto the gap, reflected from the discontinuity at the gap and traveling back toward the left, and transmitted across the gap and traveling to the right, respectively. I_i, I_r, and I_t are the corresponding current waves, and Z_0 is the characteristic impedance which gives the voltage-to-current amplitude ratio for electromagnetic waves supported by the transmission line. The transmission lines are assumed to extend sufficiently far that reflections from the ends may be neglected, except at the gap. This well-known transmission-line formulation may be considered as a voltage–current analog of dielectric mirror analysis (see Section 4.8.2) in terms of forward- and backward-traveling plane wave E and H fields.

These equations are solved by applying boundary conditions at the gap, $x = 0$. The voltage and current boundary conditions are

$$V_i(t) + V_r(t) = V_g(t) + V_t(t) \tag{10.8}$$

and

$$\frac{V_i(t) - V_r(t)}{Z_0} = I_g(t) = \frac{V_t(t)}{Z_0} \tag{10.9}$$

These equations state that the voltage just to the right of the gap must equal the voltage just to the left of the gap, reduced by the voltage drop across the gap, and that the current across the gap must be continuous. Manipulation of the boundary condition equations yields

$$V_r(t) = \frac{V_g(t)}{2} \qquad \text{and} \qquad V_t(t) = V_i(t) - \frac{V_g(t)}{2} \tag{10.10}$$

Finally, by combining these expressions with eq. (10.1), we obtain an equation governing the gap voltage,

$$\frac{dV_g}{dt} + \frac{1 + 2Z_0G(t)}{2Z_0C_g} V_g(t) = \frac{V_i(t)}{Z_0C_g} \tag{10.11}$$

We gain insight by considering the ideal case of a constant bias V_b, with the conductance stepping abruptly from zero at time $t = 0$ to value G_0. This corresponds to absorption of an ultrashort pulse by a photoconductor with long carrier lifetime. For $t < 0$ the gap is an open circuit and $I_g = 0$. Therefore, initial conditions at $t = 0$ are $V_g = V_b$, $V_r = V_i = V_b/2$, and $V_t = 0$. Since there are no reflections from the source, the value of V_i remains constant, $V_i(t) = V_b/2$ for all t. With these conditions eq. (10.11) has the solution

$$V_g(t > 0) = V_b - \frac{2Z_0G_0V_b}{1 + 2Z_0G_0}\left(1 - e^{-t/\tau_{\text{circ}}}\right) \tag{10.12a}$$

where

$$\tau_{\text{circ}} = \frac{2Z_0C_g}{1 + 2Z_0G_0} \tag{10.12b}$$

is the time constant associated with the circuit response. The voltage wave coupled onto the output transmission line is given by

$$V_t(t > 0) = \frac{Z_0G_0V_b}{1 + 2Z_0G_0}\left(1 - e^{-t/\tau_{\text{circ}}}\right) \tag{10.13}$$

and is steplike with rise time proportional to τ_{circ}. This result clearly demonstrates the importance of circuit effects in limiting electrical response times. We distinguish two limiting cases. When the resistance across the gap remains large compared to the characteristic impedance ($Z_0 G_0 \ll 1$), the voltage coupled onto the output line is proportional to the conductance step but remains small compared to the bias voltage. In this regime the response is linear in the photoconductance, and the time constant is simply $2Z_0 C_g$. For $C_g = 20$ fF and $Z_0 = 50\Omega$, this works out to 2 ps. On the other hand, when $Z_0 G_0 \gg 1$, the output voltage saturates at $V_t = V_b/2$. In this regime the large conductance helps to discharge the gap more quickly, resulting in faster response.

Photoconductive gates can be used not only to generate ultrafast electrical transients, but also to measure them [579,580]. Generally, a correlation approach similar to that used in characterization of ultrashort optical pulses and in time-resolved pump–probe spectroscopy is employed: A first optical pulse generates the electrical transient (perhaps using a photoconductive switch), and a second time-delayed optical pulse is used to sample it. Figure 10.3 shows two representative photoconductive correlation setups. In either case the strategy is to measure the total charge Q_s coupled through the second photoconductive gate (the sampling gate) as a function of the delay τ between the optical pulse pair. Q_s may be written

$$Q_s(\tau) = \int dt\, I_{t,s}(t) = \int dt\, \frac{V_{t,s}(t)}{Z_0} \tag{10.14}$$

where $I_{t,s}$ and $V_{t,s}$ are, respectively, the (transmitted) current and voltage transients on the output side of the sampling gate. $V_{t,s}$ may be analyzed using a formulation analogous to that described above for photoconductive generation and depends in general on the voltage transient $V_{i,s}$ incident on the sampling gate, its time-dependent photoconductance $g_s(t-\tau)$, and associated circuit effects. In a correlation experiment $V_{i,s}(t)$ is a delayed version of the waveform $V_t(t)$ generated by the first photoconductive switch. When the photoconductance is sufficiently small to be treated as a perturbation (similar to the condition $Z_0 G_0 \ll 1$

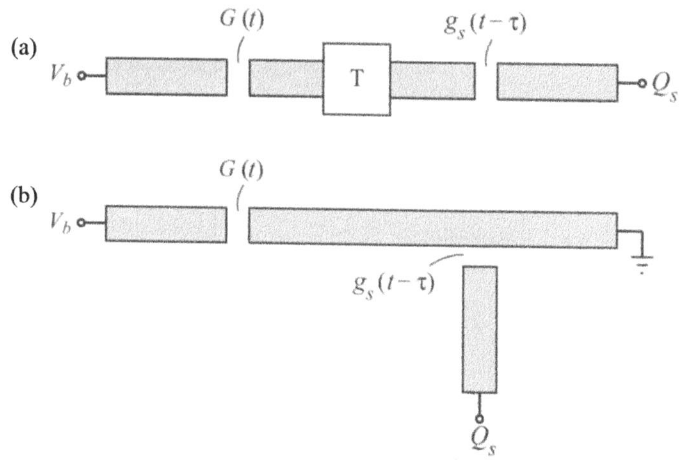

Figure 10.3 Representative circuit geometries for photoconductive electrical generation and sampling. Such setups yield a correlation measurement of the electrical response of the two photoconducting gaps. (a) In-line generation and measurement (the box labeled "T" represents a matched attenuator to prevent reflections and leakage of the bias onto the sampling gap); (b) in-line generation with sampling by a side gap. Adapted from [579]. Copyright © 1983, IEEE.

described earlier), the collected charge resulting from photoconductive sampling is given, to first order in g_s, by

$$Q_s(\tau) \sim \int dt\, g_s(t-\tau) \int dt'\, V_{i,s}(t')h_{\mathrm{sc}}(t-t') \tag{10.15}$$

Here $h_{\mathrm{sc}}(t)$ is the impulse response function representing the sampling gate circuit response (in the absence of induced photoconductance, $g_s = 0$). In this equation the rightmost integral represents the time-dependent voltage across the sampling gap resulting from the incident voltage waveform and the unperturbed circuit response. The instantaneous voltage across the sampling gap is multiplied by the time-dependent photoconductance and then integrated to give first-order current and charge, respectively. The collected charge follows the same basic formula for the in-line and side-gap correlation geometries, except that the circuit response functions are different. In either case both the carrier lifetime of the photoconductor and circuit effects associated with gap capacitance limit the time resolution of the sampling operation.

To generate (and measure) the shortest pulses possible, materials with very short carrier lifetimes should be used, and limitations due to circuit effects should be minimized. Figure 10.4a shows the *sliding contact scheme*, which essentially eliminates circuit limitations, both in generation and in measurement [584,585]. Here the transmission line has the coplanar microstrip structure, consisting simply of two closely spaced metal lines, both on the same (top) surface of a semiconductor wafer. Focusing a laser pulse to a spot between the two (biased) lines generates a transient conductance, leading to generation of a symmetric pair of electrical transients propagating along the transmission line in both forward and backward directions. Because there are no breaks in the transmission line and hence no parasitic capacitance, the electrical pulse duration mirrors the time-dependent conductance.

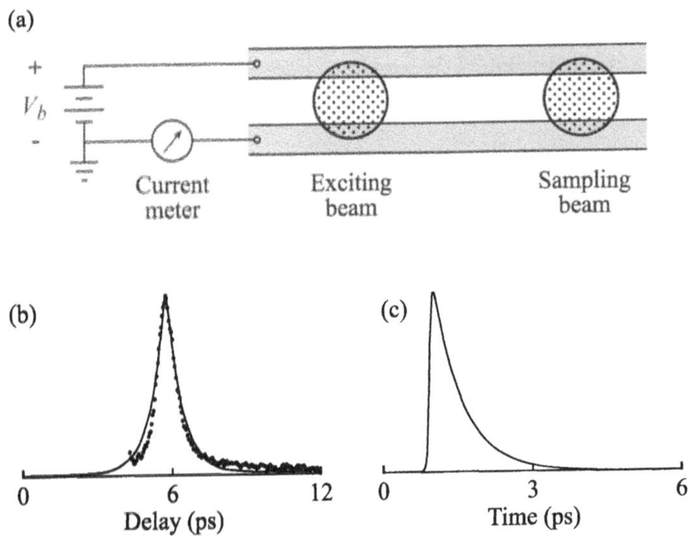

Figure 10.4 (a) Sliding contact geometry for photoconductive electrical generation and measurement; (b) electrical autocorrelation data (symbols) and theoretical autocorrelation (solid line), assuming 130-fs excitation pulse and 600-fs carrier lifetime; (c) electrical pulse calculated based on a theoretical model with parameters as in (b). Adapted from [585]. Copyright © 1988, IEEE.

Photoconductive sampling is performed by focusing a second laser pulse onto another spot between the metal lines and recording the total collected charge as a function of time delay between laser pulses. Practically, this measurement can be performed by chopping the two laser beams at different frequencies and using a lock-in amplifier to detect the modulation in the collected charge at the sum of the chopping frequencies. This implementation eliminates background signals due to transients generated by either of the laser beams acting alone. Figure 10.4b shows data for an electrical pulse generated from a very compact coplanar microstrip sample (linewidths 1.2 μm with 2.4 μm of separation) fabricated on ion-implanted silicon-on-sapphire (carrier lifetime approximately 600 fs) [585]. These data, which constitute an electrical autocorrelation, have a width of 0.85-ps FWHM and are in reasonable agreement with theoretical results based on a 130-fs excitation pulse, a 600-fs carrier lifetime, and zero excess capacitance. The time-dependent voltage pulse calculated from the same theoretical model is shown in Fig. 10.4c. The voltage has a fast rise, reflecting the rapid onset of photoconductance followed by a relatively slower decay on the scale of the carrier lifetime. The extreme speed of this electrical transient may be attributed to several factors: the fast carrier lifetime, the parasitic capacitance-free geometry, the very compact transmission-line structure which leads to low electrical dispersion, and the flexibility provided by the sliding contact scheme to generate and sample at small lateral separations, which minimizes both dispersion and frequency-dependent electrical loss.

10.1.2 Electro-optic Sampling

An alternative means for probing ultrafast electrical transients relies on the electro-optic effect. The electro-optic effect is based on a second-order nonlinear susceptibility (see Chapter 5) of the form $\chi^{(2)}(\omega : \omega, 0)$, in which a low-frequency electric field and an optical field interact to produce a second-order polarization at optical frequency. This process is commonly described in terms of an induced change in the index ellipsoid that is linear in the low-frequency field, with the proportionality specified by an electro-optic tensor. Detailed treatment of the electro-optic effect in general is given in [228], and specific discussions focused on its application to electro-optic sampling of fast electrical signals may be found in [586–589]. Here we simply assume that the low-frequency electric field induces a birefringence (or a change in birefringence) in a suitable electro-optic material. A short optical probe pulse is focused into the material and incurs a change in its polarization state as a result of the induced birefringence. By passing the pulse through a polarizer, the change in polarization is converted into a change in transmitted intensity. As in pump–probe and other ultrashort-pulse measurement techniques, the change in the transmitted probe energy is recorded using a time-integrating detector. This is repeated as the delay between the low-frequency electrical signal and the probe pulse is varied. Thus, in this electro-optic sampling method, the low-frequency field is sampled by a variable delay optical probe pulse via the electro-optic interaction.

Electro-optic sampling has several attractive features. First, it is intrinsically fast and operates well into the terahertz regime. Therefore, in the context of our discussion, subpicosecond electrical transients are considered to be "low-frequency fields." Second, there is substantial flexibility for probing signals at various positions, since no special test structure (such as a photoconductive sampling gap) needs to be incorporated into the circuit. Third, because no charge is extracted, loading of the device under test is usually small. Chief limitations include spatial resolution limited to optical wavelength scale and the need to bring an electro-optic material into proximity with the circuit to be probed.

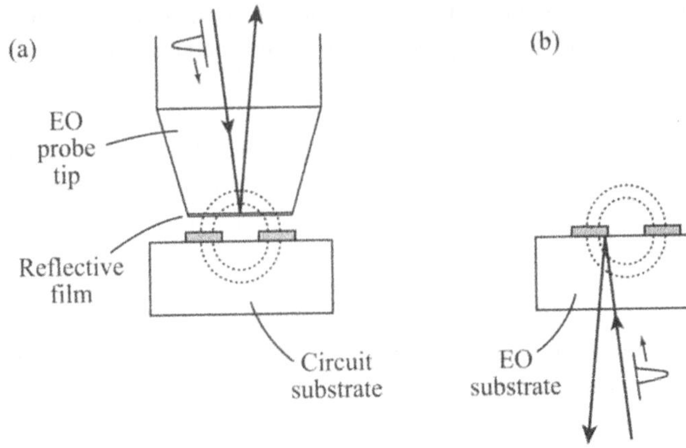

Figure 10.5 Representative geometries for electro-optic sampling of voltage transients in circuits: (a) an external electro-optic probe tip is brought into proximity with the circuit; (b) for certain materials the electro-optic response of the substrate itself may be exploited for electro-optic sampling.

A number of geometries have been employed for electro-optic sampling. Two illustrative examples are depicted in Fig. 10.5. In both cases a pair of metal interconnect lines, as in a coplanar microstrip, are shown on top of a circuit substrate. A fast electrical transient (not shown) is presumed to propagate along the metal lines in a direction normal to the page. In Fig. 10.5a an external probe, with an electro-optic crystal at its tip, is positioned just above the circuit of interest. Electric field lines from the circuit penetrate into the electro-optic tip and induce a birefringence. An optical probe pulse is focused into the tip to sense the birefringence. After reflection the change in probe polarization is analyzed to infer the voltage at the sampling instant. In Fig. 10.5b the circuit substrate itself is assumed to be electro-optic. The probe pulse passes through the back side of the wafer, reflects off one of the metal lines, and emerges from the substrate for analysis. In this case voltage is inferred from the birefringence induced within the circuit substrate. The latter approach has the obvious advantage of being minimally invasive, since no external probe tip is needed. On the other hand, its use is restricted to electro-optically active circuit materials. These include GaAs and other compound semiconductors frequently used for high-speed radio-frequency applications but not silicon, the most important electronic material.

For a simple analysis of electro-optic sampling, we take the direction in which the optical probe beam propagates as $\hat{z}$ and assume that the low-frequency field induces a birefringence in the transverse plane, with principal axes and refractive indices $\hat{x}$ and $\hat{y}$ and n_x and n_y, respectively. The input probe field is adjusted to sense both principal axes and is written

$$a(t-\tau)\frac{\hat{x}+\hat{y}e^{j\phi_0}}{\sqrt{2}} \tag{10.16}$$

where $a(t)$ is a slowly varying amplitude function. After propagation through the electro-optic material, the field becomes

$$\frac{a(t-\tau)}{\sqrt{2}}\left(\hat{x}e^{-j\phi_x}+\hat{y}e^{j\phi_0}e^{-j\phi_y}\right) \tag{10.17}$$

Here $\phi_x = (\omega_0/c) \int n_x(z)\,dz$ and $\phi_y = (\omega_0/c) \int n_y(z)\,dz$ are the phases accumulated for field components polarized along $\hat{x}$ and $\hat{y}$ axes, respectively; and the dependence of n_x and n_y on z reflects the spatial distribution of the low-frequency field inducing the birefringence. Furthermore, since the low frequency field varies according to the time dependence of the electrical transients on the circuit lines, all quantities are implicitly functions of time. The average power in the field transmitted through an analyzer oriented along $(\hat{x} - \hat{y})/\sqrt{2}$ is proportional to

$$\int dt\, |a(t-\tau)|^2 \sin^2\left(\frac{(\phi_x - \phi_y) + \phi_0}{2}\right) \tag{10.18}$$

As usual, the integral over time takes into account the use of a slow (time-integrating) photodetector.

In electro-optic sampling one usually desires that the output signal is linear in the voltage waveform being sampled. To achieve linear response, two conditions should be satisfied. First, an appropriate bias point is needed. Optimum is to bias the setup for 50% transmission, which is achieved for circular input polarization (e.g., $\phi_0 = \pi/2$). Second, the differential phase shift must remain small, $|\phi_x - \phi_y| \ll 1$. When these conditions are satisfied, the average power transmitted through the analyzer is approximately

$$\frac{1}{2}\int dt\, |a(t-\tau)|^2 + \frac{1}{2}\int dt\, |a(t-\tau)|^2 \left[\phi_x(t) - \phi_y(t)\right] \tag{10.19}$$

Provided that the induced phase retardance is itself linear, the second term above is linear in the voltage being sampled, as desired. Such a strategy to achieve linear operation is well known in electro-optic modulator practice. It is, however, in contrast to usual practice in optical characterization via polarization-gating FROG (Section 3.6.1). In PG-FROG the setup is biased at the zero transmission point, and the gated output power is proportional to the square of the gating intensity.

The time resolution of an electro-optic sampling experiment is influenced by several factors. These include:

- *Duration of the optical sampling pulse.* With femtosecond duration pulses readily available, this is usually not the primary limitation.
- *Intrinsic electro-optic response.* Here vibrational resonances, which typically occur in the range 5 to 10 THz in electro-optic crystals, are the main issue. Such resonances lead to frequency-dependent absorption and distortion of electrical transients propagating in the electro-optic material and are linked to noninstantaneous electro-optic response. These effects are usually not serious for measurement of transients on transmission lines and in circuits, where the electrical bandwidths remain below the resonance frequencies. However, such resonances can be a significant issue in electro-optic sampling of higher-frequency, freely propagating terahertz transients, explored in the following section.
- *Velocity mismatch between electrical and optical signals.* Just as mismatch in the group velocities of two optical pulses degrades temporal resolution in an optical intensity cross-correlation measurement, mismatch between electrical phase velocity and optical group velocity limits temporal resolution in electro-optic sampling. Velocity

matching may sometimes be possible. However, in geometries where optical and electrical signals propagate in orthogonal directions, as in Fig. 10.5, velocity mismatch is intrinsic. The situation is related to that in single-shot optical autocorrelation experiments (Section 3.4.2), which exploit variations in delay across the transverse beam profile that occur when beams cross at a large angle. However, unlike the pure optical case, in electro-optic sampling electrical and optical signals propagate at different speeds and have different spatial distributions. Therefore, limits to temporal resolution are described individually in terms of electrical and optical transit times [587,588]. To understand these limits, we consider electrical and optical signals that are both delta functions in time.

—If we further assume that the propagating electric field is completely localized in space, temporal resolution will be limited by the *electrical transit time* through the finite optical intensity distribution. Assuming a focused Gaussian beam with field radius w, the electro-optic sampling response is itself a Gaussian of the form $\exp[-2(c\tau/n_{\text{eff}}w)^2]$, where c/n_{eff} is the electrical phase velocity in the transmission-line structure. The FWHM of the electro-optic sampling response is $(2\ln 2)^{0.5}wn_{\text{eff}}/c$.

—Conversely, if we assume that the optical spot size is negligible, temporal resolution will be limited by the *optical transit time* through the spatial region in which electric field lines are localized. The optical transit time is given approximately by $2hn_g/c$, where h represents the spatial extent of the electric field, n_g is the optical group velocity in the electro-optic medium, and the factor of 2 accounts for a double pass through the electric field region.

Representative numbers for probing of a GaAs microstrip line, in a configuration similar to that of Fig. 10.5b but adapted for the microstrip geometry, are given in [588]. Assuming that $w = 5\ \mu\text{m}$, $n_{\text{eff}} = 3$, $n_g = 3.5$, and that the wafer thickness $h = 100\ \mu\text{m}$, the electrical transit time is only 60 fs. The optical transit time is 2.3 ps, which is a more serious limitation. However, the optical transit time can be significantly smaller for transmission-line structures with more closely spaced electrodes, such as the coplanar microstrip. The point here is that quasi-static electric field distributions, which are characteristic of transmission lines, are confined over a spatial dimension (h) comparable to the electrode spacing.

A final limitation, although not intrinsic to electro-optic sampling, is that ultrashort electric pulses propagating on transmission lines are subject to extremely strong dispersion. Therefore, measurements of the very fastest electrical transients must be performed very close to the generation point. Alternatively, such dispersion may be avoided for ultrafast electromagnetic transients propagating in free space, which is discussed in the next section.

10.2 ULTRAFAST ELECTROMAGNETICS: TERAHERTZ BEAMS

In Section 10.1 we discussed propagation of ultrafast voltage–current transients confined in simple circuits. However, time-varying currents also radiate, especially when aided by antenna structures. The resulting freely propagating terahertz waves are of considerable interest for applications in sensing and spectroscopy. In the following we cover the basics of pulsed terahertz radiation, including approaches to generation and measurement and terahertz beam systems for time-domain spectroscopy. A monograph treating pulsed terahertz radiation in detail may be found in [590].

10.2.1 Generation and Measurement of Terahertz Pulses

Photoconductive Dipole Antennas Figure 10.6 shows the setup of an early experiment in which subpicosecond electromagnetic pulses were generated and detected via photoconductive antennas. The antenna structures consist of a long coplanar strip transmission line connected at one end to a pair of contact pads, a dipole antenna at the other end of the transmission line, and a photoconducting film underneath the dipole. The photoconductor is a radiation-damaged silicon film on a transparent substrate and has a 500-fs estimated carrier lifetime. The dipole antennas consist of metal stubs at the end of the transmission line, which are fabricated in lengths from 50 to 200 μm. A pair of such antenna chips are placed a few millimeters apart with the antenna sides of the substrates facing each other; the photoconductors are illuminated from the back side by optical pulses passing through the substrates. The first (transmitter) chip is biased with a dc voltage. When its photoconductor is triggered by a laser pulse, a time-varying current flows along the metal dipole, resulting in electromagnetic radiation. According to elementary antenna theory [5], in the far-field zone the radiated electric field for a short dipole antenna of length $d \ll \lambda$ is given by

$$E_{\text{rad}} = \frac{\mu_0 d}{4\pi r} \frac{\partial I}{\partial t} e^{-jkr} \sin\theta \tag{10.20}$$

where I is the current and r and θ are the distance from the dipole and the radiation angle with respect to the direction of current flow, respectively. When this radiation arrives at the second (receiving) antenna, it provides a transient bias across the corresponding photoconductive gap. A laser probe pulse illuminating the receiver photoconductor results in a current pulse with amplitude proportional to the transient bias at the time of illumination. Recording the time-integrated current at the output of the receiving chip as a function of optical delay provides a measurement of the time-dependent radiated electric field. This scheme is similar to the photoconductive generation and sampling of Section 10.1.1, except that now

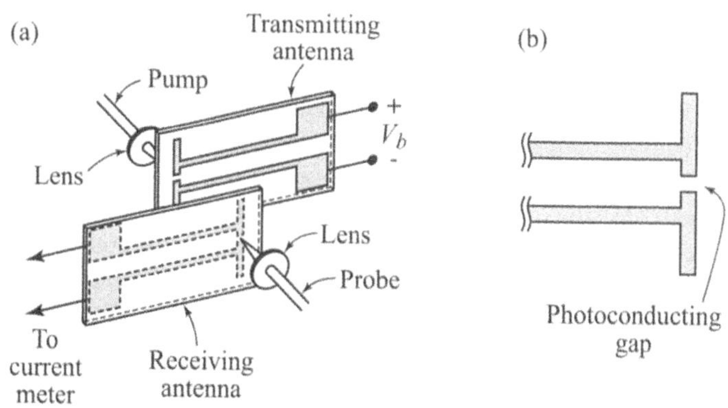

Figure 10.6 (a) Apparatus for observation of THz radiation freely propagating between a pair of photoconducting dipole antennas; (b) expanded view of dipole antenna with photoconducting gap. (a) Adapted from [591]. Copyright © 1988, IEEE.

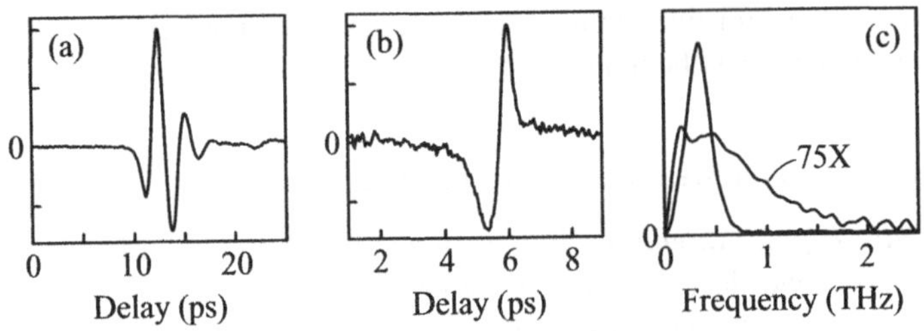

Figure 10.7 THz radiation data from apparatus of Fig. 10.6: (a) time-domain trace for 200 μm dipole antennas; (b) time-domain trace for 50-μm dipole antennas (the peak average current is about ninefold less than in the 200-μm case); (c) frequency response (spectral amplitude) obtained by Fourier transforming the time-domain traces. Adapted from [591]. Copyright © 1988, IEEE.

the photoconductive generator and sampler are connected via radiation in free space rather than by a physical transmission line.

Figure 10.7 shows terahertz pulse data from these experiments. The response obtained from the longer (200-μm) dipole antenna chips exhibits a ringing characteristic, with a 2.7-ps delay between positive peaks. The spectral response, obtained via Fourier transform, exhibits a clear peak at 340 GHz. Principally, this reflects the response of the antennas, which are expected to show resonant behavior when the effective length of the dipole is equal to one-half of the wavelength. Figure 10.7b shows an essentially single-cycle response obtained from the shorter (50-μm) dipole antenna chips. The corresponding spectrum, although lower in amplitude, is now much broader, with frequency content extending well beyond 1.5 THz. Here the waveform is determined principally by the time dependences of the transmitter photocurrent and receiver photoconductance; these quantities are governed in turn by the laser pulse duration, by carrier dynamics that may limit the photocurrent risetime, and by the carrier lifetime. The antenna resonance is shifted to a frequency higher than that supported by the carrier lifetime and does not play a significant role. The bipolar nature of the recorded signal arises because the radiated field depends on the time derivative of the current in the transmitting antenna, while the exponential tail apparent on the front edge of the waveform may be attributed to the photoconductance response function of the receiving antenna.[1]

An important point is that experiments such as these provide a measure of the coherent time-dependent electric field. Accordingly, terahertz pulse experiments retain phase information. This is in contrast to pure optics experiments, where intensity is usually measured.

Figure 10.8 depicts an experiment in which such terahertz radiation is now coupled to collection and focusing optics. Here a 9.5-mm-diameter hemispherical sapphire mirror is contacted to the sapphire side of a radiation-damaged silicon-on-sapphire chip. A coplanar transmission line (consisting of a pair of 5-μm-wide lines spaced by 10 μm) and various photoconductive sampling gaps (not shown) are fabricated onto the bottom (silicon) side of the chip. An electrical pulse is generated by biasing the transmission line and exciting it with a laser pulse, using the sliding contact excitation geometry of Fig. 10.4a. The photocurrent flowing between the metal lines at the position of the laser spot acts as a time-varying dipole that both launches a current–voltage transient down the transmission line and radiates a

[1] According to eq. (10.15), the photoconductive sampling trace may be expressed as the correlation between the time-dependent photoconductance and the time-dependent voltage induced across the sampling gap.

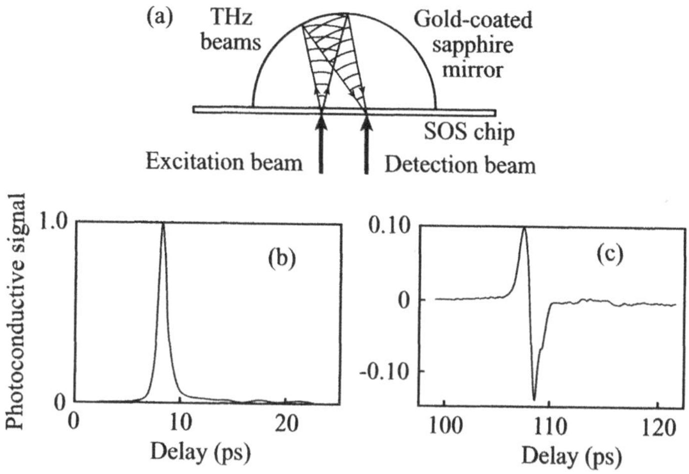

Figure 10.8 (a) Focusing optics for pulsed terahertz radiation. A terahertz pulse is generated from a biased coplanar transmission line (not shown) on the surface of a silicon-on-sapphire chip. The radiated terahertz field is collected by a spherical reflector and refocused onto a photoconductive gap detector (also not shown). (b) Measured electrical pulse on the transmission line. (c) Measured terahertz radiation waveform after collimation and refocusing. Adapted from [592], with permission. Copyright © 1988, American Institute of Physics.

transient electromagnetic field into space. Because its spatial dimensions are much smaller than the minimum wavelength of the detected signal, this dipole is essentially a point source for terahertz radiation. The transient launched onto the transmission line is measured using a side-gap photoconductive sampler fabricated a short distance down the line. The radiated terahertz field propagates into the sapphire hemisphere, reflects from a gold coating at the sapphire–air interface, and is refocused onto a photoconductive gap detector fabricated at the silicon plane but offset from the transmission line. This scheme allows direct comparison of the freely propagating electromagnetic transient radiated into the sapphire with the transient guided along the transmission line. The comparison reveals the following:

- The radiated signal occurs approximately 100 ps later than the signal traveling along the transmission line. This matches the propagation delay expected for the ca. 9.5-mm round-trip propagation distance of the terahertz radiation in the sapphire hemisphere with refractive index equal to 3.06 and confirms observation of the freely propagating terahertz radiation.
- The signal on the transmission line is unipolar, while the radiated field exhibits bipolar character similar to the derivative of the transmission-line signal. Equation (10.20) predicts a derivative relationship between the radiation signal and the time-varying current responsible for the radiation. The data of Fig. 10.8 provide direct confirmation of this prediction.
- The peak amplitude of the detected radiation signal is about 10% as large as that of the detected signal on the transmission line. Given that the radiated field refocuses to a minimum spot size of a wavelength (100 μm for a 1-THz frequency in a medium with index of 3), which is large compared to the photoconductive gap (10 μm) used for measurement, this indicates that the terahertz radiation is collected and refocused with high efficiency.

In the reflective optics system of Fig. 10.8, the terahertz field remains within the sapphire substrate and hemisphere. However, analogous refractive optics systems can collimate terahertz beams for propagation outside the lens material. This is important for spectroscopy and related applications where access to external samples is required. Such terahertz beam systems are discussed later.

Large-Aperture Emitters Because of their small size, the dipole emitters discussed above are limited in terms of the bias voltages and optical powers they can withstand. This limits the amplitudes of radiated terahertz fields. Increasing the size of the photoconductive gap allows both larger bias voltages and greater optical power, as from amplified femtosecond systems, and results in stronger terahertz pulses. Accordingly, in this section we discuss large-aperture antennas [593–595]. Here the term *large aperture* means that transverse dimensions of the emitter are large compared to the wavelength of the terahertz beam (referenced to the center of the terahertz frequency distribution).

As shown in Fig. 10.9, the large-aperture antenna consists simply of a piece of photoconductor with a bias applied parallel to the photoconductor surface. An optical pulse illuminates the photoconductor, leading to transient photoconductance and transient current. Because the current distribution is spread over an area large compared to the wavelength, the radiation it emits emerges as a beam in a well-defined direction, with angular width that depends on the ratio of aperture size to wavelength. Beams are radiated in both forward ("transmitted") and backward ("reflected") directions. Interestingly, the direction in which the terahertz beams are emitted depends on the direction from which the optical beam is incident. For an optical pulse at oblique incidence, the current transient sweeps across the photoconductor at a rate that depends on the angle of incidence θ_{in}. This results in terahertz emission in directions determined by a generalized Fresnel law,

$$\sin\theta_r = n_{\text{THz}}\sin\theta_t = \sin\theta_{\text{in}} \tag{10.21}$$

where n_{THz} and θ_t are the refractive index and propagation angle of the terahertz beam inside the photoconductive medium (transmitted direction), and θ_r is the angle of the terahertz beam in the reflected direction. Incidence from free space is assumed. The terahertz beam in the backward direction is emitted in the same direction as the reflected optical beam.

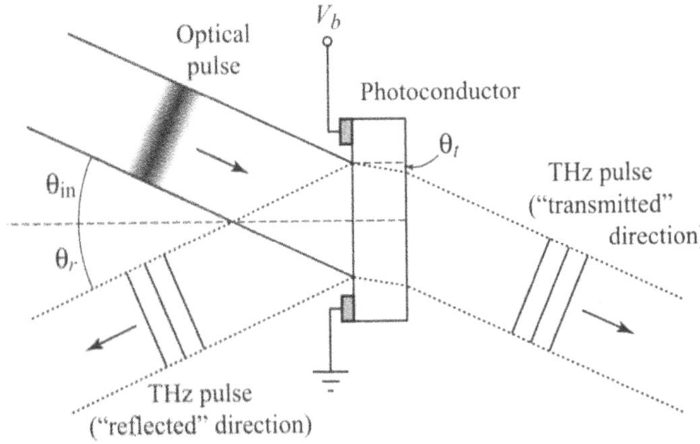

Figure 10.9 Large-aperture terahertz photoconductive antenna.

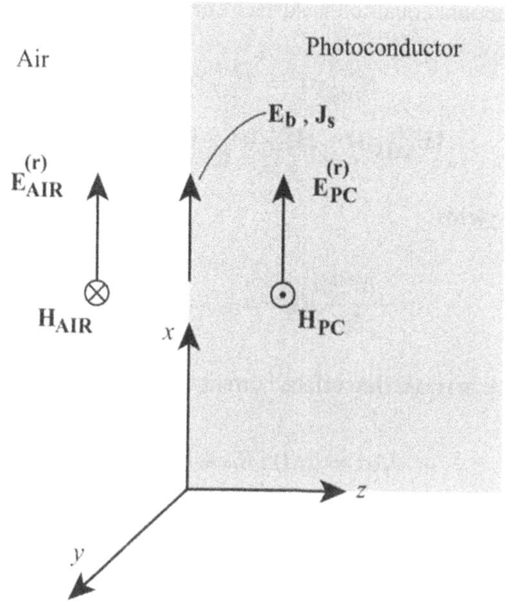

Figure 10.10 Geometry for plane-wave analysis of a large-aperture terahertz photoconductive antenna.

Because the emitter is taken to be much larger than a wavelength, an approximate analysis of large-aperture antennas is possible using a plane-wave framework. For simplicity we consider normal incidence excitation [595]. The geometry is sketched in Fig. 10.10. A time-dependent surface current $\mathbf{J_s} = J_s(t)\hat{x}$, taken to be independent of transverse spatial coordinates (x, y), is assumed to be present at the photoconductor–air boundary $(z = 0)$. The surface current drives uniform plane waves that propagate in $\pm z$ directions. The E and H fields of the waves propagating into air and into the photoconductor are written, respectively,

$$\mathbf{E_{AIR}} = \hat{x}E_b + \hat{x}E_{\text{AIR}}^{(r)}\left(t + \frac{z}{c}\right)$$
$$\mathbf{H_{AIR}} = -\hat{y}\left[E_{\text{AIR}}^{(r)}\left(t + \frac{z}{c}\right)\right]\Big/ Z_0 \tag{10.22}$$

and

$$\mathbf{E_{PC}} = \hat{x}E_b + \hat{x}E_{\text{PC}}^{(r)}\left(t - \frac{zn_{\text{THz}}}{c}\right)$$
$$\mathbf{H_{PC}} = \hat{y}\left[n_{\text{THz}}E_{\text{PC}}^{(r)}\left(t - \frac{zn_{\text{THz}}}{c}\right)\right]\Big/ Z_0 \tag{10.23}$$

In these equations $\mathbf{E_b} = \hat{x}E_b$ is the constant bias field, $Z_0 = \sqrt{\mu_0/\varepsilon_0}$ is the impedance of free space, and n_{THz} is assumed to be frequency independent within the bandwidth of the terahertz pulse. Radiating fields are designated with superscript $^{(r)}$. Boundary conditions for Maxwell's equations require that the transverse component of the electric field be continuous across the $z = 0$ boundary and that the transverse component of the magnetic field be discontinuous by an amount equal to the surface current. Mathematically, we write

$$\begin{aligned} E^{(r)}_{\mathrm{AIR}}(t) &= E^{(r)}_{\mathrm{PC}}(t) = E^{(r)}(t) \\ \left[\mathbf{H}^{(\mathbf{r})}_{\mathbf{AIR}}(t) - \mathbf{H}^{(\mathbf{r})}_{\mathbf{PC}}(t)\right] &\cdot \hat{y} = J_s(t) \end{aligned} \tag{10.24}$$

Combining the above yields

$$\frac{-E^{(r)}(t)}{Z_0}(1 + n_{\mathrm{THz}}) = J_s(t) \tag{10.25}$$

On the other hand, we can relate the surface current to the electric field through Ohm's law:

$$J_s(t) = \sigma_s(t)\left[E_b + E^{(r)}(t)\right] \tag{10.26}$$

Here surface current J_s and surface conductivity σ_s, with units A/m and Ω^{-1} respectively, are obtained by integrating current density and conductivity [with units A/m^2 and $(\Omega \cdot \mathrm{m})^{-1}$] over the thickness of the thin-film photoconductor. Inserting eq. (10.26) into eq. (10.25) yields

$$E^{(r)}(t) = \frac{-\sigma_s(t) Z_0 E_b}{(1 + n_{\mathrm{THz}}) + \sigma_s(t) Z_0} \tag{10.27}$$

This result points to several interesting observations:

- At low optical power ($\sigma_s Z_0 \ll 1$), the plane-wave electric field $E^{(r)}(t)$ evaluated at the source plane is proportional to the instantaneous surface conductivity $\sigma_s(t)$. Furthermore, the electric field amplitudes of the plane waves radiated in the two directions are the same. However, due to its stronger magnetic field, the intensity of the wave radiated into the substrate is larger compared to the wave propagating into air by a factor of n_{THz}. A substantially stronger asymmetry in radiation pattern (favoring radiation into the substrate) is seen for small integrated-circuit antennas, such as dipoles, for which the plane-wave analysis does not apply [591].
- $E^{(r)}(t)$ saturates at high optical power and reaches a limiting value equal to $-E_b$ for $\sigma_s Z_0 \gg 1$. This behavior has been confirmed experimentally [595]. A similar saturation equation was obtained in our analysis of electrical pulse generation using photoconductors and transmission-line structures in Section 10.1.1. In the large-aperture antenna case, we may understand saturation as arising because photocurrent is driven by the full electric field [i.e., the superposition of the bias field and the plane-wave field generated (evaluated at the source plane)]. When the field generated is strong, the electric field remaining to drive additional photocurrent is reduced.
- In contrast to the case of short dipole antennas, for which the radiated field depends on the time derivative of the photocurrent, here the field generated is predicted to follow the instantaneous photocurrent (this is particularly clear for the low-power case). The resolution of this paradox is that for short dipoles, our discussion centered on radiation in the far field, while for large-aperture antennas, the analysis was performed at the

source (i.e., in the near field). Finite-size electromagnetic beams propagating into the far field are subject to frequency-dependent diffraction. This frequency-dependent propagation introduces a factor of ω into the expression for the on-axis radiated field, after which the on-axis field exhibits the expected time derivative dependence.[2]

Other Terahertz Emitters So far we have only discussed externally biased, photoconductive sources of transient terahertz radiation. Although our list is not exhaustive, here we describe two additional mechanisms useful for terahertz pulse generation. These are commonly used with large-aperture excitation, although this is not fundamentally required.

Built-in Fields at Semiconductor Surfaces It has been found that femtosecond optical illumination of direct gap semiconductor surfaces can lead to terahertz radiation even without an external bias field [596]. The mechanism for this radiation relies on built-in electric fields within the depletion layer that occurs close to the semiconductor–air interface. Absorption of photons from a short pulse results in generation of electrons and holes, which drift in opposite directions in the built-in field. This provides the transient photocurrent responsible for terahertz emission. Because the built-in field and the photocurrent are normal to the semiconductor surface, terahertz emission is observed only when the laser is incident at oblique angles, so that the radiated electric field has a nonvanishing component along the photocurrent direction. Note that the photocurrent shuts off as soon as the photocarriers travel through the built-in field region; therefore, the photocurrent lifetime can be much shorter than the carrier lifetime. This method has the practical advantage of extreme simplicity: only a femtosecond laser and a suitable semiconductor wafer are needed, and terahertz pulse generation is achieved without special material preparation or fabrication. Terahertz radiation generated in this way also serves as a useful probe of semiconductor surface fields and fast carrier dynamics within these fields [597]. As one example, the polarity of the radiated terahertz pulse is found to depend on the sign of the built-in field (directed toward or away from the surface), which in turn depends on semiconductor doping.

Optical Rectification Terahertz radiation can also be generated by femtosecond illumination of electro-optic crystals [598,599]. Here, in a second-order nonlinear process (see Chapter 5) called *optical rectification*, the input pulse excites a nonlinear polarization term of the form

$$P_{\mathrm{NL}}(t) \sim |a(t)|^2 \tag{10.28}$$

Here $a(t)$ is the complex amplitude function of the input pulse, and the nonlinear susceptibility is assumed to be constant within the bandwidth of the optical rectification signal. The nonlinear polarization follows the time-dependent optical intensity and propagates within the electro-optic crystal at the optical group velocity. The terahertz radiation resulting from the nonlinear polarization, in the far-field zone, is proportional to

$$E_{\mathrm{rad}} \sim \frac{\partial^2 P_{\mathrm{NL}}}{\partial t^2} \tag{10.29}$$

[2] The appearance of a factor of ω is easily verified for the special case of a Gaussian beam. Generally, frequency-dependent diffraction is described by the space–frequency wave equation, eq. (6.97), which reduces to $\partial A/\partial z + (j\omega_0/2\beta_0\omega)\nabla_T^2 A = 0$ for propagation without dispersion in a source-free region.

The second derivative operating on the nonlinear polarization is consistent with the form of the forced wave equation, eq. (5.12). Since $\partial P/\partial t$ represents the current associated with motion of bound charge carriers, $\partial^2 P/\partial t^2$ may be understood as the bound current equivalent of the $\partial I/\partial t$ term responsible for radiation from time-varying free currents. Like many bound electron nonlinear optical processes, optical rectification can be very fast. Terahertz radiation may be generated up to frequencies limited by the laser pulse duration and (in part) by the lattice resonances that are prominent in many electro-optic crystals.

It is worth noting that the nonlinear polarization spectrum may be written

$$\mathcal{P}_{\mathrm{NL}}(\Omega) \sim \int d\omega\, A(\omega) A^*(\omega - \Omega) \tag{10.30}$$

where $A(\omega)$ is the complex spectral amplitude of the input pulse and ω and Ω refer to optical and baseband (terahertz) frequencies, respectively. Equation (10.30) is of the same form as the source term responsible for difference frequency mixing in eq. (5.154). Accordingly, optical rectification may be viewed as difference frequency mixing of optical frequency components contained within a single excitation pulse. A distinction is that the term *difference frequency generation* is often used to refer to experiments in which special care is taken to achieve phase matching, while the term *optical rectification* typically refers to situations far from phase matching. In the latter case, only the nonlinear polarization corresponding to a material thickness of one coherence length contributes constructively to the terahertz emission.

Free-Space Electro-optic Sampling Earlier we discussed electro-optic sampling for measurement of ultrafast electrical transients guided on metal interconnect lines. Electro-optic sampling may also be used for characterization of terahertz electromagnetic radiation. Like its guided-wave counterpart, free-space electro-optic sampling is capable of providing very high speed. Furthermore, similar to photoconductive sampling, electro-optic sampling resolves the coherent, time-dependent electric field, so phase information is retained.

Figure 10.11 shows a typical setup for free-space electro-optic sampling [600–602]. The terahertz radiation and a short laser pulse are superimposed for collinear propagation using a thin beamsplitter and then directed into an electro-optic (EO) crystal. The operation is then similar to that in Section 10.1.2. The polarizations of the terahertz and laser fields and the orientation of the EO crystal are chosen such that the terahertz radiation induces a

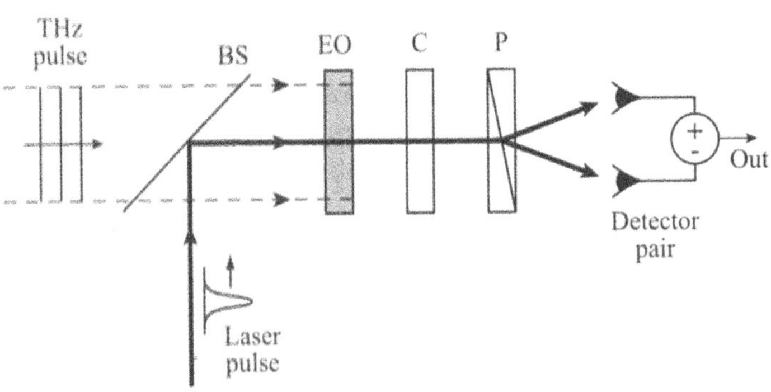

Figure 10.11 Setup for free-space electro-optic sampling. BS, beamsplitter; EO, electro-optic crystal; C, compensator; P, polarization splitter.

birefringence in the EO crystal, which results in modification of the laser polarization. A polarizing beamsplitter separates optical polarization components and directs them onto a differential photodetector pair. A compensator adjusts the laser polarization for quadrature bias, so that in the absence of the terahertz field, the laser beam is split equally onto the two photodetectors. This results in an EO sampling signal that is linear in the terahertz field. As usual, the temporal profile of the terahertz field is recorded by repeating the measurement as the delay of the laser pulse is varied. With reference to eq. (10.19), which gives the EO sampling signal for the case of a single photodetector, the differential detector pair yields an output signal of the form

$$\int dt\, |a(t-\tau)|^2 \left[\phi_x(t) - \phi_y(t)\right] \tag{10.31}$$

The differential detection scheme doubles the amplitude of the sampling output, while eliminating the delay-independent background term and suppressing laser intensity noise. This permits high detection sensitivity near the shot noise limit.[3]

Several factors affect the detection bandwidth of free-space electro-optic sampling. These include the laser pulse duration as well as frequency-dependent aspects of the electro-optic interaction. Obviously, the laser pulse should be significantly shorter than the oscillation period of any terahertz frequencies of interest; for terahertz frequencies sufficiently high that this requirement is violated, the measurement response rolls off. Frequency-dependent aspects of the electro-optic interaction have been discussed in [601,603–605]. Assuming an infinitely short probe pulse, the complex electro-optic response function $R(\Omega)$, which gives the sampling signal resulting from a unit amplitude terahertz frequency component at angular frequency Ω, may be written

$$R(\Omega) = t(\Omega)H(\Omega)r(\Omega) \tag{10.32}$$

where $t(\Omega)$, $H(\Omega)$, and $r(\Omega)$ represent the terahertz transmission coefficient into the EO crystal, propagation and phase-matching effects, and the frequency-dependent electro-optic coefficient, respectively. This formulation ignores dispersion acting on the optical sampling pulse and neglects multiple reflections of terahertz and optical fields. Assuming that the EO crystal is sufficiently thick, signal components associated with multiple-reflections in actual data may be removed in software by truncating the EO signal prior to the beginning of the first multiple-reflection contribution. With these stipulations the factors appearing in eq. (10.32) may be understood as follows:

- $t(\Omega) = 2/[n(\Omega) + 1]$ is simply the transmission coefficient [5] for a plane-wave terahertz field incident from air (the refractive index is 1) into a medium with frequency-dependent index $n(\Omega)$.
- The $H(\Omega)$ term is analyzed by considering electro-optic polarization rotation as a three-wave interaction in which fields at optical frequency ω and terahertz frequency Ω mix to give a second-order optical field at frequency $\omega + \Omega$. The second-order field mixes with the zero-order field at $\omega + \Omega$ present in the input sampling pulse to determine the change in power through the analyzer. Accordingly, the sampling

[3] Although not described earlier, the differential detection scheme is also applicable to electro-optic sampling of signals guided on interconnect lines.

signal is proportional to the generated second-order field, the amplitude of which is determined by phase matching. Referring to Chapter 5 [e.g., eq. (5.17) or (5.155], we find that the second-order field amplitude is proportional to

$$\int_0^L dz \exp\left\{j\left[k_{\text{opt}}(\omega+\Omega) - k_{\text{opt}}(\omega) - k_{\text{THz}}(\Omega)\right]z\right\} \tag{10.33}$$

where L is the crystal thickness. By expanding k_{opt} to first order in Ω, we obtain

$$H(\Omega) \sim \frac{j}{\Delta k(\Omega)}\left(1 - e^{j\Delta k(\Omega)L}\right) \tag{10.34a}$$

where

$$\Delta k(\Omega) = \frac{\Omega[n_g - n(\Omega)]}{c} \tag{10.34b}$$

and c/n_g is the optical group velocity. Due to phase-matching effects, the response rolls off at high frequencies, with a bandwidth that depends on the mismatch between the optical group velocity and terahertz phase velocity. The response drops to zero for terahertz frequencies such that the temporal walk-off is equal to an integer number of oscillation periods of the terahertz field. Velocity matching is desirable and is possible for certain EO crystals and certain specific combinations of optical and terahertz frequencies. However, in general, velocity matching cannot be achieved simultaneously over a broad terahertz bandwidth.

- The electro-optic coefficients of most commonly used EO crystals, usually denoted r_{ij} in tensor format, are flat in the low-terahertz range. However, they can vary significantly at higher frequencies due to lattice resonances. We discuss this further below.

Electro-optic sampling detectors can provide response into the midinfrared, provided that velocity mismatch is minimized by using a very thin crystal [253,602,604]. Practically, the very thin sensing crystal may be bonded to a thicker substrate made of the same material but oriented such that the substrate electro-optic response is suppressed. In addition to providing mechanical support, the extra thickness provided by the substrate increases the time separation of signals associated with multiple reflections from the air–EO crystal interfaces. An example of electro-optic sampling data in the midinfrared is given in Fig. 5.16, which shows short-pulse electric field waveforms at 13.7-μm wavelength generated via phase-matched difference frequency generation and measured using a 10-μm-thick ZnTe sensor crystal [253].

Figure 10.12 shows calculations of the electro-optic frequency response of a 13-μm-thick GaP sensor crystal. Figure 10.12a is the calculated result for $|t(\Omega)H(\Omega)|$, which provides insight into effects associated with the first two terms in eq. (10.32). The response shows pronounced dips for frequencies in the vicinity of lattice resonances around 11 THz. These dips are attributed both to high material absorption[4] and large velocity mismatch near the lattice resonances. Otherwise, the response rolls off gradually as frequency increases,

[4] Terahertz absorption is included in the formulation given above by allowing $n(\Omega)$ to be complex.

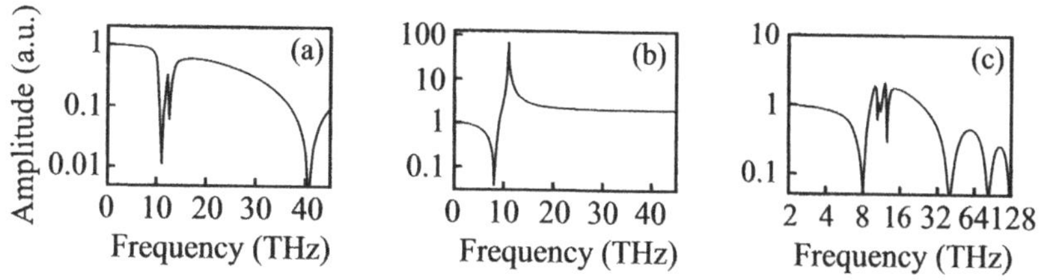

Figure 10.12 Calculated electro-optic sampling spectral amplitude response for 13-μm-thick GaP sensor crystal (amplitude units are arbitrary): (a) spectral response assuming frequency-independent electro-optic coefficient; (b) frequency-dependent amplitude of the electro-optic coefficient; (c) overall EO sampling response plotted on a logarithmic frequency axis. In the language of eq. (10.32), parts (a), (b), and (c) correspond to $t(\Omega)H(\Omega)$, $r(\Omega)$, and $R(\Omega)$, respectively. Adapted from [604] with permission. Copyright © 1999, American Institute of Physics.

dropping to zero at 40 THz and then partially recovering for even higher frequencies. This rolloff is attributed to velocity mismatch. Figure 10.12b shows a calculation of the frequency dependence of the electro-optic coefficient $|r(\Omega)|$, which is not included in Fig. 10.12a. For frequencies below the lattice resonances, both nuclear and electronic effects contribute to the electro-optic response, while for frequencies above the resonances, the response is purely electronic. Because nuclear and electronic responses in GaP contribute with opposite signs, the EO coefficient at low frequencies is actually less than at high frequencies. The nuclear component increases as resonance is approached. Initially, this leads to nearly complete destructive interference of nuclear and electronic contributions around 8 THz. At slightly higher frequency the nuclear component dominates, leading to a strong peak near 11 THz. Above resonance the EO coefficient recovers to the optical value, which is frequency independent over a broad range. Figure 10.12c shows the calculation for the full response function $|R(\Omega)|$. For GaP it turns out that the resonant peak in $|r(\Omega)|$ at 11 THz and the corresponding dip in $|t(\Omega)H(\Omega)|$ approximately cancel. The result is an overall electro-optic sampling signal which is flat to 6 THz or so, with a single strong dip at 8 THz due to cancellation of electronic and lattice responses, and recovered response at higher frequencies, ultimately limited by velocity matching. Experiments using such a 13-μm GaP sensor crystal resulted in observation of frequency components up to 70 THz, corresponding to wavelengths as short as 4 μm [604].

10.2.2 Terahertz Spectroscopy and Imaging

THz Time-Domain Spectroscopy Optical generation and coherent detection of terahertz pulses has found extensive application to spectroscopy in a frequency region difficult to reach by other means. A terahertz beam system widely adopted for terahertz time-domain spectroscopy is depicted in Fig. 10.13 [606–608]. In this setup a short laser pulse triggers a biased photoconductive antenna chip to emit terahertz radiation. The terahertz radiation is collimated by a combination of a spherical lens contacted to the back side of the transmitting chip and an off-axis paraboloidal mirror. The resulting terahertz beam may be relayed with high efficiency over meter path lengths. A similar combination of paraboloidal mirror and substrate lens couples the collimated terahertz beam efficiently onto the receiving antenna, contributing to high signal-to-noise ratio detection. Samples may be placed into the collimated region between paraboloidal mirrors for transmission spectroscopy.

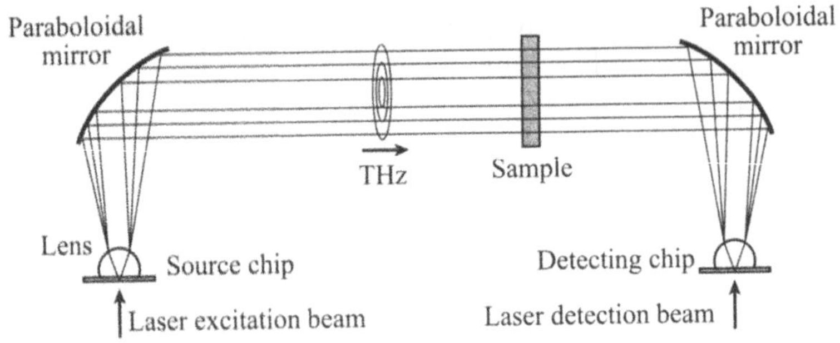

Figure 10.13 Setup for terahertz time-domain spectroscopy. Photoconductive antennas for generation and measurement of pulsed terahertz electric fields are fabricated onto the source chip and detector chip, respectively. Adapted from [608].

Note that in this setup, substrate lenses should be fabricated from materials with low terahertz absorption, such as high-resistivity silicon. A detailed description of the optics of substrate lenses and their effect on broadband terahertz beams is given in [609]. Note also that the path between transmitter and receiver chips is preferably located in an airtight enclosure, which may be filled with gases such as dry nitrogen to avoid absorption due to water vapor. Similar setups adapted for other means of terahertz generation (e.g., large-aperture antennas, optical rectification) or for detection by free-space electro-optic sampling may also be employed.

To acquire linear transmission spectra via terahertz time-domain spectroscopy, one measures the terahertz waveform with the sample in place, $E_{\text{sample}}(t)$, as well as a reference waveform with the sample removed, $E_{\text{ref}}(t)$. The time domain traces are then Fourier-transformed to give $\tilde{E}_{\text{sample}}(\Omega)$ and $\tilde{E}_{\text{ref}}(\Omega)$, respectively. The frequency-dependent refractive index $n(\Omega) = n'(\Omega) + jn''(\Omega)$ may then be found through the relation

$$\tilde{E}_{\text{sample}}(\Omega) = \tilde{E}_{\text{ref}}(\Omega)e^{-j\Omega n'(\Omega)L/c}e^{\Omega n''(\Omega)L/c} \tag{10.35}$$

where L is the sample thickness and $n'' < 0$ for absorption. Similar to eq. (9.98), spectra may also be expressed in terms of absorption coefficient and propagation constant as

$$\tilde{E}_{\text{sample}}(\Omega) = \tilde{E}_{\text{ref}}(\Omega)e^{-\alpha(\tilde{\Omega})L/2}e^{-j\beta(\tilde{\Omega})L} \tag{10.36}$$

In both cases it is assumed that any multiple-reflection effects are common to sample and reference fields or are suppressed via additional software processing. It is clear from the above that terahertz time-domain spectroscopy directly yields both amplitude and phase information. This is a consequence of the fact that terahertz techniques provide access to the coherent electric field, in contrast to optical spectroscopies that usually deal with intensity.

Figure 10.14 shows an example of terahertz time-domain spectroscopy data for water vapor. Another example of spectroscopy data, for a quantum well intersubband absorber, was shown in Fig. 9.21. For the time-domain data of Fig. 10.14, the reference and sample traces correspond to propagation through an enclosure filled, respectively, either with dry nitrogen only or with an additional 1.5 torr of water vapor. The main effect of the water vapor is to introduce additional low-amplitude oscillations onto the tail of the original

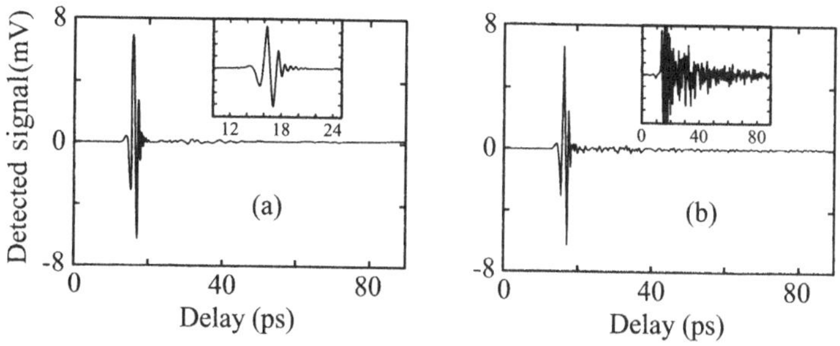

Figure 10.14 Measured terahertz electrical waveforms after propagation in (a) dry nitrogen and (b) with addition of 1.5 torr of water vapor. Insets show waveforms (a) on an expanded time scale and (b) on a 20× expanded vertical scale, respectively. From [606].

terahertz waveform. These may be understood as a superposition of the free induction decay signals from multiple water vapor absorption lines, or equivalently, as the result of the corresponding absorption and dispersion profiles. The decay of these additional oscillations, which occurs over several tens of picoseconds, may be understood qualitatively as an averaged T_2 dephasing time. Figure 10.15 shows the data in the frequency domain after Fourier analysis. The amplitude spectra of data from Fig. 10.14 are shown in Fig. 10.15a from 250 GHz to 1.5 THz. Now the effect of the water vapor is evident as a series of sharp dips in the amplitude spectrum. Figure 10.15b and c show the frequency-dependent amplitude absorption coefficient and phase obtained by dividing $\tilde{E}_{\text{sample}}(\Omega)$ by $\tilde{E}_{\text{ref}}(\Omega)$. A series of absorption lines is clearly revealed; and as expected, for each absorption line there is a corresponding derivative-like feature in the spectral phase. Such terahertz time-domain spectroscopy methods have been applied to characterize the constituents and temperatures of flames [610] and to investigate a wide range of additional materials, including dielectric and semiconducting solids, superconductors, and liquids.

Other spectroscopy techniques involving terahertz pulses are also well known. For example, in terahertz emission spectroscopy a sample of interest emits a terahertz transient upon absorption of an optical pulse. Measurement of the time profile of the radiated transient reveals information about ultrafast current dynamics induced by the laser pulse. As noted

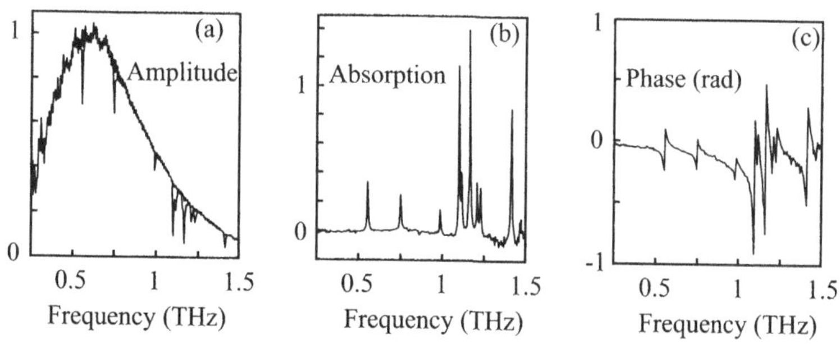

Figure 10.15 (a) Spectral amplitudes obtained by Fourier-transforming terahertz waveforms of Fig. 10.14, corresponding to propagation in nitrogen and in nitrogen–water vapor mixture, respectively. Resulting amplitude absorption spectrum (b) and phase spectrum (c) of water vapor. Adapted from [606].

previously, such terahertz emission spectroscopy has been used, for example, to investigate built-in fields at semiconductor surfaces [597] and wave packet dynamics in coupled quantum well structures (Fig. 9.18) [537]. Nonlinear pump–probe spectroscopies, in which absorption of an optical pump pulse induces a time-dependent change in the optical properties sensed by a delayed terahertz probe, have also been used for time-resolved studies of ultrafast dynamics in a variety of materials.

Terahertz Imaging Optically generated terahertz pulses may also be used to generate images, using either photoconductive [611,612] or electro-optic detection [613]. Figure 10.16 illustrates the schematic experimental configuration for terahertz imaging using photoconductive detection. The setup is related to that used for terahertz time-domain spectroscopy, but has several key differences. These include the following:

- The optics are modified in order to focus the terahertz beam to a diffraction-limited spot at the sample and to obtain spatially resolved measurements. The sample (or the focused terahertz spot) is then scanned in the transverse dimension to build up an image.
- Data acquisition is enhanced so that time-domain terahertz traces may be obtained at different transverse sample positions with sufficient rapidity. This is achieved by using a rapid scanning optical delay line and by digitizing the signal directly at the output of the terahertz receiver (thereby dispensing with lock-in amplifier detection common in spectroscopy applications).

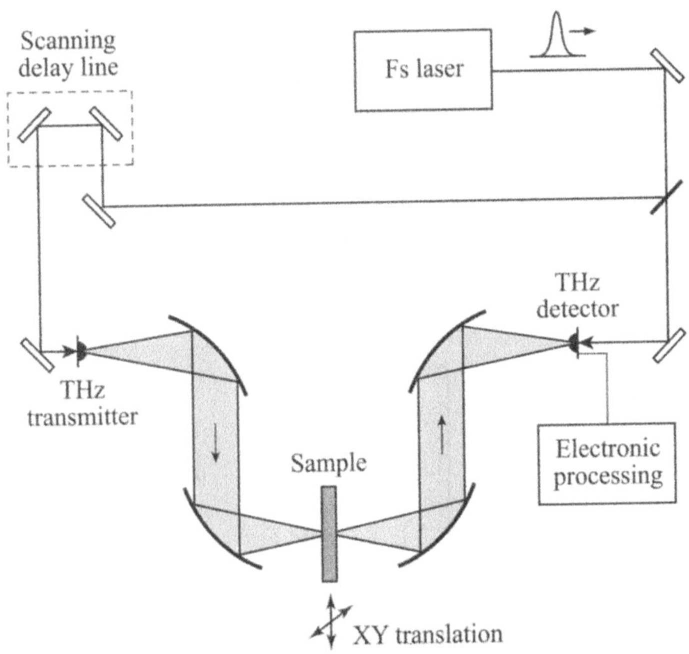

Figure 10.16 Experimental setup for terahertz transmission imaging. Generated terahertz radiation is collimated, focused onto the sample, recollimated, and finally focused onto the terahertz detector using substrate lenses and a series of paraboloidal mirrors. An image is acquired by scanning the beam across the sample. Adapted from [612]. Copyright © 1996, IEEE.

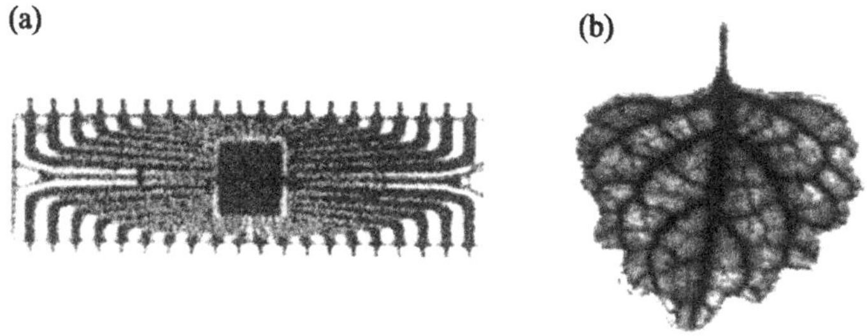

Figure 10.17 Examples of terahertz transmission images: (a) image of packaged integrated circuit; (b) image of a leaf. (a) From [611]. (b) From [612]. Copyright © 1996, IEEE.

- Digital signal processing is employed for rapid extraction of information from time-domain traces for display in image format. There are many choices of algorithms for such information extraction; different choices are appropriate for different samples. Simple examples include searching for the peak of the time-domain waveform, which may provide a measure of overall transmission and propagation time, or Fourier-transforming the data to identify the frequencies and strengths of sharp absorption lines or to determine the transmission integrated over a specified spectral region.

Examples of terahertz images are shown in Fig. 10.17. Each of these images consists of tens of thousands of pixels with spatial resolution of a few hundred micrometers. Figure 10.17a shows a terahertz image of a packaged integrated circuit, constructed by displaying the spatially resolved terahertz intensity integrated over a frequency range of 1 to 3 THz. In this example the plastic packaging material is nearly transparent in the terahertz spectral range, whereas the partially absorbing doped semiconductor chip and the opaque metal interconnect lines are clearly visible in the image. The ability to construct images of structures concealed below packaging layers is relevant for a range of industrial inspection and quality control applications. Figure 10.17b shows a terahertz transmission image of the leaf of a household plant. This example takes advantage of the strong absorption of liquid water in the terahertz range and displays the image in a gray scale where darkness correlates with water content. An important point is that such terahertz imaging studies are nondestructive, providing a unique tool for assessment of water transport dynamics in live leaves as well as for certain biomedical applications.

PROBLEMS

10.1. Consider a photoconductive switch as pictured in Fig. 10.1 fabricated using a 1-μm-thick GaAs photoconducting film and a 50-Ω microstrip transmission line. The photoconducting gap has dimensions 20 μm by 20 μm. Estimate the absorbed pulse energy for which the switch crosses from the linear to the saturated regime. You may take electron and hole mobilities as 8500 and 400 $cm^2/V\cdot s$, respectively, and may assume a long carrier lifetime.

10.2. Analyze the response of a photoconductive switch in the side-gap geometry, similar to Fig. 10.3b but used for electrical waveform generation rather than sampling. Assume that the side-gap is connected to a constant bias voltage V_b, which drops completely across the gap prior to the arrival of a light pulse that switches the conductance abruptly from zero to G_0.

10.3. Consider a photoconductive sampling gate in the in-line geometry. Assume that the conductance of the gate may be taken as $G_0 + g_s(t)$, where G_0 is the dark conductance and $g_s(t)$ is small. Formally derive eq. (10.15) by using a perturbation approach to analyze eq. (10.11) to first order in g_s.

10.4. Although not shown in the data of Fig. 10.4b, the sliding contact scheme for photoconductive generation and sampling is symmetrical: The collected charge vs. delay data show a pair of peaks, one when the optical "generation" beam precedes the "sampling" beam by a delay equal to the electrical propagation time between generation and sampling locations, and one when the generation beam follows the sampling beam by the same amount. Qualitatively, discuss the specific mechanism by which each of the peaks is generated.

10.5. Assume that the impulse response of an electro-optic sampling system limited by velocity mismatch is a rectangular pulse of duration Δt. Work out the frequency response of the electro-optic sampling system in response to electrical sinusoids of frequency f (i.e., what is the amplitude of the waveform recorded by electro-optic sampling as a function of frequency)? Assuming a velocity mismatch of 2 ps, at what frequency does the response drop to 50% of its low-frequency value?

10.6. Consider a large-aperture photoconducting antenna excited by an optical pulse at normal incidence. Assume that the optical beam has Gaussian spatial distribution with a field radius $w_0 = 5$ mm, defined as in Section 1.3.3. Estimate the full spreading angle of the emitted terahertz radiation (in air) for frequencies of 100 GHz, 500 GHz, and 1 THz.

10.7. Consider a Gaussian beam, as in Section 1.3.3, with a waist at $z = 0$ with frequency-independent beam radius w_0. Show that upon propagation into the far field, the on-axis electric field amplitude acquires a multiplicative factor proportional to ω.

10.8. Consider a free-space electro-optic sampling system that is perfectly velocity matched. Assume that the detection bandwidth is limited only by the laser pulse duration. If the laser pulse is Gaussian with 12-fs FWHM duration, at what terahertz frequency does the response drop to 50% of its low-frequency value? How about to 1% of its low-frequency value? What are the corresponding wavelengths?

REFERENCES

[1] M. Hentschel, R. Kienberger, C. Spielmann, G. A. Reider, N. Milosevic, T. Brabec, P. Corkum, U. Heinzmann, M. Drescher, and F. Krausz. Attosecond metrology. *Nature*, 414(6863):509–513, 2001.

[2] H. E. Edgerton. *Stopping Time: The Photographs of Harold Edgerton*. H.N. Abrams, New York, 1987.

[3] S. Kawanishi. Ultrahigh-speed optical time-division-multiplexed transmission technology based on optical signal processing. *IEEE J. Quantum Electron.*, 34(11):2064–2079, 1998.

[4] H. G. Weber, R. Ludwig, S. Ferber, C. Schmidt-Langhorst, M. Kroh, V. Marembert, C. Boemer, and C. Schubert. Ultrahigh-speed OTDM-transmission technology. *J. Lightwave Technol.*, 24(12):4616–4627, 2006.

[5] S. Ramo, J. R. Whinnery, and T. Van Duzer. *Fields and Waves in Communication Electronics*. Wiley, New York, 3rd edition, 1994.

[6] J. D. Jackson. *Classical Electrodynamics*. Wiley, New York, 3rd edition, 1999.

[7] A. Yariv. *Quantum Electronics*. Wiley, New York, 3rd edition, 1989.

[8] A. E. Siegman. *Lasers*. University Science Books, Mill Valley, CA, 1986.

[9] H. A. Haus. *Waves and Fields in Optoelectronics*. Prentice-Hall, Englewood Cliffs, NJ, 1984.

[10] M. Abramowitz and I. A. Stegun. *Handbook of Mathematical Functions with Formulas, Graphs, and Mathematical Tables*. U.S. Govt. Printing Office, Washington, DC, 1964.

[11] R. N. Bracewell. *The Fourier Transform and Its Applications*. McGraw-Hill, New York, 2nd edition, 1986.

[12] A. V. Oppenheim, A. S. Willsky, and H. Nawab. *Signals and Systems*. Prentice Hall, Upper Saddle River, NJ, 2nd edition, 1997.

[13] W. H. Press, B. P. Flannery, S. A. Teukolsky, and W. T. Vetterling. *Numerical Recipes*. Cambridge University Press, Cambridge, U.K., 1986.

Ultrafast Optics, By Andrew M. Weiner

[14] D. J. Kuizenga and A. E. Siegman. FM and AM mode locking of the homogeneous laser. Part I: Theory. *IEEE J. Quantum Electron.*, 6:694–708, 1970.

[15] H. A. Haus. A theory of forced mode locking. *IEEE J. Quantum Electron.*, 11:323–330, 1975.

[16] O. P. McDuff and S. E. Harris. Nonlinear theory of the internally loss modulated laser. *IEEE J. Quantum Electron.*, 3:101–111, 1967.

[17] J. R. Fontana. Theory of spontaneous mode locking in lasers using a circuit model. *IEEE J. Quantum Electron.*, 8:699–703, 1972.

[18] D. J. Kuizenga and A.E. Siegman. FM and AM mode locking of the homogeneous laser. Part II: Experiment. *IEEE J. Quantum Electron.*, 6:709–715, 1970.

[19] J. D. Kafka, M. L. Watts, and J. W. J. Pieterse. Picosecond and femtosecond pulse generation in a regeneratively mode-locked Ti:sapphire laser. *IEEE J. Quantum Electron.*, 28(10):2151–2162, 1992.

[20] T. F. Carruthers and I. N. Duling. 10-GHz, 1.3-ps erbium fiber laser employing soliton pulse shortening. *Opt. Lett.*, 21(23):1927–1929, 1996.

[21] J. P. Heritage and R. K. Jain. Subpicosecond pulses from a tunable CW mode-locked dye-laser. *Appl. Phys. Lett.*, 32(2):101–103, 1978.

[22] R. K. Jain and J. P. Heritage. Generation of synchronized CW trains of picosecond pulses at 2 independently tunable wavelengths. *Appl. Phys. Lett.*, 32(1):41–44, 1978.

[23] P. J. Delfyett, L. T. Florez, N. Stoffel, T. Gmitter, N. C. Andreadakis, Y. Silberberg, J. P. Heritage, and G. A. Alphonse. High-power ultrafast laser-diodes. *IEEE J. Quantum Electron.*, 28(10):2203–2219, 1992.

[24] H. A. Haus. Theory of mode locking with a fast saturable absorber. *J. Appl. Phys.*, 46:3049–3058, 1975.

[25] H. A. Haus. Theory of mode locking with a slow saturable absorber. *IEEE J. Quantum Electron.*, 11:736–746, 1975.

[26] A. Dienes, E. P. Ippen, and C. V. Shank. A mode locked CW dye laser. *Appl. Phys. Lett.*, 19:258–260, 1971.

[27] C. V. Shank and E. P. Ippen. Subpicosecond kilowatt pulses from a mode-locked CW dye laser. *Appl. Phys. Lett.*, 24:373–375, 1974.

[28] E. P. Ippen and C. V. Shank. Dynamic spectroscopy and subpicosecond pulse compression. *Appl. Phys. Lett.*, 27:488–490, 1975.

[29] R. L. Fork, B. I. Greene, and C. V. Shank. Generation of optical pulses shorter than 0.1 psec by colliding pulse mode-locking. *Appl. Phys. Lett.*, 38:671–673, 1981.

[30] J. A. Valdmanis, R. L. Fork, and J. P. Gordon. Generation of optical pulses as short as 27 femtoseconds directly from a laser balancing self-phase modulation, group-velocity dispersion, saturable absorption, and saturable gain. *Opt. Lett.*, 10(3):131–133, 1985.

[31] G. H. C. New. Pulse evolution in mode locked quasi-continuous lasers. *IEEE J. Quantum Electron.*, 10:115–124, 1974.

[32] O. E. Martinez, R. L. Fork, and J. P. Gordon. Theory of passively mode-locked lasers for the case of a nonlinear complex-propagation coefficient. *J. Opt. Soc. Am. B*, 2(5):753–760, 1985.

[33] A. J. DeMaria. Picosecond laser pulses. In E. Wolf, editor, *Progress in Optics*, volume 9, pages 33–71. North-Holland, Amsterdam, 1971.

[34] R. Paschotta, J. A. der Au, G. J. Spuhler, F. Morier-Genoud, R. Hovel, M. Moser, S. Erhard, M. Karszewski, A. Giesen, and U. Keller. Diode-pumped passively mode-locked lasers with high average power. *Appl. Phys. B*, 70:S25–S31, 2000.

[35] U. Morgner, F. X. Kärtner, S. H. Cho, Y. Chen, H. A. Haus, J. G. Fujimoto, E. P. Ippen, V. Scheuer, G. Angelow, and T. Tschudi. Sub-two-cycle pulses from a Kerr-lens mode-locked Ti:sapphire laser. *Opt. Lett.*, 24(6):411–413, 1999.

[36] D. H. Sutter, G. Steinmeyer, L. Gallmann, N. Matuschek, F. Morier-Genoud, U. Keller, V. Scheuer, G. Angelow, and T. Tschudi. Semiconductor saturable-absorber mirror-assisted Kerr-lens mode-locked Ti:sapphire laser producing pulses in the two-cycle regime. *Opt. Lett.*, 24(9):631–633, 1999.

[37] G. P. Agrawal. *Nonlinear Fiber Optics*. Academic Press, San Diego, CA, 2nd edition, 1995.

[38] R. H. Stolen and C. Lin. Self-phase-modulation in silica optical fibers. *Phys. Rev. A*, 17(4):1448–1453, 1978.

[39] L. F. Mollenauer and R. H. Stolen. The soliton laser. *Opt. Lett.*, 9(1):13–15, 1984.

[40] K. J. Blow and D. Wood. Mode-locked lasers with nonlinear external cavities. *J. Opt. Soc. Am. B*, 5(3):629–632, 1988.

[41] K. J. Blow and B. P. Nelson. Improved mode-locking of an F-center laser with a nonlinear nonsoliton external cavity. *Opt. Lett.*, 13(11):1026–1028, 1988.

[42] P. N. Kean, X. Zhu, D. W. Crust, R. S. Grant, N. Langford, and W. Sibbett. Enhanced mode-locking of color-center lasers. *Opt. Lett.*, 14(1):39–41, 1989.

[43] J. Mark, L. Y. Liu, K. L. Hall, H. A. Haus, and E. P. Ippen. Femtosecond pulse generation in a laser with a nonlinear external resonator. *Opt. Lett.*, 14(1):48–50, 1989.

[44] E. P. Ippen, H. A. Haus, and L. Y. Liu. Additive pulse mode locking. *J. Opt. Soc. Am. B*, 6:1736–1745, 1989.

[45] C. P. Yakymyshyn, J. F. Pinto, and C. R. Pollock. Additive-pulse mode-locked NaCl:OH-laser. *Opt. Lett.*, 14(12):621–623, 1989.

[46] J. Goodberlet, J. Wang, J. G. Fujimoto, and P. A. Schulz. Femtosecond passively mode-locked Ti:Al_2O_3 laser with a nonlinear external cavity. *Opt. Lett.*, 14(20):1125–1127, 1989.

[47] J. Goodberlet, J. Jacobson, J. G. Fujimoto, P. A. Schulz, and T. Y. Fan. Self-starting additive-pulse mode-locked diode-pumped Nd:YAG laser. *Opt. Lett.*, 15(9):504–506, 1990.

[48] L. Y. Liu, J. M. Huxley, E. P. Ippen, and H. A. Haus. Self-starting additive-pulse mode-locking of a Nd:YAG laser. *Opt. Lett.*, 15(10):553–555, 1990.

[49] J. M. Liu and J. K. Chee. Passive mode locking of a CW Nd:YLF laser with a nonlinear external coupled cavity. *Opt. Lett.*, 15:685–687, 1990.

[50] F. Krausz, C. Spielmann, T. Brabec, E. Wintner, and A. J. Schmidt. Subpicosecond pulse generation from a Nd:glass laser using a nonlinear external cavity. *Opt. Lett.*, 15(13):737–739, 1990.

[51] H. A. Haus. Short pulse generation. In Irl N. Duling III, editor, *Compact Sources of Ultrashort Pulses*, pages 1–56. Cambridge University Press, New York, 1995.

[52] D. E. Spence, P. N. Kean, and W. Sibbett. 60-fsec pulse generation from a self-mode-locked Ti:sapphire laser. *Opt. Lett.*, 16(1):42–44, 1991.

[53] L. Spinelli, B. Couilland, N. Goldblatt, and D. K. Negus. Paper CPDP7. In *Conference on Lasers and Electro-optics*, volume 10 of OSA Technical Digest Series. Optical Society of America, Washington, DC, 1991.

[54] F. Salin, J. Squier, and M. Piche. Mode-locking of Ti:Al_2O_3 lasers and self-focusing—a Gaussian approximation. *Opt. Lett.*, 16(21):1674–1676, 1991.

[55] L. Xu, G. Tempea, A. Poppe, M. Lenzner, C. Spielmann, F. Krausz, A. Stingl, and K. Ferencz. High-power sub-10-fs Ti:sapphire oscillators. *Appl. Phys. B*, 65(2):151–159, 1997.

[56] C. P. Huang, M. T. Asaki, S. Backus, M. M. Murnane, H. C. Kapteyn, and H. Nathel. 17-fsec pulses from a self-mode-locked Ti:sapphire laser. *Opt. Lett.*, 17:1289, 1992.

[57] M. T. Asaki, C.-P. Huang, D. Garvey, J. Zhou, H. C. Kapteyn, and M. M. Murnane. Generation of 11-fs pulses from a self-mode-locked Ti:sapphire laser. *Opt. Lett.*, 18:977–979, 1993.

[58] A. Stingl, C. Spielmann, F. Krausz, and R. Szipocs. Generation of 11-fs pulses from a Ti:sapphire laser without the use of prisms. *Opt. Lett.*, 19(3):204–206, 1994.

[59] A. Stingl, M. Lenzner, C. Spielmann, F. Krausz, and R. Szipocs. Sub-10-fs mirror-dispersion-controlled Ti:sapphire laser. *Opt. Lett.*, 20(6):602–604, 1995.

[60] D. H. Sutter, I. D. Jung, F. X. Kärtner, N. Matuschek, F. Morier-Genoud, V. Scheuer, M. Tilsch, T. Tschudi, and U. Keller. Self-starting 6.5-fs pulses from a Ti:sapphire laser using a semiconductor saturable absorber and double-chirped mirrors. *IEEE J. Sel. Top. Quantum Electron.*, 4(2):169–178, 1998.

[61] M. Piche and F. Salin. Self-mode locking of solid-state lasers without apertures. *Opt. Lett.*, 18(13):1041–1043, 1993.

[62] D. Georgiev, J. Herrmann, and U. Stamm. Cavity design for optimum nonlinear absorption in Kerr-lens mode-locked solid-state lasers. *Opt. Commun.*, 92(4-6):368–375, 1992.

[63] F. Krausz, E. Wintner, A. J. Schmidt, and A. Dienes. Continuous wave thin plate Nd:glass-laser. *IEEE J. Quantum Electron.*, 26(1):158–168, 1990.

[64] T. Brabec, C. Spielmann, P. F. Curley, and F. Krausz. Kerr lens mode-locking. *Opt. Lett.*, 17(18):1292–1294, 1992.

[65] H. A. Haus, J. G. Fujimoto, and E. P. Ippen. Analytic theory of additive pulse and Kerr lens mode-locking. *IEEE J. Quantum Electron.*, 28(10):2086–2096, 1992.

[66] H. Kogelnik. Imaging of optical modes: resonators with internal lenses. *Bell Syst. Tech. J.*, 44:455–494, 1965.

[67] T. Brabec, P. F. Curley, C. Spielmann, E. Wintner, and A. J. Schmidt. Hard-aperture Kerr-lens mode-locking. *J. Opt. Soc. Am. B*, 10(6):1029–1034, 1993.

[68] J. H. Marburger. Self-focusing: theory. *Prog. Quantum Electron.*, 4:35–110, 1975.

[69] Y. R. Shen. *The Principles of Nonlinear Optics*. Wiley, New York, 1984.

[70] S. A. Akhmanov, A. D. Sukhorokov, and R. V. Khoklov. Self-focusing and diffraction of light in a nonlinear medium. *Sov. Phys. Uspekhi*, 10:609, 1968.

[71] A. Yariv and P. Yeh. The application of Gaussian beam formalism to optical propagation in nonlinear media. *Opt. Commun.*, 27:295–298, 1978.

[72] M. Sheik-bahae, A. A. Said, D. J. Hagan, M. J. Soileau, and E. W. Van Stryland. Nonlinear refraction and optical limiting in thick media. *Opt. Eng.*, 30(8):1228–1235, 1991.

[73] D. Huang, M. Ulman, L. H. Acioli, H. A. Haus, and J. G. Fujimoto. Self-focusing-induced saturable loss for laser mode-locking. *Opt. Lett.*, 17(7):511–513, 1992.

[74] V. Magni, G. Cerullo, and S. De Silvestri. *ABCD* matrix analysis of propagation of Gaussian beams through Kerr media. *Opt. Commun.*, 96(4–6):348–355, 1993.

[75] V. Magni, G. Cerullo, and S. De Silvestri. Closed-form Gaussian-beam analysis of resonators containing a Kerr medium for femtosecond lasers. *Opt. Commun.*, 101(5–6):365–370, 1993.

[76] G. Cerullo, S. De Silvestri, V. Magni, and L. Pallaro. Resonators for Kerr-lens mode-locked femtosecond Ti:sapphire lasers. *Opt. Lett.*, 19(11):807–809, 1994.

[77] M. Sheik-bahae, A. A. Said, T. H. Wei, D. J. Hagan, and E. W. Van Stryland. Sensitive measurement of optical nonlinearities using a single beam. *IEEE J. Quantum Electron.*, 26:760–769, 1990.

[78] H. W. Kogelnik, E. P. Ippen, A. Dienes, and C. V. Shank. Astigmatically compensated cavities for CW dye lasers. *IEEE J. Quantum Electron.*, 8(3):373–379, 1972.

[79] R. E. Bridges, R. W. Boyd, and G. P. Agrawal. Effect of beam ellipticity on self-mode locking in lasers. *Opt. Lett.*, 18(23):2026–2028, 1993.

[80] V. Magni, G. Cerullo, S. De Silvestri, and A. Monguzzi. Astigmatism in Gaussian-beam self-focusing and in resonators for Kerr-lens mode-locking. *J. Opt. Soc. Am. B*, 12(3):476–485, 1995.

[81] H. A. Haus, J. G. Fujimoto, and E. P. Ippen. Structures for additive pulse mode-locking. *J. Opt. Soc. Am. B*, 8(10):2068–2076, 1991.

[82] A. Hasegawa and F. Tappert. Transmission of stationary nonlinear optical pulses in dispersive dielectric fibers. I: Anomalous dispersion. *Appl. Phys. Lett.*, 23:142–144, 1973.

[83] L. F. Mollenauer, R. H. Stolen, and J. P. Gordon. Experimental observation of picosecond pulse narrowing and solitons in optical fibers. *Phys. Rev. Lett.*, 45:1095–1098, 1980.

[84] L. F. Mollenauer, M. J. Neubelt, S. G. Evangelides, J. P. Gordon, J. R. Simpson, and L. G. Cohen. Experimental study of soliton transmission over more than 10,000 km in dispersion-shifted fiber. *Opt. Lett.*, 15:1203–1205, 1990.

[85] M. S. Stix and E. P. Ippen. Pulse shaping in passively mode-locked ring dye lasers. *IEEE J. Quantum Electron.*, 19:520–525, 1983.

[86] E. P. Ippen, L. Y. Liu, and H. A. Haus. Self-starting condition for additive-pulse mode-locked lasers. *Opt. Lett.*, 15(3):183–185, 1990.

[87] F. Krausz, T. Brabec, and C. Spielmann. Self-starting passive-mode locking. *Opt. Lett.*, 16(4):235–237, 1991.

[88] F. Krausz and T. Brabec. Passive-mode locking in standing-wave laser resonators. *Opt. Lett.*, 18(11):888–890, 1993.

[89] H. A. Haus and E. P. Ippen. Self-starting of passively mode-locked lasers. *Opt. Lett.*, 16(17):1331–1333, 1991.

[90] C. Spielmann, F. Krausz, T. Brabec, E. Wintner, and A. J. Schmidt. Experimental-study of additive-pulse mode-locking in an Nd:glass laser. *IEEE J. Quantum Electron.*, 27(5):1207–1213, 1991.

[91] G. Cerullo, S. De Silvestri, and V. Magni. Self-starting Kerr-lens mode-locking of a Ti:sapphire laser. *Opt. Lett.*, 19(14):1040–1042, 1994.

[92] J. Herrmann. Starting dynamic, self-starting condition and mode-locking threshold in passive, coupled-cavity or Kerr-lens mode-locked solid-state lasers. *Opt. Commun.*, 98(1–3):111–116, 1993.

[93] S. Namiki, E. P. Ippen, H. A. Haus, and C. X. Yu. Energy rate equations for mode-locked lasers. *J. Opt. Soc. Am. B*, 14(8):2099–2111, 1997.

[94] E. Sorokin, G. Tempea, and T. Brabec. Measurement of the root-mean-square width and the root-mean-square chirp in ultrafast optics. *J. Opt. Soc. Am. B*, 17(1):146–150, 2000.

[95] J.-C. Diels and W. Rudolph. *Ultrashort Laser Pulse Phenomena*. Academic Press, San Diego, CA, 1996.

[96] M. Born and E. Wolf. *Principles of Optics: Electromagnetic Theory of Propagation, Interference, and Diffraction of Light*. Cambridge University Press, New York, 7th edition, 1999.

[97] K. Naganuma, K. Mogi, and H. Yamada. Group-delay measurement using the Fourier-transform of an interferometric cross-correlation generated by white-light. *Opt. Lett.*, 15(7):393–395, 1990.

[98] K. Naganuma and Y. Sakai. Interferometric measurement of wavelength dispersion on femtosecond laser cavities. *Opt. Lett.*, 19(7):487–489, 1994.

[99] F. Reynaud, F. Salin, and A. Barthelemy. Measurement of phase-shifts introduced by nonlinear optical phenomena on subpicosecond pulses. *Opt. Lett.*, 14(5):275–277, 1989.

[100] K. Minoshima, M. Taiji, and T. Kobayashi. Femtosecond time-resolved interferometry for the determination of complex nonlinear susceptibility. *Opt. Lett.*, 16(21):1683–1685, 1991.

[101] L. Lepetit, G. Cheriaux, and M. Joffre. Linear techniques of phase measurement by femtosecond spectral interferometry for applications in spectroscopy. *J. Opt. Soc. Am. B*, 12(12):2467–2474, 1995.

[102] H. T. Shang. Chromatic dispersion measurement by white-light interferometry on meter-length single-mode optical fibers. *Electron. Lett.*, 17(17):603–605, 1981.

[103] J. Stone and D. Marcuse. Direct measurement of 2nd-order dispersion in short optical fibers using white-light interferometry. *Electron. Lett.*, 20(18):751–752, 1984.

[104] D. Huang, E. A. Swanson, C. P. Lin, J. S. Schuman, W. G. Stinson, W. Chang, M. R. Hee, T. Flotte, K. Gregory, C. A. Puliafito, and J. G. Fujimoto. Optical coherence tomography. *Science*, 254(5035):1178–1181, 1991.

[105] A. F. Fercher, W. Drexler, C. K. Hitzenberger, and T. Lasser. Optical coherence tomography: principles and applications. *Rep. Prog. Phys.*, 66(2):239–303, 2003.

[106] W. Drexler, U. Morgner, F. X. Kärtner, C. Pitris, S. A. Boppart, X. D. Li, E. P. Ippen, and J. G. Fujimoto. In vivo ultrahigh-resolution optical coherence tomography. *Opt. Lett.*, 24(17):1221–1223, 1999.

[107] Z. P. Chen, T. E. Milner, S. Srinivas, X. J. Wang, A. Malekafzali, M. J. C. vanGemert, and J. S. Nelson. Noninvasive imaging of in vivo blood flow velocity using optical Doppler tomography. *Opt. Lett.*, 22(14):1119–1121, 1997.

[108] J. A. Izatt, M. D. Kulkami, S. Yazdanfar, J. K. Barton, and A. J. Welch. In vivo bidirectional color Doppler flow imaging of picoliter blood volumes using optical coherence tomography. *Opt. Lett.*, 22(18):1439–1441, 1997.

[109] H. P. Weber. Method for pulsewidth measurement of ultrashort light pulses generated by phase-locked lasers using nonlinear optics. *J. Appl. Phys.*, 38:2231–2234, 1967.

[110] H. P. Weber. Generation and measurement of ultrashort light pulses. *J. Appl. Phys.*, 39:6041–6044, 1968.

[111] J. A. Armstrong. Measurement of picosecond laser pulse widths. *Appl. Phys. Lett.*, 10:16, 1967.

[112] M. Maier, W. Kaiser, and J. A. Giordmaine. Intense light bursts in the stimulated Raman effect. *Phys. Rev. Lett*, 17:1275–1277, 1966.

[113] E. P. Ippen and C. V. Shank. Techniques for measurement. In S. L. Shapiro, editor, *Ultrashort Light Pulses: Picosecond Techniques and Applications*, volume 18, pages 83–122. Springer-Verlag, New York, 2nd edition, 1977.

[114] J. C. M. Diels, J. J. Fontaine, I. C. McMichael, and F. Simoni. Control and measurement of ultrashort pulse shapes (in amplitude and phase) with femtosecond accuracy. *Appl. Opt.*, 24(9):1270–1282, 1985.

[115] K. Naganuma, K. Mogi, and H. Yamada. General-method for ultrashort light-pulse chirp measurement. *IEEE J. Quantum Electron.*, 25(6):1225–1233, 1989.

[116] J. H. Chung and A. M. Weiner. Ambiguity of ultrashort pulse shapes retrieved from the intensity autocorrelation and the power spectrum. *IEEE J. Sel. Top. Quantum Electron.*, 7(4):656–666, 2001.

[117] E. W. Van Stryland. Effect of pulse to pulse variation on ultrashort pulsewidth measurements. *Opt. Commun.*, 31(1):93–96, 1979.

[118] J. Jansky, G. Corradi, and R.N. Gyuzalian. On a possibility of analyzing the temporal characteristics of short light pulses. *Opt. Commun.*, 23:293–298, 1977.

[119] G. Taft, A. Rundquist, M. M. Murnane, I. P. Christov, H. C. Kapteyn, K. W. DeLong, D. N. Fittinghoff, M. A. Krumbugel, J. N. Sweetser, and R. Trebino. Measurement of 10-fs laser pulses. *IEEE J. Sel. Top. Quantum Electron.*, 2(3):575–585, 1996.

[120] J. A. Giordmaine, P. M. Rentzepis, S. L. Shapiro, and K. W. Wecht. Two-photon excitation of fluorescence by picosecond light pulses. *Appl. Phys. Lett.*, 11:216–218, 1967.

[121] J. R. Klauder, M. A. Duguay, J. A. Giordmaine, and S. L. Shapiro. Correlation effects in the display of picosecond pulses by two-photon techniques. *Appl. Phys. Lett.*, 13:174–176, 1968.

[122] Y. Takagi, T. Kobayashi, K. Yoshihara, and S. Imamura. Multiple-shot and single-shot autocorrelator based on 2-photon conductivity in semiconductors. *Opt. Lett.*, 17(9):658–660, 1992.

[123] F. R. Laughton, J. H. Marsh, D. A. Barrow, and E. L. Portnoi. The 2-photon absorption semiconductor wave-guide autocorrelator. *IEEE J. Quantum Electron.*, 30(3):838–845, 1994.

[124] J. K. Ranka, A. L. Gaeta, A. Baltuska, M. S. Pshenichnikov, and D. A. Wiersma. Autocorrelation measurement of 6-fs pulses based on the two-photon-induced photocurrent in a GaAsP photodiode. *Opt. Lett.*, 22(17):1344–1346, 1997.

[125] D. T. Reid, M. Padgett, C. McGowan, W. E. Sleat, and W. Sibbett. Light-emitting diodes as measurement devices for femtosecond laser pulses. *Opt. Lett.*, 22(4):233–235, 1997.

[126] W. Rudolph, M. Sheik-Bahae, A. Bernstein, and L. F. Lester. Femtosecond autocorrelation measurements based on two-photon photoconductivity in ZnSe. *Opt. Lett.*, 22(5):313–315, 1997.

[127] Z. Zheng, A. M. Weiner, J. H. Marsh, and M. M. Karkhanehchi. Ultrafast optical thresholding based on two-photon absorption GaAs waveguide photodetectors. *IEEE Photon. Technol. Lett.*, 9(4):493–495, 1997.

[128] D. H. Auston. Measurement of picosecond pulse shape and background level. *Appl. Phys. Lett.*, 18:249, 1971.

[129] D. Meshulach, Y. Barad, and Y. Silberberg. Measurement of ultrashort optical pulses by third-harmonic generation. *J. Opt. Soc. Am. B*, 14(8):2122–2125, 1997.

[130] P. Langlois and E. P. Ippen. Measurement of pulse asymmetry by three-photon-absorption autocorrelation in a GaAsP photodiode. *Opt. Lett.*, 24(24):1868–1870, 1999.

[131] R. W. Boyd. *Nonlinear Optics*. Academic Press, San Diego, CA, 2nd edition, 2003.

[132] R. A. Altes. Detection, estimation, and classification with spectrograms. *J. Acoust. Soc. Am.*, 67(4):1232–1246, 1980.

[133] L. Cohen. Time frequency-distributions: a review. *Proc. IEEE*, 77(7):941–981, 1989.

[134] J. Paye. The chronocyclic representation of ultrashort light-pulses. *IEEE J. Quantum Electron.*, 28(10):2262–2273, 1992.

[135] J. Paye. How to measure the amplitude and phase of an ultrashort light- pulse with an autocorrelator and a spectrometer. *IEEE J. Quantum Electron.*, 30(11):2693–2697, 1994.

[136] I. A. Walmsley and V. Wong. Characterization of the electric field of ultrashort optical pulses. *J. Opt. Soc. Am. B*, 13(11):2453–2463, 1996.

[137] R. Trebino, editor. *Frequency-Resolved Optical Gating: The Measurement of Ultrashort Laser Pulses*. Kluwer Academic Publishers, Norwell, MA, 2000.

[138] R. Trebino and D. J. Kane. Using phase retrieval to measure the intensity and phase of ultrashort pulses: frequency-resolved optical gating. *J. Opt. Soc. Am. A*, 10(5):1101–1111, 1993.

[139] K. W. Delong, R. Trebino, and D. J. Kane. Comparison of ultrashort-pulse frequency-resolved-optical-gating traces for 3 common beam geometries. *J. Opt. Soc. Am. B*, 11(9):1595–1608, 1994.

[140] R. Trebino, K. W. DeLong, D. N. Fittinghoff, J. N. Sweetser, M. A. Krumbugel, B. A. Richman, and D. J. Kane. Measuring ultrashort laser pulses in the time–frequency domain using frequency-resolved optical gating. *Rev. Sci. Instrum.*, 68(9):3277–3295, 1997.

[141] D. J. Kane and R. Trebino. Characterization of arbitrary femtosecond pulses using frequency-resolved optical gating. *IEEE J. Quantum Electron.*, 29(2):571–579, 1993.

[142] J. Paye, M. Ramaswamy, J. G. Fujimoto, and E. P. Ippen. Measurement of the amplitude and phase of ultrashort light pulses from spectrally resolved autocorrelation. *Opt. Lett.*, 18(22):1946–1948, 1993.

[143] K. W. Delong, R. Trebino, J. Hunter, and W. E. White. Frequency-resolved optical gating with the use of 2nd-harmonic generation. *J. Opt. Soc. Am. B*, 11(11):2206–2215, 1994.

[144] E. T. J. Nibbering, M. A. Franco, B. S. Prade, G. Grillon, J. P. Chambaret, and A. Mysyrowicz. Spectral determination of the amplitude and the phase of intense ultrashort optical pulses. *J. Opt. Soc. Am. B*, 13(2):317–329, 1996.

[145] H. R. Lange, M. A. Franco, J. F. Ripoche, B. S. Prade, P. Rousseau, and A. Mysyrowicz. Reconstruction of the time profile of femtosecond laser pulses through cross-phase modulation. *IEEE J. Sel. Top. Quantum Electron.*, 4(2):295–300, 1998.

[146] J. M. Dudley, L. P. Barry, P. G. Bollond, J. D. Harvey, R. Leonhardt, and P. D. Drummond. Direct measurement of pulse distortion near the zero-dispersion wavelength in an optical fiber by frequency-resolved optical gating. *Opt. Lett.*, 22(7):457–459, 1997.

[147] H. Stark. *Image Recovery: Theory and Applications*. Academic Press, Orlando, FL, 1987.

[148] J. R. Fienup. Phase retrieval algorithms: a comparison. *Appl. Opt.*, 21(15):2758–2769, 1982.

[149] J. R. Fienup. Reconstruction of a complex-valued object from the modulus of its Fourier-transform using a support constraint. *J. Opt. Soc. Am. A*, 4(1):118–123, 1987.

[150] R. P. Millane. Phase retrieval in crystallography and optics. *J. Opt. Soc. Am. A*, 7(3):394–411, 1990.

[151] R. W. Gerchberg and W. O. Saxton. A practical algorithm for the determination of phase from image and diffraction plane pictures. *Optik*, 35(2):237–246, 1972.

[152] K. W. DeLong, D. N. Fittinghoff, and R. Trebino. Practical issues in ultrashort-laser-pulse measurement using frequency-resolved optical gating. *IEEE J. Quantum Electron.*, 32(7):1253–1264, 1996.

[153] D. J. Kane. Recent progress toward real-time measurement of ultrashort laser pulses. *IEEE J. Quantum Electron.*, 35(4):421–431, 1999.

[154] D. N. Fittinghoff, J. L. Bowie, J. N. Sweetser, R. T. Jennings, M. A. Krumbugel, K. W. DeLong, R. Trebino, and I. A. Walmsley. Measurement of the intensity and phase of ultraweak, ultrashort laser pulses. *Opt. Lett.*, 21(12):884–886, 1996.

[155] A. M. Weiner. Femtosecond optical pulse shaping and processing. *Prog. Quantum Electron.*, 19:161–238, 1995.

[156] J. L. A. Chilla and O. E. Martinez. Analysis of a method of phase measurement of ultrashort pulses in the frequency-domain. *IEEE J. Quantum Electron.*, 27(5):1228–1235, 1991.

[157] J. L. A. Chilla and O. E. Martinez. Direct determination of the amplitude and the phase of femtosecond light-pulses. *Opt. Lett.*, 16(1):39–41, 1991.

[158] D. T. Reid. Algorithm for complete and rapid retrieval of ultrashort pulse amplitude and phase from a sonogram. *IEEE J. Quantum Electron.*, 35(11):1584–1589, 1999.

[159] T. K. Rhee, T. S. Sosnowski, T. B. Norris, J. A. Arns, and W. S. Colburn. Chirped-pulse amplification of 85-fs pulses at 250 kHz with 3rd-order dispersion compensation by use of holographic transmission gratings. *Opt. Lett.*, 19(19):1550–1552, 1994.

[160] J. K. Rhee, T. S. Sosnowski, A. C. Tien, and T. B. Norris. Real-time dispersion analyzer of femtosecond laser pulses with use of a spectrally and temporally resolved upconversion technique. *J. Opt. Soc. Am. B*, 13(8):1780–1785, 1996.

[161] J. P. Heritage, A. M. Weiner, R. J. Hawkins, and O. E. Martinez. Stabilized pulse-compression by multiple-order stimulated Raman-scattering with group-velocity dispersion. *Opt. Commun.*, 67(5):367–372, 1988.

[162] K. C. Chu, J. P. Heritage, R. S. Grant, K. X. Liu, A. Dienes, W. E. White, and A. Sullivan. Direct measurement of the spectral phase of femtosecond pulses. *Opt. Lett.*, 20(8):904–906, 1995.

[163] K. C. Chu, J. P. Heritage, R. S. Grant, and W. E. White. Temporal interferometric measurement of femtosecond spectral phase. *Opt. Lett.*, 21(22):1842–1844, 1996.

[164] J. E. Rothenberg and D. Grischkowsky. Measurement of optical-phase with subpicosecond resolution by time-domain interferometry. *Opt. Lett.*, 12(2):99–101, 1987.

[165] J. E. Rothenberg and D. Grischkowsky. Measurement of the phase of a frequency-swept ultrashort optical pulse. *J. Opt. Soc. Am. B*, 2(4):626–633, 1985.

[166] C. Iaconis and I. A. Walmsley. Spectral phase interferometry for direct electric-field reconstruction of ultrashort optical pulses. *Opt. Lett.*, 23(10):792–794, 1998.

[167] C. Iaconis and I. A. Walmsley. Self-referencing spectral interferometry for measuring ultrashort optical pulses. *IEEE J. Quantum Electron.*, 35(4):501–509, 1999.

[168] J. W. Goodman. *Introduction to Fourier Optics*. McGraw-Hill, New York, 1968.

[169] J. R. Birge, R. Ell, and F. X. Kärtner. Two-dimensional spectral shearing interferometry for few-cycle pulse characterization. *Opt. Lett.*, 31(13):2063–2065, 2006.

[170] D. von der Linde. Characterization of the noise in continuously operating mode-locked lasers. *Appl. Phys. B*, 39(4):201–217, 1986.

[171] D. Eliyahu, R. A. Salvatore, and A. Yariv. Noise characterization of a pulse train generated by actively mode-locked lasers. *J. Opt. Soc. Am. B*, 13(7):1619–1626, 1996.

[172] H. A. Haus and A. Mecozzi. Noise of mode-locked lasers. *IEEE J. Quantum Electron.*, 29(3):983–996, 1993.

[173] S. Namiki, C. X. Yu, and H. A. Haus. Observation of nearly quantum-limited timing jitter in an all-fiber ring laser. *J. Opt. Soc. Am. B*, 13(12):2817–2823, 1996.

[174] D. Eliyahu, R. A. Salvatore, and A. Yariv. Effect of noise on the power spectrum of passively mode-locked lasers. *J. Opt. Soc. Am. B*, 14(1):167–174, 1997.

[175] I. G. Fuss. An interpretation of the spectral measurement of optical pulse-train noise. *IEEE J. Quantum Electron.*, 30(11):2707–2710, 1994.

[176] F. Jenkins and H. White. *Fundamentals of Optics*. McGraw-Hill, New York, 4th edition, 1976.

[177] Z. Bor and B. Racz. Dispersion of optical-materials used for picosecond spectroscopy. *Appl. Opt.*, 24(21):3440–3441, 1985.

[178] O. E. Martinez, J. P. Gordon, and R. L. Fork. Negative group-velocity dispersion using refraction. *J. Opt. Soc. Am. A*, 1(10):1003–1006, 1984.

[179] Z. Bor and B. Racz. Group-velocity dispersion in prisms and its application to pulse-compression and traveling-wave excitation. *Opt. Commun.*, 54(3):165–170, 1985.

[180] O. E. Martinez. Pulse distortions in tilted pulse schemes for ultrashort pulses. *Opt. Commun.*, 59(3):229–232, 1986.

[181] J. Hebling. Derivation of the pulse front tilt caused by angular dispersion. *Opt. Quantum Electron.*, 28(12):1759–1763, 1996.

[182] S. Szatmari, G. Kuhnle, and P. Simon. Pulse-compression and traveling-wave excitation scheme using a single dispersive element. *Appl. Opt.*, 29(36):5372–5379, 1990.

[183] A. E. Treacy. Optical pulse compression with diffraction gratings. *IEEE J. Quantum Electron.*, 5:454, 1969.

[184] M. Stern, J. P. Heritage, and E. W. Chase. Grating compensation of 3rd-order fiber dispersion. *IEEE J. Quantum Electron.*, 28(12):2742–2748, 1992.

[185] S. Kane and J. Squier. Grating compensation of 3rd-order material dispersion in the normal dispersion regime: sub-100-fs chirped-pulse amplification using a fiber stretcher and grating-pair compressor. *IEEE J. Quantum Electron.*, 31(11):2052–2057, 1995.

[186] A. M. Johnson, R. H. Stolen, and W. M. Simpson. 80 $\times$ single-stage compression of frequency doubled Nd-yttrium aluminum garnet laser-pulses. *Appl. Phys. Lett.*, 44(8):729–731, 1984.

[187] C. Froehly, B. Colombeau, and M. Vampouille. Shaping and analysis of picosecond light pulses. In E. Wolf, editor, *Progress in Optics*, volume 20, pages 65–153. North-Holland, Amsterdam, 1983.

[188] O. E. Martinez. 3000 times grating compressor with positive group-velocity dispersion: application to fiber compensation in 1.3–1.6 μm region. *IEEE J. Quantum Electron.*, 23(1):59–64, 1987.

[189] R. L. Fork, O. E. Martinez, and J. P. Gordon. Negative dispersion using pairs of prisms. *Opt. Lett.*, 9(5):150–152, 1984.

[190] R. L. Fork, C. H. B. Cruz, P. C. Becker, and C. V. Shank. Compression of optical pulses to 6 femtoseconds by using cubic phase compensation. *Opt. Lett.*, 12(7):483–485, 1987.

[191] C. H. B. Cruz, P. C. Becker, R. L. Fork, and C. V. Shank. Phase correction of femtosecond optical pulses using a combination of prisms and gratings. *Opt. Lett.*, 13(2):123–125, 1988.

[192] B. Proctor and F. Wise. Quartz prism sequence for reduction of cubic phase in a mode-locked Ti:Al_2O_3 laser. *Opt. Lett.*, 17(18):1295–1297, 1992.

[193] B. E. Lemoff and C. P. J. Barty. Generation of high-peak-power 20-fs pulses from a regeneratively initiated, self-mode-locked Ti:sapphire laser. *Opt. Lett.*, 17(19):1367–1369, 1992.

[194] Z. Bor. Distortion of femtosecond laser-pulses in lenses and lens systems. *J. Mod. Opt.*, 35(12):1907–1918, 1988.

[195] M. Kempe, U. Stamm, B. Wilhelmi, and W. Rudolph. Spatial and temporal transformation of femtosecond laser-pulses by lenses and lens systems. *J. Opt. Soc. Am. B*, 9(7):1158–1165, 1992.

[196] Z. Bor, Z. Gogolak, and G. Szabo. Femtosecond-resolution pulse-front distortion measurement by time-of-flight interferometry. *Opt. Lett.*, 14(16):862–864, 1989.

[197] M. Kempe and W. Rudolph. Impact of chromatic and spherical-aberration on the focusing of ultrashort light-pulses by lenses. *Opt. Lett.*, 18(2):137–139, 1993.

[198] F. Gires and P. Tournois. Interferometre utilisable pour la compression d'impulsions lumineuses modulees en frequence. *C.R. Hebd. Acad. Sci.*, 258:6112–6115, 1964.

[199] J. Kuhl and J. Heppner. Compression of femtosecond optical pulses with dielectric multilayer interferometers. *IEEE J. Quantum Electron.*, 22(1):182–185, 1986.

[200] K. D. Li, W. H. Knox, and N. M. Pearson. Broad-band cubic-phase compensation with resonant Gires–Tournois interferometers. *Opt. Lett.*, 14(9):450–452, 1989.

[201] H. A. Macleod. *Thin-Film Optical Filters*. Macmillan, New York, 2nd edition, 1986.

[202] J. A. Dobrowolski. Optical properties of films and coatings. In Michael Bass, editor, *Handbook of Optics*, volume 1. McGraw-Hill, New York, 2nd edition, 1995.

[203] S. De Silvestri, P. Laporta, and O. Svelto. Analysis of quarter-wave dielectric-mirror dispersion in femtosecond dye-laser cavities. *Opt. Lett.*, 9(8):335–337, 1984.

[204] A. M. Weiner, J. G. Fujimoto, and E. P. Ippen. Femtosecond time-resolved reflectometry measurements of multiple-layer dielectric mirrors. *Opt. Lett.*, 10(2):71–73, 1985.

[205] W. H. Knox. Dispersion measurements for femtosecond-pulse generation and applications. *Appl. Phys. B*, 58(3):225–235, 1994.

[206] P. Laporta and V. Magni. Dispersive effects in the reflection of femtosecond optical pulses from broad-band dielectric mirrors. *Appl. Opt.*, 24(13):2014–2020, 1985.

[207] W. H. Knox, N. M. Pearson, K. D. Li, and C. A. Hirlimann. Interferometric measurements of femtosecond group delay in optical-components. *Opt. Lett.*, 13(7):574–576, 1988.

[208] N. Matuschek, F. X. Kärtner, and U. Keller. Theory of double-chirped mirrors. *IEEE J. Sel. Top. Quantum Electron.*, 4(2):197–208, 1998.

[209] R. Szipöcs, K. Ferencz, C. Spielmann, and F. Krausz. Chirped multilayer coatings for broadband dispersion control in femtosecond lasers. *Opt. Lett.*, 19(3):201–203, 1994.

[210] R. Szipöcs and A. Kohazikis. Theory and design of chirped dielectric laser mirrors. *Appl. Phys. B*, 65(2):115–135, 1997.

[211] N. Matuschek, F. X. Kärtner, and U. Keller. Exact coupled-mode theories for multilayer interference coatings with arbitrary strong index modulations. *IEEE J. Quantum Electron.*, 33(3):295–302, 1997.

[212] N. Matuschek, F. X. Kärtner, and U. Keller. Analytical design of double-chirped mirrors with custom-tailored dispersion characteristics. *IEEE J. Quantum Electron.*, 35(2):129–137, 1999.

[213] L. G. Cohen. Comparison of single-mode fiber dispersion measurement techniques. *J. Lightwave Technol.*, 3(5):958–966, 1985.

[214] K. Naganuma. Semiconductor-laser cavity dispersion measurement based on interferometric cross-correlation of amplified spontaneous emission. *Appl. Phys. Lett.*, 64(3):261–263, 1994.

[215] A. P. Kovacs, K. Osvay, Z. Bor, and R. Szipöcs. Group-delay measurement on laser mirrors by spectrally resolved white-light interferometry. *Opt. Lett.*, 20(7):788–790, 1995.

[216] D. Meshulach, D. Yelin, and Y. Silberberg. Real-time spatial-spectral interference measurements of ultrashort optical pulses. *J. Opt. Soc. Am. B*, 14(8):2095–2098, 1997.

[217] W. H. Knox. In situ measurement of complete intracavity dispersion in an operating Ti:sapphire femtosecond laser. *Opt. Lett.*, 17(7):514–516, 1992.

[218] W. H. Knox and J. P. Gordon. Frequency-domain dispersion measurements in tunable mode-locked lasers. *J. Opt. Soc. Am. B*, 10(11):2071–2079, 1993.

[219] N. Bloembergen. *Nonlinear Optics*, volume 23. Addison-Wesley, Redwood City, CA, 1992.

[220] V. G. Dmitriev, G. G. Gurzadyan, and D. N. Nikogosyan. *Handbook of Nonlinear Optical Crystals*. Springer-Verlag, New York, 2nd edition, 1997.

[221] T. Brabec and F. Krausz. Intense few-cycle laser fields: frontiers of nonlinear optics. *Rev. Mod. Phys.*, 72(2):545–591, 2000.

[222] D. Umstadter. Review of physics and applications of relativistic plasmas driven by ultra-intense lasers. *Phys. Plasmas*, 8(5):1774–1785, 2001.

[223] G. A. Mourou, C. P. J. Barty, and M. D. Perry. Ultrahigh-intensity lasers: physics of the extreme on a tabletop. *Phys. Today*, 51(1):22–28, 1998.

[224] C. J. Joshi and P. B. Corkum. Interactions of ultra-intense laser-light with matter. *Phys. Today*, 48(1):36–43, 1995.

[225] F. Zernike and J. E. Midwinter. *Applied Nonlinear Optics*. Wiley, New York, 1973.

[226] J. A. Giordmaine. Mixing of light beams in crystals. *Phys. Rev. Lett.*, 8:19–20, 1962.

[227] P. D. Maker, R. W. Terhune, M. Nisenoff, and C. M. Savage. Effects of dispersion and focusing on the production of optical harmonics. *Phys. Rev. Lett.*, 8:21–22, 1962.

[228] A. Yariv and P. Yeh. *Optical Waves in Crystals: Propagation and Control of Laser Radiation*. Wiley, New York, 1984.

[229] G. D. Boyd and D. A. Kleinman. Parametric interaction of focused Gaussian light beams. *J. Appl. Phys.*, 39:3597–3639, 1968.

[230] J. Comly and E. Garmire. Second harmonic generation from short pulses. *Appl. Phys. Lett.*, 12:7–9, 1968.

[231] S. A. Akhmanov, A. P. Sukhorukov, and A. S. Chirkin. Nonstationary phenomena and space–time analogy in nonlinear optics. *Sovi. Phys. JETP*, 28:748–757, 1969.

[232] W. H. Glenn. Second-harmonic generation by picosecond optical pulses. *IEEE J. Quantum Electron.*, 5:281–290, 1969.

[233] S. A. Akhmanov, V. A. Vysloukh, and A. S. Chirkin. *Optics of Femtosecond Laser Pulses*. American Institute of Physics, New York, 1992.

[234] H. Wang and A. M. Weiner. Efficiency of short pulse type I second harmonic generation with simultaneous spatial walk-off, temporal walk-off and pump depletion. *IEEE J. Quantum Electron.*, 39:1600–1618, 2003.

[235] A. M. Weiner, A. M. Kanan, and D. E. Leaird. High-efficiency blue generation by frequency doubling of femtosecond pulses in a thick nonlinear crystal. *Opt. Lett.*, 23:1441–1443, 1998.

[236] J. A. Armstrong, N. Bloembergen, J. Ducuing, and P. S. Pershan. Interactions between light waves in a nonlinear dielectric. *Phys. Rev.*, 127:1918–1939, 1962.

[237] P. A. Franken and J. F. Ward. Optical harmonics and nonlinear phenomena. *Rev. Mod. Phys.*, 35:23–39, 1963.

[238] M. M. Fejer, G. A. Magel, D. H. Jundt, and R. L. Byer. Quasi-phase-matched 2nd harmonic-generation: tuning and tolerances. *IEEE J. Quantum Electron.*, 28(11):2631–2654, 1992.

[239] G. Imeshev, M. A. Arbore, M. M. Fejer, A. Galvanauskas, M. Fermann, and D. Harter. Ultrashort-pulse second-harmonic generation with longitudinally nonuniform quasi-phase-matching gratings: pulse compression and shaping. *J. Opt. Soc. Am. B*, 17(2):304–318, 2000.

[240] R. L. Byer. Quasi-phase-matched nonlinear interactions and devices. *J. Nonlinear Opt. Phys.*, 6(4):549–592, 1997.

[241] G. Imeshev, M. A. Arbore, S. Kasriel, and M. M. Fejer. Pulse shaping and compression by second-harmonic generation with quasi-phase-matching gratings in the presence of arbitrary dispersion. *J. Opt. Soc. Am. B*, 17(8):1420–1437, 2000.

[242] M. A. Arbore, O. Marco, and M. M. Fejer. Pulse compression during second-harmonic generation in aperiodic quasi-phase-matching gratings. *Opt. Lett.*, 22(12):865–867, 1997.

[243] M. A. Arbore, A. Galvanauskas, D. Harter, M. H. Chou, and M. M. Fejer. Engineerable compression of ultrashort pulses by use of second-harmonic generation in chirped-period-poled lithium niobate. *Opt. Lett.*, 22(17):1341–1343, 1997.

[244] G. Imeshev, A. Galvanauskas, D. Harter, M. A. Arbore, M. Proctor, and M. M. Fejer. Engineerable femtosecond pulse shaping by second-harmonic generation with Fourier synthetic quasi-phase-matching gratings. *Opt. Lett.*, 23(11):864–866, 1998.

[245] A. M. Weiner. Effect of group velocity mismatch on the measurement of ultrashort optical pulses via second harmonic generation. *IEEE J. Quantum Electron*, 19:1276–1283, 1983.

[246] A. Laubereau. Optical nonlinearities with ultrashort pulses. In W. Kaiser, editor, *Ultrashort Laser Pulses and Applications*, volume 60 of Topics in Applied Physics. Springer-Verlag, Berlin, 1988.

[247] J. M. Manley and H. E. Rowe. Some general properties of nonlinear elements. Part I: general energy relations. *Proc. IRE*, 44:904–914, 1956.

[248] J. Shah. Ultrafast luminescence spectroscopy using sum frequency generation. *IEEE J. Quantum Electron.*, 24(2):276–288, 1988.

[249] J. Shah. *Ultrafast Spectroscopy of Semiconductors and Semiconductor Nanostructures*. Springer-Verlag, New York, 1999.

[250] H. F. Wang and A. M. Weiner. A femtosecond waveform transfer technique using type II second harmonic generation. *IEEE J. Quantum Electron.*, 40(7):937–945, 2004.

[251] R. A. Kaindl, D. C. Smith, M. Joschko, M. P. Hasselbeck, M. Woerner, and T. Elsaesser. Femtosecond infrared pulses tunable from 9 to 18 μm at an 88-MHz repetition rate. *Opt. Lett.*, 23:861–863, 1998.

[252] R. A. Kaindl, F. Eickemeyer, M. Woerner, and T. Elsaesser. Broadband phase-matched difference frequency mixing of femtosecond pulses in GaSe: experiment and theory. *Appl. Phys. Lett.*, 75(8):1060–1062, 1999.

[253] F. Eickemeyer, R. A. Kaindl, M. Woerner, T. Elsaesser, and A. M. Weiner. Controlled shaping of ultrafast electric field transients in the mid-infrared spectral range. *Opt. Lett.*, 25(19):1472–1474, 2000.

[254] R. Huber, A. Brodschelm, F. Tauser, and A. Leitenstorfer. Generation and field-resolved detection of femtosecond electromagnetic pulses tunable up to 41 THz. *Appl. Phys. Lett.*, 76(22):3191–3193, 2000.

[255] W. S. Pelouch, P. E. Powers, and C. L. Tang. Ti:sapphire-pumped, high-repetition-rate femtosecond optical parametric oscillator. *Opt. Lett.*, 17(15):1070–1072, 1992.

[256] Q. Fu, G. Mak, and H. M. van Driel. High-power, 62-fs infrared optical parametric oscillator synchronously pumped by a 76 MHz Ti:sapphire laser. *Opt. Lett.*, 17(14):1006–1008, 1992.

[257] G. M. Gale, M. Cavallari, T. J. Driscoll, and F. Hache. Sub-20-fs tunable pulses in the visible from an 82-MHz optical parametric oscillator. *Opt. Lett.*, 20(14):1562–1564, 1995.

[258] G. M. Gale, M. Cavallari, and F. Hache. Femtosecond visible optical parametric oscillator. *J. Opt. Soc. Am. B*, 15(2):702–714, 1998.

[259] T. Wilhelm, J. Piel, and E. Riedle. Sub-20-fs pulses tunable across the visible from a blue-pumped single-pass noncollinear parametric converter. *Opt. Lett.*, 22(19):1494–1496, 1997.

[260] A. Shirakawa and T. Kobayashi. Noncollinear phase- and group-velocity matching of optical parametric amplifier for ultrashort pulse generation. *IEICE Trans. Electron.*, E81C(2):246–253, 1998.

[261] G. Cerullo, M. Nisoli, S. Stagira, and S. De Silvestri. Sub-8-fs pulses from an ultrabroadband optical parametric amplifier in the visible. *Opt. Lett.*, 23(16):1283–1285, 1998.

[262] E. Riedle, M. Beutter, S. Lochbrunner, J. Piel, S. Schenkl, S. Sporlein, and W. Zinth. Generation of 10 to 50 fs pulses tunable through all of the visible and the NIR. *Appl. Phys. B*, 71(3):457–465, 2000.

[263] A. Baltuska, T. Fuji, and T. Kobayashi. Visible pulse compression to 4 fs by optical parametric amplification and programmable dispersion control. *Opt. Lett.*, 27(5):306–308, 2002.

[264] M. Zavelani-Rossi, G. Cerullo, S. De Silvestri, L. Gallmann, N. Matuschek, G. Steinmeyer, U. Keller, G. Angelow, V. Scheuer, and T. Tschudi. Pulse compression over a 170-THz bandwidth in the visible by use of only chirped mirrors. *Opt. Lett.*, 26(15):1155–1157, 2001.

[265] P. Di Trapani, A. Andreoni, P. Foggi, C. Solcia, R. Danielius, and A. Piskarskas. Efficient conversion of femtosecond blue pulses by traveling-wave parametric generation in noncollinear phase-matching. *Opt. Commun.*, 119(3–4):327–332, 1995.

[266] Y. Kodama and A. Hasegawa. Nonlinear pulse-propagation in a monomode dielectric guide. *IEEE J. Quantum Electron.*, 23(5):510–524, 1987.

[267] K. J. Blow and D. Wood. Theoretical description of transient stimulated Raman-scattering in optical fibers. *IEEE J. Quantum Electron.*, 25(12):2665–2673, 1989.

[268] A. W. Snyder and J. D. Love. *Optical Waveguide Theory*. Chapman & Hall, London, 1983.

[269] R. H. Stolen. Nonlinear properties of optical fibers. In S. E. Miller and A. G. Chynoweth, editors, *Optical Fiber Telecommunications*. Academic Press, San Diego, CA, 1979.

[270] G. P. Agrawal. *Fiber-Optic Communication Systems*. Wiley-Interscience, New York, 3rd edition, 2002.

[271] J. K. Ranka, R. S. Windeler, and A. J. Stentz. Visible continuum generation in air–silica microstructure optical fibers with anomalous dispersion at 800 nm. *Opt. Lett.*, 25(1):25–27, 2000.

[272] P. Russell. Photonic crystal fibers. *Science*, 299(5605):358–362, 2003.

[273] E. A. J. Marcatili and R. A. Schmeltzer. Hollow metallic and dielectric waveguides for long distance optical transmission and lasers. *Bell Syst. Tech. J.*, 43(4P2):1783–1809, 1964.

[274] M. Nisoli, S. De Silvestri, and O. Svelto. Generation of high energy 10 fs pulses by a new pulse compression technique. *Appl. Phys. Lett.*, 68(20):2793–2795, 1996.

[275] M. Nisoli, S. De Silvestri, O. Svelto, R. Szipocs, K. Ferencz, C. Spielmann, S. Sartania, and F. Krausz. Compression of high-energy laser pulses below 5 fs. *Opt. Lett.*, 22(8):522–524, 1997.

[276] S. Sartania, Z. Cheng, M. Lenzner, G. Tempea, C. Spielmann, F. Krausz, and K. Ferencz. Generation of 0.1-TW 5-fs optical pulses at a 1-kHz repetition rate. *Opt. Lett.*, 22(20):1562–1564, 1997.

[277] J. A. Fleck, J. R. Morris, and M. D. Feit. Time-dependent propagation of high-energy laser-beams through atmosphere. *Appl. Phys. B*, 10(2):129–160, 1976.

[278] M. D. Feit and J. A. Fleck. Light-propagation in graded-index optical fibers. *Appl. Opt.*, 17(24):3990–3998, 1978.

[279] D. Grischkowsky and A. C. Balant. Optical pulse-compression based on enhanced frequency chirping. *Appl. Phys. Lett.*, 41(1):1–3, 1982.

[280] W. J. Tomlinson, R. H. Stolen, and C. V. Shank. Compression of optical pulses chirped by self-phase modulation in fibers. *J. Opt. Soc. Am. B*, 1(2):139–149, 1984.

[281] W. J. Tomlinson, R. H. Stolen, and A. M. Johnson. Optical-wave breaking of pulses in nonlinear optical fibers. *Opt. Lett.*, 10(9):457–459, 1985.

[282] H. Nakatsuka, D. Grischkowsky, and A. C. Balant. Non-linear picosecond-pulse propagation through optical fibers with positive group-velocity dispersion. *Phys. Rev. Lett.*, 47(13):910–913, 1981.

[283] B. Nikolaus and D. Grischkowsky. 12x pulse-compression using optical fibers. *Appl. Phys. Lett.*, 42(1):1–2, 1983.

[284] B. Nikolaus and D. Grischkowsky. 90-fs tunable optical pulses obtained by 2-stage pulse-compression. *Appl. Phys. Lett.*, 43(3):228–230, 1983.

[285] C. V. Shank, R. L. Fork, R. Yen, R. H. Stolen, and W. J. Tomlinson. Compression of femtosecond optical pulses. *Appl. Phys. Lett.*, 40(9):761–763, 1982.

[286] J. G. Fujimoto, A. M. Weiner, and E. P. Ippen. Generation and measurement of optical pulses as short as 16 fs. *Appl. Phys. Lett.*, 44(9):832–834, 1984.

[287] J. M. Halbout and D. Grischkowsky. 12-fs ultrashort optical pulse-compression at a high repetition rate. *Appl. Phys. Lett.*, 45(12):1281–1283, 1984.

[288] W. H. Knox, R. L. Fork, M. C. Downer, R. H. Stolen, C. V. Shank, and J. A. Valdmanis. Optical pulse compression to 8 fs at a 5-kHz repetition rate. *Appl. Phys Lett.*, 46(12):1120–1121, 1985.

[289] A. Baltuska, Z. Y. Wei, M. S. Pshenichnikov, and D. A. Wiersma. Optical pulse compression to 5 fs at a 1-MHz repetition rate. *Opt. Lett.*, 22(2):102–104, 1997.

[290] J. D. Kafka, B. H. Kolner, T. Baer, and D. M. Bloom. Compression of pulses from a continuous-wave mode-locked Nd:YAG laser. *Opt. Lett.*, 9(11):505–506, 1984.

[291] J. P. Heritage, R. N. Thurston, W. J. Tomlinson, A. M. Weiner, and R. H. Stolen. Spectral windowing of frequency-modulated optical pulses in a grating compressor. *Appl. Phys. Lett.*, 47(2):87–89, 1985.

[292] A. S. L. Gomes, W. Sibbett, and J. R. Taylor. Generation of subpicosecond pulses from a continuous-wave mode-locked Nd:YAG laser using a 2-stage optical compression technique. *Opt. Lett.*, 10(7):338–340, 1985.

[293] C. Rolland and P. B. Corkum. Compression of high-power optical pulses. *J. Opt. Soc. Am. B*, 5(3):641–647, 1988.

[294] D. H. Reitze, A. M. Weiner, and D. E. Leaird. High-power femtosecond optical pulse-compression by using spatial solitons. *Opt. Lett.*, 16(18):1409–1411, 1991.

[295] A. M. Johnson, R. H. Stolen, and W. M. Simpson. Generation of 0.41-picosecond pulses by the single-stage compression of frequency doubled Nd:YAG laser pulses. In D. H. Auston and K. B. Eisenthal, editors, *Ultrafast Phenomena IV*, Monterey, CA. Springer-Verlag, New York, 1984.

[296] A. M. Weiner, J. P. Heritage, and R. N. Thurston. Synthesis of phase-coherent, picosecond optical square pulses. *Opt. Lett.*, 11(3):153–155, 1986.

[297] A. M. Weiner and J. P. Heritage. Picosecond and femtosecond Fourier pulse shape synthesis. *Rev. Phys. Appl.*, 22(12):1619–1628, 1987.

[298] A. M. Weiner. Femtosecond pulse shaping using spatial light modulators. *Rev. Sci. Instrum.*, 71(5):1929–1960, 2000.

[299] L. Xu, N. Nakagawa, R. Morita, H. Shigekawa, and M. Yamashita. Programmable chirp compensation for 6-fs pulse generation with a prism-pair-formed pulse shaper. *IEEE J. Quantum Electron.*, 36(8):893–899, 2000.

[300] L. Xu, L. M. Li, N. Nakagawa, R. Morita, and M. Yamashita. Application of a spatial light modulator for programmable optical pulse compression to the sub-6-fs regime. *IEEE Photon. Technol. Lett.*, 12(11):1540–1542, 2000.

[301] R. N. Thurston, J. P. Heritage, A. M. Weiner, and W. J. Tomlinson. Analysis of picosecond pulse shape synthesis by spectral masking in a grating pulse compressor. *IEEE J. Quantum Electron.*, 22(5):682–696, 1986.

[302] A. Hasegawa and W. F. Brinkman. Tunable coherent IR and FIR sources utilizing modulational instability. *IEEE J. Quantum Electron.*, 16(7):694–697, 1980.

[303] A. Hasegawa and F. Tappert. Transmission of stationary nonlinear optical pulses in dispersive dielectric fibers. II: Normal dispersion. *Appl. Phys. Lett.*, 23(4):171–172, 1973.

[304] A. M. Weiner. Dark optical solitons. In J. R. Taylor, editor, *Optical Solitons: Theory and Experiment*. Cambridge University Press, Cambridge, UK, 1992.

[305] P. Emplit, J. P. Hamaide, F. Reynaud, C. Froehly, and A. Barthelemy. Picosecond steps and dark pulses through nonlinear single-mode fibers. *Opt. Commun.*, 62(6):374–379, 1987.

[306] D. Krokel, N. J. Halas, G. Giuliani, and D. Grischkowsky. Dark-pulse propagation in optical fibers. *Phys. Rev. Lett.*, 60(1):29–32, 1988.

[307] A. M. Weiner, J. P. Heritage, R. J. Hawkins, R. N. Thurston, E. M. Kirschner, D. E. Leaird, and W. J. Tomlinson. Experimental-observation of the fundamental dark soliton in optical fibers. *Phys. Rev. Lett.*, 61(21):2445–2448, 1988.

[308] J. R. Taylor, editor. *Optical Solitons: Theory and Experiment*. Cambridge University Press, Cambridge, UK, 1992.

[309] L. F. Mollenauer, J. P. Gordon, and P. V. Mamyshev. Solitons in high bit-rate transmission systems. In I. P. Kaminow and T. L. Koch, editors, *Optical Fiber Telecommunications IIIA*. Academic Press, San Diego, CA, 1997.

[310] V. E. Zakharov and A. B. Shabat. Exact theory of 2-dimensional self-focusing and one-dimensional self-modulation of waves in nonlinear media. *Sov. Phys. JETP*, 34(1):62, 1972.

[311] J. Satsuma and N. Yajima. Initial value-problems of one-dimensional self-modulation of nonlinear-waves in dispersive media. *Suppl. Prog. Theor. Phys.*, 55:284–306, 1974.

[312] H. A. Haus and W. S. Wong. Solitons in optical communications. *Rev. Mod. Phys.*, 68:423–444, 1996.

[313] K. J. Blow, N. J. Doran, and D. Wood. Trapping of energy into solitary waves in amplified nonlinear dispersive systems. *Opt. Lett.*, 12(12):1011–1013, 1987.

[314] K. Tajima. Compensation of soliton broadening in nonlinear optical fibers with loss. *Opt. Lett.*, 12(1):54–56, 1987.

[315] H. H. Kuehl. Solitons on an axially nonuniform optical fiber. *J. Opt. Soc. Am. B*, 5(3):709–713, 1988.

[316] E. M. Dianov, P. V. Mamyshev, A. M. Prokhorov, and S. V. Chernikov. Generation of a train of fundamental solitons at a high repetition rate in optical fibers. *Opt. Lett.*, 14(18):1008–1010, 1989.

[317] V. A. Bogatyrev, M. M. Bubnov, E. M. Dianov, A. S. Kurkov, P. V. Mamyshev, A. M. Prokhorov, S. D. Rumyantsev, V. A. Semenov, S. L. Semenov, A. A. Sysoliatin, S. V. Chernikov, A. N. Guryanov, G. G. Devyatykh, and S. I. Miroshnichenko. A single-mode fiber with chromatic dispersion varying along the length. *J. Lightwave Technol.*, 9(5):561–566, 1991.

[318] M. D. Pelusi and H. F. Liu. Higher order soliton pulse compression in dispersion-decreasing optical fibers. *IEEE J. Quantum Electron.*, 33(8):1430–1439, 1997.

[319] P. V. Mamyshev, S. V. Chernikov, and E. M. Dianov. Generation of fundamental soliton trains for high-bit-rate optical fiber communication lines. *IEEE J. Quantum Electron.*, 27(10):2347–2355, 1991.

[320] S. V. Chernikov, J. R. Taylor, and R. Kashyap. Integrated all-optical fiber source of multigigahertz soliton pulse-train. *Electron. Lett.*, 29(20):1788–1789, 1993.

[321] S. V. Chernikov, E. M. Dianov, D. J. Richardson, and D. N. Payne. Soliton pulse-compression in dispersion-decreasing fiber. *Opt. Lett.*, 18(7):476–478, 1993.

[322] K. R. Tamura and M. Nakazawa. A polarization-maintaining pedestal-free femtosecond pulse compressor incorporating an ultrafast dispersion-imbalanced nonlinear optical loop mirror. *IEEE Photon. Technol. Lett.*, 13(5):526–528, 2001.

[323] T. Brabec and F. Krausz. Nonlinear optical pulse propagation in the single-cycle regime. *Phys. Rev. Lett.*, 78(17):3282–3285, 1997.

[324] R. W. Hellwarth. 3rd-order optical susceptibilities of liquids and solids. *Prog. Quantum Electron.*, 5:1–68, 1977.

[325] M. D. Levenson. *Introduction to Nonlinear Laser Spectroscopy*. Academic Press, New York, 1982.

[326] R. W. Hellwarth, A. Owyoung, and N. George. Origin of nonlinear refractive index of liquid CCl_4. *Phys. Rev. A*, 4(6):2342, 1971.

[327] R. Hellwarth, J. Cherlow, and T. T. Yang. Origin and frequency-dependence of nonlinear optical susceptibilities of glasses. *Phys. Rev. B*, 11(2):964–967, 1975.

[328] J. P. Gordon. Theory of the soliton self-frequency shift. *Opt. Lett.*, 11(10):662–664, 1986.

[329] R. H. Stolen, J. P. Gordon, W. J. Tomlinson, and H. A. Haus. Raman response function of silica-core fibers. *J. Opt. Soc. Am. B*, 6(6):1159–1166, 1989.

[330] R. H. Stolen, C. Lee, and R. K. Jain. Development of the stimulated Raman spectrum in single-mode silica fibers. *J. Opt. Soc. Am. B*, 1:652–657, 1984.

[331] P. V. Mamyshev. Generation and compression of femtosecond solitons in optical fibers. In J. R. Taylor, editor, *Optical Solitons: Theory and Experiment*. Cambridge University Press, Cambridge, UK, 1992.

[332] F. M. Mitschke and L. F. Mollenauer. Discovery of the soliton self-frequency shift. *Opt. Lett.*, 11(10):659–661, 1986.

[333] E. A. Golovchenko, E. M. Dianov, A. M. Prokhorov, and V. N. Serkin. Decay of optical solitons. *JETP Lett.*, 42(2):87–91, 1985.

[334] W. Hodel and H. P. Weber. Decay of femtosecond higher-order solitons in an optical fiber induced by Raman self-pumping. *Opt. Lett.*, 12(11):924–926, 1987.

[335] K. Tai, A. Hasegawa, and N. Bekki. Fission of optical solitons induced by stimulated Raman effect. *Opt. Lett.*, 13(5):392–394, 1988.

[336] G. P. Agrawal. Effect of intrapulse stimulated Raman-scattering on soliton-effect pulse-compression in optical fibers. *Opt. Lett.*, 15(4):224–226, 1990.

[337] P. Beaud, W. Hodel, B. Zysset, and H. P. Weber. Ultrashort pulse-propagation, pulse breakup, and fundamental soliton formation in a single-mode optical fiber. *IEEE J. Quantum Electron.*, 23(11):1938–1946, 1987.

[338] A. S. Gouveia-Neto, A. S. L. Gomes, and J. R. Taylor. Femtosecond soliton Raman generation. *IEEE J. Quantum Electron.*, 24(2):332–340, 1988.

[339] M. N. Islam, G. Sucha, I. Bar-Joseph, M. Wegener, J. P. Gordon, and D. S. Chemla. Broad bandwidths from frequency-shifting solitons in fibers. *Opt. Lett.*, 14(7):370–372, 1989.

[340] D. Anderson and M. Lisak. Non-linear asymmetric self-phase modulation and self-steepening of pulses in long optical-waveguides. *Phys. Rev. A*, 27(3):1393–1398, 1983.

[341] F. Demartini, C. H. Townes, T. K. Gustafson, and P. L. Kelley. Self-steepening of light pulses. *Phys. Rev.*, 164(2):312, 1967.

[342] A. L. Gaeta. Catastrophic collapse of ultrashort pulses. *Phys. Rev. Lett.*, 84(16):3582–3585, 2000.

[343] H. A. Haus and E. P. Ippen. Group velocity of solitons. *Opt. Lett.*, 26(21):1654–1656, 2001.

[344] J. E. Rothenberg. Space-time focusing - breakdown of the slowly varying envelope approximation in the self-focusing of femtosecond pulses. *Opt. Lett.*, 17(19):1340–1342, 1992.

[345] R. L. Fork, C. V. Shank, C. Hirlimann, R. Yen, and W. J. Tomlinson. Femtosecond white-light continuum pulses. *Opt. Lett.*, 8(1):1–3, 1983.

[346] A. Brodeur and S. L. Chin. Band-gap dependence of the ultrafast white-light continuum. *Phys. Rev. Lett.*, 80(20):4406–4409, 1998.

[347] R. R. Alfano and S. L. Shapiro. Emission in the region 4000 to 7000 angstroms via four-photon coupling in glass. *Phys. Rev. Lett.*, 24(11):584–587, 1970.

[348] R. R. Alfano and S. L. Shapiro. Observation of self-phase modulation and small-scale filaments in crystals and glasses. *Phys. Rev. Lett.*, 24(11):592–594, 1970.

[349] P. B. Corkum, C. Rolland, and T. Srinivasan-Rao. Supercontinuum generation in gases. *Phys. Rev. Lett.*, 57(18):2268–2271, 1986.

[350] P. B. Corkum and C. Rolland. Femtosecond continua produced in gases. *IEEE J. Quantum Electron.*, 25(12):2634–2639, 1989.

[351] A. Brodeur and S. L. Chin. Ultrafast white-light continuum generation and self-focusing in transparent condensed media. *J. Opt. Soc. Am. B*, 16(4):637–650, 1999.

[352] J. K. Ranka and A. L. Gaeta. Breakdown of the slowly varying envelope approximation in the self-focusing of ultrashort pulses. *Opt. Lett.*, 23(7):534–536, 1998.

[353] J. E. Rothenberg. Pulse splitting during self-focusing in normally dispersive media. *Opt. Lett.*, 17(8):583–585, 1992.

[354] J. K. Ranka, R. W. Schirmer, and A. L. Gaeta. Observation of pulse splitting in nonlinear dispersive media. *Phys. Rev. Lett.*, 77(18):3783–3786, 1996.

[355] A. L. Gaeta. Nonlinear propagation and continuum generation in microstructured optical fibers. *Opt. Lett.*, 27(11):924–926, 2002.

[356] I. N. Duling. Subpicosecond all-fiber erbium laser. *Electron. Lett.*, 27(6):544–545, 1991.

[357] I. N. Duling. Modelocking of all-fiber lasers. In I. N. Duling, editor, *Compact Sources of Ultrashort Pulses*, pages 140–178. Cambridge University Press, New York, 1995.

[358] D. J. Richardson, R. I. Laming, D. N. Payne, M. W. Phillips, and V. J. Matsas. 320 fs soliton generation with passively mode-locked erbium fiber laser. *Electron. Lett.*, 27(9):730–732, 1991.

[359] M. Nakazawa, E. Yoshida, and Y. Kimura. Generation of 98 fs optical pulses directly from an erbium-doped fiber ring laser at 1.57 μm. *Electron. Lett.*, 29(1):63–65, 1993.

[360] M. E. Fermann, F. Haberl, M. Hofer, and H. Hochreiter. Nonlinear amplifying loop mirror. *Opt. Lett.*, 15(13):752–754, 1990.

[361] N. J. Doran and D. Wood. Nonlinear-optical loop mirror. *Opt. Lett.*, 13(1):56–58, 1988.

[362] K. J. Blow, N. J. Doran, and B. K. Nayar. Experimental demonstration of optical soliton switching in an all-fiber nonlinear Sagnac interferometer. *Opt. Lett.*, 14(14):754–756, 1989.

[363] V. J. Matsas, T. P. Newson, D. J. Richardson, and D. N. Payne. Self-starting passively mode-locked fiber ring soliton laser exploiting nonlinear polarization rotation. *Electron. Lett.*, 28(15):1391–1393, 1992.

[364] V. J. Matsas, D. J. Richardson, T. P. Newson, and D. N. Payne. Characterization of a self-starting, passively mode-locked fiber ring laser that exploits nonlinear polarization evolution. *Opt. Lett.*, 18(5):358–360, 1993.

[365] M. E. Fermann, M. J. Andrejco, M. L. Stock, Y. Silberberg, and A. M. Weiner. Passive-mode locking in erbium fiber lasers with negative group delay. *Appl. Phys. Lett.*, 62(9):910–912, 1993.

[366] P. D. Maker, R. W. Terhune, and C. M. Savage. Intensity-dependent changes in refractive index of liquids. *Phys. Rev. Lett.*, 12(18):507–509, 1964.

[367] R. H. Stolen, J. Botineau, and A. Ashkin. Intensity discrimination of optical pulses with birefringent fibers. *Opt. Lett.*, 7(10):512–514, 1982.

[368] B. Nikolaus, D. Grischkowsky, and A. C. Balant. Optical pulse reshaping based on the non-linear birefringence of single-mode optical fibers. *Opt. Lett.*, 8(3):189–191, 1983.

[369] M. N. Islam, E. R. Sunderman, R. H. Stolen, W. Pleibel, and J. R. Simpson. Soliton switching in a fiber nonlinear loop mirror. *Opt. Lett.*, 14(15):811–813, 1989.

[370] I. N. Duling, C. J. Chen, P. K. A. Wai, and C. R. Menyuk. Operation of a nonlinear loop mirror in a laser cavity. *IEEE J. Quantum Electron.*, 30(1):194–199, 1994.

[371] M. Nakazawa, E. Yoshida, and Y. Kimura. Low threshold, 290-fs erbium-doped fiber laser with a nonlinear amplifying loop mirror pumped by InGaAsP laser diodes. *Appl. Phys. Lett.*, 59(17):2073–2075, 1991.

[372] A. B. Grudinin, D. J. Richardson, and D. N. Payne. Energy quantization in figure 8 fiber laser. *Electron. Lett.*, 28(1):67–68, 1992.

[373] R. P. Davey, N. Langford, and A. I. Ferguson. Interacting solitons in erbium fiber laser. *Electron. Lett.*, 27(14):1257–1259, 1991.

[374] D. U. Noske, N. Pandit, and J. R. Taylor. Source of spectral and temporal instability in soliton fiber lasers. *Opt. Lett.*, 17(21):1515–1517, 1992.

[375] N. Pandit, D. U. Noske, S. M. J. Kelly, and J. R. Taylor. Characteristic instability of fiber loop soliton lasers. *Electron. Lett.*, 28(5):455–457, 1992.

[376] M. L. Dennis and I. N. Duling. Role of dispersion in limiting pulse-width in fiber lasers. *Appl. Phys. Lett.*, 62(23):2911–2913, 1993.

[377] S. M. J. Kelly. Characteristic side-band instability of periodically amplified average soliton. *Electron. Lett.*, 28(8):806–807, 1992.

[378] N. J. Smith, K. J. Blow, and I. Andonovic. Side-band generation through perturbations to the average soliton model. *J. Lightwave Technol.*, 10(10):1329–1333, 1992.

[379] M. L. Dennis and I. N. Duling. Experimental-study of side-band generation in femtosecond fiber lasers. *IEEE J. Quantum Electron.*, 30(6):1469–1477, 1994.

[380] L. F. Mollenauer, S. G. Evangelides, and H. A. Haus. Long-distance soliton propagation using lumped amplifiers and dispersion shifted fiber. *J. Lightwave Technol.*, 9(2):194–197, 1991.

[381] K. J. Blow and N. J. Doran. Average soliton dynamics and the operation of soliton systems with lumped amplifiers. *IEEE Photon. Technol. Lett.*, 3(4):369–371, 1991.

[382] A. Hasegawa and Y. Kodama. Guiding-center soliton in optical fibers. *Opt. Lett.*, 15(24):1443–1445, 1990.

[383] L. F. Mollenauer, J. P. Gordon, and M. N. Islam. Soliton propagation in long fibers with periodically compensated loss. *IEEE J. Quantum Electron.*, 22(1):157–173, 1986.

[384] J. P. Gordon. Dispersive perturbations of solitons of the nonlinear Schrödinger-equation. *J. Opt. Soc. Am. B*, 9(1):91–97, 1992.

[385] K. Tamura, E. P. Ippen, and H. A. Haus. Optimization of filtering in soliton fiber lasers. *IEEE Photon. Technol. Lett.*, 6(12):1433–1435, 1994.

[386] M. Martinelli. A universal compensator for polarization changes induced by birefringence on a retracing beam. *Opt. Commun.*, 72(6):341–344, 1989.

[387] M. Nakazawa, E. Yoshida, and K. Tamura. 10 GHz, 2 ps regeneratively and harmonically FM mode-locked erbium fibre ring laser. *Electron. Lett.*, 32(14):1285–1287, 1996.

[388] J. Li, P. A. Andrekson, and B. Bakhshi. Direct generation of subpicosecond chirp-free pulses at 10 GHz from a nonpolarization maintaining actively mode-locked fiber ring laser. *IEEE Photon. Technol. Lett.*, 12(9):1150–1152, 2000.

[389] J. D. Kafka, T. Baer, and D. W. Hall. Mode-locked erbium-doped fiber laser with soliton pulse shaping. *Opt. Lett.*, 14(22):1269–1271, 1989.

[390] C. R. Doerr, H. A. Haus, E. P. Ippen, M. Shirasaki, and K. Tamura. Additive-pulse limiting. *Opt. Lett.*, 19(1):31–33, 1994.

[391] M. Nakazawa, K. Tamura, and E. Yoshida. Supermode noise suppression in a harmonically mode-locked fibre laser by self-phase modulation and spectral filtering. *Electron. Lett.*, 32(5):461–463, 1996.

[392] M. Horowitz, C. R. Menyuk, T. F. Carruthers, and I. N. Duling. Theoretical and experimental study of harmonically mode-locked fiber lasers for optical communication systems. *J. Lightwave Technol.*, 18(11):1565–1574, 2000.

[393] D. Kopf, F. X. Kärtner, K. J. Weingarten, and U. Keller. Pulse shortening in a Nd:glass laser by gain reshaping and soliton formation. *Opt. Lett.*, 19(24):2146–2148, 1994.

[394] F. X. Kärtner, D. Kopf, and U. Keller. Solitary-pulse stabilization and shortening in actively mode-locked lasers. *J. Opt. Soc. Am. B*, 12(3):486–496, 1995.

[395] I. D. Jung, F. X. Kärtner, L. R. Brovelli, M. Kamp, and U. Keller. Experimental verification of soliton mode-locking using only a slow saturable absorber. *Opt. Lett.*, 20(18):1892–1894, 1995.

[396] F. X. Kärtner, I. D. Jung, and U. Keller. Soliton mode-locking with saturable absorbers. *IEEE J. Sel. Top. Quantum Electron.*, 2(3):540–556, 1996.

[397] U. Keller, K. J. Weingarten, F. X. Kärtner, D. Kopf, B. Braun, I. D. Jung, R. Fluck, C. Honninger, N. Matuschek, and J. A. der Au. Semiconductor saturable absorber mirrors (SESAM's) for femtosecond to nanosecond pulse generation in solid-state lasers. *IEEE J. Sel. Top. Quantum Electron.*, 2(3):435–453, 1996.

[398] I. D. Jung, F. X. Kärtner, N. Matuschek, D. H. Sutter, F. Morier-Genoud, Z. Shi, V. Scheuer, M. Tilsch, T. Tschudi, and U. Keller. Semiconductor saturable absorber mirrors supporting sub-10-fs pulses. *Appl. Phys. B*, 65(2):137–150, 1997.

[399] I. S. Gradshteyn and I. M. Ryzhik. *Table of Integrals, Series, and Products*. Academic Press, New York, 4th edition, 1980.

[400] F. X. Kärtner and U. Keller. Stabilization of solitonlike pulses with a slow saturable absorber. *Opt. Lett.*, 20(1):16–18, 1995.

[401] D. J. Jones, H. A. Haus, and E. P. Ippen. Subpicosecond solitons in an actively mode-locked fiber laser. *Opt. Lett.*, 21(22):1818–1820, 1996.

[402] F. X. Kärtner, J. A. Der Au, and U. Keller. Mode-locking with slow and fast saturable absorbers: What's the difference? *IEEE J. Sel. Top. Quantum Electron.*, 4(2):159–168, 1998.

[403] K. Tamura, E. P. Ippen, H. A. Haus, and L. E. Nelson. 77-fs pulse generation from a stretched-pulse mode-locked all-fiber ring laser. *Opt. Lett.*, 18(13):1080–1082, 1993.

[404] K. Tamura, L. E. Nelson, H. A. Haus, and E. P. Ippen. Soliton versus nonsoliton operation of fiber ring lasers. *Appl. Phys. Lett.*, 64(2):149–151, 1994.

[405] H. A. Haus, K. Tamura, L. E. Nelson, and E. P. Ippen. Stretched-pulse additive-pulse mode-locking in fiber ring lasers: theory and experiment. *IEEE J. Quantum Electron.*, 31(3):591–598, 1995.

[406] R. W. Tkach, A. R. Chraplyvy, F. Forghieri, A. H. Gnauck, and R. M. Derosier. 4-photon mixing and high-speed WDM systems. *J. Lightwave Technol.*, 13(5):841–849, 1995.

[407] N. J. Smith, F. M. Knox, N. J. Doran, K. J. Blow, and I. Bennion. Enhanced power solitons in optical fibres with periodic dispersion management. *Electron. Lett.*, 32(1):54–55, 1996.

[408] N. J. Smith, N. J. Doran, F. M. Knox, and W. Forysiak. Energy-scaling characteristics of solitons in strongly dispersion-managed fibers. *Opt. Lett.*, 21(24):1981–1983, 1996.

[409] J. H. B. Nijhof, N. J. Doran, W. Forysiak, and F. M. Knox. Stable soliton-like propagation in dispersion managed systems with net anomalous, zero and normal dispersion. *Electron. Lett.*, 33(20):1726–1727, 1997.

[410] J. H. B. Nijhof, W. Forysiak, and N. J. Doran. Dispersion-managed solitons in the normal dispersion regime: a physical interpretation. *Opt. Lett.*, 23(21):1674–1676, 1998.

[411] Y. Chen and H. A. Haus. Dispersion-managed solitons in the net positive dispersion regime. *J. Opt. Soc. Am. B*, 16(1):24–30, 1999.

[412] F. Krausz, M. E. Fermann, T. Brabec, P. F. Curley, M. Hofer, M. H. Ober, C. Spielmann, E. Wintner, and A. J. Schmidt. Femtosecond solid-state lasers. *IEEE J. Quantum Electron.*, 28(10):2097–2122, 1992.

[413] R. Ell, U. Morgner, F. X. Kärtner, J. G. Fujimoto, E. P. Ippen, V. Scheuer, G. Angelow, T. Tschudi, M. J. Lederer, A. Boiko, and B. Luther-Davies. Generation of 5-fs pulses and octave-spanning spectra directly from a Ti:sapphire laser. *Opt. Lett.*, 26(6):373–375, 2001.

[414] L. Matos, D. Kleppner, O. Kuzucu, T. R. Schibli, J. Kim, E. P. Ippen, and F. X. Kaertner. Direct frequency comb generation from an octave-spanning, prismless Ti:sapphire laser. *Opt. Lett.*, 29(14):1683–1685, 2004.

[415] P. F. Moulton. Spectroscopic and laser characteristics of Ti:Al_2O_3. *J. Opt. Soc. Am. B*, 3:125–133, 1986.

[416] C. P. Huang, H. C. Kapteyn, J. W. McIntosh, and M. M. Murnane. Generation of transform-limited 32-fs pulses from a self-mode-locked Ti:sapphire laser. *Opt. Lett.*, 17(2):139–141, 1992.

[417] J. Zhou, G. Taft, C.-P. Huang, M.M. Murnane, and H.C. Kapteyn. Pulse evolution in a broad-bandwidth Ti:sapphire laser. *Opt. Lett.*, 19:1149–1151, 1994.

[418] L. Xu, C. Spielmann, F. Krausz, and R. Szipocs. Ultrabroadband ring oscillator for sub-10-fs pulse generation. *Opt. Lett.*, 21(16):1259–1261, 1996.

[419] I. D. Jung, F. X. Kärtner, N. Matuschek, D. H. Sutter, F. Morier-Genoud, G. Zhang, U. Keller, V. Scheuer, M. Tilsch, and T. Tschudi. Self-starting 6.5-fs pulses from a Ti:sapphire laser. *Opt. Lett.*, 22(13):1009–1011, 1997.

[420] F. X. Kärtner, U. Morgner, R. Ell, T. Schibli, J. G. Fujimoto, E. P. Ippen, V. Scheuer, G. Angelow, and T. Tschudi. Ultrabroadband double-chirped mirror pairs for generation of octave spectra. *J. Opt. Soc. Am. B*, 18(6):882–885, 2001.

[421] Y. Chen, F. X. Kärtner, U. Morgner, S. H. Cho, H. A. Haus, E. P. Ippen, and J. G. Fujimoto. Dispersion-managed mode locking. *J. Opt. Soc. Am. B*, 16(11):1999–2004, 1999.

[422] S. T. Cundiff, W. H. Knox, E. P. Ippen, and H. A. Haus. Frequency-dependent mode size in broadband Kerr-lens mode locking. *Opt. Lett.*, 21(9):662–664, 1996.

[423] I. P. Christov, V. D. Stoev, M. M. Murnane, and H. C. Kapteyn. Mode-locking with a compensated space–time astigmatism. *Opt. Lett.*, 20(20):2111–2113, 1995.

[424] I. P. Christov, H. C. Kapteyn, M. M. Murnane, C. P. Huang, and J. P. Zhou. Space-time focusing of femtosecond pulses in a Ti:sapphire laser. *Opt. Lett.*, 20(3):309–311, 1995.

[425] http://nobelprize.virtual.museum/nobel_prizes/physics/.

[426] L. Xu, C. Spielmann, A. Poppe, T. Brabec, F. Krausz, and T. W. Hänsch. Route to phase control of ultrashort light pulses. *Opt. Lett.*, 21(24):2008–2010, 1996.

[427] H. R. Telle, G. Steinmeyer, A. E. Dunlop, J. Stenger, D. H. Sutter, and U. Keller. Carrier-envelope offset phase control: a novel concept for absolute optical frequency measurement and ultrashort pulse generation. *Appl. Phys. B*, 69(4):327–332, 1999.

[428] J. Reichert, R. Holzwarth, T. Udem, and T. W. Hänsch. Measuring the frequency of light with mode-locked lasers. *Opt. Commun.*, 172(1–6):59–68, 1999.

[429] D. J. Jones, S. A. Diddams, J. K. Ranka, A. Stentz, R. S. Windeler, J. L. Hall, and S. T. Cundiff. Carrier-envelope phase control of femtosecond mode-locked lasers and direct optical frequency synthesis. *Science*, 288(5466):635–639, 2000.

[430] A. Apolonski, A. Poppe, G. Tempea, C. Spielmann, T. Udem, R. Holzwarth, T. W. Hänsch, and F. Krausz. Controlling the phase evolution of few-cycle light pulses. *Phys. Rev. Lett.*, 85(4):740–743, 2000.

[431] R. Holzwarth, T. Udem, T. W. Hänsch, J. C. Knight, W. J. Wadsworth, and P. S. J. Russell. Optical frequency synthesizer for precision spectroscopy. *Phys. Rev. Lett.*, 85(11):2264–2267, 2000.

[432] S. T. Cundiff. Phase stabilization of ultrashort optical pulses. *J. Phys. D*, 35(8):R43–R59, 2002.

[433] J. Ye and S. T. Cundiff, editors. *Femtosecond Optical Frequency Comb Technology*. Springer-Verlag, New York, 2005.

[434] S. A. Diddams, J. Ye, and L. Hollberg. Femtosecond lasers for optical clocks and low noise frequency synthesis. In J. Ye and S. T. Cundiff, editors, *Femtosecond Optical Frequency Comb Technology*. Springer-Verlag, New York, 2005.

[435] R. Ell, J. R. Birge, M. Araghchini, and F. X. Kärtner. Carrier-envelope phase control by a composite plate. *Opt. Express*, 14(12):5829–5837, 2006.

[436] K. R. Vogel, S. A. Diddams, C. W. Oates, E. A. Curtis, R. J. Rafac, W. M. Itano, J. C. Bergquist, R. W. Fox, W. D. Lee, J. S. Wells, and L. Hollberg. Direct comparison of two cold-atom-based optical frequency standards by using a femtosecond-laser comb. *Opt. Lett.*, 26(2):102–104, 2001.

[437] J. N. Eckstein, A. I. Ferguson, and T. W. Hänsch. High-resolution 2-photon spectroscopy with picosecond light-pulses. *Phys. Rev. Lett.*, 40(13):847–850, 1978.

[438] S. A. Diddams, J. C. Bergquist, S. R. Jefferts, and C. W. Oates. Standards of time and frequency at the outset of the 21st century. *Science*, 306(5700):1318–1324, 2004.

[439] R. J. Rafac, B. C. Young, J. A. Beall, W. M. Itano, D. J. Wineland, and J. C. Bergquist. Sub-dekahertz ultraviolet spectroscopy of 199Hg(+). *Phys. Rev. Lett.*, 85(12):2462–2465, 2000.

[440] G. Bonsch and E. Potulski. Measurement of the refractive index of air and comparison with modified Edlen's formulae. *Metrologia*, 35(2):133–139, 1998.

[441] D. H. Reitze, A. M. Weiner, and D. E. Leaird. Shaping of wide bandwidth 20 fsec optical pulses. *Appl. Phys. Lett.*, 61:1260–1262, 1992.

[442] J. P. Heritage, A. M. Weiner, and R. N. Thurston. Picosecond pulse shaping by spectral phase and amplitude manipulation. *Opt. Lett.*, 10:609–611, 1985.

[443] A. M. Weiner, J. P. Heritage, and E. M. Kirschner. High-resolution femtosecond pulse shaping. *J. Opt. Soc. Am. B*, 5:1563–1572, 1988.

[444] A. M. Weiner, Y. Silberberg, H. Fouckhardt, D. E. Leaird, M. A. Saifi, M. J. Andrejco, and P. W. Smith. Use of femtosecond square pulses to avoid pulse breakup in all-optical switching. *IEEE J. Quantum Electron*, 25:2648, 1989.

[445] A. M. Weiner, J. P. Heritage, and J. A. Salehi. Encoding and decoding of femtosecond pulses. *Opt. Lett.*, 13:300–302, 1988.

[446] H. P. Sardesai, C.-C. Chang, and A. M. Weiner. A femtosecond code-division multiple-access communication system testbed. *J. Lightwave Technol.*, 16:1953–1964, 1998.

[447] M. R. Schroeder. *Number Theory in Science and Communication*. Springer-Verlag, Berlin, 1986.

[448] R. Skaug and J. F. Hjelmstad. *Spread Spectrum in Communications*. Peregrinus, London, 1985.

[449] J. P. Heritage and A. M. Weiner. Advances in spectral optical code-division multiple-access communications. *IEEE J. Sel. Top. Quantum Electron.*, 13(5):1351–1369, 2007.

[450] A. M. Weiner and D. E. Leaird. Generation of terahertz-rate trains of femtosecond pulses by phase-only filtering. *Opt. Lett.*, 15:51–53, 1990.

[451] A. M. Weiner, D. E. Leaird, J. S. Patel, and J. R. Wullert. Programmable shaping of femtosecond pulses by use of a 128-element liquid-crystal phase modulator. *IEEE J. Quantum Electron.*, 28:908–920, 1992.

[452] A. M. Weiner, D. E. Leaird, J. S. Patel, and J. R. Wullert. Programmable femtosecond pulse shaping by using a multielement liquid crystal phase modulator. *Opt. Lett.*, 15:326–328, 1990.

[453] M. M. Wefers and K. A. Nelson. Generation of high-fidelity programmable ultrafast optical waveforms. *Opt. Lett.*, 20:1047–1049, 1995.

[454] T. Brixner and G. Gerber. Femtosecond polarization pulse shaping. *Opt. Lett.*, 26:557–559, 2001.

[455] M. Akbulut, R. Nelson, A. M. Weiner, P. Cronin, and P. J. Miller. Broadband polarization correction with programmable liquid-crystal modulator arrays. *Opt. Lett.*, 29(10):1129–1131, 2004.

[456] R. D. Nelson, D. E. Leaird, and A. M. Weiner. Programmable polarization-independent spectral phase compensation and pulse shaping. *Opt. Express*, 11(15):1763–1769, 2003.

[457] C.-C. Chang, H. P. Sardesai, and A. M. Weiner. Dispersion-free fiber transmission for femtosecond pulses using a dispersion-compensating fiber and a programmable pulse shaper. *Opt. Lett.*, 23:283–285, 1998.

[458] Z. Jiang, S. D. Yang, D. E. Leaird, and A. M. Weiner. Fully dispersion-compensated 500 fs pulse transmission over 50 km single-mode fiber. *Opt. Lett.*, 30(12):1449–1451, 2005.

[459] D. Yelin, D. Meshulach, and Y. Silberberg. Adaptive femtosecond pulse compression. *Opt. Lett.*, 22(23):1793–1795, 1997.

[460] D. Meshulach, D. Yelin, and Y. Silberberg. Adaptive real-time femtosecond pulse shaping. *J. Opt. Soc. Am. B*, 15:1615–1619, 1998.

[461] T. Baumert, T. Brixner, V. Seyfried, M. Strehle, and G. Gerber. Femtosecond pulse shaping by an evolutionary algorithm with feedback. *Appl. Phys. B*, 65:779–782, 1997.

[462] R. S. Judson and H. Rabitz. Teaching lasers to control molecules. *Phys. Rev. Lett.*, 68:1500–1503, 1992.

[463] A. Assion, T. Baumert, M. Bergt, T. Brixner, B. Kiefer, V. Seyfried, M. Strehle, and G. Gerber. Control of chemical reactions by feedback-optimized phase-shaped femtosecond laser pulses. *Science*, 282:919–922, 1998.

[464] O. E. Martinez. Grating and prism compressors in the case of finite beam size. *J. Opt. Soc. Am. B*, 3(7):929–934, 1986.

[465] M. M. Wefers and K. A. Nelson. Space–time profiles of shaped ultrafast optical waveforms. *IEEE J. Quantum Electron.*, 32:161–172, 1996.

[466] B. Colombeau, M. Vampouille, and C. Froehly. Shaping of short laser pulses by passive optical Fourier techniques. *Opt. Commun.*, 19(2):201–204, 1976.

[467] Ph. Emplit, J.-P. Hamaide, and F. Reynaud. Passive amplitude and phase picosecond pulse shaping. *Opt. Lett.*, 17:1358–1360, 1992.

[468] D. E. Leaird and A. M. Weiner. Femtosecond optical packet generation via a direct space-to-time pulse shaper. *Opt. Lett.*, 24:853–855, 1999.

[469] D. E. Leaird and A. M. Weiner. Chirp control in the direct space-to-time pulse shaper. *Opt. Lett.*, 25:850–852, 2000.

[470] D. E. Leaird and A. M. Weiner. Femtosecond direct space-to-time pulse shaping. *IEEE J. Quantum Electron.*, 37:494–504, 2001.

[471] P. Tournois. Acousto-optic programmable dispersive filter for adaptive compensation of group delay time dispersion in laser systems. *Opt. Commun.*, 140(4–6):245–249, 1997.

[472] F. Verluise, V. Laude, J. P. Huignard, P. Tournois, and A. Migus. Arbitrary dispersion control of ultrashort optical pulses with acoustic waves. *J. Opt. Soc. Am. B*, 17(1):138–145, 2000.

[473] S. E. Harris and R. W. Wallace. Acousto-optic tunable filter. *J. Opt. Soc. Am.*, 59(6):744–747, 1969.

[474] D. A. Smith, J. E. Baran, J. J. Johnson, and K. W. Cheung. Integrated-optic acoustically-tunable filters for WDM networks. *IEEE J. Se. Areas Commun.*, 8(6):1151–1159, 1990.

[475] M. E. Fermann, V. da Silva, D. A. Smith, Y. Silberberg, and A. M. Weiner. Shaping of ultrashort optical pulses by using an integrated acousto-optic tunable filter. *Opt. Lett.*, 18(18):1505–1507, 1993.

[476] F. Verluise, V. Laude, Z. Cheng, C. Spielmann, and P. Tournois. Amplitude and phase control of ultrashort pulses by use of an acousto-optic programmable dispersive filter: pulse compression and shaping. *Opt. Lett.*, 25(8):575–577, 2000.

[477] S. A. Akhmanov, A. S. Chirkin, K. N. Drabovic, A. I. Kovrigin, R. V. Khokhlov, and A. P. Sukhoruk. Nonstationary nonlinear optical effects and ultrashort light pulse formation. *IEEE J. Quantum Electron.*, 4(10):598–605, 1968.

[478] B. H. Kolner. Space–time duality and the theory of temporal imaging. *IEEE J. Quantum Electron.*, 30(8):1951–1963, 1994.

[479] A. Papoulis. Pulse-compression, fiber communications, and diffraction: a unified approach. *J. Opt. Soc. Am. A*, 11(1):3–13, 1994.

[480] J. Chou, Y. Han, and B. Jalali. Adaptive RF-photonic arbitrary waveform generator. *IEEE Photon. Technol. Lett.*, 15(4):581–583, 2003.

[481] I. S. Lin, J. D. McKinney, and A. M. Weiner. Photonic synthesis of broadband microwave arbitrary waveforms applicable to ultra-wideband communication. *IEEE Microwave Compon. Lett.*, 15(4):226–228, 2005.

[482] J. A. Valdmanis. Real time picosecond optical oscilloscope. In *Proceedings of Ultrafast Phenomena V*, pages 82–85. Springer-Verlag, Berlin, 1986.

[483] Z. P. Jiang and X. C. Zhang. Electro-optic measurement of THz field pulses with a chirped optical beam. *Appl. Phys. Lett.*, 72(16):1945–1947, 1998.

[484] F. Coppinger, A. S. Bhushan, and B. Jalali. Photonic time stretch and its application to analog-to-digital conversion. *IEEE Trans. Microwave Theory Tech.*, 47(7):1309–1314, 1999.

[485] F. G. Sun, Z. P. Jiang, and X. C. Zhang. Analysis of terahertz pulse measurement with a chirped probe beam. *Appl. Phys. Lett.*, 73(16):2233–2235, 1998.

[486] B. H. Kolner and M. Nazarathy. Temporal imaging with a time lens. *Opt. Lett.*, 14(12):630–632, 1989.

[487] M. T. Kauffman, W. C. Banyai, A. A. Godil, and D. M. Bloom. Time-to-frequency converter for measuring picosecond optical pulses. *Appl. Phys. Lett.*, 64(3):270–272, 1994.

[488] C. V. Bennett and B. H. Kolner. Upconversion time microscope demonstrating 103 × magnification of femtosecond waveforms. *Opt. Lett.*, 24(11):783–785, 1999.

[489] C. V. Bennett and B. H. Kolner. Principles of parametric temporal imaging. Part I: System configurations. *IEEE J. Quantum Electron.*, 36(4):430–437, 2000.

[490] S. P. Dijaili, A. Dienes, and J. S. Smith. *ABCD* matrices for dispersive pulse-propagation. *IEEE J. Quantum Electron.*, 26(6):1158–1164, 1990.

[491] G. A. Mourou, T. Tajima, and S. V. Bulanov. Optics in the relativistic regime. *Rev. Mod. Phys.*, 78(2):309–371, 2006.

[492] W. Koechner. *Solid-State Laser Engineering*. Springer-Verlag, New York, 6th edition, 2006.

[493] S. Backus, C. G. Durfee, M. M. Murnane, and H. C. Kapteyn. High power ultrafast lasers. *Rev. Sci. Instrum.*, 69(3):1207–1223, 1998.

[494] F. Salin. Ultrafast solid-state amplifiers. In M.E. Fermann, A. Galvanauskas, and G. Sucha, editors, *Ultrafast Lasers: Technology and Applications*. Marcel Dekker, New York, 2003.

[495] L. M. Frantz and J. S. Nodvik. Theory of pulse propagation in a laser amplifier. *J. Appl. Phys.*, 34(8):2346–2349, 1963.

[496] W. H. Knox. Femtosecond optical pulse amplification. *IEEE J. Quantum Electron.*, 24(2):388–397, 1988.

[497] M. D. Perry, D. Pennington, B. C. Stuart, G. Tietbohl, J. A. Britten, C. Brown, S. Herman, B. Golick, M. Kartz, J. Miller, H. T. Powell, M. Vergino, and V. Yanovsky. Petawatt laser pulses. *Opt. Lett.*, 24(3):160–162, 1999.

[498] D. Strickland and G. Mourou. Compression of amplified chirped optical pulses. *Opt. Commun.*, 56(3):219–221, 1985.

[499] P. Maine, D. Strickland, P. Bado, M. Pessot, and G. Mourou. Generation of ultrahigh peak power pulses by chirped pulse amplification. *IEEE J. Quantum Electron.*, 24(2):398–403, 1988.

[500] J. R. Klauder, A. C. Price, S. Darlington, and W. J. Albersheim. The theory and design of chirp radars. *Bell Syst. Tech. J.*, 39(4):745–808, 1960.

[501] C. P. J. Barty, C. L. Gordon, and B. E. Lemoff. Multiterawatt 30-fs Ti:sapphire laser system. *Opt. Lett.*, 19(18):1442–1444, 1994.

[502] D. T. Du, J. Squier, S. Kane, G. Korn, G. Mourou, C. Bogusch, and C. T. Cotton. Terawatt Ti:sapphire laser with a spherical reflective-optic pulse expander. *Opt. Lett.*, 20(20):2114–2116, 1995.

[503] G. Cheriaux, P. Rousseau, F. Salin, J. P. Chambaret, B. Walker, and L. F. Dimauro. Aberration-free stretcher design for ultrashort-pulse amplification. *Opt. Lett.*, 21(6):414–416, 1996.

[504] C. P. J. Barty, T. Guo, C. LeBlanc, F. Raksi, C. Rose-Petruck, J. Squier, K. R. Wilson, V. V. Yakovlev, and K. Yamakawa. Generation of 18-fs, multiterawatt pulses by regenerative pulse shaping and chirped-pulse amplification. *Opt. Lett.*, 21(9):668–670, 1996.

[505] G. R. Fleming. *Chemical Applications of Ultrafast Spectroscopy*. International Series of Monographs on Chemistry. Oxford University Press, New York, 1986.

[506] S. Mukamel. *Principles of Nonlinear Optical Spectroscopy*. Oxford Series on Optical and Imaging Sciences. Oxford University Press, New York, 1995.

[507] D. M. Jonas, P. Corkum, R. J. Dwayne Miller, and A. M. Weiner, editors. *Ultrafast Phenomena XV: Proceedings of the 15th International Conference, Pacific Grove, CA, July 30–August 4, 2006*. Springer-Verlag, Berlin, 2007.

[508] T. F. Heinz, S. L. Palfrey, and K. B. Eisenthal. Coherent coupling effects in pump–probe measurements with collinear, copropagating beams. *Opt. Lett.*, 9(8):359–361, 1984.

[509] C. Flytzanis. Theory of nonlinear optical susceptibilities. In H. Rabin and C.L. Tang, editors, *Quantum Electronics: A Treatise*, volume 1, part A. Academic Press, New York, 1976.

[510] B. S. Wherrett, A. L. Smirl, and T. F. Boggess. Theory of degenerate 4-wave mixing in picosecond excitation-probe experiments. *IEEE J. Quantum Electron.*, 19(4):680–690, 1983.

[511] Z. Vardeny and J. Tauc. Picosecond coherence coupling in the pump and probe technique. *Opt. Commun.*, 39(6):396–400, 1981.

[512] S. L. Palfrey and T. F. Heinz. Coherent interactions in pump–probe absorption-measurements: the effect of phase gratings. *J. Opt. Soc. Am. B*, 2(4):674–679, 1985.

[513] C. V. Shank and E. P. Ippen. Anisotropic absorption saturation with picosecond pulses. *Appl. Phys. Lett.*, 26(2):62–63, 1975.

[514] E. P. Ippen. Quantum electronics, picosecond laser pulses, and applications. In K. Young, editor, *Opto-electronics*, pages 155–245. Chinese University Press, Hong Kong, 1982.

[515] J. I. Steinfeld. *Molecules and Radiation: An Introduction to Modern Molecular Spectroscopy*. MIT Press, Cambridge, MA, 2nd edition, 1985.

[516] A. M. Weiner and E. P. Ippen. Femtosecond excited-state relaxation of dye molecules in solution. *Chem. Phys. Lett.*, 114(5-6):456–460, 1985.

[517] C. H. B. Cruz, R. L. Fork, W. H. Knox, and C. V. Shank. Spectral hole burning in large molecules probed with 10 fs optical pulses. *Chem. Phys. Lett.*, 132(4-5):341–344, 1986.

[518] A. Mokhtari, J. Chesnoy, and A. Laubereau. Femtosecond time-resolved and frequency-resolved fluorescence spectroscopy of a dye molecule. *Chem. Phys. Lett.*, 155(6):593–598, 1989.

[519] P. O. J. Scherer, A. Seilmeier, and W. Kaiser. Ultrafast intra-molecular and intermolecular energy-transfer in solutions after selective infrared excitation. *J. Chem. Phys.*, 83(8):3948–3957, 1985.

[520] C. Kittel. *Introduction to Solid State Physics*. Wiley, Hoboken, NJ, 8th edition, 2005.

[521] N. W. Ashcroft and D. N. D. Mermin. *Solid State Physics*. Saunders College, Philadelphia, 1976.

[522] R. W. Schoenlein, W. Z. Lin, E. P. Ippen, and J. G. Fujimoto. Femtosecond hot-carrier energy relaxation in GaAs. *Appl. Phys. Lett.*, 51(18):1442–1444, 1987.

[523] W. Z. Lin, R. W. Schoenlein, J. G. Fujimoto, and E. P. Ippen. Femtosecond absorption saturation studies of hot carriers in GaAs and AlGaAs. *IEEE J. Quantum Electron.*, 24(2):267–275, 1988.

[524] D. W. Bailey, C. J. Stanton, and K. Hess. Numerical studies of femtosecond carrier dynamics in GaAs. *Phys. Rev. B*, 42(6):3423–3434, 1990.

[525] T. Elsaesser, J. Shah, L. Rota, and P. Lugli. Initial thermalization of photoexcited carriers in GaAs studied by femtosecond luminescence spectroscopy. *Phys. Rev. Lett.*, 66(13):1757–1760, 1991.

[526] J. Shah, B. Deveaud, T. C. Damen, W. T. Tsang, A. C. Gossard, and P. Lugli. Determination of intervalley scattering rates in GaAs by subpicosecond luminescence spectroscopy. *Phys. Rev. Lett.*, 59(19):2222–2225, 1987.

[527] K. Kash and J. Shah. Carrier energy relaxation in $In_{0.53}Ga_{0.47}As$ determined from picosecond luminescence studies. *Appl. Phys. Lett.*, 45(4):401–403, 1984.

[528] R. H. Pantell and H. E. Puthoff. *Fundamentals of Quantum Electronics*. Wiley, New York, 1969.

[529] M. Sargent, M. O. Scully, and W. E. Lamb. *Laser Physics*. Addison-Wesley, Reading, MA, 1974.

[530] J. A. Yeazell and C. R. Stroud. Observation of spatially localized atomic electron wave-packets. *Phys. Rev. Lett.*, 60(15):1494–1497, 1988.

[531] A. ten Wolde, L. D. Noordam, A. Lagendijk, and H. B. van Linden van den Heuvell. Observation of radially localized atomic electron wave-packets. *Phys. Rev. Lett.*, 61(18):2099–2101, 1988.

[532] M. J. Rosker, F. W. Wise, and C. L. Tang. Femtosecond relaxation dynamics of large molecules. *Phys. Rev. Lett.*, 57(3):321–324, 1986.

[533] J. Chesnoy and A. Mokhtari. Resonant impulsive-stimulated Raman-scattering on malachite green. *Phys. Rev. A*, 38(7):3566–3576, 1988.

[534] H. L. Fragnito, J. Y. Bigot, P. C. Becker, and C. V. Shank. Evolution of the vibronic absorption-spectrum in a molecule following impulsive excitation with a 6 fs optical pulse. *Chem. Phys. Lett.*, 160(2):101–104, 1989.

[535] T. S. Rose, M. J. Rosker, and A. H. Zewail. Femtosecond real-time observation of wave packet oscillations (resonance) in dissociation reactions. *J. Chem. Phys.*, 88(10):6672–6673, 1988.

[536] M. J. Rosker, T. S. Rose, and A. H. Zewail. Femtosecond real-time dynamics of photofragment-trapping resonances on dissociative potential-energy surfaces. *Chem. Phys. Lett.*, 146(3–4):175–179, 1988.

[537] H. G. Roskos, M. C. Nuss, J. Shah, K. Leo, D. A. B. Miller, A. M. Fox, S. Schmitt-Rink, and K. Kohler. Coherent submillimeter-wave emission from charge oscillations in a double-well potential. *Phys. Rev. Lett.*, 68(14):2216–2219, 1992.

[538] E. J. Heller. The semi-classical way to molecular spectroscopy. *Acc. Chem. Res.*, 14(12):368–375, 1981.

[539] W. T. Pollard, H. L. Fragnito, J. Y. Bigot, C. V. Shank, and R. A. Mathies. Quantum-mechanical theory for 6-fs dynamic absorption-spectroscopy and its application to Nile blue. *Chem. Phys. Lett.*, 168(3–4):239–245, 1990.

[540] A. H. Zewail. Laser femtochemistry. *Science*, 242(4886):1645–1653, 1988.

[541] A. H. Zewail. Femtochemistry: Atomic-scale dynamics of the chemical bond. *J. Phys. Chem. A*, 104(24):5660–5694, 2000.

[542] L. Allen and J. H. Eberly. *Optical Resonance and Two-Level Atoms*. Wiley, New York, 1975.

[543] R. L. Shoemaker. Coherent transient infrared spectroscopy. In J. I. Steinfeld, editor, *Laser and Coherence Spectroscopy*, pages 197–270. Plenum Press, New York, 1978.

[544] F. Eickemeyer, M. Woerner, A. M. Weiner, T. Elsaesser, R. Hey, and K. H. Ploog. Coherent nonlinear propagation of ultrafast electric field transients through intersubband resonances. *Appl. Phys. Lett.*, 79(2):165–167, 2001.

[545] J. E. Rothenberg, D. Grischkowsky, and A. C. Balant. Observation of the formation of the 0π pulse. *Phys. Rev. Lett.*, 53(6):552–555, 1984.

[546] R. Kubo. A stochastic theory of line-shape and relaxation. In D. Ter Haar, editor, *Fluctuation, Relaxation and Resonance in Magnetic Systems*. Oliver & Boyd, Edinburgh, UK, 1962.

[547] R. Kubo. A stochastic theory of line shape. *Adv. Chem. Phys.*, 15:101–127, 1969.

[548] J. W. Goodman. *Speckle Phenomena in Optics: Theory and Applications*. Roberts and Co., Englewood, CO, 2007.

[549] A. M. Weiner, S. De Silvestri, and E. P. Ippen. 3-pulse scattering for femtosecond dephasing studies: theory and experiment. *J. Opt. Soc. Am. B*, 2(4):654–662, 1985.

[550] T. Yajima and Y. Taira. Spatial optical parametric coupling of picosecond light-pulses and transverse relaxation effect in resonant media. *J. Phys. Soc. Jpn.*, 47(5):1620–1626, 1979.

[551] H. J. Eichler, P. Günter, and D. W. Pohl. *Laser-Induced Dynamic Gratings*. Springer-Verlag, New York, 1986.

[552] I. D. Abella, N. A. Kurnit, and S. R. Hartmann. Photon echoes. *Phys. Rev.*, 141(1):391–406, 1966.

[553] P. C. Becker, H. L. Fragnito, C. H. B. Cruz, R. L. Fork, J. E. Cunningham, J. E. Henry, and C. V. Shank. Femtosecond photon-echoes from band-to-band transitions in GaAs. *Phys. Rev. Lett.*, 61(14):1647–1649, 1988.

[554] V. O. Lorenz and S. T. Cundiff. Non-Markovian dynamics in a dense potassium vapor. *Phys. Rev. Lett.*, 95(16):163601, 2005.

[555] S. Mukamel and R. F. Loring. Nonlinear response function for time-domain and frequency-domain 4-wave-mixing. *J. Opt. Soc. Am. B*, 3(4):595–606, 1986.

[556] T. Mossberg, A. Flusberg, R. Kachru, and S. R. Hartmann. Total scattering cross-section for Na on He measured by stimulated photon-echoes. *Phys. Rev. Lett.*, 42(25):1665–1669, 1979.

[557] S. De Silvestri, A. M. Weiner, J. G. Fujimoto, and E. P. Ippen. Femtosecond dephasing studies of dye molecules in a polymer host. *Chem. Phys. Lett.*, 112(3):195–199, 1984.

[558] G. Mourou. Spectral hole burning in dye solutions. *IEEE J. Quantum Electron.*, 11(1):1–8, 1975.

[559] A. M. Weiner. Femtosecond optical pulse generation and dephasing measurements in condensed matter. Ph.D. dissentation, Massachusetts Institute of Technology, 1984.

[560] J. Y. Bigot, M. T. Portella, R. W. Schoenlein, C. J. Bardeen, A. Migus, and C. V. Shank. Non-Markovian dephasing of molecules in solution measured with 3-pulse femtosecond photon-echoes. *Phys. Rev. Lett.*, 66(9):1138–1141, 1991.

[561] M. H. Cho, J. Y. Yu, T. H. Joo, Y. Nagasawa, S. A. Passino, and G. R. Fleming. The integrated photon echo and solvation dynamics. *J. Phys. Chem.*, 100(29):11944–11953, 1996.

[562] W. P. deBoeij, M. S. Pshenichnikov, and D. A. Wiersma. On the relation between the echo-peak shift and Brownian-oscillator correlation function. *Chem. Phys. Lett.*, 253(1–2):53–60, 1996.

[563] W. P. deBoeij, M. S. Pshenichnikov, and D. A. Wiersma. System-bath correlation function probed by conventional and time-gated stimulated photon echo. *J. Phys. Chem.*, 100(29):11806–11823, 1996.

[564] R. W. Schoenlein, D. M. Mittleman, J. J. Shiang, A. P. Alivisatos, and C. V. Shank. Investigation of femtosecond electronic dephasing in CdSe nanocrystals using quantum-beat-suppressed photon-echoes. *Phys. Rev. Lett.*, 70(7):1014–1017, 1993.

[565] R. R. Ernst, G. Bodenhausen, and A. Wokaun. *Principles of Nuclear Magnetic Resonance in One and Two Dimensions*. Clarendon Press, Oxford, UK, 1987.

[566] D. M. Jonas. Two-dimensional femtosecond spectroscopy. *Annu. Rev. Phys. Chem.*, 54:425–463, 2003.

[567] R. M. Hochstrasser. Two-dimensional spectroscopy at infrared and optical frequencies. *Proc. Natl. Acad. Sci. U.S.A.*, 104(36):14190–14196, 2007.

[568] J. R. Zheng, K. Kwak, J. Asbury, X. Chen, I. R. Piletic, and M. D. Fayer. Ultrafast dynamics of solute-solvent complexation observed at thermal equilibrium in real time. *Science*, 309(5739):1338–1343, 2005.

[569] Y. X. Yan, E. B. Gamble, and K. A. Nelson. Impulsive stimulated scattering: general importance in femtosecond laser-pulse interactions with matter, and spectroscopic applications. *J. Chem. Phys.*, 83(11):5391–5399, 1985.

[570] A. Nazarkin, G. Korn, M. Wittmann, and T. Elsaesser. Generation of multiple phase-locked Stokes and anti-Stokes components in an impulsively excited Raman medium. *Phys. Rev. Lett.*, 83(13):2560–2563, 1999.

[571] S. De Silvestri, J. G. Fujimoto, E. P. Ippen, E. B. Gamble, L. R. Williams, and K. A. Nelson. Femtosecond time-resolved measurements of optic phonon dephasing by impulsive stimulated

Raman-scattering in alpha-perylene crystal from 20-K to 300-K. *Chem. Phys. Lett.*, 116(2–3):146–152, 1985.

[572] A. M. Weiner, D. E. Leaird, G. P. Wiederrecht, and K. A. Nelson. Femtosecond pulse sequences used for optical manipulation of molecular-motion. *Science*, 247(4948):1317–1319, 1990.

[573] A. M. Weiner, D. E. Leaird, G. P. Wiederrecht, and K. A. Nelson. Femtosecond multiple-pulse impulsive stimulated Raman-scattering spectroscopy. *J. Opt. Soc. Am. B*, 8(6):1264–1275, 1991.

[574] R. A. Bartels, T. C. Weinacht, N. Wagner, M. Baertschy, C. H. Greene, M. M. Murnane, and H. C. Kapteyn. Phase modulation of ultrashort light pulses using molecular rotational wave packets. *Phys. Rev. Lett.*, 88(1), 2002.

[575] M. Y. Shverdin, D. R. Walker, D. D. Yavuz, G. Y. Yin, and S. E. Harris. Generation of a single-cycle optical pulse. *Phys. Rev. Lett.*, 94(3), 2005.

[576] M. Wittmann, A. Nazarkin, and G. Korn. Synthesis of periodic femtosecond pulse trains in the ultraviolet by phase-locked Raman sideband generation. *Opt. Lett.*, 26(5):298–300, 2001.

[577] N. Zhavoronkov and G. Korn. Generation of single intense short optical pulses by ultrafast molecular phase modulation. *Phys. Rev. Lett.*, 88(20), 2002.

[578] D. H. Auston. Ultrafast optoelectronics. In W. Kaiser, editor, *Ultrashort Laser Pulses and Applications*, pages 183–233. Springer-Verlag, Berlin, 1988.

[579] D. H. Auston. Impulse-response of photoconductors in transmission-lines. *IEEE J. Quantum Electron.*, 19(4):639–648, 1983.

[580] D. H. Auston. Picosecond photoconductivity: High-speed measurements of devices and materials. In R. B. Marcus, editor, *Measurements of High-Speed Signals in Electric Materials and Devices*, volume 28 of *Semiconductors and Semimetals*, pages 85–134. Academic Press, Orlando, FL, 1990.

[581] F. E. Doany, D. Grischkowsky, and C. C. Chi. Carrier lifetime versus ion-implantation dose in silicon on sapphire. *Appl. Phys. Lett.*, 50(8):460–462, 1987.

[582] S. Gupta, M. Y. Frankel, J. A. Valdmanis, J. F. Whitaker, G. A. Mourou, F. W. Smith, and A. R. Calawa. Subpicosecond carrier lifetime in GaAs grown by molecular-beam epitaxy at low-temperatures. *Appl. Phys. Lett.*, 59(25):3276–3278, 1991.

[583] A. C. Warren, J. M. Woodall, J. L. Freeouf, D. Grischkowsky, D. T. McInturff, M. R. Melloch, and N. Otsuka. Arsenic precipitates and the semi-insulating properties of GaAs buffer layers grown by low-temperature molecular-beam epitaxy. *Appl. Phys. Lett.*, 57(13):1331–1333, 1990.

[584] M. B. Ketchen, D. Grischkowsky, T. C. Chen, C. C. Chi, I. N. Duling, N. J. Halas, J. M. Halbout, J. A. Kash, and G. P. Li. Generation of subpicosecond electrical pulses on coplanar transmission-lines. *Appl. Phys. Lett.*, 48(12):751–753, 1986.

[585] D. R. Grischkowsky, M. B. Ketchen, C. C. Chi, I. N. Duling, N. J. Halas, J. M. Halbout, and P. G. May. Capacitance free generation and detection of subpicosecond electrical pulses on coplanar transmission-lines. *IEEE J. Quantum Electron.*, 24(2):221–225, 1988.

[586] J. A. Valdmanis and G. Mourou. Subpicosecond electrooptic sampling: principles and applications. *IEEE J. Quantum Electron.*, 22(1):69–78, 1986.

[587] J. A. Valdmanis. Electro-optic measurement techniques for picosecond materials, devices, and integrated circuits. In R. B. Marcus, editor, *Measurements of High-Speed Signals in Electric Materials and Devices*, volume 28 of *Semiconductors and Semimetals*, pages 135–219. Academic Press, Orlando, FL, 1990.

[588] B. H. Kolner and D. M. Bloom. Electro-optic sampling in GaAs integrated-circuits. *IEEE J. Quantum Electron.*, 22(1):79–93, 1986.

[589] T. Nagatsuma and M. Tsuchiya. Optical measurement technologies for high-frequency electronics. In C. H. Lee, editor, *Microwave Photonics*, pages 111–155. CRC Press, Boca Raton, FL, 2007.

[590] Y.-S. Lee. *Principles of Terahertz Science and Technology*. Springer, New York, 2009.

[591] P. R. Smith, D. H. Auston, and M. C. Nuss. Subpicosecond photoconducting dipole antennas. *IEEE J. Quantum Electron.*, 24(2):255–260, 1988.

[592] C. Fattinger and D. Grischkowsky. Point-source terahertz optics. *Appl. Phys. Lett.*, 53(16):1480–1482, 1988.

[593] J. T. Darrow, B. B. Hu, X. C. Zhang, and D. H. Auston. Subpicosecond electromagnetic pulses from large-aperture photoconducting antennas. *Opt. Lett.*, 15(6):323–325, 1990.

[594] B. B. Hu, J. T. Darrow, X. C. Zhang, D. H. Auston, and P. R. Smith. Optically steerable photoconducting antennas. *Appl. Phys. Lett.*, 56(10):886–888, 1990.

[595] J. T. Darrow, X. C. Zhang, D. H. Auston, and J. D. Morse. Saturation properties of large-aperture photoconducting antennas. *IEEE J. Quantum Electron.*, 28(6):1607–1618, 1992.

[596] X. C. Zhang, B. B. Hu, J. T. Darrow, and D. H. Auston. Generation of femtosecond electromagnetic pulses from semiconductor surfaces. *Appl. Phys. Lett.*, 56(11):1011–1013, 1990.

[597] X. C. Zhang and D. H. Auston. Optoelectronic measurement of semiconductor surfaces and interfaces with femtosecond optics. *J. Appl. Phys.*, 71(1):326–338, 1992.

[598] D. H. Auston and M. C. Nuss. Electro-optic generation and detection of femtosecond electrical transients. *IEEE J. Quantum Electron.*, 24(2):184–197, 1988.

[599] L. Xu, X. C. Zhang, and D. H. Auston. Terahertz beam generation by femtosecond optical pulses in electro-optic materials. *Appl. Phys. Lett.*, 61(15):1784–1786, 1992.

[600] Q. Wu and X. C. Zhang. Ultrafast electro-optic field sensors. *Appl. Phys. Lett.*, 68(12):1604–1606, 1996.

[601] Q. Wu and X. C. Zhang. 7 terahertz broadband GaP electro-optic sensor. *Appl. Phys. Lett.*, 70(14):1784–1786, 1997.

[602] Q. Wu and X. C. Zhang. Free-space electro-optics sampling of mid-infrared pulses. *Appl. Phys. Lett.*, 71(10):1285–1286, 1997.

[603] A. Nahata, A. S. Weling, and T. F. Heinz. A wideband coherent terahertz spectroscopy system using optical rectification and electro-optic sampling. *Appl. Phys. Lett.*, 69(16):2321–2323, 1996.

[604] A. Leitenstorfer, S. Hunsche, J. Shah, M. C. Nuss, and W. H. Knox. Detectors and sources for ultrabroadband electro-optic sampling: Experiment and theory. *Appl. Phys. Lett.*, 74(11):1516–1518, 1999.

[605] G. Gallot and D. Grischkowsky. Electro-optic detection of terahertz radiation. *J. Opt. Soc. Am. B*, 16(8):1204–1212, 1999.

[606] M. van Exter, C. Fattinger, and D. Grischkowsky. Terahertz time-domain spectroscopy of water vapor. *Opt. Lett.*, 14(20):1128–1130, 1989.

[607] M. van Exter and D. R. Grischkowsky. Characterization of an optoelectronic terahertz beam system. *IEEE Trans. Microwave Theory Tech.*, 38(11):1684–1691, 1990.

[608] D. Grischkowsky, S. Keiding, M. van Exter, and C. Fattinger. Far-infrared time-domain spectroscopy with terahertz beams of dielectrics and semiconductors. *J. Opt. Soc. Am. B*, 7(10):2006–2015, 1990.

[609] J. Van Rudd and D. M. Mittleman. Influence of substrate-lens design in terahertz time-domain spectroscopy. *J. Opt. Soc. Am. B*, 19(2):319–329, 2002.

[610] R. A. Cheville and D. Grischkowsky. Far-infrared terahertz time-domain spectroscopy of flames. *Opt. Lett.*, 20(15):1646–1648, 1995.

[611] B. B. Hu and M. C. Nuss. Imaging with terahertz waves. *Opt. Lett.*, 20(16):1716–1718, 1995.

[612] D. M. Mittleman, R. H. Jacobsen, and M. C. Nuss. T-ray imaging. *IEEE J. Sel. Top. Quantum Electron.*, 2(3):679–692, 1996.

[613] Q. Wu, T. D. Hewitt, and X. C. Zhang. Two-dimensional electro-optic imaging of THz beams. *Appl. Phys. Lett.*, 69(8):1026–1028, 1996.

INDEX

Ultrafast Optics, By Andrew M. Weiner

Printed and bound by CPI Group (UK) Ltd, Croydon, CR0 4YY

16/09/2026

14963447-0008